BFI Film and Television Handbook

1999

 British Film Institute

Editor: Eddie Dyja
Commentary: Nick Thomas

Information Services
Statistics Manager: Peter Todd
Statistics Research: Phil Wickham, Erinna Mettler
Statistics Tabulation: Ian O'Sullivan

Additional Research/Editorial Assistance:
Tom Brownlie, Sean Delaney, Jan Dyja, Allen Eyles, David Fisher, Alan Gregory, Michael Henry, Anastasia Kerameos, David Sharp, Linda Wood, Tony Worron

Production: Tom Cabot
Marketing: John Atkinson, Rebecca Watts, Sarah Prosser
Cover design: ketchup

Advertisment Managers: The Gate Partnership
4th Floor, 17 Shaftesbury Avenue,
Piccadilly, London W1V 7RL
Tel: 0171 439 3334
Fax: 0171 439 2935

With many thanks to those who assisted with photographs: BFI, Carlton, Channel 5, Electric, Entertainment, Film Four Distributors, Granada, Hollywood Pictures, Miramax, PolyGram Filmed Entertainment, 20th Century-Fox, Universal & Amblin, The Walt Disney Company

© **British Film Institute 1998**
21 Stephen Street
London WIP 2LN
Printed in Great Britain by Bath Press, Bath

A catalogue record for this book is available from the British Library.

ISBN 0 85170 682 7

Price: £18.99

Cover images:
The Full Monty (courtesy of Fox)
Love is The Devil (courtesy of BFI)
Velvet Goldmine (courtesy of Film Four Distributors)

CONTENTS

Buena Vista International UK Ltd

Forthcoming Releases

Little Voice

Studio 54

A Bug's Life

La Vita E Bella

Beloved

Mighty Joe Young

My Favourite Martian

Instinct

Tarzan

Runaway Bride

Buena Vista International (UK) Ltd
Beaumont House, Kensington Village, Avonmore Road, London W14 8TS

Acknowledgements

I am indebted to all the organisations and individuals who have supported the *BFI Film and Television Handbook* by keeping me informed of movements and changes within the industry.

Special thanks go to Peter Todd and his team in the Information Service of the BFI for their substantial input into this year's book. In particular, I'd like to thank Phil Wickham, Erinna Mettler and Ian O'Sullivan for their painstaking work on the statistical section.

Staff in the BFI National Library have also made their usual significant contributions to the Handbook. The Further Reading list was refurbished by Sean Delaney; the Books section was put together by Anastasia Keraemos with back up from Andrea King; the CD-Rom section was updated by Tony Worron, the Careers section belonged to David Sharp.

Other useful support and encouragement came from John Atkinson, Melissa Bromley, Tom Cabot, Natasha Fairbairn, Tessa Forbes, Eugene Finn. Matt Ker, Lavinia Orton, Sarah Prosser, Mandy Rosencrown, Markku Salmi, Rebecca Watts and Linda Wood.

As ever, *Screen Finance, Screen International, Screen Digest* and *Variety* have been exceptionally accommodating in sharing their information - this is much appreciated.

Thanks also go to the following organisations and individuals: The BBC, The British Film Commission (BFC), British Screen Finance, British Videogram Association (BVA), Central Statistical Office (CSO), Cinema Advertising Association (CAA), Entertainment Data Inc. (EDI), The Department for Culture, Media and Sport (DCMS), Independent Television Commission (ITC), Kate Allen, Tom Brownlie, Lavinia Carey, Peter Cowie, Allen Eyles, David Fisher, Allan Hardy, Michael Henry, Neil McCartney, Boyd Farrow, Barrie MacDonald and Steve Perrin.

Eddie Dyja, Handbook Editor
August 1998

FOREWORD

Alan Parker, Chairman of the British Film Institute

When I took up my post as Chairman of the BFI in January, 1998, I was determined that the Institute should refocus on its core activity of education based upon film and television.

The departure of Jeremy Thomas as Chairman, Wilf Stevenson as Director and Jane Clarke as Deputy Director, marked the end of an important phase in the BFI's history and effectively provided a clean break with the past. With the clear support of Chris Smith, Secretary of State for Culture, Media and Sport, and the BFI Governors, I immediately initiated a full scale review of BFI's activities.

The changes stemming from this fundamental review (which are detailed in BFI Director John Woodward's report) are being currently implemented and will fit comfortably with the Government's well-publicised plans to create an all-encompassing body to deliver film policy in the UK.

The BFI is looking forward to playing a major role in this new organisation, which will bring together the disparate government structures currently in existence for the commercial film industry, but which will also allow the BFI to reinvent itself as a cultural and educational organisation for the new century.

I am delighted at this important time in British cinema to write the foreword for this new edition of the BFI's Film and Television Handbook, which is an indispensable reference work, highly regarded and constantly used by the film and television industry.

INTRODUCTION

John Woodward, Director of the British Film Institute

Over the years the BFI's work has earned a worldwide reputation for excellence and has added considerably to the public's understanding of contemporary and classic film and television. Yet with decreasing funding available from central government, it has become apparent that the BFI has been engaged in too many activities without sufficient resources to make a real difference.

At the beginning of 1998, a comprehensive review started of every BFI activity, with the stated aim of returning to our core activities of education and access to film culture. We were faced with the dichotomy of an Institute which has to be both popular and elitist - reaching out to new audiences and connecting with the millions of children and adults who may be unaware of the BFI's existence, while at the same time maintaining and building on our reputation for intellectual and academic rigour. The intention was to be clear about what the BFI stands for and to prioritise our activities against the overriding principle of our key role which is one of education about the moving image.

During the summer of 1998, Chris Smith, the Secretary of State for Culture, Media and Sport, also announced his intention to introduce a number of changes aimed at rationalising the present machinery for delivering film policy. A new organisation is planned which will oversee the activities of the BFI, British Screen (the Government-backed film financing company), the British Film Commission (which exists to attract film production into the UK), and the Arts Council of England's Lottery-funded film production activities. The new structure for the BFI, developed from our own internal review, is designed to fit smoothly into whatever form the new body takes.

Our review also took into account the report of the Department for Culture, Media and Sport (DCMS) Film Policy Review Group, published in March 1998, which was intended to draw up an action plan for government and industry based on a number of objectives set out by the Secretary of State in May 1997. The Review recommended that the BFI give priority to stimulating interest in, and widening understanding of, film among the general public and develop a strategy to increase and broaden the audience for film.

The BFI's new structure - which is currently being implemented - will see our activities grouped into four departments - Collections, Education, Exhibition and Production, all of which will meet the twin aims of education and accessibility.

Collections

Films, television programmes, computer games, museum artefacts, stills, posters and designs will become part of the Collections Department, which will, in future, place as much emphasis on access as the National Film and Television Archive has on the process of preservation in the past.

Education

The new Education Department is working towards the creation of a coherent education policy aimed at complementing not competing with other education providers. The DCMS has asked the BFI to take a co-ordinating role in a Film Education Working Group which will set a new agenda for film education. Much of the work of the BFI Education Department will be defined by the conclusions of this report which will be published during 1999.

Exhibition

The creation of a national exhibition strategy reaching out to film festivals, regional cinemas, the under-valued film society movement, but also encompassing the National Film Theatre will be the aim of the Exhibition Department. This department will develop opportunities for cinemagoers everywhere to choose from, and learn about, the broadest possible range of British and world cinema.

Production

Although we will not for the moment finance feature films, the Production Department will administer a substantial regional production fund and continue the highly-successful New Directors Fund which has seen a steady stream of shorts produced since it was launched a decade ago.

Last year saw many successes for the BFI, in particular the 41st London Film Festival during which BFI Fellowships were awarded to Thelma Schoonmaker Powell and Sir Sydney Samuelson; the phenomenally successful Mizoguchi season, which not only played to sell-out NFT houses, but also delighted audiences at a whole host of regional cinemas; and John Maybury's *Love Is the Devil* which is now the most successful film ever to come out of BFI Production.

I am confident that the reorganisation now under way will not only enable us to achieve our objectives but also make the BFI a more useful place for the public, our clients and customers. The BFI's Governors and managers have devoted much time to ensuring our new structure will enable us to play a major role in the new film body which looks likely to start work later in 1999.

BFI: HIGHLIGHTS OF THE YEAR

BFI National Film and Television Archive (NFTVA)

Three previously thought missing, believed lost, British features from the 1930s - *Bella Donna, The Mayor's Nest*, and *His Lordship* - were acquired during the year.

The Archive Cricket Evening continued into its 16th year, with tickets for the August event selling out so fast, that a second evening in December was programmed by popular demand. David Meeker, the Keeper of Feature Films, was awarded the MBE in the New Year's Honours List for services to the NFTVA.

BFI National Film Theatre (NFT)

The NFT, one of the world's leading cinémathèques, saw more than 200,000 attendances during the year, with membership rising by 10 per cent to over 35,000.

The February season devoted to the Japanese master director Kenji Mizoguchi was one of the most successful of the year, playing to packed houses and to much critical acclaim. The highly popular Guardian Interviews brought famous faces such as Quentin Tarantino, Martin Scorsese, Sigourney Weaver, Ralph Fiennes and Kevin Spacey to the NFT's stage.

In 1997 more people than ever before visited the three BFI-run festivals - with attendances not only increasing at the London Film Festival, but also at the Lesbian and Gay, and the Jewish Film Festivals. Record amounts of sponsorship were achieved for the 41st LFF, the 12th LGFF, and for film seasons at the NFT.

Ralph Fiennes and Sigourney Weaver at the London Film Festival

Vic Reeves and Jarvis Cocker were invited on to a panel of film critics and producers, which decided to award the 1997 Sutherland Trophy for the best first feature screened during the LFF to Bruno Dumont's *La Vie de Jésus*.

BFI Museum Division

Visitors of all ages flocked to the Museum for three very different exhibitions - Re-Play (on the history of computer games), Hercules: Gerald Scarfe Meets Walt Disney and Hammer Horror.

The construction of the BFI London IMAX Cinema began in November, 1997, and when it opens this spring, visitors to the South Bank will be able to experience the magic of big screen 2D and 3D IMAX cinema on the biggest screen in Europe.

BFI Films

Two films from Distribution were big at the

box office - *Ugetsu Monogatari* earned a record-breaking £14,000 in one week at the NFT, and *Plein Soleil's* £90,000 gross made it the most successful BFI Films' re-release of the year.

BFI archival footage of native Americans visiting the London grave of Chief Lone Wolf appeared in news bulletins around the world, as the Chief's remains were returned to South Dakota.

BFI Information and Education
Use of the BFI National Library hit an all time high. Not only did more people than ever visit the Reading Room, but telephone calls to the Information Line also broke records, as did the demand for items from the Special Collections.

The year saw a higher profile and increased sales for all BFI Publishing's titles, but particularly for the first six books in the new Modern Classics strand. The BFI's *Sight and Sound* magazine more than met its target for subsidy reduction, and is on course to reach zero subsidy within two years.

In the Queen's Birthday Honours, Gillian Hartnoll, the former Head of Library and Information Services, who had left the BFI the previous year, received an MBE for services to film and television culture.

BFI Research
BFI TV's *Howard Hawks: American Artist* had its American premiere at the Museum of Modern Art in New York.

Andrew Kötting's Gallivant was released in 1997

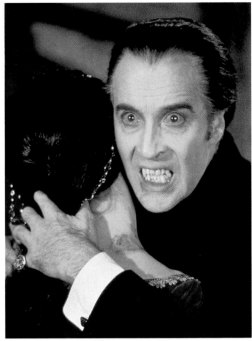

The Hammer Horror exhibition was a scream

One hundred audio tapes featuring some of the many BFI lectures, interviews and debates from the past 40 years, which were in danger of deterioration, were transferred on to CD-Rom and made available to the public for the first time.

BFI Production
Having been selected for numerous film festivals, three titles produced by the BFI - *Under the Skin*, *Gallivant* and *Stella Does Tricks* were released in the UK during the year.

Three writers whose screenplays featured in the 1996 and 1997 Script Factory seasons received full production funding - *What Rats Won't Do* (Steve Coombes and Dave Robinson), *Feeney's Rainbow* (Peter Mulryan) and *Fanny and Elvis* (Kay Mellor).

Regional Development Unit
Through the Regional Development Unit, the BFI distributed just over £2.1 million to regional arts boards and media development agencies to enable them to fund film, video and media projects throughout the UK.

BFI: FACTS AND FIGURES

BFI Membership
35,110 (including 3,816 senior citizens who are given a 10-year free membership)

BFI Information and Education
21 Book titles published
26,299 *Sight and Sound* circulation

BFI Distribution
18 theatrical releases including *Under the Skin*, *Ugetsu Monogatari* and *Plein Soleil*

Regional Film Theatres
28 Regional Film Theatres (Serviced and/or funded by the BFI); **42** screens

BFI National Library
17,551 visitors
38,590 telephone enquiries

South Bank
2,000 film, television and video programmes

'Titanic': a bigger smash
Real bullets fly in Pakistan's 'Zar Gul'
Carnal knowledge: Ang Lee's 'The Ice Storm'
Laura Mulvey on Douglas Sirk
Scorsese on 'Kundun'

Sight and Sound achieved a circulation of over 26,000

screened at The National Film Theatre (NFT)
16 Guardian-sponsored events
125,208 Admissions: NFT 1 (450 seats)
75,453 Admissions NFT 2 (162 seats)
24,459 Admissions Museum Cinema (135 seats)
329,496 Admissions to the Museum of the Moving Image

110,784 Audience for the 41st London Film Festival

BFI National Film and Television Archive
350,000 titles dating from 1894 to the present day
2 million ft of film preserved
2.15 million ft of original material duplicated
1.2 million ft of new viewing copies generated
The Viewing Service provided bookings and loans as follows:
1,536 shorts;
1,260 features;
1,807 television programmes

BFI Production
500 scripts received each year
Two features (*Love Is the Devil* and *Speak Like a Child*) and five short films completed
18 films were supported by the Projects Fund. Average budget for feature films between **£750,000** and **£1 million**

BFI Income

Grant-in-aid	£ 16,000,000
Other grants	£ 1,502,000
Lottery funding	£ 3,260,000
Investment income	£ 394,000
Income from activities	£ 10,965,000
Fund raising activities	£ 833,000
Total income	**£ 32,954,000**

Figures cover the period 01.04.1997 to 31.03.1998

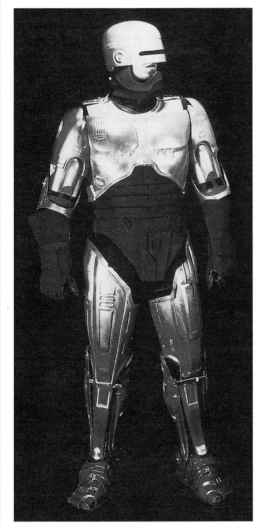

ROUTLEDGE
....for the COMPLETE picture

UK FILM, TELEVISION AND VIDEO: OVERVIEW

by Nick Thomas

Introduction

In many ways, 1997 marked a high point in the history of the film industry in Britain. Audience admissions were at their highest for a quarter of a century, at nearly 140 million for the year, while the multiplex revolution took another major step, with 47 new multi-screen sites opening in 1997.

For good measure, a British film, *The Full Monty*, (albeit made with American money) topped the UK box office, taking a record-breaking £46

million, with another British film, *Bean*, also making into the top five. Films with British input accounted for over 27 per cent of the total UK box office, well ahead of Culture Secretary Chris Smith's target of 20 per cent. There was a rise in the number and overall value of wholly British films, too, although the UK's studios were still attracting major projects from Hollywood and beyond.

The new Labour government made the right noises about the film industry and even appointed a minister with special responsibility for the film

The Full Monty caught the whole film industry with its pants down

for the film industry, who in turn set up the Film Policy Review Group, a welcome development after years of neglect from Whitehall.

Looked at in isolation, then, 1997 was to some extent a year of triumphs, and of great promise for a future of sustained growth across the film industry. But looked at from the perspective of 1998, all in the garden is not quite so rosy.

In terms of exhibition, the boom in the building of cinemas has not reached saturation point yet, but many multiplexes are now facing direct competition for the first time, and all the circuits are having to work hard to sustain their market share, despite the continued growth in admissions.

Distribution is still dominated by US companies and US product. Although UIP's future in the UK may yet be affected by a ruling from the European Commission, the US majors still accounted for over 78 per cent of the UK box office. The demise of two British distributors – Carlton and First Independent – did not help, while the future of PolyGram, the most successful domestic distributor, was undermined by uncertainty about its future following its purchase by Seagram.

That has also affected the production sector, where PolyGram has been the major force behind the revival in British filmmaking, finding audiences at home and abroad for the likes of *Bean*, *Spice World* and *The Borrowers*. As welcome as the success of *The Full Monty* was, it is these kind of accessible and commercially successful films which the UK industry needs in order to sustain its recent progress. There has certainly been much activity in the production sector, much of it helped by the influx of lottery funding. But whether the films being produced will find a British audience remains to be seen.

It remains unclear also whether the plethora of new television channels soon to be available will find an audience sufficient to sustain them. Television in the UK is on the brink of a major change, with the digital services supplied via cable, satellite and terrestrial, set to create a truly multi-channel environment for those who choose to subscribe.

Some rather over-eager City analysts have offered great claims about the massive impact the new digital services will make on UK television, and the huge amounts of money to be made. Maybe they were not around a decade ago when the launch of satellite was accompanied by claims about 50 per cent penetration by the year 2000. That was never going to happen, and the claims made for digital should similarly be taken with a large pinch of salt.

Some people will undoubtedly make fortunes from the new era of television, but they are more likely to be those who own the intellectual property – the all-important TV rights –than those who are investing billions in persuading viewers to subscribe to their particular means of accessing it. As the battle for supremacy kicks off among the different providers of digital television, the only guaranteed winners will be the rights owners, as Mr Murdoch, with his takeover of Manchester United, has clearly grasped.

What some analysts also fail to grasp is the conservatism of the British viewing public, as the 1997 ratings once again show. The digital revolution will not happen until there is a perceived confidence in the added value of the new technology among the sceptical public, and a critical mass of suitably equipped receivers. That could take some years.

Just look at the way video maintains its presence in the marketplace despite its obvious shortcomings as a format. The long-awaited introduction of DVD (Digital Versatile Disc) – a markedly superior system – in 1998 may signal the end for video, but don't hold your breath. Old technology – whether video, analogue terrestrial television, or even cinema – is set to rule the roost for a few years yet.

Film Production

Though there was drop in the number of films produced in the UK in 1997 – 116, compared to 1996's bumper crop of 128 – the upward trend of the last decade continued (Table 1).

£562.8 million was spent on these films, compared to the £741.4 million spent in 1996. The 25 per cent fall in this figure does not indicate a crisis so much as the particularly expensive crop of US co-productions made in the UK in 1996. Though the UK in 1997 housed several movies with budgets of over £30 million, these were fewer in number, creating a disproportionate dent in the overall figure. Broadly speaking, investment in film production in the UK continues to grow.

In terms of wholly UK product, indeed, there are welcome signs that not only are more films being produced, but that their average budgets are increasing (Table 2).

Previous *BFI Handbooks* have argued strongly that in order to compete at the box office both here and abroad, British films need to increase their budgets to ensure the kind of production values that today's cinemagoers expect. This is still the case, despite the great success of the occasional low-budget crossover like *The Full Monty*, whose massive success during the year should not distract us completely from the broader picture.

There was a notable increase in the number of wholly UK films made in 1997 – 65, compared with 53 in 1996 and 28 in 1995. The total value of these was £148.2 million, up from £84.9 million in 1996, with the average budget up from £1.6 million to a healthier £2.28 million. The financial input from the National Lottery has made some difference, but only 14 of the 65 films here received Lottery funding.

If we adopt the formula of previous *Handbooks* and eliminate the 19 films with a budget of less

❶   Number and Value of UK Films
1981-1997

Year	Titles produced	Current prices (£m)	Production cost (£m) (1998 prices)
1981	24	61.2	130.7
1982	40	141.1	277.7
1983	51	251.1	472.7
1984	53	270.4	480.5
1985	54	269.4	455.3
1986	41	165.8	271.0
1987	55	195.3	305.8
1988	48	175.2	264.0
1989	30	104.7	145.2
1990	60	217.4	271.8
1991	59	243.2	285.3
1992	47	184.9	208.6
1993	67	224.1	252.9
1994	84	455.2	502.7
1995	78	402.4	441.0
1996	128	741.4	785.0
1997	116	562.8	581.9

Source: BFI/Screen Finance

Lock, Stock and Two Smoking Barrels - shooting began in 1997

UK Film Production 1997 - Category A

Films where the cultural and financial impetus is from the UK and the majority of personnel are British.

A feature film is defined here as being made on celluloid, over 72 minutes long and with the intention of a theatrical release.

Title	Production Company(ies)	Production cost (£m)
Alien Blood	West Coast Films	1.00
All the Little Animals	Recorded Picture Co/BBC/British Screen/Isle of Man Film Commission/J&M	3.50
Among Giants	Among Giants Ltd/Kudos Prods/BBC/British Screen/Arts Council-England/Yorkshire Media Prod Fund	2.50
Babymother	Formation Films/C4/Arts Council-England	2.00
Beach Boys	Ragged Prods	0.10
Bedrooms & Hallways	Berwin & Dempsey Prods/BBC	3.00
Buskers	Lyndania Films	0.06
Captain Jack *	Viva Films/Granada/Winchester Films/Arts Council-England	3.50
Comic Act	Spider Pictures	1.20
Crossmaheart	Lexington Films/Arts Council-Northern Ireland	0.52
Dad Savage	Sweet Child Films/PolyGram/Dad Savage Ltd	3.50
Day Release	Bolt-On Media/Liscombe Holdings	0.25
The Devil's Snow	Eye-Cue Prods	0.03
Dirty British Boys	Firestorm Pictures	1.00
Elizabeth	Working Title/C4	13.00
The Final Cut	Fugitive Features	1.00
Fast Food	Fast Food Films	1.00
Get Real	Graphite Films/B Sky B/Distant Horizon/British Screen	1.80
Girls' Night	Granada/ITV/Showtime Networks Inc	3.00
The Governess	Parallax Pictures/BBC/British Screen/Arts Council-England/Pandora Prods	2.80
Guru in Seven*	Balhar Prods/Ratpack Films Ltd	0.10
Hard Edge	DMS Films	0.85
Heart	Granada	5.00
I Want You	Revolution Films	3.00
Jackie	Oxford Film Co/C4/British Screen/Arts Council-England	4.90
Jilting Joe	Warner Sisters/BBC Scotland/British Screen	1.00
The Jolly Boys' Last Stand	Jolly Prods/Bigger Picture Co/Function Films	0.10
Keep the Aspidistra Flying	Bonaparte Films/UBA/Sentinel Films/Arts Council-England, Overseas Filmgroup	4.00
A Kind of Hush	The First Film Co/British Screen/Arts Council-England	2.20
Laid Up	Rented Films	0.02
The Last Seduction 2	Specific Films/Ty Cefn/PolyGram	1.50
The Life of Stuff	Prairie Pictures/BBC/Scottish Arts Council/Glasgow Film Fund	2.00
Love Is the Devil	State Films/BFI/BBC/Arts Council-England/Première Heure/Uplink/Partners in Crime	1.00

Title	Production Company(ies)	Production cost (£m)
Lucia	Lexington Films	1.00
Martha - Meet Frank, Daniel and Laurence	Banshee/C4	3.00
Middleton's Changeling*	High Time Pictures	1.60
Mistress of the Craft	Armadillo Films/Vista Street Entertainment	0.80
Orphans	Antonine Green Bridge Prods/C4/Scottish Arts Council/Glasgow Film Fund	1.70
Parting Shots	Scimitar Films	4.00
Plunkett and MacLeane	Working Title	10.00
Prometheus	Holmes Associates/Michael Kustow Prods/C4/Arts Council-England	1.60
Red Mercury	Red Mercury Ltd	0.02
Resurrection Man	Revolution Films/PolyGram	3.50
Rogue Trader	David Paradine Prods/Granada/Newmarket Capitol Group	7.27
Shadow Run	Shadow Run	1.70
A Soldier's Daughter Never Cries	Merchant Ivory/British Screen/Soldiers Daughter Prods	6.20
Speak Like a Child	Leda Serene/BBC/BFI/Arts Council-England	1.00
Spice World	Fragile Films/PolyGram/Icon Entertainment Inc/Brackmont Film	4.00
Sugar, Sugar	Sweet Tooth Films	0.60
Table 5	Raw Talent Prods	0.01
Talisman	Ealing Touch Film Prods	0.40
The Theory of Flight	BBC/Distant Horizon	3.00
The Tichborne Claimant	The Bigger Picture Co/ Merseyside Film Prod Fund	3.10
Time Enough	Time Films	0.20
Titanic Town	Company Pictures/BBC/British Screen/Arts Council-Northern Ireland/Pandora Cinema	3.00
Tom's Midnight Garden	Eastern and Hyperion/BBC/Jadeinn	3.30
24 7 TwentyFourSeven	Scala/BBC	1.40
Underground	Creative Film Services	0.07
Understanding Jane	DMS Films/Flashpoint Pictures	0.50
Up 'N' Under	Touchdown Films/Colour Features/Llaniau Lliw Cyf/Entertainment Film Distributors	2.00
Urban Ghost Story	Living Spirit	0.50
Us Begins with You	Bill Kenwright Films	5.00
What Rats Won't Do	Working Title Films/PolyGram/Rat Productions Ltd	3.00
The Wisdom of Crocodiles	Zenith Prods/Arts Council-England	3.50
The Wolves of Kramer	Discodog Prods	0.80

Total number of films   65
Total Cost   £148.20m
Average Cost   £2.28m

* These films were filmed prior to 1997 but post-production was completed in that year.
Source: BFI/Screen Finance

③   **UK Film Production 1997 - Category B**

Majority UK Co-Productions. Films in which, although there are foreign partners, there is a UK cultural content and a significant amount of
British finance and personnel.

Title	Production Company(ies)/Participating Country(ies)	Production cost (£m)
Appetite	TheAlternative Cinema Co/Loudmouse/101 Films/Schlemmer Films **(Germany)**	1.53
The Croupier	Little Bird/TATfilm/C4/WDR/Filmstiftung Nordrhein Westfalen **(Germany)**	3.50
Dangerous Obsession	Working Title/Film DevCorp/Alberto Ardissone/ZDF/Isle of Man Film Commission	1.90
Divorcing Jack	Scala/IMA Films/BBC/Arts Council-England/Arts Council-Northern Ireland/	
	Winchester Films **(France)**	2.70
Hideous Kinky	Greenpoint/L Films/BBC/Arts Council-England/ Film Consortium **(France)**	3.00
The Land Girls	Greenpoint Films/West 11 Films/Camera One/Arena Film/C4/	
	Arts Council-England **(France)**	5.00
Los Angeles Without a Map	Dan Films/Marianna Films/Euro America Films/Arts Council-England/	
	Yorkshire Media Prod Agency/Finnish Film Foundation **(France/Finland)**	4.90
The Misadventures of Margaret	Lunatics and Lovers/Granada/Mandarin/TF1/Canal Plus/ECF/Film 50	4.50
My Name Is Joe	Parallax Pictures/Road Movies Vierte Producktionen/C4/Arte/Scottish Arts Council/	
	Glasgow Film Fund/Filmstiftung Nordrhein Westfalen **(Germany/France)**	2.50
The Nutcracker Prince	Sands Films/IMAX Corporation **(Canada)**	5.00
The Sea Change	Winchester Films/Granite Films/Sogedesa **(Spain)**	2.00
The Secret Laughter of Women	Elba Films/Paragon Ent./Septieme/ECF/Arts Council-England/HandMade Films	
	(France)	3.40
Sunset Heights	Scorpio Prods/Northlands Film/Arts Council-Northern Ireland/Irish Film Board/	
	Section 35 tax scheme **(Ireland)**	2.00
Vigo: A Passion For Life	Impact Pictures/Nitrate Films/Little Magic/Mact Films/Road Movies/Tomasol/	
	C4/Canal Plus/Arte/European Script Fund **(France/Germany/Spain/Japan)**	3.40
Waking Ned	Tomboy Films/Gruber Bros/Mainstream/Bonaparte/Canal Plus/Overseas Film Group/	
	Isle of Man Film Commission **(France)**	3.10

Total number of films	**15**
Total Cost	**£48.43m**
Average Cost	**£3.23m**

Source: Screen Finance/BFI

than £1 million, and add the 15 majority UK productions (Table 3), the signs are that 1996's growth has been sustained. This solid core of British films numbers 61, with an average budget of £3.13 million (compared to 56 at £3.2 million in 1996). The shift within this core to wholly UK

productions is welcome, but it remains to be seen whether this crop will be any more competitive at the box office than their predecessors.

There was a significant decrease in the number of minority UK co-productions in 1997 – 15 with

Saving Private Ryan was one of a number of big budget productions to use UK studios

an average budget of £3.11 million compared to 30 at £3.7 million in 1996 (Table 4). Perhaps the fundamental problem of this kind of film – making a commercially viable film while balancing the requirements of a number of international backers – is proving a formidable obstacle, with very few of 1996's films in this category making any impact at the box office.

While the number and value of US/UK co-productions fell – 19 in total, with an average budget of £16.48 million, (down from 25 at £17.71 million in 1996) – the UK still hosted a number of major productions during the year. Brand new studios were created for George Lucas' *Star Wars* prequel (at Leavesden) and Steven Spielberg's *Saving Private Ryan* (at Hatfield), while two large-scale film versions of 60s TV shows – *Lost in Space* and *The Avengers* – were also based here. The attractiveness of the UK as a production base for Hollywood Studios continued through 1997, with our technicians, notably in the field of special effects and CGI

(computer generated images), much in demand. It remains to be seen whether the subsequent strength of sterling in 1998 has had an adverse effect on this sector.

The good news for those making movies in the UK is that the demand for feature films here, whether at the cinema or on the small screen, continues to grow at an impressive rate (Table 6). And while video rental is a sector in decline, the increase in movie channel subscriptions more than makes up the difference, without adversely affecting cinemagoing, it would seem.

Whether British producers can consistently deliver the kind of films audiences want to see remains a moot point, but there was welcome evidence in 1997 that people will willingly pay to see a British film, if it is good enough, and if it is marketed well enough.

The strength of the UK's production sector, at least in relative terms, becomes apparent when

❹   **UK Film Production 1997 - Category C**

Minority UK Co-Productions. Foreign (non-US) films in which there is a small UK involvement in finance or personnel.

Title	Production Company(ies)/Participating Country(ies)	Production cost (£m)
Aarzoo	Dayavanti Pictures **(India)**	2.00
The Commissioner	New Era/Metropolis/Saga/Eurimages/Canal Plus/Filmboard Berlin Brendenburg/ Samuelson Production	3.70
The Dance	Oxford Film Co/Isfilm/ECF/Hamburger Kino-Kompanie/Nordisk Film **(Iceland/Germany)**	1.00
Dancing at Lughnasa	Ferndale/Pandora/Samson Films/C4 **(Ireland/France/Germany)**	7.00
The General	Nattore Ltd/Merlin/J&M **(Ireland)**	*
Himalaya - A Chief's Childhood	Antelope/Galatee' Films/JMH/Canal Plus/ECF/CNC **(France/Switzerland)**	3.00
Hooligans	Woodline Films/Liquid Films/Continent Films/ECF/ Bord Scannán na hÉireann **(Ireland/Germany)**	3.00
If Only	CLT-UFA/Escima/Mandarin/Paragon Ent Cor/HandMade Films/Wildrose **(Spain/France/Germany)**	2.20
Jinnah	Petra Films **(Pakistan)**	3.00
On connaît la chanson	Greenpoint Films/Arena Films/ECF **(France/Switzerland)**	5.00
Owd Bob	Allied Vision/Kingsborough/Greenlight Films **(Canada)**	3.10
The Real Howard Spitz	Metrodome/Imagex/Telefilm Canada/Nova Scotia Dev Corp **(Canada)**	4.60
The Red Violin	Rhombus Media/Mikado/C4/Telefilm Canada/Dor Film/New Line Cinema **(Canada/Italy/Austria)**	6.20
St Ives	Little Bird/TATfilm/ BBC/WDR/Compagnie des Phares et Balises/ Arts Council-Northern Ireland	3.40
Such a Long Journey	Amy Intl/Filmworks/British Screen **(Canada/UK/Germany)**	1.96
Treasure Island	Allied Vision/Kingsborough Greenlight Pictures **(Canada)**	2.50
Vent de Colere	Millennium Films/CNC/CLC Prods/ ECF/Rhones-Sples/France 3 **(France)**	1.30

Total number of films   **17**

Total Cost   **£52.96m**

Average Cost   **£ 3.11m**

Source: Screen Finance/BFI   * budget cannot be verified

we look at the figures from other EC countries (Table 7). France and Italy both experienced an increase in the number and value of productions, but still lag behind the UK in terms of overall level of investment. Indeed, the UK alone accounts for 30 per cent of the sum invested in European film production in 1997.

It would be nice if that share were reflected in the box office returns for British movies. But that remains a pipe dream as long as so many of the films made in the UK fail to make it into the cinema. For all the optimism in the production sector, it is sobering to see that two-thirds of UK films made in 1996 remained unreleased a year

Twentieth Century Fox

Fox Searchlight Pictures

Fox Animation Studios

Fox 2000

A NEWS CORPORATION COMPANY

Twentieth Century House, 31-32 Soho Square, London W1V 6AP
Telephone: 0171 437 7766   Fax: 0171 734 3187

UK Film Production 1997 - Category D

American films with a UK creative and/or minor financial involvement

1) American financed or part financed films made in the UK. Most titles have a British cultural content.

Title	Production Company(ies)/Participating Country(ies)	Production cost (£m)
The Avengers	Jerry Weintraub Prods/Warner Bros	45.00
Basil	Showcase/ Kushner Locke	9.00
Death and the Loss of Sexual Innocence	Red Mullet/New Line	2.40
Little Voice	Miramax/Scala	4.20
Lock, Stock and Two Smoking Barrels	Ska Prods/Steve Tisch/Paragon Entertainment/HandMade Films	2.60
Lost in Space	New Line/Prelude Pictures/Irwin Allen Pictures	45.00
My Life So Far	Miramax/Enigma Films/Scottish Arts Council	3.70
The Parent Trap	Meyers-Schyer Prods/Walt Disney/Stansbury Productions	24.00
Saving Private Ryan	Dreamworks SKG/Paramount/Cloud 9/Section 35	50.00
Sliding Doors	Mirage Enterprises/Miramax/British Screen/Sliding Doors Project	1.80
Star Wars: Episode One	Lucasfilm/Jak Prods	25.00
Tomorrow Never Dies	United Artists/Eon Prods/Danjaq	40.00
Velvet Goldmine	Killer Films/Zenith Prods/C4	4.50
Woundings	Stone Canyon Ent./Muse Prods.	2.00

Number of Films   14

Total Cost   £259.2m

Average Cost   £18.51m

2) American Films with some UK financial involvement.

The Big Lebowski	Working Title	13.00
The HiLo Country	Working Title	10.00
Les Miserables	Sarah Radclyffe Prods/Etic/Mandalay Entertainment	25.00
Palookaville	Redwave Films/ Playhouse International/Goldwyn	2.50
A Price Above Rubies	C4/Miramax	3.50

Number of Films   5

Total Cost   £54m

Average Cost   £10.80m

Source: Screen Finance/BFI

⑥ UK Consumer Spending on Feature Films 1984-1997

Year	UK Box Office (£m)	Video Rental (£m)	Video Retail (£m) *	Movie Channel Subscription (£m)**	Total (£m)
1984	103	425	40	-	568
1985	123	300	50	-	473
1986	142	375	70	-	587
1987	169	410	100	-	679
1988	193	470	175	-	838
1989	227	555	320	-	1,102
1990	273	550	365	47	1,235
1991	295	540	444	121	1,400
1992	291	511	400	283	1,485
1993	319	528	643	350	1,840
1994	364	438	698	540	2,040
1995	385	789	457	721	2,352
1996	426	491	803	1,003***	2,723
1997	506	369	858	1,290	3,023

* only a proportion of the UK video retail market is accounted for by feature films

** based on subscriptions to BSkyB of £24.99 per month which includes all movie channels but excludes sport

*** figures for Movie Channel subscriptions for 1996 have been readjusted from last year

Source: BVA/EDI/BSkyB

after their production (Table 8). The fact that several of these were eventually given a theatrical release in 1998 (Table 9) does not diminish the seriousness of the problem. Most British films produced still do not get a wide theatrical release.

There are two obvious approaches to this hardy perennial. The first is to decry the failure of the distribution sector to give these movies a chance to find an audience, and to suggest that support for the production of British films (from the Arts Councils via the National Lottery) is not necessarily helpful unless similar support is offered in both the marketing and distribution of those films. These problems were noted by the Film Policy Review Group set up by the new Labour government's first films minister, Tom Clarke, whose report was published in 1998. We await the implementation of its proposals, which include a voluntary levy of 0.5 per cent of turnover across all sectors of the industry, with interest.

From the distributors point of view, it is easy enough to argue that British producers are still generally incapable of producing films which can generate sufficient interest at the box office to justify an investment. That argument may have been dented a little by the success of *The Full Monty*. What began life as yet another low-budget British movie soon showed it 'had legs' (and a lot more besides!). This led to a clever marketing campaign, which culminated in this modest comedy becoming not only the highest grossing film ever at the UK box office (until *Titanic* overtook it), but also, given its small budget, the most profitable.

That might be a more encouraging precedent if there were more distributors willing and able to promote British films. PolyGram's emergence as both the major producer and distributor of British films was undermined by the uncertainty over its future following the sale of the company to Seagram in 1998. The closure of First Independent, one of the diminishing number of independent distributors, left the US majors with an even bigger share of the UK market than ever.

⑦ EU Film Production 1997

Country	No of Films (inc. Co-Prods)	Investment ($m)
Austria	14	18.4
Belgium	6	19.1
Denmark	23	59.7
Finland	10	16.5
France	163	902.1
Germany	61	346.3
Greece	20	7.1
Ireland	22	111.7
Italy	110	299.1
Luxembourg	5	3.3
Netherlands	15	63.0
Portugal	8	5.7
Spain	80	204.4
Sweden	29	76.3
UK	116	929.8
Total EU	**682**	**3,052.50**

Source: Screen Digest

⑧ Types of Release for UK Films 1984-1996

Proportions of UK films and UK co-productions made in 1996 which achieved:

a) a wide release - opening on 30 or more screens around the country within a year of production.

b) a limited release mainly in arthouse cinemas or a limited West End release within a year of production.

c) were unreleased a year after production. (ie 1st Jan 1998)

Year	(a)%	(b)%	(c)%
1984	50.00	44.00	6.00
1985	52.80	35.90	11.30
1986	55.80	41.90	2.30
1987	36.00	60.00	4.00
1988	29.50	61.20	9.30
1989	33.30	38.90	27.80
1990	29.40	47.10	23.50
1991	32.20	37.30	30.50
1992	38.30	29.80	31.90
1993	25.40	22.40	52.20
1994	31.00	22.60	46.40
1995	23.10	34.60	42.30
1996	19.00	14.00	67.00

Source: BFI/ Screen Finance/EDI

It is true to say that only a proportion of the feature films produced could ever realistically expect a decent theatrical release. This is certainly the case in the US. And with the planned expansion of pay-TV over the next few years, new outlets for premiering British feature films will emerge. But it is hard to see how the current production boom can be sustained while fewer than 20 per cent of British movies make it onto the big screen within a year of being produced.

With the funds now available from the National Lottery, it is arguably easier than it has ever been to produce a film in the UK. Whether there is a sufficiently deep talent base among producers, writers and directors to justify more than 100 feature film projects each year is debatable, to say the least. Though the quantity of films produced is higher than for 40 years, there is little evidence of a comparable rise in the overall quality.

Sliding Doors opened in May 1998

⑨   # What Happened to 1996's UK Films?

Distribution of 1996 UK Productions and foreign films made in the UK.

Released theatrically in 1996/7

Bean

Career Girls

Event Horizon

Extreme Measures

Fargo

The Fifth Element

The Gambler

Hard Men

Incognito

The Jackal

Lawn Dogs

A Life Less Ordinary

Mrs Brown

Photographing Fairies

Preaching to the Perverted

Prince Valiant

Regeneration

Roseanna's Grave

The Saint

The Scarlet Tunic*

Shooting Fish

The Slab Boys**

The Tango Lesson

True Blue

Twin Town

Under the Skin

Up on the Roof

Welcome to Sarajevo

Wilde

Straight to Video

Pervirella

Distribution deal but no release date

B Monkey

The Education of Little Tree

Eyes Wide Shut

Food of Love

Mandragora

The Revengers Comedy

The Scar

Spanish Fly

No current distribution deal

The Apocalypse Watch

Bogwoman

Cameleon

Diana and Me

Driven

Dust

The Eye of the Eagle

The Fifth Province

Harald

The Harpist

I Love Paris

The Island on Bird Street

Julie and the Cadillacs

Kini and Adams

Let's Stick Together

Long Hot Summer

Made in Japan

Miss Monday

Monarch

The Opium War

Peggy Su

Prairie Doves

Rhinocerous Hunting in Budapest

Romance and Rejection

The Serpent's Kiss

The Sixth Happiness

Stone Man

Thon

Unknown Things

Vicious Circles

Vol au Vent

Welcome to Woop Woop

You Can Keep the Animals

Released 1998

Amy Foster

Bent

The Big Swap

Bring Me the Head of Mavis Davis

Cousin Bette

Dance of the Wind

Downtime

Fairytale: A True Story

The Girl with Brains in her Feet

In Love and War

The James Gang

Love and Death on Long Island

The Man Who Knew Too Little (Watch That Man)

Metroland

Mojo

Monk Dawson

Mortal Kombat 2: Annihilation

Mrs Dalloway

My Son the Fanatic

Oscar and Lucinda

Razor Blade Smile

Something to Believe In

Stella Does Tricks

Stiff Upper Lips

The Wings of the Dove

The Winter Guest

Short run at National Film Theatre

Darklands

Remember Me?

The Designated Mourner

Straight to TV

The Brave

Follow the Moonstone

Hostile Waters

MacBeth

The Matchmaker

The Tribe

The Wrong Guy

Unfinished

Deadline

Deadly Wake

Hysteria

Lucan

Pulse

The Stringer

As of June 1998.
Source: BFI/EDI

⑩   Number of UK Films Produced
1912-1997

Year	No.	Year	No.	Year	No.
1912	2	1945	39	1978	54
1913	18	1946	41	1979	61
1914	15	1947	58		
1915	73	1948	74	1980	31
1916	107	1949	101	1981	24
1917	66			1982	40
1918	76	1950	125	1983	51
1919	122	1951	114	1984	53
		1952	117	1985	54
1920	155	1953	138	1986	41
1921	137	1954	150	1987	55
1922	110	1955	110	1988	48
1923	68	1956	108	1989	30
1924	49	1957	138		
1925	33	1958	121	1990	60
1926	33	1959	122	1991	59
1927	48			1992	47
1928	80	1960	122	1993	67
1929	81	1961	117	1994	84
		1962	114	1995	78
1930	75	1963	113	1996	128
1931	93	1964	95	1997	116
1932	110	1965	93		
1933	115	1966	82		
1934	145	1967	83		
1935	165	1968	88		
1936	192	1969	92		
1937	176				
1938	134	1970	97		
1939	01	1071	06		
		1972	104		
1940	50	1973	99		
1941	46	1974	88		
1942	39	1975	81		
1943	47	1976	80		
1944	35	1977	50		

Source: Screen Digest/BFI

We would do well to glance back at the 1940s, which saw British films take their highest ever share of the UK box office, not least because of the quality of the product on offer. Yet in 1946, the last year in which British films took more at the UK box office than Hollywood, we made just 41 films. It is not just a coincidence.

One welcome development in 1997, then, was the setting up of the three commercial film franchises comprising some of the most talented and experienced producers in the UK (Pathé Productions, The Film Consortium, DNA Film Ltd). A move away from the project-by-project, cottage-industry approach that has characterised the UK production sector in recent years is a step in the right direction.

Indeed, the whole process of tendering for these franchises created a number of alliances between UK producers which bode well for the future. In order for the UK to compete both at home and abroad, the evidence suggests we need fewer, better films. And to make them, we need fewer, better production entities.

My Name Is Joe

Cinema

Cinema admissions continued to rise in 1997 (Table 11), with a total of 139.3 million, a year-on-year increase of 12.5 per cent, proving the highest for a quarter of a century.

UK cinemas took a total of £506.27 million at the box office in 1997, up 18.8 per cent on 1996's total of £426 million. That box office revenue has grown faster than admissions is due to an increase in the average ticket price as exhibitors seek to recoup the cost of their huge investment in the new generation of multiplexes.

1997 was the year of the 'plex, with no fewer than 47 new multiscreen sites opening in the year. Although some of these were conversions of existing sites, the net effect was the creation of a further 217 screens in the UK. The boom looks

Bean beamed its way to box office success

Cinema Admissions 1983-1997

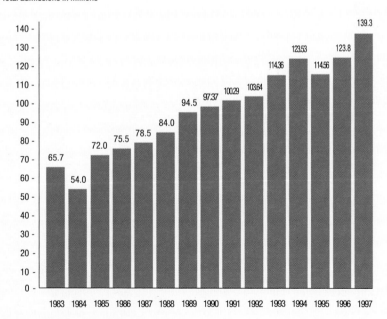

Total admissions in millions

Year	Admissions
1983	65.7
1984	54.0
1985	72.0
1986	75.5
1987	78.5
1988	84.0
1989	94.5
1990	97.37
1991	100.29
1992	103.64
1993	114.36
1994	123.53
1995	114.56
1996	123.8
1997	139.3

Source: EDI/Screen Digest/Screen Finance

⑫   UK Box Office 1997

Admissions	139.3 million
Total Cinema Sites	747
Total Cinema Screens	2383
Total Multiplex Sites	142
Total Multiplex Screens	1222
Box Office Gross	£506.27 million
Average Ticket Price	£3.68

Source: Screen Finance

set to continue, too, with a number of megaplexes, some boasting as many as 30 screens, now on the horizon.

This increase in the number of multiplexes should be good news for the cinemagoer. The resurgence of the cinema industry effectively began with the launch of the first UK multiplex in 1985, and the environment in which films are now enjoyed has improved beyond recognition since then. But now, as the number of multiplexes increases, there is real competition for customers among the different operators.

The cinema circuits are working hard to sell their product and to preserve their share of the market. If that means better seats, a better sound system and easier ways to book tickets, then the cinemagoing public can only benefit. But the cost of these improvements may lead some operators to be more conservative in deciding which films to book. With competition intense, and operating margins tight, the danger is that smaller, offbeat product cannot find a place in multiplex programmes, despite the increased number of screens.

That said, the demographic breakdown of cinemagoing in 1997 seems to suggest that more people are going more often (table 14). Across all sectors, there has been an increase in the number who go to the cinema at least once a month.

A touch of the old Spice Girls in Spice World

**⑬   UK Sites and Screens
1984-1997**

Year	Total Sites	Total Screens
1984	660	1,271
1985	663	1,251
1986	660	1,249
1987	648	1,215
1988	699	1,416
1989	719	1,559
1990	737	1,685
1991	724	1,789
1992	735	1,845
1993	723	1,890
1994	734	1,969
1995	743	2,019
1996	742	2,166
1997	747	2,383

Source: Screen Finance

⑭   Frequency of Cinema-going 1997

Age Group	7 to 14	15 to 24	25 to 34	35+	ABC1	C2DE
No. of People (million)	8.14	7.12	9.22	30.6	27.01	28.07
Once a month or more	26%	52%	34%	10%	25%	19%
Less than once a month but at least twice a year	47%	34%	43%	24%	36%	27%
Once a year or less	15%	9%	16%	26%	21%	21%
Total who ever go to the cinema	88%	95%	93%	60%	82%	67%

Source: Screen Finance/CAVIAR

The Borrowers made it big at the UK Box Office

Over half of all 15-24 year olds, and a third of all 25-34 year olds come into this category, representing in both cases a steep increase from 1996. These two groups represent the core cinemagoing audience, and they clearly see the cinema as an essential part of their leisure activity. Not surprisingly, most films are aimed at this 15-34 age group.

It is interesting to note, however, that there has been a slight decline among the proportion of 7-14 year olds who go to the cinema, even though there are now over 8 million potential cinemagoers in this age group. This remains an underserved sector, although PolyGram's success with films like *Bean*, *The Borrowers* and *Spice World* showed that it is a potentially lucrative one.

(15)

UK Box Office for UK Feature Films 1997

UK Films

Title	Distributor	£
Bean	PolyGram	17,902,161
Spice World *	PolyGram	8,532,981
The Borrowers *	PolyGram	6,504,140
Shooting Fish	Entertainment	4,023,825
Fever Pitch	Film Four	1,836,141
Face	UIP	1,158,803
Twin Town	PolyGram	707,265
Career Girls	Film Four	492,772
Quadrophenia (re-issue)	Feature Film	266,461
Photographing Fairies	Entertainment	81,609
Keep the Aspidistra Flying*	First Independent	75,397
Cold Comfort Farm	Feature Film	49,045
The Leading Man	Pathé	46,087
Preaching to the Perverted	Entertainment	43,536
The Scarlet Tunic *	RIT	31,623
The Near Room	Metrodome	29,135
Under the Skin	BFI	19,807
Alive and Kicking	Film Four	19,701
Heat & Dust (re-issue)	First Independent	18,542
Gallivant	Alliance	16,137
Slab Boys	Film Four	15,911
Remember Me	Metrodome	8,467
Railway Children (re-issue)	BFI	6,736
Up on the Roof	Carlton	5,982
La Passione	Warner	4,137
Small Time	Alliance	2,854
Hell Is a City (re-issue)	Barbican	1,122

* Films still on general release in the UK after Jan 13th 1998

27 Titles   41,900,377

US/UK Co-Productions

Title	Distributor	£
The Full Monty *	Fox	46,252,459
Tomorrow Never Dies *	UIP	16,620,181
Fierce Creatures	UIP	4,310,638
Event Horizon	UIP	3,591,607
A Life Less Ordinary	PolyGram	3,474,816
G.I. Jane	First Independent	2,331,514
Seven Years in Tibet	Entertainment	1,660,580
Extreme Measures	Carlton	1,194,157
Portrait of a Lady	PolyGram	636,188
Hamlet	Carlton	605,000
Welcome to Sarajevo	Film Four	330,650
Blood and Wine	Fox	305,976
Palookaville	Metrodome	269,373

US/UK Co-Productions (cont)

Title	Distributor	£
Roseanna's Grave	PolyGram	171,844
8 Heads in a Duffel Bag	Carlton	130,888
Walking and Talking	Alliance	118,916
Lawn Dogs	Carlton	63,476
Head Above Water	Warner	32,017
Kiss Me Guido	UIP	12,940
Snow White	UIP	2,637
The Disappearance of Kevin Johnson	DDA	580

21 Titles   82,116,437

Other UK Co-production

Title	Distributor	Prod Countries	£
Shine	BVI	AU/UK	4,415,599
Mrs. Brown	BVI	IE/US/UK	2,647,037
Wilde	PolyGram	UK/US/JP/DL	1,524,346
An American Werewolf in Paris	Entertainment	UK/NL/LU	1,234,316
Nil by Mouth	Fox	UK/FR	699,214
I Went Down	BVI	UK/IE/ES	565,113
Kolya	BVI	CZR/UK/FR	536,392
Ma vie en rose	Bluelight	FR/UK/BE/CH	325,318
Carla's Song	PolyGram	UK/ES/DL	322,661
Kama Sutra	Film Four	UK/IN/JP/DL	289,372
Regeneration	Artificial Eye	UK/CA	206,246
The Tango Lesson	Artificial Eye	UK/FR/AR/JP/PL	155,875
Intimate Relations	Fox	UK/CA	144,419
Portraits Chinois	Film Four	FR/UK/SP	76,785
Trojan Eddie	Film Four	IE/UK	60,414
Prince Valiant	Entertainment	UK/DL/IE/US	56,493
Jump the Gun	Film Four	UK/ZA	22,539
Margaret's Museum	Metrodome	UK/CA	22,454
House of America	First Independent	UK/NL	22,407
The Gambler	Film Four	UK/NL/HU	22,291
Hard Men	Entertainment	FR/UK	22,219
Disappearance of Finbar	BVI	IE/UK/FR/SE	17,478
Total Eclipse	Feature Film	FR/UK/BE/US	14,953
The Proprietor	Warner	FR/UK/TR/US	14,071
Swann	Pathé	UK/CA	11,785
Madame Butterfly	Blue Dolphin	UK/FR/JP/DE	10,506
Gold in the Streets	Carlton	IE/UK	10,336
My Mother's Courage	Clarence	DL/UK/AT	5,852
The Boy from Mercury	Blue Dolphin	IE/UK/FR	4,223
A Further Gesture	Film Four	IE/UK/DL/JP	4,207
North Star	Warner	NO/FR/UK/IT	2,627
Someone Else's America	Film Four	UK/FR/DL	2,425
Franz Fanon	BFI	UK/FR	1,684
Driftwood	Blue Dolphin	UK/IE	1,288

34 Titles   13,472,945
Total Box Office   137,489,759

Source: EDI/BFI

⑯   **Top 20 Films at the UK Box Office 1997**

	Title	Country	Distributor	Box Office (£m)
1	The Full Monty*	US/UK	20th Century Fox	46.2
2	Men in Black	US	Columbia TriStar	35.8
3	The Lost World Jurassic Park	US	UIP	25.8
4	Bean	UK	PolyGram	17.9
5	Star Wars (re-issue)	US	20th Century Fox	16.3
6	Batman & Robin	US	Warner	14.6
7	Tomorrow Never Dies*	US/UK	UIP	14.5
8	Ransom	US	Buena Vista	12.8
9	The English Patient	US	Buena Vista	12.7
10	Liar Liar	US	UIP	11.7
11	Space Jam	US	Warner	11.6
12	Hercules	US	Buena Vista	11.2
13	Sleepers	US	PolyGram	10.1
14	Jerry Maguire	US	Columbia TriStar	9.4
15	My Best Friend's Wedding	US	Columbia TriStar	8.7
16	Scream	US	Buena Vista	8.3
17	William Shakespeare's Romeo + Juliet	US	20th Century Fox	7.4
18	Empire Strikes Back (re-issue)	US	20th Century Fox	7.2
19	The Fifth Element	FR	Pathé	7.1
20	Mars Attacks	US	Warner	7.0

Source EDI/BFI

*Films still playing at cinemas after Jan 3rd 1998

NB:Box Office figures have been rounded up.

Also underserved, incidentally, are the over 35s. As circuits try to establish their points of difference in the marketplace, maybe somebody will wonder why only 10 per cent of those over 35 go to the cinema at least once a month.

With all the funds available to British producers, it is rather ironic that the film of the year in 1997 was produced by British talent but made with American money. This was the year of *The Full Monty*, which became the highest grossing film ever in the UK (a title subsequently taken by *Titanic* in 1998), accounting for over 9 per cent of the total annual box office on its own.

It spearheaded a pretty successful year for British productions and co-productions at the box office (Table 15). It helped account for a 33 per cent increase in the money taken by US/UK co-productions, but it was the success of wholly British films, whose box office increased by 110 per cent year-on-year, that was the most striking

Titanic passed The Full Monty as top UK Box Office film of all time

⑰   **Breakdown of UK Box Office by Country of Origin 1997**

Country	No. of Titles	Box Office	%
US	153	347,319,194	68.61
US/UK	21	82,116,437	16.22
UK	27	41,900,377	8.27
Other UK Co.	34	13,472,945	2.66
EU (including co-prods)	25	9,657,225	1.91
Other Co.	14	10,602,416	2.09
Rest Of World (Foreign)	4	24,419	0.01
Rest Of World (English)	6	1,164,815	0.23

Total: **284**   **506,257,828**

NB. Does not include films released in 1996 and still showing in 1997

Source: BFI/EDI

⑱   **US Box Office Revenue for UK Films Released in the US in 1997**

Title	US Distributor	Box Office ($m)
Bean	Gramercy	44.87
The Full Monty	Fox Searchlight	35.01
Mrs Brown	Miramax	8.49
The Wings of the Dove	Miramax	8.19
Secrets and Lies*	October	7.53
A Life Less Ordinary	Fox Searchlight	4.33
The Matchmaker^	Gramercy	3.40
Breaking the Waves*	October	3.08
Career Girls	October	2.63
Brassed Off	Miramax	2.59
The Pillow Book	CFP	2.37
The Van	Fox Searchlight	0.71

* Film went on release in 1995 (amount shown cumulative total)
^ Minority UK Co-Production.
Source: © Variety Inc/BFI

⑲   Top 20 Foreign Language Films Released in 1997

	Title	Distributor	Production Country(ies)	Box Office (£)
1	Ridicule	Alliance	FR	729,436
2	Kolya	BVI	CZR/GB/FR	536,392
3	L'Appartement	Artificial Eye	FR/ES/IT	352,342
4	Ma vie en rose	Bluelight	FR/GB/BE/CH	325,318
5	Beyond the Clouds	Artificial Eye	FR/IT/	286,901
6	A Self Made Hero	Artificial Eye	FR	265,847
7	Tango Lesson	Artificial Eye	GB/AR/FR/JP/DL	155,875
8	Temptress Moon	Artificial Eye	HK/CN	141,126
9	Flirt	Artificial Eye	US/DL/JP	112,498
10	Starm aker	Fox	IT	71,002
11	Will it Snow for Christmas?	Artificial Eye	FR	60,670
12	Drifting Clouds	Metro/Tartan	FI/DL/FR	55,981
13	Mon Homme	Artificial Eye	FR	50,860
14	Ma vie sexuelle	Pathé	FR	43,124
15	Tierra	Metro/Tartan	ES	33,150
16	Kitchen	Alliance	HK/JP	25,292
17	See How They Fall	Artificial Eye	FR	14,442
18	Men/Women...	Gala	FR	13,892
19	Plein Soleil	BFI	FR/IT	12,082
20	Madame Butterfly	Blue Dolphin	FR/JP/DE/GB	10,506

Source: EDI/BFI

Ridicule (left) and Kolya (right) – most foreign language films continue to struggle at the UK Box Office

⑳   Top 20 Films at the Worldwide Box Office 1997

	Film	Int. Distributor	Country of Origin	Box Office Gross ($m)
1	The Lost World Jurassic Park	UIP	US	605.0
2	Men in Black	Columbia TriStar	US	533.0
3	101 Dalmatians	BVI	US/UK	315.0
4	Liar Liar	UIP	US	285.0
5	Air Force One	BVI	US	278.9
6	Jerry Maguire	Columbia TriStar	US	269.0
7	Star Wars (re-issue)	Fox	US	255.0
8	The Fifth Element	Various	FR	235.0
9	My Best Friend's Wedding	Columbia TriStar	US	233.3
10	Ransom	BVI	US	233.0
11	The English Patient	BVI	US	228.0
12	Face/Off	BVI	US	223.0
13	Batman & Robin	Warner	US	208.0
14	Bean	PolyGram	UK	198.3
15	Con Air	BVI	US	183.0
16	Hercules	BVI	US	178.8
17	Dante's Peak	UIP	US	178.0
18	Contact	Warner	US	175.5
19	Scream	Various	US	171.0
20	Space Jam	Warner	US	163.0

Source: Screen International

The Lost World Jurassic Park – on top of the world in 1997

㉑ Top 20 of Admissions of Films Distributed in Europe 1997

Based on analysis of 70% of admissions in European Union in 1997

	Original Title	Country	Admissions
1	Men in Black	US	29,240,389
2	The Lost World Jurassic Park	US	24,718,969
3	Bean	UK	21,500,270
4	The Fifth Element	FR	18,519,222
5	The Full Monty	UK	18,100,481
6	101 Dalmatians	US	15,284,938
7	The English Patient	US/UK	14,673,687
8	Hercules	US	13,845,009
9	Ransom	US	12,745,790
10	Space Jam	US	11,933,922
11	My Best Friend's Wedding	US	11,639,364
12	Liar Liar	US	9,447,714
13	Tomorrow Never Dies	UK/US	8,368,416
14	Air Force One	US	8,306,568
15	Alien: Resurrection	US	7,867,623
16	Batman & Robin	US	7,528,968
17	Con Air	US	7,419,680
18	Star Wars	US	6,932,916
19	Face/Off	US	6,906,129
20	Seven Years in Tibet	US	6,785,589

Source: European Audiovisual Observatory

㉒ Top 20 of Admissions of European Films Distributed in Europe 1997

Based on analysis of 70% of admissions in European Union in 1997

	Original Title	Country	Admissions
1	Bean	UK	21,500,270
2	The Fifth Element	FR	18,519,222
3	The Full Monty	UK	18,100,481
4	The English Patient	US/UK	14,673,687
5	Tomorrow Never Dies	UK/US	8,368,416
6	Evita	UK/US	5,700,294
7	Fuochi D'Artifichio	IT	5,444,136
8	La Verite si Je Mens	FR	4,955,037
9	Le Pari	FR	4,017,576
10	Knockin' On Heaven's Door	DE	3,544,820
11	Rossini-Oder Die Morderische Fraage Wer Mit Wem	DE	3,376,252
12	Kleines Arschlock	DE	3,311,284
13	Didier	FR	3,055,471
14	Fraulein Smillas Gespur Fur Schnee	DE/SE/DK	2,634,030
15	Fierce Creatures	US/UK	2,549,804
16	Ballerman 6	DE	2,482,304
17	La Vita E'Bella	ES/PT/DE	2,308,238
18	Airbag	IT	2,073,524
19	On connait la chanson	FR/IT/CH	1,791,614
20	Le Bossu	FR/IT/DE	6,572,622

National sources: Austria: Columbia; Belgium: Le Moniteur du film belge; Germany: FFA/SPIO; France: Le Film Francais; Italy: Cinetel; Spain: ICAA; UK: X25/Screen Finance

For Italy, Spain and UK, data has been extrapolated from box-office revenues on the basis of average ticket price

* films supported by Eurimages
** data includes admissions in UK after 01.01.98

Source: European Audiovisual Observatory

in 1997. The lion's share was taken by the trio of PolyGram films mentioned above, with an honourable mention going to *Shooting Fish*, the first film to pay back its National Lottery investment.

Bean, a notably successful transfer to the big screen, ended up the fourth most popular film of the year, with the US/UK co-production *Tomorrow Never Dies* also figuring in the top 10 for the year (Table 16). Otherwise, the top 20 belonged as usual to Hollywood.

These 20 films accounted for some £296 million, just over 58 per cent of the total box office, the same percentage as 1996. This would seem to confirm that the increase in screens does not mean that the revenue is spread across more films: although regular cinemagoing is becoming more common, most people still go to the cinema to see the big movies rather than for the sake of it.

101 Dalmatians made it into the top 10 admissions in Europe

㉓   Breakdown Of UK Box Office for Releases by Distributor 1997

Distributor	Titles	Box Office
20th Century Fox	25	110,708,734
Buena Vista	26	89,666,351
UIP	20	82,416,777
Columbia	19	64,630,151
Warner	21	48,552,167
Total US Majors	**111**	**395,974,180**
PolyGram	22	55,753,927
Entertainment	29	22,568,012
Pathé	13	13,966,790
Carlton	11	5,925,751
Film Four	13	3,208,675
1st Independent	12	3,126,409
Artificial Eye	13	1,965,654
Alliance Releasing	9	1,391,227
Feature Film Co.	6	849,633
Metrodome	6	786,767
Bluelight	3	340,809
Metro/Tartan	4	99,499
Island Jamaica	1	98,808
BFI	11	71,232
Blue Dolphin	6	43,217
Gala	6	40,543
RIT	1	31,623
Clarence	1	5,852
Barbican	3	4,842
Hong Kong	1	2,379
Island	1	1,419
DDA	1	580
Total (Independents)	**173**	**110,283,648**
Total	**284**	**506,257,828**

Source EDI/BFI

Culture Secretary Chris Smith's stated desire to see British films accounting for 20 per cent of the UK box office was, as we pointed out in last year's *Handbook*, actually achieved in 1996. In 1997, this share was increased further: films with a UK involvement took 27.15 per cent of the total box office, up from 20.99 in 1996 (Table 17). Wholly American films accounted for 68.6 per cent of the total, down from 77.4 per cent in 1996.

There was real success for UK films abroad, too, with *Bean* and *The Full Monty* taking nearly $45 million and $35 million respectively in the US. What was particularly impressive about *Bean* was not just its relative success in the States, but its performance around the globe in countries where the TV show is an established favourite. The movie was a big hit in places as diverse as Japan, Germany and Australia, reviving a tradition of internationally popular British comedy established by Norman Wisdom in the 1950s.

Two more obviously traditional films, the period drama *Mrs Brown* and the literary adaptation *The Wings of the Dove*, also did well on the back of Miramax's campaigns to garner Oscars for them.

Foreign language films once again had difficulty making any impact on our screens, with only the French costume drama *Ridicule* and the Czech drama *Kolya* making over £500,000 at the box office. Moreover, if we go further down the list, only 13 films here managed to reach even £50,000 at the box office, compared to 19 in 1996, suggesting that this already small market is getting even smaller.

And while the UK might be fighting back against US domination locally, a glance at the top twenty films worldwide in 1997 (Table 20) shows how strong Hollywood's grip is on the global industry. This is also the case if we look at European admissions (Table 21), although *Bean*, *The Full Monty*, and the French-backed *The Fifth Element* mean that Europe provides three of the top five titles here (but only five of the top 20).

UK Cinema Circuits 1984-1997
s (sites) scr (screens)

Year	ABC		Virgin		Cine UK		Odeon		Showcase		UCI		Warner Village		Small Chains		Independents	
	s	scr	s	scr	s	scr	s	scr	s	scr	s	scr	s	scr	s	scr	s	scr
1984	-	-	-	318	-	-	-	205	-	-	-	-	-	-	-	-	-	-
1985	-	-	158	403	-	-	76	194	-	-	3	17	1	5	-	-	-	-
1986	-	-	173	443	-	-	74	190	-	-	3	17	1	5	-	-	-	-
1987	-	-	154	408	-	-	75	203	-	-	5	33	1	5	-	-	-	-
1988	-	-	140	379	-	-	73	214	10	127	12	99	1	5	-	-	-	-
1989	-	-	142	388	-	-	75	241	7	85	18	156	3	26	-	-	-	-
1990	-	-	142	411	-	-	75	266	7	85	21	189	5	48	-	-	-	-
1991	-	-	136	435	-	-	75	296	8	97	23	208	6	57	-	-	-	-
1992	-	-	131	422	-	-	75	313	9	109	25	219	7	64	-	-	-	-
1993	-	-	125	408	-	-	75	322	10	127	25	219	9	84	55		-	-
1994	-	-	119	402	-	-	76	327	11	141	26	232	10	93	143	57	437	631
1995	-	-	116	406	-	-	71	320	11	143	26	232	12	110	130		469	716
1996	92	244	24	162	2	24	73	362	14	181	26	232	16	143	58	139	437	679
1997	80	225	29	213	5	66	73	362	15	197	26	263	17	152	68	166	434	739

Source: Screen Finance

The success of *The Full Monty* was a triumph for 20th Century Fox, who claimed the largest share of the UK market in 1997. Their take of £110.7 million represents a 21.9 per cent market share (up from 13.5 per cent in 1996), putting them ahead of Buena Vista (17.7 per cent, down from 21.1 per cent in 1996) and UIP (16.3 per cent, down from 22.2 per cent in 1996).

These are now the big three distributors in the UK, with PolyGram, which took 11 per cent of the market in 1997, now joining Warner Bros, Columbia and Entertainment on the second tier. Alas, life for the rest of the distribution sector is pretty hard, with First Independent following Carlton into oblivion.

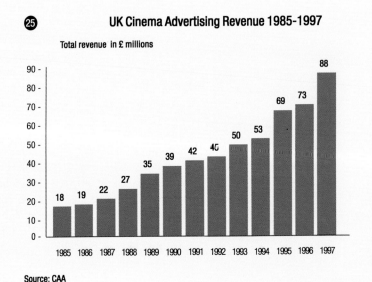

UK Cinema Advertising Revenue 1985-1997

Total revenue in £ millions

Year	1985	1986	1987	1988	1989	1990	1991	1992	1993	1994	1995	1996	1997
Revenue	18	19	22	27	35	39	42	45	50	53	69	73	88

Source: CAA

26 **Most Profitable Films of 1997**

Title	Production Company	Worldwide Box Office($m)	Budget ($m)	Profit to cost ratio
The Full Monty	Fox Searchlight/Redwave	205.4	3.5	58.7
Chasing Amy	Miramax	14.7	0.4	36.7
When the Cat's Away..	Sony Classics	7.6	0.3	25.3
In the Company of Men	Sony Classics/Alliance	4.2	0.2	21.0
Shine	New Line/Pandora	89.1	5.5	16.2
Scream	Miramax	173.0	14.0	12.4
Star Wars (reissue)	Fox	253.7	24.0	10.6
Bean	PolyGram	221.2	23.0	9.6
The Lost World Jurrasic Park	Universal	614.4	70.0	8.8
The English Patient	Miramax	229.9	27.0	8.5
Air Bud	Keystone	25.0	3.0	8.3
William Shakespeare's Romeo + Juliet	Fox	135.7	19.0	7.1
Men in Black	Sony	586.2	85.0	6.9
Liar Liar	Universal	300.0	45.0	6.7
Jerry Maguire	Sony	270.6	43.0	6.3
My Best Friend's Wedding	Sony	286.9	46.0	6.2
Soul Food	Fox	44.4	7.5	5.9
I Know What You Did Last Summer	Entertainment	98.1	17.0	5.8
Karma Sutra	Trimark/Ciby	15.7	3.0	5.2
Beavis and Butt-head Do America	Paramount	80.1	16.0	5.1
George of the Jungle	Buena Vista	165.3	33.0	5.0

Least Profitable Films of 1997

Title	Production Company	Worldwide Box Office($m)	Budget ($m)	Profit to cost ratio
The Shadow Conspiracy	Summit Ent.	5.4	45	0.120
McHale's Navy	Universal	5.5	42	0.131
Steel	Warner Bros	2.2	16	0.137
Cats Don't Dance	Warner Bros	5.2	32	0.162
'Til There Was You	Lakeshore Ent.	4.8	23	0.209
The Pest	Sony	4.1	17	0.241
A Smile Like Yours	Rysher Ent.	4.4	18	0.244
Blood and Wine	Fox Searchlight	7.3	26	0.281
Switchback	Rysher Ent.	9.3	29	0.321
Warriors of Virtue	MGM	11.1	34	0.326
Gone Fishin'	Buena Vista	19.8	53	0.373
Gattaca	Sony	13.5	36	0.375
Ghosts of Mississippi	Castle Rock	13.9	36	0.386
One Night Stand	New Line	9.4	24	0.392
A Simple Wish	Universal	11.1	28	0.396
Hamlet	Castle Rock	11.3	24	0.471
Rosewood	Warner Bros	15.2	31	0.490
Turbulence	Rysher Ent.	28.6	55	0.520
U-Turn	Sony	10.2	19	0.537
Father's Day	Warner Bros	36.2	65	0.557

Source: Source: © Variety Inc

㉗   Cinema Admissions (millions)
1933-1997

Year	Admissions	Year	Admissions
1933	903.00	1966	288.80
1934	950.00	1967	264.80
1935	912.33	1968	237.30
1936	917.00	1969	214.90
1937	946.00		
1936	987.00	1970	193.00
1939	990.00	1971	176.00
		1972	156.60
1940	1,027.00	1973	134.20
1941	1,309.00	1974	138.50
1942	1,494.00	1975	116.30
1943	1,541.00	1976	103.90
1944	1,575.00	1977	103.50
1945	1,585.00	1978	126.10
1946	1,635.00	1979	111.90
1947	1,462.00		
1948	1,514.00	1980	101.00
1949	1,430.00	1981	86.00
		1982	64.00
1950	1,395.80	1983	65.70
1951	1,365.00	1984	54.00
1952	1,312.10	1985	72.00
1953	1,284.50	1986	75.50
1954	1,275.80	1987	78.50
1955	1,181.80	1988	84.00
1956	1,100.80	1989	94.50
1957	915.20		
1958	754.70	1990	97.37
1959	581.00	1991	100.29
		1992	103.64
1960	500.80	1993	114.36
1961	449.10	1994	123.53
1962	395.00	1995	114.56
1963	357.20	1996	123.80
1964	342.80	1997	139.30
1965	326.60		

Source: Screen Digest/BFI

In the exhibition sector, meanwhile, as companies like Virgin and Cine UK have been building brand new sites, others like Odeon and UCI, the two biggest chains, have embarked on expensive programmes of refurbishment of existing sites. None of the players in the exhibition sector is prepared to stand still as the pressure to maintain that all-important share of the market increases.

With extra costs to worry about, the circuits will have been pleased that cinema advertising revenue rose in 1997 by 20.5 per cent to £88 million a year (Table 24), a reflection perhaps of the increasingly high profile of cinema in general.

The Full Monty not only topped the UK box office, it topped the list of most profitable films of 1997 (Table 25), covering its production budget, a mere $3.5 million, some 58 times over.

Yet if you want to know how and why Hollywood continues to dominate the global film industry, look at the list of least profitable movies. Most of these films made a net loss of $20 or $30 million dollars. One movie like that would bankrupt the UK industry; in the US it's just another gamble that failed. But it goes to show that, if you want to make big money, you have to be able to lose big money.

One Night Stand

Video

As we postpone video's obituary for another year, it is worth noting that more money is spent in the UK on buying videos than on going to the cinema. What is more, that figure – £858 million in 1997 (Table 28) – is still on the rise.

Rental activity may be slowing down, but even on the new readjusted figures (previous ones had been overestimated) that sector was still worth £369 million in 1997. By comparison, the market for video games (Table 33) was worth £623 million in 1997, again a bigger market than cinemagoing.

It appears that the competition from cable and satellite, with their dedicated movie channels, is taking audiences away from video rental, but not as quickly as some had thought. Channel 5's evening schedules, based around a movie slot at 9pm, have not made a big impact, but the proposed removal of ITN's 10pm news slot from the ITV schedules, freeing up the evening for movie presentations on primetime terrestrial television, surely will, as will the introduction of digital satellite television and of DVD (Digital Versatile Disc) at the end of 1998.

28   BBFC Censorship of Videos 1997

Certificate	Number of Films passed after cuts
U	1
PG	19
12	9
15	13
18	165
R18	17
Rejected	5

Source: BBFC/BVA

Yet the video rental sector moves in mysterious ways. A glance at the top ten rental titles shows that action titles such as *The Rock* and *Ransom*, perform even better on the small screen than they do at the cinema. This chart (Table 29) suggests that video rental remains a more downmarket and more male-oriented activity than cinemagoing.

That contrasts with the top 20 for retail video, which is distinctly family-friendly. Although *Independence Day* was the most popular single

29   The UK Video Market 1986-1997

Year	Retail Transactions (millions)	Value (£m)	Rental Transactions (millions) *	Value (£m) *
1986	6	55	233	284
1987	12	110	251	326
1988	20	184	271	371
1989	38	345	289	416
1990	40	374	277	418
1991	45	440	253	407
1992	48	506	222	389
1993	60	643	184	350
1994	66	698	167	339
1995	73	789	167	351
1996	79	803	175	382
1997	87	858	161	369

Source: BVA

* Readers will note that retrospective figures given for Rental transactions and Rental Value are lower than in previous volumes of the Handbook. This is due to a new Rental Montior system quoted by the BVA which discovered that the previous figures had been overestimated. Figures have been adjusted accordingly.

(30) Top Ten Rental Videos in the UK 1997	
Title	Distributor
1 The Rock	Buena Vista
2 Ransom	Buena Vista
3 Independence Day	Fox Pathe
4 Twister	CIC
5 Mission Impossible	CIC
6 The Nutty Professor	CIC
7 Jerry Maguire	Columbia TriStar
8 Eraser	Warner
9 Phenomenon	Buena Vista
10 Sleepers	PolyGram

Source: BVA

(31) Distributors' Share of UK Rental Transactions (%)	
Company	% share
1 CIC	22.2
2 Buena Vista	20.0
3 Warner/MGM	13.7
4 Columbia TriStar	13.1
5 Fox Pathé	10.7
6 EV	9.3
7 PolyGram	7.0
8 Film Four	1.4
9 High Fliers	1.1
10 First Independent	0.7

Source: BVA

video, 14 of the top 20 could be categorised as children's titles, with a notable double appearance by the ubiquitous *Teletubbies*. Video's role as the electronic babysitter, capable of mollifying small children who never seem to tire of repeat viewings, has translated into major sales figures. DVD may plan to offer us the best of Hollywood but until it enters Teletubbyland, it will struggle to challenge video's domestic supremacy.

DVD undoubtedly offers superior quality, and with anticpated sell-through prices of between £15 and £20, it is far more likely to become the new standard for home entertainment than, say, Laserdisc, which has never really caught on. But it is unclear how DVD will fit into British viewing habits, and it could be a few years before it makes the expected impact.

Video may be a prehistoric form of technology, but most people have a VCR player at home, and have access to a video rental outlet. With the imminent introduction of digital TV, viewers are being asked to spend money on set-top decoders, on top of cable and satellite subscriptions, not to

The Rock – solid at the top of the video rental chart

mention the TV license fee. Many may feel that spending several hundred pounds on a DVD player is not high on their agenda for a while. This wait and see approach could prove difficult for the DVD industry.

It is also unclear whether DVD will cater for both the retail and rental sectors. While much of DVD's pre-publicity focuses on owning movies, industry wisdom argues that most consumers, in

㉜ Top 20 UK Retail Videos in the UK 1997

	Title	Distributor
1	Independence Day	Fox
2	Star Wars Trilogy	Fox
3	101 Dalmatians	Buena Vista
4	The Hunchback of Notre Dame	Buena Vista
5	Spice-The Official Video	Virgin
6	Roald Dahl's Matilda	Columbia
7	Oliver and Company	Buena Vista
8	Here Come the Teletubbies	BBC
9	Dance with the Teletubbies	BBC
10	Space Jam	Warner Home Video
11	Cinderella	Buena Vista
12	Evita	EV
13	Batman and Robin	Warner Home Video
14	The Many Adventures of Winnie the Pooh	Buena Vista
15	The Black Cauldron	Buena Vista
16	Beauty and the Beast's Enchanted Christmas	Buena Vista
17	Billy Connolly-Two Night Stand	VVL
18	Toy Story	Buena Vista
19	Winnie the Pooh's Most Grand Adventure	Buena Vista
20	Brookside-The Lost Weekend	PolyGram

Source: BVA

㉞ The UK Video Games Market 1997 Consumer Software Sales

Value	£623 million
Units	18.7 million

Source: Computer Trade Weekly/Trade Chart

㉝ Video Sell-Through Company Market Share by Volume 1997 (%)

	Company	% share
1	Buena Vista	15.6
2	PolyGram	14.2
3	Warner/MGM	13.4
4	Twentieth Century Fox	9.1
5	BBC	8.3
6	VCI	8.1
7	CIC	7.7
8	Columbia TriStar	6.2
9	EV	1.9
10	Virgin	1.5

Source: BVA/CIN

most cases, prefer to watch a film once, and pay a couple of pounds for the privilege. Only certain titles, such as those for children, are seen to warrant a £15 purchase.

This is reflected in the distributors' market share for video in 1997. For while the US majors dominate the rental market (Table 30), the retail sector saw strong performances from the likes of the BBC and PolyGram, providers of family fare, with Disney's distribution arm, Buena Vista, unsurprisingly at the top.

Independence Day – top of the video retail chart

Television

Television executives anxious to attract viewers to the growing number of channels will be pleased to see that we are all watching more television – an extra 25 minutes per week. This would appear to defy the commonly-held notions that viewers' TV watching is both finite and in decline.

Television, then, is increasingly important to people's lives despite notional opposition from other media such as the internet and the cinema. Indeed, the number of TV households now stands at 23.9 million, out of a total of 24 million.

Whether viewer loyalty will hold firm is one of the key questions as we move into a period of great uncertainty in the UK sector. Indeed, many of the industry's key executives are still settling

---

㉟   Trends in Television Viewing 1997

Average daily hours of viewing	3.59
Number of TV households	23.9 million

Source: TBI

㊱   Average TV Audience Share (%) of TV Channels

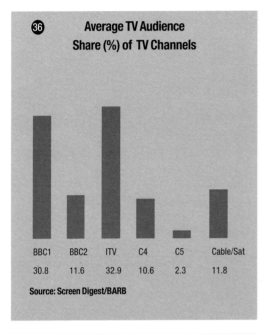

BBC1	BBC2	ITV	C4	C5	Cable/Sat
30.8	11.6	32.9	10.6	2.3	11.8

Source: Screen Digest/BARB

㊲   Cable and Satellite Penetration 1997

	Cable	Satellite
No. of subscribers	2.3m	4.3m
Penetration %	10*	18

* represents national figure. There is a 22.4% take up in cabled areas. 44% of TV households have been passed by broadband cable

Source: TBI

㊳   ITV Companies Programme Supply to the Network

Company	Hours	Minutes
Anglia	252	-
Border	2	-
Carlton	214	-
Central	200	-
Channel	3	-
Grampian	1	41
Granada	665	-
HTV	24	-
LWT	277	-
Meridian	57	-
STV	94	-
Tyne Tees	65	-
UTV	-	-
Westcountry	-	25
Yorkshire	207	-

Source: ITC

39   **Top 20 Programmes for all Terrestrial Channels 1997**

Only top rated episodes of each series are included

	Title	Channel	TX Date	Audience (m)
1	Funeral of Princess Diana	BBC1	6-Sep	19.3
2	Heartbeat	ITV	16-Nov	18.4
3	A Touch of Frost	ITV	16-Feb	18.2
4	Eastenders	BBC1	2-Jan	18.1
5	Coronation Street	ITV	17-Nov	18.0
6	Casualty	BBC1	22-Feb	16.4
7	Men Behaving Badly	BBC1	25-Dec	16.3
8	One Foot in the Grave (Xmas)	BBC1	25-Dec	15.8
9	Before They Were Famous	BBC1	31-Mar	15.3
10	It'll be Alright on the Night 8	ITV	4-Jan	14.9
11	Emmerdale	ITV	20-Feb	14.2
12	The Bill	ITV	17-Jan	14.2
13	London's Burning	ITV	2-Feb	14.1
14	Police, Camera, Action	ITV	7-Jan	13.7
15	National Lottery Live	BBC1	15-Mar	13.6
16	The Killing at Badger's Drift	ITV	23-Mar	13.5
17	BallyKissangel	BBC1	5-Jan	13.5
18	Ronnie Barker: A Life in Comedy	BBC1	1-Jan	13.2
19	Peak Practice	ITV	18-Feb	13.1
20	National TV Awards	ITV	8-Oct	13.0

Source: BARB

into their new jobs after a protracted bout of musical chairs in 1997.

The introduction of digital channels (with terrestrial, satellite and cable operators all competing for viewers), with their set-top decoders, not to mention widescreen and interactive television, is likely to change the face of television in the UK in the long-term. But it could prove a hard sell in the shorter term to the millions of viewers who are still coming to terms with cable and satellite, and for whom the existing free terrestrial services appear to be doing a good job.

Cable and satellite penetration continued to grow in 1997, but it was not spectacular (Table 36). Indeed cable is still only in 10 per cent of TV households, the same as in 1996. But overall the audience share for cable and satellite

Kirsty Young from Channel 5 News

40   **Top 20 Original Drama Productions 1997**

Including soap operas, serials and TV Movies.

Ratings are for highest rated episode of each drama during 1996.

	Title	Producer/Sponser	TX date	Audience (m)
1	Heartbeat	Yorkshire	16-Nov	18.4
2	A Touch of Frost	Yorkshire	16-Feb	18.2
3	Eastenders	BBC	2-Jan	18.1
4	Coronation Street	Granada	17-Nov	18.0
5	Casualty	BBC	22-Feb	16.4
6	Emmerdale	YTT	20-Feb	14.2
7	The Bill	Pearson/Carlton	17-Jan	14.2
8	London's Burning	LWT	2-Feb	14.1
9	The Killing at Badger's Drift	Bentley Prods/YTT	23-Mar	13.5
10	Ballykissangel	World Prods/BBC	5-Jan	13.5
11	Peak Practice	Central	18-Feb	13.0
12	Where the Heart Is	United Film and Tv/Anglia	20-Apr	12.6
13	Jane Eyre	LWT	9-Mar	12.5
14	Touching Evil	United film and Tv/Anglia	29-Apr	12.3
15	The Wingless Bird	Festival TV/YTT	26-Jan	12.3
16	The Student Prince	BBC	29-Nov	12.3
17	Inspector Morse	Zenith/Central	19-Nov	12.1
18	Reckless	Granada	13-Mar	11.9
19	The Vanishing Man	Harbour Pictures/Meridian	2-Apr	11.9
20	Taggart	STV	16-Jan	11.8

Source: BARB/AGB/BFI

programming (Table 35) has increased to 11.8 per cent (up from 10.2 per cent in 1996).

That increase, together with the 2.3 per cent share for newcomers Channel 5, has been achieved at the expense of the two main channels, BBC1 and ITV. BBC1's share was 30.8 per cent (down from 32.5 per cent in 1996), and although ITV continued to be the largest single channel, its share was reduced to 32.9 per cent (down from 35.1 per cent in 1996). BBC2 and Channel 4 both defied expectation to maintain their market shares in 1997.

ITV installed a new management team at its Network Centre in 1997, who have attempted to

41   **Television Advertising 1997**

	£m
Net TV Advertising Revenue	2,600
Consists of:	
ITV	1,716
C4	520
C5	28
Cable and Satellite	286
In addition total programme sponsorship income	36

Source: ITC

The Bill continued to capture audience figures

UK TV Films Premiered 1997

BBC1

Title	Tx date	BARB Rating (m)
Gobble	15-Feb	5.8
Deacon Brodie	8-Mar	6.2
Hostile Waters	26-Jul	8.5
The Fix	4-Oct	4.6
Sex and Chocolate	26-Oct	6.7
The Student Prince	29-Nov	12.3
Cold Enough for Snow	31-Dec	5.1

BBC2

Title	Tx date	BARB Rating (m)
*Stonewall	17-May	1.3
*I.D	1-Jun	3.0
Stone, Scissors, Paper	7-Jun	2.4
*Butterfly Kiss	14-Jun	2.0
*Brothers in Trouble	21-Jun	0.9
Eight Hours from Paris	16-Nov	1.0
Bumping the Odds	7-Dec	1.7
In Your Dreams	14-Dec	2.4
Perfect Blue	21-Dec	2.3
Mothertime	28-Dec	4.4

ITV

Title	Tx date	BARB Rating (m)
The Place of the Dead	18-Jan	6.9
Supply and Demand	5-Feb	10.3
No Child of Mine	25-Feb	5.1
Jane Eyre	9-Mar	12.5
The Killings at Badger's Drift	23-Mar	13.5
The Vanishing Man	2-Apr	11.9

ITV (cont)

Title	Tx date	BARB Rating (m)
The Ebb Tide	24-Sep	6.9
Into the Blue	15-Oct	10.9
Inspector Morse: Death is now my Neighbour	19-Nov	12.1
Agatha Christie's The Pale Horse	23-Dec	6.8
The Black Velvet Band	24-Dec	8.6

Channel Four

Title	Tx date	BARB Rating (m)
*Death and the Maiden	2-Jan	1.9
*Shallow Grave	9-Jan	2.9
*The Baby of Macon	23-Jan	0.5
The Investigator	6-May	2.7
*Trainspotting	26-Nov	4.6
*Blue Juice	3-Dec	1.6
*Sister, My Sister	10-Dec	2
*Beautiful Thing	17-Dec	1.3
*A Midsummer Night's Dream	26-Dec	0.5
*Le Confessional	27-Dec	0.3

Channel Five

Title	Tx date	BARB Rating (m)
Beyond Fear	30-Mar	1.7
One Deadly Summer	30-Nov	1.8

* denotes previous cinema release.

Source: BFI/BARB

---

arrest the freefall in the channel's share. The jury is still out, though, on whether they will be able to defy what some see as an inevitable decline. Advertising revenue for ITV dropped slightly in 1997, with more money flowing into rival channels (Table 40).

Interestingly, the advertising revenue generated by cable and satellite is still less than that of Channel 4, despite a larger market share. Presumably the fragmented, multi-channeled nature of cable and satellite makes it difficult to deliver the kinds of well-defined audiences advertisers are looking for. If that is the case, the

Coronation Street one of the top five programmes in the UK

43	Top 20 Feature Films Shown on Terrestrial TV 1997				
Title		**Country**	**Year**	**Channel**	**Rating (m)**
1	Mrs Doubtfire	US	1993	BBC1	12.3
2	Hocus Pocus	US	1993	ITV	10.4
3	The Bodyguard	US	1992	ITV	10.3
4	Airport	US	1969	BBC1	10.2
5	Sister Act 2	US	1993	ITV	10.1
6	True Lies	US	1994	BBC1	10.1
7	In the Line of Fire	US	1993	ITV	9.7
8	Made in America	US	1993	BBC1	9.7
9	Addams Family Values	US	1993	BBC1	9.7
10	Father of the Bride	US	1991	ITV	9.6
12	Tightrope	US	1984	ITV	9.5
13	Die Hard	US	1988	ITV	9.2
14	Beethoven's Second	US	1993	BBC1	9.1
15	Pretty Woman	US	1990	ITV	8.8
16	Clear and Present Danger	US	1993	BBC1	8.7
17	The Chase	US	1994	BBC1	8.4
18	Another Stakeout	US	1993	ITV	8.4
19	Lethal Weapon	US	1987	ITV	8.3
20	Home Alone 2	US	1992	ITV	8.2

Source: BARB

possibilities for revenue generation are not likely to be enhanced by adding 200 new channels.

The guaranteed revenue for small channels from cable and satellite subscribers, moreover, is threatened by new legislation brought in by the ITC which demands an end to the current system of bundling a number of channels into packages. As an attempt to kick-start the slow take-up for cable by allowing consumers to subscribe only to those channels which interest them, the ITC ruling is a good idea. But it will make life very difficult for those channels who currently receive money from subscribers mainly as part of a pre-negotiated package.

A look at the top 20 programmes in the UK in 1997 shows the conservative nature of the audience, with old stagers like *Heartbeat, A Touch of Frost, Eastenders, Casualty* and *Coronation Street* again dominating the chart, and proving once more that the most popular type of programme among British audiences remains the original drama series.

The BBC's coverage of the funeral of Diana, Princess of Wales, was the most-watched single programme in 1997, despite being shown simultaneously on ITV. It proved once again the unfashionable but indestructible virtue of public service broadcasting. In moments of national importance, the British public turn to the BBC.

This may infuriate Rupert Murdoch, but while BSkyB continues to make large amounts of money, its contribution to original programming has been distinctive in just two areas: sport and movies.

Much is being invested in Sky Movies, which was relaunched in 1998, complete with ex-BBC man Barry Norman as its star presenter. BSkyB has certainly led the way in the field, building an audience with its range of big movies, imaginatively presented.

Neither the BBC nor ITV have been able or willing to compete, and it is little wonder that the ratings for feature films on terrestrial TV have been so feeble. It is a bizarre indictment of the BBC's lack of commitment to films that *Airport,* an unexceptional potboiler from 1969, should end up being its second-highest rating film in 1997 (Table 42). No movie appeared in the year's top 20 programmes.

Nor did any sporting event (although the big games in Euro 96 and the 1998 World Cup did make it). BSKyB's attempts to win viewers by buying the rights to major sporting events – with sport being described by Mr Murdoch as his 'battering ram' – has been a mixed success. Its coverage of Premier League football was overshadowed by the BBC's triumphant coverage of the World Cup in 1998. BSkyB's takeover of Manchester United in 1998 may prove a cash cow, but could be seen as an acknowledgement that the real beneficiaries of Sky's investment in football so far have been the clubs themselves.

Sports rights remain a cheap way of providing programming that appeals to young male audiences, but it's a risky business. Boxing, for example, could still pull terrestrial audiences of 20 million in the mid 1980s, but Sky's exclusive coverage of the sport seems to have coincided with the British public's growing distaste for it. Rugby League, another BSkyB acquisition, is also in crisis. When the UK TV rights for Cricket test matches were put out to tender in 1998, the size of Sky's cheque book was no longer enough, as the sport's governing body realised that without continued exposure on free terrestrial television, a sport stands to lose a generation of fans and players.

ITV's proposed decision to scrap *News at Ten* in order to offer uninterrupted primetime movies may work if they have the right product to show. They had some successes with original TV films in 1997, but replacing a flagship news programme with cheap imported made-for-TV movies will be popular with neither audiences nor TV's regulating bodies.

Yet the terrestrial channels should not be too anxious about their future within a multi-channel environment. While the focus in UK television over the past decade has been on the technological means of delivering programmes, we are moving into an age where content is going to be king. On that score, with their massive libraries and flagship shows, BBC1 and ITV still hold the aces.

44   General Statistics 1997

Population	59 million
Number of Households	24 million
Annual rate of inflation	3.4%
GDP (current prices)	£786,308 million
Total TV Licences in force	22 million
Licence fee income	£1,915.2 million

Source: CSO/BBC

National Lottery

Funding from the National Lottery in 1997 failed to matched 1996's contribution, with £23.7 million (as opposed to £32 million) being awarded to feature and short film projects via the Arts Councils of England, Scotland, Wales and Northern Ireland.

The vast majority of this, £20.5 million, was distributed via the Arts Council of England (ACE), although that figure includes a £4.5 million grant to improve the facilities at the National Film and Television School. ACE gave nearly £16 million to 18 feature projects with a total value of around just under £70 million, although this included the first films to come out of The Film Consortium, one the three newly established production franchises.

In addition in 1997 the Arts Council's allocated development funding as well as production funding, a welcome gesture given the common criticism that many low-budget films are deficient in this area.

The desire to create a UK (as opposed to an English) film industry was not helped by a significant drop in Lottery funding in Scotland and Wales, and while Northern Ireland's total budget increased, it still amounted to less than £1 million.

The Acid House

Shooting Fish – a big fish among Lottery-funded tiddlers

In 1997 *Shooting Fish* became the first movie to pay back its Lottery funding, but it looks like remaining in a very small minority, with few of those funded so far making any impact at the box office.

Only half a dozen or so of the 39 features supported by Lottery funding in 1996 had released a decent theatrical release in the UK by the end of 1998. How the 1997 crop fares remains to be seen, but things will need to change, as the new board of the Arts Council will no doubt confirm.

Given the criterion established for Lottery funding – that the films have real commercial potential – it is difficult to be positive about such a performance. Indeed, the decision in 1998 to remove funding from the BFI's Production Board for its development of feature films seems even more unfortunate.

It is to be hoped that the proposed Alpha fund will be able to continue the tradition of what is described as 'non-commercial' cinema. The films supported by the BFI in recent years may not have been commercial hits (although many have sold widely internationally), but they at least had the merit of being artistically innovative. The danger is that the Lottery will be supporting films which manage to be neither.

Nick Thomas is Editor of Flicks

BEST
OF BRITISH

Film Images presents the *COI Footage File* – a unique visual record of Britain's culture, heritage, way of life and aspirations covering the last 60 years.

The collection begins with John Grierson's documentary masterpiece "Drifters". It moves on to include a wealth of material shot by the Crown Film Unit in the 40s and 50s and now encompasses an outstanding and constantly updated selection of Government commissioned material covering every decade.

Footage File is a kaleidoscope of imagery – social, historical, geographical, scientific and political. Whether you're looking for a single shot or a whole programme, landscapes or laser surgery, churches or Churchill, forestry or forensic science, our experienced staff will make sure that you get the very best of British every time.

fourth floor, 184-192 drummond street, london nw1 3hp. tel: 0171 383 2288

Funding of Film Productions by National Lottery Awards 1997

Arts Council of England

Title	Amount of Award(£)	Total Budget (£)
Features		
Among Giants	665,000	2,300,000
Babymother	1,000,000	2,000,000
A Christmas Carol		
(Animation)^	50,000	149,500
Divorcing Jack	800,000	2,700,000
From a View to a Death	1,000,000	2,700,000
Hideous Kinky^	1,000,000	2,000,000
HOLD BACK THE NIGHT^	560,000	1,700,000
Jackie	950,000	4,900,000
Jack Sheppard and		
Jonathan Wild*	1,000,000	5,500,000
JANICE BEARD 45 WPM^	12,000	2,500,000
Kin	1,000,000	4,500,000
Los Angeles Without a Map	870,000	4,900,000
The Lost Son^	2,000,000	6,000,000
Mike Leigh Project 98	2,000,000	13,500,000
Plunkett and Macleane	950,000	9,300,000
Still Crazy	1,890,000	7,000,000
Two Bad Mice (Post-Production)	117,500	405,960
Awards Sub-Total	**24,304,500**	
Shorts		
Appeal	23,694	48,592
Insomnia	30,000	60,000
Stiletto	38,000	76,000
Stung	5,974	11,974
Temenos	55,448	
Awards Sub-Total	**153,116**	
Projects		
National Film & Television School improvements in production facilities	**4,492,000**	
Screen Network	**5,000**	
Awards Sub-Total	**4,497,000**	
Total Awards	**£28,954,616**	

Scottish Arts Council

Title	Amount of Award(£)	Total Budget (£)
Features		
THE ACID HOUSE	170,000	900,000
Blood Relative	500,000	1,500,000
Daybreak	370,000	900,000
My Name Is Joe	500,000	2,500,000
A Pale View of the Hills*	450,000	2,600,000
Awards Sub-Total	**1,990,000**	
Shorts		
Lay of the Land	15,000	50,000
Little Sisters	10,000	23,000
The Proposal	15,000	30,000
Awards Sub-Total	**40,000**	
Projects		
West Highland Animation New TV Series of Films	52,975	105,950
Awards Sub-Total	**52,975**	
Total Awards	**£2,082,975**	

Arts Council of Wales

Title	Amount of Award(£)	Total Budget (£)
Features		
LAWRENCE OF ARABIA WAS WELSH	10,000	
ONE OF THE HOLLYWOOD TEN	28,500	3,100,000
Y Mynydd grug (Post Production)	6,000	600,000
Awards Sub-Total	**44,500**	
Shorts		
Motor Driving Made Easy	13,955	25,000
Too Old to Dream	25,619	
Awards Sub-Total	**39,574**	
Projects		
Lluniau Lliw Script Development for a number of features		120,000
Teliesyn Development for 2 feature projects.	36,000	
Awards Sub-Total	**156,000**	
Total Awards	**£240,074**	

Arts Council of Northern Ireland

Title	Amount of Award (£)	Total Budget (£)
Features		
Cycle of Violence	150,000	520,000
Divorcing Jack	200,000	2,700,000
The King's Wake (Animation)	60,000	
Mad About Mambo	17,525	900,000
ON HOLY GROUND	13,000	
A PROFUSION OF BLOOD	15,000	
St.Ives	100,000	3,500,000
Awards Sub-Total	**555,525**	

Television

Title	Amount of Award (£)	Total Budget (£)
Amazing Grace (Documentary)	20,000	
THE DREAM TEAM	12,000	
Flip Sides (Animation)	38,500	
MARY ANN	15,150	
The Stranger	200,000	
Awards Sub-Total	**285,650**	
Total Awards	**£841,175**	

Key

Upper case - Money given for development of this title.

* Production postponed after lottery award announced.

^ Money given through franchise for THE FILM CONSORTIUM, one of the three lottery franchise holders for film production announced in May 1997.

Babymother received £1 million of Arts Council of England Lottery funding

Further Reading

AGB Cable & Satellite Yearbook.
London: Taylor Nelson AGB Publications, 1996.

ADVISORY COMMITTEE ON
FILM FINANCE (Chair Peter Middleton)
Report to secretary of State.
[Department of National Heritage], July 1996.

ARTHUR ANDERSEN
The European Film Production Guide: finance, tax, legislation
London; New York: Routledge, 1996.

The ARTS COUNCIL OF ENGLAND and SPEC-TRUM
Lottery Film Franchising: a feasibility study.
London: The Arts Council of England, 1996.

BRITISH BOARD OF
FILM CLASSIFICATION
BBFC Annual Report.
London: BBFC, 1996/97.

BRITISH VIDEO ASSOCIATION
British Video Association Yearbook.
London: BVA, 1998.

BROADCASTING STANDARDS
COMMISSION
Annual report.
London: Broadcasting Standards Council, 1997/98.

BROADCASTING STANDARDS
COMMISSION
The Bulletin (Monthly).
London: Broadcasting Standards Council.

Business Ratio Plus: The film & TV industry. (13th ed.).
Hampton, Middlesex: ICC Information Group, 1998.

The Cable TV and Telecom Yearbook.
Dunstable: WOAC Communications
Company, 1996.

Cable and Satellite Yearbook
FT Media & Telecoms 1998

CAINES, Richard (editor)
Broadcasting in the UK: 1995 market review.
Hampton, Middlesex: Key Note, 1995.

CASEY, Bernard & ECKSTEIN, Jeremy et. al.
Cultural Trends: cultural trends in the '90s, part 1
(Issue 25 1995): film, cinema and video; television and radio; employment in the cultural sector.
London: Policy Studies Institute, 1996.

CASSON BECKMAN
Film 2000: an insight into the future of the UK film industry.
London: Casson Beckman, 1996

CAVIAR Report
(n.12) (3 vol set)
CAVIAR Consortium, 1995

CENTRAL STATISTICAL OFFICE
GB Cinema Exhibitors.
London: CSO, 1997.

CENTRAL STATISTICAL OFFICE
Overseas Transactions of the Film and TV industry.
London: CSO, 1997.

COULING, Katherine
Cinemagoing Europe (5 vol set) Leicester: Dodona
Research, 1998

DUNLOP, Rachael & MOODY, Dominic et. al.
Cultural Trends: cultural trends in the '90s, part 2
(Issue 25 1995): the performing arts
London: Policy Studies Institute, 1996.

DURIE, John (editor)
The Film Marketing Handbook: a practical guide to marketing strategies for independent films.
Madrid: Media Business School, 1993.

ENTERTAINMENT DATA INTERNATIONAL
EDI Database Reports.
London: EDI, 1997/98.

ENTERTAINMENT DATA INTERNATIONAL
EDI Release Schedule UK.
London: EDI, 1997/98.

European Cinema Yearbook,
Milan: Media Salles, 1997.

Film and Television Industry: ICC financial survey.
(25th ed)
Hampton, Middlesex: ICC Business Publications, 1997.

FILM POLICY REVIEW GROUP
A bigger picture: the report of the Film Policy
Review Group
London: Department of Culture Media and Sport,
(DCMS), 1998

GREAT BRITAIN House of Commons National
Heritage Committee
The British Film Industry, Second Report, Volume I,
Session 1994-95: report and minutes of proceedings.
London: HMSO, 1995 [HC 57-I]

GREAT BRITAIN Monopolies and Mergers Commission
Films: a report on the supply of films for exhibition in cinemas in the UK.
London: HMSO, 1994. [Cm 2673]

GREAT BRITAIN Statutes
Broadcasting Act, 1996
London: HMSO, 1996.

GRUMMITT, Karsten-Peter
Cinemagoing 5.
Leicester: Dodona Research, 1996.

HARBORD, Jane and WRIGHT, Jeff
40 Years of British Television.
London: Boxtree, 1995.

HART-WILDEN, Paul
A practical guide to film financing
London: FT Media & Telecoms, 1997

HILL, John & McLOONE, Martin (editors)
Big Picture, Small Screen: the relations between film and television.
John Libbey Media/University of Luton Press, 1996.

HOPEWELL, John and De MOL, Gerry
European Film, a Cultural Industry: ways of maximizing revenues.
Brussels: Media Business School, 1994.

HOWITT, Simon (ed.)
The Film Industry:1995 market report.
(1st ed.).
Hampton, Middlesex: Key Note, 1995.
A Key Note Market Report.

HYDRA ASSOCIATES
Scotland on Screen: the development of the film and television industry in Scotland.
Glasgow: Scott Stern Associates, 1996.
Commissioned by Scottish Enterprise and Highlands & Islands Enterprise.

ILOTT, Terry
Budgets and Markets: a study of the budgeting of European film.
London; New York; Routledge, 1996

INDEPENDENT TELEVISION COMMISSION
ITC Annual Report.
London: ITC, 1995.

INDEPENDENT TELEVISION COMMISSION
1995 Performance Reviews: issued by the Independent Television Commission.

London: ITC, 1997.
Released on the 24th April 1996.

INDEPENDENT TELEVISION COMMISSION
Television: the public's view.
London: ITC, 1997.

JONES, Graham (edited by Lucy Johnson)
Talking Pictures: interviews with contemporary British film-makers.
London: BFI, 1997.

MEDIA
Data on the grid of the European Audiovisual and Media Programme: statistical study of the implementation of positive discrimination principle.
[London?]: [MEDIA?], [ca. 1995].

The Media Map of Eastern Europe
The Media Map of Western Europe
Exeter: CIT Publications, 1998

Radio and Television Systems in Europe
(4 vol set) European Audiovisual Observatory, 1998

Scottish Screen Data '96.
Glasgow: Scottish Film Council, 1996.

Statistical Yearbook: cinema, television, video and new media in Europe, 1998.
Strasbourg: European Audiovisual Observatory, 1996.

Taris UK Television Book (formerly AGB Television Yearbook)
London, Taylor Nelson, 1998 AGB plc

Video and Audiovisual Industry: ICC financial survey. (22nd ed.).
Hampton, Middlesex: ICC Business Publications, 1997.

WOOLF, Myra and HOLLY, Sara
Employment Patterns and Training Needs: freelance and set crafts.
London: Skillset, 1994.

The World Television and Film Market.
Montpellier: Institute de l'audiovisuel et des télécommunications en Europe (IDATE), 1997.

ZENITH MEDIA WORLDWIDE
Television in Europe to 2004.
London: Zenith Media Worldwide, 1994.

ARCHIVES AND FILM LIBRARIES

INTERNATIONAL ORGANISATIONS

FIAF (International Federation of Film Archives)
rue Franz Merjay 190
1180 Brussels
Belgium
Tel: (32) 2 343 06 91
Fax: (32) 2 343 76 22
Christian Dimitziu
FIAF, which has over 50 member archives and many Provisional members and Associates from 60 countries, exists to develop and maintain the highest standards of film preservation and access. It also publishes handbooks on film archiving practice which can be obtained from the above address

FIAT/IFTA (International Federation of Television Archives)
National Film and Television Archive
21 Stephen Street
London W1P 2LN
Tel: 0171 957 8940
Fax: 0171 580 7503
Steve Bryant
FIAT membership is mainly made up of the archive services of broadcasting organisations. However it also encompasses national archives and other television-related bodies. It meets annually and publishes its proceedings and other recommendations concerning television archiving

EUROPEAN ARCHIVES

Below are selected European Film Archives of countries in the EC. For more specialised information consult *Film and Television Collections in Europe - The MAP-TV Guide* published by Blueprint

AUSTRIA
Österreichishes Filmarchiv (Austrian Film Library)
Rauhensteingasse 5
1010 Wien
Austria
Tel: 43 1 512 99 36
Fax: 43 1 513 53 30
Dr Josef Schuchnig

BELGIUM
Cinémathèque Royale/Koninklijk Filmarchief (Royal Film Archives)
Palais des Beaux Arts 23
Rue Ravenstein 1000
Bruxelles
Belgium
Gabrielle Claes

DENMARK
Det Danske Filmmuseum
Store Søndervoldstraede 4
1419 København
Denmark
Tel: 45 31 57 65 00
Fax: 45 31 54 12 12
Ib Monty, Director

FINLAND
Suomen Elokuva-Arkisto (Finnish Film Archive)
PO Box 177
00151 Helsinki
Finland
Tel: 358 0 171 417
Fax: 358 0 171 544
Matti Lukkarila

FRANCE
Les Archives du Film du Centre National de la Cinématographie - CNC
7 bis rue Alexandre-Turpault, 78390 Bois-D'Arcy
France
Tel: 33 1 30 14 80 00
Fax: 33 1 34 60 52 25
Eric Le Roy

GERMANY
Deutsche Rundfunkarchiv, Standort Berlin, Fernseharchiv (German Broadcasting Service, Berlin, Television Archive)
Rudower Chaussee
12489 Berlin
Germany
Tel: 49 30 63 81 60 55
Fax: 49 30 6 77 40 07
Sigrid Ritter, Head Archivist

GREECE
Teniothiki Tis Elladas (Greek Film Archives)
1 Kanari Street
Athens 1067
Greece
Tel: 30 1 3612046
Fax: 30 1 3628468
Theodoros Adamopoulos, Director

IRELAND
Irish Film Archive
Film Institute of Ireland
6 Eustace Street
Dublin 2
Republic of Ireland
Tel: 353 1 679 5744
Fax: 353 1 677 8755
email: s.howard@ucrysj.ac.uk
Sunniva O'Flynn, Curator

ITALY
Cineteca Nazionale (National Film Archive)
Centro Sperimentale di Cinematografia
Via Tuscolana
1524, 00173 Roma
Italy
Tel: 39 6 722941
Fax: 39 6 7223131
Angelo Libertini, General Director

LUXEMBOURG
Cinémathèque Municipale de Luxembourg/Villo de Luxembourg (Luxembourg National Film Archive/City of Luxembourg
10 rue Eugene Ruppert
2453 Luxembourg
Luxembourg
Tel: 352 47 96 26 44
Fax: 352 40 75 19
Fred Junck

THE NETHERLANDS
Nederlands Filmmuseum, Stichting (Netherlands Film Museum)
Vondelpark 3
1071 AA Amsterdam
The Netherlands
Tel: 31 20 5891400
Fax: 31 20 6833401
Peter Westervoorde, Head of
Cataloguing Department

PORTUGAL
Cinemateca Portuguesa - Museum do Cinema (Portuguese Film Archive - Museum of Cinema)
Rua Barata Salgueiro, No 39
1200 Lisboa
Portugal
Tel: 351 1 546279
Fax: 351 1 3523180
José Manuel Costa, Head of Film
Archive

SPAIN
Filmoteca Española (Spanish National Film Theatre)
Carretera de la Dehesa de la Villa
s/n 28040
Madrid
Spain
Tel: 34 1 549 00 11
Fax: 34 1 549 73 48

SWEDEN
Svenska Filminstitutet (Swedish Film Institute)
PO Box 27 126
102 52 Stockholm
Sweden
Tel: 46 8 665 11 00
Fax: 46 8 661 18 20
Rolf Lindfors, Head of Archive

NATIONAL ARCHIVES

Imperial War Museum Film and Video Archive
Lambeth Road
London SE1 6HZ
Tel: 0171 416 5000
Fax: 0171 416 5379
The national museum of twentieth century conflict, illustrating and recording all aspects of modern war. The Archive reflects these terms of reference with an extensive collection of film and video material, which is widely used by historians and by film and television companies

National Film and Television Archive
21 Stephen Street
London W1P 2LN
Tel: 0171 255 1444
Fax: 0171 580 7503
The NFTVA preserves and makes permanently available a national collection of moving images which have value as examples of the art and history of cinema and television, or as a documentary record of the twentieth century. The collection now holds over 300,000 titles dating from 1895 to the present. The Archive also preserves and makes accessible the BFI's collection of stills, posters and designs **(see p8)**

Scottish Screen Film and Television Archive
Dowanhill
74 Victoria Crescent Road
Glasgow G12 9JN
Tel: 0141 302 1700
Fax: 0141 302 1713
Janet McBain: Curator
Anne Docherty: Enquiries
Almost exclusively non-fiction film, the collection dates from 1896 to the present day and concerns aspects of Scottish social, cultural and industrial history. Access charges and conditions available on request

Wales Film and Television Archive
Unit 1, Aberystwyth Science Park
Cefn Llan, Aberystwyth
Dyfed SY23 3AH
Tel: 01970 626007
Fax: 01970 626008
Director: Iola Baines
The Archive locates, preserves and catalogues film and video material relating to Wales. The collection is made accessible where possible for research and viewing. The Archive is part of Sgrín, Media Agency for Wales
Chief Executive: J Berwyn Rowlands

REGIONAL COLLECTIONS

East Anglian Film Archive
University of East Anglia
Norwich NR4 7TJ
Tel: 01603 592664
Fax: 01603 458553
David Cleveland, Patrick Russell, Jane Alvey
Preserving non-fiction films, both amateur and professionally made, showing life and work in Norfolk, Suffolk, Essex and Cambridgeshire

Northeast Television Film Archives
University of Teeside
Borough Road
Middlesborough TS1 3BA
Tel: 01642 342115
Fax: 01642 342190
A.F. Gaw
Comprises of television newsfilms 1958-1972 from BBC Northeast and documentary film from Tyne Tees Television 1960-1980

North West Film Archive
Manchester Metropolitan University
Minshull House
47-49 Chorlton Street
Manchester M1 3EU
Tel: 0161 247 3097
Fax: 0161 247 3098
email: n.w.filmarchive@mmu.ac.uk
Maryann Gomes: Director
Rachael Holdsworth: Enquiries
Preserves moving images showing life in the North West and operates as a public regional archive. Urban and industrial themes are particularly well illustrated

Northern Film and Television Archive
36 Bottle Bank
Gateshead
Tyne and Wear NE8 2AR
Tel: 0191 477 3601/5532
Fax: 0191 478 3681
Bob Davis
A northern regional collection with an emphasis on industry, especially coal mining, and its community

South East Film & Video Archive
University of Brighton
68 Grand Parade
Brighton BN2 1RA
Tel: 01273 600900
Established in 1992 the function of this regional film and video archive is to locate, collect, preserve and promote films and video tapes made in the four counties of Surrey, Kent, East Sussex and West Sussex

Wessex Film and Sound Archive
Hampshire Record Office
Sussex Street
Winchester SO23 8TH
Tel: 01962 847742
Fax: 01962 878681
email: sadedm@hants.gov.uk.
David Lee
Preserves and makes publicly
accessible for research, films, video
and sound recordings of local
interest to central southern England

Yorkshire Film Archive
University College of Ripon and
York St John
College Road
Ripon HG4 2QX
Tel: 01765 602691
Fax: 01765 600516
email: s.howard@ucrysj.ac.uk
Sue Howard
The Yorkshire Film Archive exists to
locate, preserve and show film about
the Yorkshire region. Material dates
from 1897 and includes newsreels,
documentaries, advertising and
amateur films

NEWSREEL, PRODUCTION AND STOCK SHOT LIBRARIES

Adventure & Wildlife Stock Shot Library
Church House,
18 Park Mount, Leeds
Yorkshire LS5 3HE
Tel: 0113 2307150
Fax: 0113 2745387
Chris Lister
A wide range of adventure sports
and wildlife footage, most shot on
16mm or Super 16mm and available
on Beta SP or film

Archive Film Agency
21 Lidgett Park Avenue
Roundhay
Leeds LS8 1EU
Tel: 0113 2662454
Fax: 0113 2662454
Agnèse Geoghegan
Film from 1898 to present day,
including a current worldwide stock
shot library. Specialists in early
fiction, newsreel, documentary,
Music Hall, Midlands, Yorkshire,
British 1930s stills. Cassette services

Archive Films
4th Floor,
184 Drummond Street
London NW1 3HP
Tel: 0171 383 0033
Fax: 0171 383 2333
email: 100657.1661.compuserve.com.
Angela Saward
Stock film footage collection
covering silent comedies, feature
films, documentaries, TV
programmes, rare music footage,
industrial films and newsreels. Same
day service of a research cassette for
almost any request

BBC Information & Archives - Television Archive
Reynard Mills Industrial Estate
Windmill Road
Brentford TW8 9NF
Tel: 0181 576 9211/9212
Fax: 0181 569 9374
Margaret Kirby
The largest collection of broadcast
programmes in the world reflecting
the whole range of BBC output

Boulton-Hawker Films
Hadleigh, Ipswich
Suffolk IP7 5BG
Tel: 01473 822235

Fax: 01473 824519
Peter Boulton
Educational films produced over 50
years. Subjects include: health,
biology, botany, geography, history,
archaeology, and the arts

The British Defence Film Library
SSVC, Chalfont Grove
Chalfont St. Peter
Gerrards Cross
Bucks SL9 8TN
Tel: 01494 878278/878252
Fax: 01494 878007
The British Defence Film Library
(BDFL) is an independent
department within the SSVC group
of companies with experience in
providing entertainment and
producing military training films for
the armed forces. The newly formed
library is the official supplier of
Ministry of Defence footage to the
film and television industry. A wide
range of military material is
available covering many aspects of
life in the army, navy or air force

British Movietonews
North Orbital Road
Denham
Middx UB9 5HQ
Tel: 01895 833071
Fax: 01895 834893
Barbara Heavens
One of the world's major film
archives featuring high quality
cinema newsreels from the turn of
the century, with an emphasis on
1929-1979. the library now
represents on an exclusive basis the
TV-AM News Library with over
1,100 hours of British and World
news covering the period 1983-
1991. This material is available on
re-mastered digital tape

British Pathé Plc
60 Charlotte Street
London W1P 2AX
Tel: 0171 323 0407
Fax: 0171 436 3232
Larry McKinna: Chief Librarian
50 million feet of newsreel and
social documentary from 1896 to
1970. Rapid research and sourcing
through computerised catalogue
Pinewood Studios
Pinewood Road
Iver, Bucks SL0 0NH
Tel: 01753 630 361
Fax: 01753 655365
Ron Saunders

Canal + Image UK
Pinewood Studios
Pinewood Road, Iver
Bucks SL0 0NH
Tel: 01753 631111

Fax: 01753 655813
John Herron
Feature films, TV series, stock shot and stills, b/w and colour, 35mm, 1925 to present day

Channel Four Clip Library
124 Horseferry Road
London SW1P 2TX
Tel: 0171 306 8490/8155
Fax: 0171 306 8362
email: caustin@channel4.co.uk
Claire Austin/Eva Kelly
An ever growing portfolio of programmes and a diverse collection of library material

Contemporary Films
24 Southwood Lawn Road
Highgate
London N6 5SF
Tel: 0181 340 5715
Fax: 0181 348 1238
Documentaries on China, USSR, Cuba, Nazi Germany, South Africa. The library also covers areas like the McCarthy witch hunts in the 50s , the civil rights movement of the 60s, hippie culture, feminism

Editions Audiovisuel Beulah
66 Rochester Way
Crowborough TN6 2DU
Tel: 01892 652413
Fax: 01892 652413
email: iainlogan@enterprise.net
Beulah publish the following videos Vintage Music, Inland Waterways, Royal Navy, Military Transport, Yesterday's Britain. It also incorporates Film Archive Management (FAME)

Educational and Television Films (ETV)
247a Upper Street
London N1 1RU
Tel: 0171 226 2298
Fax: 0171 226 8016
Documentaries on Eastern Europe, USSR, China, Vietnam, Cuba and British Labour movement, b/w and colour, 16mm and 35mm, 1896 to present day

Environmental Investigation Agency
15 Bowling Green Lane
London EC1R 0BD
Tel: 0171 490 7040
Fax: 0171 490 0436
email: eiauk@gn.apc.org
Extensive and exclusive library of video and stills showing the exploitation of wildlife and the environment worldwide. Subjects include dolphin and whale slaughter, the bird trade, bear farms, animal products illegally on sale in shops

and to undercover investigators, and other aspects of endangered species trade. All film sales help to fund future investigations and campaigns

Film Consultancy Services
19 Penfields House
Market Road
London N7 9PZ
Tel: 0171 700 7055
Fax: 0171 700 7055
Distribute the Associated-Rediffusion Television Library

Film Images
4th Floor
184-192 Drummond Street
London NW1 3HP
Tel: 0171 383 2288
Fax: 0171 383 2333
email: filmimages@compuserve.com.
Angela Saward
Thousands of hours of classic and contemporary film images from hundreds of different sources around the world. All fully catalogued and immediately available for viewing on VHS or U-Matic. Suppliers include Central Office of Information and Overseas Film and Television

Film Research & Production Services Ltd
Suite 211-213, Mitre Street
177-183 Regent House
London W1R 8LA
Tel: 0171 734 1525
Fax: 0171 734 8017
Amanda Dunne, James Webb
Film research and copyright clearance facilities, also third party clearance. Film holding of space footage

GB Associates
80 Montalt Road
Woodford Green
Essex IG8 9SS
Tel: 0181 505 1850
Fax: 0181 504 6340
email: filmview@aol.com
Malcolm Billingsley
An extensive collection, mainly on 35mm, of fact and fiction film from the turn of the century. The collection is particularly strong in vintage trailers, the early sound era, early colour systems and adverts

Fred Goodland Film, Video & Record Collections
81 Farmilo Road
Leyton
London E17 8JN
Tel: 0181 539 4412
Fax: 0181 539 4412
Fred Goodland MBKS

Actuality and entertainment material from the 1890s through to the 1990s. Specialist collections include the early sound period and a wide range of musical material. Extensive film research facilities available on tape

Ronald Grant Archive
The Cinema Museum
The Master's House
The Old Lambeth Workhouse
2 Dugard Way (off Renfrew Road, Kennington)
London SE11 4TH
Tel: 0171 840 2200
Fax: 0171 840 2299
email: martin@cinemamuseum.org.uk
Martin Humphries
15 million feet of fact and fiction film, mainly 35mm, from 1896 on. Also 700,000 film stills, posters, programmes, scripts and information. The museum is a FIAF subscriber

Huntley Film Archives
78 Mildmay Park
Newington Green
London N1 4PR
Tel: 0171 923 0990
Fax: 0171 241 4929
Amanda Huntley, John Huntley
Archive film library for broadcast, corporate and educational purposes, specialising in documentary footage 1900-1980. Phone to make an appointment or write for brochure detailing holdings. Now also 50,000 stills from films and film history

Index Stock Shots
12 Charlotte Mews
London W1P 1LN
Tel: 0171 631 0134
Fax: 0171 436 8737
email: index@msn.com
Philip Hinds
Unique stock footage on 35mm film and tape. Including extremes of nature and world climate, time-lapse and aerial photography, cities, landmarks, aviation, wildlife

ITN Television News Archive
200 Gray's Inn Road
London WC1X 8XZ
Tel: 0171 430 4480
Fax: 0171 430 4453
John Flewin
Worldwide TV news coverage on film and video tape, from 1955 to the present day. Complete library archive on site including multi-format transfer suite. Also an on-line stills from video service, available via ISDN. A newspaper cuttings reference library with cuttings back to 1955 is available

London Electronic Arts
2-4 Hoxton Square
London N1 6NU
Tel: 0171 284 4588
Fax: 0171 267 6078
email: lea@easynet.co.uk
Britain's national centre for video and new media art, housing the most extensive collection of video art in the country. Artists' work dating from the 1970s to the present

London Film Archive
c/o 78 Mildmay Park
Newington Green
London N1 4PR
Tel: 0171 923 4074
Fax: 0171 241 4929
Dedicated to the acquisition and preservation of film relating to the Greater London region. The collection consists of material from 1895 to the present day and represents professional and amateur produced features and documentary films

London Weekend TV Images
London Television Centre
Upper Ground
London SE1 9LT
Tel: 0171 261 3690/3771
Fax: 0171 261 3456
email: images@lwt.co.uk
Julie Lewis
Clips and stockshots available from London Weekend Television's vast programme library, dating from 1968. Drama, entertainment, music, arts and international current affairs. Plus London's news, housing, transport, politics, history, wildlife etc

Medi Scene
32-38 Osnaburgh Street
London NW1 3ND
Tel: 0171 387 3606
Fax: 0171 387 9693
Aurora Salvador-Bennett
Wide range of accurately catalogued medical and scientific shots available on film and video. Part of the Medi Cine Group

Moving Image
The Basement
2-4 Dean Street
London W1V 5RN
Tel: 0171 437 5688
Fax: 0171 437 5649
Michael Maloney
Over 11,000 hours of contemporary and archive images. Collections include British Tourist Authority, British Airways, Vintage Slapstick, Leisure Sports, TVAM 1983-1992, Travelogue Classics, Medical Technology, Subaqua Films, Space Exploration, World Destinations, The Cuban Archive. Showreels and information packs

Nova Film and Video Library
11a Winholme
Armthorpe
Doncaster DN3 3AF
Tel: 01302 833422
Fax: 01302 833422
email: index@msn.com
An extensive collection of unique archive material. The library holds a huge selection of amateur cine film documenting the changing social life of Britain, dating back to 1944 and has a dedicated collection of transport footage, from 1949 to the present day. The library also holds a selection of specially shot footage and interviews. A catalogue and showreel is available

The Olympic Television Archive Bureau
TWI House
23 Eyot Gardens
London W6 9TR
Tel: 0181 233 5353
Fax: 0181 233 5354
David Williams
The International Olympic Committee owns a unique collection of film and television material covering the entire history of the Olympic Games from 1896 to 1994. Now it can be accessed via the Olympic Television Archive Bureau, which is administered by Trans World International

Oxford Scientific Films
Long Hanborough
Oxford OX8 8LL
Tel: 01993 881881
Fax: 01993 882808 or 01993 883969
Jane Mulleneux, Rachel Roberts, Sandra Berry
Stock footage on 16mm, 35mm film and video. Wide range of wildlife, special fx, timelapse, slow motion, scenics, agriculture, traffic, macro, micro etc. Catalogue and showreel available. Extensive stills library

Pearson Television International Ltd
1 Stephen Street
London W1P 1PJ
Tel: 0171 691 6732/6733
Fax: 0171 691 6080
Len Whitcher
Over 20,000 films and videotapes of a wide range of TV programmes including all Grundy, SelecTV (Alomo and Witzend) and ACI programming

PolyGram Television International
76 Oxford House
London W1N OHQ
Tel: 0171 307 7500
Fax: 0171 307 7501

Brendan Kelly
Includes the entire output from Associated Television 1955-1982. Also contains feature films, TV-movies and mini-series produced by ITC Entertainment

Post Office Film and Video Library
PO Box 145, Sittingbourne
Kent ME10 1NH
Tel: 01795 426465
Fax: 01795 474871
email: poful@edist.co.uk
Barry Wiles, Linda Gates
Holds a representative selection of documentary programmes made under the GPO Film Unit, including the classic Night Mail. Catalogue available

Reuters Television Library
40 Cumberland Avenue
London NW10 7EH
Tel: 0171 510 6444
Fax: 0171 510 8568
Pam Turner
Newsreel, TV news, special collections. Colour and b/w, 16mm, 35mm, 1896 to present day and all material pre-1951 and post-July 1981 on 1" video

RSPB Film and Video Unit
The Lodge
Sandy
Bedfordshire SG19 2LN
Tel: 01767 680551
Fax: 01767 692365
Colin Skevington: Head of Film & Video
Lesley Norman: Film Library Sales:
Over one million feet of 16mm film covering a wide variety of wildlife subjects and their habitats, particularly European birds

Sky News Library Sales
British Sky Broadcasting Ltd
Grant Way
Isleworth
Middlesex TW7 5QD
Tel: 0171 705 2872/3132
Fax: 0171 705 3201
Sue Stewardson, Nic Peters
Extensive round the clock news and current affairs coverage since 1989. Entire library held on Beta SP on site. Library operates 24 hours a day

The Sports Video Library
Trans World International
5th Floor Axis Centre Building
3 Burlington Lane
Chiswick
London W4 2TH
Tel: 0181 233 5500/5300
Fax: 0181 233 5301
Rita Costantinou
Includes golf, tennis, World Cup

rugby, America's Cup, Test cricket, skating, snooker, gymnastics, yachting, motorsport, adventure sport, many minor and ethnic sports plus expanding catalogue of worldwide stockshots

TCB Releasing
Stone House, Rudge
Frome
Somerset BA11 2QQ
Tel: 01373 830769
Fax: 01373 831028
Angus Trowbridge
Sales of jazz and blues music programmes to broadcast television and the home-video media

World Backgrounds Film Production Library
Imperial Studios
Maxwell Road
Borehamwood, Herts
Tel: 0181 207 4747
Fax: 0181 207 4276
Ralph Rogers
Locations around the world. Fully computerised. All 35mm including 3,000 back projection process plates. Numerous video masters held. Suppliers to TV commercials, features, pop promos, TV series, corporate videos etc

Worldwide Television News (WTN)
The Interchange
Oval Road
Camden Lock
London NW1 7DZ
Tel: 0171 410 5353
Fax: 0171 413 8327
email: wtnlib@abc.com
Website: http://www.wtnlibrary.com
David Simmons
Newsfilm and video from 1896 - and adding every day up to 100 new items from around the world. Hard news, stock footage, features, personalities, annual compilations, background packages etc. Story details and shotlists are stored on full-text easy-to-search database. Database also available on CD-Rom and online (see website address above)

PHOTOGRAPHIC LIBRARIES

BBC Photograph Library
B116 Television Centre
Wood Lane
London W12 7RJ
Tel: 0181 225 7193
Fax: 0181 746 0353
The BBC's unique archive of radio and television programme stills, equipment, premises, news and personalities dating from 1922. B/w and colour. Visits by appointment

BFI Stills, Posters and Designs
21 Stephen Street
London W1P 2LN
Tel: 0171 255 1444
Fax: 0171 323 9260
A visual resource of around seven million images, illustrating every aspect of the development of world cinema and television **(see p8)**

The Bridgeman Art Library
17-19 Garway Road
London W2 4PH
Tel: 0171 727 4065
Fax: 0171 792 8509
email: info@bridgeman.co.uk
Website: http//www.bridgeman.co.uk
From cave paintings to contemporary design The Bridgeman Art Library provides a one stop-source for the world's art. Representing over 750 collections and contemporary artists worldwide, the library offers large format colour transparencies, a picture research service, CD-Rom catalogues, image and copyright databases

Central Office of Information Footage File
4th Floor
184-192 Drummond Street
London NW1 3HP
Tel: 0171 383 2292
Fax: 0171 383 2333
email: filmimages.compuserve.com.
Tony Dykes
40,000 Crown Copyright titles from the Government's News and Information archives spanning over 75 years of British social and business history. Most of the collection has been thoroughly shot listed and is available on VHS viewing cassettes

The Cinema Museum
The Master's House
The Old Lambeth Workhouse
2 Dugard Way (off Renfrew Road,
Kennington)
London SE11 4TH
Tel: 0171 840 2200
Fax: 0171 840 2299
email: martin@cinemamuseum.org.uk
Martin Humphries
(See Ronald Grant Archive)

Hulton Getty Picture Collection
21-31 Woodfield Road
London W9 2BAD
Tel: 0171 266 2660
Fax: 0171 266 2414
One of the world's largest stills archives with over 15 million photographs, prints and engravings covering the entire history of photojournalism

The Image Bank
17 Conway Street
London W1
Tel: 0171 312 0300
Fax: 0171 391 9111

image.net
18 Vine Hill
London EC1
Tel: 0541 522 333
Fax: 0171 729 5098
Simon Townsley

Image Diggers Picture and Tape Library
618b Finchley Road
London NW11 7RR
Tel: 0181 455 4564
Fax: 0181 455 4564
35mm slides, stills, postcards, sheet music, magazine and book material for hire. Cinema, theatre and literature clippings archive. Audio/visual tape resources in performing arts and other areas, plus theme research

Imperial War Museum
Photograph Archive
All Saints Annexe
Austral Street
London SE11 4SL
Tel: 0171 416 5333/8
Fax: 0171 416 5355
Within an overall photographic collection of some 6 million images, the Department has several thousand film stills, primarily from material in the Museum's own film archive

Institute of Contemporary History & Wiener
Library Limited
4 Devonshire Street
London W1N 2BH
Tel: 0171 636 7247
Fax: 0171 436 6428
email: lib@wl.u-net.com
Rosemarie Nief: Head Librarian
Ben Barkow: Photo Archive,
Christine Patel: Video Collection

The Wiener Library is a private research library and institute specialising in contemporary European and Jewish history, especially the rise and fall of the Third Reich, Nazism and fascist movements, anit-Semitism, racism, the Middle East and post-war Germany. It holds Britain's largest collection of documents, testimonies, books and videos on the Holocaust. The photographic archive contains stills, postcards, posters and portraits, illustrated books, approx. 2,000 videos and recordings

Kobal Collection
4th Floor, 184 Drummond Street
London NW1 3HP
Tel: 0171 383 0011
Fax: 0171 383 0044
David Kent
One of the world's leading film photo archives in private ownership. Film stills and portraits, lobby cards and posters, from the earliest days of the cinema to modern times

Mckenzie Heritage Picture Archive
90 Ardgowan Road
London SE6 1UU
Tel: 0181 697 0147
Fax: 0181 697 0147
email: MkHeritage@aol.com
The Mckenzie Heritage Picture Archive specialises in pictures of black communities from Britain and abroad. The images span the 19th and 20th centuries

Retrograph Nostalgia Archive Ltd
164 Kensington Park Road
London W11 2ER
Tel: 0171 727 9378
Fax: 0171 229 3395
email: mbreese999@aol.com
Jilliana Ranicar-Breese
Hiring out of transparencies or colour laser prints from original labels, packaging, advertising, posters, prints. Commercial and fine art material from 1860-1960. Supply to publishers, film and television companies, record and CD companies and gift manufacturers. Visual research service and photography service. Most subjects available but travel, food and drink a speciality

Undercurrents Productions
16b Cherwell Street
Oxford OX4 1BG
Tel: 01865 203 663
Fax: 01865 243 562
email: underc@gn.apc.org
Website: http://www.undercurrents.org
Roddy Mansfield

'Undercurrents' (formerly known as Small World Media) is a quarterly VHS magazine of micro-documentaries of social and environmental issues. Filmed on domestic camcorders the films offer insights into the struggles for social justice

This section features some of the principal festival prizes and awards
from January 1997 to June 1998

1997

BAFTA CRAFT AWARDS
Awarded in London 21st April 1997 at the London Hilton

AWARDS IN THE GIFT OF COUNCIL
Fellowship Award: Steve Bochco
Michael Balcon Award for Outstanding British Contribution to Cinema: Channel Four
Alan Clarke Award for Outstanding Creative Contribution to Television: Michael Wearing
FILM AWARDS
Best Cinematography: John Seale for THE ENGLISH PATIENT (USA) Dir Anthony Minghella
Best Production Design: Tony Burrough for RICHARD III (UK) Dir Richard Loncraine
Best Costume Design: Shuna Harwood for RICHARD III (UK) Dir Richard Loncraine
Best Editing: Walter Murch for THE ENGLISH PATIENT (USA) Dir Anthony Minghella
Best Film Music: Gabriel Yared for THE ENGLISH PATIENT (USA) Dir Anthony Minghella
Best Sound: Toivo Lember for SHINE (Australia/UK) Dir Scott Hicks
Best Achievement in Special Effects: Stefen Fangmeier/Industrial Light and Magic for TWISTER (USA) Dir Jan de Bont
Best Make-up: Rick Baker and Geri. B. Oppenheim for THE NUTTY PROFESSOR (USA) Dir Tom Shadyac
Best Hair: Deborah Ann Piper for THE NUTTY PROFESSOR (USA) Dir Tom Shadyac
TELEVISION AWARDS
Best Original Television Music: Jim Parker for MOLL FLANDERS: PART ONE (Granada Television)
Best Make-up/Hair: Jean Speak for THE TENANT OF WILDFELL HALL (BBC)
Best Photography and Lighting (Fiction/Entertainment): Nic Knowland for THE FINAL PASSAGE (Channel Four)
Best Costume Design: Shirley Russell for GULLIVER'S TRAVELS (Channel Four)
Best Sound (Fiction/Entertainment): Phil Smith for HILLSBOROUGH (Granada Television)
Best Editing (Fiction/Entertainment): Barrie Vince for HILLSBOROUGH (Granada Television)
Best Design: Roger Hall for GULLIVER'S TRAVELS (Channel Four)

BAFTA PRODUCTION AND PERFORMANCE AWARDS
Awarded in London 29th April 1997 at the Royal Albert Hall

Fellowship of the Academy Award: Woody Allen, Julie Christie
Alexander Korda Award for the Outstanding British Film of the Year: SECRETS AND LIES (UK/France) Dir Mike Leigh
Dennis Potter Award: Peter Flannery
FILM AWARDS
Best Film: THE ENGLISH PATIENT (USA) Dir Anthony Minghella
David Lean Award for the Best Achievement in Direction: Joel Coen for FARGO (USA)
Best Film Not in the English Language: RIDICULE (France) Dir Patrice Leconte
Best Screenplay (Original): Mike Leigh for SECRETS AND LIES (UK/France) Dir Mike Leigh
Best Screenplay (Adapted): Anthony Minghella for THE ENGLISH PATIENT (USA) Dir Anthony Minghella
Best Actress: Brenda Blethyn for SECRETS AND LIES (UK/France)
Best Actor: Geoffrey Rush for SHINE (Australia/UK) Dir Scott Hicks
Best Supporting Actress: Juliette Binoche for THE ENGLISH PATIENT (USA) Dir Anthony Minghella
Best Supporting Actor: Paul Schofield for THE CRUCIBLE (USA) Dir Nicholas Hytner
TELEVISION AWARDS
Best Single Drama: HILLSBOROUGH (Granada for ITV)
Best Drama Series: EASTENDERS (BBC1)
Best Drama Serial: Charles Pattinson and Peter Flannery for OUR FRIENDS IN THE NORTH (BBC2)
Best Factual Series: THE HOUSE, (Double Exposure for BBC2)
Best Light Entertainment (Programme or Series): SHOOTING STARS (BBC2)
Best Comedy (Programme or Series): Gareth Gwenlan Tony Dow for ONLY FOOLS AND HORSES (Christmas Special) (BBC1)
Best News Coverage: NEWSNIGHT (BSE coverage);
Best Sports/Events Coverage in Real Time: Euro '96 (BBC1)

Best Talk Show: MRS MERTON CHRISTMAS SHOW (BBC1)
Best Actress: Gina McKee for OUR FRIENDS IN THE NORTH (BBC2)
Best Actor: Nigel Hawthorne for THE FRAGILE HEART (Carnival Films, C4)
Best Light Entertainment Performance: John Bird and John Fortune RORY BREMNER...WHO ELSE? (Vera Productions, C4)
Best Comedy Performance: David Jason for ONLY FOOLS AND HORSES (BBC1)
Huw Wheldon Award for Best Arts Programme or Series: LEAVING HOME (RHYTHM) (LWT, C4)
Best Children's Programme (Fiction/Entertainment): SHAKESPEARE SHORTS: ROMEO AND JULIET (BBC)
Flaherty Documentary Award: FERMAT'S LAST THEOREM (HORIZON), BBC2
Richard Dimbleby Award for Best Factual Television: Robert Hughes
Lew Grade Special Award for a Significant and Popular Television Programme: CORONATION STREET (Granada)
Best Foreign Television Programme: MURDER ONE, 20th Century Fox, BBC2

BFI FELLOWSHIPS
Awarded at TV97: A Festival of the Small Screen 6th-9th February 1997
Michael Parkinson, Verity Lambert, Lynda La Plante, Alan Yentob
Awarded at the NFT at the end of May 1997
Sir David Puttnam

47th BERLIN FESTIVAL
Held in Berlin 13th-24th February 1997

Golden Bear: THE PEOPLE VS. LARRY FLYNT (USA) Dir Milos Foreman
Silver Bear Special Jury Prize: HE LIU (Taiwan) Dir Ming-liang Tsai
Silver Bear (Best Director): Eric Heumann for PORT DJEMA (France/Italy/Greece)
Silver Bear (Best Actor): Leonardo DiCaprio for WILLIAM SHAKESPEARE'S ROMEO + JULIET (USA/Canada) Dir Baz Luhrmann
Silver Bear (Best Actress): Juliette Binoche for THE ENGLISH PATIENT (USA) Dir Anthony Minghella
Silver Bear for an Outstanding Single Achievement: Zbingniew Preisner (composer) for ØEN I FUGLEGADEN (Denmark/UK/Germany) Dir Søren Kragh-Jacobsen
Silver Bear for His Lifetime Contribution to the Art of Cinema: Raúl Ruiz (on the occasion of the screening of his film GÉNÉALOGIES D'UN CRIME (France) Dir Raúl Ruiz
ECUMENICAL JURY PRIZES
Best Film (In Competition): LENEGED ENAYIM MAARAVIOT (Israel) Dir Joseph Pitchhadze
Best Film (International Forum): NOBODY'S BUSINESS (USA) Dir Alan Berliner
Best Film (Panorama): BRASSED OFF (UK) Dir Mark Herman
Special Prize: MUTTER UND SOHN (Germany/Russia) Dir Aleksander Sokurov
Special Jury Award for the International Forum of New Cinema: THEY TEACH US HOW TO BE HAPPY (Switzerland) Dir Peter von Gunten
FIPRESCI (INTERNATIONAL CRITICS') PRIZES
Best Film (In Competition): HE LIU (Taiwan) Dir Ming-liang Tsai
Best Film (International Forum of Young Cinema): NOBODY'S BUSINESS (USA) Dir Alan Berliner
Best Film (11th International Panorama): SREDA (Russia/Germany/Brazil/Finland) Dir Victor Kossakovsky
UNICEF PRIZES
Best Children's Film: KAN DU VISSLA, JOHANNA (Sweden) Dir Rumle Hammerich
[Special Mention]: KRATKA - Dir Pawe Oziski
WO JE YOU BABA Dir Huang Shuquinn
Best Children's Short Film: DINNER FOR TWO (Canada) Dir Janet Perlman
BERLIN CHILDREN'S JURY PRIZES
First Prize: DER FLUG DES ALBATROS (New Zealand/Germany) Dir Werner Meyer
[Special Mention]: THE WHOLE OF THE MOON (New Zealand/Canada) Dir Ian Mune
Best Short Film: LA GRANDE MIGRATION (France) Dir Iouri Tcherenkov
[Special Mention]: OZONFISK (Norway) Dir Ingebjørg Torgesen
BLAUE ENGEL (BLUE ANGEL) PRIZE OF THE EUROPEAN ACADEMY OF FILM AND TELEVISION (SPONSORED BY EASTMAN KODAK):
LOS SECRETOS DEL CORAZÓN (Spain/France/Portugal) Dir Montxo Barrios Armendariz
FIPRESCI (INTERNATIONAL CRITICS') PRIZES:
Alfred Bauer Prize: WILLIAM SHAKESPEARE'S ROMEO + JULIET (USA/Canada) Dir Baz Luhrmann
GUILD OF GERMAN ART HOUSE CINEMAS PRIZE: GET ON THE BUS (USA) Dir Spike Lee
DAS LEBEN IST EINE BAUSTELLE (Germany) Dir Wolfgang Becker
'BERLINER MORGENPOST' NEWSPAPER READERS' JURY PRIZE: THE PEOPLE VS. LARRY FLYNT (USA) Dir Milos Foreman
[Second Public Favourite]: THE ENGLISH PATIENT (USA) Dir Anthony Minghella
[Third Public Favourite]: WILLIAM SHAKESPEARE'S ROMEO + JULIET (USA/Canada) Dir Baz Luhrmann
'BERLINER ZEITUNG' NEWSPAPER READERS' JURY PRIZE: FOCUS (Japan) Dir Satoshi Isaka

WOLFGANG STAUDTE PRIZE: SERTÃO DAS MEMORIAS (Brazil) Dir José Arrújo
[Special Mention]: PICADO FINO (Argentina) Dir Esteban Sapir
C.I.C.A.E. (INTERNATIONAL CONFEDERATION OF ART CINEMAS) PRIZE: MUTTER UND SOHN (Germany/Russia) Dir Aleksander Sokurov
NETPAC (NETWORK FOR THE PROMOTION OF ASIAN CINEMA) AWARD [Joint]: FOCUS (Japan) Dir Satoshi Isaka
GAY TEDDY BEARS (GAY AND LESBIAN FILM AWARDS)
Best Feature: ALL OVER ME (USA) Dir Alexandra Sichel
Best Documentary: HELDINNEN DER LIEBE (Germany) Dir Natualie Percillier and Lily Beilly
Best Essay Film: MURDER AND MURDER (USA) Dir Yvonne Rainer

BROADCASTING PRESS GUILD AWARDS FOR 1996
Awarded on 18th April 1997

Best Single Drama: HILLSBOROUGH (Granada for ITV)
Best Drama Series/Serial: OUR FRIENDS IN THE NORTH (BBC2)
Best Documentary Series: THE HOUSE (Double Exposure for BBC2)
Best Single Documentary: QUALITY TIME (Mosaic Pictures for Modern Times, BBC2)
Best Entertainment: TFI FRIDAY (Ginger Television Productions for C4)
Best Actor: Chris Eccleston for OUR FRIENDS IN THE NORTH (BBC2)
Best Actress: Gina McKee for OUR FRIENDS IN THE NORTH (BBC2)
Best Performer (Non-Acting): Desmond Lynam (BBC1)
TV Journalist of the Year: Charles Wheeler (WHEELER ON AMERICA for BBC2)
Writer's Award: Peter Flannery (OUR FRIENDS IN THE NORTH for BBC2)
Harvey Lee Award for Outstanding Contribution to Broadcasting: Michael Grade (Chief Executive, C4)

50th CANNES FESTIVAL
Held 7th-19th May 1997

Palme d'Or [Joint]: TA'M-E-GUILASS (Iran) Dir Abbas Kiarostami
UNAGI (Japan) Dir by Shohei Imamura
Grand Prix: THE SWEET HEREAFTER (Canada) Dir Atom Egoyan
Prix du 50emme Anniversaire: Youssef Chahine for his life work
Best Female Performance: Kathy Burke for NIL BY MOUTH (UK) Dir Gary Oldman
Best Male Performance: Sean Penn for SHE'S SO LOVELY (USA) Dir Nick Cassavetes
Best Director: Wong Kar-Wai for HAPPY TOGETHER (Hong Kong)
Best Script: James Schamus for THE ICE STORM (USA) Dir Ang Lee
Jury Award: WESTERN (France) Dir Manuel Poirier
Grand Prix Technique de la Commission Superieure Technique de L'Image et du Son: Thierry Arbogast (director of photography) for THE FIFTH ELEMENT (France) Dir Luc Besson and SHE'S SO LOVELY (USA) Dir Nick Cassavetes
Camera d'Or for First Film: SUZAKU (Japan) Dir Naomi Kawase
Special Mention Camera d'Or: Bruno Dumont for LA VIE DE JESUS (France)
Best Short Film: IS IT THE DESIGN OR THE WRAPPER (UK) Dir Tessa Sheridan
Jury Prize for Short Film [Joint]: LEONIE (Belgium) Dir Lieven Debrauwer
LES VACANCES (France) Dir Emmanuelle Bercot

22nd CÉSARS
Awarded in Paris 8th February 1997

Best Actor: Philippe Torréton for CAPITAINE CONAN (France) Dir Bertrand Tavernier
Best Actress: Fanny Ardant for PEDALE DOUCE (France) Dir Gabriel Achion
Best Supporting Actor: Jean-Pierre Darroussin for UN AIR DE FAMILLE (France) Dir Cédric Klapisch
Best Supporting Actress: Catherine Frot for UN AIR DE FAMILLE (France) Dir Cédric Klapisch
Most Promising Young Actor: Mathieu Amalric for COMMENT JE ME SUIS DISPUTÉ...(MA VIE SEXUELLE) (France) Dir Arnaud Desplechin
Most Promising Young Actress: Laurence Cote for LES VOLEURS (France) Dir André Techiné
Best Director [Joint]: Bertrand Tavernier for CAPITAINE CONAN (France)
Patrice Leconte for RIDICULE (France)
Best Producer: Jacques Perrin for MICROCOSMOS: LE PEUPLE DE L'HERBE (France/Switzerland/Italy) Dir Claude Nuridsany
Best French Film: RIDICULE (France) Dir Patrice Leconte
Best Foreign Film: BREAKING THE WAVES (Denmark/Sweden/France/Netherlands/Norway) Dir Lars von Trier
Best Original/Adapted Script: Agnés Jaoui, Jean-Pierre Bacri and Cédric Klapicsh for UN AIR DE FAMILLE (France) Dir Cédric Klapisch
Best Music: Bruno Coulais for MICROCOSMOS: LE PEUPLE DE L'HERBE (France/Switerzland/Italy) Dir Claude Nuridsany
Best First Feature: Y AURA-T-IL DE LA NEIGE A NOEL? (France) Dir Sandrine Veysset

Best Short Feature: MADAME JACQUES SUR LA CROISETTE (France) Dir Emmanuel Kinkiel
Best Cinematography: Claude Nuridsany, Marie Pérennou, Thierry Macado, Hugues Ryffel for MICROCOSMOS: LE PEUPLE DE L'HERBE (France/Switzerland/Italy) Dir Claude Nuridsany
Best Costumes: CHRISTIAN GASC for RIDICULE (France) Dir Patrice Leconte
Best Set Design: Ivan Maussion for RIDICULE (France) Dir Patrice Leconte
Best Sound: Philippe Barbeau, Bernard Leroux MICROCOSMOS: LE PEUPLE DE L'HERBE (France/Switerzland/Italy) Dir Claude Nuridsany
Best Editing: MICROCOSMOS: LE PEUPLE DE L'HERBE (France/Switzerland/Italy) Dir Claude Nuridsany

EMMY AWARDS - 25th INTERNATIONAL EMMY AWARDS
Awarded in New York 24th November 1997

Best Drama: CROSSING THE FLOOR (UK - A Hat Trick Production for BBC2)
Best Documentary: GERRIE AND LOUISE (Canada - Blackstock Pictures Inc. and Eurasia Motion Pictures Inc. in Association with the Canadian Broadcasting Corporation)
Best Arts Documentary: DANCING FOR DOLLARS: THE BOLSHOI IN VEGAS (UK - NVC Arts for Channel 4)
Best Performing Arts: ENTER ACHILLES (UK - A DV8 Films Production for the BBC and RM Arts)
Best Popular Arts: LIBERG ZAPPT (the Netherlands - NOS/TROS - Ivo Niehe Productions)
Best Children and Young People Category: WISE UP, series 4, episode one (UK - A Carlton Production for Channel 4)

EMMY AWARDS - NATIONAL ACADEMY FOR TELEVISION ARTS AND SCIENCES (49th PRIMETIME AWARDS)
Awarded in Pasedena 14th September 1997

Outstanding Directing for a Comedy Series: FRASIER - TO KILL A TALKING BIRD (Grub Street Productions in association with Paramount)
Outstanding Directing for a Drama Series: NYPD BLUE - WHERE'S SWALDO? (Steven Bochco Productions)
Outstanding Directing for a Variety or Music Program: CENTENNIAL OLYMPIC GAMES: OPENING CEREMONIES (Don Mischer Productions)
Outstanding Directing for a MIniseries or a Special: THE ODYSSEY, PART I & II (Hallmark Entertainment in association with American Zoetrope)
Outstanding Lead Actor in a Comedy Series: John Lithgow for 3RD ROCK FROM THE SUN (Carsey-Werner Productions, LLC)
Outstanding Lead Actor in a Drama Series: Dennis Franz for NYPD BLUE (Steven Bochco Productions)
Outstanding Lead Actor in a Miniseries or Special: Armand Assante for GOTTI (A Gary Lucchesi Production in association with HBO Pictures)
Outstanding Lead Actress in a Comedy Series: Helen Hunt for MAD ABOUT YOU (Infront Productions and Nuance Productions in association with Tristar Television)
Outstanding Lead Actress in a Drama Series: Gillian Anderson for THE X-FILES (Ten Thirteen Productions in association with 20th Century Fox Television)
Outstanding Lead Actress in a Miniseries or a Special: Alfre Woodard for MISS EVERS' BOYS (An Anasazi Production in association with HBO NYC)
Outstanding Supporting Actor in a Comedy Series: Michael Richards for SEINFELD (Castle Rock Entertainment)
Outstanding Supporting Actor in a Drama Series: Hector Elizondo for CHICAGO HOPE (David E. Kelley Productions in association with 20th Century Fox)
Outstanding Supporting Actor in a Miniseries or a Special: Beau Bridges for THE SECOND CIVIL WAR (Baltimore Pictures)
Outstanding Supporting Actress in a Comedy Series: Kristen Johnston for 3rd Rock From The Sun (Carsey-Werner Productions, LLC)
Outstanding Supporting Actress in a Drama Series: Kim Delaney for NYPD Blue (Steve Bochco Productions)
Outstanding Supporting Actress in a Miniseries or Special: Diana Rigg for REBECCA (Portman Productions for Carlton co-produced with WGDII/Booton)
Outstanding Performance in a Variety or Music Program: Bette Midler for BETTE MIDLER: DIVA LA VEGAS (A Miss M Production in association with Cream Cheese Films and HBO Original Planning)
The Presidents Award: MISS EVERS' BOYS (An Anasazi Production in association with HBO NYC)
Outstanding Comedy Series: FRASIER (Grub Street Productions in association with Paramount)
Outstanding Drama Series: LAW & ORDER (Wolf Films in association with Universal Television)
Outstanding Miniseries: PRIME SUSPECT 5: ERRORS OF JUDGEMENT (Granada Television in co-production with WGBH/Boston)
Outstanding Made for Television Movie: MISS EVERS' BOYS (Anasazi Productions in association with HBO NYC)
Outstanding Variety, Music or Comedy Series: TRACEY TAKES ON (Takes on Productions Inc)
Outstanding Variety, Music or Comedy Special: Chris Rock for BRING THE PAIN (Productions Partners Inc. in association with HBO Original Programming)
Outstanding Writing For a Comedy Series: ELLEN - THE PUPPY EPISODE (Black/Marlens Co. in association with

Touchstone Television)
Outstanding Writing for a Drama Series: NYPD BLUE - WHERE'S SWALDO? (Steven Bochco Productions)
Outstanding Writing for a Variety or Music Program: Chris Rock for BRING THE PAIN (Production Partners Inc. in association with HBO Original Programming)
Outstanding Writing for a Miniseries or a Special: William Faulkner's OLD MAN (Hallmark Hall Of Fame Presentation)

10th EUROPEAN FILM AWARDS
12th December 1997, Berlin
6 Segitzdam 2
10969 Berlin,
Tel: (49) 30 615 3091

Best Actor: Javier Bardem for PERDITA DURANGO (Spain/Mexico) Dir Alejandro de la Iglesia
Best Actress: Jodie Foster for CONTACT (US) Dir Robert Zemeckis
Best Cinematographer: John Seale for THE ENGLISH PATIENT (US) Dir Anthony Minghella
European Actor of the Year: Bob Hoskins for 24 7 (UK) Dir Shane Meadows
European Actress of the Year: Juliette Binoche for THE ENGLISH PATIENT (US) Dir Anthony Minghella
European Film of the Year: THE FULL MONTY (US/UK) Dir Peter Cattaneo
Best Film: THE FULL MONTY (US/UK) Dir Peter Cattaneo
Best Script: Alain Berliner and Chris Vander Stappen for MA VIE EN ROSE (France/Belgium/UK/ Switzerland) Dir Alain Berliner
Discovery of the Year: Bruno Dumont for LA VIE DE JÉSUS (France) DirBruno Dumont
European Achievement in World Cinema: THE PEOPLE VS. LARRY FLYNT (US) Dir Milos Forman
FIPRESCI Critics Prize: VIAEM AO PRINCIPIO DO MUNDO (Portugal/France) Dir Manuel de Oliveira
Life Achievement Award: Jeanne Moreau
Screen International Award: Takeshi Kitano for HANA-BI

EVENING STANDARD FILM AWARDS 1997
Awarded in London 2nd February 1997

Best Film: RICHARD III (UK) Dir Richard Loncraine
Best Actor: Liam Neeson for MICHAEL COLLINS (USA)
Best Actress: Kate Winslet for SENSE AND SENSIBILITY (UK/USA) Dir Ang Lee
Best Screenplay [Joint]: Emma Thompson for SENSE AND SENSIBILITY (UK/USA) Dir Ang Lee
John Hodge for TRAINSPOTTING (UK) Dir Danny Boyle
Technical Achievement: Tony Burrough for RICHARD III (UK) Dir Richard Loncraine
Most Promising Newcomer: Emily Watson for BREAKING THE WAVES (Denmark/Sweden/France/Netherlands/ Norway) Dir Lars von Trier
Peter Sellers Award for Comedy: Mark Herman for BRASSED OFF (UK/USA) Dir Mark Herman
Special Award for his Contribution to British Filmmaking: Leslie Phillips

54th GOLDEN GLOBE AWARDS
Awarded 20th January 1997

FILM
Best Drama: THE ENGLISH PATIENT (USA) Dir Anthony Minghella
Best Comedy/Musical: EVITA (USA) Dir Alan Parker
Best Foreign Language Film: KOLYA (Czech Republic/UK/France) Dir Jan Sverák
Best Director: Milos Forman for THE PEOPLE VS. LARRY FLYNT (USA)
Best Actor (Drama): Geoffrey Rush for SHINE (Australia/UK) Dir Scott Hicks
Best Actor (Comedy/Musical): Tom Cruise for JERRY MAGUIRE (USA) Dir Cameron Crowe
Best Actress (Drama): Brenda Blethyn for SECRETS AND LIES (UK/USA) Dir Mike Leigh
Best Actress (Comedy/Musical): MADONNA for EVITA (USA) Dir Alan Parker
Best Supporting Actor: Edward Norton for PRIMAL FEAR (USA) Dir Gregory Hoblit
Best Supporting Actress: Lauren Bacall for THE MIRROR HAS TWO FACES (USA) Dir Barbara Streisand
Best Screenplay: Scott Alexander and Larry Karaszewski for THE PEOPLE VS. LARRY FLYNT (USA) Dir Milos Foreman
Best Original Score: Gabriel Yared for THE ENGLISH PATIENT (USA) Dir Anthony Minghella
Best Original Song: Alexander Scott and Larry Karaszewski for THE PEOPLE VS. LARRY FLYNT (USA) Dir Milos Forman
Cecil B. De Mille Lifetime Achievement Award: Dustin Hoffman
TELEVISION
DRAMA
Best Series: X FILES (USA) (Fox)
Best Actor: David Duchovny for X FILES (USA) (Fox)
Best Actress: Gillian Anderson for X FILES (USA) (Fox)
MUSICAL/COMEDY
Best Series: 3RD ROCK FROM THE SUN (USA) (NBC)

Best Actor: John Lithgow for 3RD ROCK FROM THE SUN (USA) (NBC)
Best Actress: Helen Hunt for MAD ABOUT YOU (USA) NBC
MINI-SERIES/TELEFILM
Best Program: RASPUTIN (Hungary/Canada) Dir Ulrich Edel
Best Actor: Alan Rickman for RASPUTIN
Supporting Actor: Ian McKellen for RASPUTIN
Best Actress: Helen Mirren for LOSING CHASE
Best Supporting Actress: Kathy Bates for THE LATE SHIFT

37th GOLDEN ROSE OF MONTREUX TV FESTIVAL
Held 24th-29th April 1997

Golden Rose: COLD FEET (UK) (Granada)
Silver Rose (Music): THE TONY FERRINO PHENONMENON (UK)(Pozzitive)
Silver Rose (Sitcom): THE LARRY SANDERS SHOW (USA) (HBO)
Silver Rose/Special Prize of the City of Montreux: COLD FEET (UK) (Granada)
Silver Rose (Game Show): WANTED (UK) (Hewland International)
Silver Rose (Variety): NATIONAL PARODY (Spain) (Gestmusic)
Bronze Rose (Music): JAEL (Switzerland) (TSR)
Bronze Rose (Sitcom): THE NEWSROOM (Canada) (CBC)
Bronze Rose (Comedy): THE JOYBOYS STORY (Finland) (MTV3)
Bronze Rose (Game Show): TALKING TELEPHONE NUMBERS (UK) (Celador)
Bronze Rose (Variety): FRIDAY NIGHT ARMISTICE (UK) (BBC)
Press Prize: THE WAITING ROOM (Germany) (Regina Ziegler)

32nd KARLOVY VARY INTERNATIONAL FILM FESTIVAL
Awarded 4th July 1997

Best Actor: Boleslav Plívka for ZAPOMENUTE SVETLO (Czech Republic) Dir Dir Vladimír Michálek
Best Actress: Lena Endre for JULORATORIET (Sweden/New Zealand) Kjell-Åha Andersson
Honourable Mention For Direction: Martine Dugowson for PORTRAITS CHINOIS (France/UK)
Best Documentary: NESPATRENE (Czech Republic) Dir Janek Miroslav
NOEL FIELD DER ERFUNDENE SPION (Switzerland/Germany) Dir Werner Schweizer
Ecumenical Jury Special Mention: NESPATRENE (Czech Republic) Dir Janek Miroslav
Ecumenical Jury Special Mention: JUGOFILM (Austria) Dir Goran Rebic
Ecumenical Jury Prize: ZAPOMENUTE SVETLO (Czech Republic) Dir Dir Vladimír Michálek
Fédération Internationale des Ciné-Clubs Prize: DJÖFLAEYJAN (Iceland/Norway/Germany/Denmark) Dir Fridrik Thor Fridriksson
Jury Special Mention: DESADANAM (India) Dir Jayaraaj
Public Award: PRIVATE PARTS (US) Dir Betty Thomas
Public Award: ZAPOMENUTE SVETLO (Czech Republic) Dir Dir Vladimír Michálek
Special Jury Award: LA BUENA VIDA (Spain) Dir David Trueba
Lifetime Achievement Award: Milos Forman

50th LOCARNO FESTIVAL
Awarded in Locarno, Switzerland 6th August 1997

Audience Award: The Full Monty (US/UK) Dir Peter Cattaneo
Best Actor: Valerio Mastandrea for Tutti Giù Per Terra (Italy) Dir Davide Ferrario
Best Actress: Rona Hartner for Gadjo Dilo (France) Dir Tony Gatlif
Ecumenical Jury Special Mention: Ramedan Suleman for Fools (France/Zambia/Mozambique/Zibabwe) Dir Ramedan Suleman
Ecumenical Jury Prize: Gadjo Dilo (France) Dir Tony Gatlif
FIPRESCI Award: Tutti Giù Per Terra (Italy) Dir Davide Ferrario
Golden Leopard: Ayneth (Iran) Dir Jafar Panahi
Silver Leopard: Gadjo Dilo (France) Dir Tony Gatlif
Silver Leopard: Fools (France/Zambia/Mozambique/Zimbabwe) Dir Ramedan Suleman
Youth Jury Prize: Gadjo Dilo (France) Dir Tony Gatlif
Jury C.I.C.A.E: Xianggang Zhizao (Hong Kong) Dir Fruit Chan
Prize of FICC/IFFS: Clanestins (Switzerland/Canada/France/Belgium) Dir Nicolas Wadimoff and Denis Chouinard

LONDON FILM CRITICS' CIRCLE AWARDS
Awarded 2nd March 1997

Film of the Year: FARGO (USA) Dir Joel Coen
Foreign Language Film of the Year: LES MISÉRABLES (France) Dir Claude Lelouch
Screenwriter of the Year: Joel and Ethan Coen for FARGO (USA) Dir Joel Coen

Actor of the Year: Morgan Freeman for SEVEN (USA) Dir David Fincher
Actress of the Year: Frances McDormund for FARGO (USA) Dir Joel Coen
British Film of the Year: SECRETS AND LIES (UK/USA) Dir Mike Leigh
British Producer of the Year: TRAINSPOTTING (UK) Dir Danny Boyle
British Director of the Year: Mike Leigh for SECRETS AND LIES (UK/USA)
British Actor of the Year [Joint]: Sir Ian McKellen for RICHARD III (USA)
Ewan McGregor for BRASSED OFF (UK/USA) Dir Mark Herman, THE PILLOW BOOK (Netherlands/France/UK/Luxembourg)
Dir Peter Greenaway, EMMA (UK/USA) Dir Douglas McGrath and TRAINSPOTTING (UK) Dir Danny Boyle
British Actress of the Year: Brenda Blethyn for SECRETS AND LIES (UK/USA) Dir Mike Leigh
Special Lifetime Achievement Award: Norman Wisdom, John Mills

37th MONTE CARLO TV FESTIVAL
Held in Monte Carlo 7th-13th Feb 1997
Since 1996 incorporates the IMAGINA Conference

FILMS FOR TV
Gold Nymph (Best Film): L'ORANGE DE NOEL (King Movies) (France)
Silver Nymph (Best Director): Jean-Louis Lorenzi for L'ORANGE DE NOEL (King Movies) (France)
Silver Nymph (Best Script): Stuart Urban for DEADLY VOYAGE (BBC2)
Silver Nymph (Best Actor): Omar Epps for DEADLY VOYAGE (BBC2)
Silver Nymph (Best Actress): Meredith Baxter for AFTER JIMMY (Worldvision Enterprises, USA)
Special Mention: CORRERE CONTRO (SACIS Spa, Italy)
MINI-SERIES
Gold Nymph (Best Mini-Series): CLEAN SHEET (Danmarks Radio/TV) (Denmark)
Silver Nymph (Best Director): Daniel Alfredson for CLEAN SHEET (Danmarks Radio/TV) (Denmark)
Silver Nymph (Best Script): John Brown for CLEAN SHEET (Danmarks Radio/TV) (Denmark)
Silver Nymph (Best Actor): Ole Ernst for CLEAN SHEET (Danmarks Radio/TV) (Denmark)
Silver Nymph (Best Actress): Alex Kingston for MOLL FLANDERS (Granada Television) (UK)
Special Mention: Edward Klosinski, director of photography for GRAND AVENUE (HBO Enterprises) (USA)
NEWS PROGRAMS
Gold Nymph: LIZENZ ZUM QUALEN (ZDF) (Germany)
Silver Nymph: THE SELLING OF INNOCENTS (Associated Producers Inc.,) (Canada)
Silver Nymph: ENVOYE SPECIAL: ENQUETE SUR UN MASSACRE (France 2/Point du jour) (France)
Silver Nymph: LE POINT: LES NOUVEAUX GOULAGS (Société Radio Canada) (Canada)

69th OSCARS® - ACADEMY OF MOTION PICTURE ARTS AND SCIENCES
Awarded 24th March 1997 for 1996 Films

Best Film: THE ENGLISH PATIENT (USA) Dir Anthony Minghella
Best Foreign Language Film: KOLYA (Czech Republic/UK/France) Dir Jan Sverák
Best Director: Anthony Minghella for THE ENGLISH PATIENT (USA)
Best Actor: Geoffrey Rush for SHINE (Australia/UK) Dir Scott Hicks
Best Actress: Frances McDormand for FARGO (USA) Dir Joel Coen
Best Supporting Actor: Cuba Gooding Jr. for JERRY MAGUIRE (USA) Dir Cameron Crowe
Best Supporting Actress: Juliette Binoche for THE ENGLISH PATIENT (USA) Dir Anthony Minghella
Best Original Screenplay: Joel Coen and Ethan Coen for FARGO (USA) Dir Joel Coen
Best Screenplay Adaptation: Billy Bob Thornton for SLING BLADE (USA) Dir Billy Bob Thornton
Best Cinematography: John Seale for THE ENGLISH PATIENT (USA) Dir Anthony Minghella
Best Editing: Walter Murch for THE ENGLISH PATIENT (USA) Dir Anthony Minghella
Best Original Dramatic Score: Gabriel Yared for THE ENGLISH PATIENT (USA) Dir Anthony Minghella
Best Original Musical or Comedy Score: Rachel Portman for EMMA (UK/USA) Dir Douglas McGrath
Best Original Song: 'You Must Love Me' Andrew Lloyd Webber and Tim Rice EVITA (USA) Dir Alan Parker
Best Art Direction/Set Decor: Stuart Craig and Stephanie McMillan for THE ENGLISH PATIENT (USA) Dir Anthony Minghella
Best Costume Design: Ann Roth for THE ENGLISH PATIENT (USA) Dir Anthony Minghella
Best Make-up: Rick Baker and David Leroy for THE NUTTY PROFESSOR (USA) Dir Tom Shadyac
Best Sound: Mark Berger, Walter Murch, Chris Newman and David Parker for THE ENGLISH PATIENT (USA) Dir Anthony Minghella
Best Sound Effects Editing: Bruce Stambler for THE GHOST AND THE DARKNESS (USA) Dir Stephen Hopkins
Best Short Film (Animated): Tyron Montgomery, Thomas Stellmach for QUEST (Germany) Dir Tyron Montgomery
Best Short Film (Live Action): David Frankel, Barry Jossen for DEAR DIARY (USA) Dir David Frankel
Best Documentary Feature: WHEN WE WERE KINGS (USA) Dir Leon J. Gast
Best Documentary Short: BREATHING LESSONS: THE LIFE AND WORK OF MARK O'BRIEN (USA) Dir Jessica Yu
Best Visual Effects: Clay Pinney, Joe Viskocil, Douglas Smith and Volker Engel for INDEPENDENCE DAY (USA) Dir Ronald Emmerich
Scientific and Technical Award: IMAX Corp

Irving G Thalberg Award: Saul Zaentz
Life Achievement Award: Michael Kidd

ROYAL TELEVISION SOCIETY AWARDS
CRAFT & DESIGN AWARDS 1995/1996
Presented in London Hilton, 19 November 1996

Camera - Drama and Entertainment: Howard Atherton - GULLIVER'S TRAVELS (Jim Henson Productions for Channel Four Television)
Camera - Documentary and Factual: Rod Clarke and Kevin Flay - ALIEN EMPIRE: HARDWARE, (BBC)
Costume Design - Drama: Les Lansdown - GREAT MOMENTS IN AVIATION (BBC)
Special Commendation for: - SHAKESPEARE SHORTS (BBC)
Costume Design - Any Other Programme: Sarah Burns - FRENCH AND SAUNDERS (BBC)
Graphics Design - Channel Idents: Luis Cook - CHANNEL FOUR STINGS, Aardman Animations for Channel Four Television
Graphic Design - Programme Content Sequences: Andy Royston O'Conner - NEWSROUND, BBC News Resources
Graphic Design - Titles: John Durrant - OLDIE TV (Burrell Durrant Hifle for BBC)
Visual Effects: Tim Webber - GULLIVER'S TRAVELS (FrameStore for Jim Henson Productions for Channel Four)
Lighting - Drama: Nic Knowland - THE FINAL PASSAGE (Passage Productions for Channel Four Television)
Lighting - Any Other Programme: Peter Greenyer - PURCELL TERCENTENARY CONCERT (BBC Outside Broadcasts)
Production Design - Drama: Roger Hall - GULLIVER'S TRAVELS, (Jim Henson Productions for Channel Four Television)
Production Design - Any Other Programme: Phil Lewis and Trisha Budd - A CLOSE SHAVE (Aardman Animations for BBC)
Sound - Drama: Phil Smith, John Rutherford, John Senior, John Whitworth and Martin Berisford - SOME KIND OF LIFE (Granada Television)
Sound - Any Other Programme: Adrian Rhodes and Paul Hamblin - A CLOSE SHAVE (Aardman Animations for BBC)
Tape and Film Editing - Drama: Edward Marshall - CRACKER - BROTHERLY LOVE (Granada Television)
Tape and Film Editing - Documentary and Factual: Graham Shrimpton - TRUE STORIES - CRIME OF THE WOLF (Yorkshire Television)
Make Up Design - Drama: Deanne Turner - A ROYAL SCANDAL (BBC)
Make Up Design - Any Other Programme: Glenda Wood - STARS IN THEIR EYES (Granada Television)
Music - Original Music: Zbigniew Preisner - PEOPLE'S CENTURY (Zbigniew Preisner Productions for BBC)
Music - Original Score: Dashiell Rae - YN GYMSYG OLL I GYD (ALL MIXED UP) (HTV for S4C)
Team Award: GLADIATORS (London Weekend Television)
Judges' Award: Nick Park and Aardman Animations

EDUCATIONAL AWARDS
Presented in London 14th April 1997

SCHOOLS TELEVISION
Pre School & Infants: STOP LOOK LISTEN: STORIES OF FAITH - SIKHISM (A Tetra TV Production for Channel Four Television)
Junior: 7-9 Years: LIVING PROOF: BRITAIN IN THE SECOND WORLD WAR (An Open Mind Production for Channel Four Television)
Junior: 9-11 Years: HISTORY THROUGH ART: THE RAILWAY STATION - (CASE Television for Channel 4 Schools)
Secondary Science: SHORT CIRCUIT: HOMEOSTASIS (An Invincible Films Production for BBC Schools Programmes)
Secondary Arts: SHAKESPEARE SHORTS: ROMEO AND JULIET (BBC Schools Programmes)

ADULT EDUCATIONAL TELEVISION
Adult Education & Training: DEUTSCH PLUS (A Learning Media Production for BBC Education)
Personal Education: TALES FROM THE WASTELAND: FAMILY FORTUNES (October Films for Channel Four Television)
General: NETWORK FIRST: THE CONNECTION (Carlton Television)
Campaigns & Seasons: THE TROUBLE WITH MEN (BBC Bristol and BBC Education)

MULTIMEDIA AWARD
BBC SHAKESPEARE on CD-Rom: Macbeth on CD-Rom (HarperCollins Publishers and BBC Education)

TELEVISION JOURNALISM and SPORTS AWARDS
Presented at the London Hilton on 27th February 1997

JOURNALISM AWARDS
Young Journalist of the Year (new for 1996): Donal McIntyre - WORLD IN ACTION (Granada TV)
Regional Daily News Magazine: Tyne Tees Television for NORTH EAST TONIGHT WITH MIKE NEVILLE
Regional Current Affairs: BBC Scotland for FRONTLINE SCOTLAND - OPEN TO ABUSE
News Award - Home: ITN for ITV News for Dunblane
Current Affairs Award - Home: Granada Television for WORLD IN ACTION - THE UNTOUCHABLES & WAYNE'S WORLD
News Award International: ITN for IYV for ISRAEL BOMBS THE WEST BANK
Current Affairs Award - International: BBC for NEWSNIGHT - AFGHANISTAN
News Event Award: ITN for ITV News for Dunblane

Production Award: Vera, Dennis Woolf & Fulcrum Productions for Channel Four Television for SCOTT OF THE ARMS ANTICS
TV News Technician of the Year: Darren Conway, BBC
Interview of the Year: BBC for NEWSNIGHT - Neil Hamilton
Television Journalist of the Year: Colin Baker, ITN
Judges' Award: Charles Wheeler

SPORTS AWARDS
Sports News: ITN for Channel Four News: Olympic Money
Live Sports Coverage: BBC for European Football Championships
Sports Documentary: BBC for DICKIE BIRD: A RARE SPECIES
Regional Sports Awards: Carlton UK Television for 24 HOURS: BARRY'S BLUES
Sports Presenter and Commentator Award: Andy Gray, Sky Sport
Judges' Award: Peter O'Sullivan

PROGRAMME AWARDS
Presented 22nd May 1997 at the Grosvernor House Hotel, London

Situation Comedy and Comedy Drama: ONLY FOOLS AND HORSES (BBC Television)
Entertainment: THE FAST SHOW (BBC Television)
Children's Drama: RETRACE (A Perx Production for ITV Network)
Children's Factual: WISE UP (A Carlton Production for Channel Four Television)
Children's Entertainment: THE ANT & DEC SHOW (Zenith North for BBC Television)
Regional Documentary: HOME TRUTHS - A WOMAN IN TWELVE (BBC Northern Ireland)
Single Documentary: TRUE STORIES - CRIME OF THE WOLF (A Yorkshire TV Production for Channel Four Television)
Documentary Series/Strands: THE SYSTEM (BBC Television)
Regional Programme: TARTAN SHORTS - THE STAR (Renegade Films for BBC Scotland)
Regional Presenter: Kaye Adams (Scottish Television)
Presenter: Cilla Black (BLIND DATE) (London Weekend Television)
Drama Serial: OUR FRIENDS IN THE NORTH (BBC Television in association with Primetime)
Single Drama: HILLSBOROUGH (Granada Television)
Drama Series: BALLYKISSANGEL (A Ballykea Production for World Productions for BBC Television)
Actor Female: Stella Gonet (TRIP TRAP) (BBC Television)
Actor Male: David Jason (ONLY FOOLS AND HORSES) (BBC Television)
Live Event: CHRISTMAS WITH THE ROYAL NAVY (Two Four Productions for Westcountry Television)
Arts: ARENA - THE BURGER AND THE KING (BBC Television)
Network Newcomer: Francesca Joseph (Director, PICTURE THIS - FOUR TARTS AND A TENOR) (BBC Television)
Television Performance: Paul Whitehouse (THE FAST SHOW) (BBC Television)
Team: GULLIVER'S TRAVELS (A Jim Henson Production for Channel Four Television)
Writer's Award: Peter Flannery (OUR FRIENDS IN THE NORTH) (BBC Television in association with Primetime)
Cyril Bennett Judges' Award: Tony Garnett
Gold Medal: Michael Grade

54th VENICE FESTIVAL
Awarded 26th August - 6th September 1997

Golden Lion: Takeshi Kitano for HANA-BI (Japan) Dir Takeshi Kitano
Special Award of Jury: Paolo Virzi for OVOSODO (Italy) Dir Paolo Virzi
Coppa Volpi for Best Actor: Wesley Snipes for ONE NIGHT STAND (USA) Dir Mike Figgis
Coppa Volpi for Best Actress: Robin Tunney for NIAGARA, NIAGARA (USA) Dir Bob Gosse
Osella d'oro for the Best Original Screenplay: Gilles Taurand and Anne Fontaine for NETTOYAGE A SEC (France/Spain) Dir Anne Fontaine
Osella d'oro for Best Photographer: Emmanuel Machuel for OSSOS (Portugal/France/Denmark) Dir Pedro Costa
Osella d'oro for Best Original Soundtrack: Graeme Revell for CHINESE BOX (France/Japan) by Wayne Wang
Medaglia d'oro Della Presidenza del Sanato: VOR (Russia/France) Dir Pavel Churchraj
FIPRESCI Awards: HISTORIE MILOSNE Dir Jerzy Stuht; 24 7 (UK) Dir Shane Meadows
Special Mention: GUMMO (USA) Dir Harmony Korine
Film Critics Award: ALORS VOILA (France) Dir Michel Piccoli
OCIC Jury Award: THE WINTER GUEST (USA/UK) Dir Alan Rickman
BENT FAMIGLIA Dir Nouri Bouzid
Prix du festival: LA STRANA STORIA DI BANDA SONORA (Italy) Dir Francesca Arcibugi

1998

BAFTA FILM AWARDS
19th April 1998, London
195 Piccadilly
London W1V OLN
Tel: 0171 734 1792

Alexander Korda Award for Outstanding British Film of the Year: NIL BY MOUTH (UK) Dir Gary Oldman
Audience Award: THE FULL MONTY (US/UK) Dir Peter Cattaneo
Best Actor: Robert Carlyle for THE FULL MONTY (US/UK) Dir Peter Cattaneo
Best Actress: Judi Dench for MRS BROWN (UK/US/Ireland) Dir John Madden
Best Adapted Screenplay: Baz Luhrmann, Craig Pearce for William Shakespeare's ROMEO AND JULIET (US/UK) Dir Baz Luhrmann
Best Cinematography: Eduardo Serra for WINGS OF THE DOVE (US/UK) Dir Iain Softley
Best Costume Design: Deirdre Clancy for MRS BROWN (UK/US/Ireland) Dir John Madden
Best Director: Baz Luhrmann for WILLIAM SHAKESPEARE'S ROMEO AND JULIET (US/UK) Dir Baz Luhrmann
Best Film: THE FULL MONTY (US/UK) Dir Peter Cattaneo
Best Editing: Peter Honess for L.A. CONFIDENTIAL (US) Dir Curtis Hanson
Best Foreign Language Film: L'APPARTEMENT (France/Spain/Italy) Dir Gilles Mimouni
Best Make-Up/Hair: Sallie Jaye, Jan Archibald for WINGS OF THE DOVE (US/UK) Dir Iain Softley
Anthony Asquith Award for Achievement in Film Music: for WILLIAM SHAKESPEARE'S ROMEO AND JULIET (US/UK) Dir Baz Luhrmann
Best Original Screenplay: Gary Oldman for NIL BY MOUTH (UK) Dir Gary Oldman
Best Production Design: Catherine Martin for WILLIAM SHAKESPEARES ROMEO AND JULIET (US/UK) Dir Baz Luhrmann
Best Supporting Actor: Tom Wilkinson for THE FULL MONTY (US/UK) Dir Peter Cattaneo
Best Supporting Actress: Sigourney Weaver for THE ICE STORM (US) Dir Ang Lee
Best Short Film: THE DEADNESS OF DAD (UK) Dir Philippa Cousins
Best Short Animated Film: STAGE FRIGHT (UK) Dir Steve Box
Best Special Visual Effects: LE CINQUIÈME ÉLÉMENT (France) Dir Luc Besson
Best Sound: L.A. CONFIDENTIAL (US) Dir Curtis Hanson
Fellowship: Sean Connery
Outstanding British Contribution to Cinema: Michael Robert

BAFTA TELEVISION AWARDS
18th May 1998, London
195 Piccadilly
London W1V OLN
Tel: 0171 734 1792

Best Single Drama: NO CHILD OF MINE (Stonehenge Films and United Film & TV Productions for ITV)
Best Drama Series: JONATHAN CREEK (BBC entertainment for BBC1)
Best Drama Serial: HOLDING ON (BBC drama for BBC2)
Best Factual Series: THE NAZIS - A WARNING FROM HISTORY (BBC documentary & history for BBC2)
Best Light Entertainment (programme or series): THE FAST SHOW (BBC entertainment for BBC2)
Best Comedy (programme or series): I'M ALAN PARTRIDGE (Talkback Productions for BBC2)
Best Actress: DANIELA NARDINI - THIS LIFE (World Productions for BBC2)
Best Actor: SIMON RUSSELL BEALE - A DANCE TO THE MUSIC OF TIME (Table Top for Channel 4)
Best Light Entertainment Performance: PAUL WHITEHOUSE - THE FAST SHOW (BBC entertainment for BBC2)
Best Comedy Performance: STEVE COOGAN - I'M ALAN PARTRIDGE (Talkback Productions for BBC2)
The Huw Wheldon Award (for best arts programme or series): GILBERT AND GEORGE - THE SOUTH BANK SHOW (LWT for ITV)
Sports/events Coverage in Real Time: RUGBY UNION (Sky Sports)
News and Current Affairs Journalism: VALENTINA'S STORY - PANORAMA (BBC1)
The Flaherty Documentary Award: THE GRAVE - TRUE STORIES (Soul Purpose for C4)
Original Television Music: TOM JONES (BBC drama for BBC1)
Best Make-Up/Hair: TOM JONES (BBC drama for BBC1)
Best Photography (factual): POLAR BEAR WILDLIFE SPECIAL (BBC Natural History for BBC1)
Best Photography and Lighting (fiction/entertainment) THE WOMAN IN WHITE (BBC/Carlton for BBC1)
Best Costume Design: TOM JONES (BBC drama for BBC1)
Best Graphic Design: ELECTION '97 (BBC Resources for BBC1)
Best Sound: (factual/entertainment): AIRPORT (BBC documentaries and history for BBC1)

Best Sound: (fiction/entertainment): THE LAKES (BBC drama for BBC1)
Best Editing (factual): THE NAZIS - A WARNING FROM HISTORY (BBC documentary & history for BBC2)
Best Editing (fiction/entertainment): THE LAKES (BBC drama for BBC1)
Best Design: THE WOMAN IN WHITE (BBC/Carlton for BBC1)
The Academy Fellowship: Bill Cotton
The Alan Clarke award (for outstanding creative contribution to television): Ted Childs
The Richard Dimbleby Award (for the year's most important personal contribution on the screen in factual television): David Dimbleby
The Dennis Potter Award: Kay Mellor
The International Television Programme: FRIENDS
The Special Award: Roger Cook
The Lew Grade Award (for a significant and popular programme): A TOUCH OF FROST (Yorkshire TV for ITV)

BFI FELLOWSHIPS
Awarded at the NFT at the end of the London Film Festival, November 1997
Sir Sydney Samuelson, Thelma Schoonmaker Powell

48TH BERLIN INTERNATIONAL FILM FESTIVAL
11th-22nd February 1998, Berlin
Internationale Filmfestspiele Berlin
Budapester Strasse 50
10787 Berlin
Tel: (49) 30 25 48 92 25

Golden Berlin Bear (Grand Prix): CENTRAL DO BRASIL (Brazil/France) Dir Walter Salles
Silver Berlin Bear - Special Jury Prize: WAG THE DOG (US) Dir Barry Levinson
Silver Berlin Bear - Lifetime Contribution to Arts of Cinematography: Alain Resnais for ON CONNAÎT LA CHANSON (France/Switzerland/UK) Dir Alain Resnais
Silver Berlin Bear - Best Director: Neil Jordan for THE BUTCHER BOY (Ireland/US) Dir Neil Jordan
Silver Berlin Bear - Best Actress: Fernanda Montenegro for CENTRAL DO BRASIL (Brazil/France) Dir Walter Salles
Silver Berlin Bear - Best Actor: Samuel L. Jackson for JACKIE BROWN (US) Dir Quentin Tarantino
Silver Berlin Bear - Best Single Achievement: Matt Damon as scriptwriter and actor for GOOD WILL HUNTING (US) Dir Gus Van Sant
The Blue Angel: Jeroen Krabbé for LEFT LUGGAGE (Netherlands/Belgium/US) Dir Jeroen Krabbé
Alfred Bauer Prize: YUE KUAI LE, YUE DUO LUO (Hong Kong) Dir Stanley Kam-Pang Kwan
A Special Mention for the Promising Performances - actress: Isabella Rossellini for LEFT LUGGAGE (Netherlands/Belgium/US) Dir Jeroen Krabbé
A Special Mention for the Promising Performances - actor: Eamonn Owens for THE BUTCHER BOY (Ireland/US) Dir Neil Jordan
A Special Mention: Slawomir Idziak for I WANT YOU (UK) Michael Winterbottom
Prizes for Short Films
Golden Berlin Bear: I MOVE SO I AM (Netherlands) Dir Gerrit van Dijk)
Silver Berlin Bear: CINEMA ALCAZAR (Nicaragua) Dir Florence Jaugey

BROADCASTING PRESS GUILD TELEVISION AND RADIO AWARDS 1997
2nd April 1998, London
c/o Richard Last
Tiverton, Thhe Ridge
Woking
Surrey GU22 7EQ
Tel: 01483 764895

Best Single Drama: BREAKING THE CODE (The Drama House for BBC1)
Best Drama Series/Serial: HOLDING ON (BBC2)
Best Documentary Series: THE NAZIS - A WARNING FROM HISTORY (BBC2)
Best Single Documentary: CUTTING EDGE: THE DINNER PARTY
Best Entertainment: I'M ALAN PARTRIDGE (Talkback Productions for BBC2)
Best Actor: Simon Russell Beale (A DANCE TO THE MUSIC OF TIME, C4)
Best Actress: Helen Baxendale (COLD FEET, AN UNSUITABLE JOB FOR A WOMAN)
Best Performer (Non-Acting): Jeremy Paxman (UNIVERSITY CHALLENGE, NEWSNIGHT, ELECTION NIGHT 1997)
Writer's Award: David Renwick (ONE FOOT IN THE GRAVE, Jonathan Creek)
Radio Programme of the Year: I'M SORRY I HAVEN'T A CLUE (Radio 4)
Radio Broadcaster of the Year: Susan Sharpe (MIDWEEK CHOICE, R3)
Harvey Lee Award for Outstanding Contribution to Broadcasting: Michael Wearing (Head of Drama Serials, BBC TV)

51st CANNES FESTIVAL
13th-24th May 1998, Cannes
99 Boulevard Malesherbes
75008 Paris
Tel: (33) 1 45 61 66 00

Palme d'Or: MIA EONIOTITA KE MIA MERA (An Eternity and a Day) (Greece/France) Dir Theo Angelopoulos
Grand Jury Prize: LA VITA E BELLA (Italy) Dir Roberto Benigni
Best Actress: Elodie Bouchez and Natacha Regnier for LA VIE REVEE DES ANGES (France) Dir Eric Zonca
Best Actor: Peter Mullan for MY NAME IS JOE (UK/Germany) Dir Ken Loach
Best Director: John Boorman for THE GENERAL (Ireland/UK)
Best Screenplay: Hal Hartley for HENRY FOOL (US)
Jury Technical Prize: Vittorio Storaro for TANGO (Spain/Argentina/France) Dir Carlos Saura
Special Jury Prize:
LA CLASSE DE NEIGE (France) Dir Claude Miller
FESTEN (Denmark) Dir Thomas Vinterberg
Best Aristic Contribution Prize: VELVET GOLDMINE (UK/US) Dir Todd Haynes
Camera d'Or: SLAM (US) Dir Marc Levin
Best Short Film: L'INTERVIEW (France) Dir Xavier Giannoli
Short Film Second Prize:
GASMAN (UK) Dir Lynne Ramsay
HORSESHOE (UK) Dir David Lodge

23rd CÉSARS
Awarded in Paris 28th February 1998

Best Film: ON CONNAÎT LA CHANSON (France/Switzerland/UK) Dir Alain Resnais
Best Director: Luc Besson for LE CINQUIÈME ÉLÉMENT (France)
Best Actress: Ariane Ascaride for MARIUS ET JEANNETTE (France) Dir Robert Guédiguian
Best Actor: André Dussolier for ON CONNAÎT LA CHANSON (France/Switzerland/UK) Dir Alain Resnais
Best First Film: DIDIER (France) Dir Alain Chabat
Best Screenplay: Agnès Jaoui et Jean-Pierre Bacri for ON CONNAÎT LA CHANSON (France/Switzerland/UK) Dir Alain Resnais
Best Supporting Actor: Jean-Pierre Bacri for ON CONAÎT LA CHANSON (France/Switzerland/UK) Dir Alain Resnais
Best Supporting Actress: Agnès Jaoui for ON CONNAÎT LA CHANSON (France/Switzerland/UK) Dir Alain Resnais
Best Promising Young Actor: Stanislas Merhar for NETTOYAGE À SEC (France/Spain) Dir Anne Fontaine
Best Promising Young Actress: Emma de Caunes for UN FRÈRE (France) Dir Sylvie Verheyde
Best Foreign Film: BRASSED OFF (UK/S) Dir Mark Herman
Best Photography: Thierry Arbogast for LE CINQUIÈME ÉLÉMENT (France) Dir Luc Besson
Best Sound: Pierre Lenoir et Jean-Pierre Laforce for ON CONNAÎT LA CHANSON (France/Switzerland/UK) Dir Alain Resnais
Best Decor: Dan Weil for LE CINQUIÈME ÉLÉMENT (France) Dir Luc Besson
Best Costumes: Christian Gasc for LE BOSSU (France/Italy/Germany) Dir Phiilippe De Broca
Best Editing: Hervé de Luze for ON CONNAÎT LA CHANSON (France/Switzerland/UK) Dir Alain Resnais
Best Music: Bernardo Sandoval for WESTERN (France) Dir Manuel Poirer
Best Short Film: DES MAJORETTES DANS L'ESPACE (France) Dir David Fournier
Best Honor Awards:
Michael Douglas
Clint Eastwood
Jean-Luc Godard (Histoire(s) du Cinéma)

EVENING STANDARD AWARDS 1998
Awarded in London 1st February 1998

Best Actor: Robert Carlyle for THE FULL MONTY (US/UK) Dir Peter Cattaneo
Best Actress: Katrin Cartlidge for CAREER GIRLS (UK) Dir Mike Leigh
Best Film: THE FULL MONTY (US/UK) Dir Peter Cattaneo
Best Screenplay: Jeremy Brock for MRS BROWN (UK/US/Ireland)
Best Technical Achievement: Maria Djurkovic for WILDE (UK,/US/Japan/Germany) Dir Brian Gilbert
Most Promising Newcomer: Jude Law for WILDE (UK/US/Japan/Germany) Dir Brian Gilbert
Peter Sellers Award for Comedy: MRS BROWN (UK/US/Ireland)
Special Jury Award: Kenneth Branagh for HAMLET (US/UK)
Special Award: Roy Boulting

55th GOLDEN GLOBE AWARDS
19th January 1998, Los Angeles
292 South La Cinega Blvd
Suite 316, Beverly Hills
CA 90211

FILM
Drama: TITANIC (US) Dir James Cameron
Actress (drama): Judi Dench for MRS BROWN (UK/US/Ireland) Dir John Madden
Actor (drama): Peter Fonda for ULEE'S GOLD (US) Dir Victor Nunez
Musical or comedy: AS GOOD AS IT GETS (US) Dir James L. Brooks
Actress (musical or comedy): Helen Hunt for AS GOOD AS IT GETS (US) Dir James L. Brooks
Actor (musical or comedy): Jack Nicholson for AS GOOD AS IT GETS (US) Dir James L. Brooks
Foreign Language Film: MA VIE EN ROSE (France/Belgium/UK/Switzerland) Dir Alain Berliner
Supporting actress (drama, musical or comedy): Kim Basinger for L.A. CONFIDENTIAL (US) Dir Curtis Hanson
Supporting actor (drama, musical or comedy): Burt Reynolds for BOOGIE NIGHTS (US) Dir Paul Thomas Anderson
Best Director: James Cameron for TITANIC (US)
Best Screenplay: Matt Damon and Ben Affleck for GOOD WILL HUNTING (US) Dir Gus Van Sant
Original Score: James Horner for TITANIC (US) Dir James Cameron
Original Song: James Horner and Will Jennings for MY HEART WILL GO ON, TITANIC (US) Dir James Cameron

TELEVISION
Drama Series: THE X-FILES
Actress (drama series): Christine Lahti for CHICAGO HOPE
Actor (drama series): Anthony Edwards for ER
Musical or comedy series: ALLY MCBEAL
Actress (musical or comedy series): Calista Flockhart for ALLY MCBEAL
Actor (musical or comedy series): Michael J. Fox for SPIN CITY
Miniseries or movie made for television: GEORGE WALLACE
Actress in miniseries or movie made for television: Alfre Woodard for MISS EVERS' BOYS
Actor in miniseries or movie made for television: Ving Rhames for DON KING: ONLY IN AMERICA
Supporting Actress in miniseries or movie made for television: Angelina Jolie for GEORGE WALLACE
Supporting Actor in miniseries or movie made for television: George C.Scott for 12 ANGRY MEN
Cecil B. DeMille Award: Shirley MacLaine

38th GOLDEN ROSE OF MONTREUX
23rd-28th April 1998, Montreux
Télévision Suisse Romande
PO Box 234
1211 Geneva 8
Switzerland
Tel: (41) 22 708 8599

Golden Rose: YO-YO MA INSPIRED BY BACH: SIX GESTURES (UK/Canada) (Rhombus)
Silver Rose (Music): THE CANADIAN BRASS: A CHRISTMAS EXPERIMENT (UK/Canada) (CDN)
Silver Rose (Sitcom): OPERATION GOOD GUYS; FRISK 'EM, (UK) (BBC2)
Silver Rose/Special Prize of the City of Montreux: HARRY ENFIELD & CHUMS, (UK) (Tiger Aspect)
Silver Rose (Variety): CRONICAS MARCIANAS, (Spain) (Gestmusic Zepp)
Silver Rose (Special): QUEEN - "BÉJART: BALLET FOR LIFE", (UK) (Queen Productions)
Bronze Rose (Music): GAEL FORCE, (Ireland) (Tyrone/RTE)
Bronze Rose (Sitcom): FATHER TED - ARE YOU RIGHT THERE? (UK) (Channel 4)
Bronze Rose (Comedy): THE CHAMBER QUINTET, (Israel) (MATAR ARTS +)
Bronze Rose (Special): FAME AND FORTUNE: OZZY OSBOURNE, (UK) (Channel 5)
Bronze Rose (Variety): DAVID BLAINE: STREET MAGIC, (USA) (David Blaine Productions)
Press Prize: DAVID BLAINE: STREET MAGIC, (USA) (David Blaine Productions)
Special Prize: IL SEGRETO DI PULCINELLA, (Switzerland) (RTSI)
Ex-Aequo: OPERATION GOOD GUYS; FRISK 'EM, (UK) (BBC2)
UNDA Award: YO-YO MA INSPIRED BY BACH: SIX GESTURES, (UK/Canada) (Rhombus)

GRIERSON AWARD
The Grierson Memorial Trust
Ivan Sopher & Co
5 Elstree Gate
Elstree Way
Borehamwood, Herts WD6 1JD

Tel: 0181 207 0602
THE SYSTEM - THE NATURE OF THE BEAST (Peter Dale/BBC2)

LONDON CRITICS' CIRCLE FILM AWARDS
Awarded in London 5th March 1998

Best British Actor: Robert Carlyle for The Full Monty (UK/US) Dir Peter Cattaneo, Face (UK) Dir Antonia Bird,
Carla's Song (UK/Germany/Spain) Dir Ken Loach
Best British Actress: Judi Dench for Mrs Brown (UK/US/Ireland) Dir John Madden
Best British Film: The Full Monty (UK/US) Dir Peter Cattaneo
Best British Producer: Uberto Pasolini for The Full Monty (UK/US) Dir Peter Cattaneo
Best British Screenwriter: Simon Beaufoy for The Full Monty (UK/US) Dir Peter Cattaneo
Best English-language Film: L.A. Confidential (US) Dir Curtis Hanson
Best Newcomer of the Year: Peter Cattaneo The Full Monty (UK/US) Dir Peter Cattaneo
Lifetime Achievement Award: Paul Scofield
Lifetime Achievement Award: Woody Allen
Lifetime Achievement Award: Michael Caine
Lifetime Achievement Award: Martin Scorsese

38th MONTE CARLO TELEVISION FESTIVAL
Awarded in Monte Carlo 20th-26th February 1998

FILMS FOR TV
Gold Nymph (Best Film): L'ORANGE DE NOEL (King Movies) (France)
Silver Nymph (Best Director): Jean-Louis Lorenzi for L'ORANGE DE NOEL (King Movies) (France)
Silver Nymph (Best Script): Stuart Urban for DEADLY VOYAGE (BBC2)
Silver Nymph (Best Actor): Omar Epps for DEADLY VOYAGE (BBC2)
Silver Nymph (Best Actress): Meredith Baxter for AFTER JIMMY (Worldvision Enterprises, USA)
Special Mention: CORRERE CONTRO (SACIS Spa, Italy)
MINI-SERIES
Gold Nymph (Best Mini-Series): CLEAN SHEET (Danmarks Radio/TV) (Denmark)
Silver Nymph (Best Director): Daniel Alfredson for CLEAN SHEET (Danmarks Radio/TV) (Denmark)
Silver Nymph (Best Script): John Brown for CLEAN SHEET (Danmarks Radio/TV) (Denmark)
Silver Nymph (Best Actor): Ole Ernst for CLEAN SHEET (Danmarks Radio/TV) (Denmark)
Silver Nymph (Best Actress): Alex Kingston for MOLL FLANDERS (Granada Television) (UK)
Special Mention: Edward Klosinski, director of photography for GRAND AVENUE (HBO Enterprises) (USA)
NEWS PROGRAMS
Gold Nymph: LIZENZ ZUM QUALEN (ZDF) (Germany)
Silver Nymph: THE SELLING OF INNOCENTS (Associated Producers Inc.,) (Canada)
Silver Nymph: ENVOYE SPECIAL: ENQUETE SUR UN MASSACRE (France 2/Point du jour) (France)
Silver Nymph: LE POINT: LES NOUVEAUX GOULAGS (Société Radio Canada) (Canada)

70th OSCARS - ACADEMY OF MOTION PICTURE ARTS AND SCIENCES
Awarded in Los Angeles 23rd March 1998 for 1997 Films

FILM
Best Film: TITANIC (US) Dir James Cameron
Best Foreign Language Film: KARAKTER (Netherlands/Belgium) Dir Mike van Diem
Best Director: James Cameron for TITANIC (US)
Best Actor: Jack Nicholson for AS GOOD AS IT GETS (US) Dir James L. Brooks
Best Actress: Helen Hunt for AS GOOD AS IT GETS (US) Dir James L. Brooks
Best Supporting Actor: Robin Williams for GOOD WILL HUNTING (US) Gus Van Sant
Best Supporting Actress: Kim Basinger for L.A. CONFIDENTIAL (US) Dir Curtis Hanson
Best Original Screenplay: Matt Damon and Ben Affleck for GOOD WILL HUNTING (US) Dir Gus Van Sant
Best Screenplay Adaptation: Brian Helgeland & Curtis Hanson for L.A. CONFIDENTIAL (US) Dir Curtis Hanson
Best Cinematograpy: Russell Carpenter for TITANIC (US) Dir James Cameron
Best Editing: Conrad Buff, James Cameron and Richard A. Harris for TITANIC (US) Dir James Cameron
Best Original Dramatic Score: James Horner for TITANIC (US) Dir James Cameron
Best Original Musical or Comedy Score: Anne Dudley for THE FULL MONTY (UK/US) Dir Peter Cattaneo
Best Original Song: James Horner and Will Jennings for MY HEART WILL GO ON, Titanic (US) Dir James Cameron
Best Art Direction/Set Decor: Peter Lamont and Michael Ford for TITANIC (US) Dir James Cameron
Best Costume Design: Deborah L. Scott for TITANIC (US) Dir James Cameron
Best Make-Up: Rick Baker and David LeRoy Anderson for MEN IN BLACK (US) Dir Barry Sonnenfeld
Best Sound: Gary Rydstrom, Tom Johnson, Gary Summers and Mark Ulano for TITANIC (US) Dir James Cameron
Best Sound Effects Editing: Tom Bellfort and Christopher Boyes for TITANIC (US) Dir James Cameron

Best Short Film (Animated): GERI'S GAME Dir Jan Pinkava
Best Short Film (Live Action): VISAS AND VIRTUE (US) Dir Chris Tashima
Best Documentary Feature: THE LONG WAY (US) Dir Mark Jonathan Harris
Best Documentary Short: A STORY OF HEALING (US) Dir Donna Dewey
Best Visual Effects: Robert Legato, Mark Lasoff, Thomas L. Fisher and Michael Kanfer for TITANIC (US) Dir James Cameron
Life Achievement Award: Stanley Donen

ROYAL TELEVISION SOCIETY AWARDS
Awarded in London 17th March 1998

Best Actress: Sinead Cusack for HAVE YOUR CAKE AND EAT IT; PART ONE (BBC1)
Best Drama Series: THIS LIFE (BBC2)
Best Single Drama: GRANTON STAR CAUSE (C4)
Best Television Performance: Chris Morris for BRASS EYE (C4)
Best Children's Entertainment Programme: TELETUBBIES (BBC2)
Best Entertainment Programme: HARRY ENFIELD AND CHUMS (BBC1)
Gold Medal: Trevor McDonald
Judges Award: Michael Wearing
Best Presenter: Jeremy Clarkson for TOP GEAR (BBC2)
Best Situation Comedy: VICAR OF DIBLEY (BBC1)

BRITISH SUCCESSES IN THE ACADEMY AWARDS 1927-1997

The following list chronicles British successes in the Academy Awards. It includes individuals who were either born, and lived and worked, in Britain into their adult lives, or those who were not born here but took on citizenship. Compiled by Erinna Mettler

1927/28 (1st) held in 1930

Charles Chaplin - Special Award (acting, producing, directing and writing):
The Circus

1928/29 (2nd) held in 1930

Frank Lloyd - Best Direction:
The Divine Lady

1929/30 (3rd) held in 1930

George Arliss - Best Actor:
The Green Goddess

1932/33 (6th) held in 1934

William S. Darling - Best Art Direction:
Cavalcade
Charles Laughton - Best Actor:
The Private Life of Henry VIII
Frank Lloyd - Best Direction:
Cavalcade

1935 (8th) held in 1936

Gaumont British Studios - Best Short Subject:
Wings Over Mt. Everest
Victor Mclaglen - Best Actor:
The Informer

1938 (11th) held in 1939

Ian Dalrymple, Cecil Lewis & W.P. Lipscomb - Best Screenplay:
Pygmalion

1939 (12th) held in 1940

Robert Donat - Best Actor:
Goodbye Mr. Chips
Vivien Leigh - Best Actress:
Gone With The Wind

1940 (13th) held in 1941

Lawrence Butler & Jack Whitney - Special Visual Effects:
The Thief Of Bagdad
Vincent Korda - Best Colour Set Design:
The Thief Of Bagdad

1941 (14th) held in 1942

British Ministry of Information - Honorary Award:
Target For Tonight
Donald Crisp - Best Supporting Actor:
How Green Was My Valley
Joan Fontaine - Best Actress:
Suspicion
Jack Whitney & The General Studios Sound Department - Best Sound:
That Hamilton Woman

1942 (15th) held in 1943

Noel Coward - Special Award:
In Which We Serve
Greer Garson - Best Actress:
Mrs. Miniver

1943 (16th) held in 1944

British Ministry of Information - Best Documentary:
Desert Victory
William S. Darling - Best Art Direction:
The Song of Bernadette

1945 (18th) held in 1946

The Governments of the United States & Great Britain - Best Documentary:
The True Glory
Ray Milland - Best Actor:
The Lost Weekend
Harry Stradling - Best Cinematography (b/w):
The Picture of Dorian Gray

1946 (19th) held in 1947

Muriel & Sydney Box - Best Original Screenplay:
The Seventh Veil
Clemence Dane - Best Original Story:
Vacation From Marriage
Olivia de Havilland - Best Actress:
To Each His Own
Laurence Olivier - Special Award:
Henry V
Thomas Howard - Best Special Effects:
Blithe Spirit
William S. Darling - Best Art Direction (b/w):
Anna And the King Of Siam

1947 (20th) held in 1948

John Bryan - Best Art Direction:
Great Expectations

Jack Cardiff - Best Cinematography (col):
Black Narcissus
Ronald Colman - Best Actor:
A Double Life
Guy Green - Best Cinematography (b/w):
Great Expectations
Edmund Gwen - Best Supporting Actor:
Miracle On 34th Street

1948 (21st) held in 1949

Carmen Dillon & Roger Furse - Best Art Direction (b/w):
Hamlet
Brian Easdale - Best Score:
The Red Shoes
Roger Furse - Best Costume Design:
Hamlet
Laurence Olivier - Best Picture:
Hamlet
Laurence Olivier - Best Actor:
Hamlet

1949 (22nd) held in 1950

British Information Services - Best Documentary:
Daybreak In Udi
Olivia de Havilland - Best Actress:
The Heiress

1950 (23rd) held in 1951

George Sanders - Best Supporting Actor:
All About Eve

1951 (24th) held in 1952

James Bernard & Paul Dehn - Best Motion Picture Story:
Seven Days To Noon
Vivien Leigh - Best Actress:
A Streetcar Named Desire

1952 (25th) held in 1953

T.E.B. Clarke - Best Story & Screenplay:
The Lavender Hill Mob

London Films Sound Dept. - Best Sound:
The Sound Barrier

1954 (26th) held in 1954

British Information Services - Best Documentary Short Subject:
Thursday's Children
S. Tyne Jule - Best Song:
Three Coins In The Fountain
Jon Whitely & Vincent Winter - Special Award (Best Juvenile Performances):
The Kidnappers

1956 (29th) held in 1957

George K. Arthur - Best Short Subject:
The Bespoke Overcoat

1957 (30th) held in 1958

Malcolm Arnold - Best Musical Score:
The Bridge On The River Kwai
Alec Guinness - Best Actor:
The Bridge On The River Kwai
Jack Hildyard - Best Cinematography:
The Bridge On The River Kwai
David Lean - Best Director:
The Bridge On The River Kwai
Pete Taylor - Best Editing:
The Bridge On The River Kwai

1958 (31st) held in 1959

Cecil Beaton - Best Costumes:
Gigi
Wendy Hiller - Best Supporting Actress:
Separate Tables
Thomas Howard - Special Visual Effects:
Tom Thumb
David Niven - Best Actor:
Separate Tables

1959 (32nd) held in 1960

Hugh Griffith - Best Supporting Actor:
Ben Hur
Elizabeth Haffenden - Best

Costume Design (col.):
Ben Hur

1960 (33rd) held in 1961

Freddie Francis - Best Cinematography (b/w):
Sons & Lovers
James Hill - Best Documentary:
Giuseppina
Hayley Mills - Special Award (Best Juvenile Performance):
Pollyanna
Peter Ustinov - Best Supporting Actor:
Spartacus

1961 (34th) held in 1962

Vivian C. Greenham - Best Visual Effects:
The Guns Of Navarone

1962 (35th) held in 1963

John Box & John Stoll - Best Art Direction:
Lawrence Of Arabia
Anne V. Coates - Best Editing:
Lawrence Of Arabia
Jack Howells (Janus Films) - Best Documentary:
Dylan Thomas
David Lean - Best Director:
Lawrence Of Arabia
Shepperton Studios Sound Dept. (John Cox Sound Director) - Best Sound:
Lawrence Of Arabia
Freddie Young - Best Cinematography:
Lawrence Of Arabia

1963 (36th) held in 1964

John Addison - Best Score:
Tom Jones
John Osborne - Best Adapted Screenplay:
Tom Jones
Tony Richardson - Best Director:
Tom Jones
Tony Richardson (Woodfall Films) - Best Picture:
Tom Jones
Margaret Rutherford - Best Supporting Actress:
The V.I.P.s

1964 (37th) held in 1965

Julie Andrews - Best Actress:
Mary Poppins
Cecil Beaton - Best Art Direction (col):
My Fair Lady
Cecil Beaton - Best Costume Design (col):
My Fair Lady
Rex Harrison - Best Actor:
My Fair Lady
Walter Lassally - Best Cinematography (b/w):
Zorba The Greek
Harry Stradling - Best Cinematography (col):
My Fair Lady
Peter Ustinov - Best Supporting Actor:
Topkapi
Norman Wanstall - Best Sound Effects:
Goldfinger

1965 (38th) held in 1966

Julie Christie - Best Actress
Darling
Robert Bolt - Adapted Screenplay
Doctor Zhivago
Frederic Raphael - Original Screenplay
Darling
Freddie Young - Colour Cinematography
Doctor Zhivago
John Box, Terence Marsh - Best Art Direction (colour)
Doctor Zhivago
Julie Harris - Costume (b/w)
Darling
Phyllis Dalton - Costume (col)
Doctor Zhivago
John Stears - Special Visual Effects
Thunderball

1966 (39th) held in 1967

John Barry - Best Original Score:
Born Free
John Barry & Don Black - Best Song:
Born Free
Robert Bolt - Best Adapted Screenplay:
A Man For All Seasons
Joan Bridge & Elizabeth Haffenden - Best Costume (col):
A Man For All Seasons

Gordon Daniel - Best Sound:
Grand Prix
Ted Moore - Best Cinematography (col):
A Man For All Seasons
Ken Thorne - Best Adapted Score:
A Funny Thing Happened On The Way To The Forum
Peter Watkins - Best Documentary Feature:
The War Game

1967 (40th) held in 1968

Leslie Bricusse - Best Song:
Doctor Dolittle (Talk To The Animals)
Alfred Hitchcock - Irving Thalberg Memorial Award
John Poyner - Best Sound Effects:
The Dirty Dozen

1968 (41st) held in 1969

John Barry - Best Original Score:
The Lion In Winter
Vernon Dixon & Ken Muggleston - Best Art Direction:
Oliver!
Carol Reed - Best Director:
Oliver!
Shepperton Sound Studio - Best Sound:
Oliver!
Charles D. Staffell - Scientific, Class I Statuett -
for the development of a successful embodiment of the reflex background projection system for composite cinematography
John Woolf - Best Picture:
Oliver!

1969 (42nd) held in 1970

Margaret Furfe - Best Costume:
Anne Of The Thousand Days
Cary Grant - Honorary Award
John Schlesinger - Best Director:
Midnight Cowboy
Maggie Smith - Best Actress:
The Prime Of Miss Jean Brodie

1970 (43rd) held in 1971

The Beatles - Best Original Score:
Let It Be
Glenda Jackson - Best Actress:

Women In Love
John Mills - Best Supporting Actor:
Ryan's Daughter
Freddie Young - Best Cinematography:
Ryan's Daughter

1971 (44th) held in 1972

Robert Amram - Best Short:
Sentinels Of Silence
Ernest Archer, John Box, Vernon Dixon & Jack Maxsted - Best Art Direction:
Nicholas & Alexandra
Charles Chaplin - Honorary Award
David Hildyard & Gordon K. McCallum - Best Sound:
Fiddler On The Roof
Oswald Morris - Best Cinematography:
Fiddler On The Roof

1972 (45th) held in 1973

Charles Chaplin - Best Original Score:
Limelight
David Hildyard - Best Sound:
Cabaret
Anthony Powell - Best Costume Design:
Travels With My Aunt
Geoffrey Unsworth - Best Cinematography:
Cabaret

1973 (46th) held in 1974

Glenda Jackson - Best Actress:
A Touch Of Class

1974 (47th) held in 1975

Albert Whitlock - Special Achievement In Visual Effects:
Earthquake

1975 (48th) held in 1976

Ben Adam, Vernon Dixon & Roy Walker - Best Art Direction:
Barry Lyndon

John Alcott - Best Cinematography:
Barry Lyndon
Bob Godfrey - Best Animated Short:
Great
Albert Whitlock - Special Achievement In Visual Effects:
The Hindenberg

1976 (49th) held in 1977

Peter Finch - Best Actor:
Network

1977 (50th) held in 1978

John Barry, Roger Christians & Leslie Dilley - Best Art Direction:
Star Wars
John Mollo - Best Costume Design:
Star Wars
Vanessa Redgrave - Best Supporting Actress:
Julia
John Stears - Best Visual Effects:
Star Wars

1978 (51st) held in 1979

Les Bowie, Colin Chilvers, Denys Coop, Roy Field & Derek Meddings - Special Achievement In Visual Effects:
Superman
Michael Deeley, John Peverall & Barry Spikings - Best Picture:
The Deer Hunter
Laurence Oilvier - Lifetime Achievement Award
Anthony Powell - Best Costume Design:
Death On The Nile
Maggie Smith - Best Supporting Actress:
California Suite

1979 (52nd) held in 1980

Nick Allder, Denis Ayling & Brian Johnson - Special Achievement In Visual Effects: *Alien*
Alec Guinness - Honorary Award
Tony Walton - Best Art Direction:
All that Jazz

1980 (53rd) held in 1981

Brian Johnson - Special Achievement In Visual Effects:
The Empire Strikes Back
Lloyd Phillips - Best Live Action Short:
The Dollar Bottom
Anthony Powell - Best Costume Design:
Tess
David W. Samuelson - Scientific and Engineering Award -
for the engineering and development of the Louma Camera Crane and remote control system for motion picture production
Jack Stevens - Best Art Direction:
Tess
Geoffrey Unsworth - Best Cinematography:
Tess

1981 (54th) held in 1982

Leslie Dilley & Michael Ford - Best Art Direction:
Raiders Of The Lost Ark
John Gielgud - Best Supporting Actor:
Arthur
Nigel Nobel - Best Documentary Short:
Close Harmony
David Puttnam - Best Picture:
Chariots Of Fire
Arnold Schwartzman - Best Documentary Feature:
Close Harmony
Colin Welland - Best Original Screenplay:
Chariots Of Fire
Kit West - Special Achievement In Visual Effects:
Raiders Of The Lost Ark

1982 (55th) held in 1983

Richard Attenborough - Best Picture:
Gandhi
Richard Attenborough - Best Director:
Gandhi
John Briley - Best Original Screenplay:
Gandhi

Stuart Craig, Bob Laing & Michael Seirton - Best Art Direction:
Gandhi
Ben Kingsley - Best Actor:
Gandhi
John Mollo - Best Costume Design:
Gandhi
Sarah Monzani - Best Achievement In Make Up:
Quest For Fire
Colin Mossman & Rank Laboratories - Scientific and Engineering Award -
for the engineering and implementation of a 4,000 meter printing system for motion picture laboratories
Christine Oestreicher - Best Live Action Short:
A Shocking Accident
Ronnie Taylor & Billy Williams - Best Cinematography:
Gandhi

1983 (56th) held in 1984

Gerald L. Turpin (Lightflex International) - Scientific And Engineering Award
- for the design, engineering and development of an on-camera device providing contrast control, sourceless fill light and special effects for motion picture photography

1984 (57th) held in 1985

Peggy Ashcroft - Best Supporting Actress:
A Passage To India
Jim Clark - Best Editing:
The Killing Fields
George Gibbs - Special Achievement In Visual Effects:
Indiana Jones And The Temple Of Doom
Chris Menges - Best Cinematography:
The Killing Fields
Peter Shaffer - Best Adapted Screenplay:
Amadeus

1985 (58th) held in 1986

John Barry - Best Original Score:
Out Of Africa

Stephen Grimes - Best Art Direction:
Out Of Africa
David Watkin - Best Cinematography:
Out Of Africa

1986 (59th) held in 1987

Brian Ackland-Snow & Brian Saregar - Best Art Direction:
A Room With A View
Jenny Beavan & John Bright - Best Costume Design:
A Room With A View
Michael Caine - Best Supporting Actor:
Hannah & Her Sisters
Simon Kaye - Best Sound:
Platoon
Lee Electric Lighting Ltd. - Technical Achievement Award
Chris Menges - Best Cinematography:
The Mission
Peter D. Parks - Technical Achievement Award
William B. Pollard & David W. Samuelson - Technical Achievement Award
John Richardson - Special Achievement In Visual Effects:
Aliens
Claire Simpson - Best Editing:
Platoon
Don Sharpe - Best Sound Effects Editing:
Aliens
Vivienne Verdon-Roe - Best Documentary Short:
Women - For America, For The World

1987 (60th) held in 1988

James Acheson - Best Costume Design:
The Last Emperor
Sean Connery - Best Supporting Actor:
The Untouchables
Mark Peploe - Best Adapted Screenplay:
The Last Emperor

Ivan Sharrock - Best Sound:
The Last Emperor
Jeremy Thomas - Best Picture:
The Last Emperor

1988 (61st) held in 1989

James Acheson - Best Costume Design:
Dangerous Liaisons
George Gibbs - Special Achievement In Visual Effects:
Who Framed Roger Rabbit
Christopher Hampton - Best Adapted Screenplay:
Dangerous Liaisons

1989 (62nd) held in 1990

Phyllis Dalton - Best Costume:
Henry V
Daniel Day-Lewis - Best Actor:
My Left Foot
Freddie Francis - Best Cinematography:
Glory
Brenda Fricker - Best Supporting Actress:
My Left Foot
Anton Furst - Best Art Direction:
Batman
Richard Hymns - Best Sound Effects Editing:
Indiana Jones And The Last Crusade
Jessica Tandy - Best Actress:
Driving Miss Daisy
James Hendrie - Best Live Action Short:
Work Experience

1990 (63rd) held in 1991

John Barry - Best Original Score:
Dances With Wolves
Jeremy Irons - Best Actor:
Reversal Of Fortune
Nick Park - Best Animated Short:
Creature Comforts

1991 (64th) held in 1992

Daniel Greaves - Best Animated Short:
Manipulation
Anthony Hopkins - Best Actor:
Silence Of The Lambs

1992 (65th) held in 1993

Simon Kaye - Best Sound:
The Last Of The Mohicans
Tim Rice - Best Original Song:
Aladdin (A Whole New World)
Emma Thompson - Best Actress:
Howards End
Ian Whittaker - Best Art Direction:
Howards End

1993 (66th) held in 1994

Richard Hymns - Best Sound Effects Editing:
Jurassic Park
Nick Park - Best Animated Short:
The Wrong Trousers
Deborah Kerr - Career Achievement Honorary Award

1994 (67th) held in 1995

Ken Adam & Carolyn Scott - Best Art Direction:
The Madness Of King George
Peter Capaldi & Ruth Kenley-Letts - Best Live Action Short:
Franz Kafka's It's A Wonderful Life
Elton John & Tim Rice - Best Song:
The Lion King (Can You Feel The Love Tonight)
Alison Snowden & David Fine - Best Animated Short:
Bob's Birthday

1995 (68th) held in 1996

James Acheson - Best Costume Design:
Restoration
Jon Blair - Best Documentary Feature:
Anne Frank Remembered
Lois Burwell & Peter Frampton - Special Achievement In Make Up:
Braveheart
Emma Thompson - Best Adapted Screenplay:
Sense & Sensibility
Nick Park - Best Animated Short:
A Close Shave

1996 (69th) held in 1997

Anthony Minghella - Best Director:
The English Patient
Rachel Portman - Best Original Score Musical or Comedy:
Emma
Tim Rice & Andrew Lloyd Webber - Best Original song:
Evita (You Must Love Me)
Stuart Craig & Stephanie McMillan - Best Art Direction:
The English Patient

1997 (70th) held in 1998

Peter Lamont and Michael Ford - Best Achievement In Art Direction:
Titanic
Anne Dudley - Best Original Score Musical or Comedy:
The Full Monty
Jan Pinkava - Best Animated Short:
Geri's Game

BOOKS

Below is a list of some books published recently in the UK on film and television, all of which can be found at the BFI National Library. A short description is provided when the content of the book is not obvious.
Compiled by Anastasia Kerameos

THE ACT OF SEEING: ESSAYS AND CONVERSATIONS
Wenders, Wim
[Transl.] Hofman, Michael
Faber and Faber: London.
057117843X

ADAPTATIONS AS IMITATIONS: FILMS FROM NOVELS
Griffith, James
University of Delaware Press;
Associated University Presses:
Newark; London.
0-87413-633-4

THE ADDRESS BOOK: HOW TO REACH ANYONE WHO IS ANYONE
Levine, Michael
Perigree Books: New York.
0399522743

Series: *BFI Film Classics*
L'AGE D'OR
Hammond, Paul
British Film Institute: London.
0851706428

Series: *Cinema Voices*
AGENT OF CHALLENGE AND DEFIANCE: THE FILMS OF KEN LOACH
[Ed.] McKnight, George
Flicks Books: Trowbridge.
0948911948

Series: *National Film and Television Archive Filmographies Series, 6*
AIDS: THE HOLDINGS OF THE NATIONAL FILM AND TELEVISION ARCHIVE
[Compilers] Terris, Olwen; Baker, Simon
British Film Institute: London.

ALIEN: THE SPECIAL EFFECTS
Shay, Don; Norton, Bill
Titan Books: London.
1-85286-695-0

ALLEGORIES OF UNDERDEVELOPMENT: AESTHETICS AND POLITICS IN MODERN BRAZILIAN CINEMA
Xavier, Ismail
University of Minnesota Press:
Minneapolis.
0-8166-2676-6

ALL OR NOTHING AT ALL: A LIFE OF FRANK SINATRA.
Clark, Donald
Macmillan: London.
0333643208

ALL PALS TOGETHER: THE STORY OF CHILDREN'S CINEMA
Staples, Terry
Edinburgh University Press:
Edinburgh.
0-7486-0718-8

ALL THE WAY: A BIOGRAPHY OF FRANK SINATRA
Freedland, Michael
Weidenfeld & Nicolson: London.
029781723X

AMERICA ON FILM: HOLLYWOOD AND AMERICAN HISTORY
Cameron, Kenneth M.
Continuum: New York.
0-8264-1033-2

AMERICAN FILM INSTITUTE CATALOG: WITHIN OUR GATES: ETHNICITY IN AMERICAN FEATURE FILMS, 1911-1960
[Ed.] Gevinson, Alan
[Corp. Au.] American Film Institute
University of California Press:
Berkeley.
0520209648

AMERICAN FILMS ABROAD: HOLLYWOOD'S DOMINATION OF THE WORLD'S MOVIE SCREENS FROM THE 1890S TO THE PRESENT.
Segrave, Kerry
McFarland & Company: Jefferson, NC
0-7864-0346-2

AN ANAGRAM OF THE IDEAS OF FILMMAKER MAYA DEREN: CREATIVE WORK IN MOTION PICTURES
Sullivan, Moira
University of Karlstad Press:
Karlstad.
9171535829

THE ART OF LOOKING IN HITCHCOCK'S REAR WINDOW
Sharff, Stefan

Limelight Editions: New York.
0879100877

THE ARTS COUNCIL OF ENGLAND NATIONAL LOTTERY FILM PROGRAMME: CONSULTATION PAPER ON PROJECT DEVELOPMENT
Arts Council of England
Arts Council of England: London.

ASIA PRODUCER'S GUIDE: THE ESSENTIAL GUIDE TO PARTNERING IN PRODUCTION: VOLUME ONE
Walker, Sarah
FT Telecoms & Media Publishing:
London.

AUDIENCE ANALYSIS
McQuail, Denis
Sage Publications: Thousand Oaks.
0761910026

AUDIO FOR TELEVISION
Watkinson, John
Focal Press: Oxford.
0240514645

AUDIOVISUAL POLICY OF THE EUROPEAN UNION 1998: THE NEW ERA OF THE PICTURE INDUSTRY, TELEVISION WITHOUT FRONTIERS, GREATER EUROPE IN THE YEAR 2000
Baer, Jean-Michel
Office for the Official Publications of the European Commission:
Brussels.
9282821552

AUSTRALIAN FILM: A BIBLIOGRAPHY
Reis, Brian
Cassell: London.
0-7201-2315-1

THE AUTHORISED BIOGRAPHY OF DUDLEY MOORE
Paskin, Barbra
Sidgwick & Jackson: London.
HARDBACK 0-283-06264-9

AN AUTOBIOGRAPHY OF BRITISH CINEMA: AS TOLD BY THE FILMMAKERS AND ACTORS WHO MADE IT
McFarlane, Brian

Methuen: London.
041370520X

 THE AVENGERS
Miller, Toby
British Film Institute: London.
PAPERBACK 0-85170-558-8

THE AVENGERS AND ME
Macnee, Patrick; Rogers, Dave
Titan Books: London.
PAPERBACK 1-85286-801-5

THE AVENGERS COMPANION
Carraze, Alain; Putheaud, Jean-Luc
Titan Books: London.
PAPERBACK 1-85286-728-0

BACKSTORY 3: INTERVIEWS WITH SCREENWRITERS OF THE 1960S
[Ed.] McGilligan, Patrick
University of California: Berkeley, Ca.
0520204271

 BACKTRACKS
BFI/Illuminations
British Film Institute: London.
1-86215-157-1

THE BARRY DILLER STORY
Mair, George
John Wiley and Sons: New York.
HARDBACK 0-471-13082-6

BAZIN AT WORK: MAJOR ESSAYS & REVIEWS FROM THE FORTIES & FIFTIES
Bazin, Andre
[Transl.] Piette, Alain
[Ed.] Cardullo, Bert
Routledge: New York.
0415900182

Series: *House of Commons Papers - Session 1996-97 (147-I)*
THE BBC AND THE FUTURE OF BROADCASTING, FOURTH REPORT: VOLUME I, REPORT AND PROCEEDINGS OF THE COMMITTEE.
Great Britain: House of Commons:
National Heritage Committee
[Chairman] Kaufman, Gerald
The Stationary Office: London.
0102288976

Series: *House of Commons Papers - Session 1996-97 (147-II)*
THE BBC AND THE FUTURE OF BROADCASTING, FOURTH REPORT: VOLUME II, MINUTES OF EVIDENCE AND APPENDICES
Great Britain: House of Commons:
National Heritage Committee
[Chairman] Kaufman, Gerald
The Stationary Office: London.
0102187975

Series: *Rethinking British Cinema*
THE BEATLES' MOVIES
Neaverson, Bob
Cassell: London.
HARDBACK 0-304-33796-X
PAPERBACK 0-304-33797-8

BECOMING MAE WEST
Leider, Emily Wortis
Farrar Straus & Giroux: New York.
0-374-10959-1

Series: *Filmmakers, 56*
BEFORE, IN AND AFTER HOLLYWOOD: THE AUTOBIOGRAPHY OF JOSEPH E. HENABERY
Henabery, Joseph E.
[Ed] Slide, Anthony
Scarecrow Press: Lanham, MD.
0-8108-3200-3

THE BEGINNING FILMAKER'S BUSINESS GUIDE TO A SUCCESSFUL FIRST FILM
Harmon, Renee; Lawrence, Jim
Walker and Company: .
0-8027-7409-1

THE BEGINNINGS OF THE CINEMA IN ENGLAND, 1894-1901: 1900: Vol 5
Barnes, John
University of Exeter Press: Exeter.
0-85989-522-X

BENEATH MULHOLLAND: THOUGHTS ON HOLLYWOOD AND ITS GHOSTS
Thomson, David
Alfred A. Knopf: New York.
0-679-45115-3

THE BENT LENS: A WORLD GUIDE TO GAY AND LESBIAN FILM
[Eds.] Jackson, Claire; Tapp, Peter
Australian Catalogue Company: St. Kilda Victoria.
PAPERBACK 0-646-30818-1

BE SEEING YOU: DECODING THE PRISONER
Gregory, Chris
John Libbey Media/University of Luton Press: Luton, Bedford.
PAPERBACK 1-86020-521-6

BETWEEN THE SHEETS, IN THE STREETS: QUEER, LESBIAN, AND GAY DOCUMENTARY
Holmlund, Chris; Fuchs, Cynthia
University of Minnesota Press: Minneapolis, MN.
HARDBACK 0-8166-2774-6
PAPERBACK 0-8166-2775-4

 THE BFI COMPANION TO CRIME
[Ed.] Hardy, Phil
Cassell; British Film Institute:

London.
HARDBACK 0-304-33211-9
PAPERBACK 0-304-33215-1

 BFI FILM AND TELEVISION HANDBOOK 1997
[Ed] Dyja, Eddie
British Film Institute: London.
PAPERBACK 0-85170-637-1

THE BIG FIVE: (LEWISHAM'S SUPER CINEMAS)
George, Ken
Ken George: London.
0953209806

THE BIG LEBOWSKI: THE MAKING OF A COEN BROTHERS FILM
Cook, Trisia
W.W. Norton & Co.: New York; London.
0393317501

Series: *BFI Film Classics*
THE BIG SLEEP
Thomson, David
British Film Institute: London.
9851706320

THE BIRDCAGE
May, Elaine
Newmarket Press: New York.
1557042772

A BIOGRAPHICAL HANDBOOK OF HISPANICS AND UNITED STATES FILM
Keller, Gary D.
Bilingual Press: Tempe, AZ.
0-927534-65-7

THE BIOGRAPHY OF GEORGE CLOONEY
Dougan, Andrew
Boxtree: London.
PAPERBACK 0-7522-1146-3

BISEXUAL CHARACTERS IN FILM: FROM ANAïS TO ZEE
Bryant, Wayne M
Haworth Press: Binghamton, New Hampshire.
078900142X

THE BLACKBOARD JUNGLE
Hunter, Evan
Bloomsbury Publishing: London.
0747531846

Series: *BFI Modern Classics*
 BLADE RUNNER
Bukatman, Scott
British Film Institute: London.
0851706231

Series: *BFI Modern Classics*
 BLUE VELVET
Atkinson, Michael
British Film Institute: London.
0851705596

BOGART
Sperber, A. M.; Lax, Eric
Weidenfeld & Nicolson: London.
0297812750

BOGART: A LIFE IN HOLLY-WOOD
Meyers, Jeffrey
Andre Deutsch: London.
0-233-99144-1

BOOKING HAWAII FIVE-O: AN EPISODE GUIDE AND CRITI-CAL HISTORY OF THE 1968-1980 TELEVISION DETECTIVE SERIES
Rhodes, Karen
McFarland & Company: Jefferson, NC.
HARDBACK 0-7864-0171-0

THE BOOK OF FILM BIOGRA-PHIES: A PICTORIAL GUIDE OF 1000 MAKERS OF THE CINEMA
[Ed] Morgan, Robin; Perry, George
Fromm International: New York.
0-88064-185-1

BOTTICELLI IN HOLLYWOOD: THE FILMS OF ALBERT LEWIN
Felleman, Susan
Twayne Publishers; Prentice Hall International: New York; London.
PAPERBACK 0-8057-1625-4

BOX OF MOONLIGHT & NOTES FROM OVERBOARD
Di Cillo, Tom
Faber and Faber: London.
0-571-19169-X

BRAD PITT
Mundy, Chris
Little, Brown and Company: Boston, Ma.
0316893609

BRAND IDENTITY FOR TELEVISION: WITH KNOBS ON
Lambie-Nairn, Martin
[Ed] Myerson, Jeremy
Phaidon Press: London.
HARDBACK 0-7148-3447-5

Series: BFI Film Classics
 **BRIDE OF FRANKENSTEIN**
Manguel, Alberto
British Film Institute: London.
0851706088

Series: *Film Studies*
BRIGHT DARKNESS: THE LOST ART OF THE SUPER-NATURAL HORROR FILM
Dyson, Jeremy
Cassell: London.
PAPERBACK 0-304-34038-3
HARDBACK 0-304-70037-1

THE BRITISH CINEMA BOOK
[Ed.] Murphy, Robert
British Film Institute: London.
HARDBACK 0-85170-640-1
PAPERBACK 0-85170-641-X

THE BRITISH CO-OPERATIVE MOVEMENT FILM CATALOGUE
[Ed.] Burton, Alan
Trowbridge: Flicks Books
HARDBACK 0-948911-77-8

BRITISH FILM INSTITUTE TELEVISION INDUSTRY TRACKING STUDY: SECOND INTERIM REPORT
British Film Institute; Pettigrew, Nick
British Film Institute: London.

Series: *National Cinema Series*
BRITISH NATIONAL CINEMA
Street, Sarah
Routledge: London.
HARDBACK 0-415-06735-9
PAPERBACK 0-415-06736-7

Series: *Studies in Film, Television and the Media*
BRITISH POPULAR FILMS 1929-39: THE CINEMA OF REASSURANCE
Shafer, Craig
Routledge: London.
HARDBACK 0-415-00282-6

BROADCAST/CABLE PRO-GRAMMING: STRATEGIES AND PRACTICES
Eastman, Susan Tyler; Ferguson, Douglas A.
Wadsworth Publishing Company: Belmont, Ca.
0534507441

THE BROADCAST CENTURY: A BIOGRAPHY OF AMERICAN BROADCASTING
Hilliard, Robert L.; Keith, Michael C.
[2nd ed.] Focal Press: Boston, Ma.
0240802624

BROADCASTING: THE BROAD-CASTING ACT 1996, CHAPTER 55
Great Britain: Statutes
HMSO: London.
0105455962

BROADCAST JOURNALISM: TECHNIQUES OF RADIO & TV NEWS
Boyd, Andrew
[4th ed.] Focal Press: Oxford.
0240514653

Series: *Bloomsbury Film Classics*
BULLITT
Pike, Robert L.
Bloomsbury Publishing: London
0747531854

Series: *Cambridge Film Handbooks Series*
BUSTER KEATON'S SHERLOCK JR.
Horton, Andrew
Cambridge University Press: Cambridge.
0521485665

CAREER GIRLS
Leigh, Mike
Faber and Faber: London.
0571194044

CARLA'S SONG
Laverty, Paul
Faber and Faber: London.
0571191622

CELEBRITY AND POWER: FAME IN CONTEMPORARY CULTURE
Marshall, P. David
University of Minnesota Press: Minneapolis.
0816627258

CELLULOID MIRRORS: HOLLYWOOD AND AMERICAN SOCIETY SINCE 1945
Davis, Ronald L.
Harcourt Brace College Publication: Fort Worth, Tx.
0155015680

CENSORED SCREAMS: THE BRITISH BAN ON HOLLY-WOOD HORROR IN THE THIRTIES
Johnson, Tom
McFarland & Company: Jefferson, NC.
HARDBACK 0-7864-0394-2

CHANGING CHANNELS: TELEVISION AND THE STRUGGLE FOR POWER IN RUSSIA
Mickiewicz, Ellen
Oxford University Press: New York; Oxford.
0195101634

CHANNEL SURFING: RACE TALK AND THE DESTRUC-TION OF TODAY'S YOUTH
Giroux, Henry A.; Clark, Larry
Macmillan: Basingstoke.
0333720253

CHAOS AS USUAL: CONVER-SATIONS ABOUT RAINER WERNER FASSBINDER
[Ed.] Lorenz, Juliane
Schmid, Marion; Gehr, Herbert
Applause: New York.
1557832625

CHARLIE CHAPLIN: AND HIS TIMES
Lynn, Kenneth S.

Simon & Schuster: New York.
HARDBACK 0-684-80851-X

CHILDREN & TELEVISION
Gunter, Barrie; Mcaleer, Jill
Routledge: London.
0415144523

Series: *BFI Film Classics*
 CHINATOWN
Eaton, Michael
British Film Institute: London.
0851705324

**THE CHINESE FILMOGRAPHY:
THE 2,444 FEATURE FILMS
PRODUCED BY STUDIOS IN
THE PEOPLE'S REPUBLIC OF
CHINA FROM 1949 THROUGH
1995**
Marion, Donald J.
McFarland & Company: Jefferson,
NC.
HARDBACK 0-7864-0305-5

**CHINESE MODERNISM IN THE
ERA OF REFORMS: CULTURAL
FEVER, AVANT-GARDE
FICTION AND THE NEW
CHINESE CINEMA**
Zhang, Xudong
Duke University Press: Durham,
NC; London.
0822318466

**CHINESE SILENT FILM
HISTORY**
Li, Suyuan; Hu, Jubin
[Ed.] Wang, Rvi
China Film Press: Beijing.
7106012599

 **CINEMA & ARCHITECT-
URE: MÉLIÈS, MALLET-
STEVENS, MULTIMEDIA**
[Eds.] Penz, Francois; Thomas,
Maureen
British Film Institute: London.
0851705782

**CINEMA AND THE GREAT
WAR**
Kelly, Andrew
Routledge: London; New York.
0415052033

**THE CINEMA OF MARTIN
SCORSESE**
Friedman Lawrence S.
Roundhouse Publishing: Oxford.
PAPERBACK 1-85710-027-1

Series: *Contemporary Film and
Television Series*
**THE CINEMA OF WIM
WENDERS: IMAGE, NARRA-
TIVE AND THE POSTMODERN
CONDITION**
[Ed] Cook, Roger F.; Gemuenden,
Gerd
Wayne State University Press:

Detroit, MI.
PAPERBACK 0-8143-2578-5

**CINEMA, THEORY, AND
POLITICAL RESPONSIBILITY
IN CONTEMPORARY CUL-
TURE**
McGee, Patrick
Cambridge University Press:
Cambridge.
0521581303

**CINEMAGOING EUROPE:
SCANDINAVIA & FINLAND**
Couling, Katherine; Grummitt,
Karsten-Peter
Dodona Research: Leicester.
1872025757

**CINEMAS AND FILMS: THE
EUROPEAN CONVENTION ON
CINEMATOGRAPHIC CO-
PRODUCTION (AMENDMENT)
ORDER 1997**
Great Britain: Statutory Instruments
The Stationery Office: London.
011065297X
0110637151

**CINEMAS IN BRITAIN: ONE
HUNDRED YEARS OF CINEMA
ARCHITECTURE**
Gray, Richard
Lund Humphries: London.
0853316856

**THE CINEMAS OF CAMDEN: A
SURVEY AND HISTORY OF
THE CINEMA BUILDINGS OF
CAMDEN, PAST AND PRESENT**
Aston, Mark
London Borough of Camden:
London.
0901389889

**CINEMATERNITY: FILM,
MOTHERHOOD, GENRE**
Fischer, Lucy
Princeton University Press:
Princeton, NJ.
0691037744

THE CINEMATIC CITY
Clarke, David B.
Routledge: London.
0415127467

**CINEMATIC USES OF THE
PAST**
Landy, Marcia
University of Minnesota Press:
Minneapolis.
0816628254

**CITY OF DREAMS : THE
MAKING AND REMAKING OF
UNIVERSAL PICTURES**
Dick, Bernard F.
University Press of Kentucky:
Lexington, KY.
HARDBACK 0-8131-2016-0

**CLERKS AND CHASING AMY:
TWO SCREENPLAYS**
Smith, Kevin
Miramax Books/Hyperion: New
York.
0786882638

**CLIMBING THE MOUNTAIN:
MY SEARCH FOR MEANING**
Douglas, Kirk
Simon & Schuster: New York.
068484415X

**COLLECTING MOVIE
POSTERS: AN ILLUSTRATED
REFERENCE GUIDE TO
COLLECTING MOVIE ART -
POSTERS, PRESS KITS AND
LOBBY CARDS**
Poole, Ed; Poole, Susan
McFarland & Company: Jefferson, NC.
PAPERBACK 0-7864-0169-9

**COMIC VISIONS: TELEVISION
COMEDY AND AMERICAN
CULTURE**
Marc, David
[2nd ed.] Blackwell Publishers: Oxford.
HARDBACK 1-57718-002-X
PAPERBACK 1-57718-003-8

**THE COMPLETE FILM
DICTIONARY**
Konigsburg, Ira
[2nd ed.] Penguin Reference: New York.
HARDBACK 0-6701-0009-9

**THE COMPLETE FILM
PRODUCTION HANDBOOK**
Honthaner, Eve Light
Focal Press: Boston.
0240802365

A CONFESSION IN WRITING
Shaugnessy, Alfred
Tabb House: Padstow.
HARDBACK 1-873951-30-2

CONRAD ON FILM
Moore, Gene M.
Cambridge University Press: Cambridge.
0521554489

**CONSUMING TELEVISION:
TELEVISION AND ITS
AUDIENCES**
Mullan, Bob
Blackwell Publishers: Oxford.
HARDBACK 0-631-20233-1
PAPERBACK 0-631-20234-X

COPLAND & HEAVY
Mangold, James
Faber and Faber: London.
0571194257

**COPYRIGHTS AND
TRADEMARKS FOR MEDIA
PROFESSIONALS**
Lutzker, Arnold P.
Focal Press: Boston, Ma.
0240802764

COURTNEY LOVE: THE REAL STORY
Brite, Poppy Z.
Orion Books: London.
HARDBACK 0-7528-1245-9
PAPERBACK 0-7528-1337-4

CREATING LOCAL TELEVISION: LOCAL AND COMMUNITY TELEVISION UNDER THE RESTRICTED SERVICES LICENCE
Rushton, Dave
Institute of Local Television: Edinburgh.
1899405011

THE "CRIME TIME" FILMBOOK: THE YEAR IN CRIME FILMS
[Ed.] Ashbrook, John
No Exit Press: Harpenden.
PAPERBACK 1-87406-184-X

A CRITICAL HISTORY AND FILMOGRAPHY OF TOHO'S GODZILLA SERIES
Kalat, David
McFarland & Co: Jefferson, NC.
0-7864-0300-4

CRONENBERG ON CRONENBERG
Rodley, Chris
[Rev. Ed.] Faber and Faber: London.
0571191371

CROSS-CULTURAL FILMMAKING: A HANDBOOK FOR MAKING DOCUMENTARY AND ETHNOGRAPHIC FILMS & VIDEOS
Barbarsh, Ilisa; Taylor, Lucien
University of California Press: Berkeley, Ca.
0520087607

THE CROWDED PRAIRIE: AMERICAN NATIONAL IDENTITY IN THE HOLLYWOOD WESTERN
Coyne, Michael
I.B. Tauris: London.
1860640400

Series: *BFI Modern Classics*
THE CRYING GAME
Giles, Jane
British Film Institute: London.
0851705561

Series: *Depth of Field Series*
DEFINING CINEMA
[Ed.] Lehman, Peter
Rutgers University Press: New Brunswick.
PAPERBACK 0-8135-2302-8

DEFINING VISION: THE BATTLE FOR THE FUTURE OF TELEVISION
Brinkley, Joel

Harcourt Brace & Co.: New York.
0151000875

DENZEL WASHINGTON: HIS FILMS AND CAREER
Brode, Douglas
Birch Lane Press: Secaucus, NJ.
PAPERBACK 1-55972-381-5

DESIGNING DISNEY'S THEME PARKS: THE ARCHITECTURE OF REASSURANCE
Marling, Karal Ann
Flammarion: Paris.
2080136399

DIAMONDS IN THE DARK: AMERICA, BASEBALL, AND THE MOVIES
Good, Howard
Scarecrow Press: Langham, MD.
HARDBACK 0-8108-3047-7

A DICTIONARY OF COMMUNICATION AND MEDIA STUDIES
Watson, James; Hill, Ann
Arnold: London.
0340676353

DIGITAL CINEMATOGRAPHY
Leeuw, Ben de
AP Professional: London.
0122088751

DIGITAL TELEVISION AND INTERACTIVE MULTIMEDIA SERVICES: MARKET TRENDS AND INFRASTRUCTURES COMPETITION
Uzah, Laurence; IDATE
IDATE: Montpellier.

DIRECTING: FILM TECHNIQUES AND AESTHETICS
Rabinger, Michael
[2nd ed.] Focal Press: Boston.
0240802233

THE DIRECTOR'S JOURNEY : THE CREATIVE COLLABORATION BETWEEN DIRECTORS, WRITERS AND ACTORS
Travis, Mark W.
Michael Wiese Film Productions: Studio City, CA.
PAPERBACK 0-941188-59-0

THE DISNEY TOUCH: DISNEY, ABC AND THE QUEST FOR THE WORLD'S GREATEST MEDIA EMPIRE
Grover, Ron
Irwin Publishing: Burr Ridge, IL.
0786310022

DIVA
Delacorta
Bloomsbury Publishing: London.
074753182X

DOCUMENTARY FILM CLASSICS
Rothman, William

Cambridge University Pres:
Cambridge.
0521456819

DOCUMENTARY FOR THE SMALL SCREEN
Kriwaczek, Paul
Focal Press: Oxford.
PAPERBACK 0-240-51472-6

DRACULA IN THE DARK: THE DRACULA FILM ADAPTATIONS
Holte, James Craig
Greenwood Press: Westport, Ct.
0313292159

DREAM MERCHANTS, POLITICIANS AND PARTITION: MEMOIRS OF AN INDIAN MUSLIM
Masud, Iqbal
HarperCollins Publishing India: New Delhi.
8172232624

ECHOES OF AN AUTOBIOGRAPHY
Mahfouz, Naguib
[Transl.] Johnson-Davies, Denys
Doubleday: London.
0385408390

Series: *Biobibliographies in the Performing Arts, no. 73*
EDDIE CANTOR: A BIO-BIBLIOGRAPHY
Fisher, James
Greenwood Press: Westport, CT.
0313295565

EISENSTEIN: A LIFE IN CONFLICT
Bergan, Ronald
Little, Brown and Company: London.
0316877085

ELECTRONIC MOVIEMAKING
Gross, Lynne S.; Ward, Larry W.
[3rd ed.] Wadsworth Publishing Co.: London.

THE ELVIS FILM ENCYCLOPEDIA
Braun, Eric
B.T. Batsford: London.
PAPERBACK 0-7134-8128-5

AN EMBARRASSMENT OF TYRANNIES: TWENTY-FIVE YEARS OF INDEX ON CENSORSHIP
[Eds.] Webb, W.L; Bell, Rose
Victor Gollancz: London.
0575065389

EMPIRE BUILDING: THE REMARKABLE, REAL LIFE STORY OF STAR WARS
Jenkins, Garry
Simon & Schuster: London.
HARDBACK 0-684-82091-9
PAPERBACK 0-8065-1941-X

**ENCYCLOPEDIA OF
TELEVISION**
[Ed.] Newcomb, Horace
Fitzroy Dearborn: Chicago; London.
HARDBACK 1-884964-26-5

**THE ENCYCLOPEDIA OF TV
SCIENCE FICTION**
Fulton, Roger
[3rd ed.] Boxtree: London.
PAPERBACK 0-7522-1150-1

Series: *BFI Film Classics*
LES ENFANTS DU PARIDIS
Forbes, Jill
British Film Institute: London.
0851703658

Series: *Modern French Writers*
**THE EROTICS OF PASSAGE:
PLEASURE, POLITICS, AND
FORM IN THE LATER WORK
OF MARGUERITE DURAS**
Williams, James S.
Liverpool University Press:
Liverpool.
0853239908

Series: *BFI Modern Classics*
THE EXORCIST
Kermode, Mark
British Film Institute: London.
0851706223

**EXPERIMENTAL TELEVISION,
TEST FILMS, PILOTS AND
TRIAL SERIES, 1925 THROUGH
1995: SEVEN DECADES OF
SMALL SCREEN ALMOSTS**
Terrace, Vincent
McFarland & Co.: Jefferson, NC.
0786401788

**EXPLORATIONS IN
THEOLOGY AND FILM**
[Eds.] Marsh, Clive; Ortiz, Gaye
Blackwell Publishing: Oxford.
0631203559

**EXPLORING SPACE 1999: AN
EPISODE GUIDE AND
COMPLETE HISTORY OF THE
MID 1970'S SCIENCE FICTION
TELEVISION SERIES**
Muir, John Kenneth
McFarland & Company: Jefferson,
NC.
HARDBACK 0-7864-0165-6

**EYE ON THE WORLD:
CONVERSATIONS WITH
INTERNATIONAL FILMMAKERS**
Stone, Judy
Silman James Press: Beverley Hills, CA.
PAPERBACK 1-879505-36-3

**FADE IN: THE
SCREENWRITING PROCESS**
Berman, Robert A.
Michael Wiese Productions: Studio City.
0941188582

**FAMILY FICTIONS : REPRE-
SENTATIONS OF THE FAMILY
IN 1980S HOLLYWOOD
CINEMA**
Harwood, Sarah
Macmillan: Houndmills, Berkshire.
0-3334-4844-7

**FASSBINDER: THE LIFE AND
WORK OF A PROVOCATIVE
GENIUS**
Thomsen, Christian Brand
Faber and Faber: London.
HARDBACK 0-571-17842-1

**THE FEATURE FILM DISTRI-
BUTION DEAL: A CRITICAL
ANALYSIS OF THE SINGLE
MOST IMPORTANT FILM
INDUSTRY AGREEMENT**
Cones, Jones
Southern Illinois University Press:
Carbondale, IL.
0809320827

FEMINISM AND FILM
Humm, Maggie
Edinburgh University Press:
Edinburgh.
0-253-21146-8

Series: *Oxford Television Studies*
**FEMINIST TELEVISION
CRITICISM: A READER**
[Eds.] Brunsdon, Charlotte; D'Acci,
Julie; Spigel, Lynn
Oxford University Press: Oxford.
HARDBACK 0-19-871152-2

Series: *Cassell Film Studies*
**FILM AND CENSORSHIP: THE
INDEX READER**
[Ed.] Petrie, Ruth
Cassell: London.
HARDBACK 0-304-33936-9
PAPERBACK 0-304-33937-7

**FILM ART: AN
INTRODUCTION**
Bordwell, David
[5th ed.] McGraw-Hill: New York.
0070066345

**FILM COMPOSERS GUIDE
1997-1998 (4TH ED)**
[Ed.] Francillion, Vincent-Jacquet
Lone Eagle: Los Angeles, CA.
PAPERBACK 0-943728-93-2

**FILM DIRECTORS GUIDE: 1997
(12TH ED)**
[Ed.] Singer, Michael
Lone Eagle: Los Angeles, CA.
PAPERBACK 0-943728-85-1

FILM ESSAYS AND CRITICISM
Arnheim, Rudolf
[Transl.] Benthien, Brenda
University of Wisconsin Press:
Madison.
0299152642

**FILMING T.E. LAWRENCE:
KORDA'S LOST EPICS**
[Eds.] Kelly, Andrew; Richards,
Jeffrey; Pepper, James
I.B. Tauris: London.
HARDBACK 1-86064-048-6

**bfi FILM IN VICTORIAN
BRITAIN: A TEACHING
PACK FOR PRIMARY
SCHOOLS**
Stapels, Terry
[Illus.] Ecuyer, Barry
British Film Institute: London.
0851706053

**FILMMAKERS IN THE
MOVING PICTURE WORLD:
AN INDEX OF ARTICLES**
D'Agostino, Annette M.
McFarland & Company: Jefferson,
NC.
HARDBACK 0-7864-0290-3

**FILM NATION: HOLLYWOOD
LOOKS AT US HISTORY**
Burgoyne, Robert
University of Minnesota Press:
Minneapolis.
HARDBACK 0-8166-2070-9
PAPERBACK 0-8166-2071-7

FILM REVIEW 1997-98
Cameron-Wilson, James
Virgin Publishing: London.
PAPERBACK 0-7535-0108-2

Series: *Studies in Popular Culture*
**FILMS AND BRITISH NA-
TIONAL IDENTITY: FROM
DICKENS TO "DAD'S ARMY"**
Richards, Jeffrey
Manchester University Press:
Manchester.
HARDBACK 0-7190-4742-0
PAPERBACK 0-7190-4743-9

**bfi FILM SOCIETY HAND-
BOOK**
Cargin, Peter
[3rd ed.] British Film Institute:
London.

**THE FILMS OF HARRISON
FORD**
Pfeiffer, Lee; Lewis, Michael
Carol Publishing Group: Secaucus,
NJ.
0806516585

Series: *Suny Series, Cultural
Studies in Cinema/Video*
**THE FILMS OF JEAN-LUC
GODARD**
Dixon, Wheeler Winston
State University of New York Press:
Albany, NY.
HARDBACK 0-7914-3285-8
PAPERBACK 0-7914-3286-6

THE FILMS OF MEL GIBSON
McCarty, John
Citadel Press: Secaucus, NJ.
PAPERBACK 0-8065-1918-5

THE FILMS OF MERCHANT IVORY
Long, Robert Emmet
[Rev. ed.] Harry N Abrams: New York.
HARDBACK 0-8109-3618-6

Series: *Filmmakers, 53*
THE FILMS OF MICHAEL POWELL AND THE ARCHERS
Salwolke, Scott
Scarecrow Press: Lanham, MD.
HARDBACK 0-8108-3183-X

Series: *Filmmakers, 55*
THE FILMS OF OLIVER STONE
Kunz, Don
Scarecrow Press: Lanham, MD.
HARDBACK 0-8108-3297-6

THE FILMS OF THEO ANGELOPOULOS: A CINEMA OF CONTEMPLATION
Horton, Andrew
Princeton University Press: Princeton, NJ.
0691011419

FILM TECHNOLOGY IN POST PRODUCTION
Case, Dominic
Focal Press: Oxford.
0240514637

FILM THEORY AND PHILOSOPHY
[Eds.] Allen, Richard; Smith, Murray
Oxford University Press: Oxford.
0198159218

THE FINEST CREW IN THE FLEET: THE NEXT GENERATION CAST ON SCREEN AND OFF
Shrager, Adam
Wolf Valley Books: New York.
1888149035

FLATLINING ON THE FIELD OF DREAMS: CULTURAL NARRATIVES IN THE FILMS OF PRESIDENT REAGAN'S AMERICA
Nadel, Alan
Rutgers University Press: New Brunswick.
HARDBACK 0-8135-2439-3
PAPERBACK 0-8135-2440-7

FLICKERING SHADOWS: A LIFETIME IN FILM
Mitchell, John
Harold Martin & Redman Ltd.: Malvern Wells.
1901394018

 41ST LONDON FILM FESTIVAL: SOUVENIR PROGRAMME
London Film Festival Office
British Film Institute: London.
0-85170-654-1

FOSTER CHILD: A BIOGRAPHY OF JODIE FOSTER
Foster, Buddy; Wagener, Leon
Dutton: New York.
HARDBACK 0-525-94143-6

400 VIDEOS YOU'VE GOT TO RENT!: GREAT MOVIES YOU'VE PROBABLY MISSED
Sillick, Ardis; McCormick, Michael
Carroll & Graf: New York.
0-7867-0397-0

FRAME BY FRAME II: A FILMOGRAPHY OF THE AFRICAN AMERICAN IMAGE, 1978-1994.
Klotman, Phyllis Rauch; Gibson, Gloria J.
Indiana University Press: Bloomington.
025333280X

 FRAMED: INTERROGATING DISABILITY IN THE MEDIA
[Eds.] Pointon, Ann; Davies, Chris
British Film Institute: London.
0851706002

Series: *Hispanic Issues*
FRAMING LATIN AMERICAN CINEMA: CONTEMPORARY CRITICAL PERSPECTIVES
[Ed] Stock, Ann Marie
University of Minnesota Press: Minneapolis.
HARDBACK 0-8166-2972-2
PAPERBACK 0-8166-2973-0

FRANK'S 500: THE THRILLER FILM GUIDE
Frank, Alan
B.T.Batsford: London.
PAPERBACK 0-7134-2728-0

Series: *Bio-Bibliographies in the Performing Arts, No 76*
FRED ASTAIRE: A BIO-BIBLIOGRAPHY
Billman, Larry
Greenwood Press: Westport, Connecticut.
HARDBACK 0-313-29010-5

FRENCH CINEMA IN THE 1980S: NOSTALGIA AND THE CRISIS OF MASCULINITY
Powrie, Phil
Clarendon Press: Oxford.
HARDBACK 0-19-871118-2
PAPERBACK 0-19-871119-0

FRITZ LANG: THE NATURE OF THE BEAST
McGilligan, Patrick
St. Martin's Press: New York.
HARDBACK 0-3121-3247-6

FROM HEADLINE HUNTER TO SUPERMAN: A JOURNALISM FILMOGRAPHY
Ness, Richard R.
Scarecrow Press: Langham, MD.
HARDBACK 0-8108-3291-7

FROM LIVERPOOL TO LOS ANGELES
Ansorge, Peter
Faber and Faber: London
PAPERBACK 0-571-17912-6

Series: *Filmmakers, 53*
FROM OZ TO ET: WALLY WORSLEY'S HALF-CENTURY IN HOLLYWOOD: A MEMOIR IN COLLABORATION WITH SUE DWIGGINS
Worsley, Wally; Dwiggins, Sue
Scarecrow Press: Langham, MD.
HARDBACK 0-8108-3277-1

FROM SOCIOLOGY TO CULTURAL STUDIES: NEW PERSPECTIVES
Long, Elizabeth
Blackwell: Malden, MA.
1577180135

FULL CIRCLE
Palin, Michael
BBC Books: London.
0771069073

GENDER AND REPRESENTATION IN THE FILMS OF INGMAR BERGMAN
Blackwell, Marilyn Johns
Camden House Inc.: Columbia, SC.
1571130942

GEN X TV: THE BRADY BUNCH TO MELROSE PLACE
Owen, Rob
Syracuse University Press: New York.
0815604432

GEORGE LUCAS: THE CREATIVE IMPULSE: LUCASFILM'S FIRST TWENTY-FIVE YEARS
Champlin, Charles
[Rev. and updated Ed.] Virgin Publishing: London.
HARDBACK 1-85227-721-1

Series: *Contemporary interventions*
GILLES DELEUZE'S TIME MACHINE
Rodowick, D. N.
Duke University Press: Durham, Nc.
HARDBACK 0-8223-1962-4
PAPERBACK 0-8223-1970-5

GIRLS' OWN STORIES: AUSTRALIAN AND NEW ZEALAND WOMEN'S FILMS
Robson, Jocelyn; Zalcock, Beverley
Scarlet Press: London.
PAPERBACK 1-85727-053-3

GLOBALIZATION, LIBERALIZATION AND POLICY CHANGE: A POLITICAL ECONOMY OF INDIA'S COMMUNICATIONS SECTOR
McDowell, Stephen D.
MacMillan Press: Houndsmills, Berkshire.
0333657624

GLOBAL SPOTLIGHTS ON LILLEHAMMER: HOW THE WORLD VIEWED NORWAY DURING THE 1994 WINTER OLYMPICS
Puijx, Roel
John Libbey Media/University of Luton Press: Luton.
1860205208

GLOBAL TELEVISION: AN INTRODUCTION
Barker, Christopher John
Blackwell Publishers: Oxford.
HARDBACK 0-631-20149-1

GLUED TO THE SET: THE 60 TELEVISION SHOWS AND EVENTS THAT MADE US WHO WE ARE TODAY
Stark, Steven D.
The Free Press: New York.
HARDBACK 0-684-82817-0

THE GODFATHER BOOK
Cowie, Peter
Faber and Faber: London.
PAPERBACK 0-571-19011-1

THE GODFATHER LEGACY
Lebo, Harlan
Fireside Books: New York.
PAPERBACK 0-684-83647-5

GRACIE FIELDS: A BIOGRAPHY
Moules, Joan
Summerdale: Chichester.
1840240016

GRAMOPHONE FILM MUSIC GOOD CD GUIDE: 1997
[Ed.] Walker, Mark
Gramophone Publications: Harrow.
PAPERBACK 0-902470-83-3

GREAT AFRICAN AMERICANS IN FILM
Parker, Janice
Crabtree Publishing Co.: New York.
0865058229

GRETA GARBO: A LIFE APART
Swenson, Karen

Scribner: New York.
HARDBACK 0-684-80725-4

Series: *Bibliographies and Indexes in American History, No 35*
A GUIDE TO FILMS ON THE KOREAN WAR
Edwards, Paul M.
Greenwood Press: Westport, CT.
HARDBACK 0-313-30316-9

GUIDE TO THE CINEMA OF SPAIN
D'Lugo, Marvin
Greenwood Press: Westport, CT.
0313294747

HAL'S LEGACY: 2001'S COMPUTER AS DREAM AND REALITY
Stork, David G.
MIT Press: Cambridge, MA.
0262193787

THE HAMMER STORY
Barnes, Alan; Hearn, Marcus
Titan Books: London.
HARDBACK 1-85286-790-6

HARRISON FORD: IMPERFECT HERO
Jenkins, Garry
Simon & Schuster: New York.
HARDBACK 0-684-81694-6

Series: *BFI Film Classics*
HIGH NOON
Drummond, Philip
British Film Institute: London.
0851704948

HOLDING MY OWN IN NO MAN'S LAND : WOMEN AND MEN AND FILM AND FEMINISTS
Haskell, Molly
Oxford University Press: Oxford.
HARDBACK 0-19-505309-5

HOLLYWOOD CONFIDENTIAL: AN INSIDE LOOK AT PUBLIC CAREERS AND PRIVATE LIVES OF HOLLYWOOD'S RICH AND FAMOUS
Amende, Coral
Plume: New York.
0452277914

Series: *The American Moment*
HOLLYWOOD'S HIGH NOON: MOVIEMAKING AND SOCIETY BEFORE TELEVISION
Cripps, Thomas
The Johns Hopkins University Press
HARDBACK 0-8018-5315-X
PAPERBACK 0-8018-5316-8

Series: *Culture and the Moving Image Series*
HOLLYWOOD'S NEW DEAL
Muscio, Giuliana
Temple University Press:

Philadelphia, PA.
HARDBACK 1-56639-495-3
PAPERBACK 1-56639-496-1

HONG KONG BABYLON: AN INSIDER'S GUIDE TO THE HOLLYWOOD OF THE EAST
Dannen, Frederic; Long, Barry
Faber and Faber: London.
PAPERBACK 0-571-19040-5

 HONG KONG CINEMA: THE EXTRA DIMENSIONS
Teo, Stephen
British Film Institute: London.
HARDBACK 0-85170-496-4
PAPERBACK 0-85170-514-6

HOWARD HAWKS : THE GREY FOX OF HOLLYWOOD
McCarthy, Todd
Grove Press: New York.
HARDBACK 0-8021-1598-5

HOWARD STERN, A TO Z: THE STERN FANATIC'S GUIDE TO THE KING OF ALL MEDIA
Lucaire, Luigi
St. Martin's Press: New York.
0312151446

HOW TO ENTER SCREENPLAY CONTESTS - AND WIN!: AN INSIDER'S GUIDE TO SELLING YOUR SCREENPLAY IN HOLLYWOOD
Joseph, Erik
Lone Eagle Publications: Los Angeles.
0943728886

HOW TO SLEEP ON A CAMEL: ADVENTURES OF A DOCUMENTARY FILM DIRECTOR
Webster, Nicholas
McFarland & Company: Jefferson, NC.
HARDBACK 0-7864-0349-7

THE ICE STORM: THE SHOOTING SCRIPT
Schamus, James
Newmarket Press: New York.
1557043094

ILL EFFECTS: THE MEDIA/ VIOLENCE DEBATE
Barker, Martin; Petley, Julian
Routledge: London.
0415146739

THE IMAGE
Aumont, Jacques
[Transl.] Pajackowska, Claire
British Film Institute: London.
0851704107

IMAGING THE DIVINE : JESUS AND CHRIST-FIGURES IN FILM
Baugh, Lloyd

Sheed & Ward: Kansas City, MS.
PAPERBACK 1-55612-863-0

INSIDE BROADCASTING
Newby, Julian
Routledge: London.
0415151120

THE INSIDER'S GUIDE TO WRITING FOR SCREEN AND TELEVISION
Tobias, Ronald B.
Writers Digest Books: Cincinnati, OH.
0898797179

INSIDE THE MAGIC KINGDOM: SEVEN KEYS TO DISNEY'S SUCCESS
Connellan, Tom
Bard Press: Austin, TX.
1885167237

INTERNATIONAL DICTIONARY OF FILMS AND FILMMAKERS: Vol 1-4
[Ed.] Hillstrom, Laurie Collier
[3rd ed.] St James Press: Detroit MI.
HARDBACK 1-55862-199-7

INTERNATIONAL MOTION PICTURE ALMANAC 1997 (68TH ED)
[Eds.] Moser, James D.; Stevens, Tracy; Pay, William
Quigley Publishing: New York.
HARDBACK 0-900610-57-3

INTERNATIONAL TELEVISION & VIDEO ALMANAC 1997 (42ND ED)
[Ed] Moser, James D.; Stevens, Tracy; Pay, William
Quigley Publishing: New York.
HARDBACK 0-900610-58-1

IN THE COMPANY OF MEN
Labute, Neil
Faber and Faber: London.
PAPERBACK 0-571-19931-3

IN THE ZONE : THE TWILIGHT WORLD OF ROD SERLING
Wolfe, Peter
Bowling Green State University Popular Press: Bowling Green.
HARDBACK 0-87972-729-2

AN INTRODUCTION TO DIGITAL MEDIA
Feldman, Tony
Routledge: London.
0415154235

AN INTRODUCTION TO TELEVISION DOCUMENTARY: CONFRONTING REALITY
Kilborn, Richard; Izod, John
Manchester University Press: Manchester.
0719048923

AN INTRODUCTORY HISTORY OF BRITISH BROADCASTING
Crisell, Andrew
Routledge: London.
0415128021

ITALIAN MOVIE GODDESSES: OVER 80 OF THE GREATEST WOMEN IN ITALIAN CINEMA
Masi, Stefano; Lancia, Enrico
Gremese International: Rome.
HARDBACK 88-7301-071-7

THE ITC CODE OF PROGRAMME SPONSORSHIP, SPRING 1997
Independent Television Commission
Independent Television Commission: London.

I'VE STARTED SO I'LL FINISH
Magnusson, Magnus
Little, Brown & Company: London.
HARDBACK 0-316-64132-4

IVOR NOVELLO
Harding, James
Welsh Academic Press.
1860570194

I WENT DOWN: THE SHOOTING SCRIPT
McPherson, Conor
Nick Hern Books: London.
1854593935

JAMES STEWART: THE HOLLYWOOD YEARS
Pickard, Roy
Robert Hale: .
PAPERBACK 0-7090-6113-7

JEREMY BRETT: THE MAN WHO BECAME SHERLOCK HOLMES
Manners, Terry
Virgin: London.
1852276169

Series: *Cambridge Studies in Film*
JOHN HUSTON'S FILMMAKING
Brill, Lesley
Cambridge University Press: Cambridge.
HARDBACK 0-521-58359-4
PAPERBACK 0-521-58670-4

JOYCE & GINNIE: THE LETTERS OF JOYCE GRENFELL & VIRGINIA GRAHAM
Grenfell, Joyce; Graham, Virginia
[Ed.] Hampton, Janie
Hodder & Stoughton: London.
0340671920

JUMP CUT!: MEMOIRS OF A PIONEER TELEVISION EDITOR
Schneider, Arthur
McFarland & Company: Jefferson,

NC.
0786403454

JUST KEEP TALKING: THE STORY OF THE CHAT SHOW
Wright, Steve
Simon & Schuster: New York.
PAPERBACK 0-684-81699-7

KEMP'S FILM, TV AND VIDEO YEARBOOK: 1997
Variety Media Publications: London.
HARDBACK 0-611-00928-6

KISS KISS BANG BANG!: THE UNOFFICIAL JAMES BOND FILM COMPANION
Barnes, Alan; Hearn, Marcus
B.T.Batsford: London.
PAPERBACK 0-7134-8182-X

THE LADY VANISHES
White, Ethel Lina
Bloomsbury Publishing: London.
0747531889

THE LAST MODERNIST: THE FILMS OF THEO ANGELOPOULOS
[Ed] Horton, Andrew
Flicks Books: Trowbridge.
HARDBACK 0-275-96119-2
PAPERBACK 0-313-30564-1

L.A. STORY AND ROXANNE: TWO SCREENPLAYS
Martin, Steve
Grove Press: New York.
0-8021-3512-9

LATIN AMERICAN BROADCASTING: FROM TANGO TO TELENOVELA
Fox, Elizabeth
John Libbey Media/University of Luton Press: Luton.
1860205151

LATIN AMERICAN FILMS, 1932-1994: A CRITICAL FILMOGRAPHY
Schwartz, Ronald
McFarland & Company: Jefferson, NC.
0786401745

LEE: A ROMANCE
Marvin, Pamela
Faber and Faber: London.
HARDBACK 0-571-19028-6

LEGENDARY LOVE STORIES
Guttmacher, Peter
MetroBooks: New York.
156799489X

LET THERE BE LIFE!: ANIMATING WITH THE COMPUTER
Baker, Christopher W.
Walker and Company: New York.
0802784739

Series: *BFI Film Classics*
 THE LIFE AND DEATH OF COLONEL BLIMP
Kennedy, A. L.
British Film Institute: London.
851705685

A LIFE LESS ORDINARY
Hodge, John
Faber and Faber: London.
0571192815

LIVE, DIRECT AND BIASED?: MAKING TELEVISION NEWS IN THE SATELLITE AGE
Macgregor, Brent
Arnold: London.
0340662247

LOOKING FOR THE OTHER: FEMINISM, FILM AND THE IMPERIAL GAZE
Kaplan, E. Ann
Routledge: New York.
HARDBACK 0-415-91016-1
PAPERBACK 0-415-91017-X

LOST HIGHWAY
Lynch, David; Gifford, Barry
Faber and Faber: London.
0571191509

LUGOSI: HIS LIFE IN FILMS, ON STAGE, AND IN THE HEARTS OF HORROR LOVERS
Rhodes, Gary Don
McFarland & Company: Jefferson, NC.
HARDBACK 0-7864-0257-1

A MAD, MAD, MAD, MAD WORLD: A LIFE IN HOLLY-WOOD
Kramer, Stanley; Coffey, Thomas M.
Harcourt Brace: New York.
HARDBACK 0-15-154958-3

MAKE ACTING WORK
Salt, Chrys
Bloomsbury Publishing: London.
0747535957

MAKING DOCUMENTARY FILMS AND REALITY VIDEOS: A PRACTICAL GUIDE TO PLANNING, FILMING, AND EDITING DOCUMENTARIES OF REAL EVENTS
Hampe, Barry
Henry Holt and Co.: New York.
0805044515

MAKING IMAGES MOVE : PHOTOGRAPHERS AND AVANT-GARDE CINEMA
Horak, Jan-Christopher
Smithsonian Institute Press: Washington.
HARDBACK 1-56098-744-8

THE MAKING OF ALIEN RESURRECTION
Murdock, Andrew; Aberly, Rachel
Titan Books: London.
PAPERBACK 1-85286-867-8

THE MAN IN THE SHADOWS: FRED COE AND THE GOLDEN AGE OF TELEVISION
Krampner, Jon
Rutgers University Press: New Brunswick.
HARDBACK 0-8135-2359-1

MARSHALL MCLUHAN: ESCAPE INTO UNDERSTAND-ING: A BIOGRAPHY
Gordon, W. Terence
Stoddart: Toronto.
0773730451

MASKED MEN: MASCULINITY AND THE MOVIES IN THE FIFTIES
Cohan, Steven
Indiana University Press: Bloomington.
0253332974

MEDIA AND INTRA-STATE CONFLICT IN NORTHERN IRELAND
Wilson, Robin
European Institute for the Media: Dusseldorf.

MEDIA AND POLITICAL CONFLICT: NEWS FROM THE MIDDLE EAST
Wolfsfeld, Gadi
Cambridge University Press: Cambridge.
0521580455

MEDIA SEMIOTICS: AN INTRODUCTION
Bignell, Jonathon
Manchester University Press: Manchester.
0719045010

THE MEDIA STUDIES READER
[Eds.] O'Sullivan, Tim; Jewkes, Yvonne
Arnold: London.
0340645261

MEMORIES OF HOLLYWOOD 1948: ERICH MARIA REMARQUE'S ARCH OF TRIUMPH: A TRIBUTE
Lazarou, George A.
George A. Lazarou: (10 Lamias St.) Athens, Greece.

MEN IN BLACK: THE SCRIPT AND STORY BEHIND THE FILM
Solomon, Ed; Landau, Diana; Margolis, Barbara
Penguin Books: New York.
PAPERBACK 0-1402-6889-8

MICHAEL PALIN: A BIOGRAPHY
Margolis, Jonathan
Orion Media: London.
HARDBACK 0-7528-0504-5

MISS SMILLA'S FEELING FOR SNOW: BY BILLE AUGUST: ADAPTED FROM THE NOVEL BY PETER HOEG
August, Bille; Hoeg, Peter
[Ed.] Trolle, Karin
Harvill Press: London.
HARDBACK 1-86046-371-1

MONDO MACABRO: WEIRD AND WONDERFUL WORLD CINEMA AROUND THE WORLD
Tombs, Pete
Titan Books: London.
PAPERBACK 1-85286-865-1

MONSTER: LIVING OFF THE BIG SCREEN
Dunne John Gregory
Random House: New York.
0-679-45579-5

MONSTERS IN THE CLOSET: HOMOSEXUALITY AND THE HORROR FILM
Benshoff, Harry M.
Manchester University Press: Manchester.
0719044723

MONTGOMERY CLIFT
Leonard, Maurice
Hodder & Stoughton: London.
0340653574

MONTY PYTHON ENCYCLOPEDIA
Ross, Robert
B.T. Batsford: London.
0713482796

THE MOTION PICTURE GUIDE: 1997 ANNUAL (THE FILMS OF 1996)
[Eds.] Grant, Edmond; Fox, Ken
Cinebooks: New York.
HARDBACK 0-933997-39-6

Series: *Film Studies*
THE MOVIE GAME: THE FILM BUSINESS IN BRITAIN, EUROPE AND AMERICA
Dale, Martin
Cassell: London.
HARDBACK 0-304-33386-7
PAPERBACK 0-304-33387-5

MOVIES ABOUT THE MOVIES: HOLLYWOOD REFLECTED
Ames, Christopher
University Press of Kentucky: Lexington, KY.
HARDBACK 0-8131-2018-7

MOVIES AND MEANING: AN INTRODUCTION TO FILM
Prince, Stephen
Allyn and Bacon: Boston, MA.
0023968060

MOVIES AS POLITICS
Rosenbaum, Jonathan
University of California Press: Berkeley.
0520206142

MOVING PICTURES: A NEW THEORY OF FILM GENRES, FEELINGS AND COGNITION
Grodal, Torben
Oxford University Press: Oxford.
HARDBACK 0-19-815941-2

MUSIC AND THE SILENT FILM: CONTEXTS AND CASE STUDIES, 1895-1924
Marks, Martin Miller
Oxford University Press: Oxford.
HARDBACK 0-19-506891-2

MY SON THE FANATIC
Kureishi, Hanif
Faber and Faber: London.
0571192343

MYTHICAL EXPRESSIONS OF SIEGE IN ISRAELI FILMS
Ben-Shaul, Nitsan S.
Edwin Mellen Press: London.
0773486089

THE NAKED LENS: AN ILLUSTRATED HISTORY OF BEAT CINEMA
Sargeant, Jack
Creation Books: London.
PAPERBACK 1-871592-67-4

NATIONAL TELEVISION VIOLENCE STUDY: VOLUME 1
U.C. Santa Barbara
Sage Publications: Thousand Oaks.
0761908013

Series: *Culture and Moving Image*
THE NEW CENSORS: MOVIES AND THE CULTURE WARS
Lyons, Charles
Temple University Press: Philadephia.
HARDBACK 1-56639-511-9
PAPERBACK 1-56639-512-7

NEW IMAGE OF RELIGIOUS FILM
[Ed.] May, John R.
Sheed and Ward: Kansas City, KS.
1556127618

NEW LATIN AMERICAN CINEMA: VOLUME ONE: THEORY, PRACTICES, AND TRANSCONTINENTAL ARTICULATIONS
[Ed.] Martin, Michael T.
Wayne State University Press:

Detroit.
0814327052

NEW LATIN AMERICAN CINEMA: VOLUME TWO: STUDIES OF NATIONAL CINEMAS
[Ed.] Martin, Michael T.
Wayne State University Press: Detroit.
0814327060

NEWSZAK AND NEWS MEDIA
Franklin, Bob
Arnold: London.
0340614161

NEW ZEALAND FILM, 1912-1996
Martin, Helen; Edwards, Sam
Oxford University Press: Oxford.
PAPERBACK 0-19-558336-1

NIL BY MOUTH
Oldman, Gary
ScreenPress Books: Eye, Suffolk.
1-901680-03-7

NO STRINGS ATTACHED: THE INSIDE STORY OF JIM HENSON'S CREATURE SHOP
Bacon, Matt
Virgin Books: London.
HARDBACK 1-85227-669-X

NOTORIOUS: THE LIFE OF INGRID BERGMAN
Spoto, Donald
HarperCollins: London.
HARDBACK 0-0601-8702-6

OFFENSIVE FILMS: TOWARD AN ANTHROPOLOGY OF CINÉMA VOMITIF
Brottman, Mikita
Greenwood Press: Westport, CT.
031330033X

OF JOY AND SORROW: A FILMOGRAPHY OF DUTCH SILENT FICTION
Donaldson, Geoffrey
Stichting Nederlands Filmmuseum: Amsterdam.
907133810X

OLD MEDIA, NEW MEDIA: MASS COMMUNICATIONS IN THE INFORMATION AGE
Dizard, Wilson Jnr.
[2nd ed.] Longman: White Plains, NY.
0801317436

ONCE UPON A TIME IN AMERICA
Grey, Harry
Bloomsbury Publishing: London.
0747531862

ONCE WAS ENOUGH: CELEBRITIES (AND OTHERS) WHO APPEARED A SINGLE TIME ON SCREEN
Brode, Douglas
Carol Publishing Group: Secaucus, NJ.
0806517352

ONE FINE DAY
Simon, Ellen; Seltzer, Terrel
ScreenPress Books: Eye, Suffolk.
1-901680-01-0

ONE NIGHT STAND
Figgis, Mike
Faber and Faber: London.
0571194079

THE ON PRODUCTION BUDGET BOOK
Koster, Robert
Focal Press: Boston, MA.
0-240-80298-5

ON TELEVISION
Hood, Stuart; Tabary-Peterssen, Thalia
[4th ed.] Pluto Press: London.
HARDBACK 0-7453-1110-5
PAPERBACK 0-7453-1111-3

ON THE HISTORY OF FILM STYLE
Bordwell, David
Harvard University Press: Cambridge, MA.
0674634284

OUR MOVIE HERITAGE
McGreevey, Tom; Yeck, Joanne
Rutgers University Press: New Brunswick.
0813524318

OVITZ: THE INSIDE STORY OF HOLLYWOOD'S MOST CONTROVERSIAL POWER BROKER
Slater, Robert
McGraw-Hill: New York.
0070581037

OZU'S TOKYO STORY
[Ed.] Desser, David
Cambridge University Press: Cambridge.
0521484359

Series: *Jubilee Series*
PAKISTAN CINEMA: 1947-1997
Gazdar, Mushtaq
Oxford University Press Pakistan: Karachi.
HARDBACK 0-19-577817-0

PARALLEL TRACKS: THE RAILROAD AND SILENT CINEMA.
Kirby, Lynne
University of Exeter Press: Exeter.
0859895300

PARTICIPATORY VIDEO: A PRACTICAL APPROACH TO USING VIDEO CREATIVELY IN GROUP DEVELOPMENT WORK
Shaw, Jackie; Robertson, Clive
Routledge: London.
0415141052

PASSIONATE DETACHMENTS: AN INTRODUCTION TO FEMINIST FILM THEORY
Thornham, Sue
Arnold: London.
340652268

THE PASSION OF DAVID LYNCH : WILD AT HEART IN HOLLYWOOD
Nochimson, Martha P.
University of Texas Press: Austin, TX.
HARDBACK 0-292-75566-X

PAUL BLAISDELL, MONSTER MAKER: A BIOGRAPHY OF THE B MOVIE MAKEUP AND SPECIAL EFFECTS ARTIST
Palmer, Randy
McFarland and Company: Jefferson, NC.
0786402709

PAUL VERHOEVEN
Scheers, Rob van
Faber and Faber: London.
PAPERBACK 0-571-17479-5

PECKINPAH: THE WESTERN FILMS: A RECONSIDERATION
Seydor, Paul
University of Illinois Press: Chicago.
0252022688

PERSONA GRANADA: SOME MEMORIES OF SIDNEY BERNSTEIN AND THE EARLY DAYS OF INDEPENDENT TELEVISION
Forman, Denis
Andre Deutsch: London.
0233989870

A PERSONAL JOURNEY WITH MARTIN SCORSESE THROUGH AMERICAN MOVIES
Scorsese, Martin; Wilson, Michael Henry
Faber and Faber: London.
0-5711-9242-4

Series: Contributions to the Study of Popular Culture, No 61
PERSONALITY COMEDIANS AS GENRE: SELECTED PLAYERS
Gehring, Wes D.
Greenwood Press: Westport, CT.
HARDBACK 0-313-26185-7

PETER COOK - A BIOGRAPHY
Thompson, Harry
Hodder & Stoughton: London.
HARDBACK 0-340-64968-2

PICKFORD: THE WOMAN WHO MADE HOLLYWOOD
Whitfield, Eileen
University Press of Kentucky: Lexington, KY.
HARDBACK 0-8131-2045-4

PIECES OF TIME: THE LIFE OF JAMES STEWART
Fishgall, Gary
Scribner: New York.
068482454X

PIERCE BROSNAN: THE BIOGRAPHY
Membery, York
Virgin Books: London.
HARDBACK 1-85227-672-X

Series: Bloomsbury Film Classics
THE POSTMAN ALWAYS RINGS TWICE
Cain, James M.
Bloomsbury: London.
PAPERBACK 0-7475-3778-X

POSTMODERN AFTER-IMAGES: A READER IN FILM, TELEVISION AND VIDEO
Brooker, Will
[Ed.] Brooker, Peter
Arnold: London.
0340676914

POSTNEGRITUDE VISUAL AND LITERARY CULTURE
Reid, Mark A.
State University of New York: Albany.
0791433021

POVERTY ROW STUDIOS, 1929-1940 : AN ILLUSTRATED HISTORY OF 53 INDEPENDENT FILM COMPANIES, WITH A FILMOGRAPHY FOR EACH
Pitts, Michael R.
McFarland: Jefferson, NC.
HARDBACK 0-7864-0168-0

POWER IN THEY EYE: AN INTRODUCTION TO CONTEM-PORARY IRISH FILM
Byrne, Terry
Scarecrow Press: Lanham, MD.
0810832968

A PRACTICAL GUIDE TO FILM FINANCING
Hart-Wilden, Paul
FT Media & Telecoms Publishing: London.
1853347817

PRETTY IN PINK: THE GOLDEN AGE OF TEENAGE MOVIES
Bernstein, Jonathan

St. Martin's Griffin: New York.
0312151942

PRIME TIME NETWORK SERIALS: EPISODE GUIDE, CAST AND CREDITS FOR 37 CONTINUING TELEVISION DRAMAS, 1964-1993.
Morris, Bruce B.
McFarland & Company: Jefferson, NC.
0786401648

PRODUCER TO PRODUCER: INSIDER TIPS FOR ENTER-TAINMENT MEDIA
Wiese, Michael
[Ed.] McKernan, Brian
Michael Wiese Productions: Studio City.
0941188612

PRODUCTION DESIGN IN THE CONTEMPORARY AMERICAN FILM: A CRITICAL STUDY OF 23 MOVIES AND THEIR DESIGNERS
Heisner, Beverly
McFarland & Company: Jefferson, NC.
HARDBACK 0-7864-0267-9

PROJECTING THE PAST: ANCIENT ROME, CINEMA, AND HISTORY
Wyke, Maria
Routledge: London.
HARDBACK 0-415-90613-X
PAPERBACK 0-415-90614-8

PROJECTIONS 7: FILM-MAKERS ON FILM-MAKING: IN ASSOCIATION WITH CAHIERS DU CINÉMA
[Eds.] Boorman, John; Donohue, Walter
Faber and Faber: London.
0571190332

PROPAGANDA AND DEMOC-RACY: THE AMERICAN EXPERIENCE OF MEDIA AND MASS PERSUASION
Sproule, J. Michael
Cambridge University Press: Cambridge.
0521470226

PUBLIC SERVICE BROAD-CASTING: THE CHALLENGES OF THE TWENTY-FIRST CENTURY
Atkinson, Dave
[Ed.] Raboy, Marc
Unesco Publications.
9231034219

THE QUEST FOR GRAHAM GREENE
West, W.J.
Weidenfeld & Nicolson: London.
0297818228

RAINER WERNER FASSBINDER
[Ed.] Lorenz, Juliane; Kardish, Laurence
Museum of Modern Art: New York.
0-8707-0109-6

RAYMOND CHANDLER: A BIOGRAPHY
Hiney, Tom
Chatto & Windus: London.
0701163100

Series: *Suny Series, Interruptions - Border Testimony*
RECREATIONAL TERROR : WOMEN AND THE PLEASURES OF HORROR FILM VIEWING
Pinedo, Isabel Cristina
State University of New York Press: Albany, NY.
HARDBACK 0-7914-3441-9

REEL BLACK TALK: A SOURCEBOOK OF 50 AMERICAN FILMMAKERS
Moon, Spencer
Greenwood Press: Westport, CT.
HARDBACK 0-313-29830-0

REEL CONVERSATIONS: READING FILMS WITH YOUNG ADULTS
Teasley, Alan; Wilder, Ann
Boynton/Cook: Portsmouth, NH.
0867093773

Series: *Wisconsin Studies in Film*
REEL PATRIOTISM: THE MOVIES AND WORLD WAR I
DeBauche, Leslie Midkiff
University of Wisconsin Press: Madison, WI.
HARDBACK 0-299-15400-9
PAPERBACK 0-299-15404-1

REFIGURING SPAIN: CINEMA/ MEDIA/REPRESENTATION
[Ed.] Kinder, Marsha
Duke University Press: Durham, NC.
HARDBACK 0-8223-1932-2

REGULATING FOR CHANGING VALUES: A REPORT FOR THE BROADCASTING STANDARDS COMMISSION
Kiernan, Matthew; Morrison, David; Svennevig, Michael
Broadcasting Standards Commission: London.

Series: *Rutgers Depth of Field Series*
REPRESENTING BLACKNESS: ISSUES ON FILM AND VIDEO
[Ed.] Smith, Valerie
Rutgers University Press: New Brunswick.
HARDBACK 0-485-30081-8

REPRESENTING THE WOMAN: CINEMA AND PSYCHOANALYSIS
Cowie, Elizabeth

Macmillan: Basingstoke.
0333660137

THE RESEARCHER'S GUIDE TO BRITISH FILM & TELEVISION COLLECTIONS
[Ed.] Kirchner, Daniela
British Universities Film & Video Council
British Universities Film & Video Council: London.
0901299685

THE REVOLUTION WASN'T TELEVISED: SIXTIES TELEVISION AND SOCIAL CONFLICT
[Eds.] Spiegal, Lynn; Curtis, Michael
Routlege: New York.
0415911222

RHETORIC AND REPRESENTATION IN NONFICTION FILM
Planting, Carl R.
Cambridge University Press: Cambridge.
0521573262

Series: *BFI Modern Classics*
THE RIGHT STUFF
Charity, Tom
British Film Institute: London.
085170624X

THE ROAD MOVIE BOOK
Hark, Ina Rae
[Ed.] Cohan, Steven
Routlege: London.
0415149363

ROAD TO BOX OFFICE: THE SEVEN FILM COMEDIES OF BING CROSBY, BOB HOPE AND DOROTHY LAMOUR, 1940-1962
Mielke, Randall G.
McFarland & Company: Jefferson, NC.
HARDBACK 0-7864-0162-1

ROGER EBERT'S BOOK OF FILM
Ebert, Roger
W.W. Norton & Co.: New York.
HARDBACK 0-393-04000-3

Series: *Bio-Bibliographies in the Performing Arts, No 74*
RONALD COLMAN: A BIO-BIBLIOGRAPHY
Frank, Sam
Greenwood Press: Westport, Connecticut.
HARDBACK 0-313-26433-3

SCARFACE
Trail, Armitage
Bloomsbury Publishing: London.
0747531838

SCIENCE FICTION, FANTASY AND HORROR FILM SEQUELS, SERIES AND REMAKES: AN

ILLUSTRATED FILMOGRAPHY, WITH PLOT SYNOPSES AND CRITICAL COMMENTARY
Holston, Kim R.;Winchester, Tom
McFarland & Company: Jefferson, NC.
HARDBACK 0-7864-0155-9

SCI-FI ON TAPE: A COMPREHENSIVE GUIDE TO SCIENCE FICTION AND FANTASY FILMS ON VIDEO
O'Neill, James
Billboard Books: New York.
0823076598

SCREEN DREAMS: FANTASISING LESBIANS IN FILM
Whatling, Clare
Manchester University Press: Manchester.
0719050669

SCREENING THE LOS ANGELES "RIOTS": RACE, SEEING AND RESISTANCE
Hunt, Darnell M.
Cambridge University Press: Cambridge.
0521570875

SCREEN TASTES: FROM SOAP OPERA TO SATELLITE DISHES
Brunsdon, Charlotte
Routledge: New York.
HARDBACK 0-415-12154-X
PAPERBACK 0-415-12155-8

SCREEN WORLD: 1996: Vol 47
Willis, John
Applause Books: New York; London.
HARDBACK 1-55783-252-8
PAPERBACK 1-55783-253-6

SECRETS AND LIES
Leigh, Mike
Faber and Faber: London.
0571192912

THE SECRET WOMAN: A LIFE OF PEGGY ASHCROFT
O'Connor, Garry
Weidenfeld & Nicolson: London.
0297815865

SELZNICK'S VISION: GONE WITH THE WIND AND HOLLYWOOD FILMMAKING
Vertrees, Alan David
Univesity of Texas Press: Austin, TX.
0292787294

SERIOUSLY FUNNY: FROM THE RIDICULOUS TO THE SUBLIME
Jacobson, Howard
Viking: London.
0670855464

Series: *Bloomsbury Film Classics*
SERPICO
Maas, Peter
Bloomsbury: London.
PAPERBACK 0-7475-3779-8

SET LIGHTING TECHNICIAN'S HANDBOOK: FILM LIGHTING EQUIPMENT, PRACTICE, AND ELECTRICAL DISTRIBUTION
Box, Harry C.
[2nd ed.] Focal Press: Boston, MA.
0-240-80257-8

SET VISITS: INTERVIEWS WITH 32 HORROR AND SCIENCE FICTION FILMMAKERS
Warren, Bill
McFarland & Company: Jefferson, NC.
HARDBACK 0-7864-0247-4

SEX AND ZEN & A BULLET IN THE HEAD: THE ESSENTIAL GUIDE TO HONG KONG'S MIND-BENDING MOVIES
Hammond, Stefan; Wilkins, Mike
Titan Books: London.
PAPERBACK 1-85286-775-2

Series: *Bloomsbury Film Classics*
SHAFT
Tidyman, Ernest
Bloomsbury: London.
PAPERBACK 0-7475-3777-1

SHAKESPEARE, THE MOVIE: POPULARIZING THE PLAYS ON FILM, TV, AND VIDEO
Boose, Lynda E.; Burt, Richard
Routledge: London.
0415165849

SHINE
Sardi, Jan
Bloomsbury: London.
0747531730

SHOCKING ENTERTAINMENT: VIEWER RESPONSE TO VIOLENT MOVIES
Hill, Annette
John Libbey Media: Luton.
1860205259

SHORT ORDERS: FILM WRITING
Romney, Jonathon
Serpent's Tail: London.
1852425121

SILENT CINEMA: WORLD CINEMA BEFORE THE COMING OF SOUND
Finler, Joel W.
B.T. Batsford: London.
PAPERBACK 0-7134-8072-6

SIRK ON SIRK: CONVERSA-TIONS WITH JON HALLIDAY
Sirk, Douglas
[Rev. ed.] Faber and Faber: London.
PAPERBACK 0-571-19098-7

SIX SCREENPLAYS
Riskin, Robert
[Ed.] McGilligan, Patrick
University of California Press: Berkeley, Ca.
0520203054

SKY HIGH: THE INSIDE STORY OF BSKYB
Horsman, Mathew
Orion Business Books: London.
0752811967

THE SLAB BOYS
Byrne, John
Faber and Faber: London.
0571192548

SMOKE AND MIRRORS: VIOLENCE, TELEVISION AND OTHER AMERICAN CULTURES
Leonard, John
New Press: New York.
156584226X

SOME OF ME
Rossellini, Isabella
Random House: New York.
HARDBACK 0-679-45252-4

SOMEWHERE IN THE NIGHT: FILM NOIR AND THE AMERICAN CITY
Christopher, Nicholas
Free Press: New York.
0684828030

SOUND ASSISTANCE
Talbot-Smith, Michael
Focal Press: Oxford.
0240514394

SOUND FOR FILM AND TELEVISION
Holman, Tomlinson
Focal Press: Boston, MA.
PAPERBACK 0-240-80291-8

SOUNDS OF MOVIES: INTER-VIEWS WITH THE CREATORS OF FEATURE SOUNDTRACKS
Pasquariello, Nicolas
Port Bridge Books: San Francisco.
0965311473

SPECIAL AGENT SCULLY: THE GILLIAN ANDERSON FILES
Butt, Malcom
Plexus Publishing: London.
0859652548

SPECIAL EDITION: A GUIDE TO NETWORK TELEVISION DOCUMENTARY SERIES AND SPECIAL NEWS REPORTS 1980-1989
Einstein, Daniel
Scarecrow Press: Lanham, MD.
0810832208

THE SPEED OF SOUND: HOLLYWOOD AND THE TALKIE REVOLUTION
Eyman, Scott
Simon & Schuster: New York.
HARDBACK 0-684-81162-6

SPIELBERG'S HOLOCAUST
Loshitzky, Y. Osefa
Indiana University Press: Bloomington.
HARDBACK 0-253-33232-X
PAPERBACK 0-253-21098-4

SPIKE LEE'S DO THE RIGHT THING
[Ed.] Reid, Mark A.
Cambridge University Press: Cambridge.
0521559545

STANLEY KUBRICK: A BIOGRAPHY
Baxter, John
HarperCollins: London.
HARDBACK 0-00-255588-3

STANLEY KUBRICK : A BIOGRAPHY
Lobrutto, Vincent
Donald I Fine: New York.
HARDBACK 1-55611-492-3

STANLEY KUBRICK AND THE ART OF ADAPTATION: THREE NOVELS, THREE FILMS
Jenkins, Greg
McFarland & Company: Jefferson, NC.
0786402814

THE STARS OF HOLLYWOOD REMEMBERED: CAREER BIOGRAPHIES OF 82 ACTORS AND ACTRESSES OF THE GOLDEN ERA, 1920S -1950S
Ellrod, J.G.
McFarland & Company: Jefferson, NC.
PAPERBACK 0-7864-0294-6

STARS, STARS, STARS...OFF THE SCREEN
[Phot.] Quinn, Edward
Scalo: Zurich.
HARDBACK 3-931141-28-4

Series: *Star Trek (trade/hardcover)*
STAR TREK PHASE II: THE LOST SERIES
Reeves-Stevens, Judith; Reeves-Stevens, Garfield
Pocket Books: New York.
PAPERBACK 0-671-56839-6

STAR WARS: A NEW HOPE
Lucas, George
Faber and Faber: London.
057119236X

STAR WARS: RETURN OF THE JEDI
Kasdan, Lawrence; Lucas, George
Faber and Faber: London.
0571192386

STAR WARS: THE EMPIRE STRIKES BACK
Brackett, Leigh; Kasdan, Lawrence
Faber and Faber: London.
0571192378

STEVEN SPIELBERG : A BIOGRAPHY
McBride, Joseph
Simon & Schuster: New York.
HARDBACK 0-684-81167-7

THE STORY OF THE FIFTH ELEMENT: THE ADVENTURE AND DISCOVERY OF A FILM
Besson, Luc
[Illus.] Wildman, Andrew
Titan Books: London.
PAPERBACK 1-85286-863-5

SUBJECT TO CHANGE: GUERRILLA TELEVISION REVISITED
Boyle, Deirdre
Oxford University Press: Oxford.
HARDBACK 0-19-504334-0
PAPERBACK 0-19-511054-4

SUBURBIA
Bogosian, Eric
St. Martin's Griffin: New York.
031216615X

Series: *A Bob Adelman Book*
THE SUNDAY TIMES 1000 MAKERS OF CINEMA
[Eds.] Morgan, Robin; Perry, George
Thames and Hudson: London.
PAPERBACK 0-500-27994-2

SURVIVING PRODUCTION: THE ART OF PRODUCTION MANAGEMENT FOR FILM AND TELEVISION
Patz, Deborah S.
Michael Wiese Productions: Studio City.
0941188604

Series: *History of the American Cinema Series, no.4*
THE TALKIES: AMERICAN CINEMA'S TRANSITION TO SOUND, 1926-1931
Crafton, Donald
Charles Scribner's Sons: New York.
HARDBACK 0-684-19585-2

THE TALKING CURE: TV TALK SHOWS WITH WOMEN
Shattuc, Jane
Routledge: London.
0415910889

 TALKING PICTURES: INTERVIEWS WITH YOUNG BRITISH FILM-MAKERS
Jones, Graham
[Ed] Johnson, Lucy
British Film Institute: London.

HARDBACK 0-85170-603-7
PAPERBACK 0-85170-604-5

THE TANGO LESSON
Potter, Sally
Faber and Faber: London.
0571191665

THE TECHNIQUE OF FILM AND VIDEO EDITING
Dancyger, Ken
Focal Press: Boston.
0240802551

TELEVISION AND ETHNIC MINORITIES: PRODUCERS' PERSPECTIVES: A STUDY OF BBC IN-HOUSE, INDEPENDENT AND CABLE TV PRODUCERS
Cottle, Simon
[Contributor] Ismond, Patrick
Avebury: Aldershot.
1859725023

TELEVISION FICTION AND IDENTITIES: AMNERICA, EUROPE, NATIONS
[Eds.] Bechelloni, Giovanni; Buonanno, Milly
Ipermedium: Naples.
8886908113

THE TELEVISION HANDBOOK
Holland, Patricia
Routledge: London; New York.
HARDBACK 0-415-12731-9

TELEVISION IN AMERICA: LOCAL STATION HISTORY FROM ACROSS THE NATION
[Eds.] Murray, Michael D.; Godfrey, Donald G.
Iowa State University Press: Ames, IA.
0813829690

TELEVISION NEWS AND THE ELDERLY: BROADCAST MANAGERS' ATTITUDES TOWARDS OLDER ADULTS
Hilt, Michael L.
Garland Publishing: New York.
0815326270

TELEVISION WESTERNS EPISODE GUIDE: ALL UNITED STATES SERIES, 1949-1996
Lentz, Harris M.
McFarland & Company: Jefferson, NC.
0786403772

TENDER COMRADES: A BACKSTORY OF THE HOLLY-WOOD BLACKLIST
McGilligan, Patrick; Buhle, Paul
St. Martin's Press: New York.
0312170467

THEORIZING VIDEO PRACTICE
Wayne, Mike
Lawrence and Wishart: London.
0853158274

Series: *BFI Modern Classics*
 THE THING
Billson, Anne
British Film Institute: London.
0851705669

Series: *Filmmakers, 54*
THOROLD DICKINSON AND THE BRITISH CINEMA
Richards, Jeffrey
[2nd ed.] Scarecrow Press: Lanham, MD.
HARDBACK 0-8108-3279-8

Series: *"Time Out" Guides*
THE TIME OUT FILM GUIDE: 1998
[Ed.] Pym, John
[6th ed.] Penguin Books: London.
PAPERBACK 0-14-026564-3

THE TITANIC: THE EXTRAORDINARY STORY OF THE UNSINKABLE SHIP
Tibballs, Geoff.
Carlton Books: London.
HARDBACK 1-85868-291-6

TOASTING CHEERS: AN EPISODE GUIDE TO THE 1982-1993 COMEDY SERIES, WITH CAST BIOGRAPHIES AND CHARACTER PROFILES
Bjorklund, Dennis A.
McFarland & Company: Jefferson, NJ.
0899509622

TOM CRUISE: A BIOGRAPHY
Sellers, Robert
Robert Hale: London.
HARDBACK 0-7090-5441-6

TOM HANKS: A CAREER IN ORBIT
Quinlan, David
B.T. Batsford: London.
PAPERBACK 0-7134-8073-4

TO THE RESCUE: HOW IMMIGRANTS SAVED THE AMERICAN FILM INDUSTRY, 1896-1912
Weiss, Ken
Austin & Winfield: San Francisco.
1572920505

Series: *Film/fiction Series, v. 2*
TRASH AESTHETICS: POPU-LAR CULTURE AND ITS AUDIENCE
[Eds.] Cartmell, Deborah; Kaye, Heidi; Hunter, Ian; Whelehan, Imelda
Pluto Press: London.
PAPERBACK 0-7453-1202-0
HARDBACK 0-7453-1203-9

A TREASURE HARD TO ATTAIN: IMAGES OF ARCHAE-OLOGY IN POPULAR FILM, WITH A FILMOGRAPHY
Day, David Howard

Scarecrow Press: Lanham, MD.
HARDBACK 0-8108-3171-6

TROPICAL MULTICULTURALISM: A COMPARATIVE HISTORY OF RACE IN BRAZILIAN CINEMA AND CULTURE
Stam, Robert
Duke University Press: Durham;
London.
HARDBACK 0-8223-2035-5
PAPERBACK 0-8223-2048-7

Series: *Current debates in broadcasting, 6*
TUNE IN OR BUY IN?: PAPERS FROM THE 27TH UNIVERSITY OF MANCHESTER BROAD-CASTING SYMPOSIUM, 1996.
[Eds.] Ralph, Sue; Brown, Jo
Langham; Lees, Tim
John Libby Media: Luton.
1860205283

TV DRAMA IN TRANSITION: FORMS, VALUES AND CUL-TURAL CHANGE
Nelson, Robin
Macmillan Press: Houndmills.
HARDBACK 0-333-67753-6
PAPERBACK 0-333-67754-4

T.V. SCENIC DESIGN HAND-BOOK
Millerson, Gerald
[2nd ed.] Focal Press: Oxford.
PAPERBACK 0-240-51493-9

TV WRITERS GUIDE (5TH ED)
[Ed] Naylor, Lynne
Lone Eagle: .
PAPERBACK 0-943728-86-X

TWENTIETH CENTURY'S FOX: DARRYL F. ZANUCK AND THE CULTURE OF HOLLYWOOD
Custen, George F.
Basic Books: New York.
HARDBACK 0-465-07619-X

TWENTY FOUR FRAMES UNDER: A BURIED HISTORY OF FILM MUSIC
Lack, Russell
Quartet Books: London.
0704380455

TWILIGHT ZONESl THE HIDDEN LIFE OF CULTURAL IMAGES FROM PLATO TO O.J.
Bordo, Susan
University of California Press:
Berkeley, Ca.
0520211014

TWIN TOWN
Allen, Kevin; Durden, Paul
ScreenPress Books: Eye, Suffolk.
1901680002

THE UNDECLARED WAR: THE STUGGLE FOR CONTROL OF THE WORLD'S FILM INDUS-TRY
Puttnam, David; Watson, Neil
HarperCollins: London.
HARDBACK 0-00-255675-8
PAPERBACK 0-00-638744-6

UNDERSTANDING THE FILM: AN INTRODUCTION TO FILM APPRECIATION
Bone, Jan; Johnson, Ron
NTC Publishing Group:
Lincolnwood.
0844257974

UNDRESSING CINEMA: CLOTHING AND IDENTITY IN THE MOVIES
Bruzzi, Stella
Routledge: London.
0415139562

THE UNKNOWN JAMES DEAN
Tanitch, Robert
B.T. Batsford: London.
PAPERBACK 0-7134-8034-3

THE UNCOMMON WISDOM OF OPRAH WINFREY : A POR-TRAIT IN HER OWN WORDS
Winfrey, Oprah
[Ed.] Adler, Bill
Birch Lane Press (Carol Pub Group):
Secaucus, NJ.
HARDBACK 1-55972-419-6

THE VAMPIRE FILM : FROM NOSFERATU TO INTERVIEW WITH THE VAMPIRE
Silver, Alain; Ursini; James
[3rd ed.] Limelight Editions: New York.
PAPERBACK 0-87910-266-7

VARIETY INTERNATIONAL FILM GUIDE 1998
[Ed.] Cowie, Peter
Andre Deutsch; Samuel French:
London; Hollywood.
PAPERBACK 0-233-99183-2

VERTICAL INTEGRATION IN CABLE TELEVISION
Waterman, David; Weiss, Andrew A.
MIT Press: Cambridge, MA.
0262231905

VERTIGO
Boileau, Pierre; Narcejac, Thomas
Bloomsbury Publishing: London.
074753442X

THE VIDEO SOURCE BOOK (19TH ED)
Gale: Detroit. (2 vols)
HARDBACK 0-7876-0199-3

VIEWS FROM THE EDGE OF THE WORLD: NEW ZEALAND FILM
Conrich, Ian; Davy, Sarah
Kakapo Books: London.
0953017702

VISIONS OF THE EAST: ORIENTALISM IN FILM
[Ed.] Bernstein, Matthew; Studlar, Gaylyn
Rutgers University Press: New Brunswick.
0813522951

WAR OF THE BLACK HEAVENS: THE BATTLES OF WESTERN BROADCASTING IN THE COLD WAR
Nelson, Michael
Syracuse University Press: Syracuse, NY.
0815604793

WELCOME TO SARAJEVO
Boyce, Frank Cottrell
Faber and Faber: London.
0571193854

WES CRAVEN'S LAST HOUSE ON THE LEFT: THE MAKING OF A CULT CLASSIC
Szulkin, David A.
FAB Press: Guildford
0952926008

WHAT FALLS AWAY: A MEMOIR
Farrow, Mia
Nan A. Talese: New York.
HARDBACK 0-385-40488-3

WHITE HOUSE TO YOUR HOUSE: MEDIA AND POLITICS IN VIRTUAL AMERICA
Diamond, Edwin; Silverman, Robert A.
MIT Press: Cambridge.
026254086X

bfi **WHO IS ANDY WARHOL?**
[Eds.] MacCabe, Colin;
Francis, Mark; Wollen, Peter
British Film Institute: London.
0851705898

WHO ON EARTH IS TOM BAKER?: AN AUTOBIOGRAPHY
Baker, Tom
HarperCollins: London.
HARDBACK 0-00-255834-3

WHO THE DEVIL MADE IT
Bogdanovich, Peter
Alfred A. Knopf: London.
HARDBACK 0-679-44706-7

WHOOPI GOLDBERG: HER JOURNEY FROM POVERTY TO MEGA STARDOM
Parish, James Robert
Birch Lane Press (Carol Pub Group):
Secaucus, NJ.
HARDBACK 1-55972-431-5

WILDE
Mitchell, Julian
Orion: London.
0752810421

**WILLIAM GOLDMAN: FIVE
SCREENPLAYS**
Goldman, William
Applause Books: New York.
HARDBACK 1-55783-266-8

**WILLIAM GOLDMAN: FOUR
SCREENPLAYS**
Goldman, William
Applause Books: New York.
PAPERBACK 1-55783-265-X

**THE WINONA RYDER
SCRAPBOOK**
Siegel, Scott; Siegel, Barbara
Citadel Press (Carol Pub Group):
Secaucus, NJ.
HARDBACK 0-8065-1883-9

**WITHOUT LYING DOWN:
FRANCES MARION AND THE
POWERFUL WOMEN OF
EARLY HOLLYWOOD**
Beauchamp, Cari
Scribner (Lisa Drew Book): New
York.
HARDBACK 0-684-80213-9

**WOMEN AND SOAP OPERA: A
CULTURAL FEMINIST PER-
SPECTIVE**
Blumenthal, Dannielle
Praeger Publishers: Westport, CT.
0275960390

**WOMEN BEHIND THE
CAMERA: CONVERSATIONS
WITH CAMERAWOMEN**
Krasilovsky, Alexis
Praeger Publishers: Westport, CT.
HARDBACK 0-275-95744-6
PAPERBACK 0-275-95745-4

**WOMEN FILMMAKERS OF
THE AFRICAN AND ASIAN
DIASPORA: DECOLONIZING
THE GAZE, LOCATING
SUBJECTIVITY**
Foster, Gwendolyn Audrey
Southern Illinois University Press:
Carbondale, IL.
HARDBACK 0-8093-2119-X
PAPERBACK 0-8093-2120-3

**WOMEN PIONEERS IN
TELEVISION: BIOGRAPHIES
OF FIFTEEN INDUSTRY
LEADERS**
O'Dell, Cary
McFarland & Company: Jefferson, NC.
HARDBACK 0-7864-0167-2

**WOODY ALLEN'S ANGST:
PHILOSOPHICAL
COMMENTARIES ON HIS
SERIOUS FILMS**
Lee, Sander H.
McFarland & Company: Jefferson,
NC.
0786402075

**THE WORLD ACCORDING TO
HOLLYWOOD: 1918-1939**
Vasey, Ruth
University of Wisconsin Press:
Madison.
HARDBACK 0-85989-553-X
PAPERBACK 0-85989-554-8

**WORLDS WITHOUT END: THE
ART AND HISTORY OF THE
SOAP OPERA**
Museum of TV and Radio
Harry N Abrams: New York.
HARDBACK 0-8109-3997-5

**WRITING SHORT FILMS:
STRUCTURE AND CONTENT
FOR SCREENWRITERS**
Cowgill, Linda J.
Lone Eagle Publications: Los
Angeles.
0943728800

**WRITING TREATMENTS THAT
SELL: HOW TO CREATE AND
MARKET YOUR STORY IDEAS
TO THE MOTION PICTURE
AND TV INDUSTRY**
Atchity, Kenneth; Wong, Chi-Li
Henry Holt and Co.: New York.
0805042830

**THE X FACTORY: INSIDE THE
AMERICAN HARDCORE FILM
INDUSTRY**
Petkovich, Anthony
Headpress: Manchester.
PAPERBACK 0-9523288-7-9

BOOKSELLERS

Arnolfini Bookshop
16 Narrow Quay
Bristol BS1 4QA
Tel: 0117 9299191
Fax: 0117 9253876
Open: 10.00-7.00 Mon-Sat,
12.00-6.30 Sun
Stock: A, B, F
Based in the Arnolfini Gallery,
concentrating on the visual arts. No
catalogues issued. Send requests for
specific material with SAE

At the Movies
9 Cecil Court
London WC2N 4EZ
Tel: 0171 240 7221
Stock includes books, stills and
memorabilia. No catalogue
Open: 11.00-6.00 Mon-Fri
11.30-6.00 Sat

Bath Old Books
9c Margaret's Buildings
Bath
Somerset BA1 2LP
Tel: 01225 422244
Second hand books.
Open 10.00-5.00 Mon-Sat

Blackwell's
48-51 Broad Street
Oxford OX1 3BQ
Tel: 01865 792792
Fax: 01865 794143
email: blackwells.extra@blackwell.co.uk
Website: http://www.blackwell.co.uk/
bookshops/
Open: 9.00-6.00 Mon, Wed-Sat,
9.30-6.00 Tue, 11.00-5.00 Sun
Stock: A
Literature department has sections
on cinemas and performing arts,
sociology department has a Media
Studies section 'and performing arts.
International charge and send service
available

Blackwell's Art & Poster Shop
27 Broad Street
Oxford OX1 2AS
Tel: 01865 792792
Open: 9.00-6.00 Mon, Wed-Sat,
9.30-6.00 Tues, 11.00-5.00 Sun
Stock: A, B, C, F
A wide selection of books, posters,
cards, calendars and gift items, all
available by mail order

Brockwell Books
5 Old School House Court
High Street
Honiton
Devon EX14 8NZ
Tel: 01404 42628

Stock: A, B, C, D
Film books offered by mail order
only. Mainly deal in out of print
books but also some new titles.
Catalogue 'Serious about Cinema'
produced three times a year

Cinegrafix Gallery
4 Copper Row
Shad Thames
Tower Bridge Piazza
London SE1 2LH
Tel: 0171 234 0566
Fax: 0171 234 0577
Gallery open Tues-Sat 11.00-7.00
Stock: A, B, C, D, F
Specialist in rare film posters.
Poster catalogues available at £5.
Fully illustrated Portrait catalogue
available at £4

The Cinema Bookshop
13-14 Great Russell Street
London WC1B 3NH
Tel: 0171 637 0206
Fax: 0171 436 9979
Open: 10.30-5.30 Mon-Sat
Stock: A, B, C, D
Comprehensive stock of new, out-of-
print and rare books. Posters,
Pressbooks and stills etc. No
catalogues are issued. Send requests
for specific material with SAE

The Cinema Store
Unit 4B, Orion House
Upper Saint Martin's Lane
London WC2H 9EJ
Tel: 0171 379 7838 (general enquiries)
Fax: 0171 240 7689
email: cinemastor@aol.com
Website: http//www. atlasdigital.com/
cinemastore
Tel: 0171 379 7865 (laserdiscs, mail
order books, cd's)
Tel: 0171 379 7895 (trading cards, VHS)
Open: 10.00-6.00 Mon-Wed, Sat,
10.00-7.00 Thu-Fri
12-6 Sun
Stock: A, B, C, D, E, F
Mail order available worldwide.
Latest and vintage posters/stills,
magazines, models and laser discs,
new/rare VHS, soundtracks and
trading cards

Cornerhouse Books
70 Oxford Street
Manchester M1 5NH
Tel: 0161 228 7621
Fax: 0161 236 7323
Stock: A, B, F
Open: 12.00-8.30 daily
No catalogues issued. Send requests
for specific material with SAE

A E Cox
21 Cecil Road
Itchen
Southampton SO2 7HX
Tel: 01703 447989
Stock: A, B, C, D
Telephone enquiries and orders
accepted at any time. Mail order
only. A catalogue, including rare
items, is published at least six times
yearly. Send two first-class stamps or
three international reply vouchers
overseas to receive the current issue

Culture Vultures Books
329 St Leonard's Road
Windsor SL4 3DS
Tel: 01753 851 693
Fax: 01923 224714
Stock: A
Mail order only. Periodic catalogues
issued (separate catalogues for
cinema, theatre, music). SAE
appreciated. Comprehensive stock of
out-of-print titles

Ray Dasilva Puppet Books
63 Kennedy Road
Bicester
Oxfordshire OX6 8BE
Tel: 01869 245793
Mail order (visitors by appointment).
New and second hand books on
puppetry and animation including
film and television. Catalogue
available

Decorum Books
24 Cloudsley Square
London N1 0HN
Tel: 0171 278 1838
Mail order only. Secondhand books
on film and theatre; music and art.
Also secondhand scores and sheet
music

Dress Circle
57-59 Monmouth Street
Upper St Martin's Lane
London WC2H 9DG
Tel: 0171 240 2227
Fax: 0171 379 8540
Open: 10.00-7.00 Mon-Sat
Stock: A, B, C, D, E, F
Specialists in stage music and
musicals. Catalogue of the entire
stock issued annually with updates
twice yearly. Send SAE for details

David Drummond at Pleasures of Past Times
11 Cecil Court
Charing Cross Road
London WC2H 0AA
Tel: 0171 836 1142

Open: 11.00-2.30, 3.30-5.45 Mon-Fri. First Sat in month 11.00-2.30
Stock: A, D, F
Extended hours and other times by arrangement. No catalogue

Anne FitzSimons
62 Scotby Road
Scotby
Carlisle CA4 8BD
Tel: 01228 513815
Stock: A, B, C, D, F
Mail order only. Antiquarian and out-of-print titles on cinema, broadcasting and performing arts. A catalogue is issued twice a year. Send three first-class postage stamps for current issue

Flashbacks
6 Silver Place
Beak Street
London W1R 3LJ
Tel: 0171 437 8562
Fax: 0171 437 8562
email: Flashbacks@compuserve.com
Website: http//ourworld.compuserve.com/homepages/Flashbacks
Stock: C, D
Shop and mail order service. Send SAE and 'wanted' list for stock details

Forbidden Planet
71 New Oxford Street
London WC1A 1DG
Tel: 0171 836 4179
Fax: 0171 240 7118
Open: 10.00-6.00 Mon-Wed, Sat, 10.00-7.00 Thur, Fri
Stock: A, B, C, D, E, F
Science fiction, horror, fantasy and comics specialists. Mail order service available on 0171 497 2150

Hay Cinema Bookshop (including Francis Edwards)
The Old Cinema
Castle Street
Hay-on-Wye
via Hereford HR3 5DF
Tel: 01497 820071
Large second hand stock. Open 9.00-7.00 Mon-Sat, 11.30-5.30 Sun

Heffers Booksellers
20 Trinity Street
Cambridge CB2 3NG
Tel: 01223 568568
Fax: 01223 568591
Open: 9.00-5.30 Mon-Sat 11.00-5.00 Sun
Stock: A, E
Catalogues of videocassettes and spoken word recordings issued. Copies are available on request.

David Henry
36 Meon Road

London W3 8AN
Tel: 0181 993 2859
email: filmbook@Netcomuk.co.uk
Stock: A,B
Mail order only. A catalogue of out-of-print and second hand books is issued once a year and is available on request

LV Kelly Books
6 Redlands
Blundell's Road
Tiverton
Devon EX16 4DH
Tel: 01884 256170
Fax: 01884 242550
email: lenkelly@eclipse.co.uk
Stock: A, B, E
Principally mail order but visitors welcome by appointment. Catalogue issued regularly on broadcasting and mass communications. Occasional lists on cinema, music, journalism

Ed Mason
Room 301
Third Floor
River Bank House Business Centre
1, Putney Bridge Approach
London SW6 3JD
Tel: 0171 736 8511
Stock: A, B, C, D
Specialist in original film memorabilia from the earliest onwards. Also organises the Collectors' Film Convention six times a year
Office only - all memorabilia stock is re-located to Rare Discs (see entry)

MOMI Bookshop
BFI South Bank
London SE1 8XT
Tel: 0171 815 1343
Open: 10.30 -6.30 daily
Stock: A, B, C, D, E, F
Based in the Museum of the Moving Image. Mail order available with special orders on request **(see p8)**

NFT Bookshop
BFI South Bank
London SE1 8XT
Tel: 0171 815 1343
Open: 10.00-9.00 daily
Stock: A, B, C, E, F
Based in the main NFT foyer. Mail order available on request. Book/video orders taken. Comprehensive range of film/media magazines **(see p8)**

National Museum of Photography, Film & Television
Pictureville
Bradford BD1 1NQ
Tel: 01274 203300
Fax: 01274 723155
(Closed until February 1999)
Open: 11.00-5.00 Tue-Fri, 11.00-6.00 Sat-Sun

Stock: A, C, D, F
Mail order available. Send SAE with requests for information

Offstage Theatre & Film Bookshop
37 Chalk Farm Road
London NW1 8AJ
Tel: 0171 485 4996
Fax: 0171 916 8046
Free cinema and media catalogues available. Send SAE. Open 7 days a week

C D Paramor
25 St Mary's Square
Newmarket
Suffolk CB8 0HZ
Tel: 01638 664416
Fax: 01638 664416
Stock: A, B, C, F, E
Mail order only. Visitors welcome strictly by appointment. Catalogues and most of the performing arts issued regularly free of charge

Rare Discs
18 Bloomsbury Street
London WC1B3 QA
Tel: 0171 580 3516
Open: 10.00-6.30 Mon-Sat
Stock: E
Retail shop with recorded mail order service. Over 7,000 titles including soundtracks, original cast shows, musicals and nostalgia. Telephone for information

Spread Eagle Bookshop (Incorporates Greenwich Gallery)
9 Nevada Street
London SE10 9JL
Tel: 0181 305 1666
Fax: 0181 305 1666
Open: 10.00-5.30 daily
Stock: A, B, C, D
All second-hand stock. Memorabilia, ephemera. Large stock of books on cinema, theatre, posters and photos

Stable Books
Holm Farm
Coldridge
Crediton
Devon EX17 6DR
Tel: 01363 83227
Mail order (visitors by appointment only). Second hand stock concerning theatre, cinema and puppetry

Stage Door Prints
9 Cecil Court
London WC2N 4EZ
Tel: 0171 240 1683
General stock of performing arts titles including antiquarian prints, ephemera and movie memorobilia.
Open 11.00-6.00 Mon-Fri, 11.30-6.00 Sat

Treasure Chest

61 Cobbold Street
Felixstowe
Suffolk 1P11 7BH
Tel: 01394 270717
Second hand stock specialising in cinema and literature. Open 9.30-5.30 Mon-Sat

Tyneside Cinema Shop

10 Pilgrim Street
Newcastle upon Tyne NE1 6QG
Tel: 0191 232 5592
Fax: 0191 221 0535
Open: 10.30-6.00 Mon-Sat
Stock: A, B, C, D, E, F
Send requests for specific material with SAE

Vintage Magazine Co

39-43 Brewer Street
London W1R 3FD
Tel: 0171 439 8525
Open: 10.00-8.00 Mon-Sat,
12.00-7.00 Sun
Stock: B, C, D, F
247 Camden High Street
London NW1
Tel: 0171 482 0587
Open: 10.00-6.00
Mon-Fri, 10.00-7.00 Sat, Sun
Stock: B,C,D,F

Peter Wood

20 Stonehill Road
Great Shelford
Cambridge CB2 5JL
Tel: 01223 842419
Stock: A, D, F
Mail order. Visitors are welcome by appointment. A free catalogue is available of all books in stock

A Zwemmer

80 Charing Cross Road
London WC2H 0BB
Tel: 0171 240 4157
Open: 9.30-6.00 Mon-Fri,
10.00-6.00 Sat
Stock: A, B
A catalogue of new and in-print titles on every aspect of cinema is available on request. Mail order service for all books available through Mail Order Department

STOCK

A	Books
B	Magazines
C	Posters
D	Memorabilia eg Stills
E	Cassettes, CDs, Records and Videos
F	Postcards and Greeting Cards

CABLE AND SATELLITE

We gratefully acknowledge the continuing support of **Screen Digest** in providing information about cable and satellite from its database, and thank David Fisher, Editor of **Screen Digest** for compiling this section

All broadband cable franchises to date were granted by the Cable Authority (apart from 11 previously granted by the Department of Trade and Industry), the role of which was taken over by the Independent Television Commission (ITC) in January 1991, under the Broadcasting Act 1990.

The Act empowered the ITC to grant fifteen-year 'local delivery licences', which can include use of microwave distribution. Licences must be awarded to the highest bidder on the basis of an annual cash bid in addition to forecasts of the sums that will be paid to the Exchequer as a percentage of revenue earned in the second and third five-year periods of the licence.

The franchises are arranged in alphabetical order of area. Where appropriate the principal towns in the area are identified under the area name; cross references are provided for these and other principal towns at the appropriate alphabetical point. 'Homes in area' is the number of homes in the franchise area at the time of the last census before award of the franchise. 'Build completed' is indicated where all homes nominally in the area are passed by cable. 'Homes passed' is the number of homes to which a cable service is available. 'Subscribers' (abbreviated to 'Subs') are those taking at least the basic service, with the percentage this represents of homes passed. Unless stated, services have not yet begun.

Extensive consolidation of ownership has occurred since the last edition of the *Handbook*. To reflect this, the name of the parent franchise or local delivery licence holder is given, with a cross-reference to the listing under the Group Ownership heading, where any significant local trading names are given. However, for marketing reasons such local variations are expected to disappear in time.

In some towns an older cable system still exists. These are not franchised but are licensed by the ITC to provide limited services. They are gradually being superseded by new broadband networks

Aberdeen
Franchise holder: Aberdeen Cable Services = Atlantic Telecom Group (see GO)
Homes in area: 91,000 (build complete)
Awarded: 29 Nov 83
Service start: 1 May 85
Homes passed: 97,288
Subs: 15,174 = 15.6% (1 Jul 98)

Abingdon see **Oxford**
Accrington see **Lancashire, East**
Airdrie see **Cumbernauld**

Alconbury
Narrowband upgrade system operated by Cablecom Investments (see GO)

Aldershot see **Guildford**
Amersham see **Aylesbury**

Andover/Salisbury/Romsey
Franchise holder: NTL (see GO)
Homes in area: 84,500
Awarded: Andover Apr 89, Salisbury/Romsey 6 Apr 90
Service start: Andover Mar 90, Salisbury/Romsey Jun 95
Homes passed: 50,726 (1 Jul 98)
Subs: 9,785 = 19.3% (1 Jul 98)
Franchises awarded separately but amalgamated

Ashford see **Kent, South-east**

Avon
Bristol, Bath, Weston-super-Mare, Frome, Melksham etc
Franchise holder: Telewest (see GO)
Homes in area: 300,000 (build complete)
Awarded: 16 Nov 88
Service start: 14 Sept 90
Homes passed: 299,356 (1 Jul 98)
Subs: 66,563 = 22.3% (1 Jul 98)

Aylesbury/Amersham/Chesham/Wendover
Franchise holder: Cable & Wireless Communications (see GO)
Homes in area: 89,000
Awarded: 31 May 90; acquired Jul 94
Service start: Apr 96
Homes passed: 34,344 (1 Jul 98)
Subs: 4,330 = 12.6% (1 Jul 98)

Ayr
Local delivery franchise holder: Cable & Telecoms (see GO)
Homes in area: 155,000
Awarded: July 1997

Baldock see **Hertfordshire, Central**
Banstead see **East Grinstead**
Barking/Dagenham, London Borough of see **Greater London East**
Barnet, London Borough of see **London, North West**

Barnsley
Franchise holder: Telewest (see GO)
Homes in area: 82,000
Awarded: 14 Jun 90; acquired Apr 93
Homes passed: 30,210 (1 Jul 98)
Subs: 5,046 = 16.7% (1 Jul 98)

Barrow in Furness see **Cumbria, South**
Basildon see **Thames Estuary North**
Basingstoke see **Thames Valley**

Bassetlaw
Local delivery licence holder: NTL (see GO)
Homes in area: 32,800
Date awarded: 13 Jul 95

Bath see **Avon**

Bearsden/Milngavie
Franchise holder: CableTel Glasgow (see GO)
Homes in area: 16,000
Awarded: 7 Jun 90
Homes passed: 12,834 (1 Jul 98)
Subs: 3,940 = 30.7% (1 Jul 98)

Bedford
Franchise holder: CableTel Bedfordshire

(see GO)
Homes in area: 55,000 (build complete)
Awarded: 14 Jun 90
Service start: Nov 1994
Homes passed: 53,173 (1 Jul 98)
Subs: 15,675 = 29.5% (1 Jul 98)

Bedworth see Nuneaton

Beith
Narrowband upgrade system operated by A Thomson (Relay)

Belfast
Franchise revoked from Ulster Cablevision
Homes in area: 136,000
Awarded: 29 Nov 83
see Northern Ireland

Belper see Derbyshire, East
Berkhamsted see Hertfordshire, West
Bexley, London Borough of see Greater London East

Birmingham/Solihull
Franchise holder: Birmingham Cable (see GO)
Homes in area: 465,000
Awarded: 19 Oct 88
Service start: Dec 89
Homes passed: 436,997 (1 Jul 98)
Subs: 116,893 = 26.7% (1 Jul 98)

Bishops Stortford see Harlow

Black Country
Dudley, Sandwell, Walsall, Wolverhampton, urban parts of Bromsgrove, Cannock, Kidderminster
Franchise holder: Telewest (see GO)
Homes in area: 470,000
Awarded: 14 Jul 89
Service start: Sept 91
Homes passed: 375,420 (1 Jul 98)
Subs: 80,729 = 21.5% (1 Jul 98)

Blackburn see Lancashire, East

Blackpool and Fylde
Local delivery franchise holder: Telewest (see GO)
Homes in area: 101,000
Date awarded: Sept 94
Homes passed: 38,310 (1 Jul 98)
Subs: 7,508 = 19.6% (1 Jul 98)

Blaenau Ffestiniog
Narrowband upgrade system operated by John Sulwyn Evans

Bognor see Chichester
Bolsover see Derbyshire, East

Bolton
Franchise holder: Cable & Wireless

Communications (see GO)
Homes in area: 135,000
Awarded: 13 Aug 85; acquired 22 Mar 93
Service start: Jul 90
Homes passed: 126,947 (1 Jul 98)
Subs: 24,947 = 19.7% (1 Jul 98)

Bootle see Liverpool, North
Borehamwood see Hertfordshire, South
Bosworth see Hinkley

Bournemouth/Poole/Christchurch
Franchise holder: Cable & Wireless Communications (see GO)
Homes in area: 143,000
Awarded: 6 Apr 90
Service start: mid 94
Home passed: 117,984 (1 Jul 98)
Subs: 26,095 = 22.1% (1 Jul 98)

Bracknell see Thames Valley

Bradford
Franchise holder: Telewest (see GO)
Homes in area: 175,000
Awarded: 14 Jun 90
Service start: Jul 92
Homes passed: 142,439 (1 Jul 98)
Subs: 25,436 = 17.9% (1 Jul 98)

Braintree see East Anglia, South

Brecon
Narrowband upgrade system operated by Metro Cable

Brent, London Borough of see London, North West
Brentwood see Thames Estuary North
Brighouse see Calderdale

Brighton/Hove/Worthing
Franchise holder: Cable & Wireless Communications (see GO)
Homes in area: 160,000 (build complete)
Awarded: 20 Oct 89; acquired 22 Mar 93
Service start: Apr 1992
Homes passed: 156,853 (1 Jul 98)
Subs: 33,912 = 21.6% (1 Jul 98)
Separate upgrade system in Brighton operated by CDA Communications

Bristol see Avon
Broadstairs see Thanet, Isle of

Bromley, London Borough of
Franchise holder: Cable & Wireless Communications (see GO)
Homes in area: 117,000 (build complete)
Awarded: 16 Mar 90
Service start: Jan 93
Homes passed: 116,187 (1 Jul 98)

Subs: 26,442 = 22.8% (1 Jul 98)

Burgess Hill see Haywards Heath
Burnley see Lancashire, East
Burntwood see Litchfield
Bury St Edmunds see East Anglia, South

Burton-on-Trent/Swadlincote/Ashby-de-la-Zouch/Coalville/Uttoxeter
Local delivery franchise holder: NTL (see GO)
Homes in area: 77,675
Awarded: Jun 95
Original franchise revoked from N-Com Cablevision

Bury/Rochdale
Franchise holder: Cable & Wireless Communications (see GO)
Homes in area: 143,000
Awarded: 17 May 90; acquired 4 May 93
Homes passed: 69,847 (1 Jul 98)
Subs: 15,959 = 22.8% (1 Jul 98)
Separate upgrade system in Rochdale operated by CDA Communications

Bushey see Hertfordshire, South

Calderdale
Halifax, Brighouse
Franchise holder: Telewest (see GO)
Homes in area: 75,000
Awarded: 14 Jun 90
Homes passed: 26,248 (1 Jul 98)
Subs: 4,084 = 15.6% (1 Jul 98)

Camberley see Guildford

Cambridge and district
Cambridge, Newmarket, Ely, Saffron Walden, Huntingdon, St Ives, St Neots, Royston, etc
Franchise holder: Cambridge Cable = Comcast Europe (see GO)
Homes in area: 132,000
Awarded: 4 Jun 89
Service start: Jul 91
Homes passed: 116,397 (1 Jul 98)
Subs: 26,547 = 22.8% (1 Jul 98)

Camden, London Borough of
Franchise holder: Cable London (see GO)
Homes in area: 70,000 (build complete)
Awarded: 1 Feb 86
Service start: Dec 89
Homes passed: 82,207 (1 Jul 98)
Subs: 19,511 = 23.7% (1 Jul 98)

Canterbury/Thanet
No applicants for local delivery franchise (January 1997)
Upgrade systems in Canterbury and

Isle of Thanet operated by CDA Communications

Cardiff/Penarth
Franchise holder: CableTel South Wales (see GO)
Homes in area: 103,000
Awarded: 5 Feb 86
Service start: Sept 94
Homes passed: 74,147 (1 Jul 98)
Subs: 28,176 = 38.0% (1 Jul 98)

Carlisle
Local delivery licence holder: Cable & Telecoms (UK) (see GO)
Homes in area: 35,000
Awarded: Nov 95
Original franchise surrendered by Carlisle Cablevision (awarded: 21 Jun 90)

Carmarthen
Narrowband upgrade system operated by Metro Cable

Castleford see **Wakefield**
Chatham see **Thames Estuary South**
Chelmsford see **Thames Estuary North**

Cheltenham/Gloucester
Franchise holder: Telewest (see GO)
Homes in area: 90,000
Awarded: 13 Aug 85
Service start: Aug 94
Homes passed: 55,802 (1 Jul 98)
Subs: 13,206 = 23.7% (1 Jul 98)

Cheshire, North
Chester, Ellesmere Port, Warrington, Widnes, Runcorn
Franchise holder: Cable & Wireless Communications (see GO)
Homes in area: 175,000
Awarded: 12 Jan 90; acquired 21 Apr 93
Homes passed: 111,983 (1 Jul 98)
Subs: 23,933 = 21.4% (1 Jul 98)

Chesham see **Aylesbury**
Cheshunt see **Hertfordshire, East**
Chesterfield see **Derbyshire, East**

Chichester/Bognor
Local delivery franchise holder: Cable & Wireless Communications (see GO)
Homes in area: 67,100
Awarded: Nov 95

Chigwell see **Epping Forest**
Chorley see **Southport**
Chorleywood see **Hertfordshire, South**
Christchurch see **Bournemouth**
Clacton on Sea see **East Anglia, South**

Cleethorpes see **Grimsby**
Clydebank see **Glasgow, North West**
Coalville see **Burton-on-Trent**
Coatbridge see **Cumbernauld**

Colchester/Ipswich/Felixstowe/Harwich/Woodbridge
Franchise holder: East Coast Cable = Comcast Europe (see GO)
Homes in area: 126,000
Awarded: 21 Jul 89
Service start: late 94
Homes passed: 31,233 (1 Jul 98)
Subs: 7,730 = 24.8% (1 Jul 98)

Colne see **Lancashire, East**

Consett/Stanley (Derwentside)
Local delivery franchise holder: Telewest (see GO)
Homes in area: 37,000
Awarded: 27 Jul 98

Corby see **Northampton, North-east**

Coventry
Franchise holder: Coventry Cable = NTL (see GO)
Homes in area: 119,000 (build complete)
Awarded: 29 Nov 83
Service start: 1 Sept 85
Homes passed: 112,983
Subs: 17,925 = 15.9% (1 Jul 98)

Crawley/Horley/Gatwick
Franchise holder: Eurobell (see GO)
Homes in area: 44,000 (build complete)
Awarded: 27 Apr 89
Service start: Jun 93
Homes passed: 42,582 (1 Jul 98)
Subs: 8,912 = 20.9% (1 Jul 98)

Crosby see **Liverpool, North**

Croydon, London Borough of
Franchise holder: Telewest (see GO)
Homes in area: 115,000 (build complete)
Awarded: 1 Nov 83
Service start: 1 Sept 85
Homes passed: 124,068
Subs: 27,842 = 22.4% (1 Jul 98)

Cumbernauld/Kilsyth/Airdrie/Coatbridge
Franchise holder: Telewest (see GO)
Homes in area: 55,000
Awarded: 27 Apr 89
Service start: May 95

Cumbria, Central
Local delivery franchise holder: Cable & Telecoms (UK) (see GO)
Homes in area: 84,000
Awarded: Oct 96

Cumbria, South
Barrow-in-Furness, South Lakeland District
Local delivery franchise holder: South Cumbria Cable & Telecoms = Cable & Telecoms (UK) (see GO)
Homes in area: 61,500
Awarded: May 97

Cwmbran see **Newport**
Dagenham see **Greater London East**

Darlington
Franchise holder: Comcast Teesside (see GO)
Homes in area: 34,000
Awarded: 21 Jun 90
Service start: Jun 95
Homes passed: 28,822 (1 Jul 98)
Subs: 8,712 = 30.2% (1 Jul 98)

Dartford/Swanley
Franchise holder: Cable & Wireless Communications (see GO)
Homes in area: 35,000
Awarded: 16 Mar 90
Service start: Dec 94
Homes passed: 23,708 (1 Jul 98)
Subs: 3,096 = 13.1% (1 Jul 98)

Daventry
Local delivery franchise holder: NTL (see GO)
Homes in area: 8,710
Awarded: Nov 96

Deal see **Kent, South East**

Derby/Spondon
Franchise holder: Cable & Wireless Communications (see GO)
Homes in area: 83,000 (build complete)
Awarded: 16 Feb 90; acquired 22 Mar 93
Service start: Oct 91
Homes passed: 92,771 (1 Jul 98)
Subs: 17,088 = 18.4% (1 Jul 98)

Derbyshire, East
Chesterfield, Bolsover, Matlock, Belper
Local delivery franchise holder: NTL (see GO)
Homes in area: 89,000
Awarded: Jun 96

Devon, South
Exeter, Plymouth, Torbay
Franchise holder: Eurobell (see GO)
Homes in area: 236,000
Awarded: 15 Dec 89
Service start: May 1996
Homes passed: 50,260 (1 Jul 98)
Subs: 5,044 = 10.0% (1 Jul 98)

Dewsbury see **Kirklees**
Diss see **East Anglia, South**

Doncaster/Rotherham

Franchise holder: Telewest(see GO)
Homes in area: 192,000
Awarded: 10 May 90
Homes passed: 65,212 (1 Jul 98)
Subs: 9,905 = 15.2% (1 Jul 98)

Dorset, West

Dorchester, Weymouth, Portland
Homes in area: 35,000
Awarded: 10 Feb 90
Franchise revoked from Coastal
Cablevision = Leonard
Communication (US)

Dover see **Kent, South East**
Droitwich see **Worcester**
Dudley see **Black Country**

Dumbarton/Vale of Leven

Franchise holder: Telewest (see GO)
Homes in area: 17,000
(build complete)
Awarded: 27 Apr 89
Service start: Nov 96
Homes passed: 19,898 (1 Jul 98)
Subs: 8,884 = 44.6% (1 Jul 98)

Dumfries and Galloway

Local delivery franchise holder:
Dumfries and Galloway Cable &
Telecoms
Ownership: US Cable Group,
McNicholas Construction, Morgan
Cable
PO Box 319
Whipsnade
Bedfordshire LU6 2LT
Tel: 01582 873006
Fax: 01582 873003
Homes in area: 155,000
Awarded: Dec 97

Dundee/Broughty Ferry/ Monifieth/Carnoustie

Franchise holder: Telewest (see GO)
Homes in area: 95,000
Awarded: 19 Jan 90
Service start: Jan 91
Homes passed: 65,632 (1 Jul 98)
Subs: 11,201 = 17.1% (1 Jul 98)

Dunstable see **Luton**

Durham, South/North Yorkshire

Local delivery franchise holder:
Cable & Wireless Communications
(see GO)
Homes in area: 155,000
Awarded: Apr 96

Durham see **Wearside**

Ealing, London Borough of

Franchise holder: Cable & Wireless
Communications (see GO)
Homes in area: 105,000
Awarded: 8 Nov 83

Service start: 1 Nov 86
Homes passed: 80,865 (1 Jul 98)
Subs: 20,306 = 25.1% (1 Jul 98)

East Anglia, South

Bury St Edmunds, Sudbury,
Braintree, Clacton on Sea,
Stowmarket, Thetford, Diss
Local delivery licence holder:
Southern East Anglia Cable =
Comcast Europe (see GO)
Homes in area: 205,000
Date awarded: Jan 95

East Grinstead

East Grinstead, Crowborough, parts
of Banstead and Reigate
Local delivery licence holder:
Convergence (East Grinstead)
(see GO)
Homes in area: 30,300
Date awarded: 11 Jul 96

East Kilbride see **Motherwell**

Eastbourne/Hastings

Local delivery franchise holder:
Cable & Wireless Communications
(see GO)
Homes in area: 150,000
Awarded: Feb 96
Separate upgrade systems in
Eastbourne and Hastings operated by
CDA Communications

Eastleigh see **Southampton**
East Lothian see **Lothian**

Edinburgh

Franchise holder: Telewest (see GO)
Homes in area: 183,000
(build complete)
Awarded: 5 Feb 86
Service start: May 92
Homes passed: 220,018 (1 Jul 98)
Subs: 50,233 = 22.8% (1 Jul 98)
see also Lothian

Ellesmere Port see **Cheshire, North**
Elmbridge see **Surrey, North**
Elstree see **Hertfordshire, South**

Enfield

Franchise holder: Cable London
(see GO)
Homes in area: 105,000 (build
complete)
Awarded: 31 May 90
Service start: Oct 91
Homes passed: 107,175 (1 Jul 98)
Subs: 20,003 = 27.1% (1 Jul 98)

Epping Forest/Chigwell/ Loughton/Ongar

Franchise holder: Cable & Wireless
Communications (see GO)
Homes in area: 45,000
Awarded: 3 May 90

Service start: Dec 94
Homes passed: 33,892 (1 Jul 98)
Subs: 4,587 = 13.5% (1 Jul 98)

Epsom see **Surrey, North East**
Exeter see **Devon, South**

Falkirk/West Lothian

Franchise holder: Telewest (see GO)
Homes in area: 30,000
Awarded: 21 Jun 90
Service start: Oct 94

Fareham see **Portsmouth**
Farnborough see **Guildford**

Faversham

Narrowband upgrade system
operated by CDA Communications
(see GO)

Felixstowe see **Colchester**

Fenland

Wisbech, March, Whittlesey
Franchise holder: Cable & Wireless
Communications (see GO)
Homes in area: 21,000
Awarded: 5 Jul 90
Service start: Apr 97

Fife

Kingdom of Fife excluding
Glenrothes and Kirkcaldy
Local delivery franchise holder:
Telewest (see GO)
Homes in area: 35,000
Awarded: Jul 97

Folkestone see **Kent, South East**
Gateshead see **Tyneside**
Gatwick see **Crawley**
Gillingham see
Thames Estuary South

Glamorgan, West

Swansea, Neath, Port Talbot
Franchise holder: CableTel South
Wales (see GO)
Homes in area: 110,000
Awarded: 16 Nov 89
Service start: Dec 90
Homes passed: 87,593 (1 Jul 98)
Subs: 38,979 = 44.5% (1 Jul 98)
Separate narrowband upgrade
systems operated in parts of the area
by Metro Cable

Glamorgan/Gwent

Franchise holder: CableTel South
Wales (see GO)
Homes in area: 230,000
Awarded: Oct 95
Service start: Apr 97 on existing
network
Homes passed: 15,179 (1 Jul 98)
Subs: 6,193 = 40.8% (1 Jul 98)
Separate narrowband upgrade

systems operated in parts of the area by Metro Cable

Glasgow, Greater
Franchise holder: CableTel Glasgow (see GO)
Homes in area: 274,000
Awarded: 7 Jun 90
Homes passed: 67,201 (1 Jul 98)
Subs: 23,708 = 35.3% (1 Jul 98)

Glasgow, North West/Clydebank
Franchise holder: CableTel Glasgow (see GO)
Homes in area: 112,000; 16,000 business premises
Awarded: 29 Nov 83
Service start: 1 Oct 85
Homes passed: 101,370 (1 Jul 98)
Subs: 21,305 = 21.0% (1 Jul 98)

Glenrothes/Kirkcaldy/Leven/Buckhaven/Methil
Franchise holder: Telewest (see GO)
Homes in area: 60,000
Awarded: 21 Jun 90
Service start: Oct 91
Homes passed: 50,145 (1 Jul 98)
Subs: 11,368 = 22.7% (1 Jul 98)

Gloucester see **Cheltenham**
Godalming see **Guildford**
Gosport see **Portsmouth**
Gourock see **Inverclyde**

Grantham
Franchise holder: NTL (see GO)
Homes in area: 30,000
Awarded: 26 Apr 90
Service start: Oct 95

Gravesend see **Thames Estuary South**

Great Yarmouth/Lowestoft/Caister
Franchise holder: Cable & Wireless Communications (see GO)
Homes in area: 64,000
Awarded: 5 Jul 90
Service start: Jun 96
Homes passed: 18,995 (1 Jul 98)
Subs: 3,409 = 17.9% (1 Jul 98)

Greater London East Boroughs of Barking/Dagenham, Bexley, Redbridge
Franchise holder: Cable & Wireless Communications (see GO)
Homes in area: 229,000
Awarded: 15 Dec 88
Service start: Dec 90
Homes passed: 193,839 (1 Jul 98)
Subs: 44,321 = 22.9% (1 Jul 98)

Greenock see **Inverclyde**

Greenwich/Lewisham, London Boroughs of
Franchise holder: Cable & Wireless Communications (see GO)
Homes in area: 175,000
Awarded: 7 Apr 89
Service start: Jan 91
Homes passed: 139,199 (1 Jul 98)
Subs: 30,843 = 22.2% (1 Jul 98)

Grimsby/Immingham/Cleethorpes
Franchise holder: NTL (see GO)
Homes in area: 63,000
Awarded: 5 Jul 90
Service start: Jun 95
Homes passed: 58,647 (1 Jul 98)
Subs: 11,246 = 19.2% (1 Jul 98)

Guildford/West Surrey/East Hampshire
Guildford, Aldershot, Farnborough, Camberley, Woking, Godalming
Franchise holder: CableTel UK (see GO)
Flagship House
Reading Road North
Surrey GU13 8XR
Tel: 01252 652000
Fax: 01252 652100
Homes in area: 22,000 + 115,000
Awarded: 29 Nov 83 + Aug 85
Service start: 1 Jul 87
Homes passed: 119,593 (1 Jul 98)
Subs: 39,027 = 32.6% (1 Jul 98)

Gwent see **Glamorgan/Gwent**

Hackney/Islington, London Boroughs of
Franchise holder: Cable London (see GO)
Homes in area: 150,000
Awarded: 13 Apr 90
Service start: Apr 95
Homes passed: 95,271 (1 Jul 98)
Subs: 20,976 = 22.0% (1 Jul 98)

Halifax see **Calderdale**
Hamilton see **Motherwell**

Hammersmith and Fulham, London Borough of see **London, North West**

Haringey, London Borough of
Franchise holder: Cable London (see GO)
Homes in area: 85,000
Awarded: Sept 89
Homes passed: 73,844 (1 Jul 98)
Subs: 18,626 = 25.2% (1 Jul 98)

Harlow/Bishops Stortford/Stansted Airport
Franchise holder: Anglia Cable = Comcast Europe (see GO)
Homes in area: 43,000 (build

complete)
Date awarded: 23 Mar 90
Service start: Jun 93
Homes passed: 44,305 (1 Jul 98)
Subs: 13,118 = 29.6% (1 Jul 98)

Harpenden see **Hertfordshire, West**

Harrogate/Knaresborough
Franchise holder: Cable & Wireless Communications (see GO)
Homes in area: 78,000
Awarded: 30 Mar 90, acquired Apr 94
Service start: Sep 95
Homes passed: 30,149 (1 Jul 98)
Subs: 4,349 = 14.4% (1 Jul 98)

Harrow
Franchise holder: Cable & Wireless Communications (see GO)
Homes in area: 79,000
Awarded: 24 May 90
Service start: Dec 91
Homes passed: 68,299 (1 Jul 98)
Subs: 16,171 = 23.7% (1 Jul 98)

Hartlepool see **Teesside**
Harwich see **Colchester**
Hastings see **Eastbourne**
Hatfield see **Hertfordshire, Central**
Havant see **Portsmouth**

Haverfordwest
Narrowband upgrade system operated by Metro Cable

Havering, London Borough of
Franchise holder: Cable & Wireless Communications (see GO)
Homes in area: 90,000
Awarded: 6 Apr 90
Service start: Sept 93
Homes passed: 82,498 (1 Jul 98)
Subs: 8,195 = 9.9% (1 Jul 98)

Haywards Heath
Local delivery franchise holder: Convergence Group (see GO)
Homes in area: 31,150
Original franchise revoked from N-Comm Cablevision

Heathrow see **Windsor**
Hemel Hempstead see **Hertfordshire, West**
Henley-on-Thames see **Thames Valley**

Herne Bay
Narrowband upgrade system operated by CDA Communications (see GO)

Hertford see **Hertfordshire, East**

Hertfordshire, Central
Stevenage, Welwyn, Hatfield, Hitchin, Baldock, Letchworth

Franchise holder: CableTel
Hertfordshire (see GO)
Homes in area: 100,000
Awarded: 3 Nov 89
Homes passed: 78,903 (1 Jul 98)
Subs: 41,322 = 52.4% (1 Jul 98)
Separate upgrade system operated in
Hatfield by Metro Cable

Hertfordshire, East
Hertford, Cheshunt, Ware, Lea
Valley, Hoddesdon
Franchise holder: CableTel Bedford-
shire (see GO)
Homes in area: 60,000
Awarded: 31 May 90
Homes passed: 45,785 (1 Jul 98)
Subs: 15,588 = 34.0% (1 Jul 98)

Hertfordshire, South
Watford, Chorleywood,
Rickmansworth, Bushey, Radlett,
Elstree, Borehamwood, Potters Bar
Franchise holder: Cable & Wireless
Communications (see GO)
Homes in area: 95,000
Awarded: 3 Nov 89
Service start: Apr 92
Homes passed: 87,285 (1 Jul 98)
Subs: 22,694 = 26.0% (1 Jul 98)

Hertfordshire, West
Harpenden, Hemel Hempstead, St
Albans, Berkhamsted, Tring,
Redbourne
Franchise holder: NTL (see GO)
Homes in area: 100,000
Awarded: 3 Nov 89
Service start: Mar 91
Homes passed: 87,787 (1 Jul 98)
Subs: 18,190 = 20.7% (1 Jul 98)

High Wycombe see **Thames Valley**
Hillingdon see **Middlesex**

Hinckley/Bosworth
Local delivery franchise holder: NTL
(see GO)
Homes in area: 31,200
Awarded: Jun 95
Original franchise revoked from N-
Comm Cablevision

Hitchin see **Hertfordshire, Central**
Hoddesdon see
Hertfordshire, East
Horley see **Crawley**
Hounslow see **Middlesex**
Hove see **Brighton**
Huddersfield see **Kirklees**
Hull see **Kingston upon Hull**
Immingham see **Grimsby**

Inverclyde
Greenock, Port Glasgow, Gourock,
Kilmacolm
Franchise holder: CableTel Glasgow

(see GO)
Homes in area: 32,000
Awarded: 5 Jul 90
Service start: 1995
Homes passed: 19,538 (1 Jul 98)
Subs: 9,189 = 47.0% (1 Jul 98)

Ipswich see **Colchester**
Isle of Thanet see **Thanet, Isle of**

Isle of Wight
Local delivery franchise holder: Isle
of Wight Cable and Telephone
Company
Elm Farm
Elm Lane
Calbourne
Isle of Wight PO30 4JY
Ownership: Utility Cable, Fortuna
Advanced Communications
Networks
Homes on area: 43,000
Awarded: May 1997

Islington see **Hackney**

Jersey
Franchise holder: Jersey Cable (not
ITC licensed)
3 Colomberie
St Helier, Jersey
Channel Islands JE4 9SY
Tel: 01534 66477
Fax: 01534 66681
Ownership: Carveth 50.4%, Mattbrel
30%, others 19.6%
Homes in area: 28,000
Service start: 1987
Franchise renewed: Jan 94
Homes passed: 3,500 cable, 4,600
SMATV

Kenilworth see **Warwick**

Kensington/Chelsea, London Borough of
Franchise holder: Cable & Wireless
Communications (see GO)
Homes in area: 82,000
Awarded: 4 Feb 88
Service start: Sep 89
Homes passed: 70,590 (1 Jul 98)
Subs: 14,958 = 21.2% (1 Jul 98)

Kent, South East
Ashford, Deal, Dover, Folkestone
Local delivery franchise holder:
Cable & Wireless Communications
(see GO)
Homes in area: 116,300
Awarded: May 90
Service start: Sep 96
Homes passed: 40,040 (1 Jul 98)
Subs: 4,083 = 10.2% (1 Jul 98)
Separate upgrade system in Ashford
operated by CDA Communications

Kent, West
Tunbridge Wells, Tonbridge,

Sevenoaks
Local delivery franchise holder:
Eurobell (see GO)
Homes in area: 90,600
Awarded: May 94

Kettering see
Northampton, North-east
Kidderminster see **Black Country**

Kilbirnie
Narrowband upgrade system
operated by A Thomson (Relay)

Kilsyth see **Cumbernauld**

Kingston and Richmond, London Boroughs of
Franchise holder: Telewest (see GO)
Homes in areas: 124,000
Awarded: 6 May 89
Service start: Jan 91
Homes passed: 105,839 (1 Jul 98)
Subs: 23,584 = 22.3% (1 Jul 98)

Kingston upon Hull
Narrowband upgrade system
operated by Hull Cablevision =
Atlantic Telecom Group (see GO)

Kirkcaldy see **Glenrothes**
Kirkby-in-Ashfield see **Mansfield**

Kirklees
Huddersfield, Dewsbury
Franchise holder: CableTel Kirklees
(see GO)
Homes in area: 148,000
Awarded: 14 Jun 90
Service start: Jun 95
Homes passed: 73,776 (1 Jul 98)
Subs: 24,535 = 33.3% (1 Jul 98)

Knaresborough see **York**
Knowsley see **St Helens**

Lakenheath
Narrowband upgrade system
operated by Cablecom Investments
(see GO)

Lambeth/Southwark, London Boroughs of
Franchise holder: Cable & Wireless
Communications (see GO)
Homes in area: 191,000
Awarded: 6 Jul 89
Service start: Jul 91
Homes passed: 141,756 (1 Jul 98)
Subs: 30,157 = 21.3% (1 Jul 98)

Lanark see **Motherwell**

Lancashire, Central
Preston, Leyland
Franchise holder: Telewest (see GO)
Homes in area: 114,000
Awarded: 5 Feb 86
Build start: Jun 90

Homes passed: 95,465 (1 Jul 98)
Subs: 16,206 = 17.0% (1 Jul 98)

Lancashire, East

Blackburn, Burnley, Accrington,
Nelson, Colne, Rossendale Valley
Franchise holder: Cable & Wireless
Communications (see GO)
Homes in area: 168,000
Awarded: 9 May 88; acquired 21
Apr 93
Service start: 30 Nov 89
Homes passed: 114,505 (1 Jul 98)
Subs: 21,487 = 18.8% (1 Jul 98)
Narrowband upgrade system in
Burnley operated by Cablecom
Investments (see GO)

Lancaster/Morecambe

No applications
Homes in area: 40,000

Lancashire West see Southport

Largs

Narrowband upgrade system
operated by Harris of Saltcoats
(see GO)

Lea Valley see Hertfordshire, East
Leamington Spa see Warwick

Leeds

Franchise holder: Cable & Wireless
Communications (see GO)
Homes in area: 289,000
Awarded: Mar 90
Service Start: Jun 94
Homes passed: 207,653 (1 Jul 98)
Subs: 39,911 = 19.2% (1 Jul 98)

Leicester/ Loughborough/ Shepshed

Franchise holder: NTL (see GO)
Homes in area: Leicester 170,670 +
Loughborough 30,000
Awarded: Leicester 22 Sept 89,
Loughborough 9 March 90
Service start: 1 Mar 91
Homes passed: 113,110 (1 Jul 98)
Subs: 22,546 = 19.9% (1 Jul 98)
Separate upgrade system in Leicester
operated by CDA Communications

Leighton Buzzard see Luton
Letchworth see Hertfordshire, Central

Lewes

Narrowband upgrade system
operated by CDA Communications
(see GO)

Leyland see Lancashire, Central

Lincoln

Franchise holder: NTL (see GO)
Homes in area: 42,000
Awarded: 5 Jul 90

Service start: Jul 95
Homes passed: 37,417 (1 Jul 98)
Subs: 7,105 = 19.0% (1 Jul 98)

Lincolnshire/South Humberside

Local delivery franchise holder: NTL
(see GO)
Homes in area: 144,000
Awarded: Jan 96

Litchfield/Burntwood/Rugeley

Local delivery licence holder: NTL
(see GO)
Homes in area: 39,290
Date awarded: Jun 95
Service start: Mar 97

Liverpool, North/Bootle/Crosby

Franchise holder: Telewest (see GO)
Homes in area: 119,000 (build
complete)
Awarded: 5 Jul 90
Homes passed: 121,836 (1 Jul 98)
Subs: 28,731 = 23.6% (1 Jul 98)

Liverpool, South (Merseyside)

Franchise holder: Telewest (see GO)
Homes in area: 125,000
Awarded: 29 Nov 83
Service start: Oct 90
Homes passed: 116,519 (1 Jul 98)
Subs: 26,287 = 22.6% (1 Jul 98)

Llandeilo

Narrowband system operated by
John Jones

London see also Greater London East and individual boroughs

London, North West Boroughs of Barnet, Brent, Hammersmith and Fulham

Franchise holder: Cable & Wireless
Communications (see GO)
Homes in area: 280,000
Awarded: 19 Jan 89
Service start: Jul 91
Homes passed: 137,522 (1 Jul 98)
Subs: 24,481 = 17.8% (1 Jul 98)
Separate upgrade system operated in
Brent by Sapphire

Lothian

East Lothian, Midlothian, parts of
City of Edinburgh
Local delivery franchise applicant:
Telewest (see GO)
Homes in area: 30,000

Loughborough see Leicester
Loughton see Epping Forest
Lowestoft see Great Yarmouth

Luton/Dunstable/Leighton Buzzard

Franchise holder: CableTel Bedfordshire (see GO)

Homes in area: 97,000
Awarded: Jul 86
Service start: Nov 86 on upgrade
system, Mar 90 on new build
network
Homes passed: 78,342 (1 Jul 98)
Subs: 21,998 = 28.1% (1 Jul 98)

Macclesfield/Wilmslow

Franchise holder: Cable & Wireless
Communications (see GO)
Homes in area: 45,000
Awarded: 11 Jul 90; acquired 4 May 93
Homes passed: 40,354 (1 Jul 98)
Subs: 7,268 = 18.0% (1 Jul 98)

Maidenhead see Windsor
Maidstone see Thames Estuary South

Manchester/Salford/Trafford

Franchise holder: Cable & Wireless
Communications (see GO)
Homes in area: 363,000
Awarded: 17 May 90; acquired 22
Mar 93
Service start: Oct 94
Homes passed: 209,094 (1 Jul 98)
Subs: 40,485 = 19.4% (1 Jul 98)
Separate upgrade system in Salford
operated by CDA Communications

Mansfield/Sutton/Kirkby-in-Ashfield

Franchise holder: NTL (see GO)
Homes in area: 58,000 (build
complete)
Awarded: 3 Mar 90
Homes passed: 70,772 (1 Jul 98)
Subs: 16,242 = 22.9% (1 Jul 98)

March see Fenland
Margate see Thanet, Isle of
Market Harborough see Northampton, North-east
Marlow see Thames Valley
Matlock see Derbyshire, East

Melton Mowbray

Franchise holder: NTL (see GO)
Homes in area: 30,000
Awarded: 26 Apr 90
Service start: Oct 95

Merton and Sutton, London Boroughs of

Franchise holder: Telewest (see GO)
Homes in area: 135,000
(build complete)
Awarded: 6 May 89
Service start: Mar 90
Homes passed: 135,037 (1 Jul 98)
Subs: 30,145 = 22.3% (1 Jul 98)

Middlesbrough see Teesside

Middlesex

Hillingdon, Hounslow (franchises awarded separately but since combined)
Franchise holder: Telewest (see GO)
Homes in area: 186,886
Awarded: 24 May 90
Service start: Nov 91
Homes passed: 174,363 (1 Jul 98)
Subs: 27,221 = 15.6% (1 Jul 98)

Midlothian see Lothian

Mildenhall

Narrowband upgrade system operated by Cablecom Investments (see GO)

Milford Haven

Narrowband upgrade system operated by Metro Cable

Milton Keynes

Local delivery franchise holder: BT New Towns Cable TV (see GO)
51 Alston Drive
Bradwell Abbey
Milton Keynes MK13 9HB
Tel: 01908 322522
Fax: 01908 319802
Homes in area: 114,000
Awarded: 29 May 1997
BT is upgrading its existing narrowband system

Mole Valley see Surrey, North East
Monifieth see Dundee
Morecambe see Lancaster

Motherwell/East Kilbride/ Hamilton/Wishaw/Lanark

Franchise holder: Telewest (see GO)
Homes in area: 125,000
(build complete)
Awarded: 27 Apr 89
Service start: Mar 92
Homes passed: 177,841 (1 Jul 98)
Subs: 37,410 = 21.0% (1 Jul 98)

Neath see Glamorgan, West
Nelson see Lancashire, East

Newark on Trent

Franchise holder: NTL (see GO)
Homes in area: 35,000
Awarded: 26 Apr 90
Service start: Sep 95

Newbury see Thames Valley
Newcastle-under-Lyne see Stoke-on-Trent
Newcastle-upon-Tyne see Tyneside

Newham and Tower Hamlets, London Boroughs of

Franchise holder: Cable & Wireless

Communications (see GO)
Homes in area: 127,000
Awarded: 13 Aug 85
Service start: May 87
Homes passed: 113,775 (1 Jul 98)
Subs: 22,291 = 19.6% (1 Jul 98)

Newport/Cwmbran/Pontypool

Franchise holder: CableTel South Wales (see GO)
Homes in area: 85,000
Awarded: 11 Jul 90
Homes passed: 42,655 (1 Jul 98)
Subs: 19,664 = 46.1% (1 Jul 98)

Neyland

Narrowband upgrade system operated by Metro Cable

Northampton

Franchise holder: NTL (see GO)
Homes in area: 72,000 (build complete)
Awarded: 19 Jan 89
Service start: 1988 on 13-channel upgrade network (classified as broadband), Mar 91 on new-build network
Homes passed: 72,803 (1 Jul 98)
Subs: 16,311 = 22.4% (1 Jul 98)
Separate upgrade system operated by CDA Communications

Northamptonshire, North-east Corby, Kettering, Wellingborough, Market Harborough

Franchise holder: NTL (see GO)
Homes in area: 90,000
Awarded: 21 Jun 90
Service start: Dec 94
Homes passed: 85,680 (1 Jul 98)
Subs: 13,204 = 15.4% (1 Jul 98)

Northern Ireland

Franchise holder: CableTel Northern Ireland (see GO)
Homes in area: 428,000
Date awarded: May 95
Homes passed: 114,590 (1 Jul 98)
Subs: 45,309 = 39.5% (1 Jul 98)

Northumberland

Local delivery franchise holder: Cable & Telecoms
Homes in area: 125,000
Awarded: Oct 97

Norwich

Franchise holder: Cable & Wireless Communications (see GO)
Homes in area: 83,000
Awarded: 21 Jul 89, acquired Jul 94
Service start: Jun 90
Homes passed: 68,692 (1 Jul 98)
Subs: 12,191 = 17.7% (1 Jul 98)

Nottingham

Franchise holder: NTL (see GO)
Homes in area: 160,000
(build complete)

Awarded: 22 Sept 89
Service start: 10 Sept 90
Homes passed: 177,733 (1 Jul 98)
Subs: 34,516 = 19.4% (1 Jul 98)

Nuneaton/Bedworth/Rugby

Franchise holder: Telecential Communications (see GO)
Homes in area: Nuneaton 44,000 + Rugby 23,000 (awarded as two separate franchises)
Awarded: 6 Apr 90
Service start: Feb 96

Oldham/Tameside

Franchise holder: Cable & Wireless Communications (see GO)
Homes in area: 172,000
Awarded: 17 May 90; acquired 4 May 93
Service start: Oct 94
Homes passed: 98,462 (1 Jul 98)
Subs: 20,872 = 21.2% (1 Jul 98)

Ongar see Epping Forest

Oxford/Abingdon

Franchise holder: NTL (see GO)
Homes in area: 72,000 (build complete)
Awarded: 14 Jun 90
Service start: Sept 95
Homes passed: 106,293 (1 Jul 98)
Subs: 14,259 = 13.4% (1 Jul 98)

Paisley/Renfrew

Franchise holder: CableTel Glasgow (see GO)
Homes in area: 67,000
Awarded: 7 Jun 90
Service start: Aug 94
Homes passed: 44,900 (1 Jul 98)
Subs: 21,861 = 48.7% (1 Jul 98)

Pembroke Dock

Narrowband upgrade system operated by Metro Cable

Penarth see Cardiff

Perth/Scone

Franchise holder: Telewest (see GO)
Homes in area: 18,000 (build complete)
Awarded: 19 Jan 90
Service start: 1997
Homes passed: 18,758 (1 Jul 98)
Subs: 4,201 = 22.4% (1 Jul 98)
Separate upgrade system operated in Perth by Perth Cable TV

Peterborough

Franchise holder: Cable & Wireless Communications (see GO)
Homes in area: 58,000
Awarded: 21 Jul 89
Service start: May 90
Homes passed: 55,118 (1 Jul 98)
Subs: 13,483 = 24.5% (1 Jul 98)

Plymouth see **Devon, South**
Pontefract see **Wakefield**
Pontypool see **Newport**
Poole see **Bournemouth**
Port Glasgow see **Inverclyde**
Port Talbot see **Glamorgan, West**

**Portsmouth/Fareham/Gosport/
Havant/East Hampshire**
Franchise holder: Cable & Wireless
Communications (see GO)
Homes in area: 213,000
Awarded: 2 Feb 90
Service start: Sept 91
Homes passed: 208,574 (1 Jul 98)
Subs: 55,391 = 26.6% (1 Jul 98)

Potters Bar see
Hertfordshire, South
Preston see **Lancashire, Central**
Radlett see **Hertfordshire, South**
Ramsgate see **Thanet, Isle of**

Ravenshead
Local delivery licence holder: NTL
(see GO)
Homes in area: 2,500
Date awarded: 13 July 95

Reading see **Thames Valley**
Redbourne see
Hertfordshire, West
Redbridge, London Borough of
see **Greater London East**
Reddish see **Worcester**
Redhill see **Surrey, North East**
Reigate see **Surrey, North East
and East Grinstead**
Renfrew see **Paisley**
Richmond, London Borough of
see **Kingston**
Rickmansworth see
Hertfordshire, South
Rochdale see **Bury**
Rochester see **Thames Estuary
South**
Romsey see **Andover**
Rossendale Valley see
Lancashire, East
Rotherham see **Doncaster**
Rugby see **Nuneaton**
Rugeley see **Litchfield**
Runcorn see **Cheshire, North**
Runnymede see **Surrey, North**
St Albans see **Hertfordshire, West**

St Helens/Knowsley
Franchise holder: Telewest (see GO)
Homes in area: 121,000
Awarded: 5 Jul 90

Service start: Jun 92
Homes passed: 105,951 (1 Jul 98)
Subs: 24,838 = 23.4% (1 Jul 98)

Salford see **Manchester**
Salisbury see **Andover**

Saltcoats
Narrowband upgrade system operated
by Harris of Saltcoats (see GO)

Sandwell see **Black Country**
Sefton see **Southport**
Scone see **Dundee**

Sheffield
Franchise holder: Telewest (see GO)
Homes in area: 210,000
Awarded: 31 May 90
Service start: Apr 94
Homes passed: 85,152 (1 Jul 98)
Subs: 15,086 = 17.7% (1 Jul 98)

Shepshed see **Leicester**

Shrewsbury
Local delivery licence holder: Cable
& Telecoms (see GO)
Homes in area: 90,000
Awarded: Jan 96
Licence revoked Dec 97

Sittingbourne see **Thames
Estuary South**

Skelmersdale
Narrowband upgrade system
operated by Tawd Valley Cable

Slough see **Windsor**
Solihull see **Birmingham**
South Ribble see **Southport**

Southampton/Eastleigh
Franchise holder: Cable & Wireless
Communications (see GO)
Homes in area: 119,371
Awarded: 12 Sept 86
Service start: 1 Dec 90
Homes passed: 106,720 (1 Jul 98)
includes Winchester
Subs: 25,348 = 23.8% (1 Jul 98)
includes Winchester

**Southport/Sefton/West
Lancashire/South Ribble/Chorley**
Local delivery licence holder:
Telewest
Homes in area: 90,000
Awarded: Jan 96
Service start: Jun 97

Southend see
Thames Estuary North
Southwark, London Borough of
see **Lambeth**

Stafford/Stone
Franchise holder: NTL (see GO)
Homes in area: 30,600
(build complete)
Awarded: 1 Dec 89
Service start: Sept 95
Homes passed: 66,803 (1 Jul 98)
Subs: 11,139 = 16.9% (1 Jul 98)

Staines see **Windsor**
Stanley see **Consett**
Stanwell see **Windsor**
Stevenage see **Hertfordshire,
Central**

Stockport
Franchise holder: Cable & Wireless
Communications (see GO)
Homes in area: 113,000
Awarded: 17 May 90; acquired 4
May 93
Service start: Oct 94
Homes passed: 103,102 (1 Jul 98)
Subs: 21,120 = 20.5% (1 Jul 98)

Stockton see **Teesside**

**Stoke-on-Trent/Newcastle-
under-Lyne**
Franchise holder: Cable & Wireless
Communications (see GO)
Homes in area: 156,000
Awarded: 1 Dec 89; acquired 21 Apr 93
Homes passed: 116,749 (1 Jul 98)
Subs: 19,204 = 16.4% (1 Jul 98)

Stone see **Stafford**
Stowmarket see **East Anglia,
South**
Stratford-upon-Avon see **Warwick**
Sudbury see **East Anglia, South**
Sunderland see **Wearside**

Surrey, North/North East
Banstead, Caterham, Chertsey,
Cobham, Dorking, Elmbridge,
Epsom, Ewell, Leatherhead, Reigate,
Redhill, Sunbury, Weybridge etc
Franchise holder: Cable & Wireless
Communications (see GO)
Homes in area: 71,000 + 98,000
(awarded as two franchises)
Awarded: 21 Jun 90
Service start: Apr 93
Homes passed: 149,782 (1 Jul 98)
Subs: 28,159 = 18.8% (1 Jul 98)

Sutton see **Mansfield**
Sutton, London Borough of see
Merton
Swansea see **Glamorgan, West**

Swindon
Franchise holder: Swindon Cable =
NTL (see GO)
Homes in area: 65,000

(build complete)
Service start: 1 Sep 84
Homes passed: 66,537
Subs: 21,950 = 33.0% (1 Jul 98)

Tameside see Oldham

Tamworth/North Warwickshire/Meriden

Local delivery licence holder: NTL (see GO)
Homes in area: 43,315
Awarded: Jun 95
Service start: Mar 97
Original franchise revoked from N-Com Cablevision

Taunton/Bridgwater

Local delivery licence holder: Telewest
Homes in area: 71,300
Awarded: Feb 97

Teesside
Middlesbrough, Stockton, Hartlepool

Franchise holder: Comcast Teesside (see GO)
Homes in area: 195,000
Awarded: 5 Jul 90
Service start: Jun 95
Homes passed: 147,037 (1 Jul 98)
Subs: 45,748 = 31.1% (1 Jul 98)

Telford

Franchise holder: Telewest (see GO)
Homes in area: 55,000
Awarded: 26 Apr 90
Service start: May 92
Homes passed: 52,451 (1 Jul 98)
Subs: 13,417 = 25.6% (1 Jul 98)

Thames Estuary North

Southend, Basildon, Billericay, Brentwood, Chelmsford etc
Franchise holder: Telewest (see GO)
Homes in area: 300,000
Awarded: 16 Nov 88
Service start: Jun 94
Homes passed: 172,405 (1 Jul 98)
Subs: 44,881 = 26.0% (1 Jul 98)
Separate upgrade system in Basildon operated by CDA Communications

Thames Estuary South

Gravesend, Chatham, Rochester, Gillingham, Maidstone, Sittingbourne
Franchise holder: Telewest (see GO)
Homes in area: 145,000
Awarded: 16 Nov 88
Homes passed: 92,834 (1 Jul 98)
Subs: 24,184 = 26.1% (1 Jul 98)
Separate upgrade systems in Chatham and Sittingbourne operated by CDA Communications

Thames Valley

Reading, Twyford, Henley-on-Thames, Wokingham, High Wycombe, Marlow, Bracknell, Basingstoke, Ascot, Newbury, Thatcham, Franchise holder: Telecential Communications (see GO)
Homes in area: 215,000
Awarded : 2 Dec 88
Service start: Dec 91
Homes passed: 149,747 (1 Jan 97)
Subs: 29,604 = 19.8% (1 Jan 97)

Thamesmead

Franchise holder: Videotron Corporation (see GO)
Homes in area: 11,000
Awarded: 31 May 90
Service start: Jul 91
Homes passed: 8,544 (1 Jan 97)
Subs: 2,577 = 30.2% (1 Jan 97)

Thanet, Isle of

Margate, Ramsgate, Broadstairs
Franchise revoked from Coastal Cablevision = Leonard Communications
Homes in area: 51,000
Awarded: 16 Feb 90
Separate narrowband upgrade system operated CDA Communications (see GO)

Thetford see East Anglia, South
Torbay see Devon, South

Totton/Hythe

Local delivery franchise holder: Cable & Wireless Communications (see GO)
Homes in area: 25,200
Awarded: Sep 95
Service start: Apr 97

Tower Hamlets, London Borough of see Newham
Tring see Hertfordshire, West
Twyford see Thames Valley

Tyneside

Newcastle-upon-Tyne, Gateshead, North and South Tyneside
Franchise holder: Telewest (see GO)
Homes in area: 325,000
Awarded: 14 Dec 89
Service start: Sept 90
Homes passed: 225,079 (1 Jul 98)
Subs: 55,201 = 24.5% (1 Jul 98)

Upper Heyford

Narrowband upgrade system operated by Cablecom Investments (see GO)

Uttoxeter see Burton-on-Trent

Vale of Belvoir

Local delivery franchise holder: NTL (see GO)
Homes in area: 4,545
Awarded: Jul 96

Wakefield/Pontefract/Castleford

Franchise holder: Telewest (see GO)
Homes in area: 94,000
Awarded: 2 Mar 90; acquired Apr 93
Homes passed: 43,981 (1 Jul 98)
Subs: 5,847 = 13.3% (1 Jul 98)

Walsall see Black Country

Waltham Forest, London Borough of

Franchise holder: Cable & Wireless Communications (see GO)
Homes in area: 83,000
Awarded: 28 Sept 89
Service start: Feb 94
Homes passed: 81,083 (1 Jul 98)
Subs: 17,511 = 21.6% (1 Jul 98)

Wandsworth, London Borough of

Franchise holder: Cable & Wireless Communications (see GO)
Homes in area: 100,000
Awarded: 13 Aug 85
Service start: Aug 93
Homes passed: 60,915 (1 Jul 98)
Subs: 11,455 = 18.8% (1 Jul 98)

Ware see Hertfordshire, East
Warrington see Cheshire, North

Warwick/Stratford-upon-Avon/Kenilworth/Leamington Spa

Franchise holder: NTL (see GO)
Homes in area: 50,000
(build complete)
Awarded: 30 Mar 90
Homes passed: 101,408 (1 Jul 98)
Subs: 13,389 = 13.2% (1 Jul 98)

Watford see Hertfordshire, South

Wearside

Sunderland, Durham, Washington
Franchise holder: Cable & Wireless Communications (see GO)
Homes in area: 200,000
Awarded: 14 Jun 90
Service start: Aug 96
Homes passed: 57,108 (1 Jul 98)
Subs: 11,046 = 19.3% (1 Jul 98)

Wellingborough see Northampton, North-east
Welwyn see Hertfordshire, Central
Wendover see Aylesbury
West Lothian see Falkirk

Westminster, London Borough of

Franchise holder: Westminster Cable

Company = British Telecom
(see GO)
87-89 Baker Street
London W1M 1AG
Tel: 0171 935 6699
Fax: 0171 486 9447
Homes in area: 120,000
Awarded: 29 Nov 83
Service start: Sept 85
Homes passed: 97,365 (1 Jul 98)
Subs: 21,144 = 21.7% (1 Jul 98)

Weymouth see **Dorset, West**
Whittlesey see **Fenland**
Widnes see **Cheshire, North**

Wigan

Franchise holder: Telewest (see GO)
Homes in area: 110,000
Awarded: 17 May 90
Service start: Jun 92
Homes passed: 108,525 (1 Jul 98)
Subs: 23,459 = 21.6% (1 Jul 98)

Wilmslow see **Macclesfield**

Winchester

Franchise holder: Cable & Wireless
Communications (see GO)
Homes in area: 19,000
Awarded: 6 Apr 90
Service start: 1995
Homes passed and subs included
with Southampton/Eastleigh

Windsor/Slough/Maidenhead/ Ashford/Staines/Stanwell/ Heathrow/Iver

Franchise holder: Telewest (see GO)
Homes in area: 110,000 (build
complete)
Awarded: 1 Nov 83; Iver added
subsequently to create contiguity
with Middlesex
Service start: 1 Dec 85
Homes passed: 110,934 (1 Jul 98)
Subs: 16,449 = 14.8% (1 Jul 98)

Wirral, The

Franchise holder: Cable & Wireless
Communications (see GO)
Homes in area: 120,000
Awarded: 11 Jul 90
Homes passed: 73,125 (1 Jul 98)
Subs: 13,833 = 18.9% (1 Jul 98)
Separate upgrade system operated by
CDA Communications

Wisbech see **Fenland**
Wishaw see **Motherwell**
Woking see **Guildford**
Wokingham see **Thames Valley**
Wolverhampton see **Black Country**

Worcester/Redditch/Droitwich

Franchise holder: Telewest (see GO)
Homes in area: 70,000

Awarded: 14 Jun 90
Service start: 1997
Homes passed: 26,953 (1 Jul 98)
Subs: 6,937 = 25.7% (1 Jul 98)

Worthing see **Brighton**

Wythall

Local delivery franchise holder:
Birmingham Cable (see GO)
Homes in area: 4,000
Awarded: Sep 95
Service start: Apr 97

Yeovil

Local delivery licence holder:
Convergence Group
Ownership: Convergence Group (see
GO), Orbis Trust (Guernsey)
Homes in area: 62,300
Awarded: Jul 96

York

Franchise holder: Cable & Wireless
Communications (see GO)
Homes in area: 78,000
Homes passed: 3,925
(upgrade system)
Awarded: 30 Mar 90, acquired Apr 94
Service start: Sep 95
Homes passed: 29,710 (1 Jul 98)
Subs: 4,286 = 14.4% (1 Jul 98)

Yorkshire, North see **Durham, South**

GROUP OWNERSHIP

Almost all franchises are held as part of groups of holdings. Such groups are called multiple system operators (MSOs). Extensive consolidation has taken place since 1995 and especially during the first half of 1998. As a result, three dominant groups have emerged Cable & Wireless Communications, NTL and Telewest.

Atlantic Telecom Group
303 King Street
Aberdeen AB2 3AP
Tel: 01224 646644
Fax: 01224 644601
Website: http://www.atlantic
telecom.co.uk
Areas: Aberdeen

British Telecommunications (BT)
87-89 Baker Street
London W1M 2LP
Tel: 0171 487 1254
Fax: 0171 487 1259
Areas: as BT New Towns Cable TV
Services: Milton Keynes
as Westminster Cable Company:
Westminster LB.
Also upgrade systems at Barbican (London), Brackla, Irvine, Martlesham, Walderslade, Washington
Note: BT is to sell its Milton Keynes and Westminster cable systems. It has also been authorised to deliver cable television services to the 17 per cent of UK homes outside franchised areas. From 1 January 2001 BT will be allowed to compete in delivery of television-related services with existing cable networks

Cable & Telecoms (UK)
PO Box 319
Whipsnade
Dunstable
Bedfordshire LU6 2LT
Tel: 01582 873006
Fax: 01582 873003
Ownership: US Cable Corporation, McNicholas Construction, Morgan Cable
as Ayrshire Cable & Telecoms: Ayr
as Cumbria Cable & Telecoms: Carlisle; Cumbria, Central
as Northumberland Cable & Telecoms: Northumberland
as South Cumbria Cable & Telecoms: Cumbria, South

Cable & Wireless Communications
26 Red Lion Square
London WC1R 4HQ

Tel: 0171 528 2000
Website: http://www.cwcom.co.uk
Formed 1997 by merger of cable operators Bell Cablemedia (inc Videotron), Nynex Cablecomms and telecom operator Mercury
Ownership: Cable & Wireless 52.6%, Nynex 18.5%, Bell Canada International (BCI) 14.2%, others 14.7%
Areas: Aylesbury/Amersham/ Chesham, Bolton, Bournemouth/ Poole/Christchurch, Brighton/Hove/ Worthing, Bromley, Bury/Rochdale, Cheshire North, Chichester/Bognor, Dartford/Swanley, Derby/Spondon, Durham South /North Yorkshire, Ealing, Eastbourne/Hastings, Epping Forest/Chigwell/Loughton/Ongar, Fenland, Great Yarmouth/Lowestoft/ Caister, Greater London East, Greenwich/Lewisham, Harrogate/ Knaresborough, Harrow, Havering, Hertfordshire South, Kensington/ Chelsea, Kent South East, Lambeth/ Southwark, Lancashire East, Leeds, London North West, Macclesfield/ Wilmslow, Manchester/Salford, Newham/Tower Hamlets, Norwich, Oldham/Tameside, Peterborough, Portsmouth/Fareham/Gosport/ Havant, Southampton/Eastleigh, Stockport, Stoke-on-Trent/Newcastle, Surrey North, Surrey North East, Thamesmead, Totton/Hythe, Waltham Forest, Wandsworth, Wearside, Whittlesey/March/Wisbech, Winchester, The Wirral, York

Cable London
2 Stephen Street
London W1P 2LN
Tel: 0171 911 0555
Fax: 0171 209 8459
Ownership: Comcast UK 50% [qv], Telewest 50% [qv]
Areas: Camden, Enfield, Hackney & Islington, Haringey
Note: Comcast UK and Telewest have agreed that Comcast UK (or NTL following the amalgamation of NTL and Comcast) will notify Telewest of the price at which it is willing to sell its interest in Cable London to Telewest. Telewest must then either buy Comcast's share or sell its own share to Comcast/NTL at that price

Comcast Europe
Network House
Bradfield Close
Woking
Surrey GU22 7RE
Tel: 01483 880800
Fax: 01483 880828
Ownership: Comcast Corporation [US], Warburg Pincus and Bankers

Trust, Avalon Telecommunications, others
Note: Comcast is to amalgamate with NTL (qv) by 4 November 1998
50% stake in Cable London (qv), as Anglia Cable: Harlow/Bishops Stortford/Stansted Airport
as Cambridge Cable: Cambridge/ Ely/Newmarket,
as Comcast Teesside: Darlington, Teesside
as East Coast Cable: Colchester/ Ipswich/etc,
as Southern East Anglia Cable: East Anglia South,

ComTel
Acquired by NTL [qv] June 98

Convergence Group
Premiere House
3 Betts Way
Crawley, West Sussex RH10 2GB
Tel: 01293 540444
Fax: 01293 540900
Areas: East Grinstead, Haywards Heath
50% in Taunton/Bridgewater, Yeovil (with Orbis Trust (Guernsey))

Cox Communications
US cable operator
10% stake in Telewest (23% of preference shares) [qv]

Eurobell (Holdings)
Multi-Media House
Lloyds Court, Manor Royal
Crawley, West Sussex RH10 2PT
Tel: 01293 400444
Fax: 01293 400440
Ownership: Detecon (Deutsche Telepost Consulting)
Areas: Crawley/Horley/Gatwick, Devon South, Kent West

General Cable
Acquired by Telewest June 1998

NTL
Bristol House
1 Lakeside Road
Farnborough
Hampshire GU14 6XP
Tel 01252 402662
Fax 01252 402665
Website: HYPERLINK http:// www.cabletel.co.uk http:// www.cabletel.co.uk
HQ: 110 East 59th Street, New York, NY 10022 USA
Tel +1/212 906 8440
Fax +1/212 752 1157
Formerly: International CableTel
Ownership: Rockefeller family, Capital Cities Broadcasting Company (subsidiary of Walt Disney Company)
Areas: former CableTel franchises

as CableTel Bedfordshire: Bedford
as CableTel Glasgow: Bearsden/
Milngavie, Glasgow Greater,
Glasgow North West/Clydebank,
Invercylde, Paisley/Renfrew
as CableTel Herts & Bedfordshire:
Luton/South Bedfordshire
as CableTel Hertfordshire: Hertfordshire Central, Hertfordshire East,
as CableTel Kirklees: Huddersfield/
Dewsbury
as CableTel Northern Ireland:
Northern Ireland
as CableTel South Wales: Cardiff/
Penarth, Glamorgan West, Glamor-
gan/Gwent, Newport/Cwmbran/
Pontypool
as CableTel Surrey: Guildford/West
Surrey

former ComTel franchises
Andover/Salisbury/Romsey,
Daventry, Corby/Kettering/
Wellingborough, Hertfordshire West,
Litchfield/Burntwood/Rugeley,
Northampton, Nuneaton/Bedworth/
Rugby, Oxford/Abingdon, Stafford/
Stone, Swindon, Tamworth/North
Warwickshire/Meriden, Thames
Valley, Warwick/Stratford-upon-
Avon/Kenilworth/Leamington Spa

former Diamond Cable franchises:
Bassetlaw, Burton-on-Trent,
Coventry, East Derbyshire,
Grantham, Grimsby/Immingham/
Cleethorpes, Hinckley/Bosworth,
Huddersfield/Dewsbury, Leicester,
Lincoln, Lincolnshire/South Humber-
side, Loughborough/Shepshed,
Mansfield/Sutton/Kirkby-in-Ashfield,
Melton Mowbray, Newark-on-Trent,
Northern Ireland, Nottingham,
Ravenshead, Vale of Belvoir

SBC International
Ownership: Southwestern Bell
Telecom [US telecom operator]
10% stake in Telewest (23% of
preference shares) [qv]

Telecential Communications
Acquired by NTL June 1998

Tele-Communications Inc (TCI)
(largest US cable operator)
50% share in TW Holdings, which
owns 53% of Telewest [qv]

Telewest Communications
Unit 1, Genesis Business Park
Albert Drive
Woking, Surrey GU21 5RW
Tel: 01483 750900
Fax 01483 750901
Website: http://www.telewest.co.uk
Ownership: TW Holdings (= Tele-
Communications International
(TINTA) 50% and US West

International 50%) 53%, Cox
Communications* 10%, SBC
International (= Southwestern Bell
Telecom)* 10% (*each hold 23% of
preference shares issued), Media One
48.9% stake in Cable London [qv]

Areas:
as Birmingham Cable: Birmingham/
Solihull, Wythall
as Cable Corporation: Hillingdon/
Hounslow, Windsor
as Telewest London & the South East):
Croydon, Kingston/Richmond,
Merton/Sutton, Thames Estuary
North, Thames Estuary South
as Telewest Midlands & the South
West: Avon, Black Country,
Cheltenham/Gloucester, Taunton/
Bridgewater, Telford, Worcester
as Telewest North West): Blackpool/
Fylde, Lancashire Central, Liverpool
North/Bootle/Crosby, Liverpool
South, St Helens/Knowsley,
Southport, Wigan
as Telewest Scotland & North East:
Cumbernauld, Dumbarton, Dundee,
Edinburgh, Falkirk/West Lothian,
Fife, Glenrothes/Kirkaldy/Leven,
Motherwell/East Kilbride/Hamilton/
Wishaw/Lanark, Perth/Scone,
Tyneside
as Yorkshire Cable Communica-
tions: Barnsley, Bradford,
Calderdale, Doncaster/Rotherham,
Sheffield, Wakefield/Pontefract/
Castleford

US West International
50% share in TW Holdings, which
owns 53% of Telewest [qv]

ENGLISH-LANGUAGE SATELLITE AND CABLE TELEVISION CHANNELS

All channels transmitting via cable or satellite within or to the UK, wholly or partly in the English language or intended for viewing by other linguistic groups within the UK. Services are licensed and monitored by the Independent Television Commission (ITC).

Channels not intended for reception in the UK are excluded, as are those that are licensed but not actively broadcasting (many licensed channels never materialise).

The television standard and encrypting system used are indicated after the name of the satellite.

Services for which a separate charge is made are marked 'premium' after the programming type.

The advent of digital television from the fourth quarter 1998 will create many new channels, while the ITC's ruling on bundling (removing the requirement that operators and subscribers must take a number of channels, including unwanted ones) may also affect the survival of some in this listing.

MULTIPLE SERVICE PROVIDERS (MSP)

BBC Worldwide
Woodlands
80 Wood Lane
London W12 0TT
Tel: 0181 576 2000
Services: Animal Planet 50%, BBC News 24, UK Arena 50%, UK Gold 50%, UK Horizons 50%, UK Style 50%

British Sky Broadcasting (BSkyB)
6 Centaurs Business Park
Grant Way, Syon Lane
Isleworth
Middlesex TW7 5QD
Tel: 0171 705 3000
Fax: 0171 705 3030
Website: http://www.sky.co.uk
Ownership: News International Television 39.88 %, BSB Holdings (= Pathé 30.27%, Granada 36.22%, Pearson 4.29%) 12.82 %, Pathé 12.71 %, Granada Group 6.48 %
Services: The Computer Channel, The History Channel 50%, National Geographic Channel 50%, Nickelodeon 50%, QVC 20%, Sky Box Office, Sky Cinema, Sky MovieMax, Sky News, Sky One, Sky Premier, Sky Scottish, Sky Soap, Sky Sports1 , Sky Sports 2, Sky Sports 3, Sky Travel 40% stake in Granada Sky Broadcasting

Carlton Communications
45 Fouberts Street
London W1V 2DN
Tel: 0171 432 9000
Fax: 0171 432 3151
Services: Carlton Food Network, Carlton Select

Discovery Communications
160 Great Portland Street
London W1N 5TB
Tel: 0171 462 3600
Fax: 0171 462 3700
Services: Animal Planet, Discovery Channel Europe, TLC Europe

Flextech Television
160 Great Portland Street
London W1N 5TB
Tel: 0171 299 5000
Fax: 0171 299 5400
Ownership: Tele-Communications International (TINTA)
Services: Bravo, Challenge TV, Living, Trouble, UK Arena 50%, UK Gold 50%, UK Horizons 50%, UK Style 50%
Service management: Discovery, Discovery Home & Leisure, Playboy TV, Screenshop, TV Travel Shop

Granada Sky Broadcasting
Franciscan Court
16 Hatfields
London SE1 8DJ
Tel: 0171 578 4040
Fax: 0171 578 4176
Ownership: Granada Group 60%, British Sky Broadcasting 40%
Services: Granada Breeze, Granada Plus, Granada Men & Motors

Home Video Channel
Aquis House
Station Road
Hayes
Middlesex UB3 4DX
Tel: 0181 581 7000
Fax: 0181 581 7007
Ownership: Spice Entertainment Companies
Services: The Adult Channel, HVC

Landmark Communications
64-66 Newman Street
London W1P 3PG
Tel: 0171 665 0600
Fax: 0171 665 0601
Ownership: Landmark Communications Inc
Services: Travel Channel

NBC
4th Floor
3 Shortlands
Hammersmith
London W6 8BX
Tel: 0181 600 6600
Fax: 0181 600 6278
Ownership: General Electric
Services: CNBC Europe

Portland Enterprises
Portland House
Portland Place
London E14 9TT
Tel: 0171 308 5095
Services: Gay TV, Television X The Fantasy Channel

Turner Broadcasting System (TBS)
CNN House
19-22 Rathbone Place
London W1P 1DF
Tel: 0171 637 6700
Fax: 0171 637 6768
Ownership: Time Warner
Services: Cartoon Network, CNN International, Turner Network Television

CHANNELS

The Adult Channel
Ownership: Home Video Channel [see MSP above]
Service start: Feb 1992
Satellite: Astra 1B (PAL/Videocrypt)
Programming: 'adult' entertainment (premium)

Animal Planet
Ownership: BBC Worldwide, Discovery Communications [see MSP above]
Service start: Sep 98
Satellite: Astra 1E, Hot Bird 1 (PAL/encrypted)
Programming: natural history documentaries
Website: http://www.animal.discovery.com

Asianet
Unit 1, Endsleigh Industrial Estate
Endsleigh Road
Uxbridge
Middlesex UB2 5QR
Tel: 0181 930 0930
Fax: 0181 930 0546
Cable only from videotape
Programming: movies and entertainment in Hindi, Punjabi and other languages

BBC News 24
Ownership: BBC Worldwide [See MSP above]
Programming: news

Bloomberg Television
City Gate House
39-45 Finsbury Square
London EC2A 1PQ
Tel: 0171 330 7500
Fax: 0171 256 5326
Website: http://www.bloomberg.com
Service start: 1 Nov 1995
Satellite: Astra 1E, Eutelsat II-F1
Programming: business and finance
Website: http://www.bloomberg.com/uk/

The Box
Imperial House
11-13 Young Street
London W8 5EH
Tel: 0171 376 2000
Fax: 0171 376 1313
Ownership: Emap
Service start: 2 Mar 1992
Satellite: Astra 1A (PAL/Videocrypt; cable only)
Programming: interactive pop music
Website: http://www.thebox.com

Bravo
Ownership: Flextech Television [see MSP above]
Service start: Sept 1985
Satellite: Astra 1C (PAL/Videocrypt)
Programming: old movies and television programmes
Website: http://www.bravo.co.uk

Carlton Food Network
Ownership: Carlton Communications [see MSP above]
Service start: 2 Sep 1996
Satellite: Intelsat 601 (MPEG2 encrypted)
Programming: Food
Websites: HYPERLINK http://www.carltonfoodnetwork,co.uk
http://www.cfn.co.uk

Carlton Select
Ownership: Carlton Communications [see MSP above]
Service start: 1 Jun 1995
Satellite: Intelsat 601 (MPEG2 encrypted); cable exclusive
Programming: entertainment; classic TV shows
Website: http://www.carltonselect.co.uk

Cartoon Network
Ownership: Turner Broadcasting [see MSP above]
Service start: Sept 93
Satellite: Astra 1C, Astra 1F (PAL/clear)
Programming: children's animation
Website: http://www.cartoon-network.co.uk/toonnet.app/

Challenge TV
Ownership: Flextech [see MSP above]
Service start: 3 Feb 1997
Satellite: Astra 1C (PAL/Videocrypt)
Programming: general entertainment, game shows
Website: http://www.challengetv.co.uk

The Channel Guide
1a French's Yard
Amwell End
Ware, Herts SG12 9HP
Tel: 01920 469238
Fax: 01920 468372
Ownership: Picture Applications
Service start: May 1990
Cable only (text)
Programming: programme listings

Channel One
60 Charlotte Street
London W1P 2AX
Tel: 0171 209 1234
Fax: 0171 209 1235
Ownership: Daily Mail & General Trust
Service start: 30 Nov 94
Cable only on London Interconnect, Avon, Liverpool/Merseyside

Programming: news and general features in rolling format

Chinese News and Entertainment (CNE)
Marvic House
Bishops Road, Fulham
London SW6 7AD
Tel: 0171 610 3880
Fax: 0171 610 3118
Ownership: The CNT Group
Service start: Nov 92
Satellite: Astra 1C (PAL/Clear)
Programming: news, current affairs, films, dramas, lifestyle

Christian Channel Europe
Television Centre
3rd Floor
Crown House
Borough Road
Sunderland SR1 1HW
Tel: 0191 514 4777
Fax: 0191 514 4747
Service start: 1 Oct 1995
Satellite: Astra 1B
Programming: Christian
Website: http://www.indigo.ie/spugradio/cce.html

CNBC Europe
Ownership: NBC [see MSP above]
Service start: 11 Mar 1996
Satellite: Astra 1E
Programming: business news

CNN International
Ownership: Turner Broadcasting [see MSP above]
Service start: Oct 1985
Satellite: Astra 1B, Intelsat 605 (PAL/clear)
Programming: news
Website: http://www.europe.cnn.com

The Computer Channel
Ownership: British Sky Broadcasting (see MSP)
Satellite: Astra 1D
Programming: computer topics and programs

The Discovery Channel Europe
Ownership: Discovery Communications [see MSP above]
Service start: Apr 89
Satellite: Astra 1C, Hot Bird 1 (PAL/encrypted)
Programming: documentaries
Website: http://www.discovery.com/digitnets/international/europe/europe.html

Discovery Home & Leisure
Ownership: Discovery Communications [see MSP above]
Service start: Mar 1992
Satellite: Astra 1C, Hot Bird 1 (PAL/encrypted)

Programming: lifestyle
Website: http://www.discovery.com/
digitnets/learning/learning.html

The Disney Channel UK
Beaumont House
Kensington Village
Avonmore Road
London W14 8TS
Tel: 0181 222 1000
Ownership: Walt Disney Company
Satellite: Astra 1B (PAL/Videocrypt)
Programming: children's (supplied
as bonus with The Movie Channel
and Sky Movies)

EBN: European Business News
10 Fleet Place
London EC4M 7RB
Tel: 0171 653 9300
Fax: 0171 653 9333
Website: http://www.ebn.co.uk
Ownership: Dow Jones & Co 70%,
Flextech 30%
Service start: 27 Feb 95
Satellite: Eutelsat II F6 (PAL/clear)
Programming: financial and business
news

EDTV (Emirates Dubai TV)
c/o Teleview Productions
7a Grafton Street
London W1X 3LA
Tel: 0171 493 2496
Fax: 0171 629 6207
Ownership: Dubai government
Service start: Dec 93
Satellite: Arabsat 2A, Intelsat K
Programming: news (from ITN),
entertainment, film, sports,
children's in Arabic and English
Website: http://www.edtv.com

Euronews
60 Chemin des Mouilles
69130 Ecully
France
Tel: (33) 4 72 18 80 00
Fax: (33) 4 73 18 93 71
Ownership: 18 European Broadcast-
ing Union members 51%, Générale
Occidentale 49%
Service start: 1 Jan 1993
Satellite: Hot Bird 3, Eutelsat II-F1
(PAL/clear)
Programming: news in English,
French, Spanish, German and Italian

Eurosport
55 Drury Lane
London WC2B 5SQ
Tel: 0171 468 7777
Fax: 0171 468 0024
Ownership: ESO Ltd = TF1 34%,
Canal Plus 33%, ESPN 33%
Service start: Feb 89
Satellite: Astra 1A, Hot Bird 1 (PAL/
clear)
Programming: sport

Website: http://www.eurosport-
tv.com

Fox Kids Network
Ownership: Fox Television (man-
aged by BSkyB, see MSP above)
Satellite: Astra 1A (PAL/Videocrypt)
Programming: children's
Website: http://www.lineone.net/
foxkids/

Gay TV
Ownership: Portland Enterprises [see
MSP above]
Satellite: Astra 1C (PAL/encrypted)
Programming: erotic

Granada Breeze
Ownership: Granada Sky Broadcast-
ing [see MSP above]
Satellite: Astra 1E (PAL/encrypted)
Programming: lifestyle
Website: http://www.gsb.co.uk/
breeze/home.html

Granada Men & Motors
Ownership: Granada Sky Broadcast-
ing [see MSP above]
Satellite: Astra 1A (PAL/Videocrypt)
Programming: male-oriented, motoring
Website: http://www.gsb.co.uk/men/
home.html

Granada Plus
Ownership: Granada Sky Broadcast-
ing [see MSP above]
Satellite: Astra 1A (PAL/Videocrypt)
Programming: classic TV programmes
Website: http://www.gsb.co.uk/plus/
home.html

The History Channel
Ownership: BSkyB 50%, A&E
Television Networks 50%
Service start: 1 Nov 1995
Satellite: Astra 1B (PAL/Videocrypt)
Programming: history

HVC: Home Video Channel
Ownership: Home Video Channel
[see MSP above]
Service start: Sept 1985
Satellite: Astra 1D (cable exclusive)
Programming: movies (premium)

Japan Satellite TV (JSTV)
Quick House
65 Clifton Street
London EC2A 4JE
Tel: 0171 426 7330
Fax: 0171 426 7333
Ownership: NHK, private Japanese
investors
Satellite: Astra 1E (PAL/Videocrypt)
Programming: Japanese news,
drama, documentary, entertainment,
sport
Website: http://www.jstv.co.uk

The Landscape Channel
Landscape Studios
Hye House
Crowhurst, East Sussex TN33 9BX
Tel: 01424 83688
Fax: 01424 83680
Service start: Nov 1988 (on
videotape); Apr 1993 (on satellite)
Satellite: Orion, Hispasat (PAL/
clear)
Programming: music and visual
wallpaper
Website: http://www.landscapetv.com

Live TV
24th floor
1 Canada Square
Canary Wharf
London E14 5AP
Tel: 0171 293 3900
Fax: 0171 293 3820
Ownership: Mirror Group Newspapers
Service start: 12 June 95
Programming: general entertainment
Website: http://www.livetv.co.uk

Living
160 Great Portland Street
London W1N 5TB
Tel: 0171 299 5000
Fax: 0171 306 6101
Ownership: Cox Communications,
TeleCommunications Inc, Thames
Television
Service start: Sept 93
Satellite: Astra 1C (PAL/Videocrypt)
Programming: daytime lifestyle,
evening general entertainment
Website: http://www.livingtv.co.uk

MBC: Middle East Broadcasting Centre
80 Silverthorne Road
Battersea
London SW8 3XA
Tel: 0171 501 1111
Fax: 0171 501 1110
Service start: Sept 91
Programming: general and news in
Arabic

MTV UK
180 Oxford Street
London W1N 0DS
Tel: 0171 478 6000
Ownership: Viacom
Service start: Aug 87
Satellite: Astra 1A (PAL/Videocrypt)
Programming: pop music
Website: http://www.mtv.co.uk

Muslim TV Ahmadiyyah
16 Gressenhall Road
London SW18 5QL
Tel: 0181 870 8517
Fax: 0181 870 0684
Ownership: Al-Shirkatul Islamiyyah
Service start: Jan 94

Satellite: Intelsat 601
Programming: spiritual, educational, training
Website: http://www.alislam.org/mta

Namaste Asian Television
7 Trafalgar Business Centre
77-87 River Road
Barking
Essex IG11 0EZ
Tel: 0181 507 8292
Fax: 0181 507 8292
Service start: Sept 92
Satellite: Intelsat 601
Programming: Asian entertainment

National Geographic Channel
Ownership: British Sky Broadcasting (see MSP), National Geographic
Service start: 1997
Satellite: Astra 1A (PAL/Videocrypt)
Programming: natural history documentaries
Website: HYPERLINK http://www.nationalgeographic.com   http://www.nationalgeographic.com

Nickelodeon
15-18 Rathbone Place
London W1P 1DF
Tel: 0171 462 1000
Fax: 0171 462 1030
Ownership: British Sky Broadcasting 50% [see MSP above], MTV Networks 50%
Service start: 1 Sept 93
Satellite: Astra 1C (PAL/Videocrypt)
Programming: children's
Website: http://www.nickelodeon.uk.com

The Paramount Comedy Channel
15-18 Rathbone Place
London W1P 1DF
Tel: 0171 462 1000
Fax: 0171 462 1030
Ownership: British Sky Broadcasting [see MSP above], Viacom
Service start: 1 Nov 1995
Satellite: Astra 1C (PAL/Videocrypt)
Programming: comedy
Website: http://www.paramount.com

The Parliamentary Channel
160 Great Portland Street
London W1N 5TB
Tel: 0171 299 5000
Fax: 0171 299 6000
Ownership: consortium of cable operators
Service start: Jan 92
Satellite: Intelsat 601 (PAL/clear)
Programming: coverage of British parliamentary debates
Website: http://www.parlchan.co.uk

Performance: The Arts Channel
60 Charlotte Street
London W1P 2AX

Tel: 0171 927 8808
Ownership: Harmsworth Media
Service start: Oct 92
Cable only from videotape
Programming: opera, jazz and classical concerts, drama

Playboy TV
Ownership: Flextech 51% [see MSP above], BSkyB, Playboy
Service start: 1 Nov 1995
Satellite: Astra 1B (PAL/Videocrypt)
Programming: erotic (premium)
Website: http://www.flextech.co.uk/playboytv

QVC: The Shopping Channel
Marcopolo House, Chelsea Bridge
Queenstown Road
London SW8 4NQ
Tel: 0171 705 5600
Fax: 0171 705 5602
Ownership: QVC (= Comcast, TCI) 80%, BSkyB 20%
Satellite: Astra 1C (soft scrambled)
Service start: Oct 93
Programming: home shopping
Website: http://www.qvc.com

The Racing Channel
17 Corsham Street
London N1 6DR
Tel: 0171 253 2232
Fax: 0171 696 8681
Service start: Nov 1995
Satellite: Astra 1D
Programming: horse racing

The Sci-Fi Channel Europe
77 Charlotte Street
London W1P 2DD
Tel: 0171 805 6100
Fax: 0171 805 6150
Service start: 1 Nov 1995
Satellites: Astra 1B, Hot Bird 1 (PAL/encrypted)
Programming: science fiction
Website: http://www.scifi.com/sfeurope/index.html

Sky Box Office 1-4
Ownership: British Sky Broadcasting [see MSP above]
Service start: 1 Dec 97
Satellite: Astra 1E (PAL/Videocrypt)
Programming: movies, concerts, events (pay-per-view)

Sky Cinema
Ownership: British Sky Broadcasting [see MSP above]
Service start: Oct 92
Satellite: Astra 1C (PAL/Videocrypt)
Programming: movies (premium)

Sky MovieMax
Ownership: British Sky Broadcasting [see MSP above]
Service start: Feb 89

Satellite: Astra 1A (PAL/Videocrypt)
Programming: movies (premium)

Sky News
Ownership: British Sky Broadcasting [see MSP above]
Service start: Feb 89
Satellite: Astra 1A (PAL/Videocrypt)
Programming: news

Sky One
Ownership: British Sky Broadcasting [see MSP above]
Service start: Feb 89
Satellite: Astra 1A (PAL/Videocrypt)
Programming: entertainment

Sky Premier
Ownership: British Sky Broadcasting [see MSP above]
Service start: Apr 91
Satellite: Astra 1B (PAL/Videocrypt)
Programming: movies (premium)

Sky Scottish
Ownership: British Sky Broadcasting [see MSP above] 50%, Scottish Media 50%
Service start: Feb 89
Satellite: Astra 1A (PAL/Videocrypt)
Programming: entertainment

Sky Soap
Ownership: British Sky Broadcasting [see MSP above]
Satellite: Astra 1B (PAL/Videocrypt)
Programming: entertainment

Sky Sports 1
Ownership: British Sky Broadcasting [see MSP above]
Service start: Apr 91
Satellite: Astra 1B (PAL/Videocrypt)
Programming: sport (premium)

Sky Sports 2
Ownership: British Sky Broadcasting [see MSP above]
Service start: Aug 94
Satellite: Astra 1C (PAL/Videocrypt)
Programming: sport (premium)

Sky Sports 3
Ownership: British Sky Broadcasting [see MSP above]
Service start: Aug 94
Satellite: Astra 1B (PAL/Videocrypt)
Programming: sport (premium)

Sky Travel
Ownership: British Sky Broadcasting [see MSP above]
Satellite: Astra 1C (PAL/Videocrypt)
Programming: travel documentaries

Tara Television
The Forum
74-80 Camden Street
London NW1 0EG
Tel: 0171 383 3330

Fax: 0171 383 3450
Service start: 15 Nov 1996
Satellite: Intelsat 601 (MPEG-2 encrypted)
Programming: Irish entertainment
Website: htp://www.tara-tv.co.uk

TCC
9-13 Grape Street
London WC2H 8DR
Tel: 0171 240 3422
Fax: 0171 497 9113
Ownership: Flextech [see MSP above]
Service start: Sept 1984
Satellite: Astra 1C (PAL/Videocrypt)
Programming: children's
Website: http:// www.tcc.flextech.co.uk/

Television X: The Fantasy Channel
Portland House
Portland Place
Millharbour
London E14 9TT
Tel: 0171 987 5095
Service start: 2 Jun 1995
Satellite: Astra 1C (PAL/Videocrypt)
Programming: erotic (premium)
Website:http://www.televisionx.co.uk

TNT Classic Movies
Ownership: Turner Broadcasting [see MSP above]
Service start: Sept 93
Satellite: Astra 1C, Astra 1F (PAL/clear)
Programming: movies, entertainment

Travel Channel
66 Newman Street
London W1P 3LA
Tel: 0171 636 5401
Fax: 0171 636 6424
Ownership: Landmark Communications [see MSP]
Service start: 1 Feb 94
Satellite: Astra 1E
Programming: travel
Website: http:// www.travelchannel.co.uk

Trouble
Ownership: Flextech Television [see MSP]
Service start: Sept 1985
Satellite: Astra 1C (PAL/Videocrypt)
Programming: teenagers
Website: HYPERLINK http:// www.trouble.co.uk   http:// www.trouble.co.uk

[.tv]
Victoria House
98 Victoria Street
London SW1E 5JL
Tel: 0171 705 5000
Fax: 0171 233 9110
Satellite: Astra 1E

Programming: computer-related topics
Website: http:// www.tvchannel.co.uk/dottv/

TV Travel Shop
Satellite: Astra 1C
Website: http:// www.tvtravelshop.co.uk

TVBS Europe
30-31 Newman Street
London W1P 3PE
Tel: 0171 636 8888
Satellite: Astra 1E (digital)
Programming: Chinese-language
Website: http://www.chinese-channel.co.uk

UK Arena
Ownership: BBC Worldwide, Flextech [see MSP above]
Satellite: Astra 1E
Programming: arts

UK Gold
Ownership: BBC Worldwide, Flextech [see MSP above]
Service start: Nov 92
Satellite: Astra 1B (PAL/Videocrypt)
Programming: entertainment

UK Horizons
Ownership: BBC Worldwide, Flextech [see MSP above]
Satellite: Astra 1E
Programming: documentaries

UK Style
Ownership: BBC Worldwide, Flextech [see MSP above]
Satellite: Astra 1E
Programming: lifestyle

VH-1
180 Oxford Street
London W1N 0DS
Tel: 0171 478 6000
Ownership: MTV Networks = Viacom (100%)
Satellite: Astra 1B (PAL/encrypted)
Programming: pop music
Website: http://www.vh1.com

Zee TV Europe
Unit 7
Belvue Business Centre
Belvue Road
Northolt
Middlesex UB5 5QQ
Tel: 0181 839 4000
Fax: 0181 842 3223
Ownership: Asia TV Ltd
Service start: Mar 1995
Satellite: Astra 1E (PAL/Videocrypt)
Programming: films, discussions, news, game shows in Hindi, Punjabi, Urdu, Bengali, Tamil, English, etc
Website:http://www.zeetelevision.com/

CAREERS

No one organisation gives individually-tailored advice about careers in the media industries, but it is an area much written about, and we have included in this section details of some of the books and other sources that may help. Compiled by David Sharp

There is no doubt that the media industries are perceived as being "glamorous" and young people are attracted to them. Opportunities in television appear to be increasing as the number of broadcasting (and "narrowcasting") companies and organisations continues to grow, whilst the film sector too appears healthier than for some time.

To this could be added the growth in the development of new technologies. The multimedia explosion is one very visible product and is informing film and television programme making and broadcasting at every level. It signals to anyone wanting to work in these industries, that they should expect to be open to working with new technologies and should anticipate the need to constantly update their skills. It is also the case that offering a range of skills rather than just one can be to an applicant's benefit.

Finally, it is important to recognise that this area of training and learning, like many others, has been undergoing shifts of emphasis that provide vocational alternatives to more traditional ways of obtaining qualifications and experience.

For these reasons it is important that anyone considering a career in the industry takes care to investigate what courses are available that will help prepare the way, and if possible, although this is never easy, talks to someone already doing the same kind of job to the one that interests them.

The Jobs

The media industry contains a wide range of jobs, some of which, usually of a supportive or administrative nature, (eg librarian, accountant) have equivalents in many other areas, and some of which are quite specialist and have unique, though possibly misleading titles (eg best boy; gaffer). The reading list below will help guide you.

Bibliography

Below is a selected list, based on holdings at the BFI National Library. These will give you some guidance as to the range of jobs available, the structure of the industry, and they will offer some general guidance on preparing a CV. There are publications (and short courses) devoted to writing and presenting CVs, and you should check with your nearest library about these.

Getting into Film & TV
Burder, John
Saltcoats Publishing, 1994
ISBN 0-95238-900-2

How to get into the Film & TV Business
Gates, Tudor
Alma House, 1995
ISBN 1-899830-00-6

Inside Broadcasting
Newby, Julian
Routledge, 1997
ISBN 0-415-15112-0

 ### Lights Camera Action! Careers in Film, Television, Video
Langham, Josephine
BFI, 2nd edition, 1996
ISBN 0-85170-573-1

Making Acting Work
Salt, Chrys
Bloomsbury, 1997
ISBN 07475 35957

A Woman's Guide To Jobs in Film and Television
Muir, Anne Ross
Pandora Press, 1987.
ISBN 0-86358-061-0

Working in Television, Film & Radio
Foster, Val et al
DFEE, 1997
ISBN 0861106962

Courses

The following titles are recommended for information on courses. You will need to consider what balance between theory practice and academic study you wish to undertake, what qualification and skills you will acquire, who validates the course, and what equipment is available to learn with.

 ### Media Courses UK 1999
Orton, Lavinia
BFI
A sample selection from this book will be put up on the BFI website during the currency of this edition of the *BFI Handbook*

 ### Media and multimedia short courses
Orton, Lavinia
BFI/Skillset, (Three per year)

Most of the Regional Arts Boards (see separate section of the handbook) have regularly updated versions of the above two publications on disk.

Floodlight
(for the Greater London region) and other local guides to courses may be worth checking at your local library.

Courses Abroad

Complete Guide to American Film Schools and Cinema and Television Courses
Pintoff, Ernest
Penguin, 1994
ISBN 0 1401-7226 2

Complete Guide to Animation and Computer Graphics Schools
Pintoff, Ernest
Watson-Guptill, 1995
ISBN 0-8230-2177-7
Restricted to American courses only

Variety International Film Guide
Cowie, Peter ed.
This annual publication includes an international Film Schools section.

There is also a database, **ATENA**, which stores information on about 450 institutions providing training in these subjects. The database is updated annually, and letters, faxes and phone calls can be directed to:

Christopher Poirel
Media Section
Council of Europe
F-67075 Strasbourg Cedex
France
Tel: 00 33 3 88 41 20 00
Fax: 00 33 3 88 41 27 81

Organisations

Training
Skillset
2nd Floor
91-101 Oxford Street
London W1R 1RA
Tel: 0171 534 5300
Fax: 0171 534 5333
email: info@skillset.org
Website: http//www.skillset.org
Skillset is the industry training organisation for broadcast film and video. It takes an overview and does not carry out training itself. It produces a free (and copyright free) careers pack for people interested in entering the industry, but please send SAE with £1 stamp

Training Courses
These are only a **selection** of the organisations running training courses

Broadcast Training Wales
Ty Crichton
3rd Floor,
Sgwar Mount Stuart
Caerdydd CF1 6EE
Tel: 01222 465533
Fax: 01222 463344
email: BTW@cyfle-cyf.demon.co.uk
The regional/national training consortium for Wales

Cyfle
Gronant, Penrallt Isaf
Caernarfon, Gwynedd
LL55 1NW
Tel: 01286 671000
Fax: 01286 678831
email: cyfle@cyfle-cyt.demon.co.uk
Website: http://
www.cyfle.cyf.demo.co.uk
This organisation supports the training needs of the Welsh film and television industry
Cardiff address: **Cyfle**
Crichton House
11-12 Mont Stuart Square
Cardiff CF1 6EE

ft2 - Film & Television Freelance Training
4th Floor
Warwick House
9 Warwick Street
London W1R 5RA
Tel: 0171 734 5141
Fax: 0171 287 9899
FT2 is the only UK-wide provider of new entrant training for young people wishing to enter the freelance sector of the industry in the junior construction, production and technical grades. Funded by Skillset, Freelance Training Fund, European Social Fund, the AFVPA and Channel 4, FT2 is the largest industry managed training provider in its field and has a 100 per cent record of people graduating from the scheme and entering the industry

4FIT
Managed by FT2 (see above), this is Channel 4's training programme for people from ethnic minority backgrounds wishing to train as new entrants in junior production grades

Gaelic Television Training Trust
Sabhal Mor Ostaig College
An Teanga
Isle of Skye, IV44 8RQ
Tel: 01471 844373
Fax: 01471 844383
Website: http://www.smo.uhi.ac.uk

Intermedia Film & Video
19 Heathcoat Street,
Nottingham NG1 3AF
Tel: 0115 950 5434
Fax: 0115 955 6909
Offers a range of courses

Mediaskill
Broadcasting Centre
Barrack Road
Newcastle upon Tyne
NE99 2NE
Tel: 0191 232 5484
Fax: 0191 232 8871
email: marion@ mediaskill.demon.co.uk
vicky-mediaskill@watermans.net

National Film & Television School
Short Course Training Programme
Beaconsfield Film Studios
Station Road, Beaconsfield
Bucks HP9 1LG
Tel: 01494 677903
Fax: 01494 678708
Short course training for people already working in the industry

Northern Ireland Film Commission
21 Ormeau Avenue

Belfast BT2 8HD
Tel: 01232 232444
Fax: 01232 239918
email: info@nifc.co.uk

Scottish Screen Training
74 Victoria Crescent Road
Glasgow G12 9JN
Tel: 0141 302 1700
Fax: 0141 302 1711

Skillnet South West
The Regional Training
Consortium for the South West
59 Prince Street
Bristol BS1 4QH
Tel: 0117 9254011
Fax: 0117 925 3511
Website: http://www.telescope.co.uk

Additionally all the Regional Arts Boards and Media Development Agencies are involved with or have information on training. These are listed in the separate section of this handbook under **Funding.**

Education
Film Education
Alhambra House
27-31 Charing Cross Road
London WC2H OAU
Tel: 0171 976 2291
Fax: 0171 839 5052
Website: http://www.filmeducation.org
Useful general background on how films are put together, generally as part of their study packs on particular titles

Paying Your Way

It is important to be clear on the cost of any course you embark on and sources of grants or other funding. Generally speaking short courses do not attract grants. Again the reading list should help with this, but remember that your local authority or local careers office may also be able to advise on this, or your nearest Training and Enterprise Council. Your local library may hold reference books that direct applicants to other sources of grants.

CD-ROMS

This section contains a selection of Cinema and Television related CD-Roms that are still available to purchase. Those listed are mainly reference sources, indexes and directories most of which are held in the BFI National Library. All are PC compatible unless stated. Compiled by Tony Worron

Australian Feature Films
Plymbridge Distributors Ltd
Estover Road
Plymouth PL6 7PZ
Tel: 01752 202301
Fax: 01752 202331
Information on over 1,100 feature and made for television films. Also includes stills, film excerpts, essays on different eras of Australian film making and a bibliography
PC and Macintosh compatible.

Avance Database of the British Universities Film & Video Council
BUFVC
55 Greek Street
LondonW1V 5LR
Tel: 0171 734 3687
Fax: 0171 287 3914
email: bufvc@open.ac.uk
Textual database containing over 15,000 records of audio-visual teaching aids in a variety of formats including film and video.
One subscription with BUFVC membership

Backtracks
Plymbridge Distributors Ltd.
Estover Road
Plymouth PL6 7PZ
Tel: 01752 202 301
Fax: 01752 202 331
Interactive teaching aid enabling students to edit and mix video clips with music and sound to create adverts, trailers and title sequences. Includes a reference section and detailed guidance notes for teachers. Produced by the BFI and Illuminations Interactive for Channel 4. PC and Macintosh compatible

Chinemania 95 Guide
Guide to films produced in Hong Kong , 1990-94, on two discs, with stills and some clips and trailers. Can be consulted in either English or Chinese.

Chinemania 96/97
MEI AH Laser Disc Co. Ltd
Multimedia Division
Unit 15-28
17/F Metro Centre
Phase 1
32 Lam Hing Street
Kowloon Bay

Hong Kong
Tel: (852) 2754 2855
Fax: (852) 2799 3643
Guide to films produced in Hong Kong, 1995-96. Expanded information on Hong Kong box office 1980-96, contact information, awards, biographies and filmographies

Cinema Ireland
Film Institute of Ireland
6 Eustace Street
Temple Bar
Dublin 2
Ireland
Text based guide to 1600 Irish-related films drawn from the catalogues of the Irish Film Archive and National Library. Fiction and Non-fiction films. Cast and Credits. 1896-1986. PC (Windows 95) and Macintosh compatible

Cultural Connections - Australian National Film and Sound Archive Catalogue
Reed Business Publishing Pty Ltd
Circulation Dept.
Post Office Box 5487
West Chatswood NSW 2057
Australia
Tel: (61-2) 372 5222
Fax: (61-2) 412 3317
The catalogue contains over 250,000 records for the wide range of material held in the archive

FIAF Index
FIAF (PIP)
Rue Defacqz 1
1000 Bruxelles
Belgium
Tel: (322) 534 6130
Fax: (322) 534 4774
Index to periodicals on film and television based on the International Federation of Film Archives subject index. No filmographic information but references to a wide range of world-wide film and television journals

bfi Film Index International
Chadwyck-Healy Ltd
The Quorum
Barnwell Rd
Cambridge CB5 8SW
Tel: 01223 215 512

Fax: 01223 215 514
e-mail: marketing@chadwyck.co.uk
Based on the BFI's SIFT database, containing over 1 million cast and credit references, 100,000 detailed records of entertainment films, references to over 400,000 periodical articles from the wide range of journals held by the BFI National Library and details of Awards (Oscars, BAFTA) and prize winners at the Cannes, Berlin and Venice festivals

International Film Index
Customer Services Dept
Bowker -Saur
Maypole House
Maypole Road
East Grinstead
West Sussex RH 19 1HU
Tel: 01342 330100
Fax: 01342 330 198
email: custserv@bowker-saur.co.uk
Website: http://www.bowker-saur.com/service/
Database with information on 245,000 film titles, with credit information and filmographies. Available to test on Web site

The Knowledge
Miller Freeman Information Services
Miller Freeman plc
Riverbank House
Angel Lane
Tonbridge
Kent TN9 1SE
Tel: 01732 377 586
Fax: 01732 367 301
email: ccurtis@mfplc.com
Website: http://www.mfplc.com
CD-Rom version of the standard directory of production services for the film, television and video industries

The Rebecca Project
BFI Publications
Available from:
Plymbridge Distributors Ltd.
Estover Road
Plymouth PL6 7PZ
Tel: 01752 202 301
Fax: 01752 202 331
Detailed analysis of the Hitchcock film of the Daphne Du Maurier book

and all of the main figures involved in this production. Contains essays and information on Authorship, Feminist perspectives of the film, Genre, Production and Marketing and Publicity as well as biographies, bibliographies, filmographies, film clips, stills and illustrations. Macintosh compatible only

Scotland on Location

Scottish Screen Locations
Scottish Film Council
74 Victoria Crescent Road
Glasgow G12 9JN
Tel: 0141 302 1700
email: SFC@cityscape.co.uk
A source for film and television clips, video and stills of potential locations in Scotland.
Also information on finance, joint ventures, production companies, personnel, facilities and public services.
PC and Macintosh compatible

Spotlight

Spotlight CD
7 Leicester Place
London WC2 7BP
Tel: 0171 437 7631
Fax: 0171 437 5881
Directory of British actors and actresses, with photographs, skills, credits and agents.
2 Discs, for Actors and Actresses.
PC and Macintosh compatible

Variety's Video Directory Plus

Bowker-Saur Ltd (UK)
Customer Services Dept.
Maypole House
Maypole Road
East Grinstead
West Sussex RH19 1HU
Tel: 01342 330100
Fax: 01342 330198
email: custserv@bowker-saur.co.uk
Website: http://www.bowker-saur.co.uk/service/
Lists over 100,000 video cassettes on the American market

Videolog

Trade Service Into Ltd
Cherryhold Rd
Stamford
Lincolnshire PE9 2HT
Tel: 01780 64331
Fax: 01780 57679
Lists all videos commercially available in the UK Updated monthly

CINEMAS

Listed below are the companies who control the major chains of cinemas and multiplexes in the UK, followed by the cinemas themselves listed by area, with seating capacities. The listing includes disabled access information where this is available (see below for details).
This section was compiled by Allen Eyles

Key to symbols

 BFI supported with financial and/or programming assistance (see p8)

O   Part-time or occasional screenings

●   Cinema open seasonally

Disability codes

West End/Outer London

E   Hearing aid system installed. Always check with venue whether in operation

W   Venue with unstepped access (via main or side door), wheelchair space and adapted lavatory

X   Venue with flat or one step access to auditorium

A   Venue with 2-5 steps to auditorium

G   Provision for Guide Dogs

England/Channel Islands/Scotland/Wales/Northern Ireland

X   Accessible to people with disabilities (advance arrangements sometimes necessary - please phone cinemas to check)

E   Hearing aid system installed. Always check with venue whether in operation

The help of Artsline, London's Information and Advice Service for Disabled People on Arts & Entertainment, in producing this section, including the use of their coding system for venues in the Greater London area, is gratefully acknowledged. Anyfurther information on disability access would be welcome.

CINEMA CIRCUITS

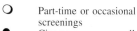

ABC Cinemas
80 Great Portland Street
London W1N 5PA
Tel: 0171 291 9000
Fax: 0171 580 1080
Operates 74 sites with 215 screens, including three multiplexes.
Further Multiplexes planned

Apollo Cinemas UK
7 Palatine Suite
Coppull Enterprises Centre
Mill Lane, Coppull
Lancs PR7 5AN
Tel: 01257 471012
Fax: 01257 794109
Operates cinemas in the North West, Wales, Yorkshire and the Midlands and has opened a nine screen multiplex at Burnley

Artificial Eye Film Company
14 King Street
London WC2E 8HN
Tel: 0171 240 5353
Fax: 0171 240 5242
Film distributors operating the Chelsea Cinema and Renoir in London's West End

Caledonian Cinemas
1st Floor, Highland Rail House
Station Square
Inverness 1V1 1LE
Tel: 01463 718888
Operates 11 screens on five sites, all in Scotland

Cine-UK Ltd
Sutherland House
5/6 Argyll Street
London W1V 1AD
Tel: 0171 494 1394
Fax: 0171 734 1443
Has opened two multiplexes at Stevenage and Wakefield with others following at Feltham, Wolverhampton, Chesterfield, Bristol and elsewhere

City Screen
86 Dean Street
London W1V 5AA
Tel: 0171 734 4342

Fax: 0171 734 4027
Operates the Picture House cinemas in Clapham, Brighton, Oxford, Exeter, Stratford upon Avon, Stratford East London and East Grinstead. Opening new three screen cinemas in York and Cambridge in 1999. The company also operates the Cambridge Arts cinema and the Curzon group of cinemas in London's West End

Film Network
23 West Smithfield
London EC1A 9HY
Tel: 0171 489 0531
Fax: 0171 248 5781
Operates nine screens on two sites at Greenwich and Peckham in South East London

Mainline Pictures
37 Museum Street
London WC1A 1LP
Tel: 0171 242 5523
Fax: 0171 430 0170
Operates Screen cinemas at Baker Street, Haverstock Hill, Islington Green, Reigate, Walton-on-Thames and Winchester with a total of 10 screens

National Amusements (UK)
200 Elm Street, Dedham
Massachusetts
02026-9126, USA
Tel: 001 617 461 1600
Fax: 001 617 326 1306
Owners and operators of 15 Showcase cinemas with 197 screens in Nottingham, Derby, Peterborough, Leeds, Liverpool, Walsall, Birmingham, Coventry, Manchester, Stockton, Bristol, Wokingham (Reading), Newham (London) and two in the Glasgow area near Coatbridge and Linwood

Oasis Cinemas
20 Rushcroft Road
Brixton
London SW2 1LA
Tel: 0171 733 8989
Fax: 0171 733 8790
Owns the Gate Notting Hill and Cameo Edinburgh, and the Ritzy Brixton which is a five-screen multiplex

Odeon Cinemas
Rank Leisure Division
Stafferton Way
Maidenhead
Bucks SL6 1AY
Tel: 01628 504000
Fax: 01628 504383
The Odeon chain totalled 361 screens on 73 sites at summer 1997, with new multiplexes opening soon at Camden, Leicester, Southampton and Wrexham and others under development

Panton Films
Coronet Cinema
Notting Hill Gate
London W11 3LB
Tel: 0171 221 0123
Fax: 0171 221 6312
Operates the Coronet circuit of 11 screens on five sites, comprising former circuit cinemas of Rank and ABC

Picturedrome Theatres
1 Duchess Street
London W1N 3DE
Tel: 01372 460 108
Independent chain of seven cinemas at Bognor, Bristol, Cannock, Chippenham, Newport (Isle of Wight), Sittingbourne and Weymouth

Robins Cinemas
Studio 3B
Highgate Business Centre
33 Greenwood Place
London NW5 1DH
Tel: 0171 482 3842
Fax: 0171 482 4141
Operates 11 buildings with 23 screens

Charles Scott Cinemas
Alexandra, Nexton Abbot
Devon
Tel: 01626 65368
West Country circuit with cinemas at Bridgwater, Exmouth, Lyme Regis, Newton Abbot, Sidmouth and Teignmouth

UCI (UK) Ltd
7th Floor, Lee House
90 Great Bridgewater Street
Manchester M1 5JW
Tel: 0161 455 4000
Fax: 0161 455 4076
Operators of 26 purpose-built multiplexes with 235 screens in the UK plus the Empire and Plaza in London's West End in summer 1997, with more multiplexes scheduled for Huddersfield, Surrey, Cardiff, Surrey Quays (South London) Pipps Hill,

Basildon The Printworks, Basildon Cardiff Bay and Trafford Centre, Manchester. Total: 278 screens

Virgin Cinemas
6th Floor, Adelaide House
626 High Road
Chiswick
London W4 5RY
Tel: 0181 987 5000
Fax: 0181 742 2998
Operates 24 muliplexes (plus one in Dublin) and 4 traditional cinemas. Has further multiplexes opening at Birmingham Great Park, Ipswich, Sheffield, Crawley and Bolton, with rapid expansion over next few years

Warner Bros Theatres (UK)
79 Wells Street
London W1P 3RD
Tel: 0171 465 4090
Fax: 0171 465 4919
Currently operating the nine-screen Warner Village West End in Leicester Square and 18 other multiplexes with 152 screens with further cinemas under construction at Leeds, Bristol, London (Finchley Road), Bolton, and 10 further sites announced

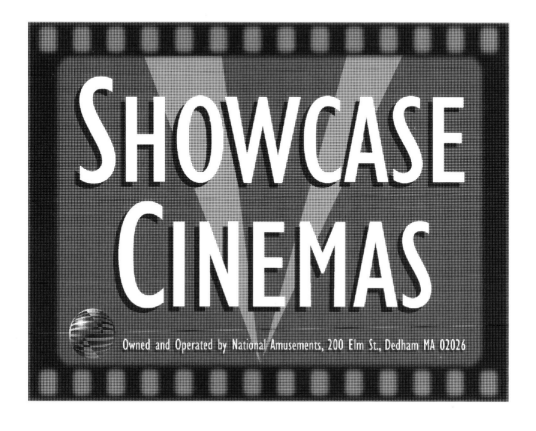

SHOWCASE CINEMAS

Owned and Operated by National Amusements, 200 Elm St, Dedham MA 02026

LONDON WEST END - PREMIERE RUN

Baker Street
ABC
Marylebone Road, NW1   A
Tel: 0171 935 9772
Seats: 1:171, 2:169

Screen on Baker Street
Baker Street, NW1
Tel: 0171 935 2772
Seats: 1:95, 2:100

Bayswater
UCI Whiteleys
Queensway, W2   WG
Tel: 0171 792 3303
Seats: 1:333, 2:284, 3:196, 4:178,
5:154, 6:138, 7:147, 8:125

Bloomsbury
Renoir
Brunswick Square, WC1
Tel: 0171 837 8402
Seats: 1:251, 2:251

Charing Cross Road
Curzon Phoenix
Phoenix Street, WC2
Tel: 0171 369 1721
Seats: 208

Chelsea
Chelsea Cinema
Kings Road, SW3
Tel: 0171 351 3742
Seats: 713

Virgin Cinemas
Kings Road, SW3
Tel: 0171 376 4744
Seats: 1:220, 2:238, 3:122, 4:111

City of London
Barbican
Silk Street, EC2   WE
Tel: 0171 638 8891/638 4141
Seats: 1:288, 2:555

Fulham Road
Virgin Cinemas
Fulham Road, SW10
Tel: 0171 370 2110
Seats: 1:348 X, 2:329 X, 3:173 X,
4:203 X, 5:218, 6:154,

Haverstock Hill
Screen on the Hill
Haverstock Hill, NW3   A

Tel: 0171 435 3366/9787
Seats: 339

Haymarket
Virgin Cinemas
Haymarket, SW1
Tel: 0171 839 1528
Seats: 1:448, 2:200, 3:201

Odeon
Haymarket SW1   A
Tel: 0181 315 4212
Seats: 566

Islington
London Filmmakers' Co-op
bfi   Lux Centre, Hoxton
        Square, N1
Seats: 121
Tel: 0171 684 0200/0201

Screen on the Green
Upper Street, Islington, N1   A
Tel: 0171 226 3520
Seats: 280

Kensington
Odeon
Kensington High Street, W8
Tel: 0181 315 4214
Seats: 1:526, 2:68, 3:85, 4:268 X,
5:173 X, 6:234 X

Knightsbridge
Curzon Minema
45 Knightsbridge, SW1X 7NL
Tel: 0171 235 4226
Seats: 68

Leicester Square
ABC Panton St
Panton Street, SW1
Tel: 0171 930 0631/2
Seats: 1:127 X, 2:144 X, 3:138,
4:136

ABC Swiss Centre
Swiss Centre, W1
Tel: 0171 439 4470/437 2096
Seats: 1:97, 2:101, 3:93, 4:108

Empire
Leicester Square, WC2
Tel: 0171 437 1234
Seats: 1:1,330 X, 2:353, 3:77

Odeon Leicester Square
Leicester Square, WC2
Tel: 0181 315 4215
Seats: 1,930 EX; Mezzanine: 1:60
W, 2:50, 3:60, 4:60, 5:60

Odeon West End
Leicester Square, WC2   E

Tel: 0181 315 4221
Seats: 1:503, 2:838

Prince Charles
Leicester Place, WC2   X
Tel: 0171 437 8181
Seats: 488

Warner Village West End
Leicester Square, WC2
Tel: 0171 437 4347/3484
Seats: 1:177, 2:126, 3:305, 4:298,
5:414, 6:263, 7:412, 8:180, 9:300

The Mall
ICA Cinema
bfi   The Mall, SW1   AG
        Tel: 0171 930 3647
Seats: 185, C'thèque: 45

Marble Arch
Odeon
Edgware Road, W1   E
Tel: 0181 315 4216
Seats: 1:254, 2:126, 3:174, 4:229,
5:239

Mayfair
Curzon Mayfair
Curzon Street, W1
Tel: 0171 369 1720
Seats: 542

Notting Hill Gate
Coronet
Notting Hill Gate, W11 A
Tel: 0171 727 6705
Seats:388

Gate
Notting Hill Gate, W11   X
Tel: 0171 727 4043
Seats: 240

Piccadilly Circus
ABC
Piccadilly, W1
Tel: 0171 437 3561
Seats: 1:124, 2:118

IMAX
Trocadero Centre
Piccadilly Circus, W1
Seats: 300

Metro
Rupert Street, W1   W
Tel: 0171 437 0757
Seats: 1:195, 2:85

Plaza
Lower Regent Street, W1
Tel: 0171 930 0144/ 437 1234
Seats: 1:732, 2:367 X, 3:161, 4:187

Virgin Cinemas
Trocadero Centre
Piccadilly Circus, W1   XE
Tel: 0171 434 0032
Seats: 1:540, 2:241, 3:137, 4:153,
5:121, 6:94, 7:89

Portobello Road
Electric
Portobello Road, W11   X
Tel: 0171 792 2020/0328
Seats: 400
(Temporarily closed)

Shaftesbury Avenue
ABC
Shaftesbury Avenue, WC2
Tel: 0171 836 6279/8606
Seats: 1:616, 2:581

Curzon West End
Shaftesbury Avenue, W1
Tel: 0171 369 1722
Seats: 624

South Kensington
Ciné Lumière, French Institute
Queensberry Place, SW7 ◯
Tel: 0171 838 2144/2146
Seats: 350

Goethe Institute
50 Princes Gate,
Exhibition Rd, SW7 ◯
Tel: 0171 411 3400
Seats: 170

Tottenham Court Road
ABC
Tottenham Court Road, W1
Tel: 0171 636 6148/6749
Seats: 1:328, 2:145, 3:137

Waterloo
National Film Theatre/Museum of the Moving Image
bfi   South Bank,
       Waterloo, SE1   WE
Tel: 0171 928 3232
Seats: 1:450, 2:160,
Museum:135

Queen Elizabeth Hall
South Bank, Waterloo, SE1 ◯   X
Tel: 0171 928 3002
Seats: 906

Royal Festival Hall
South Bank, Waterloo, SE1 ◯   X
Tel: 0171 928 3002
Seats: 2,419

OUTER LONDON

Acton
Warner Village
Royale Leisure Park, Park Royal
Tel: 0181 896 0099
Seats: 1:424, 2:155, 3:201, 4:272,
5:312, 6:272, 7:201, 8:155, 9:424

Barking
Odeon, Longbridge Road
Tel: 0181 507 8444
Seats: 1:796, 2:83, 3:131 X, 4:130,
5:132, 6:162

Barnet
Odeon, Great North Road
Tel: 0181 315 4210
Seats: 1:522 E, 2:178, 3:78, 4:190 W,
5:158

Beckenham
ABC, High Street
Tel: 0181 650 1171/658 7114
Seats: 1:478, 2:228 A, 3:127 A

Studio, Beckenham Road ◯
Tel: 0181 663 0901
Seats: 84

Bexleyheath
Cineworld The Movies, The
Broadway
Tel: 0181 303 0015
Seats: 1:157, 2:128, 3:280, 4:244,
5:88, 6:84, 7:111, 8:168, 9:221

Borehamwood
The Venue, Elstree Way ◯   WG
Tel: 0181 207 2575
Seats: 692

Brentford
Watermans Arts Centre,
High Street   WEG
Tel: 0181 568 1176
Seats: 240

Brixton
Ritzy, Brixton Oval
Coldharbour Lane SW2
Tel: 0171 737 2121
Seats: 1:354, 2:180, 3:126, 4:109, 5:84

Bromley
Odeon, High Street
Tel: 0181 315 4211
Seats: 1:392, 2:129 X, 3:105 X, 4:273

Camden Town
Odeon, Parkway
TeL: 0181 315 4229
Seats: 1:403, 2:92, 3:238, 4:90, 5:99

Catford
ABC, Central Parade SE6 2TF
Tel: 0181 698 3306/697 6579
Seats: 1:519 X, 2:259

Clapham
Picture House, Venn Street, SW4
Tel: 0171 498 3323
Seats: 1:202, 2:132 X, 3:115 X

Croydon
Safari, London Road
Tel: 0181 688 0486/5775
Seats: 1:650, 2:399 X, 3:187 X

David Lean Cinema, Clock Tower
Katherine St   X
Tel: 0181 253 1030
Seats:68

Fairfield Hall/Ashcroft Theatre
Park Lane ◯
Tel: 0181 688 9291
Seats: Fairfield: 1,552 WEG;
Ashcroft: 750

Warner Village , Purley Way
Tel: 0181 680 8090
Seats: 1:253, 2:205, 3:178, 4:396,
5:396, 6:178, 7:205, 8:253

Dagenham
Warner Village, Dagenham Leisure
Park, off Cook Road
Tel: 0181 592 1090/2020
Seats: 1:402, 2:144, 3:187, 4:250,
5:297, 6:250, 7:187, 8:144, 9:402

Dalston
Rio, Kingsland High Street, E8   WEG
Tel: 0171 254 6677/249 2722
Seats: 400

Ealing
Belle-Vue, Northfield Avenue, W13
Tel: 0181 567 1075
Seats: 1:155, 2:149

Virgin Cinemas, Uxbridge Road, W5
Tel: 0181 579 4851
Seats: 1:576, 2:371, 3:193

East Finchley
Phoenix, High Road, N2   XG
Tel: 0181 883 2233
Seats: 300

East Ham
Boleyn, Barking Road
Tel: 0181 471 4884
Seats: 800

Edgware
Belle-Vue, Station Road
Tel: 0181 381 2558
Seats: 1:700, 2:200, 3:158

Elephant & Castle
Coronet Film Centre,
New Kent Road, SE1
Tel: 0171 703 4968/708 0066
Seats: 1:546, 2:271 X, 3:211X

Feltham
Cineworld The Movies, Leisure
West Browells Lane

Tel: 0181 867 0888
Seats: 1:104, 2:116, 3:132, 4:205,
5:253, 6:351, 7:302, 8:350, 9:265,
10:90, 11:112, 12: 137, 13:124, 14:99

Finchley
Warner Village, Great North
Leisure Park, North Circular
Road, Chaplin Square, N12
Tel: 0181 446 9933
Seats: 1:367, 2:164, 3:219, 4:333,
5:333, 6:219, 7:164, 8:367

Finchley Road
Warner Village
(Multiplex under construction - 8
screens, 2,000 seats)

Golders Green
ABC, Finchley Road, NW11
Tel: 0181 455 1724/4134
Seats: 524

Greenwich
Greenwich Cinema
High Road, SE10   WEG
Tel: 01426 919 020
Seats: 1:350, 2:288, 3:144

Hammersmith
Virgin Cinemas, King Street, W6
Tel: 0181 748 2388
Seats: 1:322, 2:322, 3:268 A,
4:268 A

Riverside Studios
Crisp Road, W6   E
Tel: 0181 237 1111/1000
Seats: 200

Hampstead
ABC, Pond Street, NW3
Tel: 0171 794 4000/6603
Seats: 1:476, 2:198 X, 3:193 X

Everyman, Holly Bush Vale, NW3   X
Tel: 0171 435 1525
Seats: 285

Harringay
Curzon Frobisher Road
Tel: 0181 347 6664
Seats: 498

Harrow
Safari, Station Road
Tel: 0181 426 0303
Seats: 1:612, 2:133

Warner Village, St George's
Centre, St. Ann's Road
Tel: 0181 427 9944
Seats: 1:347, 2:288, 3:424, 4:296,
5:121, 6:109, 7:110, 8:87, 9:96

Hayes
Beck Theatre
Grange Road ○   XE
Tel: 0181 561 8371
Seats: 536

Holloway
Odeon, Holloway Road, N7
Tel: 0181 315 4213
Seats: 1:243, 2:192, 3:223, 4:328,
5:301,6:78, 7:112, 8:124

Ilford
Odeon, Gants Hill
Tel: 0181 315 4223
Seats: 1:768, 2:255 X, 3:290 X,
4:190, 5:62

Kingston
ABC Options, Richmond Road
Tel: 0181 546 0404/547 2860
Seats: 1:287 X, 2:273 X, 3:200

Lambeth
Imperial War Museum
Lambeth Road, SE1 ○   X
Tel: 0171 735 8922
Seats: 216

Lee Valley
UCI, Picketts Lock Lane
Meridian Way,
Edmonton   X
Tel: 0181 482 5280/2
Seats: 164 (6 screens), 206
(4 screens), 426 (2 screens)

Muswell Hill
Odeon, Fortis Green Road, N10
Tel: 0181 315 4216
Seats: 1:568, 2:173 X, 3:169 X

Newham
Showcase Cinemas,
Jenkins Lane, off A13   X
Tel: 0181 477 4520
Seats: 4,000 (14 screens)

Peckham
Premier, Rye Lane   X
Tel: 0171 732 1010
Seats: 1:397, 2:255, 3:275, 4:197,
5:218, 6:112

Purley
ABC, High Street
Tel: 0181 660 1212/763 1620
Seats: 1:438, 2:134 X, 3:119 X

Putney
ABC, High Street, SW15   AWG
Tel: 0181 788 3003/785 3493
Seats: 1:433, 2:313, 3:147

Richmond
Filmhouse, Water Lane   WG
Tel: 0181 332 0030
Seats: 150

Odeon, Hill Street
Tel: 0181 315 4218
Seats: 1:418, 2:186 X, 3:186 X

Odeon Studio, Red Lion Street
Tel: 0181 315 4218
Seats: 1:81, 2:78, 3:78, 4:92

Romford
ABC, South Street
Tel: 01708 743848/747671
Seats: 1:652, 2:494 A, 3:246 X

Odeon Liberty 2,
Mercury Gardens
Tel: 01708 729040
Seats: 1:412 W, 2:255, 3:150, 4:181,
5:181, 6:150, 7:331, 8:254

Sidcup
ABC, High Street
Tel: 0181 300 2539/300 3603
Seats: 1:503 A, 2:309

Staples Corner
Virgin Cinemas, Geron Way   WE
Tel: 0181 208 1367
Seats: 1:455, 2:362, 3:214, 4:210,
5:166, 6:166

Stratford
Picture House
Gerry Raffles Square
Salway Road E15
Tel: 0181 522 0043
Seats: 1:212, 2:236, 3:256, 4:157

Streatham
ABC, High Road, SW16
Tel: 0181 769 1928/6262
Seats: 1:630, 2:427 X, 3:227 X

Odeon, High Road, SW16
Tel: 0181 315 4219
Seats: 1:1,091, 2:231 X, 3:163 X,
4:253, 5:198

Surrey Quays
UCI, Mart Leisure Park
Seats: 1:411, 2:401, 3:328, 4:200,
5:198, 6:198, 7:164, 8:164, 9:164

Sutton
Secombe Centre,
Cheam Road ○   XE
Tel: 0181 661 0416
Seats: 330

UCI, St Nicholas Centre
St Nicholas Way   X
Tel: 0181 395 4400/4477/4433
Seats: 1:305, 2:297, 3:234, 4:327,
5:261, 6:327

Swiss Cottage
Odeon, Finchley Road, NW3
Tel: 0181 315 4220
Seats: 1:724, 2:114, 3:233, 4:132,
5:160, 6:162

Turnpike Lane

Coronet, Turnpike Parade, N15
Tel: 0181 888 2519/3734
Seats: 1:624, 2:417 **X**, 3:269 **X**

Walthamstow

ABC, Hoe Street, E17
Tel: 0181 520 7092
Seats: 1:592, 2:183 **A**, 3:174 **A**

Well Hall

Coronet, Well Hall Road, SE9
Tel: 0181 850 3351
Seats: 1:450, 2:131 **XG**

Willesden

Belle Vue, Willesden Green
Library Centre, NW10
Tel: 0181 830 0823
Seats: 204

Wimbledon

Odeon, The Broadway, SW19
Tel: 0181 315 4222
Seats: 1:662, 2:90, 3:190 **X**, 4:175,
5:226 **X**

Woodford

ABC, High Road, E18
Tel: 0181 989 3463/4066
Seats: 1:561, 2:199 **X**, 3:131 **X**

Woolwich

Coronet, John Wilson Street
Tel: 0181 854 2255
Seats: 1:678, 2:370 **X**

ENGLAND

Aldeburgh *Suffolk*

Aldeburgh Cinema, High Street **X**
Tel: 01728 452996
Seats: 286

Aldershot *Hants*

ABC, High Street
Tel: 01252 317223/20355
Seats: 1:313, 2:187, 3:150

West End Centre, Queens Road **X**
Tel: 01252 330040
Seats: 98

Alnwick *Northumberland*

Playhouse, Bondgate Without ○
Tel: 01665 510785
Seats: 272

Alton *Hants*

Palace, Normandy Street
Tel: 01420 82303
Seats: 111

Ambleside *Cumbria*

Zeffirelli's, Compston Road **X**
Tel: 01539 431771
Seats: 1:205, 2:63

Andover *Hants*

Savoy, London Street
Tel: 01264 352624
Seats: 350

Ardwick *Greater Manchester*

Apollo, Ardwick Green ○   **X**
Tel: 0161 273 6921
Seats: 2,641

Ashton-under-Lyne

Greater Manchester
Metro, Old Street
Tel: 0161 330 1993
Seats: 987

Aylesbury *Bucks*

Odeon, Cambridge Street
Tel: 01296 339588
Seats: 1:450, 2:108, 3:113

Banbury *Oxon*

ABC, Horsefair
Tel: 01295 262071
Seats: 1:431, 2:225

Barnsley *South Yorks*

Odeon, Eldon Street
Tel: 0122 620 5494
Seats: 1:416, 2:619   **X**

Barnstaple *Devon*

Astor, Boutport Street
Tel: 01271 42550
Seats: 360

Barrow *Cumbria*

Apollo, Abbey Road
Tel: 01229 832772
Seats: 1:531, 2:260, 3:230, 4:115

Apollo, Hollywood Park
Seats: 1,000 (6 screens)
(Scheduled to open December 1998)

Basildon *Essex*

Robins, Great Oaks
Tel: 01268 527421/527431
Seats: 1:644, 2:435, 3:90

Towngate ○
Tel: 01268 532632
Seats: 552, (Mirren Studio) 158

Basingstoke *Hants*

Anvil, Churchill Way ○   **X**
Tel: 01256 844244
Seats: 70

Warner Village,
Basingstoke, Leisure Park,
Churchill Way West   **XE**
Tel: 01256 818739/818448
Seats: 1:427, 2:238, 3:223, 4:154,
5:157, 6:157, 7:154, 8:223, 9:238,
10:427

Bath *Avon*

ABC Beau Nash,
Westgate Street   **X**
Tel: 01225 461730/462959
Seats: 727

Little Theatre, St Michael's Place
Tel: 01225 466822
Seats: 1:192, 2:74

Robins, St John's Place
Tel: 01225 461506
Seats: 1:151, 2:126 **X**, 3:49

Bedford *Beds*

Civic Theatre, Horne Lane ○
Tel: 01234 44813
Seats: 266

Virgin Cinemas, Aspect Leisure
Park, Newnham Avenue   **XE**
Tel: 01234 212826
Seats: 1:340, 2:300, 3:300, 4:300,
5:200, 6:200

Berwick *Northumberland*

Maltings Art Centre
Eastern Lane ○

Playhouse, Sandgate
Tel: 01289 307769
Seats: 650

Beverley *East Yorks*

Playhouse, Market Place
Tel: 01482 881315
Seats: 310

Bexhill *East Sussex*

Curzon, Western Road

Tel: 01424 210078
Seats: 175

Billingham *Cleveland*
Forum Theatre, Town Centre ○
Tel: 01642 552663
Seats: 494

Birmingham *West Mids*
Electric, Station Street   X
Tel: 0121 643 7277
Seats: 1:200, 2:100

Virgin Cinemas, Arcadian Centre,
Hurst Street   XE
Tel: 0121 622 3323
Seats: 1:419, 2:299, 3:275, 4:240,
5:192, 6:222, 7:210, 8:196, 9:168

MAC
Cannon Hill Park
Tel: 0121 440 3838
Seats: 1:202, 2:144

Odeon, New Street
Tel: 0121 643 6103
Seats: 1:231, 2:390, 3:298, 4:229,
5:194, 6:180, 7:130, 8:80

Piccadilly, Stratford Road
Sparkbrook
Tel: 0121 773 1658

Showcase Cinemas,
Kingsbury Road, Erdington
Tel: 0121 382 9669
Seats: 3,600 (12 screens)

Virgin Cinemas, Birmingham
Great Park, Rubery
Tel: 0870 907 0726
Seats: 1:165, 2:187, 3:165, 4:149,
5:288, 6:194, 7:523, 8:247, 9:400
10:149 11:187 12:165 13:82

Blackburn *Lancs*
Apollo Five, King William Street
Tel: 01254 51779
Seats: 1:295, 2:186, 3:114, 4:105, 5:99

Blackpool *Lancs*
ABC, Church Street
Tel: 01253 27207/24233
Seats: 1:714, 2:324, 3:225

Odeon, Dickson Road
Tel: 01253 26211
Seats: 1:1,377, 2:200, 3:200
Odeon, Rigby Road
Seats: 2,615
(10 screens)

Blyth *Northumberland*
Wallaw, Union Street
Tel: 01670 352504
Seats: 1:850, 2:150, 3:80

Bognor Regis *West Sussex*
Picturedrome, Canada Grove

Tel: 01243 823138
Seats: 1:399, 2:100

Odeon, Butlin's Southcoast World
Tel: 01243 819916
Seats: 1:240, 2:240

Boldon *Tyne and Wear*
Virgin Cinemas, Boldon Leisure
Park, Boldon Colliery
Tel: 0541 550512
Seats: 1:284, 2:197, 3:80, 4:119,
5:263, 6:529, 7:263, 8:136, 9:119,
10:197, 11:284

Bolton *Greater Manchester*
Warner Village
Middlebrook Leisure Park
Tel: 01204 66968
Seats: 1:368, 2:124, 3:124, 4:166,
5:244, 6:269, 7:269, 8:244, 9:166,
10:124, 11:124, 12:368

Virgin Cinemas
Seats: 3,557 (15 screens) (scheduled
to open December 1998)

Boston *Lincs*
Blackfriars Arts Centre,
Spain Lane ○
Tel: 01205 363108
Seats: 237

Regal, West Street
Tel: 01205 350553
Seats: 182

Bournemouth *Dorset*
ABC, Westover Road
Tel: 01202 558433/290345
Seats: 1:652, 2:585, 3:223

Odeon, Westover Road
Tel: 01202 293554
Seats: 1:757, 2:359, 3:266, 4:120,
5:121, 6:146

Bowness-on-Windermere
Cumbria
Royalty, Lake Road   X
Tel: 01539 443364
Seats: 1:400, 2:100, 3:65

Bracknell *Berks*
South Hill Park Arts Centre   X
Tel: 01344 427272/484123
Seats: 1:60, 2:200 ○

UCI, The Point
Skimpedhill Lane   X
Tel: 01344 868181/868100
Seats: 1:177, 2:205, 3:205,
4:177, 5:316, 6:316, 7:177,
8:205, 9:205, 10.177

Bradford *West Yorks*
National Museum of Photography,
Film and Television,
Prince's View X

Tel: 01274 732277/727488
Seats: 340 **(IMAX)**

Odeon, Prince's Way
Tel: 0142 691 5550
Seats: 1:466, 2:1,117, 3:244

bfi Pictureville Cinema,
NMPFTV,
Pictureville, BD1 1NQ   X
Tel: 01274 732277
Seats: 306

Priestley Centre for the Arts
Chapel Street,
Little Germany BD1 5DL   XE
Tel: 01274 820666
Seats: 280

Brentwood *Essex*
ABC, Chapel High
Tel: 01277 212931/227574
Seats: 1:300, 2:196

Bridgnorth *Shropshire*
Majestic, Whitburn Street
Tel: 01746 761815/761866
Seats: 1:500, 2:86, 3:86

Bridgwater *Somerset*
Film Centre, Penel Orlieu
Tel: 01278 422383
Seats: 1:246, 2:245

Bridlington *Humberside*
Forum, The Promenade
Tel: 01262 676767
Seats: 1:202, 2:103, 3:57

Bridport *Dorset*
Palace, South Street
Tel: 01308 22167
Seats: 420

Brighton *East Sussex*
ABC, East Street
Tel: 01273 327010/202095
Seats: 1:345, 2:271, 3:194

Cinematheque, Media Centre,
Middle Street
Tel: 01273 384 300

Duke of York's Premier Picture
House, Preston Circus
Tel: 01273 602503
Seats: 330

Gardner Arts Centre,
University of Sussex, Falmer ○
Tel: 01273 685861
Seats: 354

Virgin Cinemas, Brighton Marine
Tel: 0541 555 145
Seats: 1:351, 2:351, 3:251, 4:251,
5:223, 6:223, 7:202, 8:203

Odeon Kingswest, West Street
Tel: 01273 207977
Seats: 1:388, 2:883, 3:504, 4:273,
5:242, 6:103

Bristol *Avon*
Arnolfini, Narrow Quay   XE
Tel: 0117 929 9191
Seats: 176

Arts Centre Cinema,
King Square X
Tel: 0117 942 0195
Seats: 124

Cineworld The Movies
Hengrove Leisure Park, Hengrove
Way
Tel: 01275 831099
Seats: 1:97, 2:123, 3:133, 4:211,
5:264, 6:343, 7:312, 8:344, 9:262,
10:88, 11:113, 12:152, 13:123, 14:98

Orpheus, Northumbria Drive,
Henleaze
Tel: 0117 962 1644
Seats: 1:186, 2:129, 3:125

ABC, Whiteladies Road
Tel: 0117 973 0679/973 3640
Seats: 1:372, 2:252 X, 3:135 X

Odeon, Union Street
Tel: 0117 929 0884
Seats: 1:399, 2:244, 3:215

Showcase Cinemas, Avon Meads
off Albert Road, St Phillips
Causeway
Tel: 0117 972 4001
Seats: 3,408 (14 screens)

bfi Watershed,
1 Canon's Road,
BS1 5TX   XE
Tel: 0117 927 6444/925 3845
Seats: 1:200, 2:50

Warner Village
Seats:3,000 (12 screens)

Broadstairs *Kent*
Windsor, Harbour Street
Tel: 01843 865726
Seats: 120

Bromborough *Merseyside*
Odeon, Wirral Leisure Retail
Park, Welton Road   X
Tel: 0151 334 0777
Seats: 1:465, 2:356, 3:248, 4:203,
5:338, 6:168, 7:168, 8:86, 9:135,
10:71, 11:122

Bude *Cornwall*
Rebel, off A39, Rainbow
Trefknic Cross

Tel: 01288 361442
Seats: 120

Burgess Hill *West Sussex*
Orion, Cyprus Road
Tel: 01444 232137/243300
Seats: 1:150, 2:121

Burnley *Lancs*
Apollo, Hollywood Park
Centenary Way
Manchester Road
Tel: 01282 456222/456333
Seats: 1:61, 2:238, 3:93, 4:339,
5:93, 6:339, 7:93, 8:238, 9:93

Burnham-on-Crouch *Essex*
Rio, Station Road
Tel: 01621 782027
Seats: 1:220, 2:60

Burnham-on-Sea *Somerset*
Ritz, Victoria Street
Tel: 01278 782871
Seats: 204

Burton Latimer *Northants*
Ohio, High Street
Tel: 01536 420195
Seats: 100

Burton-on-Trent *Staffs*
Robins, Guild Street
Tel: 01283 563200
Seats: 1:502, 2:110, 3:110

Bury *Greater Manchester*
Warner Village, Pilsworth
Industrial Estate,
Pilsworth Road   X
Tel: 0161 766 2440/1121
Seats: 1:559, 2:322, 3:278, 4:434,
5:208, 6:166, 7:166, 8:208, 9:434,
10:278, 11:322, 12:573

Bury St Edmunds *Suffolk*
ABC, Hatter Street
Tel: 01284 754477
Seats: 1:196, 2:117

Camberley *Surrey*
Robins, London Road
Tel: 01276 63909/26768
Seats: 1:420, 2:114, 3:94

Globe, Hawley ○
Tel: 01252 876769
Seats: 200

Cambridge *Cambs*
Arts Cinema
8 Market Passage CB2 3PF ●   XE
Tel: 01223 578944/504444
Seats: 275

ABC, St Andrews Street
Tel: 01223 354572/645378
Seats: 1:704, 2:452

Warner Village,
Grafton Centre, East Road   XE

Tel: 01223 460442/460225
Seats: 1:163, 2:180, 3:194, 4:205,
5:175, 6:177, 7:335, 8:442

Cannock *Staffs*
Picturedrome, Walsall Road
Tel: 01543 502226
Seats: 1:368, 2:185

Canterbury *Kent*
bfi Cinema 3, Cornwallis
South, University of Kent
CT2 7NX
Tel: 01227 769075/764000 x4017
Seats: 300

ABC, St Georges Place
Tel: 01227 462022/453577
Seats: 1:536, 2:404

Carlisle *Cumbria*
Lonsdale, Warwick Road
Tel: 01228 514654
Seats: 1:375, 2:216, 3:54

City Cinemas 4 & 5, Mary Street X
Tel: 01228 514654
Seats: 4:122, 5:112

Chatham *Kent*
ABC, High Street
Tel: 01634 846756/842522
Seats: 1:520, 2:360, 3:170

Central Theatre, High Street ○
Tel: 01634 403868

Odeon, Kings Head Walk   EX
Tel: 01245 495068
Seats: 1:338, 2:110, 3:160, 4:236,
5:174, 6:152, 7:131, 8:141

Cheltenham *Glos*
Odeon, Winchcombe Street
Tel: 01242 514421
Seats: 1:252, 2:184, 3:184, 4:90, 5:129,
6:104, 7:177

Chesham *Bucks*
Elgiva Theatre, Elgiva Lane ○   XE
Tel: 01494 774759
Seats: 328

Chester *Cheshire*
Virgin Cinemas,
Chaser Court
Greyhound Park
Sealand Road XE
Tel: 01244 380459/380301/380155
Seats: 1:366, 2:366, 3:265, 4:232,
5:211, 6:211

Odeon, Northgate Street
Tel: 01244 343216
Seats: 1:408, 2:148, 3:148, 4:122,
5:122

Chesterfield *Derbyshire*
Cineworld The Movies
Derby Road
Seats: 2,000 (10 screens)

Chichester *East Sussex*
Minerva Studio Cinema
Chichester Festival Theatre
Oaklands Park ●   X
Tel: 01243 781312
Seats: 212

New Park Film Centre
New Park Road   X
Tel: 01243 786650
Seats: 120

Chippenham *Wilts*
Astoria, Marshfield Road
Tel: 01249 652498
Seats: 1:215, 2:215

Chipping Norton *Oxon*
The Theatre, Spring Street ◯
Tel: 01608 642349/642350
Seats: 195

Christchurch *Dorset*
Regent Centre, High Street ◯
Tel: 01202 499148
Seats: 485

Cirencester *Glos*
Regal, Lewis Lane
Tel: 01285 658755
Seats: 1:100, 2:100

Clacton *Essex*
Flicks, Pier Avenue
Tel: 01255 429627
Seats: 1:600, 2:187

Clevedon *Avon*
Curzon, Old Church Road
Tel: 0117 987 2158
Seats: 392

Clitheroe *Lancs*
Civic Hall, York Street
Tel: 01200 423278
Seats: 400

Colchester *Essex*
Odeon, Crouch Street
Tel: 01206 760707
Seats: 1:480, 2:237, 3:118, 4:133, 5:126, 6:177

Coleford *Glos*
Studio, High Street
Tel: 01594 833331
Seats: 1:200, 2:80

Consett *Co Durham*
Empire, Front Street   XE
Tel: 01207 506751
Seats: 535

Cosham *Hants*
ABC, High Street
Tel: 01705 376635
Seats: 1:441, 2:118, 3:107

Coventry *West Mids*
Odeon, Jordan Well
Tel: 0120 352 0923
Seats: 1:718, 2:178, 3:213, 4:390, 5:121

Showcase Cinemas, Cross Point, Hinckley Road
Tel: 01203 602555
Seats: 4,413 (14 screens)

Warwick Arts Centre, University of Warwick, CV4 7AL   X
Tel: 01203 524524/523060
Seats: 1:240

Cranleigh *Surrey*
Regal, High Street
Tel: 01483 272373
Seats: 268

Crawley *West Sussex*
Hawth, Hawth Avenue ◯   XE
Tel: 01293 553636
Seats: 800

ABC, High Street
Tel: 01293 527497/541296
Seats: 1:294, 2:212, 3:110

Virgin Cinemas
Seats: 3,158 (15 screens)
(Scheduled to open December 1998)

Crewe *Cheshire*
Apollo, High Street
Tel: 01270 255708
Seats: 1:110, 2:110, 3:95

Lyceum Theatre, Heath Street ◯
Tel: 01270 215523
Seats: 750

Victoria Film Theatre, West Street ●
Tel: 01270 211422
Seats: 180

Cromer *Norfolk*
Regal, Hans Place
Tel: 01263 513311
Seats: 1:129, 2:136, 3:66, 4:55

Crookham *Hants*
Globe, Queen Elizabeth Barracks
Tel: 01252 876769
Seats: 340

Crosby *Merseyside*
Plaza, Crosby Road North, Waterloo
Tel: 0151 474 4076
Seats: 1:600, 2:92, 3:74

Darlington *Co Durham*
Arts Centre, Vane Terrace ◯   XE
Tel: 01325 483168/483271
Seats: 100

ABC, Northgate
Tel: 01325 462745/484994
Seats: 1:578, 2:201, 3:139

Dartford *Kent*
Orchard Theatre
Home Gardens ◯   XE
Tel: 01322 343333
Seats: 930

Dartington *Devon*
bfi Barn Theatre,
Arts Society,
The Gallery, TQ9 6EJ ◯ X
Tel: 01803 865864/863073
Seats: 208

Daventry *Northants*
Regal, Bowen Square
Tel: 01327 702674
Seats: 158

Deal *Kent*
Flicks, Queen Street
Tel: 01304 361165
Seats: 1:162, 2:99

Derby *Derbyshire*
Assembly Rooms
Market Place ◯   XE
Tel: 01332 255800
Seats: 998

Guildhall Theatre, Market Place ◯
Tel: 01332 255447
Seats: 186

bfi Metro, Green Lane,
DE1 1SA   XE
Tel: 01332 340170/347765
Seats: 128

Showcase Cinemas, Foresters Park Osmaston Park Road at Sinfin Lane X
Tel: 01332 270050
Seats: 2,549 (11 screens)

UCI Meteor Centre 10, Mansfield Rd   X
Tel: 01332 295010/296000
Seats: 1:191, 2:188, 3:188, 4:191, 5:276, 6:276, 7:191, 8:188, 9:188, 10:191

Dereham *Norfolk*
Hollywood, Dereham Entertainment Centre, Market Place
Tel: 01362 693261
Seats: 1:160, 2:90, 3:108

Devizes *Wilts*
Palace, Market Place
Tel: 01380 722971
Seats: 253

Doncaster *South Yorks*
Civic Theatre, Waterdale ◯
Tel: 01302 62349
Seats: 547

Odeon, Hallgate   X
Tel: 0130 234 2523
Seats: 1:1,003, 2:155, 3:158

Warner Village, Doncaster Leisure Park, Bawtry Road
Tel: 01302 371313
Seats: 1:224, 2:212, 3:252, 4:386, 5:252, 6:212, 7:224

Dorchester *Dorset*
Plaza, Trinity Street
Tel: 01305 262488
Seats: 1:100, 2:320

Dorking *Surrey*
Premier, Dorking Halls
Tel: 01306 889694
Seats: 851

Dover *Kent*
Silver Screen, White Cliffs Experience, Gaol Lane
Tel: 01304 228000
Seats: 110

Dudley *West Mids*
Limelight Cinema, Black Country Living Museum
Tel: 0121 557 9643
Seats: 100

UCI Merryhill Shopping Centre 10   X
Tel: 01384 78244/78282
Seats: 1:350, 2:350, 3:274, 4:274, 5:224, 6:224, 7:254, 8:254, 9:178, 10:178

Durham *Co Durham*
Robins, North Road
Tel: 0191 384 3434
Seats: 1:312 **X**, 2:98, 3:96, 4:74

Eastbourne *East Sussex*
Curzon, Langney Road
Tel: 01323 731441
Seats: 1:530, 2:236, 3:236

Virgin Cinemas, Crumbles Harbour Village Pevensey Bay Road   XE
Tel: 01323 470070
Seats: 1:322, 2:312, 3:271, 4:254, 5:221, 6:221

East Grinstead *West Sussex*
King Street Picture House Atrium Building, King Street
Tel: 01342 321666
Seats: 1:240, 2:240

Elland *Yorks*
Rex, Coronation Street   X
Tel: 01422 372140
Seats: 294

Ellesmere Port *Cheshire*
Epic Cinema, Epic Leisure Centre ◯   X
Tel: 0151 355 3665
Seats: 163

Ely *Cambs*
The Maltings, Ship Lane ◯
Tel: 01353 666388
Seats: 212

Epsom *Surrey*
Playhouse, Ashley Avenue ◯   XE
Tel: 01372 742555/6
Seats: 300

Esher *Surrey*
ABC, High Street
Tel: 01372 465639/463362
Seats: 1:918 **A**, 2:117

Evesham *Hereford & Worcs*
Regal, Port Street
Tel: 01386 446002
Seats: 540

Exeter *Devon*
Northcott Theatre, Stocker Road ◯
Tel: 01392 54853
Seats: 433

Odeon, Sidwell Street
Tel: 01392 217175
Seats: 1:740, 2:121, 3:106, 4:344

Picture House, 51 Bartholomew Street West, Exeter
Tel: 01392 251341
Seats: 1:157, 2:220

Exmouth *Devon*
Savoy, Rolle Street
Tel: 01395 268220
Seats: 1:230, 2:110, 3:170

Farnham *Surrey*
Redgrave Theatre, Brightwells X
Tel: 01252 727 720
Seats: 362

Faversham *Kent*
New Royal, Market Place
Tel: 01795 591211
Seats: 448

Fawley *Hants*
Waterside, Long Lane
Tel: 01703 891335
Seats: 355

Felixstowe *Suffolk*
Cascade, Crescent Road
Tel: 01394 282787
Seats: 1:150, 2:90

Folkestone *Kent*
Silver Screen, Guildhall Street
Tel: 01303 221230
Seats: 1:435, 2:106

Frome *Somerset*
Westway, Cork Street
Tel: 01373 465685
Seats: 304

Gainsborough *Lincs*
Trinity Arts Centre Trinity Street ◯   X
Tel: 01427 810710
Seats: 210

Gateshead *Tyne & Wear*
UCI Metro 11, Metro Centre
Tel: 0191 493 2022/3
Seats: 1:200, 2:200, 3:228, 4:256, 5: 370, 6:370, 7:256, 8:228, 9:200, 10:200, 11:520

Gatley *Greater Manchester*
Tatton, Gatley Road
Tel: 0161 428 2133
Seats: 1:648, 2:247, 3:111

Gerrards Cross *Bucks*
ABC, Ethorpe Crescent
Tel: 01753 882516/883024
Seats: 1:350, 2:212

Gloucester *Glos*
Guildhall Arts Centre, Eastgate Street   X
Tel: 01452 505086/9
Seats: 1:120, 2: 120 ◯

Virgin Cinemas Peel Centre, St. Ann Way Bristol Road   XE
Tel: 01452 331181
Seats: 1:354, 2:354, 3:238, 4:238, 5:219, 6:219

Godalming *Surrey*
Borough Hall ◯
Tel: 01483 861111
Seats: 250

Gosport *Hants*
Ritz, Walpole Road
Tel: 01705 501231
Seats: 1,136

Grantham *Lincs*
Paragon, St Catherine's Road   X
Tel: 01476 570046
Seats: 1:270, 2:160

Gravesend *Kent*
ABC, King Street
Tel: 01474 356947/352470
Seats: 1:571, 2:296, 3:109

Grays *Essex*
Thameside, Orsett Road ◯
Tel: 01375 382555
Seats: 303

Great Yarmouth *Norfolk*
Hollywood, Marine Parade
Tel: 01493 842043
Seats: 1:500, 2:296, 3:250, 4:250

Grimsby *Humberside*
ABC, Freeman Street

Tel: 01472 342878/349368
Seats: 1:393, 2:236, 3:126

Screen, Crosland Road, Willows
DN37 9EH O   X
Tel: 01472 240410
Seats: 206

Guildford *Surrey*
Odeon, Bedford Road
Tel: 01483 578017
Seats: 1:430, 2:361, 3:343, 4:273,
5:297, 6:148, 7:112, 8:130, 9:130

Halifax *West Yorks*
ABC, Ward's End
Tel: 01422 352000/346429
Seats: 1:670, 2:199, 3:172

Halstead *Essex*
Empire, Butler Road
Tel: 01787 477001
Seats: 320

Halton *Bucks*
Astra, RAF Halton O
Tel: 01296 623535
Seats: 570

Hanley *Staffs*
ABC, Broad Street
Tel: 01782 212320/268970
Seats: 1:572, 2:248, 3:171

Forum Theatre, Stoke-on-Trent
City Museum, Bethesda Street O
Tel: 01782 394766
Seats: 300

Harlow *Essex*
Virgin Cinemas, Queensgate Centre
Edinburgh Way   XE
Tel: 01279 436014
Seats: 1:356, 2:260, 3:240, 4:234,
5:233, 6:230

Odeon, The High
Tel: 01279 635067
Seats: 1:450, 2:239, 3:200

Playhouse, The High O   XE
Tel: 01279 431945
Seats: 330

Harrogate *North Yorks*
Odeon, East Parade
Tel: 0142 352 0412
Seats: 1:532, 2:105 X, 3:78 X, 4:339

Harwich *Essex*
Electric Palace,
King's Quay Street O
Tel: 01255 553333
Seats: 204

Haslemere *Surrey*
Haslemere Hall, Bridge Road O
Tel: 01428 2161
Seats: 350

Hastings *East Sussex*
ABC, Queens Road

Tel: 01424 420517/431180
Seats: 1:387, 2:176, 3:129

Hatfield *Herts*
UCI, The Galleria
Comet Way
Tel: 01707 264662/270222/272734
Seats: 1:172, 2:235, 3:263, 4:167,
5:183, 6:183, 7:260, 8:378, 9:172

Havant *Hampshire*
Arts Centre, East Street O   X
Tel: 01705 472700
Seats: 130

Haverhill *Suffolk*
Cinema, Town Hall Arts
Centre, High Street O
Tel: 01440 714140
Seats: 210

Haywards Heath *West Sussex*
Clair Hall, Perrymount Road O
Tel: 01444 455440/454394
Seats: 350

Heaton Moor *Greater*
Manchester
Savoy, Heaton Manor Road
Tel: 0161 432 2114
Seats: 476

Hebden Bridge *West Yorks*
Picture House, New Road   XE
Tel: 01422 842807
Seats: 498

Hemel Hempstead *Herts*
Odeon, Leisure World,
Jarmans Park   XE
Tel: 01442 232224
Seats: 1:136, 2:187, 3:187, 4:320,
5:260, 6:435, 7:168, 8:168

Henley-on-Thames *Oxon*
Kenton Theatre, New Street O X
Tel: 01491 575698
Seats: 240

Regal, Broma Way,
off Bell Street
Tel: 01491 414150
Seats: 1:152, 2:101, 3:85

Hereford *Hereford & Worcs*
ABC, Commercial Road
Tel: 01432 272554
Seats: 378

Hereford Theatre and Arts Centre
Edgar Street O   X
Tel: 01432 359252
Seats: 364

Herne Bay *Kent*
Kavanagh, William Street   X
Tel: 01227 365676
Seats: 1:137, 2:95

Hexham *Northumberland*
Forum, Market Place

Tel: 01434 602896
Seats: 207

High Wycombe *Bucks*
UCI Wycombe 6,
Crest Road, Cressex   X
Tel: 01494 600601
Seats: 1:388, 2:388, 3:284, 4: 284,
5:202, 6:202

Hoddesdon *Herts*
Broxbourne Civic Hall
High Street O
Tel: 01992 441946/31
Seats: 564

Hollinwood *Greater Manchester*
Roxy, Hollins Road
Tel: 0161 681 1441
Seats: 1:470, 2:130, 3:260, 4:260,
5:320, 6:96

Hordern *Co Durham*
WMR Film Centre,
Sunderland Road
Tel: 01783 864344
Seats: 1:156, 2:96

Horsham *Sussex*
Arts Centre (Ritz Cinema and
Capitol Theatre), North Street
Tel: 01403 268689
Seats: 1:126, 2:450 O

Horwich *Lancs*
Leisure Centre, Victoria Road O
Tel: 01204 692211
Seats: 400

Hucknall *Notts*
Byron, High Street
Tel: 0115 963 6377
Seats: 430

Huddersfield *West Yorks*
Tudor, Queensgate, Zetland Street
Tel: 01484 530874
Two screens

UCI, McAlpine Stadium
Bradley Mills Road
Tel: 01484 469999
Seats: 1:375, 2:296, 3:296, 4:268,
5:268, 6:176, 7:176, 8:148, 9:148

Hull *Humberside*
Odeon, Kingston Street   X
Tel: 0148 258 6420
Seats: 1:172, 2:172, 3:152, 4:174,
5:168, 6:275, 7:134, 8:152, 9:110,
10:91

🅱️ Screen, Central Library
Albion Street HU1 3TF XE
Tel: 01482 226655
Seats: 247

UCI St Andrew's Quay
Clive Sullivan Way   X
Tel: 01482 587525

Seats: 1:166, 2:152, 3:236, 4:292, 5:292, 6:236, 7:152, 8:166

Hunstanton *Norfolk*
Princess Theatre, The Green ◯
Tel: 01485 532252
Seats: 467

Huntingdon *Cambs*
Cromwell Centre, Princes Street
Tel: 01480 433499
Seats: 264

Ilfracombe *Devon*
Pendle Stairway, High Street   X
Tel: 01271 863260
Seats: 460

Ilkeston *Derbyshire*
Scala, Market Place
Tel: 0115 932 4612
Seats: 500

Ipswich *Suffolk*
 Film Theatre, Corn
Exchange, King Street,
IP1 1DH   XE
Tel: 01473 255851/215544
Seats: 1:220, 2:40

Odeon, St Margaret's Street
Tel: 01473 287717
Seats: 1:509, 2:320, 3:292, 4:220, 5:220

Virgin Cinemas, Cardinal Park,
Greyfriars Road
Tel: 0870 907 0748
Seats: 1:168, 2:186, 3:168, 4:270, 5:179, 6:510, 7:238, 8:398, 9:186, 10:168, 11:83

Keighley *West Yorks*
Picture House
Tel: 01535 602561
Seats: 1:364, 2:95

Kendal *Cumbria*
Brewery Arts Centre
Highgate, LA9 4HE ●   XE
Tel: 01539 725133
Seats: 246

Keswick *Cumbria*
Alhambra, St John Street   ●
Tel: 017687 72195
Seats: 313

Kettering *Northants*
Odeon, Pegasus Court,
Wellingborough Road
Tel: 01536 485 522
Seats: 1:174, 2:125, 3:232, 4:349, 5:105, 6:83, 7:105, 8:310

King's Lynn *Norfolk*
Arts Centre, King Street
Tel: 01553 774725/773578
Seats: 359

Majestic, Tower Street
Tel: 01553 772603
Seats: 1:450, 2:130, 3:400

Kirkby-in-Ashfield *Notts*
Regent
Tel: 01623 753866
Seats: 180

Knutsford *Cheshire*
Studio, Toft Road   X
Tel: 01565 633005
Seats: 400

Lake *Isle of Wight*
Screen De Luxe, Sandown Road
Tel: 01983 404050
Seats: 150

Lancaster *Lancs*
ABC, King Street
Tel: 01524 64141/841149
Seats: 1:250, 2:244

 The Dukes,
Moor Lane,
LA1 1QE ◯ XE
Tel: 01524 66645/67461
Seats: 307

Leamington Spa *Warwicks*
Apollo, Portland Place
Tel: 01926 426106/427448
Seats: 1:309 X, 2:199 X, 3:138, 4: 108 X

Robins, Spa Centre, Newbold Terrace
Tel: 01926 887726/888997
Seats: 208

Leeds *West Yorks*
Cottage Road Cinema, Headingley
Tel: 0113 230 2562
Seats: 468

Hyde Park Cinema
Brudenell Road
Tel: 0113 275 2045
Seats: 360

Lounge, North Lane, Headingley
Tel: 0113 275 1061/258932
Seats: 691

ABC, Vicar Lane
Tel: 0113 245 1013/245 2665
Seats: 1:670, 2:483, 3:228

Odeon, The Headrow
Tel: 0142 697 7333
Seats: 1:975, 2:423 X, 3:198 X, 4:172, 5:110

Showcase Cinemas
Gelderd Road
Birstall X
Tel: 01924 420622
Seats: 3,725 (14 screens)

Warner Village
Seats: 2,800 (12 screens)

Leicester *Leics*
Bollywood, Melton Road
Tel: 0116 268 1215
Seats: 1:450, 2:150

Odeon, Aylestone Road,
Freemans Park   XE
Tel: 0116 255 7069
Seats: 1:128, 2:164, 3:154, 4;239, 5:210, 6:632, 7:332, 8:212, 9:329, 10:154, 11:164, 12:126

bfi Phoenix Arts,
11 Newarke Street,
LE1 5SS ◯ XE
Tel: 0116 255 4854/255 5627
Seats: 274

Warner Village, Meridian Leisure
Park, Brownstone
Tel: 0116 282 7733
Seats: 1:423, 2:158, 3:202, 4:266, 5:306, 6:266, 7:202, 8:158, 9:423

Leighton Buzzard *Bedfordshire*
Theatre, Lake Street ◯
Tel: 01525 378310
Seats: 170

Leiston *Suffolk*
Film Theatre, High Street
Tel: 01728 830549
Seats: 350

Letchworth *Herts*
Broadway, Eastcheap
Tel: 01462 681 223
Seats: 1:488, 2:176 X, 3:174 X

Leyburn *North Yorks*
Elite, Railway Street ◯
Tel: 01969 624488
Seats: 173

Lichfield *Staffs*
Civic Hall, Castle Dyke
Tel: 01543 254021
Seats: 278

Lincoln *Lincs*
Odeon, Valentine Road
Tel: 0152 254 2522
Seats: 1:279, 2:164, 3:181, 4:138, 5:134, 6:138

Littlehampton *West Sussex*
Windmill Theatre
Church Street ◯
Tel: 01903 724929
Seats: 252

Liverpool *Merseyside*
ABC, Allerton Road
Tel: 0151 724 3550/5095
Seats: 472

Odeon, London Road
Tel: 01426 950072
Seats: 1:967, 2:591, 3:146, 4:146, 5:145

Odeon, South Aintree
Seats: 2,484 (12 screens)

Philharmonic Hall, Hope Street ○X
Tel: 0151 709 2895/3789
Seats: 1,627

**Showcase Cinemas, East Lancashire
Road, Norris Green** X
Tel: 0151 549 2021
Seats: 3,408(12 screens)

**Virgin Cinemas, Edge Lane Retail
Park, Binns Road** XE
Tel: 0151 252 0544
Seats: 1:356, 2:354, 3:264, 4:264,
5:220, 6:220, 7:198, 8:200

Woolton, Mason Street X
Tel: 0151 428 1919
Seats: 256

Looe *East Cornwall*
Cinema, Higher Market Street
Tel: 015036 2709
Seats: 95

Loughborough *Leics*
Curzon, Cattle Market
Tel: 01509 212261
Seats: 1:420, 2:303, 3:199, 4:186,
5:140, 6:80

**Stanford Hall Cinema at the
Co-operative College** ○
Tel: 01509 852333
Seats: 352

Louth *Lincs*
Playhouse, Cannon Street
Tel: 01507 603333
Seats: 1:215, 2:158 X, 3:78 X

Lowestoft *Suffolk*
Hollywood, London Road South
Tel: 01502 564567
Seats: 1:200, 2:175, 3:40

Marina Theatre, The Marina ○
Tel: 01502 573318
Seats: 751

Ludlow *Shropshire*
**Assembly Rooms
Mill Street** ○   X
Tel: 01584 878141
Seats: 320

Luton *Beds*
ABC, George Street
Tel: 01582 27311/22537
Seats: 1:562, 2:436, 3:272

Cineworld The Movies
Seats: 2,200 (11 screens)

Artezium, Arts and Media Centre
Tel: 01582 876969
Seats: 96
(scheduled to open October 1998)

**St George's Theatre
Central Library** ○
Tel: 01582 21628
Seats: 238

Lyme Regis *Dorset*
Regent, Broad Street X
Tel: 01297 442053
Seats: 400

Lymington *Hants*
Community Centre, New Street ○
Tel: 015907 2337
Seats: 110

Lytham St. Annes *Lancs*
**Pleasure Island Cinemas,
South Promenade**
Tel: 01253 780085
Seats: 1:160, 2:88

Mablethorpe *Lincs*
Loewen, Quebec Road
Tel: 0150 747 7040
Seats: 1:203, 2:80

Maidstone *Kent*
ABC, Lower Stone Street
Tel: 01622 752628/758838
Seats: 1:272, 2:267, 3:150

Odeon, Knights Park
Seats: 1:1,646 (8 screens)

Malton *North Yorks*
Palace, The Lanes E
Tel: 01653 600008
Seats: 142 (temporarily closed in
Spring 1998)

Malvern *Hereford & Worcs*
**Cinema,
Winter Gardens Complex
Grange Road**
Tel: 01684 892277/892710
Seats: 407

Manchester *Greater
Manchester*
**Arena 7, Nynex Arena
Complex** X
Tel: 0161 839 0700
Seats: 1:138, 2:143, 3:287, 4:257,
5:221, 6:370, 7:156

Arena, Nynex Arena
Tel: 0161 930 8000 ○   X

**Cine City, Wilmslow Road,
Withington**
Tel: 0161 445 9888
Seats: 1:150, 2:130, 3:130

🅑🅕🅘 **Cornerhouse
70 Oxford Street
M1 5NH XE**
Tel: 0161 228 2467/7621
Seats: 1:300, 2:170, 3:60

Odeon, Oxford Street
Tel: 0161 236 0537
Seats: 1:629 E, 2:346 E, 3:144 X,
4:97, 5:203 E, 6:143 X, 7:86

**Showcase Cinemas
Hyde Road, Belle Vue**

Tel: 0161 220 8505
Seats: 3,191 (14 screens)

**UCI, Trafford Shopping Centre,
Barton Dock Road**
Seats: 1:427, 2:427, 3:371, 4:301,
5:243, 6:243, 7:181, 8:181, 9:181,
10:181, 11:181, 12:181, 13:152,
14:152, 15:140, 16:140, 17:112,
18:112, 19:112, 20:112

Mansfield *Notts*
**ABC, Mansfield Leisure Park
Park Lane**
Tel: 01623 422 462
Seats: 1:390, 2:390, 3:246, 4:246,
5:221, 6:221, 7:193, 8:193

Margate *Kent*
Dreamland, Marine Parade
Tel: 01843 227822
Seats: 1:378, 2:376

Market Drayton *Shropshire*
Royal Festival Centre ○
Seats: 165

Marple *Greater Manchester*
Regent, Stockport Road X
Tel: 0161 427 5951
Seats: 285

Matlock *Derbyshire*
Ritz, Causeway Lane
Tel: 01629 580 607
Seats: 1:170, 2:100

Melton Mowbray *Leics*
Regal, King Street
Tel:0166 267 3127
Seats: 226

Middlesbrough *Cleveland*
Odeon, Corporation Road
Tel: 01426 981167
Seats: 1:611, 2:129, 3:148, 4:254

Millom *Cumbria*
Palladium, Horn Hill ●
Tel: 01657 2441
Seats: 400

Milton Keynes *Bucks*
**UCI The Point 10
Midsummer Boulevard**
Tel: 01908 661662/695444
Seats: 1:156, 2:169, 3:250, 4:222, 5:222,
6:222, 7:222, 8:250

Minehead *Somerset*
**Odeon, Butlin's
Summerwest World** X
Tel: 01643 703331
Seats: 218

Monkseaton *Tyne & Wear*
ABC, Cauldwell Lane
Tel: 0191 297 2121/252 5540
Seats: 1:329, 2:117

Monton *Greater Manchester*
Princess, Monton Road
Tel: 0161 789 3426
Seats: 580

Morecambe *Lancashire*
Apollo, Central Drive
Tel: 01524 401040
Seats: 1:207, 2:207, 3:106, 4:106

Morpeth *Northumberland*
Coliseum, New Market
Tel: 01670 516834
Seats: 1:132, 2:132

Nailsea *Avon*
Cinema, Scotch Horn Leisure
Centre, Brockway ◯
Tel: 01275 856965
Seats: 250

Nantwich *Cheshire*
Civic Hall, Market Street ◯
Tel: 01270 628633
Seats: 300

Newark *Notts*
Palace Theatre, Appleton Gate ◯
Tel: 01636 71636
Seats: 351

Newbury *Berks*
Corn Exchange,
Market Place ◯   X
Tel: 01635 522733
Seats: 370

Robins, Park Way
Tel: 01635 41291
Seats: 480

Newcastle upon Tyne
Tyne & Wear
Odeon, Pilgrim Street
Tel: 01426 950527
Seats: 1:1,1171, 2:155, 3:250, 4:361

(bfi) Tyneside,
10-12 Pilgrim Street,
NE1 6QG XE
Tel: 0191 232 8289
Seats: 1:383, 2:155

Warner Village, New Bridge Street
Tel: 0191 221 0202/0545
Seats: 1:404, 2:398, 3:236, 4:244,
5:290, 6:657, 7:509, 8:398, 9:248

Newport *Isle of Wight*
Picturedrome, High Street
Tel: 01983 527169
Seats: 1:377, 2:291

Medina Theatre
Mountbatten Centre
Fairlee Road ◯   XE
Tel: 01983 527020
Seats: 419

Newton Abbot *Devon*
Alexandra, Market Street   X
Tel: 01626 65368
Seats: 1:210, 2:142

Northampton *Northants*
Virgin Cinemas, Sixfields Leisure,
Weeden Road, Upton
Tel: 01604 580700
Seats: 1:452, 2:287, 3:287, 4:207,
5:207, 6:147, 7:147, 8:147, 9:147

(bfi) Forum Cinema, Lings
Forum, Weston Favell
Centre, NN3 4JR ◯
Tel: 01604 401006/ 402 833
Seats: 270

Northwich *Cheshire*
Regal, London Road
Tel: 01606 43130
Seats: 1:797, 2:200

Norwich *Norfolk*
ABC, Prince of Wales Road
Tel: 01603 624677/623312
Seats: 1:523, 2:343, 3:186, 4:105

(bfi) Cinema City, St Andrew's
Street, NR2 4AD   X
Tel: 01603 625145/622047
Seats: 230

Odeon, Anglia Square   E
Tel: 01603 621903
Seats: 1:442, 2:197, 3:195 X

Nottingham *Notts*
ABC, Chapel Bar
Tel: 0115 941 8483/ 947 5260
Seats: 1:764, 2:436, 3:283

(bfi) Broadway, Nottingham
Media Centre
14 Broad Street, NG1 3AL
Tel: 0115 952 6600/952 6611
Seats: 1:379 E, 2:155 XE

Odeon, Angel Row
Tel: 0142 695 7022
Seats: 1:903, 2:557, 3:150, 4:150,
5:113, 6:100

Savoy, Derby Road
Tel: 0115 947 2580/941 9123
Seats: 1:386, 2:128, 3:168

Showcase Cinemas
Redfield Way
Lenton
Tel: 0115 986 2505
Seats: 3,307 (13 screens)

Okehampton *Devon*
Carlton, St James Street
Tel: 01837 52167
Seats: 380

Oldham *Lancs*
Roxy, Hollins Road

Tel: 0161 683 4759
Seats: 1:400, 2:300, 3:130

Oxford *Oxon*
ABC, George Street
Tel: 01865 244607/723911
Seats: 1:626, 2:327, 3:140

ABC, Magdalen Street
Tel: 01865 243067
Seats: 864

Phoenix, 57 Walton Street   X
Tel: 01865 512526
Seats: 1:220, 2:102

Ultimate Picture Palace,
Jeune Street   X
Tel: 01865 245288
Seats: 185

Oxted *Surrey*
Plaza, Station Road West   X
Tel: 01883 712567
Seats: 442

Padstow *Cornwall*
Capitol, Lanadwell Street
Tel: 01841 532344
Seats: 210

Paignton *Devon*
Torbay, Torbay Road
Tel: 01803 559544
Seats: 484

Penistone *South Yorks*
Metro, Town Hall
Tel: 01226 762004
Seats: 348

Penrith *Cumbria*
Alhambra, Middlegate
Tel: 01768 62400
Seats: 202

Penzance *Cornwall*
Savoy, Causeway Head
Tel: 01736 363330
Seats: 1:200, 2:50, 3:50

Peterborough *Cambs*
Showcase Cinemas
Mallory Road, Boon Gate   X
Tel: 01733 555636
Seats: 3,365 (13 screens)

Pickering *North Yorks*
Castle, Burgate
Tel: 01751 472622
Seats: 250

Plymouth *Devon*
Arts Centre, Looe Street   X
Tel: 01752 660060
Seats: 73

Odeon, Derry's Cross
Tel: 01752 668825
Seats: 1:422, 2:164, 3:164, 4:218,
5:133

ABC, Derry's Cross
Tel: 01752 663300/225553
Seats: 1:582, 2:380, 3:115

Pontefract *West Yorks*
Crescent, Ropergate
Tel: 01977 703788
Seats: 412

Poole *Dorset*
Arts Centre
Kingsland Road ○   X
Tel: 01202 670521
Seats: 143

UCI Tower Park, Mannings Heath
Tel: 01202 715010
Seats: 1:194, 2:188, 3;188, 4:194,
5:276, 6:276, 7:194, 8:188, 9:188,
10:194

Portsmouth *Hants*
ABC, Commercial Road
Tel: 01705 823538/839719
Seats: 1:542, 2:255, 3:203

Odeon, London Road, North End
Tel: 01705 651434
Seats: 1:624, 2:225, 3:173, 4:259

Rendezvous, Lion Gate Building
University of Portsmouth ●
Tel: 01705 833854
Seats: 90

UCI Port Way, Cosham X
Tel: 01705 649999
Seats: 1:214, 2:264, 3:318, 4:264,
5:257, 6:190

Potters Bar *Herts*
Wyllyotts Centre
Darkes Lane ○   X
Tel: 01707 645005
Seats: 345

Preston *Lancs*
Guild Hall, Lancaster Road   X
Tel: 01772 258858   ○

UCI Riversway,
Ashton-on-Ribble   X
Tel: 0990 888990
Seats: 1:194, 2:188, 3:188, 4:194, 5:276,
6:276, 7:194, 8:188, 9:188, 10:194

Warner, The Capitol Centre,
London Way, Walton-le-dale   X
Tel: 01772 881100
Seats: 1:180, 2:180, 3:412, 4:236,
5:236, 6:412, 7:192

Quinton *West Mids*
ABC, Hagley Road West
Tel: 0121 422 2562/2252
Seats: 1:300, 2:236, 3:232, 4:121

Ramsey *Cambs*
Grand, Great Whyte ○
Tel: 01487 813777
Seats: 173

Ramsgate *Kent*
Granville, Victoria Parade ○
Tel: 01843 591750
(conversion to two auditoria
underway in summer 1998 to seat
300 and 240)

Reading *Berks*
Film Theatre, Whiteknights ○
Tel: 01734 868497/875123
Seats: 409

ABC, Friar Street
Tel: 01734 573907/573931
Seats: 1:532, 2:222, 3:119

Odeon, Cheapside
Tel: 01734 576803
Seats: 1:410, 2:221, 3:221

Town Hall ○
Tel: 01734 591591

Redcar *Cleveland*
Regent
Tel: 01642 482094
Seats: 350

Redditch *Hereford & Worcs*
ABC, Unicorn Hill
Tel: 01527 62572
Seats: 1:203, 2:155, 3:155

Redhill *Surrey*
The Harlequin
Warwick Quadrant ○   X
Tel: 01737 765547
Seats: 494

Redruth *Cornwall*
Regal Film Centre, Fore Street
Tel: 01209 216278
Seats: 1:200, 2:128, 3:600, 4:95

Reigate *Surrey*
Screen, Bancroft Road
Tel: 01737 223213
Seats: 1:139, 2:142

Rickmansworth *Herts*
Watersmeet Theatre
High Street ○
Tel: 01923 771542
Seats: 390

Rochdale *Greater Manchester*
ABC Sandbrook Way
Sandbrook Park
Tel: 01706 719 955
Seats: 1:474, 2:311, 3:311, 4:236,
4:236, 5:236, 6:208, 7:208, 8:165,
9:165

Rochester *Kent*
Virgin Cinemas
Chariot Way, Strood
Tel: 01634 719963
Seats: 1:485, 2:310, 3:310, 4:217,
5:220, 6:199, 7:199, 8:92, 9:142

Royston *Herts*
Priory, Priory Lane
Tel: 01763 43133
Seats: 305

Rushden *Northants*
Ritz, College Street
Tel: 01933 312468
Seats: 822

Ryde *Isle of Wight*
Commodore, Star Street
Tel: 01983 564064
Seats: 1:186, 2:184, 3:180

St Albans *Herts*
Alban Arena, Civic Centre ○   XE
Tel: 01727 844488
Seats: 800

St Austell *Cornwall*
Film Centre, Chandos Place
Tel: 01726 73750
Seats: 1:276, 2:134, 3:133, 4:70, 5:70

St Ives *Cornwall*
Royal, Royal Square
Tel: 01736 796843
Seats: 1:350, 2:150, 3:63

Salford Quays *Lancs*
Virgin Cinemas, Clippers Quay
Tel: 0161 873 7279
Seats: 1:287, 2:265, 3:249, 4:249,
5:213, 6:213, 7:177, 8:177

Salisbury *Wilts*
Odeon, New Canal
Tel: 01722 335924
Seats: 1:471, 2:281 X, 3:128 X,
4:111 X, 5:70

Sandwich *Kent*
Empire, Delf Street
Tel: 01304 620480
Seats: 136

Scarborough *North Yorks*
Futurist, Forshaw Road ○   X
Tel: 01723 370742
Seats: 2,155

Hollywood Plaza
North Marine Road
Tel: 01723 365119
Seats: 275

Stephen Joseph Theatre,
Westborough   XE ○
Tel: 01723 370541
Seats: 165 (McCarthy Auditorium)

Scunthorpe *Humberside*
Majestic, Oswald Road
Tel: 01724 842352
Seats: 1:176, 2:155 X, 3:76 X,
4:55 X, 5:38

Screen, Central Library
Carlton Street, DN15 6TX ○   X

Tel: 01724 860190/860161
Seats: 253

Sevenoaks *Kent*
Stag Theatre and Majestic 1 & 2,
London Road
Tel: 01732 450175/451548
Seats: 1:470 ◯, 2:140, 3:111

Shaftesbury *Dorset*
Arts Centre, Bell Street ◯
Tel: 01747 854321
Seats: 160

Sheffield *South Yorks*
 The Showroom, Media and
Exhibition Centre,
Paternoster Row, S1 2BXX
Tel: 0114 275 7727
Seats: 1:83, 2:110, 3:178, 4:282

Odeon, Arundel Gate
Tel: 0114 272 3981
Seats: 1:253 **XE**, 2:231 **X**,
3:250 **XE**, 4:117 **XE**, 5:115 **XE**,
6:131, 7:170, 8:160, 9:161, 10:123

UCI Crystal Peaks 10
Eckington Way, Sothall **X**
Tel: 0114 248 0064/247 0095
Seats: 1:202; 2:202, 3:230, 4:226,
5:316, 6:316, 7:226, 8:230, 9:202,
10:202

Virgin Cinemas
Seats: 4,857 (20 screens) (to open
October 1998)

Warner Village, Meadowhall
Shopping Centre **X**
Tel: 0114 256 9444
Seats: 1:199, 2:198, 3:98, 4:233,
5:198, 6:362, 7:192, 8:192, 9:80,
10:190, 11:329

Shepton Mallet *Somerset*
Amusement Centre, Market Place ◯
Tel: 01749 3444688
Seats: 270

Sheringham *Norfolk*
Little Theatre, Station Road ●
Tel: 01263 822347
Seats: 198

Shipley *West Yorks*
Apollo Unit Four, Bradford Road
Tel: 01274 583429
Seats: 1:89, 2:72, 3:121, 4:94

Shrewsbury *Shropshire*
Cineworld The Movies
Seats: 1,500 (8 screens) (opening
circa December 1998)

The Film Theatre, The Music
Hall, The Square, SY1 1LH
Tel: 01743 350763/352019
Seats: 100

Sidmouth *Devon*
Radway, Radway Place
Tel: 01395 513085
Seats: 400

Sittingbourne *Kent*
Picturedrome, High Street
Tel: 01795 423984/426018
Seats: 1:300, 2:110

Skegness *Lincs*
Tower, Lumley Road
Tel: 01754 3938
Seats: 401

Skelmersdale *Lancs*
Premiere Film Centre
Tel: 01695 25041
Seats: 1:230, 2:248

Skipton *North Yorks*
Plaza, Sackville Street **X**
Tel: 01756 793417
Seats: 320

Sleaford *Lincs*
Sleaford Cinema, Southgate
Tel: 01529 303187
Seats: 60

Slough *Berks*
Virgin Cinemas, Queensmere
Centre
Tel: 01753 511299
Seats: 1,821 (10 screens)

Solihull *West Mids*
UCI 8, Highlands Road, Shirley **X**
Tel: 0121 608 7090
Seats: 286 (2 screens), 250 (2 screens),
214 (2 screens), 178 (2 screens)

South Shields *Tyne & Wear*
Customs House, Mill Dam
Tel: 0191 455 6655
Seats: 1:400, 2:160

South Woodham Ferrers *Essex*
Flix, Market Street
Tel: 01245 329777
Seats: 1:249, 2:101

Southampton *Hants*
 The Gantry,
Off Blechynden Terrace,
SO1 0GW **X**
Tel: 01703 229319/330729
Seats: 198

 Harbour Lights, Ocean
Village SO14 3TL
Tel: 01703 635335/335533
Seats: 1:350, 2:150

Mountbatten Theatre, East Park
Terrace ◯
Tel: 01703 221991
Seats: 515

Northguild Lecture Theatre
Guildhall ◯ **XE**
Tel: 01703 632601
Seats: 118

Odeon, Leisureworld
West Quay Road
Tel: 01703 333515
Seats: 1:540, 2:495, 3:169, 4:111,
5:112, 6:139, 7:270, 8:318, 9:331,
10:288, 11:502, 12:102, 13:138

Virgin Cinemas, Ocean Way
Ocean Village
Tel: 01703 232880
Seats: 1:421, 2:346, 3:346, 4:258,
5:258

Southend *Essex*
Odeon, Victoria Circus **X E**
Tel: 01702 393544
Seats: 1:200, 2:264, 3:148, 4:224,
5:394, 6:264, 7:264, 8:200

Southport *Merseyside*
Arts Centre, Lord Street ◯ **X**
Tel: 01704 540004/540011
Seats: 400

ABC, Lord Street **X**
Tel: 01704 530627
Seats: 1:504, 2:385

Spilsby *Lincs*
Phoenix, Reynard Street
Tel: 01790 753 675
Seats: 264

Stafford *Staffs*
Apollo, New Port Road
Tel: 01785 227007
Seats: 1:305, 2:170, 3:168

Staines *Middlesex*
ABC, Clarence Street
Tel: 01784 453316/459140
Seats: 1:586, 2:363 **X**, 3:174 **X**

Stalybridge *Greater Manchester*
New Palace, Market Street
Tel: 0161 338 2156
Seats: 414

Stamford *Lincs*
Arts Centre, St. Mary's Street
Tel: 01780 763203
Seats: 166

Stanley *Co Durham*
Civic Hall ◯
Tel: 01207 32164
Seats: 632

Stevenage *Herts*
Gordon Craig Theatre
Lytton Way ◯
Tel: 01438 766 866
Seats: 507

Cineworld The Movies,
Stevenage Leisure Park

Six Hills Way
Tel: 01438 740944/740310
Seats: 1:357, 2:289, 3:175, 4:148,
5:88, 6:99, 7:137, 8:112, 9:168,
10:135, 11:173, 12:286

Stockport *Greater Manchester*
Virgin Cinemas
Grand Central Square
Wellington Road South   **XE**
Tel: 0161 476 5996
Seats: 1: 303, 2:255, 3:243, 4:243,
5:122, 6:116, 7:96, 8:120, 9:84, 10:90

Stockton *Cleveland*
Dovecot Arts Centre
Dovecot Street
Tel: 01642 611625/611659
Seats: 100

Showcase Cinemas, Aintree Oval
Teeside Leisure Park
Tel: 01642 633111
Seats: 3,400 (14 screens)

Stoke-on-Trent *Staffs*
(bfi) Film Theatre,
College Road, ST4 2DE
Tel: 01782 411188/413622
Seats: 212

Odeon, Etruria Road   **X**
Tel: 0178 221 5805
Seats: 1:201, 2:216, 3:368, 4:162,
5:169, 6:185, 7:564, 8:162, 9:104,
10:75

Stourport *Hereford and Worcs*
Civic Centre, Civic Hall, New Street
Tel: 01562 820 505
Seats: 399

Stowmarket *Suffolk*
Regal, Ipswich Street ○
Tel: 01449 612825
Seats: 234

Stratford-on-Avon *Warwicks*
Picture House, Windsor Street **X**
Tel: 01789 415500
Seats: 1:249, 2:82

Street *Somerset*
(bfi) Strode Theatre, Strode
College, Church Road
BA16 0AB ○   **XE**
Tel: 01458 42846/46529
Seats: 400

Sudbury *Suffolk*
Quay Theatre, Quay Lane
Tel; 01787 374745
Seats: 129

Sunderland *Tyne & Wear*
ABC, Holmeside
Tel: 0191 565 5011/567 4148
Seats: 1:533, 2:210

Sunninghill *Berks*
Novello Theatre, High Street ○
Tel: 01990 20881
Seats: 160

Sutton Coldfield *West Mids*
Odeon, Birmingham Road
Tel: 0121 354 2714
Seats: 1:590, 2:128 **X**, 3:110 **X**,
4:330 **X**

Swanage *Dorset*
Mowlem, Shore Road
Tel: 01929 422239
Seats: 400

Swindon *Wilts*
Arts Centre, Devizes Road,
Old Town ○   **E**
Tel: 01793 614 837
Seats: 228

Cineworld The Movies
Seats: 2,200 (12 screens)

Virgin Cinemas, Shaw Ridge
Leisure Park, Whitehill Way XE
Tel: 01793 881118
Seats: 1:349, 2:349, 3:297, 4:297,
5:272, 6:166, 7:144

Wyvern, Theatre Square ○
Tel: 01793 24481
Seats: 617

Tadley *Hants*
Cinema Royal, Boundary Road ○
Tel: 01734 814617

Tamworth *Staffs*
Palace, Lower Gungate ○
Tel: 01827 57100
Seats: 325

UCI, Bolebridge Street   **X**
Tel: 0800 888980
Seats: 203 (8 screens), 327 (2 screens)

Taunton *Somerset*
Odeon, Heron Gate, Riverside **X**
Tel: 01823 443237
Seats: 1:106, 2:316, 3:218, 4:252,
5:106

Tavistock *Devon*
The Wharf, Canal Street ○
Tel: 01822 613928
Seats: 212

Teignmouth *Devon*
Riviera, Den Crescent
Tel: 01626 774624
Seats: 417

Telford *Shropshire*
UCI Telford Centre 10,
Forgegate   **X**
Tel: 01952 290606/290126
Seats: 1:194, 2:188, 3:188, 4:194, 5:276,
6:276, 7:194, 8:188, 9:188, 10:194

Tenbury Wells *Hereford & Worcs*
Regal, Teme Street
Tel: 01584 810971
Seats: 260

Tewkesbury *Glos*
Roses Theatre ○
Tel: 01684 295074
Seats: 375

Thetford *Norfolk*
Diamond Screen
Tel: 01842 750075
Seats: 1:230, 2:115

Thirsk *North Yorks*
Ritz
Tel: 01845 523484
Seats: 238

Tiverton *Devon*
Tivoli, Fore Street
Tel: 01884 252157
Seats: 364

Toftwood *Norfolk*
CBA, Shipham Road
Tel: 01362 693261
Seats: 30

Tonbridge *Kent*
Angel Centre, Angel Lane ○
Tel: 01732 359588
Seats: 306

Torquay *Devon*
English Riviera Centre
Chestnut Avenue ○   **XE**
Tel: 01803 295676
Seats: 800

Odeon, Abbey Road
Tel: 01803 295805
Seats: 1:304, 2:333

Torrington *Devon*
Plough, Fore Street
Tel: 01805 622552/3
Seats: 108

Truro *Cornwall*
Plaza, Lemon Street
Tel: 01872 272 894
5 screens

Tunbridge Wells *Kent*
ABC, Mount Pleasant
Tel: 01892 541141/523135
Seats: 1:450 **X**, 2:402, 3:130

Odeon
Seats: 2,180 (9 screens)

Uckfield *East Sussex*
Picture House, High Street
Tel: 01825 763822/764909
Seats: 1:150, 2:100

Ulverston *Cumbria*
Laurel & Hardy Museum

Upper Brook Street ○ ●
Tel: 01229 52292/86614
Seats: 50

Roxy, Brogden Street
Tel: 01229 53797/56211
Seats: 310

Urmston *Greater Manchester*
Curzon, Princess Road
Tel: 0161 748 2929
Seats: 1:400, 2:134

Uttoxeter *Staffs*
Elite, High Street
Tel: 018893 3348
Seats: 120

Uxbridge *Middx*
Odeon, High Street
Tel: 01895 813139
Seats: 1:236, 2:445

Wadebridge *Cornwall*
Regal, The Platt
Tel: 01208 812791
Seats: 1:250, 2:120

Wakefield *West Yorks*
Cineworld The Movies,
Westgate Leisure Centre   X
Tel: 01924 332114
Seats: 1:323, 2:215, 3:84, 4:114,
5:183, 6:255, 7:255, 8:183, 9:114,
10:84, 11:215, 12:323

Walkden *Greater Manchester*
Apollo, Bolton Road
Tel: 0161 790 9432
Seats: 1:118, 2:108, 3:86, 4:94

Wallasey *Merseyside*
Apollo Unit Six, Egremont
Tel: 0151 639 2833
Seats: 1:181, 2:127, 3:177, 4:105,
5:91, 6:92

Wallingford *Oxon*
Corn Exchange ○
Tel: 01491 825000
Seats: 187

Walsall *West Mids*
Showcase Cinemas,
Bentley Mill Way, Darlaston   X
Tel: 01922 22120
Seats: 2,070 (12 screens)

Walton on Thames *Surrey*
Screen, High Street
Tel: 01932 252825
Seats: 1:200, 2:140

Wantage *Oxon*
Regent, Newbury Street
Tel: 01235 771 155/767878
Seats: 1:110, 2:87

Wareham *Dorset*
Rex, West Street
Tel: 01929 552778
Seats: 239

Warrington *Cheshire*
UCI 10, Westbrook Centre,
Cromwell Avenue   X
Tel: 0990 888990
Seats: 1:186, 2:180, 3:180, 4:186,
5:276, 6:276, 7:186, 8:180, 9:180,
10:186

Washington *Tyne & Wear*
Fairworld, Victoria Road
Tel: 0191 416 2711
Seats: 1:227, 2:177

Watford *Herts*
Warner Village, Woodside Leisure
Park, Garston
Tel: 01923 682244
Seats: 1:251, 2:233, 3:264, 4:330,
5:221, 6:198, 7:215, 8:301

Wellingborough *Northants*
Palace, Gloucester Place
Tel: 01933 222184
Seats: 1:128, 2:107

Wellington *Somerset*
Wellesley, Mantle Street
Tel: 01823 666668/666880
Seats: 400

Wells *Somerset*
Film Centre, Princes Road
Tel: 01749 672036/672195
Seats: 1:150, 2:60

Welwyn Garden City *Herts*
Campus West, The Campus, AL8
6BX ○
Tel: 01707 332880
Seats: 364

West Bromwich *West Mids*
Kings, Paradise Street   X
Tel: 0121 553 0192
Seats: 1:326, 2:287, 3:462

Westgate-on-Sea *Kent*
Carlton, St Mildreds Road
Tel: 01843 832019
Seats: 303 (two additional screens
under construction in summer 1998)

Weston-super-Mare *Avon*
Odeon, The Centre
Tel: 01934 641251
Seats: 1:586, 2:109, 3:126, 4:268

Playhouse, High Street ○
Tel: 01934 23521/31701
Seats: 658

West Thurrock *Essex*
UCI Lakeside 10
Lakeside Retail Park   X

Tel: 01708 860268
Seats: 276 (2 screens), 194 (4 screens),
188 (4 screens)

Warner, Lakeside Shopping
Centre   X
Tel: 01708 891010/890567/890600
Seats: 1:382, 2:184, 3:177, 4:237,
5:498, 6:338, 7:208

Wetherby *West Yorkshire*
Film Theater, Crossley Street
Tel: 01937 580544
Seats: 156

Weymouth *Dorset*
Picturedrome
Gloucester Street
Tel: 01305 785847
Seats: 418

Whitehaven *Cumbria*
Gaiety, Tangier Street
Tel: 01946 693012
Seats: 330

Rosehill Theatre, Moresby ○   X
Tel: 01946 694039/692422
Seats: 208

Whitley Bay *Tyne & Wear*
Playhouse, Marine Avenue
Tel: 0191 252 3505
Seats: 746

Whitstable *Kent*
Imperial Oyster
Tel: 01227 770829
Seats: 144

Wigan *Greater Manchester*
Virgin Cinemas, Robin Park Road
Newtown   X
Tel: 01942 218005
Seats: 1:554, 2:290, 3:290, 4:207,
5:207, 6:163, 7:163, 8:163, 9:163,
10:207, 11:129

Wilmslow *Cheshire*
Rex, Alderley Road ○
Tel: 01625 522266
Seats: 838

Wimborne *Dorset*
Tivoli, West Borough ○
Tel: 01202 849103

Wincanton *Somerset*
Plaza, South Street
Seats: 380

Winchester *Hants*
The Screen at Winchester,
Southgate Street   X
Tel: 01962 877007
Seats: 1:214, 2:170

Theatre Royal, Jewry Street ○
Tel: 01962 842122
Seats: 405

Windsor *Berkshire*
Arts Centre, St Leonards Road ○
Tel: 01753 8593336
Seats: 108

Witney *Oxon*
Corn Exchange, Market Square ○
Tel: 01993 703646
Seats: 207

Woking *Surrey*
Ambassador Cinemas, Peacock
Centre off Victoria Way   X
Tel: 01483 761144
Seats: 1:434, 2:447, 3:190, 4:236,
5:268, 6:89

Wokingham *Berkshire*
Showcase Cinemas, Loddon
Bridge, Reading Road,
Winnersh   X
Tel: 01189 788 766
Seats: 2,980 (12 screens)

Wolverhampton *West Mids*
Cineworld The Movies,
Wednesfield Way, Wednersfield
Tel: 01902 305418
Seats: 1:103, 2:113, 3:151, 4:205,
5:192, 6:343, 7:379, 8:343, 9:184,
10:89, 11:105, 12:162, 13:143, 14:98

Light House, Chubb Buildings,
Fryer Street   XE
Tel: 01902 716055
Seats: 1:242, 2:80

Woodbridge *Suffolk*
Riverside Theatre
Quay Street
Tel: 01394 382174/380571
Seats: 280

Woodhall Spa *Lincs*
Kinema in the Woods
Coronation Road
Tel: 01526 352166
Seats: 1:290, 2:90

Worcester *Hereford & Worcs*
Odeon, Foregate Street
Tel: 01905 24006
Seats: 1:306, 2:201, 3:125, 4:99, 5:68,
6:306, 7:131

Workington *Cumbria*
Rendezvous, Murray Road
Tel: 01900 602505
Two screens

Worksop *Notts*
Regal, Carlton Road
Tel: 01909 482896
Seats: 1:326 ○, 2:154

Worthing *West Sussex*
Connaught Theatre, Union Place ○
Tel: 01903 231799/235333
Seats: 1:514, 2 (Ritz): 220

Dome, Marine Parade
Tel: 01903 200461
Seats: 600

Wotton Under Edge *Glos*
Town Cinema
Tel: 01453 521666
Seats: 200

Yeovil *Somerset*
ABC, Court Ash Terrace
Tel: 01935 413333/413413
Seats: 1:602, 2:248, 3:246

York *North Yorks*
 Film Theatre, City Screen,
Yorkshire Museum, Museum
Gardens, YO1 2DR ○   X
Tel: 01904 612940
Seats: City Screen 300,
Film Theatre 720

Odeon, Blossom Street
Tel: 0190 462 3040
Seats: 1:832, 2:115 X, 3:115 X

Warner Village, Clifton Moor
Centre, Stirling Road   X
Tel: 01904 691199/691094
Seats: 1:128, 2:212, 3:316, 4:441,
5:185, 6:251, 7:251, 8:185, 9:441,
10:316, 11:212, 12:128

CHANNEL ISLANDS

Forest *Guernsey*
Mallard Cinema, Mallard Hotel
La Villiaze
Tel: 01481 64164
Seats: 1:154, 2:54, 3:75, 4:75

St Helier *Jersey*
Odeon, Bath Street
Tel: 01534 22888
Seats: 1:409, 2:244, 3:213, 4:171

St Peter Port *Guernsey*
Beau Sejour Centre
Tel: 01481 26964
Seats: 250

St Saviour *Jersey*
Cine Centre, St Saviour's Road
Tel: 01534 871611
Seats: 1:400, 2:291, 3:85

ISLE OF MAN

Douglas *Isle of Man*
Palace Cinema
Tel: 01624 76814
Seats: 1:319, 2:120

Summerland Cinema
Tel: 01624 25511
Seats: 200

Virgin Cinemas, Queens Link,
Leisure Park, Links Road
Tel: 0541 550502
Seats: 1:160, 2:86, 3:208, 4:290,
5:560, 6:290, 7:208, 8:160, 9:160

SCOTLAND

A number of BFI-supported cinemas in Scotland also receive substantial central funding and programming/management support via the Scottish Screen

Aberdeen *Grampian*
Odeon, Justice Mill Lane
Tel: 01244 587160
Seats: 1:415, 2:123, 3:123, 4:225, 5:225

Annan *Dumfries & Gall*
Ladystreet, Lady Street ●
Tel: 01461 202796
Seats: 450

Aviemore *Highland*
Speyside, Aviemore Centre X
Tel: 01479 810624/810627
Seats: 721

Ayr *Strathclyde*
Odeon, Burns Statue Square
Tel: 01426 979722
Seats: 1:388, 2:168, 3:135, 4:371

Brodick, Arran *Strathclyde*
Brodick, Hall Cinema
Tel: 01770 302065/302375
Seats:250

Campbeltown *Strathclyde*
Picture House, Hall Street ○
Tel: 01825 553899
Seats: 265

Castle Douglas *Dumfries & Gall*
Palace, St Andrews Street ●
Tel: 01556 2141
Seats: 400

Clydebank *Strathclyde*
UCI Clydebank 10, Clyde
Regional Centre, Britannia Way
Tel: 0141 951 1949/2022
Seats: 1:202, 2:202, 3:230, 4:253, 5:390, 6:390, 7:253, 8:230, 9:202, 10:202

Coatbridge *Strathclyde*
Showcase Cinemas, Langmuir
Road, Bargeddie, Bailleston X
Tel: 01236 434 434
Seats: 3,800 (14 screens)

Dumfries *Dumfries & Gall*
ABC, Shakespeare Street
Tel: 01387 253578
Seats: 526

Robert Burns Centre Film
Theatre, Mill Road ○
Tel: 01387 264808
Seats: 67

Dundee *Tayside*
ABC, Seagate

Tel: 01382 225247/226865
Seats: 1:581, 2:301

Odeon, The Stack Leisure Park
Harefield Road X
Tel: 01382 400855
Seats: 1:574, 2:210, 3:216, 4:233, 5:192, 6:221

 Steps Theatre, Central
Library, The Wellgate
DD1 1DB ○ X
Tel: 01382 434037
Seats: 250

Dunfermline *Fife*
Robins, East Port
Tel: 01383 623535
Seats: 1:209, 2:156, 3:78

Dunoon *Strathclyde*
Studio, John Street
Tel: 01369 4545
Seats: 1:188, 2:70

East Kilbride *Strathclyde*
Arts Centre, Old Coach Road ○
Tel: 01355 261000

UCI, Olympia Shopping Centre
Rothesay Street, Town Centre
Tel: 01355 249699
Seats: 1:319, 2:206, 3:219, 4:207, 5:207, 6:219, 7:206, 8:206, 9:219

Edinburgh *Lothian*
Cameo, Home Street, Tollcross X
Tel: 0131 228 4141
Seats: 1:253, 2:75, 3:66

Dominion, Newbattle Terrace
Tel: 0131 447 2660/4771
Seats: 1:586, 2:317, 322:47, 4:67

 Filmhouse,
88 Lothian Road
EH3 9BZ XE
Tel: 0131 228 2688/6382
Seats: 1:280, 2:97, 3:73

ABC, Lothian Road
Tel: 0131 228 1638/229 3030
Seats: 1:868, 2:730 X, 3:318 X

ABC, Westside Plaza,
Wester Hailes Road
Tel: 0131 442 2200
Seats: 1:416, 2:332, 3:332, 4:244, 5:228, 6:213, 7:192, 8:171

The Lumière, Movies at the Museum
Lothian Street, Edinburgh EH1 1JF
Tel: 0131 225 7434
Seats: 280

Odeon, Clerk Street
Tel: 0131 667 7331
Seats: 1:675, 2:301 X, 3:203 X, 4:262, 5:173

UCI Kinnaird Park,
Newcraighall Road

Tel: 0800 888955
Seats: 170 (6 screens), 208 (4 screens), 312 (2 screens)

Elgin *Grampian*
Moray Playhouse, High Street
Tel: 01343 542680
Seats: 1:320, 2:220

Falkirk *Central*
ABC, Princess Street
Tel: 01324 631713/623805
Seats: 1:690, 2:140 X, 3:137

Fort William *Highlands*
Studios 1 and 2, Cameron Square
Tel: 01397 705095
Seats: 1:126, 2:76

Galashiels *Borders*
Pavilion, Market Street
Tel: 01896 752767
Seats: 1:335, 2:172, 3:147, 4:56

Girvan *Strathclyde*
Vogue, Dalrymple Street ●
Tel: 01465 2101
Seats: 500

Glasgow *Strathclyde*
ABC, Clarkston Road, Muirend X
Tel: 0141 637 2641
Seats: 1:482, 2:208, 3:90

ABC, Sauchiehall Street
Tel: 0141 332 9513/0490
Seats: 1:970, 2:872 E (rear), 3:384, 4:206 E, 5:194 E

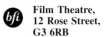 Film Theatre,
12 Rose Street,
G3 6RB XE
Tel: 0141 332 6535/8128
Seats: 1:404, 2:144

Grosvenor, Ashton Lane
Hillhead
Tel: 0141 339 4298
Seats: 1:277, 2:253

Odeon, The Quay, Paisley Road
Tel: 0141 418 0345 X
Seats: 1:428, 2:128, 3:89, 4:201, 5:200, 6:277, 7:321, 8:128, 9:89, 10:194, 11:242, 12:256

Odeon, Renfield Street X
Tel: 0141 333 9551
Seats: 1:1,142, 2:235, 3:256, 4:243, 5:271, 6:222

Virgin Cinemas, The Forge
Parkhead XE
Tel: 0141 556 4282
Seats: 1:434, 2:434, 3:322, 4:262, 5:208, 6:144, 7:132

Greenock *Strathclyde*
Waterfront,
Tel: 01475 732201
Seats: 1:258, 2:148, 3:106, 4:84

Glenrothes *Fife*
Kingsway, Church Street
Tel: 01592 750980
Seats: 1:294, 2:223

Hamilton *Strathclyde*
Odeon, Townhead Street
Tel: 01698 283802/422384
Seats: 1:466, 2:226, 3:310

Inverness *Highland*
Eden Court Theatre
Bishops Road
Tel: 01463 234234
Seats: 1:797 ◯, 2:70

La Scala, Strothers Lane
Tel: 01463 233302
Seats: 1:438, 2:255

Warner Village
Seats: 1,800 (7 screens)

Inverurie *Grampian*
Victoria, West High Street
Tel: 01467 21436
Seats: 467

Irvine *Strathclyde*
Magnum, Harbour Street   X
Tel: 01294 278381
Seats: 323

WMR Film Centre, Bank Street X
Tel: 01294 279900/276817
Seats: 252

Kelso *Borders*
Roxy, Horse Market
Tel: 01573 224609
Seats: 260

Kilmarnock *Strathclyde*
ABC, Titchfield Street
Tel: 01563 525234/571906
Seats: 1:602, 2:195, 3:152

Odeon
Seats: 1,750 (8 screens)

Kirkcaldy *Fife*
bfi   Adam Smith Theatre
Bennochy Road
KY1 1ET   ◯ XE
Tel: 01592 412929
Seats: 475

ABC, High Street
Tel: 01592 260143/201520
Seats: 1:546, 2:285 X, 3:235 X

Kirkwall *Orkney*
Phoenix, Junction Road
Tel: 01856 4407
Seats: 500

Largs *Strathclyde*
Barrfields Cinema Pavilion
Tel: 01475 689777
Seats: 470

Lockerbie *Dumfries & Gall*
Rex, Bridge Street   ●
Tel: 01576 202547
Seats: 195

Millport *Strathclyde*
The Cinema (Town Hall)
Clifton Street ●
Tel: 01475 530741
Seats: 250

Motherwell *Lanarkshire*
Civic Theatre, Civic Centre ◯
Tel: 01698 66166
Seats: 395

Newton Stewart *Dumfries & Gall*
Cinema
Tel: 01671 403 333

Oban *Strathclyde*
Highland Theatre, George Street ◯
Tel: 01631 562444
Seats: 1:277, 2:25

Paisley *Strathclyde*
Showcase Cinemas, Phoenix
Business Park, Linwood
Tel: 0141 887 0011
Seats: 3,784 (14 screens)

Perth *Tayside*
Playhouse, Murray Street
Tel: 01738 623126
Seats: 1:590, 2:227, 3:196

Peterhead *Grampian*
Playhouse, Queen Street
Tel: 01779 471052
Seats: 731

Pitlochry *Tayside*
Regal, Athal Road ●
Tel: 01796 2560
Seats: 400

Rothesay, *Isle of Bute*
Winter Gardens, Promenade,
Victoria Street
Tel: 01700 502 487
Seats: 98

St Andrews *Fife*
New Picture House, North Street
Tel: 01334 473509
Seats: 1:739, 2:94

Saltcoats *Strathclyde*
La Scala, Hamilton Street
Tel: 01294 63345/68999
Seats: 1:301, 2:142

Stirling *Central*
Carlton Cinemas, Allanpark
Tel: 01786 474137
Seats: 1:321, 2:287

   MacRobert Arts Centre,
University of Stirling,
FK9 4LA ◯   XE

Tel: 01786 461081
Seats: 495

Stornoway *Western Isles*
Twilights
Seaforth Hotel
James Street   ◯
Tel: 01851 702740

Wishaw *Strathclyde*
Arrow Cinema
Wishaw Retail Park
Tel: 01698 371 000
Seats: 1:242, 2:82,, 3:188, 4:80

WALES

Aberaman *Mid Glam*
Grand Theatre, Cardiff Road ○
Tel: 01685 872310
Seats: 950

Abercwmboi *Mid Glam*
Capitol Screen
Tel: 01443 475766
Seats: 280

Aberdare *Mid Glam*
Coliseum, Mount Pleasant Street ○X
Tel: 01685 882380
Seats: 621

Aberystwyth *Dyfed*
Arts Centre, Penglais, Campus,
University of Wales ○
Tel: 01970 623232
Seats: 125

Commodore, Bath Street
Tel: 01970 612421
Seats: 410

Bala *Gwynedd*
Neuadd Buddig ○
Tel: 01678 520 800
Seats: 372

Bangor *Gwynedd*
Plaza, High Street X
Tel: 01248 362059
Seats: 1:306, 2:163

Theatr Gwynedd, Deiniol Road X
Tel: 01248 351707/351708
Seats: 343

Barry *South Glam*
Theatre Royal, Broad Street
Tel: 01446 735019
Seats: 496

Bethesda *Gwynedd*
Ogwen, High Street ○
Tel: 01286 676335
Seats: 315

Blackwood *Gwent*
Miners' Institute, High Street ○ X
Tel: 01495 227206
Seats: 409

Blaengarw *Mid Glam*
Workmen's Hall, Blaengarw Rd ○X
Tel: 01656 871142
Seats: 250

Blaenavon *Gwent*
Welfare Hall

Brecon *Powys*
Coliseum Film Centre, Wheat Street
Tel: 01874 622501
Seats: 1:164, 2:164

Bridgend *Mid Glamorgan*
Odeon
Seats: 2,026 (9 screens)

Brynamman *Dyfed*
Public Hall, Station Road
Tel: 01269 823232
Seats: 838

Brynmawr *Gwent*
Market Hall, Market Square
Tel: 01495 310576
Seats: 320

Builth Wells *Powys*
Wyeside Arts Centre, Castle Street
Tel: 01982 552555
Seats: 210

Cardiff *South Glam*
 Chapter
Market Road
Canton, CF5 1QE   X
Tel: 01222 396061/399666
Seats: 1:194, 2:68

ABC, Queen Street
Tel: 01222 231715
Seats: 1:588, 2:310, 3:152

Monico, Pantbach Road, Rhiwbina
Tel: 01222 693426
Seats: 1:500, 2:156

Monroe Globe Centre, Albany Road
Tel: 01222 461690
Seats: 216

Odeon, Queen Street
Tel: 01222 237846
Seats: 1:424, 2:635

Odeon, Capitol Shopping Centre
Station Terrace
Tel: 01222 377410
Seats: 1:433, 2:257, 3:220, 4:183, 5:158

St David's Hall, The Hayes ○
Tel: 01222 371236/42611
Seats: 1,600

UCI
Hemingway Road, Atlantic
Wharf, Cardiff Bay
Tel: 01222 498811
Seats: 1:520, 2:353, 3:351, 4:313,
5:267, 6:267, 7:200, 8:200, 9:153,
10:153, 11:147, 12:147

Cardigan *Dyfed*
Theatr Mwldan,
Bath House Rd ○X
Tel: 01239 621200
Seats: 210

Carmarthen *Dyfed*
Lyric, King's Street ○
Tel: 01267 612200
Seats: 800

Cross Hands *Dyfed*
Public Hall
Tel: 01269 844441

Cwmaman *Mid Glam*
Public Hall, Alice Place ○
Tel: 01685 876003
Seats: 340

Cwmbran *Gwent*
Scene, The Mall
Tel: 016333 366621
Seats: 1:115, 2:78, 3:130

Ferndale *Mid Glam*
Cinema, The Hall, High Street ○

Fishguard *Dyfed*
Theatr Gwaun, West Street
Tel: 01348 873421/874051
Seats: 252

Gilfach Goch *Mid Glam*
Workmen's Hall, Glenarvon Terrace
Tel: 01443 86231
Seats: 400

Harlech *Gwynedd*
Theatr Ardudwy Coleg Harlech ○
Tel: 01766 780667
Seats: 266

Llandudno *Gwynedd*
Palladium, Gloddaeth Street
Tel: 01492 876244
Seats: 355

Llanelli *Dyfed*
Entertainment Centre, Station Rd
Tel: 01554 774057/752659
Seats: 1:516, 2:310, 3:122

Maesteg *Mid Glam*
Town Hall Cinema, Talbot Street
Tel: 01656 733269
Seats: 170

Merthyr Tydfil *Mid Glam*
Castle
Tel: 01685 386669
Seats: 1:98, 2:198

Milford Haven *Dyfed*
Torch Theatre, St Peters Road
Tel: 01646 695267
Seats: 297

Mold *Clwyd*
Theatr Clwyd, County Civic
Centre, CH7 1YA   X
Tel: 01352 756331/755114
Seats: 1:530, 2:129

Monmouth *Gwent*
Savoy, Church Street
Tel: 01600 772467
Seats: 450

Newport *Gwent*
ABC, Bridge Street

Tel: 01633 254326
Seats: 1:572, 2:190, 3:126

**Virgin Cinemas, Newport, Retail
Park Seven Styles Avenue**
Tel: 0541 550516
Seats: 1:199, 2:178, 3:123, 4:187,
5:267, 6:405, 7:458, 8:287, 9:180,
10:123, 11:211, 12:156, 13:77

Newtown *Powys*
Regent, Broad Street
Tel: 01686 625917
Seats: 210

Pontypool *Gwent*
Scala, Osborne Road
Tel: 0149 575 6038
Seats: 197

Pontypridd *Mid Glam*
Muni Screen, Gelliwastad Rd○XE
Tel: 01443 485934
Seats: 400

Port Talbot *West Glam*
Apollo, Hollywood Park
Seat: 1,000 (6 screens)
(scheduled to open December 1998)

Plaza Theatre, Talbot Road
Tel: 01639 882856
Seats: 1:725, 2:182, 3:103, 4:380

Porthcawl *Mid Glam*
Grand Pavilion ● ○
Tel: 01656 786996
Seats: 500

Portmadoc *Gwynedd*
Coliseum, Avenue Road
Tel: 01766 512108
Seats: 582

Prestatyn *Clwyd*
Scala, High Street
Tel: 01745 854365
Seats: 314

Pwllheli *Gwynedd*
Odeon, Butlin's Starcoast World
Tel: 01758 612112
Seats: 200

Town Hall Cinema ○
Tel: 01758 613371
Seats: 450

Resolven *West Glam*
Welfare Hall Cinema
Tel: 01269 592395
Seats: 541

Rhyl *Clwyd*
**Apollo, Children's Village
West Parade**
Tel: 01745 360066
Seats: 1:208, 2:208, 3:119, 4:109, 5:109

Swansea *West Glam*
**Taliesin Arts Centre,
University College,**

Singleton Park, SA2 8PZ   XE
Tel: 01792 296883/295491
Seats: 328

UCI, Quay Parade, Parc Tawe
Tel: 01792 645005
Seats: 1:180, 2:188, 3:188, 4:194,
5:276, 6:276, 7:194, 8:188, 9:188,
10:180

Tenby *Dyfed*
Royal Playhouse, White Lion Street
Tel: 01834 844809
Seats: 479

Treorchy *Mid Glam*
**Parc and Dare Theatre
Station Road**
Tel: 01443 773112
Seats: 794

Tywyn *Gwynedd*
The Cinema, Corbett Square   X
Tel: 01654 710260
Seats: 368

Welshpool *Powys*
Pola, Berriew Street
Tel: 01938 555715
Seats: 1:150, 2:40

Wrexham *Clwyd*
**Odeon Plas Coch Retail Park, Plas
Coch Road**
Tel: 01978 310656
Seats: 1:354, 2:191, 3:148, 4:254,
5:112, 6:112, 7:112

Ystradgynlais *Powys*
**Miners' Welfare and Community
Hall, Brecon Road** ○   X
Tel: 01639 843163
Seats: 345

NORTHERN IRELAND

Antrim *Antrim*
Cineplex, Fountain Hill
Tel: 01849 461 111
Seats: 1:312, 2:232, 3:132, 4:112

Armagh *Armagh*
City Film House
Tel: 01861 511033
Four screens

Ballymena *Antrim*
State, Ballymoney Road
Tel: 01266 652306
Seats: 1:215, 2:166

Banbridge *Down*
Iveagh, Huntly Road
Tel: 01820 662423
Seats: 930

Bangor *Down*
**Cineplex, Valentines Road
Castle Park**
Tel: 01247 454729/465007
Seats: 1:287, 2:196, 3:163, 4:112

Belfast *Antrim*
**Cineworld, Kennedy Centre
Falls Road**   E
Tel: 01232 600988
Seats: 1:296, 2:190, 3:182, 4:178,
5:165

Curzon, Ormeau Road
Tel: 01232 491071
Seats: 1:453, 2:360, 3:200, 4:104, 5:90

**Movie House, Yorkgate Shopping
Centre**   X
Tel: 01232 755000
Seats: 1:365, 2:266, 3:248, 4:183,
5:174, 6:97, 7:97, 8:341, 9:75,
10:68, 11:68, 12:85, 13:85, 14:481

 **Queen's Film Theatre,
25 College Gardens,
BT9 6BS**   X
Tel: 01232 244857/667687
Seats: 1:250, 2:150

Virgin Cinemas, Dublin Road
Tel: 01232 245700
Seats: 1:436, 2:354, 3:262 **X**, 4:264
X, 5:252, 6:272, 7:187 **X**, 8:187 **X**,
9:169, 10:118 **X**

The Strand, Hollywood Road
Tel: 01232 673500
Seats: 1:276, 2:196, 3:90, 4:80

Coleraine *Londonderry*
Jet Centre, Dunhill Road
Tel: 01265 58011
Seats: 1:286, 2:193, 3:152, 4:124

Cookstown *Tyrone*
Ritz, Burn Road

Tel: 016487 65182
Seats: 1:192, 2:128

Dungiven *Londonderry*
St Canice's Hall, Main Street

Enniskillen *Fermanagh*
Ardhowen Theatre
Dublin Road ◯
Tel: 01365 325440
Seats: 295

Castle Centre, Race View
Factory Road
Tel: 01365 324172
Seats: 1:302, 2:193, 3:130

Glengormley *Antrim*
Movie House, Glenville Road
Tel: 01232 833424
Seats: 1:309, 2:243, 3:117, 4:110,
5:76, 6:51

Larne *Antrim*
Regal, Curran Road
Tel: 01574 277711
Seats: 1:300, 2:220, 3:120, 4:120

Lisburn *Antrim*
Omniplex, Governors Road
Tel: 01846 663664
Seats: 1:500, 2:230, 3:160, 4:120,
5:180, 6:230, 7:150, 8:170, 9:90,
10:70, 11:70, 12:90, 13:110, 14:160

Londonderry *Londonderry*
Orchard, Orchard Street
Tel: 01504 267789
Seats: 132, 700 ◯

Strand, Strand Road
Tel: 01504 373939
Seats: 1:316, 2:252, 3:220, 4:220,
5:132, 6:126, 7:96

Lurgan *Armagh*
Centre Point Cinemas
Multi-Leisure Complex
Portadown Road
Tel: 01762 324667/321997
Seats: 1:304, 2:254, 3:160, 4:110

Maghera *Londonderry*
Movie House, St Lurach's Road
Tel: 01648 43872
Seats: 1:221, 2:117, 3:95

Magherafelt *Londonderry*
Queen Street
Tel: 01648 33172
Seats: 1:230, 2:75

Newry *Down*
Savoy 2, Merchant's Quay
Tel: 01693 67549
Seats: 1:197, 2:58

Omagh *Tyrone*
Studios 1-4, Quin Road
Seats: 1:800, 2:144, 3:300, 4:119

Portrush *Antrim*
Playhouse, Mainstreet
Tel: 01265 823917
Seats: 1:299, 2:65

COURSES

Film and TV study courses generally fall into two categories: academic and practical. Listed here are a selection of educational establishments which offer film and television as part of a course or courses. Where a course is mainly practical, this is indicated with a ●next to the course title. In the remaining courses, the emphasis is usually on theoretical study; some of these courses include a minor practical component as described. A wider range of courses and more detailed information can be found in two other BFI publications, **Media Courses UK '99** and **Media & Multimedia Short Courses**

AFECT (Advancement of Film Education Charitable Trust)
4 Stanley Buildings
Pancras Road
London NW1 2TD
Tel: 0171 837 5473
Patron: Mike Leigh
Makes professional-level, practical film education available on a part-time basis to those who may have neither the means nor the time to attend a full-time film course
● **Practical Part-time 16mm Film-making Course**
Two year course integrating learning with production. Bias is traditional narrative; despite limitations of scale, students are enabled to realise personal, artistic, social and cultural expression in this medium. Term 1: Shoot 35mm stills storyboard. Instruction/practicals camera. Interior lit sequence. Editing. Term 2: Script/shoot/edit group mute film with individual sequences, rotating crewing jobs. Term 3: Individual shot-mute three minute films. Term 4: Obtaining and adding sound. Dubbing. Sync-sound intro. Term 5: Individual six minute sync-sound film each. Term 6: Completing these. Year 3: Advanced projects; semi-independent productions

The American College in London
Department of Video Production
110 Marylebone High Street
London W1M 3DB
Tel: 0171 486 1772
Fax: 0171 935 8144
● **BA Video Production Associate Degree Video Production**
BA four years, associate degree two years. Four year course has option to concentrate on commercial, documentary or music video

Barking College
Dagenham Road
Romford RM7 0XU
Tel: 01708 766841
Fax: 01708 731067

● **B/TEC National Diploma Media**
This two year broad based media course covers video, radio and sound recording, print and journalism. Facilities include: television studio; portable video; video editing suites; sound recording studio and DTP equipment. There is a practical and vocational emphasis and students are prepared for either a career in the media industries, or for entry to higher education. Applications from mature students are welcome
RSA - Lens Based Media (Video and Photography)
This one year course provides an introduction to media work and has a 50 per cent practical 'hands on' approach to video production and photography. It is equivalent to a GNVQ Intermediate qualification - but has a more specialist emphasis. BTEC National Diploma courses in either Media or Photography would be a suitable progression
Access to Media
This ne year evening class - two evenings a week - prepares students for Higher Education in Media Studies/Production and includes some practical work

University of Bath
School of Modern Languages and International Studies
Claverton Down
Bath BA2 7AY
Tel: 01225 826482
Fax: 01225 826099
BA (Hons) Modern Languages and European Studies
First year lectures and seminars on the language and theory of film, whilst the second and fourth years offer a wide range of options on French films between the wars, the films of the Nouvelle Vague, film and television in German speaking countries and film in Italy and Russia from the 1930's through to 'glasnost'. There is a final year option in European cinema in the

70s and 80s
MPhil and PhD
Part-time or full-time research degrees in French and Russian cinema

Birkbeck College, University of London
Department of Media Studies
Centre for Extra-Mural Studies
26 Russell Square
London WC1B 5DQ
Tel: 0171 631 6667/6639
Fax: 0171 631 6683
BA Humanities
Four year part-time interdisciplinary course, including 4-6 media modules. First year offers a broad introduction to film, television or journalism studies. Second and third years build on this and develop work on theories of genre, audience, narrative and realism, communications and press history as appropriate. Fourth year is a specialist option, currently television drama or recent developments in European cinema
Contact: Penny Lazenby
Part-time Media Studies Courses
The above modules are part of a large programme of part-time courses in film, television, journalism and in areas of media practice such as screenwriting, freelance journalism, video, radio, leading to the Certificate/Diploma in Media Studies or in Media Practice. The Sessional Diploma in Media Education is designed for teachers who are having to teach media studies without having studied it as part of their degree

University of Birmingham
Department of French Studies
Birmingham B15 2TT
Tel: 0121 414 5965
Fax: 0121 414 5966
BA (Hons) French Studies
Four year course which includes options on French cinema (Year 1);

television genres; the practice of transposing works of fiction to the screen and Auteur Theory and the Star as Sign (Year 4). Also options on French television: Reading television (Year 1), packaging television programmes (Year 2). Year 3 is spent at a French university. Students are encouraged to follow courses in cinema and/or television as preparation for Year 4 studies

Bournemouth University

School of Media Arts and Communication, Poole House Talbot Campus, Fern Barrow Poole, Dorset BH12 5BB
Tel: 01202 524111
Fax: 01202 595530
email: srose@bournemouth.ac.uk

● **BA (Hons) Media Production**
A three year course covering the academic, practical, aesthetic, technical and professional aspects of work in the media. The course is divided equally between practical and theoretical studies. Students work in audio, video and interactive multimedia leading to a major project in Year 3 produced as an interactive CD Rom. In addition, students complete a piece of individual written research.

● **BA (Hons) Scriptwriting for Film and Television**
A comprehensive three year programme, taught by practising scriptwriters, comprising theoretical and practical work specifically designed to meet the needs of new writers in the industry. All graduates will have a thorough knowledge of the industry and a portfolio of work developed to a very high standard. Applications from mature students are encouraged.

● **BA (Hons) Television and Video Production**
A three year degree course which enables students to work with broadcast-quality equipment to produce video and television programmes. There are supporting courses dealing with media, communication and film theories, professional studies, and the history of cinema and broadcasting.

● **PGDip/MA in Television and Video Production**
A one year full-time course for graduates or proven practitioners who seek ultimately to become directors or producers. Using Betacam SP, the course centres on practical productions on location and in the studio, supported by theoretical and professional studies.

The MA element is a continuation period of 3-6 months part time study and includes a dissertation on a selected research topic.

● **MA in Music Design for Film and Television**
This course centres on composing for film and television. It offers tuition in and experience of the practical and theoretical aspects of combining music with moving pictures. The syllabus includes: Composition; Film Music Analysis; Film Theory; Production Theory; Law of Contract & Copyright; and Technology. Each composer spends the entire year based at his/her dedicated workstation which is designed to produce music at the highest broadcast standard

Bournemouth & Poole College of Art & Design

Faculty of Visual Communication Wallisdown, Poole, Dorset BH12 5HH
Tel: 01202 363281
Fax: 01202 537729

● **B/TEC National Diploma Audio Visual Design**
Contact: Ted May
Two year vocational course centred around the disciplines of video and audio production. These practical studies are supported by elements of design studies, drama, music, scriptwriting, animation, Contextual Studies, Business and Professional Studies. The course is recognised by BKSTS

● **B/TEC HND Design (Film and Television)**
Contact: Neil Grant
Tel: 01202 538204
Fax: 01202 537729
An intensive practical two year course based on the production process with an emphasis on drama and documentary and including critical and theoretical studies. The majority of work is student originated, with an equal emphasis on film and tape processes. The course has substantial input from, and contact with, working professionals within the industry. The course is accredited by BECTU and recognised by BKSTS

GCE A-level Critical & Historical Studies Programme
Contact: Emma Hunt
Tel: 01202 363305
Fax: 01202 537729
It is hoped that the Faculty's A-level programme will be extended so that students will be able to select A-level Film Studies, Media Studies, Photography and History of Art and

Design, for study on either a part-time or full-time basis

University of Bradford

Department of Electronic Imaging and Media Communications Richmond Road Bradford BD7 1DP
Tel: 01274 234011
Fax: 01274 233727
email: P.E.Dale@bradford.ac.uk

● **BSc Electronic Imaging and Media Communications**
A three year, full-time course, developed by a group of staff with various specialities – electronics, art and design, digital music, sociology, photography and television. The breadth of the course is unusual and offers real advantages in preparing students for a career in the media. A Foundation Year is available

● **BSc Media Technology and Production**
In this course high calibre candidates will be able to develop full media products in realistic environments

University of Bristol

Department of Drama Cantocks Close Woodland Road Bristol BS8 1UP
Tel: 0117 9303030
Fax: 0117 9288251
BA Drama
Three year course with theatre, film and TV options. Alongside theatre-based Units, critical and theoretical approaches to film and TV are part of the core syllabus in year 1. Additional critical and practical options are offered in years 2 and 3, at which point students may choose to specialise in film and TV studies. Practical work is group-based, extends and enriches critical study in a range of forms in fiction and non-fiction, and results in the production or original work for public exhibition. Theoretical work may be developed through individual dissertations as well as a range of seminar courses

● **MA/PgD in Film and Television Production**
This one-year course was the first of its kind in a British university and has produced numerous distinguished practitioners working internationally. It offers a broad grounding in practical skills in film and television production, regular consultation with professional practitioners, and a collective forum for the development of critical thinking and creative practice. Based around a core of group-based

practical work, the Diploma offers modular options in a range of practical and critical disciplines. The MA extends this with group-based production for public festival entry and/or broadcast, and individual analysis. Production platforms include broadcast-standard video and 16mm film, as well as digital media. The course enjoys widespread support from film and television organisations and leading practitioners

Brunel University
Department of Human Sciences
Uxbridge, Middx UB8 3PH
Tel: 01895 274000
Fax: 01895 232806
BSc Media and Communications Studies
Four year interdisciplinary course which aims to give an understanding of the social, intellectual and practical dimensions of the communications media, with particular reference to the new information technologies
MA Communications and Technology
This course offers detailed study of the new communications and information technologies

Canterbury Christ Church College
Dept of Radio, Film and Television
North Holmes Road
Canterbury CT1 1QU
Tel: 01227 767700
Fax: 01227 470442
email: P.Simpson@cant.ac.uk
● **BA or BSc (Hons) Radio, Film & Television with one other subject**
Contact: Philip Simpson
RFTV is one half of a three year joint honours degree and may be combined with Art, American Studies, English, Geography, History, Mathematics, Sport Science, Music, Religious Studies and Science. The course introduces students to an understanding and appreciation of radio, film and television as media of communication and creative expression, stressing their relevance to the individual and to society. It also offers an opportunity to develop and practice production skills in each of the three media
● **MA Media Production**
Contact: Dickon Reed
A one year taught MA which concentrates on production in radio, film and television. Part I of the course introduces relevant production skills; in Part II members will

fulfil a measurable major role in a production project. Course members with practical experience can update their skills and concentrate on one medium in Part I. All course members attend theory seminars through the course. Assessment will be based on the major piece of practical work and an extended essay

University of Cardiff
Centre of Film Production Studies
Bute Building
King Edward VII Avenue
Cardiff CF1 3NB
Tel: 01222 874786
Fax: 01222 238832
Contact: Steve Gough
● **Courses in Fiction, Documentary and Screen writing**
The centre's fiction, documentary and screenwriting courses are currently being reviewed, with a view to introducing a new, revised scheme in all production areas for 1998

The City College Manchester
School of Art and Design,
Manchester M23 0DD
Tel: 0161 957 1749
Fax: 0161 949 3854
● **B/TEC National Diploma Audio-Visual Design**
Multi-disciplinary course working between video, film animation, photography, graphics, sound and Digital Imaging. Assessment is continuous, based on practical projects linked to theoretical studies

Coventry University
Coventry School of Art and Design
Priory Street
Coventry CV1 5FB
Tel: 01203 838690
Fax: 01203 838667
BA (Hons) Communication Studies
Three year course which includes specialities in Cultural and Media Studies, and in Communication Management, built around a core of studies in communication, culture and media, with a range of other options from which students select. European exchange and work placement programmes are included; also options in journalism, photography and video. Projects enable students to combine theoretical and practical work according to their particular interests
Postgraduate Programme Communication, Culture and Media: MA or PgC/PgD

One full-time, two year part-time. The programme is a modular scheme with core theory and research elements, specialist options, and a selection of electives which include film theory and psychoanalysis, journalism, media policy, television culture and politics. Students may specialise in applied communications, or cultural policy, or media and culture for the MA qualification

De Montfort University Bedford
Polhill Avenue
Bedford MK41 9EA
Tel: 01234 351671
Fax: 01234 217738
Modular BA/BSc Programme (Arts Pathway) Film Studies
First, Second and Third level modules studying film as a cultural product and a social practice through lectures, screenings, seminars. Level 1: Intro to Film Studies; Level 2: double module - early to contemporary mainstream cinema; Level 3: single module - Art Cinema plus optional research module
Television and Drama
Two third level modules. 1: forms of television drama and study of key practitioners; 2: television as a cultural practice

De Montfort University Leicester
School of Arts
The Gateway
Leicester LE1 9BH
Tel: 0116 255 1551 or 0116 257 8391
Fax: 0116 257 7199
Dr Paul Wells/Tim O'Sullivan
BA (Hons) Media Studies (Single, Joint or Combined Honours Degrees)
As a Single Honours degree, Media Studies offers a range of courses which focus specifically on Film, Television/Video, Photography and Media institutions. It offers courses in both theoretical and practical work which provide students with the opportunity to develop their skills and learning through detailed analysis of media texts, through understanding the social and political processes of media industries and institutions and through practical work in video, photography, radio and journalism. With Joint Honours, it is possible to take Media Studies in conjunction with one other arts discipline; for Combined Honours, with two other disciplines

University of Derby
School of Art and Design
Green Lane,

Derby DE1 1RX
Tel: 01332 622282
Fax: 01332 622296
email: R.Fowler@derby.ac.uk
Contact: Tony Hill
● **BA (Hons) Photography & Time Based Media**
A three year full-time modular degree course. Practical work forms 75 per cent of the course work and academic studies 25 per cent. The course covers a wide spectrum of approaches and techniques and generates work in a healthy intermix of genres. The development of creative ideas alongside inventive practical experience together with a critical awareness of contemporary practice are the essential ingredients. Pathways may be followed which favour one or more of the media including digital media. The department is well-equipped with a wide range of media production facilities including new digital technology. Time Based Media facilities include 16mm film cameras and edit suites, video cameras, two and three machine video editing, sound recording, mixing and dubbing, digital post-production, animation rostrum, studio and cinema

Dewsbury College
Batley School of Art & Design
Wheelwright Campus
Birkdale Road
Dewsbury, West Yorks WF13 4HQ
Tel: 01924 451649
Fax: 01924 469491
● **B/TEC National Diploma Design (Audio-Visual)**
Two year full-time course developing creative and technical skills in audio-visual and video production, sound recording, photography and AV graphics. Supporting studies include film and business studies
BA (Hons) Moving Image Design
This course provides an opportunity to study one of the most rapidly developing areas of design. Using computer technology, the course combines live action video techniques with 3D animation to produce time-based imagery for the broadcast media, advertising and publicity. The course demands an imaginative approach with an understanding of both 2D and 3D design

University of East Anglia
School of English and American Studies,
Norwich NR4 7TJ
Tel: 01603 592283
Fax: 01603 507728

BA (Hons) Film and English Studies
A Joint Major programme which integrates Film and Television history and theory with work on English literature, history and cultural studies; the film work deals mainly with Hollywood, but also with British cinema. Course includes instruction in film and video production, and the option of submitting a practical project. All students submit an independent dissertation on a film or television topic.
BA (Hons) Film and American Studies
A four year Joint Major programme which integrates Film and Television history and theory with work on American literature, history, cultural studies and politics. Course includes instruction in film and video production, and the option of submitting a practical project. All students spend a year at a University in the USA, and submit an independent dissertation on a film or television topic.
BA (Hons) Modular System
Students admitted to the University to major in other subjects including Literature, Drama, American Studies etc have the option of taking one or more units in film and television study: together, these may comprise up to one third of the degree work. No practical element.
MA Film Studies
One year full-time taught programme. MA is awarded 50 per cent on course work and 50 per cent on dissertation. Within the School's modular system, it is possible to replace one or two of the four film seminars with others chosen from a range of topics in literary theory, creative writing, American studies and cultural studies. The film seminars deal with early cinema, British film history, film and narrative theory, screen costume and theories of the image, and research resources and methodology. Dissertation topics are freely chosen and may deal with television as well as cinema.
MA Film Studies: Film Archive option
One year full-time taught programme, run in conjunction with the East Anglian Film Archive (located in the University). Students take two of the MA film seminars, plus two more that deal with the practical and administrative aspects of film archive work. Course includes visits to other archives, and a one-month

placement at a chosen archive in Britain or overseas. Assessment is based on two essays, a video production, a placement report, and an independent dissertation (counting 50 per cent)
MPhil and PhD
Students are accepted for research degrees. Areas of special expertise include early cinema, British film history, television history, gender and cinema, classical and contemporary Hollywood, and screen costume

University of East London
Faculty of Design Built Environment
Greengate Street
London E14 0BG
Tel: 0181 590 7722
Fax: 0181 849 3694
● **BA (Hons) Fine Art, Time Based Art**
During the first year students can experiment with each of the disciplines that are available but can also specialise in film, video and video animation throughout the three years
BA Visual Theories: Film Studies
A specialist pathway within the university's modular structure and the range of options is generally extensive with theoretical work on the history of cinema and avant-garde film. Complementary options would include art and psycho-analysis, semiotics, and the history and theory of photography

Department of Cultural Studies
Longbridge Road
Dagenham
Essex RM8 2AS
Tel: 0181 590 7722 x 2741
Fax: 0181 849 3598
BA (Hons) in Media Studies
Media Studies is offered as a single honours degree or as a major, minor or joint degree in combination with other subjects (eg Cultural Studies; History; Literature and Women's Studies). All students are required to take a set of core history and theory units, and will select from approved optional units in Media Studies and related subject areas. Single honours students will be required to take production units, amounting to one-third of their course in the areas of video, audio, photography, graphics and multimedia. They will also take a working-in-the-media unit which will include short placements and visiting lecturers from culture industries. Production and placement units are not available to combined honours students

Department of Innovation Studies
Maryland House,
Manbey Park Road
London E15 1EY
Tel: 0181 590 7722 x 4216
Fax: 0181 849 3677
BSc (Hons) New Technology:
Media & Communication
This degree examines the media
industries in the context of a study
of technological change in society.
Covers the social relations of
technology, the film, recording,
newspaper, television, cable and
satellite industries. Students opt for
a practical path in either computer
graphics or video production. The
degree builds a combination of
analytical, critical, writing, research
and production skills

Edinburgh College of Art
School of Visual Communication
Lauriston Place
Edinburgh EH3 9DF
Tel: 0131 221 6138
Fax: 0131 221 6100
These courses are strongly based on
practical production work and run
for three years. Most applicants have
done either a foundation course in
art and design or an HNC in Media
Production or similar. In the film/
television option, students will
generally combine individual
projects with participation in group
productions. All kinds of work can
be tackled – drama, documentary
and experimental. Facilities include
non-linear digital editing
● **BA (Hons) Film and Television**
The course runs for three years and
most applicants have done either a
foundation course in art and design,
or a further education course in
video/audio-visual. Film/television
students will generally combine
individual projects with participation in group projects. All kinds of
work can be tackled – drama,
documentary and experimental. The
course includes possibilities of
cross-disciplinary projects with other
departments in the school -
animation, illustration, photography,
and graphic design. All students are
also encouraged to use the school's
computer workshop
● **Masters Degree**
A small number of postgraduates can
be accepted, studying either for a
diploma (three terms) or a masters
degree (four terms). In both cases
there is no formal taught course –
the programme is tailored to the
practical production proposals of the
individual student. Postgraduates
must already have appropriate skills
and experience to use the resources

available. The masters degree is
awarded on the strength of the
practical work produced

University of Exeter
School of English,
Queen's Building,
The Queen's Drive
Exeter EX4 4QH
Tel: 01392 264263
Fax: 01392 264361
BA (Hons) English Studies
Students can take up to a half of
their degree in Film Studies,
including courses on British
Cinema, Hollywood and Europe and
an introduction to Key Issues in Film
Studies. No practical component.
● **MA Programme in the History**
of Cinema and Popular Culture
The core modules in this programme
concentrate on key moments in
cinema history and the relationship
between cinema and 19th Century
optical media and popular entertainment. Optical modules cover a
variety of theoretical and historical
topics including Cult Movies and
Postcolonial Cinema.
Mphil and PhD
Applications for postgraduate study
in British Cinema, Early and Pre
Cinema History, and Cinema and
Cultural Theory will be particularly
welcome

School of Modern Languages
The Queen's Building
Queen's Drive
Exeter EX4 4QH
Tel: 01392 264231
Fax: 01392 264377
BA (Single, Combined Hons and
Modular) Italian
Italian cinema option for second/
third- and final-year students. In
general, Neo-realism to the present
day. Films are studied for their
intrinsic merit, as commercial
products, and as part of Twentieth-
century Italian culture.
BA (Single and Combined Hons)
Spanish
Option in Spanish cinema. Spanish
films 1963 to 1990. Selection
covering social, literary and war
themes, film censorship and its
circumvention

Farnborough College of Technology
Media and Visual Arts
Boundary Road, Farnborough
Hants GU14 6SB
Tel: 01252 407270
Fax: 01252 407271
email: A.Harding@Farnct.ac.uk
● **HND Media Technology and**
Business

Two year full-time course to study
media production techniques with
business studies. Course includes
television and video production,
video and audio systems, radio,
journalism and finance in the media
● **HND Design Technology**
(Multimedia, Video Graphics &
Animation)
Provides training in television and
video production, animation and
computerised video graphics. In the
first and second years, all students
undertake the following modules:
visual studies, television and video
production, animation and graphics,
historical and contextual studies
and business management. In the
second year, students select three
options from the following: video
systems, documentary and drama
production, advertising copywriting,
photography, marketing and the
media, journalism, desktop
publishing and radio production
● **BSc (Hons) Media (Production)**
Technology
This popular degree adopts a bimedia approach studying both
Television and Radio. Students will
be expected to develop and
demonstrate technical skills, as well
as an understanding of the appropriate theories and concepts. This will
be achieved by practical units in
television/video and radio, as well as
theoretical ones such as audio-
visual systems, television and film,
radio in society etc. Optional units
enable the students to create a
vocational or academic pathway of
their choice
● **Higher National Diploma in**
Media Technology (Broadcast
Engineering)
This new programme has been
created at the specific request of the
broadcast industries. As a result of
the digital revolution there is a
dramatic increase in the number of
broadcasters of both television and
radio. Through this programme of
study students learn to become
broadcast engineers using and
maintaining a range of equipment
necessary to provide television and
radio broadcasts.

University of Glasgow
Department of Theatre,
Film and Television Studies
University of Glasgow
Glasgow G12 8QQ
Tel: 0141 330 5162
Fax: 0141 330 4142
email: tfts.office@arts.gla.ac.uk
MA Joint Honours Film and
Television Studies

Four year undergraduate course. Film/Television Studies represents 50 per cent of an Honours degree or 30 per cent of a non-Honours degree. Year 1 is concerned with Film and TV as 'languages', and with the institutional, industrial and technological contexts of cinema and television. Year 2 is structured under two headings; Film and Television: Theories and Methods and Film and Television: National and Cultural Identities. Years 3 and 4 consist of a range of Honours optional courses, seven to be taken over two years in addition to a dissertation. There is also a compulsory practical course, involving either the production of a video, a contractual work placement or an applied research project

Department of French
Glasgow G12 8QQ
Tel: 0141 339 8855
Contact: Jim Steel, Ramona Fotiade
MA (Hons) French
Study of French cinema is a one year special subject comprising one two-hour seminar per fortnight and weekly screenings

Glasgow Caledonian University
Department of Language & Media
Cowcaddens Road
Glasgow G4 0BA
Tel: 0141 331 3255
Fax: 0141 331 3264
email: dhu@gcal.ac.uk
BA Communication and Mass Media
Four year course (unclassified and honours) examining the place of mass communication in contemporary society. Includes practical studies in print, television, advertising and public relations

Goldsmiths College
University of London
Lewisham Way
London SE14 6NW
Tel: 0171 919 7171
Fax: 0171 919 7509
email: admissions@gold.ac.uk
www.http://www.goldsmiths.ac.uk
● **BA Media and Communications**
This course brings together theoretical analyses in social sciences and cultural studies with practical work in creative writing (fiction), electronic graphics and animation, photography, radio, script writing or television (video and film) production. The practical element constitutes 50 per cent of the total degree course. The theoretical element includes media history and sociology, textual and cultural

studies, anthropology and psychology and media management
BA Anthropology and Communication Studies
Half of this course constitutes Communication Studies.The course is mainly theoretical but does include two short practical courses of ten weeks in length in two of the practice areas. These include television, videographics and animation, radio, print journalism, photography, creative writing and script writing. The theory component is concerned with media history, sociology, psychology, textual and cultural studies
BA Communication Studies/ Sociology
Communication Studies constitutes half this course and is split into theoretical studies and two ten week practical courses. Practical options include television, videographics and animation, radio, print journalism, photography, creative writing and script writing. The theory component is concerned with psychology, media sociology, cultural studies, semiotics and media history
● **MA Image and Communication (Photography or Electronic Graphics)**
One year full-time course combines theory and practice, specialising in either photography or electronic graphics. Practical workshops cover medium and large format cameras, flash, colour printing, lighting, computer and video graphics, design, desktop publishing, animation, animatics, two and three dimensional computer animation. Assessment by course work, practical production and viva voce
MA Television (TV Drama or Documentary)
One year full-time course specialising in either documentary or drama modes, taught by practical and theoretical sessions. Course covers script writing, programme planning, camera work, studio and location work, interviewing, sound and post production. Assessment is by course work, practical production and viva voce
MA Media and Communication Studies
This course offers an interdisciplinary approach as well as the opportunity to specialise in media and communications. The course is based around a series of compulsory courses and options drawing on theoretical frameworks from cultural studies, political economy,

sociology, anthropology, and psychology to develop a critical understanding of the role of the media and communications industries in contemporary culture. Assessment is by course work, written examinations and dissertation
MA Journalism
The course is essentially a practical introduction to journalism as a multi-media skill with the emphasis on print journalism. In addition, you will take a subsidiary course dealing, in the first term, with the Law and ethical issues and in the second term, with the history and changing structure of the media industry. There is also a course related to wide theoretical issues in the study of media and culture
Music, Image and the Word (an introduction to composing for Film and Television)
9 week course 13 January 1999 or 28 April 1999. Aimed at composers and writers/directors developing a practical and aesthetic understanding of how music image and narrative work together

Havering College of Further & Higher Education
Department of Creative Studies
Ardleigh Green Road,
Hornchurch
Essex RM11 2LL
Tel: 01708 455011
Fax: 01708 477961
● **HNC Media Technology (part time)**
National Diploma in Media and Communications (full time)
GNVQ intermediate (level two) in Media and Communications (full time)
Areas covered include Film and Video, Audio; DTP; Graphic Design; Multimedia applications; Marketing and Advertising; Journalism

University of Hertfordshire
Watford Campus, Wall Hall, Aldenham, Watford, Herts WD2 8AT
Tel: 01707 284000
Fax: 01702 285616
BA (Hons) Humanities
Full- or part-time degree. Within the Historical Studies major/minor and single honours there is a second year option, Film and History, which examines the inter-war period through film and focuses on the historian's use of film

The Hull School of Art and Design
University of Lincolnshire &

Humberside
Queens Gardens
Kingston-upon-Hull HU1 3DQ
Tel: 01482 440550
Fax: 01482 462101
Contact: Rob Gawthorp (Head of Art)
All courses in The Hull School of
Art & Design are 80 per cent
practice and 20 per cent theory and
are provided with extensive facilities
including: SVHS, DV, DAT, Digital,
Analogue, 16mm; production and
post production including non linear
editing; and current professional
programmes for Macs and PCs.
● **BA (Hons) Graphic Design**
Students may specialise in anima-
tion/film/video or graphic design or
illustration. After an introductory
period, animation and film primers
are located in Terms 1 and 2. Full
specialisation begins at the start of
Term 3. The course is essentially
practical, with a strong theoretical/
critical element and a programme of
visiting animators and film-makers.
The course has strong components
in scriptwriting, cinematography,
direction, animation and production
competence. Films are initiated by
students in narrative fiction, animation, and include sponsored public
information film and videotape
productions, computer animation and
much experimental work.
● **BA (Hons) Fine Art**
The Fine Art Course is based on
contemporary practice in all its
diversity. The Time Based Media
section includes Film, video, sound,
performance, digital media,
installation site specific and live
work. The course is specialised (ie
non-modular) and work is student
initiated. Time based work frequently
crosses other disciplines including
photography, sculpture, painting and
printmaking. Supported by a
programme of visiting artists/lectures,
screening and critical theory.
● **BA (Hons) Phonic Art**
This innovative course commences
in September 1998 and runs in
parallel to the Fine Art course. The
course is specialised (ie non-
modular) and work is student
initiated. Phonic Art is basically a
sound-based Fine Art Course where
the thinking is essentially derived
from the aural and is concerned with
the converging areas of contempo
rary art practice and experimental
music. The work produced may
include Film, Video, Performance,
Sound Sculpture and visual scores in
addition to work that is purely audio.
● **The School of Media**
BA (Hons) Documentary Production

A mixed mode honours degree, 70
per cent practical, 30 per cent
theoretical, with three production
pathways: still photography/text;
sound/radio; and video. Students
follow a multi-disciplinary first year
and then specialise in one pathway
for the following two years. The
course provides the context within
which individuals acquire the
knowledge and skills pertinent to
communicating in documentary
forms. It seeks to produce graduates
able to operate professionally in a
variety of contexts producing a
range of work within a strong
conceptual base
● **BA (Hons) European Audio-**
Visual Production
The degree will equip graduates with
producers skills for the developing
European market, including
language skills. There will be stress
on generating creative and successful ideas and having the expertise
required to get them on screen. 80
per cent practice, 20 per cent theory.
Teaching is largely project based.
The third year is spent abroad at a
university in France or Spain.
Graduates emerge with a showreel
and business plan suitable for the
European market place. Studio and
single camera video operation is
supported by projects focusing on
research and development,
programme proposals, sound,
scripting and storyboarding
● **BA (Hons) Media Production**
This course offers students technical
pathways in combinations of sound,
video, photography, graphic design
and typography and is 80 per cent
practical. It has three major strands
in fiction, journalism and alternative
practices and concentrates on
'writing the image'. It will give
students experience of working in
multi-disciplinary teams. Integrated
with the practical work is an analysis
of cultural industries

Institute of Education, University of London
Department of English,
Media and Drama
20 Bedford Way
London WC1H 0AL
Tel: 0171 612 6511/3
Fax: 0171 612 6330
MA Media Education
One year full-time or two year part-
time. Three elements: 1) Mandatory
module in The Theory and Practice
of Media Education, assessed by
final examination; 2) Optional
module, assessed by course work,
from Ideology and the Media;

Childhood, Youth and Popular
Culture; Media, Race and Gender;
Principles of Production for Media
Education; British Media: the
European Dimension; Hollywood
Cinema: Text and Context;
Television and its Audiences; Text in
a Social Semiotic Perspective.
3) Dissertation
MA Media Studies
One year full-time, two year part-
time, with three elements:
1) Mandatory Module in Ideology
and the Media (assessed by final
examination); 2) Optional module,
assessed by course work, as for the
MA Media Education with The
Theory and Practice of Media
Education replacing Ideology and
the Media; 3) Dissertation
MPhil and PhD
Supervision of research thesis in the
area of Film Studies, TV Studies,
Media Studies and Media Education
Institute Associateship
Individualised one year courses for
mature educationalists wishing to
study pedagogic and intellectual
developments in the field of Media
Education and Media Studies

Kent Institute of Art and Design
School of Visual Communication
Maidstone College
Oakwood Park
Maidstone, Kent ME16 8AG
Tel: 01622 757286
Fax: 01622 692003
email: kiadmarketing@kiad.ac.uk
● **BA (Hons) Visual Communication -**
Time Based Media
Explores all aspects of the moving
image, including video, film, sound
and animation. The emphasis is on
personal authored work, mainly in
video, but traditional production
roles and values are also taught. The
pathway encourages creative and
investigative video production, and
is a pioneer of the 'multi-skilled'
approach used today in all aspects of
professional and independent TV
and Video production
Master of Arts in Visual
Communication
This Master of Arts Programme has
been developed for designers,
illustrators, photographers, film/
video makers, and theorists who
wish to develop their practices either
as a single discipline or combination
of disciplines within an ethos of
interdisciplinary. Each application to
the programme is determined by an
individually proposed MA Project
defined within the speculative
framework of the programme. All
graduates in Visual Communication

are part of one programme but develop their Master of Arts projects within a negotiated study plan drawn from the School's principal study areas.

Time Based Media with Electronic Imaging

Time Based Media with Electronic Imaging (TBN) includes video/film production, conventional and digital animation, multi-media and sound.

Visual Theory

This aspect of the programme provides an introduction to a range of theoretical frameworks for understanding the role of visual media in contemporary society. For further information contact the Register on: 01622 757286 or Fax 01622 692003

MPhil and PHd Study

Applicants for MPhil/PhD must demonstrate an understanding of research methodologies. These are taught as part of MPhil study. Students wishing to register for PhD will normally be required to register for MPhil in the first instance. Conversion to PhD is dependent upon completion of MPhil requirements (two years) and evidence of sufficient aptitude and ability to sustain the research project through to the Doctoral award. The areas of expertise that the Institute is initially providing for research degree activity includes Electronic and Time-Based Media. For further information contact the Registrar on: 01622 757286 or Fax: 01622 692003

University of Kent
Rutherford College
Canterbury
Kent CT2 7NX
Tel: 01227 764000
BA Combined Hons
A Part 1 course on Narrative Cinema is available to all Humanities students in Year 1. The Part 2 component in Film Studies in Years 2 and 3 can vary from 25 per cent to 75 per cent of a student's programme. Courses include film theory, British cinema, non-narrative cinema, comedy, and sexual difference and cinema. The rest of a student's programme consists of courses from any other humanities subject. No practical component
BA Single Hons
This includes a practical film production option
MA, MPhil and PhD
An MA course in Film and Art History is available. Students are also accepted for MA, MPhil and PhD by thesis

King Alfred's College Winchester (Affiliated to Southampton University)
School of Community and Performing Arts
Sparkford Road
Winchester SO22 4NR
Tel: 01962 841515
Fax: 01962 842280
BA (Hons) Drama, Theatre and Television Studies
Three year course relating theories of contemporary television and drama to practical work in both media. The course looks at both the institutions and the practices of the two media from the perspectives of psychology and critical ideologies of the Twentieth Century. It includes television projects in which students work in groups to produce documentaries or drama documentaries. These projects are community based Media and Film Studies
Further details available from David Lusted: Head of Media and Film
School of Cultural Studies

Kingston University
School of Three Dimensional Design, Knights Park
Kingston-Upon-Thames
Surrey KT1 2QJ
Tel: 0181 547 2000 ext 4165
● **MA Design for Film and Television**
One year MA Course in scenic design tailored to the needs of those who wish to enter the industry with the eventual aim of becoming production designers or art directors. The course is constructed as a series of design projects to cover different types of film and television production

School of Art and Design History
Tel: 0181 547 7112
BA/BA (Hons) Combined Studies: History of Art, Architecture and Design
Five to six year part-time or three year full-time. Optional film strand: three Film Studies modules, each representing one sixth of a full-time student's yearly programme, one third of a part-time student's. Foundation level: concepts of 'Art' cinema. Intermediate level: photographic issues. Advanced level; the study of a selected artist.

School of Languages
Penrhyn Road
Kingston-Upon-Thames
Surrey KT1 2EE
Tel: 0181 547 2000
Fax: 0181 547 7392

BA (Hons) French Full and Half-field, Full and Part-time
Introduction to French Cinema. Year two on French Cinema. Year four special subject on New Wave Cinema

Kingsway College
Media & Photography Unit
Sans Walk, Clerkenwell
London EC1R 0AS
Tel: 0171 306 5700
Fax: 0171 306 5855
● **B/TEC National Diploma in Design Communications (Media Studies)**
Two year full-time course for those interested in pursuing a career in the media industry or going on to higher education. The course covers an integrated programme of practical training in photography, video, film and computer technology, and theoretical studies relating to the analysis of media texts. 70 per cent practical/30 per cent theoretical
Evening Classes
Courses available (36 wks) in Super 8 film-making and animation

LSU College of Higher Education
Department of English
The Avenue
Southampton SO17 1BG
Tel: 01703 228761/225333 (Registry)
Fax: 01703 230944
BA Combined Honours/BA English Studies
All courses contain options in Film Study for those students for whom English is a significant component of their degree

University of Leicester
Centre for Mass Communication Research
104 Regent Road
Leicester LE1 7LT
Tel: 0116 252 3863
Fax: 0116 252 3874
BSc Communications and Society
A three-year social science based undergraduate course. The modules taught cover a wide range of areas including media institutions, research methods in mass communications, film and TV forms. Students are assessed by a combination of continuous assessment and examination
MA Mass Communications
One year taught course studying the organisation and impact of the mass media both nationally and internationally and providing practical training in research methods

MA Mass Communications (by Distance Learning)
Two year part-time course by distance learning. Organized in 10 modules plus dissertation. Course materials include 60 course units, readers, set books, AV materials. Contributions from a team of international experts. The course covers media theories, history, regulation, media in global context, methodology, media industries, professional practices, audiences, texts and issues of representation. Options to include media education, management and film. Day and weekend schools are voluntary but highly recommended

Light House Media Centre
The Chubb Buildings
Fryer Street
Wolverhampton WV1 1HT
Tel: 01902 716055
Fax: 01902 717143
email: Lighthse@Waverider.co.uk
Training courses in animation computer animation, video production at beginners and intermediate level, creative editing

University of Liverpool
School of Politics and Communication Studies
Roxby Building
PO Box 147
Liverpool L69 3BX
Tel: 0151 794 2890
Fax: 0151 794 3948
BA Combined Hons (Social and Environmental Studies)
BA Joint Hons (English and Communication Studies)
BA Joint Hons (Politics and Communication Studies)
In all these programmes, students combine work in the Communication Studies Department with largely non media-related work in other Departments; Communication Studies forms up to 50 per cent of their programme. Year 1: Communication: a programme of introductory work on communication and cultural analysis. Year 2: courses on Broadcasting, Film Studies and Drama. Year 3: courses available include Documentary, exploring a range of work in literature, photography, film and television. No practical component
MA Cultural Research and Analysis
The purpose of this degree is to introduce students to current work in the area of research on popular cultural institutions, forms and behaviours. Core courses look at the

mass media and culture, at culture and national identity, and at city culture and urban life. There is a particular emphasis on research of a broadly ethnographic character, involving students in the field work within the Merseyside area

Liverpool John Moores University
School of Design and Visual Arts
68 Hope Street
Liverpool L1 9EB
Tel: 0151 231 2147
Fax: 0151 709 6408
● **BA (Hons) Graphic Design and Fine Art (modular)**
Film/Animation is a specialised option within the degree. After a general first term a number of students may specialise in Film/Animation in term 2 and continue studies throughout their second and third years. From a basic introduction to animation to an in-depth production of two and three dimensional short filmmaking the course encourages individual creative development and preparedness for the animation industry

London College of Printing & Distributive Trades
Media School, Back Hill,
Clerkenwell Road,
London EC1R 5EN
Tel: 0171 514 6500
Fax: 0171 514 6848
● **BA (Hons) Film and Video**
An autonomous course in Film and Video, part of a degree scheme in Communication Media courses, leading to the award of BA (Hons) degree. Main concerns are Women's Cinema, Third World Cinema, Popular Culture and Film. Stress on experimentation and innovation, education, independent filmmakers. Practice/theory ratio is 70:30. Course stresses integration of theory and practice. The course also includes an option in Animation. This course is accredited by BECTU
MA Screenwriting, MA Documentary Research, MA Independent Film & Video
These are part-time (1 day a week), two years, commencing in January 1998. Enquiries to the course leaders: Screenwriting - Phil Parker; Documentary Research - Michael Chanan; Independent Film & Video - Liz Wells

London Guildhall University
Sir John Cass Department of Art
133 Whitechapel High Street
London E1 7QA

Tel: 0171 320 3455/3456
Fax: 0171 320 3462
BA (Hons) Communication and Audio-Visual Production Studies (Early Specialisation)
This degree includes both practical and theoretical studies. Practical units include film television and video production, photo-journalism, radio journalism and writing for the media. Theoretical units include cultural history and cultural studies. The degree may be studied full-time or part-time. Communication Studies may also be studied as half of a joint degree or as a minor component of a degree.

The London Institute
Central St Martins College of Art and Design, School of Art
107-109 Charing Cross Road
London WC2H 0DU
Tel: 0171 753 9090
Fax: 0171 413 0586
● **BA (Hons) Fine Art, Film and Video**
Three year full-time. Students are recruited directly into the Film and Video subject of the Fine Art course. The first four terms are designed to develop technical and conceptual skills with a series of projects which explore image, sound, animation, video and 16mm film production. During the first year there will be opportunities to spend some time in another area of Fine Art and a placement during the second year. The Fine Art context encourages experimental enquiry, and the course is 80 per cent practical and 20 per cent theoretical. Details obtainable from the School of Art office, although applications are made through **the Art and Design Registry, Penn House, 9 Broad Street, Hereford HR4 9AP**
MA Fine Art, Film and Video
Contact School of Art office

London International Film School
Department F17
24 Shelton Street
London WC2H 9HP
Tel: 0171 836 9642
Fax: 0171 497 3718
● **Diploma in Film Making**
A two year full-time practical course teaching skills to professional levels. All students work on one or more films each term and are encouraged to interchange unit roles termly to experience different skill areas. Approximately half each term is spent in film making, half in practical instruction, seminars, workshops, tutorials, and script writing.

The London
International
Film School

• Training film makers for 40 years •
• Graduates now working worldwide •
• Located in Covent Garden in the heart of London •
• 16mm documentary & 35mm studio filming •
• Two year Diploma course in film making •
• Commences three times a year: January, May, September •

London International Film School,
Department F23. 24 Shelton Street, London WC2H 9HP
Tel: 0171 836 9642/0171 240 0168 Fax: 0171 497 3718
Email: lifs@dial.pipex.com Web Page: http://www.tecc.co.uk/lifs/index.html

Established for over 40 years, the school is constituted as an independent, non profit-making, educational charity and is a member of NAHEFV and CILECT – respectively the national and international federations of film schools. Graduates include Bill Douglas, Danny Huston, John Irwin, Mike Leigh, Michael Mann and Franc Roddam. The course is accredited by BECTU and widely recognised by local education authorities for grants. New courses commence each January, April and September

London School of Economics and Political Science
Department of Social Psychology
Media and Communications
Houghton Street
London WC2A 2AE
Tel: 0171 955 7710/7714
Fax: 0171 955 7565
MSc Media and Communications
One year MSc programme (two years part-time) provides an advanced understanding of the development and forms of media systems (eg text, audience, organisation, effects) in Britain and elsewhere. Students take two core courses, one inter-disciplinary theoretical approaches to media and communications, and one research methodology in media and communications. Additionally, students choose from a range of optional courses reflecting social science approaches to media and communications, and complete an original, supervised, research report on a subject of their choice

University of Manchester
Department of Drama
Oxford Road, Manchester M13 9PL
Tel: 0161 275 3347
Fax: 0161 275 3349
BA Single and Joint Honours Drama and BA in Drama and Screen Studies
Two films studies courses on Hollywood and European cinema compulsory in Year 1 and 2, with additional film studies and video production courses optional in Years 2 and 3.
MPhil/PhD
Opportunity for research theses on aspects of film and television and documentary
MA module course in Screen Studies

School of Education
Oxford Road
Manchester M13 9PL
Tel: 0161 275 3398

Fax: 0161 275 3398
MEd/Diploma Advanced Study Education and the Mass Media
Course Director: Sue Ralph
These courses are designed for educators and media practitioners from the UK and overseas who wish to explore educational uses of media and to develop effective communication techniques in their fields of work. Courses include: video, audio and tape/slide production; media policy; interpersonal communication; distance learning; mass communication theory; visits to media organisations. Courses place strong emphasis on experimental learning and collaborative group activity

Manchester Metropolitan University
Department of Communication Media, Chatham Building
Cavendish Street
Manchester M15 6BR
Tel: 0161 247 1284
Fax: 0161 247 6393
● **BA (Hons) Television Production**
A three year, full-time course based around practical projects. Typically, these include dramas, documentaries, corporate productions and magazine programmes. Students work mainly in groups and are encouraged to experience different production roles. Emphasis is placed on development of original programme material through individually based research and scripting. Equipment includes a four-camera TV studio and non-linear editing. Complimentary studies include media history, narrative studies and semiology. Subject to validation

Department of English and History
Lower Ormond Street
Manchester M15 6BX
Tel: 0161 247 1730
Fax: 0161 247 6308
BA Humanities & Social Sciences/ BA English Studies
Film and film theory. Third year optional course.
Introduction for Contemporary Cinema. Second Year optional Course: Analysing Television. Second Year optional course

Department of Interdisciplinary Studies, Cavendish Building
Cavendish Street
Manchester M15 6BG
Tel: 0161 247 3026
Dip HE
Two year course which includes an introduction to film and film theory

in Year 1 and a course on film as propaganda in Year 2. Also a second year course on Gender and the Gothic in film and literature. A third Year unit in the BA Modern Studies degree offers a study of Hollywood – Text into Film

Middlesex University
Faculty of Art, Design & Performing Arts
Cat Hill, Barnet
Herts EN4 8HT
Tel: 0181 362 5000
Fax: 0181 440 9541
email: admissions@mdx.ac.uk
BA (Hons) Visual Culture
Modular system degree. Film studies develops from a first level in Art and Design History. Critical and theoretical approaches to film are covered, including production, distribution and reception. The set allows detailed studies of different genres and modes of production in filmmaking, and raises issues of gender, nationality, representation and narrative
● **MA Video**
A one year full-time course (45 wks) emphasising the creative aspects of professional video production in the independent sector. Intended for graduate students with considerable lo-band video experience. The course covers all aspects of the production cycle, with an emphasis on scriptwriting. 50 per cent practical; 50 per cent theoretical
BA Honours Film Studies (combined with another subject-modular)
Modular system degree. Students combine Film Studies and another subject chosen from some 80 different subjects available

Faculty of Humanities
White Hart Lane, Tottenham
London N17 8HR
Tel: 0181 362 5000
Fax: 0181 362 6878
Media & Cultural Studies Set
(Major or Minor study as part of multi-disciplinary degree programme) (BA Hons Media and Cultural Studies also offered if taken as a Major). This course offers theoretical approaches to Media and Cultural Studies which contextualise processes of production and consumption. Students develop a broad range of skills relevant to the fields of new media

Napier University
Department of Photography, Film and Television
61 Marchmont Road

Edinburgh EH9 1HU
and 6/7 Coates Place
Edinburgh EH3 7AA
Tel: 0131 538 7614
Fax: 0131 538 76629
● **BA (Hons) Photography, Film and Television**
With option of specialising in Film and Television production from the start of the 3rd year. At the end of the 2nd year students take either the still image stream or the moving image stream in this four year course.
MPhil/PhD
A 2/3 year research programme with tutorial support facilitating opportunities for advanced study in creative practice in the moving image, including production of a major film or multimedia project

National Film and Television School
Beaconsfield Studios
Station Road, Beaconsfield
Bucks HP9 1LG
Tel: 01494 671234
Fax: 01494 674042
● A full-time professional training leading to an NFTS Associateship in specialisations of producing, fiction direction, documentary direction, screenwriting, cinematography, editing, animation direction, screen design, screen sound, screen music and television. Shortlisted applicants are offered induction courses prior to final selection. Previous experience in film or a related field is expected. Assistant level training is also available and a wide range of short courses for industry professionals. The School is funded by a partnership of Government and industry (film, television and video). Its graduates occupy leading roles in all aspects of film and television production

University of Newcastle upon Tyne
Centre for Research into French
Newcastle upon Tyne NE1 7RU
Tel: 0191 222 7492
Fax: 0191 222 5442
email: p.p.powrie@ncl.ac.uk
Website: http://www.ncl.ac.uk/\ncrif
Diploma/MA in Film Studies
One year full-time; two year part-time course. Obligatory research training and introduction to the study of film, followed by 4 from 19 day-time and evening options, although not all are taught in every year: 6 on Hollywood cinema (the Biblical Epic; Lubitsch; romantic comedy; film noir; the Western; gender in the action film); 3 on British Cinema (pre-50s; post-60s; class and sex); on French cinema (New Wave; cinema and conflict since 1968; the contemporary nostalgia film & postmodern cinema (1960-1979; 1980-present) 2 on Spanish Cinema (60s-70s; 80s-90s); 4 on Media and Industry (Broadcasting in France; TV Comedy; Programming and Marketing; Financial Structures). Dissertation required for the MA
MLitt in Film Studies
Research-based course tailor-made for individual students. Three/four essays followed by a dissertation on negotiated topic in British, French, Hollywood, or Spanish cinemas.
PhD in Film Studies
Supervision offered in British, French, Hollywood, and Spanish cinemas. For current and suggested postgraduate projects

Department of French Studies
Newcastle upon Tyne NE1 7RU
Tel: 0191 222 7441
Fax: 0191 222 5442
BA (Hons) French, French/Spanish, French/German
Optional modules in film studies. Stage 1: introduction to the study of film. Stage 2: introduction to film theory; Classic French cinema.

Stage 3: French cinema in the 1980s; cinema and conflict post-1968

University of North London

School of Literary and Media Studies
116-220 Holloway Road
London N7 8DB
Tel: 0171 753 5083
Fax: 0171 753 5078
BA Humanities Scheme
Three year full-time course. Six year part-time course by day or evening study. Film Studies is one of 16 subject components and may be taken as a Major, Joint or Minor. One practical unit
MA Drama and Theatre Studies
Two year part-time evening modular course with optional Film Studies and Television. No practical component
MA Literature, Representation and Modernity
Two year part-time evening modular course. Core courses in The Subject and Modernity, Postmodernism. Optional unit in Film Studies

Northern School of Film and Television

Leeds Metropolitan University
2 Queen Square
Leeds LS2 8AF
Tel: 0113 283 1900
Fax: 0113 283 1901
email: nsftv@lmu.ac.uk
Website: http://www.lmu.ac.uk/
This is run by Leeds Metropolitan University, with the support of Yorkshire Television, providing postgraduate level professional training in practical film production
● **MA/PgD Scriptwriting for Film and TV (Fiction)**
An intensive one year practical course running from February, one year full-time, and one year part-time (off site). Staffed largely by working professional writers, it covers the various forms of fiction scriptwriting for film and television – short film, feature film, television drama, soap opera, series etc. The course has a strong emphasis on professional presentation, and aims to help graduates to set up a credible freelance practice. Work consists of a short film script, a 30 minute script, a 60 minute script proposal and a full length feature script or television equivalent.
● **PgD Film Production (Fiction)**
An intensive one year practical course running from October to October. Students are admitted into specialist areas: Direction (six students per year), Production (six),

Camera (three), Art-Direction (six), Editing (three) and Sound (three). Students work in teams to produce six short films, in two batches of three. The resulting films may be broadcast on Yorkshire Television, which provide the base production funding and some facilities. Scripts are normally drawn from the product of the Scriptwriting Course and the emphasis is on team working and joint creativity under pressure. It is not a course for 'auteur' film makers. There is also a theoretical studies component
MA Film Production (Fiction)
Part-time course. Normally taken up by students who have completed the Postgraduate Diploma (see above), the MA is available via several options: 1) a 10,000 word dissertation; 2) 2 x 5,000 word extended essays; 3) exchange placement with one of NSFTV's exchange partners (Poland, Germany, Holland) and report

University of Northumbria at Newcastle

Faculty of Art & Design
Squires Building
Sandyford Road
Newcastle upon Tyne NE1 8ST
Tel: 0191 227 4935
Fax: 0191 227 3632
● **BA (Hons) Media Production**
Practical three year course with fully integrated theoretical and critical components in which students are offered the opportunity to specialise in individual programmes of study.
BA (Hons) History of Modern Art, Design and Film
Offered as a three year full-time course. Film Studies is given equal weighting with painting and design in the first year. In the second year up to 60 per cent of a student's time can be devoted to Film Studies, with this rising to nearly 100 per cent in the third year
MPhil
There are possibilities for research degrees in either film theory or practice

Nova Camcorder School

11a Winholme
Armthorpe
Doncaster DN3 3AF
Tel: 01302 833422
Fax: 01302 833422
● **Practical evening course for camcorder beginners**
A 10 week course, one night a week, for people who want to learn how to use their camcorders properly. The course explains all the features and functions of a camcorder before moving onto basic film-making

techniques and home editing and titling. The course is specifically designed for beginners, and participants receive a worksheet every week which summarises the topics covered

Plymouth College of Art and Design

School of Media and Photography
Tavistock Place
Plymouth
Devon PL4 8AT
Tel: 01752 203434
Fax: 01752 203444
● **B/TEC HND Media Production**
(in partnership with the University of Plymouth). A two year modular course with pathways in film, video, animation and electronic imaging. All areas of film, video and television production are covered and the course is well supported by visiting lecturers and workshops. Strong links with the industry have been developed and work based experience forms an important part of the course. The course has BKSTS accreditation. Opportunities exist through the ERASMUS programme to undertake a programme of exchange with European universities or polytechnics during the course. In addition suitably qualified students can progress to third level modules for the award of a BA (Hons) PhotoMedia
Advanced Diploma Photography, Film and Television leading to the BIPP Professional Qualifying Exam
A one year course post HND and postgraduate. The photography, film and television option allows students to plan their own line of study, including practical work, dissertation and an extended period of work based experience. Students from both courses have had considerable success in film and video scholarships and competitions. Students on both courses have the opportunity for three month work placements in the media industry in Europe
NCFE Foundation in Lens Based Media
A one year full time foundation course for those wishing to progress to Higher Education in one of the many exciting areas of lens based media. This practical course covers: photography, video, electronic imaging, multi-imaging and contextual studies. The course aims to help the student develop a portfolio which shows how the student has integrated technical

focus on...

Modern Apprenticeships in
photography
City & Guilds
professional photography C&G 7470
photography C&G 9231
BTEC National Diplomas in
media studies
design for multi media
photography
NCFE Foundation in
lens based media
BTEC Higher National Certificate
photography (subject to validation)
BTEC Higher National Diploma
media production
BA (Hons)
photomedia
BIPP Professional
Qualifying Examination
**photography,
film and
television**

P L Y M O U T H
COLLEGE OF ART AND DESIGN
Tavistock Place • Plymouth
*Plymouth College of Art and Design -
enriching the future through creativity*

For more details ring
01752 203434
e-mail; enquiries@pcad.plym.ac.uk

skills with creative concepts and critical analysis. (This course may also be studied part time over two academic years)
National Diploma Programmes
ND Photography; ND MultiMedia, ND Media Studies

University of Portsmouth
Department of Design
Lion Gate Building, Lion Terrace
Portsmouth PO1 3HF
Tel: 01705 843805
Fax: 01705 843808
email: lingardm@env2.enf.port.ac.uk
BA (Hons) Art, Design and Media
Three year unitised programme has six specialist pathways. All are structured around historical, cultural and theoretical analysis which form an important part of the degree. Student placements in Europe and the UK and outside projects maintain the degree's links with industry and art practice.
Media Arts - Moving Image Strand
In the first and second years,

students undertake briefs around personal and cultural identity, gender, media arts practice, documentary as intervention etc. Students work in video, sound, multimedia and photography. In the third year students work on self-directed projects
Media Arts - Photography Strand
Encourages collaborative work with moving image and sound artists
Communication Design
Design for television, video graphics and multimedia are central areas of concern
History and Cultural Theory
Aim to produce graduates with particular skills in research and communication

School of Social and Historical Studies, Milldam
Burnaby Road
Portsmouth PO1 3AS
Tel: 01705 876543
Fax: 01705 842174
BA (Hons) Cultural Studies
Year 3: options on British Cinema

1933-70, British television Drama, Avant-Garde Films and Feminism

Ravensbourne College of Design & Communication
School of Television
Walden Road, Chislehurst
Kent BR7 5SN
Tel: 0181 289 4900
Fax: 0181 325 8320
● **B/TEC HND Engineering (Television Systems)**
Two year full-time vocational course designed in consultation with the broadcasting industry leading to employment opportunities as technician engineers. Students develop skills in installing, aligning and maintaining a wide range of professional broadcast equipment using analogue and digital technologies
B/TEC HND Design Communication (Television Programme Operations)
Two year full-time vocational course designed in consultation with the TV broadcasting industry leading to employment opportunities in television and video production as members of programme making presentation and transmission teams. Students develop skills in budgeting, production management and lighting, camera operators, sound, video recording and editing, vision-mixing, Telecine, and audio-recording
BA (Hons) in Professional Broadcasting
90 week full-time vocational course designed in consultation with the broadcasting industry leading to employment opportunities for creative, team-centred individuals possessing a fundamental competence for working in the production process, business practice and technology of a changing and highly competitive industry

University of Reading
Department of Film and Drama
Bulmershe Court
Woodlands Ave
Reading RG6 IHY
Tel: 0118 931 8878
Fax: 0118 931 8873
BA Film and Drama (Single Subject)
After the first two terms in which three subjects are studied (two being in film and drama), students work wholly in film and drama. The course is critical but with significant practical elements which are designed to extend critical understanding. It does not provide professional training

BA Film and Drama with English, German, Italian
Students in general share the same teaching as Single Subject students

MA Film and Drama
One year taught course. Four taught elements, running throughout the 12 months: Text and Performance Analysis (covering both film and drama); Critical Case Studies (one in each form); Critical Practice (practical assignments in drama or video); Research methodologies in film and drama. Plus 15,000 word dissertation

MPhil and PhD
Research applications for MPhil and PhD degrees are also invited in areas of cinema, television and twentieth century theatre

Department of English Studies
Whiteknights
Reading RG6 2AA
Tel: 01734 318332
Fax: 01734 318333
BA (Hons) English
Second year optional course on film, television and literature. Third year optional course in media semiotics
PhD
Research can be supervised on the history of the BBC and other mass media topics

Department of French Studies
Whiteknights
Reading RG6 6AA
Fax: 0118 931 8122
BA (Hons) French
First year introductory course: detailed study of one film (one half-term). Final year: Two-term option: French cinema, with special emphasis on the 30s, 40s and the Nouvelle Vague. Includes introductory work on the principle of film study. Available also to students combining French with certain other subjects.

Department of Italian Studies
Whiteknights
Reading RG6 2AA
BA (Hons) Italian/French and Italian with Film Studies
First year introductory course: Post-War Italian Cinema (one half-term). Second year course: Italian Cinema (three terms). Final year course: Italian Cinema in its European and American context (two terms). Dissertation on an aspect of Italian cinema. These courses available to students reading other subjects in the Faculty.
MA Italian Cinema
One year full-time or two year part-time course on Italian cinema:

compulsory theory course, options on film and literature, Bertolucci, Italian industry and genre – the Spaghetti Western.
MPhil and PhD
Research can be supervised on Italian cinema for degree by thesis

Department of German Studies, University of Reading
Whiteknights
Reading RG6 6AA
Tel: 0118 931 8332
Fax: 01118 931 8333
email: j.e.sandford@reading.ac.uk
BA (Hons) German
Two-term Finals option: The German Cinema. Course covers German cinema from the 1920s to the present, with special emphasis on the Weimar Republic, the Third Reich, and the 'New German Cinema'.

University College of Ripon and York St John
Faculty of Creative and Performing Arts
Lord Mayor's Walk
York YO3 7EX
Tel: 01904 616672
Fax: 01904 612512
email: D.Browne@UCRYSJ.ac.uk
● **BA (Hons) Film, Television, Literature and Theatre Studies**
This specialist degree takes an integrated approach to Film, Television and Theatre complemented by the study of literature (text and script). The courses are largely practical and include: Television - forms and conventions, documentary, drama, news and current affairs, independent production. Theatre - devised theatre, performance practice. Literature - author/text/audience, fictions, text/process/adaptation. Film - forms and conventions, British cinema. Work-related modules and opportunities for placements, internships and European/US exchanges. This degree is based on the York campus. Three bursaries are available from Yorkshire Television to 1st Year Students

Ripon Campus,
College Road
Ripon HG4 2QX
Tel: 01765 602691
Fax: 01765 600516
BA (Hons) Communication Arts: Studies in Media and Performance
Head of Scheme: Alan Clarke
This interdisciplinary 'pathway' degree offers a flexible and innovative programme comprising a range of core and specialist modules

allowing students to undertake pathways in performance (music, dance, drama) or word and image (electronic imaging, desktop publishing, video and photography). There are also opportunities to study information technology and to spend a semester abroad. Students elect to study pathways from either the performance or word and image strands. This degree is based on the Ripon campus
BA (Hons) Cultural and Critical Studies
The inter-disciplinary 'pathway' degree comprises core modules which introduce key concepts in cultural and critical studies and elective modules available in two broad strands: 'Ideology, Identity and Belief', offers studies in Hollywood, race and gender studies, New Age and Post Modernism. International Contexts offers studies in American popular culture, Renaissance culture, the New Europe, Film as evidence and propaganda, Cultural imperialism. There are opportunities to stay abroad. This degree is based on the Ripon campus

Roehampton Institute, London
Faculty of Arts and Humanities
Digby Stuart College
Roehampton Lane
London SW15 5PU
Tel: 0181 392 3230
Fax: 0181 392 3289
email: j.ridgman@roehampton.ac.uk
BA Drama and Theatre Studies
The Drama and Theatre Studies modular programme may be combined with other subjects. Year 1 includes a one term module in Reading Film and Television with some practical work in portable video. A double module in Contemporary Television Drama and a practical double module in Television Drama Production are offered in Year 3. Other modules, such as Representing Women and Shakespeare on Film and Television, contain substantial study of film and television production. Students may also specialise in areas of film and television for research dissertations.
BA Film and Television Studies
A three year modular degree programme, which may be combined with a variety of other subjects. Several core courses are available (Film Narrative, British Television Drama, Cultural Theory and Media, Representing Women etc) to which may be added selected topic modules. The course includes up to

40 per cent practical work in television and video, moving from principles of single camera production in year one to sustained, independent project work in the final year

Royal College of Art
Department of Film and Television
School of the Moving Image
Kensington Gore
London SW7 2EU
Tel: 0171 584 5020
Fax: 0171 589 0178
email: j.smith@rca.ac.uk
Website:http://www.rca.ac.uk/Design
● **Cinematography**
The aim of the Cinematography Specialism is to create opportunities for students to increase their practical knowledge and creative experience. From pre-production onwards the cinematographer plays a central role in working with designers, directors and producers and a thorough understanding of all processes, equipment and creative possibilities and restrictions has to develop continuously
● **Design for Film and Television**
The Design Specialism aims to cover all aspects of production design in film and television through theoretical and practical teaching. The Course is structured to develop the creative potential and skillbase of the students and to promote an awareness of professional practice through industry placements and tutorial inputs from professional designers and allied specialists
● **Direction - Film and TV Drama**
A high-level of professional skill combined with original talent is expected of all students and the syllabus is designed to develop the practice of their creative and technical abilities. In the first year this is a achieved through a curriculum of seminars, tutorials, master classes and workshops focusing on script-writing, directing actors, mise-en-scene, technical camera, light and sound and post-production, with film screenings and seminars to develop critical analysis. In the first year this leads to the practical experience of working on the production of TV drama, documentary and commercials, and in the second term each student will direct a short film and be tutored through all aspects of production from concept to final print
● **Documentary - Film and TV**
This is a new specialism on the course and is open to a few, dedicated documentary students. They will already have a significant body of work and arrive with several projects they wish to pursue. In addition to specialised tutoring in the documentary form, students will be encouraged to develop a broad vision and will experience all aspects of film making in the first year, joining the drama directors and producers for relevant workshops.
● **Editing**
This is a new specialism on the course and offers editors the opportunity to develop technical and creative excellence and to further their skills by interweaving the study of film with maximum practical experience. The course offers a broad range of work in drama on film, video and documentary and commercial editing experience. Students are encouraged to pursue all new post-production technologies, including non-linear editing systems and multimedia post-production techniques
● **Production**
The aim of the Production Specialism is to develop the skills of creative producers and to give them the practical opportunity to take total responsibility for film and television productions from initial concept through to marketing the finished product. In addition to production and the shared curriculum, students will also be expected to complete a major development project on a subject of their choice, whether in filmed Drama, other forms of television or in Documentary
● **Sound Design**
The Sound Specialism aims to develop a full understanding of the technique and role of sound in film and video and to increase the students skill in sound recording, sound editing, post-sync recording and missing. With a thorough knowledge of all aspects of sound editing, the student can conceive and design the sound-track

St Helens College
School of Arts, Media and Design
Brook Street
St Helens
Merseyside WA10 1PZ
Tel: 01744 733766
Fax: 01744 28873
● **B/TEC National Diploma Media**
A two year full-time course aiming to provide a foundation in basic skills relevant to many areas of the media industry and the opportunity, through option selection, to examine one of more sectors in detail: television, film and video; sound and radio; print and publishing and audio-visual exhibition
● **HND Design (Multi-Disciplinary)**
A two year full-time programme offering core studies in the principles and methods of design and production, and specialist options in the following: film, television and video; graphics; 3D craft; Photography; illustration, Electronic imaging
● **B/TEC National Diploma in Design (Photography)**
A two year full-time programme providing a foundation in basic technical and design skills relevant to many areas of the photographic industries
● **BA (Hons) Media and Cultural Studies**
It is possible to study the first year of this course as part of the Integrated Credit Scheme (ICS) at St Helens College and years two and three at Liverpool John Moores University. The ICS offers a full range of academic awards. For course details, see entry under School of Media, Critical and Creative Arts at Liverpool John Moores University
● **HNC/D Media Technology**
A modular higher level programme of study in the technological concepts and practices of the media industries. Completion of the first year studies leads to the award of the Higher National Certificate, with second year students reading to the Higher National Diploma. The programme offers a broad-based, multi-skilled approach to technology, specialist options include: Television/Video Technology; Audio/Music Technology; Interactive Communications Technology
● **Women into Media Foundation**
A practical course offering City and Guilds 770 TV and Video and includes scripting, lighting, camera operations, editing and directing. Additional optional modules include animation, creative writing, music technology. No formal qualifications required
● **B/TEC National Diploma in Music Technology**
Programme of study incorporates - Music theory and practice; Technical/Production studies; Complementary studies; Analytical/Perceptual studies; Digital synthesis and Performance Technology
City and Guilds 1831 Audio Engineering. Programme of study incorporates: Acoustic Principle; Studio Design; MIDI/sequencing;

Signal processing and many more applications within the audio field
● **HNC/HND Media Production**
A modular programme available as a full-time or part-time course. Core subjects include technical and production studies, the media industry, production management and design, media and society, CAP, European Media
● **BTEC HNC/D Music Technology (Subject to validation)**
A one or two year programme. Core studies include Digital multi-track recording, Digital Hard Disc recording, practical musicianship skills and technical specialising in areas including audio post production, data systems analysis, software and design and micro electronics
● **Women into Media (Advanced)**
This one year part-time BTEC Higher National Certificate programme is for women who wish to add practical skills to their CV. Includes training in all aspects of technical and practical video production. Skills include directing, production design and management, camera, lighting and sound operations, digital communication, electronic imaging and non-linear editing. This programme is particularly useful for women graduates from academic media and cultural studies programmes

University of Salford
Faculty of Media, Music and Performance
Adelphi, Peru Street
Salford, Manchester M3 6EQ
Tel: 0161 295 6000
Fax: 0161 295 6023
International Media Centre
Director and Faculty Dean: Keith Wilson,
Adelphi House, The Crescent
Salford, Manchester M3 6EN
BA (Hons) Television and Radio
BA (Hons) Media and Performance
BA (Hons) Media, Language and Business
BSc (Hons) Media Technology
B/TEC HND Media Production
B/TEC HND Media Performance
MA Television Feature and Documentary production
MA Scriptwriting for Television and Radio (part-time)
Plus - degree courses in Band
Musicianship; Popular Music & Recording; Music, Acoustics & Recording; Composition; Performance
Visiting Professor: David Plowright.

Fellows: Richard Ellis, Ray Fitzwalter. Professional Patrons: Ken Russell, Ben Kingsley, Liz Forgan, Robert Powell, Gareth Morgan, Sir George Martin CBE, Jack Rosenthal CBE, Stuart Prebble, Leslie Woodhead OBE, Gillian Lynne. All courses are 50 per cent practical production/performance based. The Granada Education Awards are available to ethnic students. AVID Technology Bursaries also available

University of Sheffield
Department of English Literature
Shearwood Mount
Shearwood Road
Sheffield S10 2TD
Tel: 0114 222 8480
Fax: 0114 282 8481
BA (Hons) English Literature
Students may study one or two Special Subjects in Film in their second or third years

Sheffield Hallam University
Communications Subject Group
School of Cultural Studies
36 Collegiate Crescent
Sheffield S10 2BP
Tel: 0114 253 2236
Fax: 0114 253 2344
BA (Hons) Communication Studies
Course covers all aspects of human communication, one area being Mass Communication. Option course in Television Fictions in Year 3. Some practical work
MA/PgD/Certificate Communication Studies
Part-time course to gain certificate in three terms, Diploma in six terms, MA in 8 terms. Aims to develop theoretical understanding and analytical skills in relation to the processes and practices of communication in modern society. Students attend for two sessions of 2+ hours each week. Full-time/route 12 months. 8 hours per week

School of Cultural Studies
Psalter Lane
Sheffield S11 8UZ
Tel: 0114 253 2601 /272 0911
Film and Media Studies Programme
BA (Hons) Film Studies
BA (Hons) Media Studies
The Film and Media Studies Programme consists of two degree routes. The courses provide opportunities for the study of film and a range of media (including television, radio and journalism) from a variety of perspectives including historical development, social, political and economic contexts, and the artistic and

aesthetic dimensions of film and media. The courses also provide a grounding in basic media production skills with units in film, video etc and scriptwriting
● **BA (Hons) History of Art, Design and Film**
Film studies is a major component of this course. Year 1: introduction to film analysis and history. Year 2: special study on Hollywood. Year 3: critical and theoretical studies in Art, Design and Film and Contemporary Film Theory and Practice.
MA Film Studies
Two year part-time course; two evenings per week, plus dissertation to be written over two terms in a third year. Main areas of study: Problems of Method; The Classical Narrative Tradition; British Cinema 1927-45; Hollywood and Popular culture
BA (Hons) Fine Art (Combined and Media Arts)
After initial work with a range of media, students can specialise in film and/or video. Film productions range from short 8mm films to 16mm documentaries or widescreen features, to small 35mm productions

Northern Media School
The Workstation
15 Paternoster Row
Sheffield S1 2BX
Tel: 0114 272 0994
Fax: 0114 275 6816
● **PgDip Broadcast Journalism**
Main focus is on practical work. Much of the teaching is conducted through workshops and practical exercises supplemented by seminars and lectures
● **PgDip/MA Film and Television Documentary**
Provides intensive teaching in practical documentary production, backed up by a thorough grounding in its history and current trends
● **PgDip/MA Experimental Film, Video and Audio**
Aimed at self-motivated experimental artists with a background in drama, the fine arts or music, the course offers the opportunity to develop a high standard of specialist skills in film, video and audio production
● **PgDip/MA Screenwriting**
Offers a unique opportunity to develop hands-on production skills alongside writing skills. Students produce short scripts for production and develop feature scripts or material for television series
● **PgDip/MA Film and Television Drama**

The NMS offers funding to facilitate a broad range of production projects throughout the year, culminating in eight drama shorts aimed at broadcast and cinema exhibition. Each student is given the opportunity to work in all grades of their major specialism

MA Film, TV and Radio
A high-level production programme for media professionals and those possessing an NMS PgDip wishing to work within an environment of small business independent groups

South Bank University
Education, Politics and Social Science
103 Borough Road
London SE1 1AA
Tel: 0171 928 8989
Fax: 0171 815 8273
email: registry@sbu.ac.uk
● **BSc (Hons) Media and Society**
Three year full time course. Two thirds critical studies, one third practical work. This course combines units assessing the social and political significance of the mass media, together with units introducing practical production skills. Critically, the course grows from studies of the media in the Britain during year one, to studies of European and global media in years tow and three. Other units also address the understanding of media audiences, news forms and media law. Individual research leads to the completion of a dissertation thesis in year three. Practically, the course develops skills in audio, radio, video and multimedia production. These skills are then employed by students in the creation of their own final year projects

South Kent College
DASH
Maison Dieu Road
Dover CT16 1DH
Tel: 01304 204573
Fax: 01304 204573
B/TEC National Diploma Media Studies
Two year full-time course covering video, film, print/DTP, photography and radio. Students complete advertising, drama, news and documentary projects closely linked to community groups. The course is modular and work experience is offered

South Thames College
Department of Design and Media
Wandsworth High Street
London SW18 2PP

Tel: 0181 870 2241
Fax: 0181 874 6163
B/TEC HNC Design (Communication) – Television Production
Two year part-time course. BKSTS approved. All students experience every aspect of making both portable single camera (location) and multi-camera (studio) television programmes, starting with discussion of a brief with a 'client' and submitting a programme proposal, through the implementation of the programme production to the submission of the finished product. In addition to the practical work there are lectures, tutorials and visits to production companies
Evening Television Production
A series of thirty hour modules. There are two strands to the course programme, studio and location. All students do a level one course in studio and a level one course in location production before progressing to levels two and three. Specific production techniques are covered in each module, these include News and Magazine Production, Drama Production, Documentary Production and Demonstration Programmes. A record of achievement is awarded on the completion of each module. On completion of these six practical units plus additional theory units students are eligible for a City and Guilds 770 qualification

The University of Southampton
School of Education
Faculty of Educational Studies
Southampton
Hants SO9 5NH
Tel: 01703 593387
Fax: 01703 593556
Postgraduate Certificate Education
This one year initial training course for secondary/6th form teachers offers specialist work in Media Studies as an integral part of English Drama and Media Studies
Certificate and Diploma Advanced Educational Studies – Media Education
Certificate is one year course involving 60 hours of contact time; the Diploma is taken over two years, with 120 hours of contact time. Both include a range of media courses. The Certificate is also available as a distance learning package, involving 240 hours of independent study
MA (Ed) Media Education
The MA in Education is run on a modular basis as a full- or part-time

taught course. The course as a whole requires the completion of 12 15-hour units and a supervised dissertation. Included are television studies, media education, video in education, response studies and others
MPhil and PhD
Research degrees in any area of Media Education, Media Studies, Educational Broadcasting and Educational Technology are available
PgD/MA Television for Development
This one year course aims to be of value to students who wish to work or are working in development agencies in the UK and overseas as well as the staff of broadcasting institutions in developing countries. The rationale of the course is to produce an interface between development and the media, both in the North and in the South (Partnership course with King Alfred's College)

School of Research & Graduate Studies, Highfield
Southampton
Hants SO17 1BJ
Tel: 01703 593406/592248
Fax: 01703 593288/595437
email: fnl@soton.ac.uk
MA Film Studies
The course aims to equip students with the capacity to engage intellectually with significant developments in film theory and history, together with the skills required to undertake contextual and textual analysis of films and critical writing. The weight given to European cinemas, including British cinema, and to transnational perspectives, is a unique feature, and Hollywood and American independent cinema represent core elements. Tutors include Tim Bergfelder, Caroline Blinder, Pam Cook, Deniz Göktürk, Sylvie Lindeperg, Bill Marshall, Lucy Mazdon, David Vilaseca and Linda Ruth Williams

Staffordshire University
School of Humanities & Social Sciences
Field of Media and Cultural Studies
PO Box 661, College Road
Stoke on Trent ST4 2XW
Tel: 01782 294413
Fax: 01782 294760
BA (Hons) Film, Television and Radio Studies
This single honours degree provides a broad study of the media, with an emphasis on film, television and

radio, but offering opportunities to consider new technologies, journalism and advertising, together with a strand of practical work in scriptwriting and production which runs throughout the degree.

After a foundation year in which students are introduced to key problems and issues of media study, there is a wide choice of options which enable students to construct a programme to suit their own interests. Students can, for instance, choose to focus more strongly on one of the media offered - film, television or radio - or they might decide to spread their study evenly across the diversity of the media. By the final year it is possible to spend up to half their time on independent projects such as researching and writing a dissertation, producing a script, radio or video production, or developing and costing a project proposal for a media organisation or funding body

BA (Joint/Combined Hons) in Film Studies

In the half degree in Film Studies students are introduced to the diversity of practices involved in the cinema, including film-making, film-going, and film funding. A series of options enable contemporary and historical, American and British, European and non-European. Students also have the opportunity to explore some of the problems and issues involved in understanding cinema through practical work such as scriptwriting or video production.

BA (Joint/Combined Hons) in Media Studies

The half degree in Media Studies begins with an introduction to ways of studying the media, with core modules on the British Press and Broadcasting History. It then goes on to explore the operation of the mass media in society through a focus on the broadcast media, together with consideration of the press, advertising and new technologies. Within this focus students are introduced to different theories and methodologies for analysing the relation between the media and society, including such issues as: the development of new technologies, their national and global implications, changing media audiences and patterns of consumption, and public debates about media policies and practices. As well as studying media problems and issues, a series of options offer modules in broadcasting history and different radio, television or journalistic

forms. Students also have the opportunity to engage with media practices, formats and problems through practical work such as scriptwriting or audio/video production

University of Stirling
Film and Media Studies
Stirling FK9 4LA
Tel: 01786 467520
Fax: 01786 466855
BA (Hons) Film and Media Studies (Single and Joint Hons)
Four year degree in the theory and analysis of all the principal media. All students take courses in the theories of mass communication and in cultural theories, as well as problems of textual analysis and then select from a range of options, including practical courses in the problems of news reporting in radio and television and in television documentary. As a joint honours degree Film and Media Studies can be combined with a variety of other subjects

BA General Degree
Students can build a component of their degree in film and media studies ranging from as much as eight units (approximately 50 per cent of their degree) if they take a major in the subject, down to as little as three if they wish merely to complete a Part 1 major. For the most part students follow the same units as do Film and Media Studies Honours students

MSc/Diploma Media Management
One year full-time programme consisting of two taught terms (Sept-May) followed by a dissertation (May-Aug). Internationally oriented and comparative in approach, the course offers media practitioners a wider analytical perspective on the key issues affecting their work and offers graduates a rigorous foundation for a career in the media industry. Areas covered include media policy and regulation, media economics, management and marketing, analytical methods and case studies and advanced media theory

MLitt and PhD
The specialist fields of the Stirling Media Research Institute: Media and National/Cultural identity; political communication and the sociology of journalism; screen interpretation; media management and media policy; public relations. Further details of the Institute's work are obtainable on request

Msc/Diploma Public Relations

Available in full time (12 months) and distance learning (30 months) formats. The degree develops the key analytical and practical skills for a career in Public Relations. Areas covered include; Public Relations; management and organisational studies; research and evaluation; media and communication studies; marketing and political communication

Suffolk College
School of Art and Design
Rope Walk
Ipswich IP4 1LT
Tel: 01473 255885
Fax: 01473 230054
B/TEC HND Design Communication
A two year course with options in film/television graphics, animation and art direction. Students complete a period of work experience with employers in film and television companies. Facilities include two colour television studios, post-production facilities for film and video, and a film animation unit. After the two years students can enter the third year BA (Hons) Art and Design, which is part of the Suffolk Modular Degree programme
● **B/TEC National Diploma Photography and Video**
A two year full-time course which provides students with a broad knowledge of media activities and working practices in photography and video and familiarity with desk top publishing (DTP), sound recording and video production. Students choose specialist options from: photography and production for video and television

The University of Sunderland
School of Arts, Design & Communications, Forster Building
Chester Road
Sunderland SR1 3RL
Tel: 0191 515 2154
Fax: 0191 515 2178
BA (Hons) Communication and Cultural Studies
The academic study of human communication and culture. The study of natural language and mass communication common to both routes. 'Communication' route includes social psychology and information processing; 'Cultural' route includes literary and art-historical studies. Options include study of film, broadcasting and popular culture at each level. Up to 20 per cent practical study of the media is possible, including video, radio, photography, and broadcast

and print journalism

BA (Hons) Media Studies
Comprises study of social, historical and artistic aspects of the mass media and popular culture together with development of practical skills in media arts. Broadcast quality radio studios provide practical experience. Television studio, dark rooms and computer systems also in use

MA Cultural and Textual Studies
One year full-time or two year part-time MA. Postgraduate courses are constructed from a wide range of modules. The compulsory module provides students with a flexible theoretical foundation, and a multi media and comparative study of verbal and visual forms of cultural communication, representing both 'high' and 'popular' culture. Students are then asked to choose three other modules, and to write a dissertation which allows them to specialise in film studies if they wish

MA Woman, Culture and Identity
One year full-time or two year part-time MA. The compulsory module introduces students to feminist theory and criticism in the areas of film and media studies, cultural studies, literary studies and philosophy. Students are then asked to choose three other specialist modules, and to write a dissertation, which allows them to specialise in feminist film and/or media studies if they wish

Surrey Institute of Art and Design
Faculty of Arts and Media
Falkner Road, The Hart
Farnham, Surrey GU9 7DS
Tel: 01252 722441
Fax: 01252 732213
BA (Hons) Photography
BA (Hons) Film and Video
BA (Hons) Animation
The approach in each course is essentially practical, structured to encourage a direct and fundamental appraisal of photography, film, video and animation through practice and by theoretical study. 70 per cent practical, 30 per cent theoretical. Courses are BECTU accredited

BA (Hons) Media Studies
A range of theoretical approaches to the mass media are examined. Emphasis is placed on the critical application of such theories to the actual production and consumption of media, primarily visual, culture. Units on professional practice, the European context of media production, and the learning of a

modern European language prepare students for a career in the media industry

University of Sussex
School of Cultural and Community Studies
Media Studies Co-Ordinator
Essex House, Falmer
Brighton BN1 9RQ
Tel: 01273 678019
Fax: 01273 678644
email: a.m.oxley@sussex.ac.uk
BA Media Studies
The degree course in Media Studies enables students to develop a critical understanding of the press, cinema, radio television, new information technologies and of the particular character of media communications. The Major in Media Studies is taught in two Schools of Studies – Cultural and Community Studies (CCS) and European Studies (EURO): different School Courses accompany it according to the School. The course in EURO also involves study of a modern European language and an additional year abroad in Europe

BA English and Media Studies
BA Music and Media Studies
A three year full-time degree course which includes analysis of television, film and other media, together with some opportunity to be involved in practical television, video and radio production

MA in Media Studies
The MA comprises a two-term core course in media theory and research which students study the conceptual, methodological and policy related issues emerging from the study of the media. In addition, students choose, in each of the first two terms, an optional course from: European Media in Transition; Media Technology and Everyday Life; The Political Economy of the New Communications Media; Promotional Culture; Queering Popular Culture; Sexual Difference; Theories of Representation; Memories of the Holocaust

MA in Media Studies (Multimedia)
The course shares a core course, Media Theory and Research, with the MA in Media Studies. In addition, students take two dedicated courses: The Political Economy of the New Communications Media, and Theory and Practice of Interactive Multimedia. After two terms students either complete an academic dissertation, or undertake an industry placement

and a multimedia project

Institute of Education
Falmer
Brighton BN1 9RG
Tel: 01273 606755
Fax: 01273 678466
MA Language, Arts and Education
The course is designed for all teachers of the arts and for practising art-makers. Four major concerns: to establish a continuous relationship between contemporary art-makers and the teachers and custodians of all the arts; to examine some of the more recent theoretical approaches to artistic work; to offer a supportive context for the creation of original artistic work; to explore the philosophy and practice of arts teaching

Swansea Institute of Higher Education
Faculty of Art & Design
Townhill Road
Swansea SA2 0UT
Tel: 01792 481285
Fax: 01792 205305
Modular Degree Programme
Designed for those who wish to combine interests in the creative arts, art history and art theory with studies of visual communications, mass media and media production. Includes studies in: Audio-Visual, Video, Graphics, Photography, CAD. Validated by the University of Wales

Faculty of Education
Townhill Road
Swansea SA2 0UT
Tel: 01792 481285
Fax: 01792 205305
BA (Hons) Combined Studies
Three year degree with options. Modern English studies option includes film and television studies
BA (Ed) (Hons) Primary
Course includes a Literature and Media Studies main subject option

Thames Valley University, London
London College of Music and Media
St Mary's Road
Ealing
London W5 5RF
Tel: 0181 231 2304
Fax: 0181 231 2656
BA (Hons) Design & Media Management
Multi-disciplinary course. Students take design and management studies which are compulsory, and media production units including photography, sound, computer videographics, video production and multi-image

MA Cultural Studies

Part-time taught evening course of six units plus dissertation. Topics include film and television, popular culture

Trinity and All Saints College

(A College of the University of Leeds)
Brownberrie Lane, Horsforth
Leeds LS18 5HD
Tel: 0113 283 7100
Fax: 0113 283 7200
email: J.Foale@tasc.ac.uk
Website: http://www.tasc.ac.uk

Diploma/MA in BiMedia or Print Journalism

The course takes the form of a Postgraduate Diploma which, for suitable candidates, can be enhanced to Master's level. The Diploma consists of three taught modules: Basic Journalism, Journalism Skills (Bimedia or Print and Essential Knowledge). The Diploma courses run full-time for 39 weeks and include a minimum attachment at a news organisation

University of Ulster at Belfast

School of Design & Communication
Faculty of Art and Design
York Street
Belfast BT15 1ED
Tel: 01232 328515
Fax: 01232 321048

● **DipHE/BA (Hons) Visual Communication**

Practical and theoretical film/video/media studies available to all students plus a specialist pathway, Screen Based Imaging (SBI) which includes Video production, Animation and Multimedia Design

● **DipHE/BA (Hons) Combined Studies**

Students choose from modules across all courses and many specialise in a combination of Visual/Communication SBI and Fine Art Video plus media studies theory modules

School of Fine and Applied Arts
DIPHE/BA Hons Fine and Applied Arts

Students specialising in the Fine Art pathway may specialise in video from year two

University of Ulster at Coleraine

School of Media and Performing Arts, Coleraine
Co Londonderry
Northern Ireland BT52 1SA
Tel: 01265 324196
Fax: 01265 324964

● **BA (Hons) Media Studies**

Three year course integrating theoretical, critical and practical approaches to film, television, photography, radio and the press. Important practical component

MA Media Studies

A two year part-time course combining advanced study of the mass media with media practice. There are also specialist options dealing with media education and cultural identity. MA is awarded 40 per cent on course work, 60 per cent on dissertation (which may incorporate production element)

MPhil and DPhil

Students are accepted for MPhil and DPhil by thesis. Particular expertise is offered in the area of the media and Ireland, although supervision is provided in most areas of Media Studies

MA International Media Studies

A one year full-time course in association with Aichi Shukutoku University, Nagoya, Japan. The course is concerned with the mass media is an international context and students spend one semester at ASU, in Japan, and one semester at UUC, in Northern Ireland

UpGrade Media and Training Services

University of Exeter
St Luke's
Exeter EX1 2LU
Tel: 01392 264738
Fax: 01392 264736

UpGrade offers a variety of courses designed for both the educational and commercial markets. Courses and workshops based on VHS/SVHS cover pre and post production, camera techniques and off line editing. For those intending to work in the Media Industry, more advanced taster courses are offered using a full range of digital technology.

University of Wales, College of Cardiff

PO Box 908
French Section/German Section, EUROS
Cardiff CF1 3YQ
Tel: 01222 874000
Fax: 01222 874946

BA French

Study of French cinema included as part of optional courses. Small practical component

BA German

Study of contemporary German cinema forms part of optional courses

University of Wales College, Newport

University Information Centre
Caerleon Campus
PO Box 101
Newport NP6 1YH
Tel: 01633 432432
Fax: 01633 432850
email: uic@newport.ac.uk
Website: www.newport.ac.uk

BA (Hons) Film and Video

The course is intended for students wishing to explore the moving image in the broadest possible sense as an expressive and dynamic medium. It provides them with a programme of work designed to support and stimulate their personal development as creative and aware practitioners of film, regardless of their ultimate ambition. The practice of film is studied in a wider culture and intellectual context and students are encouraged to be analytical and critical. Their study acknowledges existing conventions in dominant cinema but seeks to extend them through experimentation and exploration

● **BA (Hons) Animation**

Intended for students wishing to use animation as part of a wider personal practice, as well as becoming professional animators working in independent production, advertising and design. The course is designed to develop students imaginations and ideas to explore and extend their animation technique. Therefore it is presented in a cultural context which promotes critical debate and rigorous analysis in terms of representation and expression. In Year 3 students develop their own programme for the production of major pieces of animation on high quality production equipment to broadcast standards

BA (Hons) Media and Visual Culture

At a time when our culture seems dominated as never before by the presence of media systems and images, this stimulating programme critically examines issues relating to media and visual culture. Specialist courses in the theory and history of film, photography design and contemporary art complement the central study of media culture. Practical options in subjects such as film, photography and new media can be selected, leading to major work involving practice in the final year.

● **MA Film**

This practical MA programme will offer an opportunity to: complete a

short broadcast standard film
explore and challenge the notions of the cinematic subject and language
explore developing forms and changing technologies
The course will include:
teaching through group and individual tutorials
close links with the film and television industry and media agencies in Wales
visiting masterclasses and facilities made available in professional production houses . The discursive and practical work on a short film will be in one of the following areas; Fiction, Faction, Animation, Non-genre/experimental. Acceptance of the course will be based on interview including: the submission of a treatment of a proposed film to be made during the course, the screening of a previous film, some fees/production bursaries could be available.

University College Warrington
University College (A college of the University of Manchester)
Media and Performing Arts
Padgate Campus
Crab Lane WA2 0DB
Tel: 01925 494494
Fax: 01925 816077
email: Media@Warr.ac.uk
● **BA (Joint Hons) Media with Business Management and Information Technology**
A modular degree structure allows students to undertake academic analysis of the media through core units in media forms, issues in media representation, media institutions and audiences, media professional issues, as well as a very wide range of optional units, including popular cinema, radio, media law, popular music and television, new media technologies, and European media. The course encourages students to relate media to their work in Business and IT. Students undertake extensive specialist production work chosen

from Television Production (in association with Granada TV), Radio Production, Music Recording Production, or Multimedia Journalism, as well as an extended work placement in the media industry
● **BA (Hons) Combined Studies**
A modular degree which enables students to combine academic study of the media and specialist production work in either Television, Radio, Music or Multimedia, commercial music production, Journalism, with a broad range of other modules, such as modern languages. All students undertake work experience in relevant industries
● **BA (Hons) Media and Cultural Studies**
A modular degree which enables wide-ranging study in media and cultural studies, as well as extensive production work chosen from Television Production (in association with Granada TV), Radio Production, Music Recording Production, or Multimedia Journalism. All

students undertake extensive and relevant work experience in the media industry

MA/PGDip Media and Cultural Studies

Offered on a one year full-time or two year part-time basis, this is a critical course designed to engage with key issues and debates in contemporary cultural analysis. Following core units in Classical and Post-Classical Narrative, Contemporary Theory and Critique, and Research Methods, students choose 3 optional units from a broad range on offer covering media and cultural issues. To gain the award of MA, students complete a 20,000 word dissertation

MA/PGDip Screen Studies

Offered on a one year full-time or two year part-time basis. Students take the same 3 core units as in the MA Media and Cultural Studies, but then choose 3 optional units specifically related to aspects of screen culture, including new media technologies. To gain the award of MA, students complete a 20,000 word dissertation

● **MA/PGDip Television Production** (in association with Granada TV) Run in association with Granada TV, this is a one year full-time fast track vocational course for those with serious aspirations to work in television production. Following intensive skills workshops, students are expected to produce largely, though not exclusively, factual-based programming to professional standards on Beta, using programme budgets supplied by Granada TV. The course includes 3-4 weeks work experience at Granada or BBC North. In addition to extensive production work, students develop an understanding of the contemporary broadcasting, independent and corporate sectors, and engage with aspects of television theory. To gain the award of MA, students complete a written project, which can be a conventional academic study, an industrial study or an original script. Granada TV bursaries are available for successful applicants

University of Warwick
Department of Film and Television Studies, Faculty of Arts
Coventry CV4 7AL
Tel: 01203 523523
Fax: 01203 524757

BA Joint Degree Film and Literature

Four courses offered each year, two in film and two in literature. Mainly film studies but some television included.

BA in Film with Television Studies

Four courses offered each year, three of which on film and, from year two, television. Further options available in film and/or television in year three

BA French or Italian with Film Studies

This degree puts a particular emphasis on film within and alongside its studies of French or Italian language, literature and society.

Various Degrees

Options in film studies can be taken as part of undergraduate degrees in other departments.

MA Film and Television Studies

Taught courses on Textual Analysis, Methods in Film History, Modernity and Innovation, and Issues of Representation, Introduction to Film and Television Studies for Graduates

MA for Research in Film and Television Studies

Combination of taught course and tailor-made programme of viewing and reading for students with substantial knowledge of film and television studies at BA level. For students wishing to proceed to PhD research

MA, MPhil and PhD

Students are accepted for research degrees

University of Westminster
School of Communication, Design and Media
Harrow Campus
Northwick Park HA1 3TP
Tel: 0171 911 5903
Fax: 0171 911 5955
email: hw06@wmin.ac.uk

BA (Hons) Film and Television

A modular degree course for young and mature students interested in film-making (fiction, documentary and experimental), television drama and documentary, screenwriting and film and television theory and criticism. The course emphasises creative collaboration and encourages some specialisation. It aims to equip students with understanding and competence in relevant critical ideas and the ability to work confidently and professionally in film and allied media using traditional and new technologies

BA (Hons) Media Studies

This degree studies the social context in which the institutions of mass communications operate, including film and television, and teaches the practice of print and

broadcasting journalism and video production. On levels 2 and 3 students choose one of the following pathways: radio, journalism or video production. The course gives equal emphasis to theory/criticism and practice. The video pathway is accredited by BECTU

MA PgD Film and Television Studies

Advanced level part-time course taught in Central London (evenings and study weekends) concerned with theoretical aspects of film and television. Modular credit and accumulation scheme, with exemption for work previously done. The MA is normally awarded after three years' study (120 credits). A Postgraduate Certificate can be awarded after one year (45 credits) or a Diploma after two years (75 credits)

● **BA (Hons) Contemporary Media Practice**

A modular three year full-time course offering an integrated approach to photography, film, video and digital-imaging. Students are encouraged to use a range of photographic and electronic media and theoretical studies are considered crucial to the development of ideas. In Years 1 and 2, the taught programme covers basic and applied skills on a project basis; these are complemented by a range of options. In Year 3 students are given the opportunity to develop their own programme of study, resulting in the production of major projects in practice and dissertations

Weymouth College
Creative & Performing Arts
Cranford Avenue
Weymouth
Dorset DT4 7LQ
Tel: 01305 208856
Fax: 01305 208892

● **GNVQ Intermediate and Advanced Levels Media Communication and Production BTEC)**

Intermediate level: one year course designed to give a foundation in media theory and production across audio, video and print media.

BTEC National Diploma in Media

The two year, full-time programme offers a broad foundation in media production skills, working practices and contextual knowledge. Core components include: video and audio production techniques; desk top publishing; photography, writing for the media; and research presentation skills. The second year allows for more specialisation in video or audio production but in a multi-skilled context. The programme is sponsored by Total Video Ltd, providing work experience and professional support for students

HND Video Production
Offers students multi-skilled script to screen training in video production techniques, underpinned by relevant study and research into current media issues and debates. Students undertake work on a range of productions exploring fictional and factual programme techniques. In Year 2 each student will produce, direct and crew on a variety of projects including commissioned programmes and a special one week residential location shoot

Wimbledon School of Art
Merton Hall Road
London SW19 3QA
Tel: 0181 540 0231
Fax: 0181 543 1750
● **BA (Hons) Fine Art**
Students enrol in either Painting or Sculpture. Some painting students study film and/or video

University of Wolverhampton
School of Humanities and Social Sciences, Castle View, Dudley West Midlands DY1 3HR
Tel: 01902 323400
Fax: 01902 323379
BA (Hons) Media and Cultural Studies
This is a modular programme which may be studied as a Specialist or Combined Award. Students follow a core programme covering key theoretical, historical and critical debates in both Media and Cultural Studies. A level 2 module in Research Methods prepares students for a final year project. Alongside the core students choose option modules drawn from the following themes: Film Studies; Video and Professional Communications (including Print and Broadcast Journalism, European Broadcasting and Organisational Communication); Gender and Representation (including Fashion, Style and Consumption). All students have the option of a Student Link, in which they undertake a small-scale research project in an organisation or company. This is not principally a practical programme but students taking the Video and Professional Communications Theme will undertake a range of practical work for assessment. This programme includes the option of studying a foreign language.

BA (Hons) Applied Communications
This modular programme may be studied as a Specialist or Combined Award and is concerned with issues and problems of human communication in their professional or industrial context. It draws upon media studies and planned communication, including marketing, journalism, public relations, corporate communication and interactive multimedia. The programme offers a balance between the theory and practice of different forms of planned communications. Specialists take a professional placement within an appropriate sector of the communication industry as part of their final year. This programme combines theory and practice and on completion students will have acquired a range communications skills in marketing, journalism, design for interactive multimedia and public relations. This programme includes the options of studying a foreign language.

BA (Hons) Film Studies
This modular programme will be available for the first time in 1988 as a Joint degree, which can be studied in combination with another subject. Students take a general foundation in Media Studies and then follow modules which introduce approaches to the analysis of films, studies of different genres (musicals, melodrama, film noir), theories of authorship, and the Hollywood studio system. The study of national cinemas includes options on British, French and Spanish cinema and special study of contemporary America. Film modules are taught at Wolverhampton's award-winning Light House Media Centre, which offers two purpose-built and fully equipped cinemas, library and exhibition facilities. The majority of films studied are screened in full cinema format. Additional library resources are housed on the Dudley and Wolverhampton campuses

DISTRIBUTORS (NON-THEATRICAL)

Companies here control UK rights for non-theatrical distribution (for domestic and group viewing in schools, hospitals, airlines and so on). For an extensive list of titles available non-theatrically with relevant distributors' addresses, see the **British National Film & Video Catalogue**, available for reference from BFI National Library and major public reference libraries. Other sources of film and video are listed under Archives and Film Libraries, and Workshops

Air-India
Publicity Dept
17-18 New Bond Street
London W1Y 0BD
Tel: 0181 745 1020
Fax: 0181 745 1059

Albany Video Distribution
Battersea Studios
Television Centre
Thackeray Road
London SW8 3TW
Tel: 0171 498 6811
Fax: 0171 498 1494
Education videos

Amber Films
5 Side
Newcastle upon Tyne NE1 3JE
Tel: 0191 232 2000
Fax: 0191 230 3217
email: amberside@btinternet.com

Arts Council of England
Film, Video & Broadcasting Dept
14 Great Peter Street
London SW1P 3NQ
Tel: 0171 973 6443
Fax: 0171 973 6581
Vicki Allsebrook
Support of production/distribution of films and video within two broad strands: programmes on arts subjects for television; film and video art; also access through sales/educational distribution and support of activities promoting film and video art. Also via Concord Video and Film Council

BBC for Business
Woodlands
80 Wood Lane
London W12 0TT
Tel: 0181 576 2361
Fax: 0181 576 2867

BFI Films (including Glenbuck Films)
21 Stephen Street
London W1P 2LN
Tel: 0171 957 8935
Fax: 0171 580 5830

email: bookings@bfi.org.uk
Handles non-theatrical 16mm, 35mm and video. Subject catalogues available

 BFI Production via BFI Films
21 Stephen Street
London W1P 2LN
Tel: 0171 957 8935
Fax: 0171 580 5830
email: bookings@bfi.org.uk
(see p8)

BUFVC (British Universities Film & Video Council)
77 Wells Street
London W1P 3RE
Tel: 0171 393 1500
Fax: 0171 393 1555
email: bufvc@open.ac.uk
Videocassettes and videodiscs for sale. Also off-air recording back-up service for education. Hire via Concord Video and Film Council

Big Bear Records
PO Box 944
Birmingham B16 8UT
Tel: 0121 454 7020/8100
Fax: 0121 454 9996

Black Audio Films
21b Brooksby Street
Islington
London N1 1EX
Tel: 0171 607 6161
Fax: 0171 607 7171
Lina Gopaul, Avril Johnson
Black independent films (see Distributors (Theatrical) for titles)

Boulton-Hawker Films
Hadleigh
near Ipswich
Suffolk IP7 5BG
Tel: 01473 822235
Fax: 01473 021510
Educational films and videos: health education, social welfare, home economics, P.S.E., P.E., Maths, biology, physics, chemistry, geography

John Burder Films
7 Saltcoats Road
London W4 1AR
Tel: 0181 995 0547
Fax: 0181 995 3376

CFL Vision
PO Box 35
Wetherby
Yorks LS23 7EX
Tel: 01937 541010
Fax: 01937 541083
email: euroview@compuserve.com
Website: http://www.euroview.co.uk

CTVC Video
Hillside Studio, Merry Hill Road
Bushey
Watford WD2 1DR
Tel: 0181 950 4426
Fax: 0181 950 1437
Christian, moral and social programmes

Castrol Video and Film Library
Athena Avenue
Swindon
Wilts SN2 6EQ
Tel: 01793 693402
Fax: 01793 511479

Carlton UK Television
Video Resource Unit
Broad Street
Birmingham B1 2JP
Tel: 0121 643 9898
Fax: 0121 634 4259

Central Office of Information
Tel: 0171 261 8594
Fax: 0171 261 8874
via CFL Vision

Chain Production
2 Clanricarde Gardens
London W2 4NA
0171 219 4277
0171 229 0861
Specialist in European films and world cinema, cult classics, handling European Film Libraries with all rights to over 1,000 films - also clip rights and clip search

Channel Four International

124 Horseferry Road
London SW1P 2TX
Tel: 0171 396 4444
Fax: 0171 306 8363
email: idjones@channel4.co.uk
Laurence Dawkin-Jones

Le Channel Ltd

10 Frederick Place
Weymouth
Dorset DT4 8HT
Tel: 01305 780446
Fax: 01305 780446
Agatha Rodgers
Art videos about famous paintings of
the Western World

Cinema Action

27 Winchester Road
London NW3 3NR
Tel: 0171 267 6878
Documentaries and mixed media

Cinenova (Women's Film and Video Distribution)

113 Roman Road
London E2 0QN
Tel: 0181 981 6828
Fax: 0181 983 4441
email: admin@cinenova.demon.co.uk
Website: http://www.luton.ac.uk/
cinenova/
Promotion and distribution of films
and videos by women spanning 90
years of film-making from around
the world

Concord Video and Film Council

201 Felixstowe Road
Ipswich, Suffolk IP3 9BJ
Tel: 01473 726012
Fax: 01473 274531
email: ericwalker@gn.apc.org
Lydia Vulliamy
Videos and films for hire/sale on
domestic and international social
issues - counselling, development,
education, the arts, race and gender
issues, disabilities, etc - for training
and discussion. Also incorporates
Graves Medical Audio Visual Library

Connaught Training

Gower House
Croft Road
Aldershot
Hants GU11 3HR
Tel: 01252 317700
Fax: 01252 343151
Training videos

Darvill Associates

280 Chartridge Lane
Chesham
Bucks HP5 2SG
Tel: 01494 783643
Fax: 01494 784873

Derann Film Services

99 High Street
Dudley
West Mids DY1 1QP
Tel: 01384 233191/257077
Fax: 01384 456488
D Simmonds, S Simmonds
8mm package movie distributors;
video production; bulk video
duplication; laser disc stockist

Education Distribution Service

Education House
Castle Road
Sittingbourne
Kent ME10 3RL
Tel: 01795 427614
Fax: 01795 474871
Distribution library for many clients
including film and video releases.
Extensive catalogue available

Educational and Television Films

247 Upper Street
London N1 1RU
Tel: 0171 226 2298
Fax: 0171 226 8016
Documentary films from Eastern
Europe. Archive film library

Educational Media, Film & Video

235 Imperial Drive
Rayners Lane
Harrow HA2 7HE
Tel: 0181 868 1908/1915
Fax: 0181 868 1991
Lynda Morrell
Distributors of British and overseas
educational, health, training/safety
video games as well as new CD-
ROM titles. Act as agent for the
promotion of British productions
overseas. Free catalogue

Electricity Association Video Library

30 Millbank
London SW1P 4RD
Tel: 0171 963 5827
Fax: 0171 963 5800

Film Quest Ltd

71 (b) Maple Road
Surbiton
Surrey KT6 4AG
Tel: 0181 390 3677
Fax: 0181 390 1281
Booking agents for university,
school and private film societies

Filmbank Distributors

Grayton House
498-504 Fulham Road
London SW6 5NH
Tel: 0171 386 9909/5411
Fax: 0171 381 2405
Bookings Department
Filmbank represents all of the major
film studios for the non-theatrical

market (group screenings) and
distributes titles on either 16mm
film or video

First Take Ltd

19 Liddell Road
London NW6 2EW
Tel: 0171 328 4676

Ford Film and Video Library

Ford Motor Company
Room 1/455, Eagle Way
Brentwood CM13 3BW
Tel: 01277 252442
Fax: 01277 252896

Golds

The Independent Home
Entertainment Wholesaler
Gold House, 69 Flempton Street
Leyton
London E10 7NL
Tel: 0181 539 3600
Fax: 0181 539 2176
Contact: Garry Elwood, Sales &
Marketing Director
Gold product range ever increasing
multi-format selection including:
Audio cassettes, CD's, T-Shirts,
DVD, Spoken Word Cassettes and
CDs, Video, CD Rom, CDi, Video
CD, Laserdisc, computer games and
accessories to all formats. 42 years
of service and expertise 32 years
spent in home entertainment market

Sheila Graber Animation Limited

50 Meldon Avenue
South Shields
Tyne and Wear NE34 0EL
Tel: 0191 455 4985
Fax: 0191 455 3600
email: sheila@graber.demon.co.uk
Over 70 animated shorts available -
16mm, video and computer
interactive featuring a range of 'fun'
educational shorts on art, life, the
universe and everything. Producers
of interactive CD-Roms

IAC (Institute of Amateur Cinematographers)

24c West Street, Epsom
Surrey KT18 7RJ
Tel: 01372 739672

IUN Entertainment

Centre 500
500 Chiswick High Road
London W4 5RG
Tel: 0181 956 2454
Fax: 0181 956 2339

Imperial War Museum

Film and Video Archive (Loans)
Lambeth Road
London SE1 6HZ
Tel: 0171 416 5000
Fax: 0171 416 5379

Brad King/Toby Haggith
Documentaries, newsreels and
propaganda films from the Museum's
film archive on 16mm, 35mm and video

Leeds Animation Workshop (A Women's Collective)
45 Bayswater Row
Leeds LS8 5LF
Tel: 0113 248 4997
Fax: 0113 248 4997
Milena Dragic
Producers and distributors of
animated films on social issues

London Electronic Arts
Lux Building
2-4 Hoxton Square
London N1 6NU
Tel: 0171 684 0101
Fax: 0171 684 1111
email: info@lea.org.uk
Website: http://www.lea.org.uk
Britain's national centre for video
and new media art, housing the most
extensive collection of video art in
the country

London Film-Makers' Co-op
Lux Centre
2-4 Hoxton Square
London N1 6NU
Tel: 0171 684 0202
Fax: 0171 684 2222
email: 101563.332@compuserv.com
Beth Copley
Experimental/art-based films: 2,000
classic and recent titles for hire from
1920s to current work, new
catalogue available

Melrose Film Productions
Dumbarton House
68 Oxford Street
London WIN OLH
Tel: 0171 627 8404
Fax: 0171 622 0421

National Educational Video Library
University of Bangor
Arts and Crafts Building
Hen Goleg Site
Bangor
Gwynedd LL57 2DG
Tel: 01248 353053
Fax: 01248 370144
Educational audio visual aids
consisting of videotapes, 16mm
films, slides and overhead projector
transparencies available for hire or
purchase

National Film and Television School
Beaconsfield Studios
Station Road
Beaconsfield

Bucks HP9 1LG
Tel: 01494 671234
Fax: 01494 674042
Karin Farnworth

Open University Worldwide
The Berrill Building
Walton Hall
Milton Keynes MK7 6AA
Tel: 01908 858785
Fax: 01908 858787
email: D.M.Ruault@open.ac.uk

Edward Patterson Associates
Treetops, Cannongate Road
Hythe
Kent CT21 5PT
Tel: 01303 264195
Fax: 01303 264195
Health, safety and education video
programming

Post Office Video and Film Library
PO Box 145
Sittingbourne
Kent ME10 1NH
Tel: 01795 426465
Fax: 01795 474871
Includes many video programmes
and supporting educational material
including curriculum guidelines.
Also a comprehensive range of
extension and other curriculum
linked material. TV rights available

RSPCA
Causeway
Horsham
West Sussex RH12 1HG
Tel: 01403 264181
Fax: 01403 241048

RoSPA
Film Library, Head Office
Edgbaston Park
353 Bristol Road
Birmingham B5 7ST
Tel: 0121 248 2000
Fax: 0121 248 2001
Safety films

Royal College of Art
Department of Film and Television
Kensington Gore
London SW7 2EU
Tel: 0171 584 5020 x 412
Fax: 0171 589 0178

Royal Danish Embassy
55 Sloane Street
London SW1X 9SR
Tel: 0171 333 0200
Fax: 0171 333 0270

The Short Film Bureau
68 Middle Street
Brighton BN1 1AL
Tel: 01273 235524/5

Fax: 01273 235528
email: Matt@shortfilmbureau.com
Specialising in the promotion and
distribution of short films for
theatrical and non-theatrical release
world wide

South West Arts
Bradninch Place
Gandy Street
Exeter EX4 3LS
Tel: 01392 218188
Fax: 01392 413554
Vikki Scott, Kirsty Hayes

THE (Total Home Entertainment)
National Distribution Centre
Rosevale Business Park
Newcastle-under-Lyme
Staffs ST5 7QT
Tel: 01782 566566
Fax: 01782 565400
Mike Fay, Jed Taylor
Exclusive distribution for Paradox, Pide,
Kiseki, NTV, Sportsworld, Dangerous to
Know, Labyrinth, Grosvenor, Empire
and distribution of over 3,000 other titles
(see also Video Labels)

TV Choice
22 Charing Cross Road
London WC2H 0HR
Tel: 0171 379 0873
Fax: 0171 379 0263

Team Video Productions
Canalot
222 Kensal Road
London W10 5BN
Tel: 0181 960 5536
Fax: 0181 960 9784
Chris Thomas, Billy Ridgers
Producer and distributor of
educational video resources

Television History Centre
27 Old Gloucester Street
Queen Square
London WC1N 3AF
Tel: 0171 405 6627
Fax: 0171 242 1426
Programmes for hire or purchase
about work, education, health,
women, community action,
particularly suitable for educational
discussion groups 16 plus. See also
under Production Companies

Texaco Film Library
Texaco Ltd
Public Affairs and Advertising
1 West Ferry Circus
London E14 4HA
Tel: 0171 719 3000
Fax: 0171 719 3172

Thames Television Video Sales
via Yorkshire International Thomson
Multimedia (YITM)

Training Direct

Longman House, Burnt Mill
Harlow, Essex CM20 2JE
Tel: 01279 623927
Fax: 01279 623795
Leading training resource provider,
including: multimedia, video, audio
and print based material covering all
aspects of skills development and
Health and Safety training

Training Services

Brooklands House
29 Hythegate
Werrington
Peterborough PE4 7ZP
Tel: 01733 327337
Fax: 01733 575537
email: tipton@training
services.demon.co.uk
Distribute programmes from the
following producers:

3E's Training
Aegis Healthcare
Angel Productions
Barclays Bank Film Library
CCD Product & Design
Flex Training
Grosvenor Career Services
Hebden Lindsay Ltd
Kirby Marketing Associates
McPherson Marketing
Promotions Sound & Vision
Schwops Productions
Touchline Training Group
Video Communicators Pty
John Burder Films
Easy-i Ltd

Transatlantic Films

Cabalva Studios
Whitney-on-Wye
Hereford HR3 6EX
Tel: 01497 831428
Fax: 01497 831565

The University of Westminster

School of Communications, Design
and Media
Harrow Campus
Northwick Park HA1 3TP
Tel: 0171 911 5003
Fax: 0171 911 5955
email: hwo6@wmin.ac
Applications: com.3@wmin.ac.uk

Vera Media

30-38 Dock Street
Leeds LS10 1JF
Tel: 0113 242 8646
Fax: 0113 245 1238
email: vera@vera-
media.demon.co.uk

Video Arts

Dumbarton House
68 Oxford Street
London W1N 0LH

Tel: 0171 637 7288
Fax: 0171 580 8103
Video Arts produces and exclusively
distributes the John Cleese training
films; Video Arts also distributes a
selection of meeting breaks from
Muppet Meeting Films TM as well
as Tom Peters programmes (produced by Video Publishing House
Inc) and In Search of Excellence and
other films from the Nathan/Tyler
Business Video Library

Viewtech Film and Video

7-8 Falcons Gate
Northavon Business Centre
Dean Road, Yate
Bristol BS37 5NH
Tel: 01454 858055
Fax: 01454 858056

WFA

9 Lucy Street
Manchester M15 4BX
Tel: 0161 848 9782/5
Fax: 0161 848 9783
email:wfa@timewarp.com.uk

Westbourne Film Distribution

1st Floor, 17 Westbourne Park
Road
London W2 5PX
Tel: 0171 221 1998
Fax: 0171 221 1998
Agents for broadcasting/video sales
for independent animators from
outside the UK, particularly Central
Eastern Europe. Classic children's
film *The Singing Ringing Tree*

Yorkshire International Thomson Multimedia

Television Centre
Leeds LS3 1JS
Tel: 0113 222 8369
Fax: 0113 243 4884
email: yitminfo@yitm.co.uk
Distributor for many ITV companies, including Yorkshire-Tyne Tees,
Thames, and Channel 4. Has an
educational catalogue, containing
videos and CD-Rom's, and further
industry catalogue, providing
training packages for companies

DISTRIBUTORS (THEATRICAL)

These are companies which acquire the UK rights to films for distribution to cinemas and, in many cases, also for sale to network TV, satellite, cable and video media. Listed is a selection of recently certificated features, and/or some past releases or re-releases available during this period

Alliance Releasing
(formerly Electric Pictures)
2nd Floor, 184-192 Drummond Street
London NW1 3HP
Tel: 0171 391 6900
Fax: 0171 383 0404
Film distribution. Films in joint distribution deals with PolyGram:
Before the Rain
Blood Simple
Butterfly Kiss
The Celluloid Closet
Dead Man
Death and the Maiden
Fallen Angels
The Flower of My Secret
kids
Kitchen
The Last Supper
Electric Pictures:
Albino Alligator
Arizona Dream
Big Night
Bits and Pieces/Il cielo è sempre più blu
Gallivant
Red Firecracker, Green Firecracker
The Sweet Hereafter
Trees Lounge

Apollo Film Distributors
14 Ensbury Park Road
Bournemouth BH9 2SJ
Tel: 01202 520962
Fax: 01202 539969

Arrow Film Distributors
18 Watford Road
Radlett
Herts WD7 8LE
Tel: 01923 858 306
Fax: 01923 859673
Angie Agran
L'amore molesto
Fiorile
Kaspar Hauser
Sister My Sister
Salut Cousin

Artificial Eye Film Company
14 King Street
London WC2E 8HN
Tel: 0171 240 5353
Fax: 0171 240 5242
Contact: Robert Beeson, Sam Shinton
L'Appartement
Flirt

Mon homme
Regeneration
A Self-Made Hero/Un héro très discret
The Tango Lesson
Temptress Moon/Fengyue
Unhook the Stars/Décroches les étoiles
Welcome to the Dollhouse
Will it Snow for Christmas?/Y'aura t'il de la neige à Noël
Your Beating Heart/Un Coeur qui bat

Asian Film Company
The Asian Film Library
Suite 19, 2 Lansdowne Row
Berkeley Square
London W1X 8HL
Tel: 07970 506326
Fax: 0171 493 4935
Stephen Cremin
April Story
Fried Dragon Fish
Fireworks

BFI Films
21 Stephen Street
London W1P 2LN
Tel: 0171 957 8905
Fax: 0171 580 5830
email: bookings@bfi.org.uk
Website: http//:www.bfi.org.uk
Contact: Heather Stewart (see p8)
Features, including:
A Bit of Scarlet
Cloud-capped Star/Meghe Dhaka Tara
Close-Up/Namayeh Nazdik
Frantz Fanon Black Skin White Mask
His Girl Friday
Plein soleil/Delitto in pieno sole/Blazing Sun/Purple Noon
Smalltime
Under the Skin

Black Audio Films
21b Brooksby Street
Islington
London N1 1EX
Tel: 0171 607 6161
Fax: 0171 607 7171
Black independent films
Beaton but Unbowed
Black Cab
Handsworth Songs
Mysteries of July
Seven Songs for Malcolm X
A Touch of the Tar Brush
Twilight City
Three Songs on Pain, Light and Time

Mothership Connection
The Last Angel of History
Martin Luther King - Days of Hope
Who Needs A Heart
The Darker Side of Black

Blue Dolphin Film & Video
40 Langham Street
London W1N 5RG
Tel: 0171 255 2494
Fax: 0171 580 7670
Joseph D'Morais
The Boy From Mercury
The Butterfly Effect/El efecto mariposa
Driftwood
Madame Butterfly
No Way Home
The Square Circle/Daayraa
Tokyo Fist

Blue Light
(See Made in Hong Kong)
231 Portobello Road
London W11 1LT
Tel: 0171 792 9791
Fax: 0171 792 9871

Buena Vista International (UK)
Beaumont House
Kensington Village
Avonmore Road
London W14 8TS
Tel: 0171 605 2890
Fax: 0171 605 2827
Contact: Daniel Battsek
Air Force One
Con Air
CopLand
The English Patient
Everyone Says I Love You
Face/Off
Flirting With Disaster
George of the Jungle
Grosse Pointe Blank
Hercules
Jungle 2 Jungle
Kolya
Ma vie en rose
Marvin's Room
Metro
Mrs. Brown
Nothing to Lose
The Preacher's Wife
Ransom
Romy and Michele's High School Reunion
Rumble in the Bronx/Hongfan Qu
Scream

John Burder Films

7 Saltcoats Road
London W4 1AR
Tel: 0181 995 0547
Fax: 0181 995 3376
Broadcast TV programmes
See also under Distributors (Non-Theatrical)

Carlton Film Distributors

(Formerly Rank Film Distributors)
127 Wardour Street
London W1V 4AD
Tel: 0171 224 3339
Fax: 0171 434 3689
8 Heads in a Duffel Bag
Lawn Dogs
SubUrbia

Cavalcade Films

Regent House
235-241 Regent Street
London W1R 8JU
Tel: 0171 734 3147
Fax: 0171 734 2403

Chain Production

2 Clanricarde Gardens
London W2 4NA
Tel: 0171 229 4277
Fax: 0171 229 0861
Specialist in European films and
world cinema, cult classics, handling
European Film Libraries with all
rights to over 1,000 films - also clip
rights and clip search

Cinenova

113 Roman Road
London E2 0QN
Tel: 0181 981 6828
Fax: 0181 983 4441
email: admin@cinenova.demon.co.uk
Website: http://www.luton.ac.uk/
cinenova/
Promotion and distribution of
women's film and video
B/Side
Deviant Beauty
Dialogues with Madwomen
Jodie: An Icon
A Life in a Day with Helena Goldwater
Mad, Bad and Barking
Sluts and Goddesses Video Workshop
True Blue Camper
See also under Distributors (Non-Theatrical)

Columbia TriStar Films (UK)

Europe House
25 Golden Square
London W1R 6LU
Tel: 0171 533 1111
Fax: 0171 533 1105
Feature releases from Columbia,
TriStar, and Orion Pictures
Adrenalin Fear the Rush
Anaconda
Booty Call
Crash

The Devil's Own
Excess Baggage
Fly Away Home
Fools Rush
Get on the Bus
High School High
Jerry Maguire
Maximum Risk
Men in Black
The Mirror Has Two Faces
My Best Friend's Wedding
Never Talk To Strangers
The People vs. Larry Flynt
The Swan Princess The Secret of the Castle

Contemporary Films

24 Southwood Lawn Road
Highgate
London N6 5SF
Tel: 0181 340 5715
Fax: 0181 348 1238
La Règle du jeu
Battleship Potemkin

Dangerous to Know

4th Floor
17a Newman Street
London W1P 3HB
Tel: 0171 735 8330
Fax: 0171 793 8488
email: (name)@dtk.co.uk
Website: www.dtk.co.uk
Frisk
Latin Boys Go to Hell
Lillies
Sebastian
Skin & Bone
The Watermelon Woman

Documedia International Films Ltd

Programme Sales/Acquisitions
19 Widegate Street
London E1 7HP
Tel: 0171 625 6200
Fax: 0171 625 7887
Distributors of award winning drama
specials, drama shorts and feature
films for theatrical release, also
video sales/Internet and video on
demand. Drama specials -
Soulscapes;
Telemovies/Features - *Deva's Forest,
Leaves and Thorns*
International short drama for
theatrical release, also educational/
film club and non theatrical release -
*Pile of Clothes, JoyRidden, The
Cage, Nazdrovia, Thin Lines, Late
Fred Morse, The Extinguisher, The
Summer Tree, Arch Enemy, Tea and
Bullets, Isabelle, Peregrine, Beyond
Reach, Trauma, Edge of Night,
Something Wonderful*

Double: Take

21 St Mary's Grove
London SW13 0JA
Tel: 0181 788 5743
Fax: 0181 785 3050
Maya Kemp
Distributors and producers of

children's TV and video
The Clangers
Crystal Tipps and Alistair
Fred Basset
Ivor the Engine
Willow the Wisp

Electric Avenue

191 Portobello Road
London W11 2ED
Tel: 0171 792 2020
Fax: 0171 792 2617
Brian Bonaparte

Electric Pictures

(see Alliance Releasing)

Entertainment Film Distributors

27 Soho Square
London W1V 6HU
Tel: 0171 439 1606
Fax: 0171 734 2483
An American Werewolf in Paris
The Arrival
BAPS
Dangerous Ground
Donnie Brasco
Evening Star
*First Strike/Jingcha gushi 4 Zhi Jiandan
Renwu*
Hard Men
Hard Eight
I Know What You Did Last Summer
In Love and War
It Takes Two
love jones
Love! Valor! Compassion!
Mother Night
One Night Stand
Photographing Fairies
Preaching to the Perverted
Prince Valiant/Printz Eisenherz
Private Parts
Set It Off
Seven Years in Tibet
Shadow Conspiracy
Shooting Fish
Space Truckers
Spawn
Trial and Error
Turbulence
Warriors of Virtue

Feature Film Company

68-70 Wardour Street
London W1V 3HP
Tel: 0171 734 2266
Fax: 0171 494 0309
The Blackout
Cold Comfort Farm
Das Boot; the Director's Cut
Female Perversions
Gang Related
It's A Wonderful Life
My Son the Fanatic
The Myth of Fingerprints
Total Eclipse
Ulees Gold
Quadrophenia
Wild Man Blues

Film and Video Umbrella

2 Rugby Street
London WC1N 3QZ

Tel: 0171 831 7753
Fax: 0171 831 7746
email: umbrella@cityscape.co.uk
Promoting innovation with film and
the electronic image
Bill Viola: Landscapes of the Mind
Daniel Reeves: Ongoing Obsessions
Never A Dumb Moment - new
artists' video work. *Computer World
3* - international computer animation
showcase. *The Digital Underground
VideoFile* - series of programmes
designed for use by the education
sector. *Electric Dance* - two
programme video dance package
exploring movement and the body in
digital space

Film Booking Offices
19 Penfields House
Market Road
London N7 9PZ
Tel: 0171 700 7055
Fax: 0171 700 7055
Distribute the Associated-
Rediffusion Television Library

Film Four Distributors
Castle House
75-76 Wells Street
London W1P 3RE
Tel: 0171 436 9944
Fax: 0171 436 9955
Alive and Kicking/Indian Summer
Career Girls
Fever Pitch
A Further Gesture
The Gambler
Jump the Gun
Kama Sutra A Tale of Love
*Martha - Meet Frank, Daniel and
Lawrence*
Portraits Chinois
Land Girls
Remember Me?
The Slab Boys
*Someone Else's America/L'Amerique des
autres*
Trojan Eddie
Velvet Goldmine
Welcome to Sarajevo
The Winter Guest

First Independent Film Ltd
99 Baker Street
London W1M 1FB
Tel: 0171 317 2500
Fax: 0171 317 2502/2503
Peter Naish, Managing Director
Box of Moonlight
Broken English
Eddie
G.I. Jane
House of America
Keep the Aspidistra Flying
Killer a Journal of Murder
Night Falls on Manhattan
This World, Then the Fireworks
Trigger Happy

Gala
(See Porter Frith)

*Love Lessons/Lust och fagring stor/
Laererinden*
*Men, Women: Users Manual/Hommes
femmes Mode d'emploi*

The Samuel Goldwyn Company
St George's House
14-17 Wells Street
London W1P 3FP
Tel: 0171 436 5105
Fax: 0171 580 6520
Angels and Insects
I Shot Andy Warhol
Reckless

Guild Entertainment
(See Pathé)
Kent House
Market Place
London W1N 8AR
Tel: 0171 323 5151
Fax: 0171 631 3568
The Addiction
*Austin Powers International Man of
Mystery*
Basquiat
Bound
Crying Freeman
The Fifth Element/Le Cinquiéme Elément
The Funeral
Larger Than Life
The Leading Man
*Microcosmos/Microcosmos Le Peuple de
l'herbe*
Swann
Swingers

HandMade Films
19 Beak Street
London W1R 3LB
Tel: 0171 434 3132
Fax: 0171 434 3143
Gareth Jones, Hiliary Davis, Jan
Roldanus, Juliette Gill
Complete films include:
Intimate Relations
Sweet Angel Mine
The James Gang
The Man With Rain in His Shoes
The Secret Laughter of Women
Lock, Stock & Two Smoking Barrels
Sweety Barrett.
Paragon Entertainment Corporation
acquired HandMade Films'
catalogue of 23 feature films in
August 1994. The entire catalogue
now amounts to 33 titles. HandMade

Hemdale Communications
10 Avenue Studios, Sydney Close,
Chelsea
London SW3 6HW
Tel: 0171 581 9734
Fax: 0171 581 9735
The Legend of Wolf Mountain
Little Nemo – Adventures in Slumberland
The Magic Voyage
The Mighty Kong
Quest of the Delta Knights
Savage Land

Hourglass Distribution
Charlton House

Charlton Road
London SE7 8RE
Tel: 0181 858 6870/0181 319 8949
Fax: 0181 858 6870
Contacts: John Walsh, David Walsh,
Maura Walsh
New distribution and sales wing of
Hourglass Productions. Specialising
in the theatrical and TV distribution
and sales of documentary and drama
shorts. Features include:
The Frozen Four
In From the Storm
Monarch
Shorts include:
Comedy Store
*Masque of Draperie in Presence of HRH
The Queen*
Sceptic & The Psychic
The Sleeper
Spiritual World
A State of Mind

ICA Projects
12 Carlton House Terrace
London SW1Y 5AH
Tel: 0171 930 0493
Fax: 0171 930 9686
Conspirators of Pleasure/Spiklenci slasti
Fetishes
Gabbeth
Irma Vep
Kids Return

Indy UK
Independent Feature Film
Distributors
13 Mountview
Northwood
Middlesex HA6 3NZ
Tel: 1923 820330
Fax: 1923 820518
email: indyuk@realit.demo.co.uk
The Scarlet Tunic

Made in Hong Kong/Blue Light
231 Portobello Road
London W11 1LT
Tel: 0171 792 9791
Fax: 0171 792 9871
Made in Hong Kong releases the
finest in Hong Kong cinema
Full Contact/Xia Dao Gao Fei
Heroic Trio
The Killer
Saviour of the Soul
Blue Light distributes European and
other titles

Mainline Pictures
37 Museum Street
London WC1A 1LP
Tel: 0171 242 5523
Fax: 0171 430 0170
Bandit Queen
The Diary of Lady M
Go Fish
The Premonition
Ruby in Paradise
Smoking/No Smoking
The Wedding Banquet

Manga Entertainment Ltd
40 St Peter's Road
London W6 9BD
Tel: 0181 748 9000
Fax: 0181 748 0841
Ghost in the Shell
Dancehall Queen
Razor Blade Smile
Gravesend
Perfect Blue

Mayfair Entertainment UK Ltd
110 St Martins Lane
London WC2N 4AD
Tel: 0171 304 7922
Fax: 0171 867 1121

Medusa Communications & Marketing Ltd
Regal Chambers, 51 Bancroft
Hitchin
Herts SG5 1LL
Tel: 01462 421818
Fax: 01462 420393
The Brides Guide to a Traditional
Wedding
Harry the Horny Hypnotist
Howie and the Leprechauns Buemania
The Cool Surface
Shape Up and Rave
Ticks
Come Play With Me
Playbirds
Eskimo Nell
The David Galaxy Affair
Emanuelle in Soho
Queen of the Blues
Naughty
Groupie Girl
The Love Box
The Wife Swappers

Metro Tartan Distribution Ltd
79 Wardour Street
London W1V 3TH
Tel: 0171 734 8508/9
Fax: 0171 287 2112
Un Air De Famille
Deep Crimson
Dobermann
Drifting Clouds/Kauas pilvet kerkaavat/
Au loin s'en vont les nuages
Funny Games
JunkMail
Kissed
Ponette
Prisoner of the Mountain
Tierra
To Have and To Hold
Les Voleurs

Metrodome Distribution
3rd Floor
25 Maddox Street
London W1R 9LE
Tel: 0171 408 2121
Fax: 0171 409 1935
Metrodome Distribution is part of the Metrodome Group. The distribution arm was set up in order to distribute films that Metrodome Films produces, as well as to actively

acquire and release another 8-10 films per year. 1997 films include:
Chasing Amy
Darklands
Johns
Margaret's Museum
The Near Room
Palookaville
Pusher
1998 films:
The Daytrippers
Metroland
The Real Blonde
Stiff Upper Lips

Miracle Communications
69 New Oxford Street
London WC1A 1DG
Tel: 0171 528 7767
Fax: 0171 528 7770
Michael Myers

New Line International
25-28 Old Burlington Street
London W1X 1LB
Tel: 0171 440 1040
Fax: 0171 439 6118
Camela Galano
European film distribution
Lost in Space
Mortal Kombat 2
Wag The Dog
One Night Stand
Boogie Nights

Oasis Cinemas and Film Distribution
20 Rushcroft Road
Brixton
London SW2 1LA
Tel: 0171 733 8989
Fax: 0171 733 8790
email: oasiscinemas@compuserve.co.uk
Dancehall Queen
Gravesend
Laws of Gravity
The Lunatic
The Secret Rapture

Pathé Distribution
Kent House
14-17 Market Place
Great Titchfield Street
London W1N 8AR
Tel: 0171 323 5151
Austin Powers International Man of Mystery
Le Bouffe
The Fifth Element
The Grass Harp
If Only
Live Flesh
Ma vie sexuelle (How I Got Into an
Argument)/Comment je me suis
disputé...("ma vie sexuelle")
Love and Death on Long Island
Spanish Prisoner

Pilgrim Entertainment
1 Richmond Mews
London W1
Tel: 0171 287 6314
Fax: 0181 961 6454
White Angel

PolyGram Filmed Entertainment
1 Sussex Place
Hammersmith
London W6 9QB
Tel: 0181 910 5000
Fax: 0181 910 5121
Contact: Julia Short
The Associate
Bang
Bean
The Borrowers
Carla's Song
City of Industry
Def Jam's How to be a Player
The Game
Gridlock'd
Guy
Keys to Tulsa
A Life Less Ordinary
Lost Highway
The Portrait of a Lady
The Relic
Roseanna's Grave
Snow White A Tale of Terror
Spice World The Movie
Trainspotting
Twin Town
When We Were Kings
Wilde

Porter Frith
(Formerly Gala Films)
26 Danbury Street
Islington
London N1 8JU
Tel: 0171 226 5085
Fax: 0171 226 5897
Artemisia
Men, Women: A User's manual
Marius et Jeanette

Poseidon Film Distributors
Hammer House
117 Wardour Street
London W1V 3TD
Tel: 0171 734 4441
Fax: 0171 437 0638
Autism
Dyslexia
Russian Composers - Writers
The Steal
Animation series: *The Bears*
"The Odyssey"
The Night Witches

Squirrel Films Distribution
119 Rotherhithe Street
London SE16 4NF
Tel: 0171 231 2209
Fax: 0171 231 2119
As You Like It
The Fool
The Little Gang
Tales of Beatrix Potter
Amahl and Ki Wight Viritose
The Swann Princess

Starlight
(now Metrodome)
86 Dean Street
London W1V SAA
Tel: 0171 734 6877

Glastonbury The Movie
The Grotesque
Guiltrip
Kids in the Hall Brain Candy
The Secret Agent Club
Swimming with Sharks

Supreme Film Distributors

Premier House
77 Oxford Street
London W1R 1RB
Tel: 0171 437 4415
Fax: 0171 734 0924

TKO Communications

PO Box 130
Hove, East Sussex BN3 6QU
Tel: 01273 550088
Fax: 01273 540969
Adventures of Scaramouche
Catch me a Spy
The Diamond Mercenaries
Gallavants
The Investigator
The Mark of Zorro
Swan Lake (featuring stars of the Kirov
and Bolshoi Ballets)
The Three Musketeers (animated version)

Twentieth Century Fox Film Co

20th Century House
31-32 Soho Square
London W1V 6AP
Tel: 0171 437 7766
Fax: 0171 734 3187
Alien: Resurrection
Blood and Wine
The Crucible
The Full Monty
Home Alone 3
Intimate Relations
Inventing the Abbotts
Looking For Richard
Love and Other Catastrophes
Moll Flanders
Nil By Mouth
One Fine Day
Paradise Road
Picture Perfect
She's The One
Smilla's Feeling for Snow/Fräulein
Smillas Gespür für Schnee
Speed 2 Cruise Control
The Starmaker/L'uomo delle stelle
Star Wars IV A New Hope
That Thing You Do!
Unforgettable
Volcano
White Man's Burden
William Shakespeare's Romeo + Juliet

UIP (United International Pictures (UK))

12 Golden Square
London W1A 2JL
Tel: 0171 534 5200
Fax: 0171 636 4118
Releases product from Paramount,
Universal, MGM/UA and SKG
DreamWorks
Beavis and Butt-Head Do America
The Chamber
Dante's Peak

Event Horizon
Face
Fierce Creatures
The Frighteners
Grace of My Heart
Harriet the Spy
Kiss Me, Guido
Liar Liar
The Lost World Jurassic Park
The Peacemaker
The Phantom
The Quest
The Saint
A Simple Wish
That Old Feeling
Tomorrow Never Dies

Warner Bros Distributors

135 Wardour Street
London W1V 4AP
Tel: 0171 734 8400
Fax: 0171 437 2950
Contact: Maj-Britt Kirchner
187
Addicted to Love
Anna Karenina
Batman & Robin
Conspiracy Theory
Contact
*Entertaining Angels the Dorothy Day
Story*
Eyes Wide Shut
Father's Day
Free Willy 3 The Rescue
Head Above Water
I Am Legend
Incognito
L.A. Confidential
Mars Attacks!
Murder at 1600
Lethal Weapon 4
La Passione
A Perfect Murder
Practical Magic
The Proprietor/La Propriétaire
Soldier
Space Jam
Stephen King's Thinner
You Have Mail

Westbourne Film Distribution

1st Floor,
17 Westbourne Park Road
London W2 5PX
Tel: 0171 221 1998
Fax: 0171 221 1998
Agents for broadcasting/video sales
for independent animators from
outside the UK, particularly Central
Eastern Europe. Classic children's
film *The Singing Ringing Tree*

Winstone Film Distributors

18 Craignish Avenue
Norbury
London SW16 4RN
Tel: 0181 765 0240
Fax: 0181 765 0564
Mike G. Ewin
Sub-distribution for Alliance
Releasing (library), (now Canal +
Image UK Ltd) - Library only
My Mothers Courage

Ma vie en Rose
Eskiya
The Umbrella's of Cherbourg
Rothschilds Violin
Monk Dawson
Respect
Girls Night
Sweet Hereafter
In the Company of Men
Kiss or Kill

FACILITIES

AFM Lighting Ltd
12 Alliance Road
London W3 0RA
Tel: 0181 752 1888
Fax: 0181 752 1432
Greg Humphries/Gary Wallace
(due to be moving in 1998)
Lighting equipment and crew hire;
generator hire

Abbey Road Studios
3 Abbey Road
London NW8 9AY
Tel: 0171 266 7000
Fax: 0171 266 7250
Four studios; music to picture;
35mm projection; film sound
transfer facilities; audio post-
production. Sonic Solutions
computer sound enhancement
system; residential accommodation,
restaurant and bar. Multimedia
design and authoring, dvd authoring

After Image Facilities
32 Acre Lane
London SW2 5SG
Tel: 0171 737 7300
Fax: 0171 326 1850
email: Jane@arc.co.uk
Jane Thorburn
Full broadcast sound stage - Studio
A (1,680 sq ft, black, chromakey,
blue, white cyc) and insert studio
(730 sq ft hard cyc). Multiformat
broadcast on-line post production.
Special effects - Ultimatte/blue screen

Air Studios
Lyndhurst Hall, Lyndhurst Road
Hampstead
London NW3 5NG
Tel: 0171 794 0660
Fax: 0171 794 8518
Alison Burton
Lyndhurst Hall: capacity - 500 sq m
by 18m high with daylight; 100 plus
musicians; four separation booths. Full
motion picture scoring facilities. Neve
VRP Legend 72ch console, flying
fader automation. LCRS monitoring.
Studio 1: capacity - 60 sq m with
daylight. 40 plus musicians. Neve/
Focusrite 72ch console; GML
automation; LCRS monitoring. Studio
2: Mixing Room; SSL8000G plus
series console with Ultimation; ASM
system. Film and TV dubbing
facilities; two suites equipped with
AMS Logic II consoles; 16 output;
AudioFilc spectra plus; LCRS
monitoring. Exabyte back-up. One
suite equipped with an AMS Logic III
console. Every tape machine format
available

Alphabet Communications
Parkgate Industrial Estate
Knutsford
Cheshire WA16 8DX
Tel: 01565 755678
Fax: 01565 634164
On line edit suites. Sony Digital and
SP Betacam VTR's with 9100 plus
edit controller with DVS-6000C
Digital Vision Mixer. Off-line Avid
Media Composer 800. 2D and 3D
Computer Graphics. 1,800 ft drive in
studio. VHS duplication, standards
conversion. Animation studio. Hire
department

Angel Recording Studios
311 Upper Street
London N1 2TU
Tel: 0171 354 2525
Fax: 0171 226 9624
Gloria Luck
Two large orchestral studios with
Neve desks, and one small studio.
All with facilities for recording to
picture

Anvil Post Production Ltd
Denham Studios
North Orbital Road, Denham
Uxbridge
Middx UB9 5HL
Tel: 01895 833522
Fax: 01895 835006
email: *@anvil.nildram.co.uk
Sound completion service; re-
recording, ADR, post-sync, Fx
recording, transfers, foreign version
dubbing; non-linear and film editing
rooms, neg cutting, off-line editing,
production offices

ARRI Lighting Rental
20a The Airlinks, Spitfire Way,
Heston, Middx TW5 9NR
Tel: 0181 561 6700
Fax: 0181 569 2539
Tim Ross
Lighting equipment hire

Avolites
184 Park Avenue
London NW10 7XL
Tel: 0181 965 8522
Fax: 0181 965 0290
Manufacture and sale of dimming
systems, memory and manual lighting
control consoles. Also complete range
of distribution hardware

BUFVC
77 Wells Street
London W1P 3RE
Tel: 0171 393 1500
Fax: 0171 393 1555

email: bufvc@open.ac.uk
16mm cutting room plus 35mm and
16mm viewing facilities. Betacam 2
machine edit facility for low-cost
assembly off-line work

Jim Bambrick and Associates
William Blake House
8 Marshall Street
London W1V 2AJ
Tel: 0171 434 2351
Fax: 0171 734 6362
6 x Avid Editing Suite with versions
6.5 software, 35mm Steinbeck

Barcud
Cibyn
Caernarfon
Gwynedd LL55 2BD
Tel: 01286 671671
Fax: 01286 671679
Video formats: 1"C, Beta SP, D2 OB
Unit 1: up to 7 cameras 4VTR OB
Unit 2: up to 10 cameras 6VTR,
DVE, Graphics Betacam units.
Studio 1: 6,500 sq ft studio with
audience seating and comprehensive
lighting rig. Studio 2: 1,500 sq ft
studio with vision/lighting control
gallery and sound gallery. Three edit
suites; two graphics suites, one with
Harriet. DVE: three channels
Charisma, two channels Cleo. Two
Sound post-production suites with
AudioFile and Screen Sound; BT
lines. Wales' leading broadcast
facility company can supply OB
units, studios, Betamac Kits (all
fully crewed if required) and full
post production both on and off-line

Bell Digital Facilities
Lamb House
Church Street
Chiswick Mall
London W4 2PD
Tel: 0181 996 9960
Fax: 0181 996 9966
email: sales@bel-media.co
ProTools IV sound dubbing studio
with non-linear picture. VocAlign &
other ADR and outboard tools. Voice
booth accessible from all suites.
Extensive 3D animation & 2D
graphics studio. Sound proofed, air-
conditioned. 600 sq ft video studio
available as 4-waller with
cameras. Avid off and on-line and
After Effects

Blue Post Production
58 Old Compton Street
London W1V 5PA
Tel: 0171 437 2626
Fax: 0171 439 2477

Contact: Catherine Spruce, Director of Marketing
Digital Online Editing with Axial edit controllers, GVG 4000 digital vision mixers, Kaleidoscope DVEs, disc recorders, Abekas A72, digital audio an R-Dat
Quantel Edit Box 4000 with 2 hours non-compressed storage
Sound Studio with Avid Audio Vision, 32 input MTA fully automated desk
Offline Editing on Avid Media Composer 800
Telecine Ursa Diamond System, incorporating Pogle Platinium DCP with ESR & TWiGi

CTS Studios Ltd
Engineers Way
Wembley
Middx HA9 0DR
Tel: 0181 903 4611
Fax: 0181 903 7130
Four studios ranging in musician capacity from 130 to 10. Synchronised film projection and video facilities available for recording to picture. Digital or analogue multitrack formats. Studio 2 all digital with Neve Capricorn. Digital Mastering. Restaurant and car park

Canalot Production Studios
222 Kensal Road
London W10 5BN
Tel: 0181 960 8580
Fax: 0181 960 8907
Nieves Heathcote
Media business complex housing over 80 companies, involved in TV, film, video and music production, with boardroom to hire for meetings, conferences and costings

Capital FX
21A Kingly Court
London W1R 5LE
Tel: 0171 439 1982
Fax: 0171 734 0950
Graphic design and production, subtitling, opticals and editing, non-broadcast Telecine

Capital Studios
Wandsworth Plain
London SW18 1ET
Tel: 0181 877 1234
Fax: 0181 877 0234
Central London: 3,000 and 2,000 sq ft fully equipped broadcast standard television studios. 16x9/4x3 switchable, two on line edit suites (D3, D2, D5, Digital Betacam & Beta SP). Avid on/off line editing. Multi track and digital sound dubbing facilities with commentary booth. 'Harriet' graphics suite. BT lines. All support facilities. Car park.

Expert team, comfortable surroundings, immaculate standards

Charter Broadcast
47 Theobald Street
Borehamwood
Herts WD6 4RY
Tel: 0181 905 1213
Fax: 0181 905 1424
Jo Fyfe
Broadcast equipment hire, with comprehensive range of facilities including cameras, camcorders, lenses, VTRs, mixers, matrices, DVEs, character generators, communications, audio equipment, radio camera and radio links, all of which can be hired individually. Full 'fly away' production facilities, short and long term. All systems individually designed for rapid installation and flexibility

Chromacolour International Ltd
11-16 Grange Mills
Weir Road
London SW12 0NE
Tel: 0181 675 8422
Fax: 0181 675 8499
Animation supplies/equipment

Chrysalis Mobiles
3 Chrysalis Way
Langley Bridge
Eastwood
Nottingham NG16 3RY
Tel: 01773 718111
Fax: 01773 716004
6 x OB units, PAL and component. High definition television, five outside broadcast units; two mobile edit suites; single camera units, all 1250 EBU standard

Cine Image Film Opticals
7a Langley Street
London WC2H 9JA
Tel: 0171 240 6222
Fax: 0171 240 6242
Film opticals, titles and effects on 35 and 16mm; film rostrum

Cine-Europe
7 Silver Road
Wood Lane
London W12 7SG
Tel: 0181 743 6762
Fax: 0181 749 3501
16mm, Super 16 cameras, lenses and grip equipment hire

Cine-Lingual Sound Studios
27-29 Berwick Street
London W1V 3RF
Tel: 0171 437 0136
Fax: 0171 439 2012
Mike Anscombe
Three sound studios, 16/35mm computerised ADR; Foley recording;

Dolby stereo mixing; sound transfers; cutting rooms. DAWN and AudioFile digital recording and editing facilities

Cinebuild
Studio House
Rita Road
London SW8 1JU
Tel: 0171 582 8750
Fax: 0171 793 0467
Special effects: rain, snow, fog, mist, smoke, fire, explosions; lighting and equipment hire. Studio: 200 sq m

Cinecontact
27 Newman Street
London W1P 4AR
Tel: 0171 323 0618
Fax: 0171 323 1215
Contact: Jacqui Timberlake
Documentary film-makers. Avid post production facilities

Cinesite (Europe) Ltd
9 Carlisle Street
London W1V 5RG
Tel: 0171 973 4000
Fax: 0171 973 4040
Also: Pinewood Studios
The Props Building
Pinewood Road, Iver
Bucks SL0 0NH
Tel: 01753 630285
Fax: 01753 650261
Utilising state-of-the art technology, Cinesite provides expertise in every area of resolution-free digital imaging and digital special effects for feature films. Our creative and production teams offer a full spectrum of services from the storyboard to the final composite, including digital effects, and shoot supervision. Credits include: *Mission Impossible, Pinocchio, Muppets Treasure Island, Restoration* and *A Midsummer Night's Dream*

Cinevideo
Broadcast Television Equipment Hire
7 Silver Road, White City
Industrial Park
Wood Lane
London W12 7SG
Tel: 0181 743 3839
Fax: 0181 743 8417
Video formats: Beta SP, D1, D2, D3. Cameras: Ikegami HK355, HL57, HLV55, Sony BVP370, BVP70isp, BVP90. Full range of microwave links/radio cameras. PAL/NTSC

Clark Facilities
Cavendish House
128-134 Cleveland Street
London W1P 5DN
Tel: 0171 388 7700

Fax: 0171 388 3366
Allan Hay
On-line editing. Autoconform, Avid.
Transfers: D5, D3, Beta SP, U-Matic, 1", Hi-8

Colour Film Services Group
10 Wadsworth Road
Perivale
Middx UB6 7JX
Tel: 0181 998 2731
Fax: 0181 997 8738

Complete
Slingsby Place
Off Long Acre
London WC2E 9AB
Tel: 0171 379 7739
Fax: 0171 497 9305
email: info@complete.co.uk
Richard Ireland,
Richard Ireland, Lucy Pye, Sarah
Morgan, Lisa Sweet and Holly Ryan
Henry, Flame, Harriet. Digital
editing. C-reality-Hires-Telecine. 3D
Animation with Alias wave front and
soft/maxDigital Ursa Diamond
Telecine with Russell Square DI tape
grading. Digital playouts and ISDN
links. Award-winning creative team

Connections Communications Centre
Palingswick House
241 King Street
Hammersmith
London W6 9LP
Tel: 0181 741 1766
Fax: 0181 593 9134
Betacam SP editing with computer-
ised edit controller and dynamic
tracking. SVHS on-line and off-line
facilities. Production equipment also
available

Corinthian and Synchro Sonics
5 Richmond Mews
Richmond Buildings
London W1V 5AG
Tel: 0171 734 3325
Fax: 0171 437 3502
Paul White
Paul White, Ian Pickford, Torian
Brown
Digital dubbing studio for video or
film; hard disc tracklaying;
commentary booth; syncing rushes
to Beta SP. Comprehensive digital
FX libraries; sound and video
transfer; 16mm cutting rooms;
equipment hire

Corinthian Television
87 St John's Wood Terrace
London NW8 6PY
Tel: 0171 483 6000
Fax: 0171 483 4264
OBs: Multi-camera and multi-VTR
vehicles. Post Production: 3 suites, 1

SP component, 2 multi-format with
1", D2, D3, Abekas A64, A72,
Aston and colour caption camera.
Studios: 2 fully equipped television
studios (1 in St John's Wood, 1, in
Piccadilly Circus), 1-5 camera,
multi-format VTRs, BT lines,
audience seating. Audio: SSL
Screensound digital audio editing
and mixing system

Crow Film and Television Services
12 Wendell Road
London W12 9RT
Tel: 0181 749 6071
Fax: 0181 740 0795
Cheryl Nunn
2 x on-line suites - D3, D2, Beta 1".
Non-linear - Avid. Computer
rostrum camera. Harriet Graphics.
Digital audio dubbing

Crystal Film and Video
50 Church Road
London NW10 9PY
Tel: 0181 965 0769
Fax: 0181 965 7975
Aatons, Arriflex, Nagras, radio mics,
lights and transport; Sony Beta SP;
studio 50' x 30'; crews

Cygnet
Communication Business Centre
14 Blenheim Road
High Wycombe
Bucks HP12 3RS
Tel: 01494 450541
Fax: 01494 462154
16mm and video production
company

DBA Television
7 Lower Crescent
Belfast BT7 1NR
Tel: 01232 231197
Fax: 01232 333302
Crew hire and 16mm edit facilities,
sound transfers; Aaton, Steenbeck

Dateline Productions
79 Dean Street
London W1V 5HA
Tel: 0171 437 4510
Fax: 0171 287 1072
Avid non-linear editing, 16/35mm
film editing

De Lane Lea Sound Centre
75 Dean Street
London W1V 5HA
Tel: 0171 439 1721
Fax: 0171 437 0913
2 high speed 16/35mm Dolby stereo
dubbing theatres with Dolby SR;
high speed ADR and FX theatre (16/
35mm and NTSC/PAL video);
Synclavier digital FX suite; digital
dubbing theatre with Logic 2

console; 3 x AudioFile preparation
rooms; sound rushes and transfers;
video transfers to VHS and U-Matic;
Beta rushes syncing. 24 cutting
rooms/offices. See also under studios

Denman Productions
60 Mallard Place
Strawberry Vale
Twickenham TW1 4SR
Tel: 0181 891 3461
Fax: 0181 891 6413
Video and film production,
including 3D computer animation
and web design

Despite TV
113 Roman Road
London E2 0HU
Tel: 0181 983 4278
Fax: 0181 983 4278
Training through production on 3
machine series 9 lo/hi SP edit suite
with 4A controller, Amiga 2000, 6
channel audio mixer

Digital Sound House
14 Livonia Street
London W1V 3PH
Tel: 0171 434 2928/ 437 7105
Fax: 0171 287 9110
Sound transfer; 16/35mm 1/4" DAT;
FX library; full footsteps theatre,
also available for voiceover
commentary recordings and fully
compatible for video; Avid

Diverse Production
6 Gorleston Street
London W14 8XS
Tel: 0171 603 4567
Fax: 0171 603 2148
Ray Nunney
TV post-production. Digital on-line
editing; off-line editing; comprehensive graphic design service; titles
sequences, programme graphics,
generic packaging, sets and
printwork

Document Films
8-12 Broadwick Street
London W1V 1FH
Tel: 0171 437 4526
Avid suites, 16mm Aaton crews,
sound and video transfers

Dolby Laboratories
Wootton Bassett
Wilts SN4 8QJ
Tel: 01793 842100
Fax: 01793 842101
Cinema processors for replay of
Dolby Digital, and Dolby SR
(analogue) film soundtracks; audio
noise reduction equipment. Sound
consultancy relating to Dolby film
productions and Dolby Surround
productions for television

Dubbs
25-26 Poland Street
London W1V 3DB
Tel: 0171 629 0055
Fax: 0171 287 8796
Videotape duplication: All digital
formats; Beat SP 1", BVU SP,
SVHS, VHS, U-matic, Hi-8 and
Video 8. Standards Conversion:
Alchemist Ph C, Adac, Tetra. Audio:
Tascam DA88, DAT, Audio cassette.
Full labelling, packing and despatch
service available

Edinburgh Film and Video Productions
Edinburgh Film and TV Studios
Nine Mile Burn
By Penicuik
Midlothian EH26 9LT
Tel: 01968 672131
Fax: 01968 672685
Stage: 50 sq m; 16/Super 16/35mm
cutting rooms; preview theatre; edge
numbering; lighting grip equipment
hire; scenery workshops

Edinburgh Film Workshop Trust
29 Albany Street
Edinburgh EH1 3QN
Tel: 0131 557 5242
Fax: 0131 557 3852
email: post@efwt.demon.co.uk
Website: http:\\www.efwt.demon.co.uk

Production and post-production on
tape and film. Umbrella production
for new producers and consultancy
to individuals and organisations

The Edit Works
No 1, 2nd Floor
Chelsea Harbour Design Centre
London SW10 0XE
Tel: 0171 352 5244
Fax: 0171 376 8645
Barry Noakes
Digital component and composite
digital. Full editing service to the
broadcast industry. Including Avid
Audiovision dubbing suite

Edric Audio-visual Hire
34-36 Oak End Way
Gerrards Cross
Bucks SL9 8BR
Tel: 01753 884646
Fax: 01753 887163
Audiovisual and video production
facilities

Elstree Light and Power
Millennium Studios
Elstree Way
Borehamwood
Herts WD6 1SF
Tel: 0181 236 1300
Fax: 0181 236 1333
Tony Slee
TV silent generators; Twin Sets

HMI, MSR and Tungsten Heads.
Distribution to BS 5550. Rigging
Specialists

Essential Pictures
222 Kensal Road
London W10 5BN
Tel: 0181 969 7017
Fax: 0181 960 8201
Post-production facility. 2 x
broadcast studios complete with
lines to BT Tower and Phone-in
facilities; Camera crews;
composium; highly sophisticated
digital component suite, powerful
compositing machine, dedicated
paint and graphics station, on-line
editing, multi-layered digital keying,
matte creation, large stills library;
Avid on-line and off-line suites,
Sound dubbing facilities. Comprehensive language transfer service
including translation, voice over,
subtitling and project management
to and from all languages

Eye Film and Television
The Guildhall
Church Street
Eye
Suffolk IP23 7BD
Tel: 01379 870083
Fax: 01379 870987
Betacam SP crews, Avid Non Linear
offline & online systems available

for wet & dry hire. Associated production services

Faction Films
28-29 Great Sutton Street
London EC1V 0DU
Tel: 0171 608 0654/3
Fax: 0171 608 2157
Avid MC1000 composer; Montage 22 suite; 6 plate 16mm Steenbeck edit suite; Low-band U-matic suite; Sony VX1000 digi-cam; Sony Hi-8; HHB Portadat; Nagra 4.2; Production office space; experienced Editors and Sound recordist available

The Film Centre
Leathermarket
Weston Street
London SE1
Tel: 0171 261 1115 or 0700 FILMCENTRE (345623)
Fax: 0118 961 7392
email: info@filmcentre.co.uk
35/16mm Camera Equipment hire, primarily to the independent sector. Also film editing and off-line hire, rostrum, lighting truck, audio equipment, film/video projectors, multimedia authoring. Production office and crewing facilities.
Film Centre Shop: one stop for stock and consumables

The Film Factory at VTR
64 Dean Street
London W1V 5HG
Tel: 0171 437 0026
Fax: 0171 439 9427
email: info@vtr.co.uk
The Film Factory at VTR is one of London's major feature film post production facilities specialising in high-resolution digital special effects. Creative teams cover all aspects of feature visual post production work including special effects filming, extensive CGI work, and high resolution compositing. The Domino system is an integral part of The Film Factory. Already proven for its extensive US work on Independence Day, the Domino is an ideal and versatile medium for 35mm high resolution compositing at 3,000 lines. Experienced producer teams, one of the largest CGI feature film units in the UK and highly-experienced Domino high-resolution compositing teams, make the Film Factory one of the most exciting feature film post production facilities in London

Film Work Group
Top Floor, Chelsea Reach
79-89 Lots Road
London SW10 0RN
Tel: 0171 352 0538
Fax: 0171 351 6479

Emma Plimmer
Video and film post-production facilities with graphic design. Three machine hi-band SP with effects and Beta/Hi-8 play-ins. Two machine lo-band, videographics and 6-plate Steenbeck. CD-Rom burning/writing

FinePoint Broadcast
Furze Hill
Kingswood
Surrey KT20 6EZ
Tel: 01737 370033
Fax: 01737 370088
email: hire@finepoint.co.uk
Website: http://www.finepoint.co.uk
Broadcast equipment hire. Cameras, lenses, control units, cables, VTRs, edit controllers, digital video effects, vision mixers, monitors, sound kit, full outside broadcast unit

FrameStore
9 Noel Street
London W1V 4AL
Tel: 0171 208 2600
Fax: 0171 208 2626
email:jane.white@framestore.co.uk
Digital effects for film and video. The latest technology, including Ursa Diamond Telecine, 2x inferno/flame, 4x Henry, Digital Edit Suite, 3D Computer Animation, Avid Editing for commercials, broadcast and graphic design projects. Plus digital film opticals for feature effects, repairs, pick-ups, restoration, titles and tape to film transfers. Call bookings or post-producers: Fiona, Lottie, AJ and Drew for advice on projects

Mike Fraser
Unit 6
Silver Road
White City Industrial Park
London W12 7SG
Tel: 0181 749 6911
Fax: 0181 743 3144
Mike Fraser
Mike Fraser, Rod Wheeler
Telecine transfer 35mm, 16mm and S16; rushes syncing; non-linear edit suites; film video list management, post-production through OSC/R to negative cutting. Storage

Frontline Television Services
44 Earlham Street
London WC2H 9LA
Tel: 0171 836 0411
Fax: 0171 379 5210
Charlie Sayle
Component Broadcast Editing and Multi-Format Composite suite, both with Grass Valley 200 vision mixer, Sony 9000 edit controller, Abekas A53D, Aston 4/A72 Graphics generators. Computerised video

rostrum camera, tape duplication, broadcast standards conversion via ADAC. D2, D3 transfers and editing

GBS Lighting (part of the Arri Bell Group)
169 Talgarth Road
London W14 9DA
Tel: 0181 748 0316
Fax: 0181 563 0679
Lighting equipment hire

General Screen Enterprises
Highbridge Estate
Oxford Road
Uxbridge
Middx UB8 1LX
Tel: 01895 231931
Fax: 01895 235335
Pinewood Studios
Iver Heath
Bucks SL0 0NH
Tel: 01753 650260
Fax: 01753 650259
Studio: 100 sq m. 16mm, 35mm opticals including matting, aerial image work, titling, editing, trailers, promos, special effects, graphics. Cineon Digital Fx Unit, Film Opticals, Motion Control Stage, VistaVision; computerised rostrum animation motion control; video suite; preview theatre

Goldcrest Post Production
Facilities Ltd
Entrance 1 Lexington Street
36/44 Brewer Street
London W1R 3HP
Tel: 0171 439 4177 or 0171 437 7972
Fax: 0171 437 5402
email: mailbox@goldcrest-post.co.uk
John Spirit
John Spirit, Louise Seymour
Theatre One with SSL5000 console, Dolby SRD, film + video projection; Theatre Two with Harrison Series 12 Console & Synclavier 9600 16 track direct to disk; Dolby SRD ADR + effects recording, built-in Foley surfaces and extensive props; AMS AUDIOFILE suite with Yamaha digital desk; SOUND TRANSFER BAYS all film and video formats with Dolby SRD; Rank Cintel MKIIC TELECINE enhanced 4:2:2, Pogle and secondary colour correction. Keycode + Aaton code readers, Electronic Wetgate; Video transfers to 1", Beta SP, U-Matics, VHS and D2, ADAC standards conversion. Non Linear editing Avid or Lightworks available. 40 Cutting Rooms; Production offices; Duplex apartments available

Hammonds AVS
64a Queens Road

Watford
Herts WD1 2LA
Tel: 01923 239733
Fax: 01923 221134
M F Hammond
M F Hammond, Clare Driver
ENG crews; BVU, Pro S SVHS, Hi-8, Professional editing, multi-format. DVE audio-recording facilities. SVHS on line Video Editing, SVHS off-line; Duplication and Standards Conversion via AVS System Slide Film to Video and vice versa, CD Production

Hays Information Management
320 Western Road
Merton
London SW19 2QA
Tel: 0181 640 6626
Fax: 0181 640 1297
Production material archive services. Secure, controlled environment. Rapid access

Headline Video Facilities
3 Nimrod Way
Elgar Road
Reading
Berks RG2 0EB
Tel: 0118 975 1555
Fax: 0118 986 1482
email: post@headlinevideo.demon.co.uk
Tom Street
Talented and experienced designers and editors who create and produce quality Broadcast graphics and editing. Quantel Hal Express, Harriet and 3D Studio Max with fully integrated Highly Specified Digital Betacam and Beta SP editing suites. Avid and VHS offline, duplication and standards conversion

Hillside Studios
Merry Hill Road
Bushey
Herts WD2 1DR
Tel: 0181 950 7919
Fax: 0181 421 8085
email: hillside@ctvc.co.uk
Website: http://ourworld.compuserve. com/homepages/hillside_studios
Production and Post-Production facilities to Broadcast standards. 1500 sq ft studio with 16 x 9 switchable cameras and Digital Mixer. Smaller studio and single camera location units available. Sounds Studios and Dubbing Suites, Non-Linear and Digital Editing. Graphics, Set Design and Construction. Offices, restaurant and parking

Holloway Film & TV
68-70 Wardour Street
London W1V 3HP
Tel: 0171 494 0777

Fax: 0171 494 0309
Matt Stoddart
D5, D3, D2, Digital Betacam, 3 m/c Digital Betacam suite, AVID (AVRTT) on-line/Offline. Betacam SP, 1"C, BVU, Lo-Band Hi-8, Video-8, S-VHS, VHS, Standards Conversion, Audio Laybacks/Layoffs

Hull Time Based Arts
8 Posterngate
Hull HU1 2JN
Tel: 01482 215050
Fax: 01482 589952
email: Ron@htba.demon.co.uk
Website: http://www.htba.demon.co.uk
Avid Media Composer 8000 on line digital non linear video editing suite with pro tools. Avid Media Composer 400 off line digital non linear video editing suites. DVC Pro and DV video cameras. DAT recorders, Video/Data projectors and all anciliary video equipment available. Special rates for non commercial projects

Humphries Video Services
Unit 2, The Willow Business Centre
17 Willow Lane
Mitcham
Surrey CR4 4NX
Tel: 0181 648 6111/0171 636 3636
Fax: 0181 648 5261
email: sales@hvs.bdx.co.uk
Website: http://www.hvs.co.uk
David Brown, Emma Lincoln
Video cassette duplication: all formats, any standard. Standards convertors. Macrovision anti-copy process, labelling, shrink wrapping, packaging and mail out services, free collections and deliveries in central London. Committed to industrial and broadcast work

ITN
200 Gray's Inn Road
London WC1X 8XZ
Tel: 0171 430 4134
Fax: 0171 430 4655
Martin Swain
Martin Swain, Jenny Mazzey
2400 sq ft studio; live or recorded work; comprehensive outside source ability; audience 65; crews; video transfer; Westminster studio; graphics design service using Flash Harry, Paintbox etc; Training offered; Sound and dubbing; tape recycling; experienced staff

IVPTV
16a York Place
Edinburgh EH1 3EP
Tel: 0131 558 1888
Fax: 0131 557 5465
Edit 1: Two Ampex 1" VTRs; Sony

9000 edit controller; Grass Valley 300 switcher; Kaleidoscope with Kurl and Wipe; Dubner character generator; Abekas A64 Digital Disk Recorder; two Beta SP studio players and DAT Audio with T/c. Edit 2: Sony 910 edit controller; GVG 110 CV component mixer; three Beta SP studio players; D-Vision non-linear off-line + VHS; Matisse Paint and 3-D; Symbolics XL1200 3-D Animator and video studio. Cameras: Sony BVW 400 (2), Ampex CVC 7. 2 off-line Avids and one on-line Avid. Lightworks Turbo

Interact Sound
160 Barlby Road
London W10 6BS
Tel: 0181 960 3115
Fax: 0181 964 3022
16/35mm film dubbing, (12 track Dolby Stereo, SR & A); sound transfer and digital sound edit suites. Video studio: hard disc capability, DAT, Beta, Tascam. Mixers: Aadwirtz, Lee, Taylor

International Broadcast Facilities
12 Neal's Yard
London WC2H 9DP
Tel: 0171 497 1515
Fax: 0171 379 8562
Digital audio dubbing studio with Screensound. Full film and video audio post-production services. Multi standard/format video dubbing and duplication. Fast turnaround

Terry Jones PostProductions Ltd
The Hat Factory
16-18 Hollen Street
London W1V 3AD
Tel: 0171 434 1173
Fax: 0171 494 1893
Terry Jones
Terry Jones, Paul Jones
2 x Lightworks non-linear editing suites, plus computerised Beta off-line and 35mm film editing facilities. Two experienced award-winning editors handling commercials, documentaries, features and corporate work

Lee Lighting
Wycombe Road
Wembley
Tel: 0181 900 2949
Fax: 0181 903 3012
Film/TV lighting equipment hire

Light House Media Centre
The Chubb Buildings
Fryer Street
Wolverhampton WV1 1HT
Tel: 01902 716044
Fax: 01902 717143

Contact: Technical department
Three machine U-Matic edit suite (hi-band - BVE 900, lo-band BVE 600) VHS/U-Matic/Betacam/ENG kits, also animation and chroma keying

Lighthouse

Brighton Media Centre
9-12 Middle Street
Brighton BN1 1AL
Tel: 01273 384222 Facilities: 01273 384255
Fax: 01273 384233
email: info@lighthouse.org.uk
Jane Finnis
Jane Finnis, Caroline Freeman
A training and production centre, providing courses, facilities and production advice. Avid off- and online edit suites. Apple Mac graphics and animation workstations. Digital video capture & manipulation. Output to/from Betacam SP. SVHS offline edit suite. Post Production and Digital Artists equipment bursaries offered three times a year

London Electronic Arts

Lux Centre
2-4 Hoxton Square
London N1 6NU
Tel: 0171 684 0101
Fax: 0171 684 1111
email: infor@lea.co.uk
Subsidised facilities for video and new media arts. Avid off-line & on-line, Beta SP/DVCPRO on-line with Alladin 3D DVE. Media Illusion/Matador compositing, multiformat off-line (VHS, SVHS, Hi8, lo-band, Beta SP) Digital Audio: 02R, Cubase, ProTools. Multimedia/Internet workstations, CD recorders. DVCAM, DVCPRO and Hi8 camera kits. LCD projectors, monitors and decks for exhibition

London Fields Film and Video

10 Martello Street
London E8 3PE
Tel: 0171 241 2997
Computer graphics; video editing; 16mm editing

London Filmmakers' Co-op

Lux Centre
2-4 Hoxton Square,
London N1 6NU
Tel: 0171 684 0202
Fax: 0171 684 2222
email: 101563.3332@compuserve.com
Paul Murray
Paul Murray, Patricia Diaz
Optical Printer, for gauge transfers, colouring and all kinds of optical manipulation of film; Macs with Adobe Premiere for editing and

Macromedia Director for interactive multimedia authoring; state of the art Avid editing suite; 16mm black and white film processing; Super 8, 16mm and Super 16 camera kits; Lighting including standard blonde and redhead kits and the latest Dedolight and Kino Flo Kits; Analogue editing suites with Steenbecks, plus a rare Super 8 Steenbeck; Sound transfer room; Rostrum Camera for animation, titles and effects; broadcast quality Telecine

M2 Facilities Group

The Forum
74-80 Camden Street
London NW1 0EG
Tel: 0171 387 5001
Fax: 0171 387 5025
Helen Clarke
Helen Clarke, Shula Parker
Non-linear off-line, on-line and sound post-production with full broadcast facility back-up

MAC Sound Hire

1-2 Attenburys Park
Park Road
Altrincham
Cheshire WA14 5QE
Tel: 0161 969 8311
Fax: 0161 962 9423
Professional sound equipment hire

The Machine Room

54-58 Wardour Street
London W1V 3HN
Tel: 0171 734 3433
Fax: 0171 287 3773
David Atkinson
3 wet/dry gate digital Telecine suites. VT viewing and sound layback suite. Most digital and analogue video tape formats in both PAL and NTSC. Standards conversion with Alchemist Phe and Vector Motion Compensation (VMC). Programme dubbing. VHS duplication. Macrovision anti-piracy system, Lightworks non-linear, day hire, edit suite. FACT accredited. Full range of film treatment services. See also Film Treatment Centre under Laboratories. Nitrate handling and nitrate storage vaults

Magnetic Image Production Facilities

6 Grand Union Centre
West Row
London W10 5AS
Tel: 0181 960 7337/964 5000
Fax: 0181 964 4110
Cathy Burns
On-line video editing using Integra, multi-format component digital

processing system. Standards conversions for all countries and video duplication. Beta SP shooting also available. Off-line editing and computer graphics

Mersey Film and Video

40a Bluecoat Chambers
School Lane
Liverpool L1 3BX
Tel: 0151 708 5259
Fax: 0151 707 0048
Production facilities for: BETA SP, SUHS, Hi8, 16mm. Dolly Post Production: Avid, SUHS, Hi8, FX library, music library, Jibarm, lights, mics, etc. Guidance and help for funding, finance, budgets, production

Metropolis

8-10 Neal's Yard
London WC2H 9DP
Tel: 0171 240 8423
Fax: 0171 379 6880
Sam Timms
Sam Timms, Adam Peat
2 x Avid suites; 3 machines U-matic Editmaster; 2 machine U-matic/VHS VITC Reads. Duplication - all formats standards conversion

MetroSoho

6-7 Great Chapel Street
London W1V 3AG
Tel: 0171 439 3494
Fax: 0171 437 3782
Mark Cox
Mark Cox, Nayle Kemah
Broadcast Hire: Avid and Lightworks, Ron Line editing Camera formats - Digital Beta, Beta SX, Beta SP, DVC Pro, DV Cam, Mini DV. VTRs, D2, D3, D5,, DVIV A5ooP, BV W 75. 24 hour support. Crew hire also available. Duplication: Alchemist standards conversion from/to all formats. Technical assessment. Format include: D1, D2, D3, Digital Beta, Beta SX, Ducplo, Du Cam mini DV. CD-Rom origination and replication

MetroVideo

The Old Bacon Factory
57-59 Great Suffolk Street
London SE1 0BS
Tel: 0171 928 2088
Fax: 0171 261 0685
Sales: Non-linear systems + DVE; Digital and analogue VTR's, multi-media, cameras, tape sales - most key manufacturers. Approved Sony dealer. Presentation hire; video conferencing; CD Rom + M.Peg solutions; Hi-8 cameras and editing; OHD projectors; data display panels; PCT Mac computers; sound equipment; projectors; monitors;

prowalls; videowalls; shooting crews; all formats; archives; film and video projects; design and project management

The Mill
40/41 Great Marlborough Street
London W1V 1DA
Tel: 0171 287 4041
Fax: 0171 287 8393
Andy Barmer
Andy Barmer, Liz McLean
Special effects for feature film and television commercials using Flame and Advance, Henry, Harry and digital editing

Millennium Studios
Elstree Way
Borehamwood
Herts WD6 1SF
Tel: 0181 236 1400
Fax: 0181 236 1444
Kate Tufano
Sound stage 80'x44'x24' with 6'x44'x11' balcony flying and cyc grid. In house suppliers of: lighting; generators; rigging; photography; crew catering and fully licensed bar

Mister Lighting Studios Ltd
2 Dukes Road
Western Avenue
London W3 0SL
Tel: 0181 956 5600
Fax: 0181 956 5604
Steve Smith
Lighting equipment/studio hire

Molinare
34 Fouberts Place
London W1V 2BH
Tel: 0171 439 2244
Fax: 0171 734 6813
Video formats: Digital Betacam, D1, D2, D3, 1", Beta SP, BVU, U-Matic, VHS. NTSC: 1", Beta SP, U-Matic & VHS. Editing: Editbox, three D1 serial digital suite; two component multi-format; one composite multi-format. DVEs: two A57, four A53, DME, four ADO, Encore. Storage: two A66, A64. Caption Generators: Aston Motif, A72, Aston Caption, Aston 3. Graphics: Harry with V7 Paintbox, Encore and D1. Harriet with V7 Paintbox, D1 and Beta SP. 3D graphics with Silicon Graphics and Softimage. Telecine: Ursa Gold with Pogle + DCP, A57, Rank Cintel 111 with 4.2.2 digital links, wetgate, Pogle and DCP controller and secondary colour grading, 35mm, 16mm, S16mm/S8. Audio: two digital studios, two 24 track and AudioFile studios, track-laying studio with DAWN, voice record studios, transfer room, sound Fx libraries. Duplication, standards

conversion, Matrix camera, BT landlines, satellite downlink

Morgan Broadcast Ltd
6 Indeson Court
Millharbour
London E14 9TN
Tel: 0181 908 6377
Fax: 0181 908 4211
-Large screen video protection monitor and projections videowalls

Mosaic Pictures Ltd
2nd Floor
8-12 Broadwick Street
London W1V 1FH
Tel: 0171 437 6514
Fax: 0171 494 0595
email: 75337.1233@compuserve.com
Mosaic Pictures offers help with cameras, transfers, off-line suites and on-line suites in Soho location

The Moving Picture Company
25 Noel Street
London W1V 3RD
Tel: 0171 434 3100
Fax: 0171 437 3951/287 5187
Video formats: D1, D2, Digital Betacam, Betacam SP, 1" C format, hi-/lo-band.
Editing: 3xD1/Disk based edit suites, Sony 9100 and Abekas A84 (8 layers) A57 DVE, A64, A60 and A66 Disks; A72 and Aston Motif caption generator. Video Rostrum and Colour Caption Camera
Non Linear Offline Editing: 1 x Avid 4000 with Betacam SP. 35/16mm cutting room
Telecine: 2 URSA Gold 4 x 4 with Pogle DCP/Russell Square Colour Correction Jump Free, Low Speed/Silk Scan Options, Matchbox Stills Store, Key Code, noise reduction
SFX: Discreet Logic 2 x Flame, 1 x Flint and Quantel 2 x Henry
3D: Hardware: 7 x SGI systems (3 x High Impacts and 4 x Indigo 2 Extremes). Software: Alias Poweranimator, Custom Programming and Procedural Effects, Matador, 3D Studio Paint, Elastic Reality and Pandemonium.
Rendering: SGI Challenge and Onyx (x2). Digital Film: High resolution 35mm digital film post production, comprising 7 x Kodak Cineon, 1 x Discreet Logic Inferno and Matador. Filmtel TM video tape to 35mm transfer. Mac: Disk or ISDN input of artwork. File transfer, Photoshop and Illustrator and stills output to 35mm or high resolution 5 x 4 transparencies
Studio: 47' x 30' with L cyc

Northern Light
35/41 Assembly Street

Leith
Edinburgh EH6 7RG
Tel: 0131 553 2383
Fax: 0131 553 3296
Gordon Blackburn
Stage lighting equipment hire. Mains distribution, staging, PA equipment hire. Sale of colour correction pyrotechnics etc

The OBE Partnership
16 Kingly Street
London W1R 5LD
Tel: 0171 734 3028
Fax: 0171 734 2830
16/35mm film editing, off-line and Avid editing, post-production supervision

Oasis Television
4-7 Great Pulteney Street
London W1V 3LF
Tel: 0171 434 4133
Fax: 0171 494 2843
Helen Leicester
Five large on-line edit suites, composite or component: D3, D2, D5, DGBeta, 1"C, Beta SP. Three graphics suites: Acrobat plus Matador for 3D and 2D animation, Matisse paint system. 3 Lightworks non-linear off-line suites. High quality dubbing and duplication. Digital audio dubbing

Ocean Post
5 Upper James Street
London W1R 3HF
Tel: 0171 287 2297
Fax: 0171 287 0296
email: bookings@oceanpost.co.uk
Editbox suite, Avid online suite, Avid Audio Vision 16 track sound mixing suite, Avid offline

Omnititles
28 Manor Way
London SE3 9EF
Tel: 0181 297 7877
Fax: 0181 297 7877
email: Omnititles@compuserve.com
Spotting and subtitling services for film, TV, video, satellite and cable. Subtitling in most world languages

Oxford Film and Video Makers
The Stables, North Place
Headington
Oxford OX3 7RF
Tel: 01865 741682 or 01865 60074 (course enquiries)
Fax: 01865 742901
email: ofvm@ox39hy.demon.co.uk
Film and video equipment hire - including Beta SP and non-linear editing facility (FAST). Wide range of evening and weekend courses

PMPP Facilities
69 Dean Street

London W1V 5HB
Tel: 0171 437 0979
Fax: 0171 434 0386
Off-line editing: BVW SP, lo-band
and VHS. Non-linear editing: 5
custom built Avid suites either self
drive or with editor. On-line editing:
Digital Betacam, D3, D2, Beta SP,
1", BVU SP and Hi-8 formats. Three
suites with Charisma effects Aston
or A72 cap gen and GVG mixers.
Graphics: Matisse Painting,
Softimage 3D, Acrobat 3D,
animation and T-Morph morphing
on Silicon Graphics workstations.
Sound dubbing on Avid Audiovision
or AudioFile. Voiceover studio/A-
DAT digital multi-track recording.
Full transfer, duplication and
standards conversion service. Pack
shot studio

The Palace
8 Poland Street
London W1V 3DG
Tel: 0171 439 8241
Fax: 0171 287 1741
Ray Davis
3 Beta SP/D3 on-line suites. Graphics

Panavision Grips
5-11 Taunton Road
Metropolitan Centre
Greenford
Middx UB6 8UQ
Tel: 0181 578 2382
Fax: 0181 578 1536
email: pangrip.co.uk
Grip equipment and studio hire
Also: The Greenford Studios
5-11 Taunton road
Metropolitan Centre
Greenford Middx UB6 8UQ
Tel: 0181 575 7300
Fax: 0181 839 1640
Grip equipment hire only

Panavision UK
Wycombe Road, Wembley
Middx HA0 1QN
Tel: 0181 903 7933
Fax: 0181 902 3273
Services many TV, commercial and
feature film productions

Paxvision
The Albany
Douglas Way
London SE8 4AG
Tel: 0181 692 6322 0973 416447
(mobile)
Fax: 0181 692 6322
Facilities for hire - concessionary
rates available

Picardy Television
23-27 Broughton Street Lane
Edinburgh EH1 3LY
Tel: 0131 558 1551

Fax: 0131 558 1555
Biddy Joyce
1 Park Circus
Glasgow G3 6AX
Tel: 0141 333 1200
Fax: 0141 332 6002
Digital editing, Avid non-linear off-
line, digital sound post-production,
graphics/graphic design studio. Dry/
crewed camera hire

Piccadilly Audio-visual Systems
6 Indeson Court
London E14 9TN
Tel: 0171 538 1622
Fax: 0171 538 1480
Supplies a full range of video, audio-
visual and multi-media hardware
products for conferences, exhibitions
and installations

Picture Post Productions
13 Manette Street
London W1V 5LB
Tel: 0171 439 1661
Fax: 0171 494 1661
Avid non-linear editing

Pinewood Studios
Sound Dept
Pinewood Road
Iver, Bucks SL0 0NH
Tel: 01753 656301
Fax: 01753 656014
email: graham_hartstone@rank.com
Graham Hartstone
Two large stereo dubbing theatres
with automated consoles, all digital
release formats. 35mm and Digital
dubbing, ADR & Foley recording.
Large ADR/Fx recording theatre,
35mm or AVID AUDIOVISION,
removable drives, ISDN Dolbyfax
with timecode in aux data. Digital
dubbing theatre with AMS/NEVE
Logic 2 and AudioFile Spectra 16.
Preview theatre 115 seats. Formats
35/70mm Dolby SR.D, DTS and
SDDS. Comprehensive transfer bay.
Stereo Optical Negative transfer
including Dolby SR.D, SDDS and
DTS. Cutting rooms

Post Box
8 Lower James Street
London W1R 3PL
Tel: 0171 439 0600
Fax: 0171 439 0700
Jo Smith
Jo Smith, Erica Rae
UK and international post produc-
tion. Offline, online, editbox and
more - offering the whole package
for broadcasters

Q Broadcast Ltd
1487 Melton Road
Queniborough
Leicester LE7 3FP

Tel: 0116 260 8813
Fax: 0116 260 8329
Avid MC 1000 on PCI with AVR 75
and Digital Betacam - On-Line. 3
Machine Betacam SP Component
Edit Suite with Digital Betacam,
Inscriber and Aston Captioning,
Alladin and Charisma DVE - On-
Line. Heavyworks II Non-Linear
Off-Line with 64Gb storage. Drive
in Studio, 1,350 sq ft, Cyc on two
walls. Location Crews in Midlands
and North. Comprehensive Graphics
services

The Ragged School
47 Union Street
London SE1 1SG
Tel: 0171 403 1316
Tel: 0171 403 1316
The Ragged School is a location and
studio for stills and moving pictures.
The dimensions are 40' by 25' by 14'
(high) White Plaster room + Parquet
Floor. Daylight. Facilities include
make-up, showers, free parking,
office space and equipment hire.
Second studio 30' x 40' x 12' (high)
Daylight red brick + dance floor,
Third studio 30' x 30' x 12' (high)
concrete floor paintable walls

Rank Video Services
Phoenix Park
Great West Road
Brentford
Middx TW8 9PL
Tel: 0181 568 4311
Fax: 0181 847 4032
Julie Johnson
Julie Johnson, Paul Gooderham
Europe's leading video duplication
facility with duplication centres in
London and Germany. Duplication,
standards conversion, packing and
distribution. Accredited with
BS5750/ISO9002

Red Post Production
Hammersley House
London W1R 6JD
Tel: 0171 439 1449
Fax: 0171 439 1339
email: Red-Post@Demon.co.uk
email: Redfx@Demon.co.uk
Post production company specialising in design and technical special
effects for commercials, video
promos, broadcast titles and idents,
feature film projects, broadcast
projects utilising computer animation techniques. Motion capture,
Flame, Henry, Flash Harry. Full
technical supervision

Redapple
214 Epsom Road, Merrow
Guildford
Surrey GU1 2RA

Tel: 01483 455044
Fax: 01483 455022
Video formats: Beta SP, Beta Sx,
NTSC/PAL. Cameras: Sony DNW
90WSP 4:3016:9, IKEGAMI, V-55
Camcorders; Transport; VW
Caravelle and Volvo Camera Cars

Redwood Studios Recording & Film Post Production
1-6 Falconberg Court
London W1V 5FG
Andre Jacquemin - Managing
Director - Sound Designer
Post production for features
including large f/x library and digital
audio post work

Reuters Television Ltd
40 Cumberland Avenue
London NW10 7EH
Tel: 0171 542 5810
Tel: 0171 542 5810
Kate P Charters, Sales Manager,
Location Specials
Crewing, editing, uplinking and
assistance with booking satellite
transmissions from anywhere in the
world. Offers mobile facilities ranging
from the hire of a single camera crew,
to the setting up of an SNG operation
or even the building of a complete
outside broadcast studio

Richmond Film Services
The Old School, Park Lane
Richmond
Surrey TW9 2RA
Tel: 0181 940 6077
Fax: 0181 948 8326
Sound equipment available for hire,
sales of tape and batteries, and UK
agent for Ursta recordists' trolleys
and Denecke timecode equipment

Rostrum Cameras Ltd
11 Charlotte Mews
off Tottenham Street
London W1P 1LN
Tel: 0171 637 0535
Fax: 0171 323 3892
Matthew Ferris
16/Super 16mm, 35mm and video
rostrum computerised cameras

Rushes
66 Old Compton Street
London W1V 5PA
Tel: 0171 437 8676
Fax: 0171 734 2519
Mike Uden
Mike Uden, Joce Capper
Digital and film post production:
Digital editing, 4:4:4 Ursa Gold
Telecine, 3-D animation, 2,400 sq ft
motion control studio, Nicam digital
playouts, Flame/Inferno SFX, AVID,
Henry, Flint

SVC Television
142 Wardour Street
London W1V 3AU
Tel: 0171 734 1600
Fax: 0171 437 1854
Tracey Whitaker
Video Post Production including the
following: Diamond Telecini, Flame,
3 Henrys, Computer Animation & 2
Motion Control Studios

Salon Post-Productions
10 Livonia Street
London W1V 3PH
Tel: 0171 437 0516
Fax: 0171 437 6197
16/35mm Steenbecks and editing
equipment rental, Steenbeck
Telecine, off-line suites, non-linear
editing systems including Avid and
Lightworks. Digital sound editing

Samuelson Film Service London Ltd
21 Derby Road
Metropolitan Centre
Greenford
Middx UB6 8UJ
Tel: 0181 578 7887
Fax: 0181 578 2733
email: johnfx@sammys.demon.co.uk
Cameras: Moviecam, Arriflex,
VistaVision, Mitchell and
Photosonics. Sony High Definition
Video systems. Lenses: Canon,
Cooke, Nikon, Leitz, Zeiss,
Hasselblad and Samcine. Heads,
filters, video assist, sound, editing,
stock, consumables and transport.
24hr service

Michael Samuelson Lighting
Pinewood Studios
Iver Heath
Bucks SL0 0NH
Tel: 01753 631133
Fax: 01753 630485
Unit 9 Maybrook Industrial Park
Armley Road
Leeds LS12 2EL
Tel: 0113 242 8232
Fax: 0113 245 4149
Unit K, Llantrisant Business Park
Llantrisant,
Mid Glamorgan CF7 8LF
Tel: 01443 227777
Fax: 01443 223656
Units 7/8, Piccadilly Trading Estate,
Great Ancoats Street
Manchester M1 2NP
Tel: 0161 272 8462
Fax: 0161 273 8729
Meridian Studios, Television Centre,
Northam
Southampton SO2 0TA
Tel: 01703 222555
Fax: 01703 335050
Hire of film and television lighting

equipment and generators. Six UK
depots. Largest range of MSR
discharge equipment in Europe
including the powerful SUNPAR
range from France

Sheffield Independent Film
5 Brown Street
Sheffield S1 2BS
Tel: 0114 272 0304
Fax: 0114 279 5225
Colin Pons
Colin Pons, Gloria Ward,
Alan Robinson
Aaton XTR + (S16/St 16). Vision 12
tripod S16/St 16. 6-plate Steenbeck,
Picsync. Nagra IS. SQN 45 mixer.
Microphones: 416, 816, ECM 55s.
SVHS edit suite. Avid MSP edit
suite Sony DXC537. UVW100
Betakit/Betacam (PVE 2800)/Hi-
band SP (BVU 950)/Hi-8, 2 and 3
machine edit suite. Three Chip
cameras. Lighting equipment. 1,200
ft studio. Sony DVC Digital Camcorder

Shepperton Sound
Shepperton Studios
StudiosRoad
Shepperton
Middx TW17 0QD
Tel: 01932 572676
Fax: 01932 572396
three Dubbing Theatres (16mm,
35mm, video) Post-sync, and
footsteps; effects, theatre, in-house
sound transfers

Shepperton Studios
Studios Road
Shepperton
Middx TW17 0QD
Tel: 01932 562611
Fax: 01932 568989
Cutting rooms; 16mm, 35mm
viewing theatres

Soho Images
8-14 Meard Street
London W1V 3HR
Tel: 0171 437 0831
Fax: 0171 734 9471
All 35mm/16mm and Super 16mm
processing and printing including
overnight video rushes service;
Ultrasonic cleaning with restoration
and optical printing of all types of
archive film including nitrate; 3 x
Wetgate Rank Cintel 4.2.2. Telecine
for transfer of film formats;
Broadcast Standard Conversions Pal,
NTSC and SECAM via Adac and
duplication of single/bulk quantities

The South East England Location Company
2 Dixwell Close
Gillingham

Kent ME8 9TB
Tel: 01634 262606
Fax: 01634 263606
David Neale
We provide a location research and
support service to film and video
tape production companies

SPLICE-RITE
Pinewood Studios, Pinewood Road
Iver Heath
Slough SL0 0NH
Tel: 01753 650006
Fax: 01753 650016

Brian Stevens
Animated Films Ltd
11 Charlotte Mews
off Tottenham Street
London W1P 1LN
Tel: 0171 637 0535
Fax: 0171 323 3892
Brian Stevens
Full 2D animation facilities
including rostrum cameras

Studio Pur
c/o Gargoyle Graphics
16 Chart St
London N1 6UG
Tel: 0171 490 5177
Fax: 0171 490 5177
email: sbayly@ich.ucl.ac.uk
Website: http://www.ace.mdx.ac.uk/
hyperhomes/houses/pur/index.htm
Simon Bayly, Lucy Thane
Media 100 & After Effects, Pro
Tools, Sony DSR-200 DVCAM
camcorder & tripod, Tascam
portable DAT, audio effects
processing, 500 lumens video
projector, multimedia workstation,
mics, lights & PA equipment

Studiosound
84 Wardour Street
London W1V 3LF
Tel: 0171 734 0263
Fax: 0171 434 9990
Film recording, dubbing and transfer
facilities, film recording, dubbing,
post-sync, Foleys, voice-overs and
transfer facilities

TSI
10 Grape Street
London WC2H 8DY
Tel: 0171 379 3435
Fax: 0171 379 4589
Full range of post-production facilities.
Four edit suites: One component digital
suite with Digital Betacam, Alpha Image
ELITE vision mixer; two composite
multi-format suites; one component Beta
SP suite. Graphics: HAL and Harriet
suites. Audio: DAWN digital sound
post-production, with voiceover booth.
All formats available: 1", Beta SP,
DigiBeta, D2, D3, D5 etc

TVP Videodubbing Ltd
2 Golden Square
London W1R 3AD
Tel: 0171 439 7138
Fax: 0171 434 1907
Jaqui Winston
Telecine transfer from 35mm, Super
16mm, 16mm and Super 8mm to all
video formats with full grading,
blemish concealment and image
restoration service. Video mastering,
reformatting and duplication to and
from any format; standards conversion service including motion
compensation via the Alchemist Ph.
C. digital converter. Also landlines for
feeds to the BT Tower and commercials playouts. Laserdisc pre-mastering
and full quality assessment. Packaging

Tattooist International
Westgate House
149 Roman Way
London N7 8XH
Tel: 0171 700 3555
Fax: 0171 700 4445
16mm cutting room, Super 16 and
stereo options, lo-band U-Matic off-
line, Aaton camera hire specialists,
Steadicam, time lapse library,
production offices

Tele-Cine
Video House
48 Charlotte Street
London W1P 1LX
0171 208 2000
Fax: 0171 208 2250/1
Siân Morgan
Component digital editing;
composite editing graphics 13D;
Avid non-linear editing telecine;
audio post-production laser disc
preparation multi-media applications; copy and duplication
commercials playouts; 2" quad/
archive retrieval; standards
conversion: telecom lines; satellite
downlink

Third Eye Productions
Unit 210 Canalot Studios
222 Kensal Road
London W10 5BN
Tel: 0181 969 8211
Fax: 0181 960 8790

Tiny Epic Video Co
37 Dean Street
London W1V 5AP
Tel: 0171 437 2854
Fax: 0171 434 0211
Non-linear offlining on Avid, and D/
Vision. Tape offlining on Umatic &
VHS with and without shotlister.
Rushes dubbing. Tape transfers -
most formats including Hi-8 and
DAT. EDL Generation and Translation

Todd-AO UK
Hawley Crescent
London NW1 8NP
Tel: 0171 284 7900
Fax: 0171 284 1018
Todd-AO UK serves both a local and
International production client base,
in the Film, Television, Cable and
Satellite markets. A comprehensive
post production service, in film and
video is offered. Staff work closely
together with clients to find post
production routes that enable
producers and production managers
to maximise their budgets

TVi
Film House
142 Wardour Street
London W1V 3AU
Tel: 0171 434 2141
Fax: 0171 439 3984
Mark Ottley
Mark Ottley, Joy Hancock
Post production; Telecine; sound
dubbing; copying and conversion.
Extensive integrated services
especially for film originated
programmes. Full digital compo-nent
environment. Wetgate digital
Telecine with Aaton and ARRI and
full range of gates including Super
16mm and 35mm wide aperture
wetgate transfers. Full film rushes
transfer service. Free film preparation service. Wide range of VTRs
including Digital Betacam

TV Media Services
420 Sauchiehall Street
Glasgow G2 3JD
Tel: 0141 331 1993
Fax: 0141 331 1993
Peter McNeill
Beta SP and BVU SP editing. All
formats shooting kit and crews
available. Short call out. BITC
copies of rushes. Dubbing to audio
for transcription

Twickenham Film Studios
St Margaret's
Twickenham
Middx TW1 2AW
Tel: 0181 607 8888
Fax: 0181 607 8889
Gerry Humphreys,
ISDN: 0181 744 1415
Gerry Humphreys, Caroline Tipple
Two dubbing theatres; ADR/Foley
theatre; 40 cutting rooms;
Lightworks, Avid, 16/35mm

VMTV
34 Fouberts Place
London W1V 2BH
Tel: 0171 439 4356
Fax: 0171 437 0952

OBs: Five units with 2-20 cameras, 1" or Beta SP record format, full complement ancillary equipment, plus new digital truck. Dry hire: Betacam SP and 16mm equipment with/without crews. Studios: two West End broadcast studios 45 x 33 x 217ft high and 25 x 15 x 13ft high

VTR

64 Dean Street
London W1V 5HG
Tel: 0171 437 0026
Fax: 0171 439 9427
Fiona Watson
Two Henry suites. Two D1 and Betacam Digital edit suites. One sound relay/viewing suite. Two Telecine suites with tape to tape grading facility. Three 3D graphic systems. One Harriet suite. Land line play out facilities. Teletext subtitling. Advance suite. Digital facility line

Vector Television

Battersea Road
Heaton Mersey
Stockport
Cheshire SK4 3EA
Tel: 0161 432 9000
Fax: 0161 443 1325
Martin Tetlow
Vector Graphics; Vector Digital Audio; Vector Digital editing; Vector Studios; 2D/3D design and visualisation consultancy

The Video Duplicating Co Ltd and CD Systems

VDC House, South Way
Wembley
Middx HA9 0HB
Tel: 0181 903 3345
Fax: 0181 900 1427
SanJay Mohindra
Comprehensive video services in all formats, tape to tape, high-speed bulk cassette duplication, CD replication of all formats such as CD ROM, CD Video, and CD audio

The Video Lab

Back West Crescent
St Annes on Sea
Lancs FY8 1SU
Tel: 01253 725499/712011
Fax: 01253 713094
Cintel Telecine 9.5/8/Super 8/16/35mm, slides and stills. Video formats: 2", 1"C, BVU, U-Matic, Beta SP. Cameras: Sony. Duplication, standards conversion. Specialists in transfer from discontinued videotape formats. Library of holiday videos (Travelogue), corporate and TV production, TV and cinema commercial production

Video Film & Grip Company

23 Alliance Court
Alliance Road
London W3 0RB
Tel: 0181 993 8555
Fax: 0181 896 3941
Contact: G.Stubbings
Unit 9, Orchard Street Industrial Estate, Salford
Manchester M6 6FL
Tel: 0161 745 8146
Fax: 0161 745 8161
Cardiff Studios, Culverhouse Cross, Cardiff CF5 6XT
Tel: 01222 599777
Fax: 01222 597957
Suppliers of 35mm camera equipment. 16mm camera equipment for documentaries, Digital SP and Beta SP video equipment for broadcast, and extensive range of cranes, dollies and ancillary grip equipment

Video Time

142 Wardour Street
London W1V 3AT
Tel: 0171 439 1211
Fax: 0171 287 1456
One stop facilities with over 20 different formats including D1, D2, D3, D5, Digital Betacam PAL/NTSC. Motion-compensated standards conversion, Telecine 4:2:2 16/Super 16/35mm PAL/NTSC/SECAM. Audio playback. Specialists in CAV/CLV laser disc cutting and disc mastering. CD-Rom mastering from any data source. MPEG encoding from video source

Videola (UK)

162-170 Wardour Street
London W1V 3TA
Tel: 0171 437 5413
Fax: 0171 734 5295
Video formats: 1", U-Matic, Beta SP. Camera: JVC KY35. Editing: three machine Beta SP. Computer rostrum camera. Lightworks offline

Videolondon Sound

16-18 Ramillies Street
London W1V 1DL
Tel: 0171 734 4811
Fax: 0171 494 2553
email: info@videolon.ftech.co.uk
Website: http://www.ftech.net/~videolon
Five sophisticated sound recording studios with overhead TV projection systems. 16mm, 35mm and video post-sync recording and mixing. Two Synclavier digital audio suites with four further Synclaviers, five AvidAudiovision, two StudioFrame and one AudioFile assignable to any of the studios. All sound facilities

for film or video post-production including D3, DigiBetacam, Betacam SP, 1" PAL and Dolby Surround for TV with three Lightworks non-linear editing systems

Videoscope

Cattwg Studio
Llancarfan
Barry
South Glam CF6 9AG
Tel: 01446 710963
Fax: 01446 710023
Video formats: hi-band U-Matic, Betacam, 1". Cameras: Sony 3000. Timecode transfer, computerised editing; copying bank. 3-machine, hi-band editing, SP editing; copying bank

Videosonics Cinema Sound

68a Delancey Street
London NW1 7RY
Tel: 0171 209 0209
Fax: 0171 419 4470
2 x All Digital THX Film Dubbing Theatres. Dolby Digital and SR 35 mm, 16mm and Super 16mm. All aspect ratios, all speeds. Video Projection if required Theatre I: AMS-Neve Logic II console (112 channels) with 24track Audiofile. Theatre II (Big Blue): AMS-Neve DFC console (224 channels) with 2 x 24 track Audio files. 3 x additional television Sound Dubbing Suites, 2 with AMS-Neve digital consoles, 1 x SSL console. 6 x Digital Audio Editing rooms, 35mm film editing, Facilities for Lightworks and Avid 2 x Foley and ADR Studios. A total of 14 AMS Audiofiles. Parking by arrangement. Wheelchair Access

VTR Ltd

64 Dean Street
London W1V 5HG
Tel: 0171 437 0026
Fax: 0171 439 9427
email: info@vtr.co.uk
VTR is one of London's major digital non-linear post production facilities specialising in commercials, corporates and promos. Facilities include: The Spirit, the world's first real-time high resolution film scanner for 35mm, 16mm and super 16; 2 x Ursa Gold telecines with Pogle Platinum Electric sunroof, full range of Ursa optical effects incl. Kaleidoscope. Inferno, for resolution independent special effects for TV & film. 2 x Henrys with V8 software for non-linear editing and effects. Flint RT, for visual effects work. Harriet with V series Paintbox. MAC with Adobe Photoshop. CGI, 3-D graphics,

Domino see under The Film Factory at VTR

Warwick Sound
WFS (Film Holdings)
Wardour Street
London W1V 3TD
Tel: 0171 437 5532
Fax: 0171 439 0372
Sound transfer all magnetic formats, Optical sound negative recording Stereo/Mono

Waterside Television
Waters' Close
Leith
Edinburgh EH6 6RB
Tel: 0131 555 3456
Fax: 0131 555 1346
Full serial digital edit suite. Accom RAVE non-linear on-line editing system. D1 VTRs, Accom RTDs. D75 Betacams. DAT. Fully loaded A57 DVE with Superwarp, DD20 digital mixer. Graphics/compositing suite

West One Television
10 Bateman Street
London W1V 5TT
Tel: 0171 437 5533
Fax: 0171 287 8621

West One West
186 Campden Hill Road
London W8 7TH
Tel: 0171 221 8221
Fax: 0171 792 4718
Shelley Fox, David Lale
Five multi-format on-line suites, one component suite; multi-format low volume duplication/one component suite; four Avid non-linear suite; one Lightworks non-linear suite. Sound-dubbing suite, 24 track and AudioFile multi-format low volume duplication

Windmill Lane Pictures
4 Windmill Lane
Dublin 2
Ireland
Tel: (353) 1 6713444
Fax: (353) 1 6718413
Telecine, digital off-line/on-line graphics and ENG crews

Wired For Post
5 Upper James Street
London W1R 3ED
Tel: 0171 287 2297
Fax: 0171 287 0296
Avid off-line and on-line, Audio vision 24 track mixing Editbox

Wiseman
29-35 Lexington Street
London W1R 3HQ
Tel: 0171 439 8901

Fax: 0171 437 2481
Ann Jones
Ann Jones, Jane Kellock
Component digital and analogue on-line editing. Non-linear Lightworks and Avid suites. Telecine suites with DCP secondary colour processor. Digital audio suite. Paintbox, Harriet, Harry and Softimage 3D graphics. Duplication and standards conversion. Full storyboard and design service available

World Wide Sound
21-25 St Anne's Court
London W1V 3AW
Tel: 0171 434 1121
Fax: 0171 734 0619
16/35mm, digital and Dolby recording track laying facilities specialising in post sync foreign dubbing

Worldwide Television News (WTN Facilities)
The Interchange
Oval Road
Camden Lock
London NW1
Tel: 0171 410 5410
Fax: 0171 410 5335
Anne Marie Phelan
2 TV studios (full cyc and component key); Digital Betacam and Beta SP editing (PAL and NTSC), Quantel Newsbox Non-linear online editing; Vistek VMC Digital standards conversion; Soundstation Digital Audio dubbing; UK and international satellite delivery and crews

Wrap it up
116a Action Lane,
Chiswick
London W4 5HH
Tel: 0181 333 1681/995 3357
(Mobiles 0973 198154/0976 380 781)
Fax: 0181 879 7710
Wrap it up provides production services which include transcription, post production scripts, voice scripts and logging of rushes for production companies. Recent work: Agenda for Britain, October Films - Transcription and Post Production Scripts. Horizon BBC - Transcription. Dennis and Gnasher, Tony Collingwood Productions - Voice Scripts

FESTIVALS

Listed below by country of origin are international film, television and video festivals with contact addresses and brief synopses

AUSTRALIA

Melbourne International Film Festival
July/August
PO Box 2206, Fitzroy Mail Centre
Fitzroy 3065
Victoria
Tel: (61) 3 417 2011
Fax: (61) 3 417 3804
Non-competitive showcase for Australian and International features, shorts, documentaries, animation and experimental films.

Sydney Film Festival
June
PO Box 950
Glebe NSW 2037
Tel: (61) 2 9660 3844
Fax: (61) 2 9692 8793
email: sydfilm@ozonline.com.au
A broad-based non-competitive Festival screening around 200 films not previously shown in Australia: features, documentaries, shorts, animation, video and experimental work. Competitive section for Australian short films only. Audience votes for best documentary, short and feature

AUSTRIA

Viennale - Vienna International Film Festival
October
Stiftgasse 6
A-1070 Vienna
Tel: (43) 1 526 5947
Fax: (43) 1 523 4172
email: office@viennale.or.at
Website: http//www.viennale.or.at
Non-competitive for features and documentaries. Additional categories: Twilight Zone; Tributes; Historical Retrospective

BELGIUM

Brussels International Film Festival
19-30 January 1999
Chaussée de Louvain 30
B - 1210 Brussels
Tel: (32) 2 227 39 80
Fax: (32) 2 218 18 60
email: infoffb@netcity.be
Website: http://ffb.cinebel.com
This is a competitive festival for European general interest films, annually showing about 100 features and 120 shorts. European features and shorts eligible to compete for Crystal Star Award. Belgian shorts eligible to compete for Golden Iris Award. Sections include European Competition, Kaleidoscope of the World Cinema, Belgian Focus with a National Competition, Special Events and Tributes. Feature entries should be over 60 minutes and shorts should be under 30 minutes. Formats accepted: 35 mm, 16mm. Deadline: 31st October. No entry fee

Brussels International Festival of Fantasy, Thriller and Science Fiction Films
March
144 Avenue de la Reine
1030 Brussels
Tel: (32) 2 201 17 13
Fax: (32) 2 201 14 69
Competitive for features and shorts (less than 20 mins)

Flanders International Film Festival - Ghent
October
1104 Kortrijksesteenweg
B-9051 Ghent
Tel: (32) 9 221 89 46
Fax: (32) 9 221 90 74
email: filmfestival@glo.be
Website: http://www.filmfestival.be
Contact: Jacques Dubrulle, Secretary-general, Walter Provo, Programme Director, Peter Bouckaert, Assistant Manager/Press Officer, Marian Ponnet, Guest Officer.
Belgium's most prominent yearly film event. Competitive, showing 150 feature films and 80 shorts from around the world. Deadline for entry forms mid August

BRAZIL

Mostrario - Rio de Janeiro Film Festivals
September
Rua Voluntários dá Pátria 97
CEP 22270-010
Rio de Janeiro RJ
Tel: (55) 21 539 1505
Fax: (55) 21 539 1247
Non-competitive, promoting films that would not otherwise get to Brazilian screens

Gramado International Film Festival - Latin and Brazilian Cinema
August
Avenida das Hortensias 2029
Grande 95670-000
Gramado - Rio do Sul
Tel: (55) 54 286 2335
Fax: (55) 54 286 2397
email: festival@via-rs.com.br
Website: http://www.viadigital.com.br/gramado
For exhibition of audiovisual products from Latin language speaking countries

São Paulo International Film Festival
October
Al Lorena 937, Cj 303
01424-001 São Paulo SP
Tel: (55) 11 883 5137/3064-5819
Fax: (55) 11 853 7936
email: info@mostra.org
Website: http://www.mostra.org
Two sections, international selection (for features, shorts, documentary, animation.) and a competitive section for films of new directors (first, second or third feature), produced during two years preceding the festival

BURKINA FASO

Panafrican Film and TV Festival of Ouagadougou
February/March odd years
Secrétariat Général Permanent du FESPACO
01 BP 2505
Ouagadougou 01
Tel: (226) 30 75 38
Fax: (226) 31 25 09
Competitive, featuring African diaspora and African film-makers, whose work has been produced

during the three years preceding the Festival, and not shown before at FESPACO

CANADA

The Atlantic Film Festival
September
c/o Suite 220 - 5600 Sackville Street (CBC)
Halifax
Nova Scotia B3J 3E9
Tel: (1) 902 422 3456
Fax: (1) 902 422 4006
email: festival@atlanticfilm.com
Website: http://www.atlanticfilm.com
Gordon Whittaker - Executive Director, Lia Rinaldo - Senior Programmer
Presents current films for both adults and children, several competitions (both public and juried), workshops and seminars, and a number of gala social events. Attracts films and videos from Atlantic Canada,and regions and countries bodering on the North Atlantic Rim

Banff Television Festival
June
1516 Railway Avenue
Canmore, Alberta, T1W 1P6
Tel: (1) 403 678 9260
Fax: (1) 403 678 9269
email: banff@bowest.awinc.com
Website: http://www.cochran.com/banfftv
Competitive for programmes made for television, including short and long drama, limited and continuing series, arts and social and political documentaries, children's programmes, comedy, performance specials, information and animation programmes broadcast for the first time in the previous year and popular science programmes

Festival International du Film Sur L'Art (International Festival of Films on Art)
9-14 March 1999
640 rue St Paul Ouest
Bureau 406
Montreal, Quebec H3C 1L9
Tel: (1) 514 874 1637
Fax: (1) 514 874 9929
email: Fifa@maniacom.com
Website: http://www.maniacom.com/fifa.html
The Festival encompasses all the arts, of any period or style. Films and videos must preferably be in French, otherwise in English, in their original, subtitled or dubbed version

Montreal International Festival of Cinema and New Media
June
3668 Boulevard Saint-Laurent
Montreal
Quebec H2X 2V4
Tel: (1) 514 843 4725
Fax: (1) 514 843 4631
email: montrealfest@fcmm.com
Website: http://www.fcmm.com
Claude Chamberlain
Discovery and promotion of outstanding international films, video and new media creations produced during previous two years, which have not been previously screened in Canada. Non-competitive (although some prizes in cash are awarded), indoor and outdoor screenings

Montreal World Film Festival (+ Market)
August/September
1432 de Bleury St
Montreal
Quebec H3A 2JI
Tel: (1) 514 848 3883
Fax: (1) 514 848 3886
Competitive festival recognized by the International Federation of Film Producers Associations. Categories: Official Competition, Hors Concours section, Cinema of Today, Cinema of Tomorrow, Latin American Cinema, Focus on One Country's Cinema, Panorama Canada, TV Films, Tributes. Feature films and shorts produced during the 12 months preceding the Festival, and unreleased in Canada

Ottawa International Animation Festival
September/October even years
2 Daly Avenue
Ottawa
Ontario K1N 6E2
Tel: (1) 613 232 6727
Fax: (1) 613 232 6315
Competitive

Toronto International Film Festival
September
Suite 1600 2 Carlton Street
Toronto
Ontario M5B IJ3
Tel: (1) 416 967 7371
Fax: (1) 416 967 9477
Non-competitive for feature films and shorts not previously shown in Canada. Also includes some American premieres, retrospectives and national cinema programmes. Films must have been completed within the year prior to the Festival to be eligible

Vancouver International Film Festival
September/October
Suite 410, 1008 Homer Street
Vancouver
British Columbia V6B 2X1
Tel: (1) 604 685 0260
Fax: (1) 604 688 8221
email: viff@viff.org
Webstite: http://viff.org
Third largest festival in North America, with special emphasis on East Asian, Canadian and documentary films. Also British and European cinema and 'The Screenwriter's Art'. Submission deadline mid-July - only feature film entries accepted from outside of Canada

CROATIA

World Festival of Animated Films - Zagreb
June even years
Koncertna direkcija Zagreb
Kneza Mislava 18
10000 Zagreb
Tel: (385 - 1) 46 11 808/46 11 709/46 11 598
Fax: (385 - 1) 46 11 807/46 11 808
email: kdz@zg.tel.hr
Website: http://animafest.hr
Competitive for animated films (up to 30 mins). Categories: a) films from 30 secs-6 mins, b) films from 6-15 mins, c) 15 min-30mins. Awards: Grand Prix, First Prize in each category (ABC), Best First Production (Film Debut) Best Student Film, Five Special Distinctions. Films must have been completed in two years prior to the Festival and not have been awarded prizes

CUBA

International Festival of New Latin American Cinema
December
Calle 23
1155 Vedado
Havana 4
Tel: (53) 7 34169/36072
Fax: (53) 7 33 30 78
Competitive for films and videos. Market for Latin American films

CZECH REPUBLIC

'Golden Prague' International TV Festival
May
Czech Television

Kavci Hory
140 70 Prague 4
Tel: (42) 2 6113 4405/4028
Fax: (42) 2 6121 2891
Competitive for television music
programmes and other types of
serious music, dance, jazz and world
music.music programmes. Entry
forms must be submitted by 31st
January and videos by 15th February

Karlovy Vary International Film Festival
July
Film-Festival Karlovy Vary
Foundation
Panská 1
110 00 Prague 1
Tel: (42) 0 2 2423 5412
Fax: (42) 0 2 2423 3408
Feature film competition, informa-
tion sections, Eastern Europe and
Czech panorama, retrospectives,
documentary, competition

DENMARK

Balticum Film and TV Festival
June
Skippergade 8
3740 Svaneke
Tel: (45) 7023 0024
Fax: (45) 7023 0025
Competitive for short films and
documentaries from the countries
around the Baltic Sea. Three
categories: Best Film; Best Televi-
sion Programme; Special Prizes.
Three prizes in each category in the
Balticum competition. The Film
School Competition: Only European
Film Schools can enter new projects.
2 Prizes: Best film school documen-
tary and best film school fiction

Copenhagen Film Festival
September
FSI
Vesterbrogade 35
1620 Copenhagen V
Tel: (45) 33 25 25 01
Fax: (45) 33 25 57 56
email: fside@datashopper.dk
Festival for the public. Previews of
American, European and Danish
films, both by established filmmak-
ers and those less well known.
Around 120 films, plus seminars and
exhibition

International Odense Film Festival
August
Vindegade 18
5100 Odense C
Tel: (45) 6613 1372 x4044
Fax: (45) 6591 4318

email: filmfestival@post.odkomm.dk
Website: www.filmfestival.dk
Competitive for fairy-tale and
experimental-imaginative films.
Deadline for entries 1 April

EGYPT

Cairo International Film Festival
November/December
17 Kasr El Nil Street
Cairo
Tel: (20) 2 392 3562/3962/393 3832
Fax: (20) 2 393 8979
Competitive for feature films, plus a
film, television and video market

Cairo International Film Festival for Children
September
17 Kasr el El Nil Street
Cairo
Tel: (20) 2 392 3562/3962/393 3832
Fax: (20) 2 393 8979
Competitive for children's films:
features, shorts, documentaries,
educative, cartoons, television films
and programmes for children up to
14 years

FINLAND

Midnight Sun Film Festival
June
Jäämerentie 9
99600 Sodankylä
Tel: (358) (0)16 614 524/614 522
Fax: (358) (0)16 618 646
Non-competitive for feature films,
held in Finnish Lapland

Tampere International Short Film Festival
10-14 March 1999
PO Box 305
33101 Tampere
Tel: (358) 3 2130034
Fax: (358) 3 2230121
email: film.festival@tt.tampere.fi
Competitive for short films, max. 30
mins. Categories for animated,
fiction and documentary short films,
completed on or after 1 January
1997. Videos (VHS) accepted for
selection only. 10-15 large
retrospectives and tributes of short
films from all over the world.
Competition deadline 5 Jan 1999

FRANCE

18th Amiens International Film Festival
November
MCA - 2 place Léon Gontier

F-8000 Amiens
Tel: (33) 322 713570
Fax: (33) 322 92 53 04
Films completed after 15 September
1997, and which make a contribution to the identity of people or an
ethnic minority, are eligible for
entry. They may be either full-length
or short, fiction or documentary
films

Annecy International Festival of Animation (+ Market)
May
JICA/MIFA
BP 399
74013 Annecy Cèdex
Tel: (33) 04 50 10 09 00
Fax: (33) 04 50 10 09 70
Competitive for animated short
films, feature-length films, TV films,
commercials, produced in the
previous 26 months

Cannes International Film Festival
May
99 Boulevard Maiesherbes
75008 Paris
Tel: (33) 1 45 61 66 00
Fax: (33) 1 45 61 97 60
email: festival@cannes.bull.net
Competitive section for feature films
and shorts (up to 15 mins) produced
in the previous year, which have not
been screened outside country of
origin nor been entered in other
competitive festivals, plus non-
competitive section: Un Certain
Regard. Other non-competitive
events: Directors Fortnight
(Quinzaine des Réalisateurs) and
Programme of French Cinema
(Cinémas en France)
215 rue du Faubourg St Honoré
75008 Paris
Tel: (33) 1 45 61 01 66
Fax: (33) 1 40 74 07 96
Critic's Week (Semaine de la
Critique)
73 rue de Lourmel
75015 Paris
Tel: (33) 1 45 75 68 27
Fax: (33) 1 40 59 03 99

Cinéma du Réel, (International Festival of Visual Anthropology)
March
Bibliothèque Publique
d'Information
19 rue Beaubourg
75197 Paris Cedex 04
Tel: (33) 1 44 78 44 21/45 16
Fax: (33) 1 44 78 12 24
Documentaries only (film or video).
Competitive - must not have been
released commercially or been
awarded a prize at an international

festival in France. Must have been made in the year prior to the Festival

Cognac International Thriller Film Festival

April
Le Public Systeme
36 rue Pierret
92200 Neuilly-sur-Seine
Tel: (33) 1 46 40 55 00
Fax: (33) 1 46 40 55 39
email: cognac@pobox.com
Competitive for thriller films, which have not been commercially shown in France or participated in festivals in Europe (police movies, thrillers, 'film noirs', court movies, investigations etc)

Deauville Festival of American Film

September
36 rue Pierret
92200 Neuilly-sur-Seine
Tel: (33) 1 46 40 55 00
Fax: (33) 1 46 40 55 39
email: deauville@pobox.com
Studio previews (non competitive) Independent Films Competition and panorama. US productions only

FIFREC (International Film and Student Directors Festival)

June
16 chemin de Pommier
69330 Jons/Lyon
Tel: (33) 72 02 48 64
Fax: (33) 72 02 20 36
Official film school selections (three per school) and open selection for directors from film schools, either students or recent graduates. Categories include fiction, documentaries and animation. Also best film school award. Films to be under 40 mins

Festival Cinèmatographique d'Automne de Gardanne

October/November
Cinèma 3 Casino
11 cours Forbin
13120 Gardanne
Tel: (33) (0)442 51 44 93
Fax: (33) (0) 442 58 17 86
Includes European Competition of Shorts. Aims to discover high quality European cinema. Also junior section and retrospectives. All films for competition to be submitted on VHS, and to have been produced in the year prior to the Festival

Festival des Trois Continents

November
BP 43302
44033 Nantes Cedex 1

Tel: (33) 2 40 69 74 14
Fax: (33) 2 4073 55 22
Feature-length fiction films from Africa, Asia, Latin and Black America. Competitive section, tributes to directors and actors, panoramas

Festival du Film Britannique de Dinard

September
47 boulevard Féart
35800 Dinard
Tel: (33) 99 88 19 04
Fax: (33) 99 46 67 15
Competitive, plus retrospective and exhibition; tribute meeting between French and English producers

Festival International de Films de Femmes

March
Maison des Arts
Place Salvador Allende
94000 Créteil
Tel: (33) 1 49 80 38 98
Fax: (33) 1 43 99 04 10
email: filmsfemme@wanadoo.fr
Website: http://www.gdebussac.fr/filmfem
Competitive for feature films, documentaries, shorts, retrospectives directed by women and produced in the previous 23 months and not previously shown in France

French-American Film Workshop

June
10 Montée de la Tour
30400 Villeneuve-les-Avignon
Tel: (04) 90 25 93 23
Fax: (04) 90 25 93 24
198 Avenue of the Americas
New York, NY 10013
USA
Tel: (212) 343 2675
Fax: (212) 343 1849
email: JHR2001@AOL
Contact: Jerome Henry Rudes, General Director
The Workshop brings together independent filmmakers from the United States and France at the Avignon/New York Film Festival and Rencontres Cinématographiques Franco-Américans d'Avignon (see below). French and American independent film is celebrated with new films, retrospectives, round-tables on pertinent issues and daily receptions

Avignon/New York Film Festival

April
Alliance Française/French Institute, 22 East 60th Street, New York, NY - with 'the 21st Century Filmmaker Awards'

Rencontres Cinématographiques Franco-Américans d'Avignon

June
Cinéma Vox, Place de l'Horloge, Avignon, France - with 'The Tournage Awards'

Gérardmer-Fantastic Arts International Fantasy Film Festival

January
36 rue Pierret
92200 Neuilly-sur-Seine
Tel: (33) 1 46 40 55 00
Fax: (33) 1 46 40 55 39
email: fantasticarts@pdox.com
Competitive for international fantasy feature films (science-fiction, horror, supernatural etc)

International Festival of European Cinema La Boule

October
97 Rue Raumur
75002 Paris
Tel: (33) 1 4041 0454
Fax: (33) 1 4026 5478
Categories for European feature, short film, animation and documentary. Prizes for best director, actor and actress

MIP-TV

April
Reed MIDEM Organisation
179 avenue Victor Hugo
75116 Paris
Tel: (33) 1 44 34 44 44
Fax: (33) 1 44 34 44 00
International television programme market, held in Cannes

MIPCOM

October
Reed MIDEM Organisation
179 avenue Victor Hugo
75116 Paris
Tel: (33) 1 44 34 44 44
Fax: (33) 1 44 34 44 00
International film and programme market for television, video, cable and satellite, held in Cannes

GERMANY

Berlin International Film Festival

February
Internationale Filmfestspiele Berlin
Budapester Strasse 50
10787 Berlin
Tel: (49) 30 254 890
Fax: (49) 30 254 89249
Competitive for feature films and shorts (up to 10 mins), plus a separate competition for children's films - feature length and shorts - produced in the previous year and not entered for other festivals. Also

has non-competitive programme consisting of forum of young cinema, panorama, film market and New German films

Feminale, 9th International Women's Film Festival

October
Feminale
Maybachstr, 111
50670 Cologne
Tel: (49) 221 416066/1300 225
Fax: (49) 221 1300281
email: Feminale@t-online.de
Website: www.dom.de./filmworks/Feminale
Non-competitive for films by women directors only made in last two years, all genres, formats, lengths. Retrospectives, special programmes

Femme Totale - International Frauen Film Festival

10-14 March 1999
c/o Kulturbüro der Stadt Dortmund
Kleppingstr 21-23
44122 Dortmund
Tel: (49) 231 50 25 162
Fax: (49) 231 50 22 497
1062123237@compuserve.com
Silke Johanna Räbiger
Held every two years. Women Film-makers' Festival screens features, short films, documentaries and videos. Workshops and seminars.

Filmfest Hamburg

September
Friedensallee 1
22765 Hamburg
Tel: (49) 40 398 26 210
Fax: (49) 40 398 26 211
Non-competitive, international features and shorts for cinema release (fiction, documentaries), presentation of one film country/continent, premieres of Hamburg-funded films, and other activities

International FilmFest Emden

May
An der Berufsschule 3
26721 Emden
Tel: (49) 49 21 91 55 35
Fax: (49) 49 21 91 55 91
A festival for European feature films and shorts. Competitive section for feature films and a short and animation films (not more than 20 minutes in length) (35mm and 16mm) from northwestern Europe and German speaking countries (Audience Awards 15 000 DM 6000 DM) Regular Sections: New British and New German Films. Retrospectives, tributes to directors, children's films, midnight talks

International Film Festival Mannheim - Heidelberg

October
Collini-Center, Galerie
68161 Mannheim
Tel: (49) 621 10 29 43
Fax: (49) 621 29 15 64
email: ifmh@mannheim-filmfestival.com
Additional parts of the annual event are the 'Co-Production Meetings' for producer seeking co-production partners and the 'Independent Market Service' for buyers. Deadline for entry: 25 July, 1998

Internationale Filmwochenende Würzburg

January
Gosbertsteige 2
97082 Würzburg
Tel: (49) 931 414 098
Fax: (49) 931 416 279
Competitive section for recent European and international productions, plus non-competitive section including tributes to directors as well as panoramas. Videos accepted for selection only

Internationales Leipziger Festival für Dokumentar und Animationsfilm

October/November
Box 940
04009 Leipzig
Tel: (49) 341 9 80 39 21
Fax: (49) 341 9 80 48 28
Competition, special programmes, retrospective, international juries and awards

International Festival of Animated Film Stuttgart

April
Festivalbüro
Teckstrasse 56 (Kulturpark Berg)
70190 Stuttgart
Tel: (49) 711/9254610
Fax: (49) 711/9254615
Competitive for animated short films of an artistic and experimental nature, which have been produced in the previous two years and not exceeding 35 mins. Animation, exhibitions and workshops. DM 80,000 worth of prizes

Munich Film Festival

June/July
Internationale Münchner Filmwochen
Kaiserstrasse 39
80801 Munich
Tel: (49) 89 38 19 04 0
Fax: (49) 89 38 19 04 26
Non-competitive for feature films,

shorts and documentaries which have not previously been shown in Germany

Munich International Documentary Festival

April/May
Troger Strasse 46
Munich D-81675
Tel: (49) 89 470 3237
Fax: (49) 89 470 6611

Nordic Film Days Lübeck

November
23539 Lübeck
Tel: (49) 451 122 41 05
Fax: (49) 451 122 41 06
email: filmtage@luebeck.de
Website: http://www.luebeckk.de/filmtage
Festival of Scandinavian and Baltic films. Competitive for feature, children's, documentary, and Nordic countries' films

Oberhausen International Short Film Festival

April
Grillostrasse 34
46045 Oberhausen
Tel: (49) 208 825 2652
Fax: (49) 208 825 5413
email: kurzfilmtage_oberhausen@uni-duisburg.de
Website: http://www.shortfilm.de
Competitive for documentaries, animation, experimental, short features and videos (up to 35 mins), produced in the previous 28 months; international competition and German competition; international symposia

Prix Europa

September/October
Sender Freies Berlin
Masurenallee 8-14
14046 Berlin
Tel: (49) 30 3031 1610
Fax: (49) 30 3031 1619
Competitive for fiction, non-fiction, series/serials in television. Open to all television stations and television producers in Europe. Eight awards of 6,150 ECU

Prix Futura Berlin

May/April odd years
International Radio and TV Contest
Sender Freies Berlin
14046 Berlin
Tel: (49) 30 3031 1610
Fax: (49) 30 3031 1619
Competitive, one entry per category (documentary and drama). Open to all broadcasting organisations; radio evening competition for newcomers

in drama and features; three television competitions for films from Asia, Africa and Latin America in evenings

Prix Jeunesse International
June
Bayerischer Rundfunk
80300 Munich
Tel: (49) 89 5900 2058
Fax: (49) 89 5900 3053
Competitive for children's and youth television programmes (age groups up to 7, 7-12 and 12-17), in fiction and non-fiction, produced in the previous two years. (In odd years: seminars in children's and youth television)

GREECE

40th International Thessaloniki Film Festival
November
36 Sina Street
10672 Athens
Tel: (30) 1 3610418/3620907/ 3620962
Fax: (30) 1 362 1023
Dedicated to the promotion of independent cinema from all over the world. International Competition for first or second features (Golden Alexander worth approx. $43,000, Silver Alexander $27,000). Official non competitive section for Greek films produced in 1998, informative section with the best independent films of the year, retrospectives (last year's retrospective dedicated to Claude Chabrol and Arturo Ripstein), exhibitions, special events etc

HONG KONG

Hong Kong International Film Festival
April
Level 7, Admin Building, Hong Kong Cultural Centre
10 Salisbury Road, Tsimshatsui Kowloon
Tel: (852) 2734 2892
Fax: (852) 2366 5206
Non-competitive for feature films, documentaries and invited short films, which have been produced in the previous two years, also a local short film competition

HUNGARY

Hungarian Film Week
February
Magyar Filmunio
Varosligeti fasor 38

H - 1068 Budapest
Tel: (361) 351 7760
Fax: (361) 351 7766
Competitive festival for Hungarian features, shorts and documentaries

INDIA

Bombay International Film Festival for Documentary, Short and Animation Films
(February even years)
Films Division, Ministry of Information and Broadcasting Government of India
24-Dr G Deshmukh Marg
Bombay 400 026
Tel: (91) 22 3864633/3873655/ 3861421/3861461
Fax: (91) 22 3860308
Competitive for fiction, non-fiction and animation films, plus Golden/ Silver Conch and cash awards and Information Section. Films to have been produced between Jan 1996 and Dec 1997 (competition)

International Film Festival of India (IFFI)
January
Directorate of Film Festivals
Fourth Floor, Lok Nayak Bhavan Khan Market,
New Delhi 110 003
Tel: (91) 11 4615953/4697167
Fax: (91) 11 4623430
Malti Sahai (Director)
Organised from January 10-20, each year and recognised by FIAPF. It is held in different Indian Film Cities by rotation including New Delhi, Bangalore, Bombay, Calcutta, Hyderbad and Trivandrum. IFFI'98 was organised in New Delhi and featured a specialised competition section for Asian film makers

Kerala International Film Festival
30-10 April 1999
15/63, C2 Elankom House,
Elankom Gardens, Vellayambalam, Trivandrum 695 010 Kerala
Tel: (91) 471 310323
Fax: (91) 471 310322
email: chitram@md3.vsnl.net.in
Competition for Asian, African and Latin American films. Plus student films, Hitchcock retrospective and cinema from Burma

5th Mumbai International Film Festival
(March)
24- Dr Gopalrao Deshmukh Marg
Mumbai
400 026
Tel: (91) (22) 386 46 33

Fax: (91) (22) 380 00 308
email: films@giasbm01.Vsnl.net.in

IRAN

Tehran International Market (TIM)
c/o CMI
53 Koohyari Street
Fereshteh Avenue, Tehran 19658
Tel: (98) 21 254 8032
Fax: (98) 21 255 1914
The fourth Tehran International Market is designed to provide major producers from the West a personalised arena to target regional program buyers and theatrical distributors in the lucrative Middle East market and surrounding areas. Buyers will also represent the Persian Gulf States, Asia and the Indian subcontinent, Central and Eastern Europe. More than 1,500 hours of programming was brought by Iranian TV alone during TIM'96

IRELAND

Cork International Film Festival
October
Hatfield House, Tobin Street
Cork
Tel: (353) 21 271711
Fax: (353) 21 275945
email: ciff@indigo.ie
Non-competitive, screening a broad range of features and shorts from over 40 countries. Films of every category welcomed for submission

Dublin Film Festival
March
1 Suffolk Street
Dublin 2
Tel: (353) 1 679 2937
Fax: (353) 1 679 2939

ISRAEL

Haifa International Film Festival
October
142 Hanassi Avenue
Haifa 34633
Tel: (972) 4 383424/386246
Fax: (972) 4 384327
The biggest annual meeting of professionals associated with the film industry in Israel. Competitions: 1. 'Golden Anchor' award $25,000 for mediterranean cinema. 2. Israeli Film Competition award $30,000

Jerusalem Film Festival
July
PO Box 8561, Wolfson Gardens

Hebron Road
91083 Jerusalem
Tel: (972) 2 672 4131
Fax: (972) 2 673 3076
Finest in recent international
cinema, documentaries, animation,
avant garde, retrospectives, special
tributes and homages, Mediterranean
and Israeli cinema, retrospectives,
restored classics. 150 films during
10 days. Prizes include Wolgin
Awards For Israeli Cinema; the
Lipper Award for Best Israeli
Screenplay. Three international
awards: Wim van Leer In Spirit of
Freedom focus on human rights;
Mediterranian Cinema; Jewish
Theme Awards

ITALY

Da Sodoma a Hollywood
April
**Turin Lesbian and Gay Film
Festival, Associazione Culturale
L'Altra Communicazione
Via Tasso 11
10122 Turin**
Tel: (39) 11 436 6855
Fax: (39) 11 521 3737
Specialist lesbian/gay themed
festival. Competitive for features,
shorts and documentaries. Also
retrospectives and special showcases
for both cinema and television work

Europa Cinema & TV Viareggio, Italy
September
Via XX Settembre Mo 3
Tel: (39) 6 42011184
     (39) 6 42000211
Fax: (39) 6 42010599
An international competition of
European Films; a section of films
regarding food; a retrospective
section of films about 'The
European Roots of American
Cinema'; a retrospective of
Radford's films; a section of Arte's
(a European Cultural Network) most
decent productions; a section of
films produced thanks to Eurimages;
an international conference
organised by ACE (Atelier du
Cinéma Européen); a RAI première
(opening day); screening of
Ballando, Ballando by E.Scola
(closing day)

Festival dei Popoli - International Review of Social Documentary Films
November
**Borgo Pinti 82r
50121 Firenze**

Italy
Tel: (39) 55 244 778
Fax: (39) 55 241364

Giffoni Film Festival
July/August
**Piazza Umberto I
84095 Giffoni Valle Piana**
Tel: (39) 89 868 544
Fax: (39) 89 866 111
Competitive for full-length fiction
for children 12-14 and 12-18 years.
Entries must have been produced
within two years preceding the
festival

MIFED
November
**Largo Domodossola 1
20145 Milan**
Tel: (39) 2 480 12912 -48012920
Fax: (39) 2 499 77020
International market for companies
working in the film and television
industries

Mystery & Noir Film Festival
3-9 December 1998
**Via Tirso 90
00198 Rome**
Tel: 39 6 8848030 - 8844672
Fax: 39 6 8840450
Giorgio Gosetti
Competitive for thrillers between 30-
180 mins length, which have been
produced in the previous year and
not released in Italy. Festival now
takes place at Courmayeur (at the
foot of Mount Blanc)

Pesaro Film Festival (Mostra Internazionale del Nuovo Cinema)
August/September
**Via Villafranca 20
00185 Rome**
Tel: (39) 6 4456643/491156
Fax: (39) 6 491163
email: pesarofilmfest@mclink.it
Non-competitive. Particularly
concerned with the work of new
directors and emergent cinemas, with
innovation at every level. In recent
seasons the festival has been devoted
to a specific country or culture

The Pordenone Film Fair
October
**c/o Le Giornate del Cinema Muto
c/o La Cineteca del Frioli
Via G.Biui, Polozzo Guiisotti
33013 Gemona (UD)**
Tel: 39 434 520446
Fax: 39 434 520584
email: gcm@proxima.conecto.it
Website: http://
www.cinetec@del.frivli.org/gcm
An exhibition of books and journals,

collectibles and ephemera presented
by the Giornate del Cinema Muto.
The Film Fair is held near the Verdi
Theatre which will host the film
screenings. Authors attending the
festival are invited to discuss their
latest works. An auction sale of
cinema memorabilia is part of the
program

Pordenone Silent Film Festival (La Giornate del Cinema Muto)
October
**c/o La Cineteca del Friuli
Via G.Bini, Polozzo Gurisotti
33013 Gemona (UD)**
Tel: (39) 432 98 04 58
Fax: (39) 432 97 05 42
email: gcm@proxima.conecto.it
Website: http://www.
cinetec@delfrivli.org/gcm/
Non-competitive silent film festival.
Annual award for restoration and
preservation of the silent film
heritage

Prix Italia
September
**RAI Radiotelevisione Italiana
V.le Mazzini 14
00195 Rome**
Tel: (39) 6 37514996
Fax: (39) 6 3613401
Competitive for television and radio
productions from national broadcasting organisations. In the three
categories (music and arts, fiction,
documentaries) each broadcasting
organisation may submit: Radio -
maximum four entries; Television -
maximum three entries.

Salerno International Film Festival
October
**PO Box 137
84100 Salerno**
Tel: (39) 89 223 632
Fax: (39) 89 223 632
All films - feature, documentaries,
experimental and animated films -
which are entered in the competitive
section are eligible for the "Gran
Premio Golta di Salerno"

Taormina International Film Festival
July
**Palazzo Firenze
Via Pirandello 31
98039 Taormina
Sicily**
Tel: (39) 942 21142
Fax: (39) 942 23348
Competitive for features. Recog-
nised by FIAPF, category B.
Emphasis on new directors and
cinema from developing countries

Cinema Giovani - Torino Film Festival

November
Via Monte di Pietá 1
10121 Torino
Tel: (39) 11 5623309
Fax: (39) 11 5629796
email: cinemagiovani@
torinofilmfest.org
Website: http://www.torinofilmfest.org
Competitive sections for feature and short films. Italian Space section (videos and films) open solely to Italian work. All works must be completed during 13 previous months, with no prior release in Italy

Venice Film Festival

September
Mostra Internazionale d'Arte
Cinematografica
La Biennale di Venezia
Ca' Giustinian
San Marco, 30124 Venice
Tel: (39) 41 5218711
Fax: (39) 41 5227539
Website:www.labiennale.it
Competitive for feature films competitive for shorts (up to 30 mins); has competitive sections, perspectives, night and stars, Italian section; retrospective. Non-participation at other international festivals and/or screenings outside country of origin. Submission by 30 June

JAPAN

International Animation Festival Hiroshima

August
4-17 Kako-machi
Naka-ku
Hiroshima 730-0812
Tel: (81) 82 245 0245
Fax: (81) 82 245 0246
email: hiroanim@urban.ne.jp
Website: http: www.city.hiroshima.jp
Competitive biennial festival. Also retrospective, symposium, exhibition etc. For competition, animated works under 30 mins, and completed during preceding two years are eligible on either 16mm, 35mm, 3/4" videotape (NTSC, PAL, SECAM) or Betacam (only NTSC)

Tokyo International Film Festival

October
Organising Committee
4F, Landic Ginza Bldg II
1-6-5 Ginza, Chuo-Ku
Tokyo 104-0061
Tel: (81) 3 3563 6305
Fax: (81) 3 3563 6310
Competitive for International Films

and Young Cinema sections. Also special screenings, cinema prism, Nippon cinema classics, symposium, no film market

Tokyo Video Festival

January
c/o Victor Co of Japan Ltd
1-7-1 Shinbashi
Victor Bldg, Minato-ku
Tokyo 105
Tel: (81) 3 3289 2815
Fax: (81) 3 3289 2819
Competitive for videos; compositions on any theme and in any style accepted, whether previously screened or not, but maximum tape playback time must not exceed 20 minutes

MALTA

Golden Knight International Amateur Film & Video Festival

November
PO Box 450
Valletta CMR,01
Tel: (356) 222345/236173
Fax: (356) 225047
Three classes: amateur, student, professional - maximum 30 mins

MARTINIQUE

Festival du Film Caribéen Cinéma, Vidéo: Images Caraïbes

June even
77 route de la Folie
97200 Fort-de-France
Tel: (596) 69 10 12/70 23 81
Fax: (596) 69 21 58/62 23 93
Competitive for all film and video makers native to the Caribbean Islands - features, shorts and documentary

MONACO

Monte Carlo Television Festival and Market

February
4, Boulevard du Jardin Exotique
Monte-Carlo 98000 Monaco
Tel: 337 93 10 40 60
Fax: 337 93 50 70 14
Contact: David Tomatis
Annual festival and market, includes awards for television films, mini-series and news categories. In 1996 joined with Imagina conference

THE NETHERLANDS

Cinekid

Weteringschans 249
1017 XI

Amsterdam
Tel: (31) 2 624 7110
Fax: (31) 2 620 9965
International children's film and television festival. Winning film is guaranteed distribution in the Netherlands

Dutch Film Festival

September/October
Stichting Nederlands Film Festival
PO Box 1581
3500 BN Utrecht
Tel: (31) 30 2322684
Fax: (31) 30 2313200
email: ned.filmfest@inter.nl.net
Website: http://www.nethlandfilm.nl
Annual screening of a selection of new Dutch features, shorts, documentaries, animation and television drama. Retrospectives, seminars, talkshows, Cinema Militans Lecture, Holland Film Meeting, outdoor programme. Presentation of the Grand Prix of Dutch Film: the Golden Calf Awards

International Documentary Filmfestival Amsterdam

December
Kleine-Gartmanplantsoen 10
1017 RR Amsterdam
Tel: (31) 20 6273329
Fax: (31) 20 6385388
Competition programme: compet-itive for documentaries of any length, 35mm or 16mm, produced in 15 months prior to the festival; retrospectives; Joris Ivens award; Top 10 selected by well-known filmmaker; competitive video-programme; forum for international co-financing of European documentaries. Workshop, seminar and debates

International Film Festival Rotterdam

January/February
PO Box 21696
3001 AR Rotterdam
Tel: (31) 10 411 8080
Fax: (31) 10 413 5132
Comprising 200 features, documenta-ries, shorts. Main programme presents international premiers, and a selection of the last year's festivals. Several sidebars and retrospectives. Cinemart: co-production market for films in progress. Deadline for entries 1 November. Tiger Award competition: three premiums, each US 10,000 for promising new film-makers

NEW ZEALAND

Auckland International Film Festival

July
PO Box 9544
Te Aro
Wellington 6035
Tel: (64) 4 385 0162
Fax: (64) 4 801 7304
Festival includes feature films, short films, documentaries, video and animation

Wellington Film Festival
July
C/o New Zealand Film Festival
PO Box 9544
Te Aro
Wellington
Tel: (64) 4 385 0162
Fax: (64) 4 801 7304
email: enzedff@actrix.gen.nz
Festival includes feature films, short films, documentaries, video and animation

NORWAY

Norwegian International Film Festival
August
PO Box 145
5501 Haugesund
Tel: (47) 52 73 44 30
Fax: (47) 52 73 44 20
Non-competitive film festival, highlighting a selection of films for the coming theatrical season. New Nordic films - a market presenting Nordic films with a potential outside the Nordic Countries (27-29 Aug)

POLAND

International Short Film Festival in Kraków
May/June
c/o PIF 'Apollo Film'
ul. Pychowicka 7
30-364 Kraków
Tel: (48) 12 67 23 40/67 13 56
Fax: (48) 12 67 15 52
Festival Director: Wit Dudek
Competitive for short films (up to 35 mins), including documentaries, fiction, animation, popular science and experimental subjects, produced in the previous 15 months. Deadline for submitting films 31st January 1998

PORTUGAL

Cinanima (International Animated Film Festival)
November
Apartado 743
4501 Espinho Codex

Tel: (351) 2 734 4611/734 1621
Fax: (351) 2 734 6015
Competitive for animation short films, features, advertising and institutional, didactic and information, first film, title sequences and series. Entries must have been completed from January 1998 onwards

Encontros Internacionais de Cinema Documental
November
Centro Cultural Malaposta
Rua Angola, Olival Basto
2675 Odivelas
Tel: (351) 9388570/407
Fax: (351) 9389347
email: amascultura@mail.telepac.pt
Director: Manuel Costa e Silva
Two categories: film and video (competition). Only event dedicated to documentary in Portugal, to increase awareness of the form and show work from other countries

Fantasporto - Oporto International Film Festival
26th February - 6 March 1999
Cinema Novo - Multimedia Centre
Rua da Constituição 311
4200 Porto
Tel: (351) 2 5508990/1/2
Fax: (351) 2 5508210
email: fantas@caleida.pt
Website: http://
www.caleida.pt.fantasporto
Mario Dorminsky
Competitive section for feature films and shorts, particularly fantasy and science fiction films. Includes New Directors week (non-fantasy films) and in the Retrospective Sections (general). Also planned in conjunction with the Portugese Film Institute, a programme of Portuguese film including the beginnings of Portugese cinema

Festival Internacional de Cinema da Figueira da Foz
August/September
Apartado de Correios 5407
1709 Lisbon Codex
Tel: (351) 1 812 62 31
Fax: (351) 1 812 62 28
email: jose.marques@ficff.pt
Web site: http:\\www.ficff.pt
Competitive for fiction and documentary films, films for children, shorts and video. Some cash prizes. Special programmes on different directors and countries. Also retrospective of Portuguese cinema. Entries must have been produced during 20 months preceding Festival

Festival Internacional de Cinema de Troia
June
International Film Festival
Forum Luisa Todi
Avenida Luisa Todi, 61-65
2900 Setúbal
Tel: (351) 65 52 59 08 - 53 40 59
Fax: (351) 65 52 56 81
Three categories: Official Section, First Films, Man and His Environment, also information section. The Official Section is devoted to films coming from those countries which have a limited production (less than 21 features per year). Films must not have been screened previously in Portugal and must have been produced during 12 months preceding the Festival. Also film market, retrospectives in the information section, Gay and Lesbian section, Jury selection

PUERTO RICO

Puerto Rico International Film Festival
November
70 Mayagüez Street, Suite B1
Hato Rey PR 00918
Tel: (1) 809 764 7044
Fax: (1) 809 753 5367
Non-competitive international, full-length feature event with emphasis on Latin American, Spanish and women directors. FIPRESCI jury for the Latin American selection

RUSSIA

International Film Festival of Festivals
June
10 Kamennoostrovsky Avenue
St Petersburg 197101
Tel: (7) 812 237 0304
Fax: (7) 812 233 2174
Non-competitive, aimed at promoting films from all over the world that meet the highest artistic criteria, and the distribution of non-commercial cinema

Moscow International Film Festival
July
Interfest, General Management of International Film Festivals
10 Khokholski Per
Moscow 109028
Tel: (7) 95 917 9154
Fax: (7) 95 916 0107

SERBIA

Belgrade International Film Festival
January/February
Sava Centar
Milentija Popovica 9
11070 Novi Beograd
Tel: (38) 11 222 49 61
Fax: (38) 11 222 11 56
Non-competitive for features refl-ecting high aesthetic and artistic values and contemporary trends

SINGAPORE

Singapore International Film Festival
April
29A Keong Saik Road
Singapore 089136
Tel: (65) 738 7567
Fax: (65) 738 7578
email: filmfest@pacific.net.sg
Specialised competitive festival for Best Asian Film. Non-competitive includes panorama of international film. 8mm, 16mm, 35mm and video are accepted. Films must not have been shown commercially in Singapore

SLOVAKIA

Forum - Festival of First Feature Films
October
Brectanova
833 14 Bratislava 1
Tel: (42) 7 378 8290
Fax: (42) 7 378 8290
International competition for first feature films at least 50 minutes long, and made or first shown in the 16 months preceding the Festival

SOUTH AFRICA

Cape Town International Film Festival
April/May
University of Cape Town
Private Bag
Rondebosch 7700
Tel: (27) 21 23 8257/8
Fax: (27) 21 24 2355
Annual festival which runs over a three week period during April/May. Screenings are in 35mm, 16mm as well as video with an emphasis on contemporary, international feature films

Durban International Film Festival
University of Natal
Centre for Creative Arts
4014 Durban
Tel: (27) 31 260 2594
Fax: (27) 31 260 3074
The Festival aims to showcase films of quality to local audiences, including screenings in peri-urban areas of the city

SPAIN

L'Alternativa, International Independent Film Festival of Barcelona
November
Centre de Cultura Contemporània de Barcelona
C/Montalegre 5
08001 Barcelona
Tel: 34 3 306 41 00
Fax: 34 3 306 41 04
Contact: Albert Plans
L'Alternativa is dedicated to screening creative and innovative cinema from around the world. Composed of three official sections (short, feature and documentary) all films compete in front of an international jury. A film market, retrospective screenings and seminars run parallel to the official sections

Bilbao International Festival of Documentary & Short Films
November/December
Colón de Larreátegui 37, 4o drcha
48009 Bilbao
Tel: (34) 4 248698/247860
Fax: (34) 4 245624
Competitive for animation, fiction and documentary

International Film Festival For Young People of Gijón
November/December
Maternidad 2-20
33207 Gijón
Tel: (34) 8 5343739
Fax: (34) 8 5354152
Competitive for features and shorts. Must have been produced during 18 months preceding the festival and not awarded a prize at any other major international film festival

International Short Film Contest 'Ciudad de Huesca'
June
C/Del Parque 1,2
(Circulo Oscense)
Huesca 22002
Tel: (34) 74 212582

Fax: (34) 74 210065
email: huescafest@tsai.es
Website: http://www.huesca-filmfestival.com
Competitive for short films (up to 30 mins) on any theme except tourism and promotion

Mostra de Valencia/Cinema del Mediterrani
October
Pza del Arzobispo 2 bajo
46003 Valencia
Tel: (34) 6 392 1506
Fax: (34) 6 391 5156
Competitive official section. Informative section, special events section, 'mostra' for children, and International Congress of Film Music

Ourense Film Festival
September/October
C/Cardenal Quiroga 15, 3o Of. 25
32003 Ourense
Tel: (34) 88 215885
Fax: (34) 88 215885
Festival Internacional de Cine Independiente de Ourense - Second international festival for independent cinema. Competitive sections for every independent short or long length film produced after January 1st 1996

San Sebastian International Film Festival
September
Plaza OQuendo S/N
20004 San Sebastian
Tel: (34) 943 48 12 12
Fax: (34) 943 48 12 18
email: ssiff@mail.ddnet.es
Website:http://www.ddnet.es/san_sebastian_film_festival
Diego Galan
Competitive for feature films produced in the previous year and not released in Spain or shown in any other festivals. Non-competitive for Zabaltegi/Open Zone. Also retrospective sections. New Director's Prize to the best first or second film. 300,000 ECUS given to the producer and director of the winning film as co-production for their next film

Sitges International Film Festival of Catalonia
October
Rosselló, 257, 3E
08008 Barcelona
Tel: (34) 3 415 39 38
Fax: (34) 3 237 65 21
email:cinsit@sitgesfur.com
Two official sections. One for fantasy films and others for non-

genre files. Also shorts, retrospective and animation sections

Valladolid International Film Festival
October
PO Box 646
47080 Valladolid
Tel: (34 83) 30 57 00/77/88
Fax: (34 83) 30 98 35
email: festval.ladolid@seminci.com
Website: http://www.seminic.com
Director: Fernando Lara
Competitive for 35mm features and shorts, plus documentaries, entries not to have been shown previously in Spain. Also film school tributes, retrospectives and selection of new Spanish productions

SWEDEN

Gotebörg Film Festival
29 January - 7th February 1999
Box 7079
402 32 Gothenburg
Tel: (46) 31 41 05 46
Fax: (46) 31 41 00 63
email:goteborg.filmfestival@mailbox.swipnet.se
Website: http://goteborg.filmfestival.org
Non-competitive for features, documentaries and shorts not released in Sweden

Stockholm International Film Festival
November
PO Box 7673
103 95 Stockholm
Tel: (46) 8 67 75 000
Fax: (46) 8 20 05 90
Competitive for innovative current feature films, focus on American Independents, a retrospective, summary of Swedish films released during the year, survey of world cinema. Around 100 films have their Swedish premiere during the festival. FIPRESCI jury, FIAPF accredited. 'Northern Lights' - Critics Week

Umea International Film Festival
September
PO Box 43
S 901 02
Umea
Tel: (46) 90 13 33 88
Fax: (46) 90 77 79 61
email: film.festival@ff.umea.se
Non-competitive festival with focus mainly on features but does except shorts and documentaries. About 150 films are screened in several sections. The festival also organises seminars

Uppsala International Short Film Festival
October
Box 1746
751 47 Uppsala
Tel: (46) 18 12 00 25
Fax: (46) 18 12 13 50
email: uppsala@shortfilmfestival.com
Website: http://www.shortfilmfestival.com
Competitive for shorts (up to 60 mins), including fiction, animation, experimental films, documentaries, children's and young people's films. 16 and 35mm only

SWITZERLAND

Festival International de Films de Fribourg
7-14 March 1999
Rue de Locarno 8
1700 Fribourg
Tel: (41) (26) 322 22 32
Fax: (41) (26) 322 79 50
email: info@fiff.ch
Ingrid Kramer
Competitive for films from Africa, Asia and Latin America (16/35mm, video). Films (16/35mm) may be circulated throughout Switzerland after the Festival

Golden Rose of Montreux TV Festival
April
Télévision Suisse Romande
PO Box 234
1211 Geneva 8
Tel: (41) 22 708 8599
Fax: (41) 22 781 5249
email: gabrielle.bucher@tsr.ch
Competitive for television productions (24-60 mins) of light entertainment, music and variety, first broadcast in the previous 14 months

International Film - Video - Multimedia Festival Lucerne - VIPER
26-31 October 1999
PO Box 4929
6002 Lucerne
Tel: (41) 1-4506262
Fax: (41) 1-4506261
Competitive for innovative experimental films and videos (two awards), plus retrospective, special programmes and 'Swiss Videoworkshop' (two awards programme of current Swiss video) and Multimedia Projects (internet and CD-Rom)

Locarno International Film Festival
August

Via della Posta 6
CP 844
6601 Locarno
Tel: (41) 91 751 02 32
Fax: (41) 91 751 74 65
Programme includes: a) Competition reserved for fiction features representative of Young Cinema (first or second features) and New Cinema (films by more established filmmakers who are innovating in film style and content and works by directors from emerging film industries). b) A (non-competitive) selection of films with innovative potential in style and content. c) A retrospective designed to enlarge perspectives on film history

Nyon International Documentary Film Festival - Visions du Réel
April
PO Box 593
CH -1260 Nyon
Tel: (41) 22 361 60 60
Fax: (41) 22 361 70 71
International competition

Semaine Internationale du Vidéo/ International Video Week
November odd years
Saint Gervais Gèneve, Centre pour l'image contemporaine
5 rue du Temple
1201 Geneva
Tel: (41) 22 908 20 60
Fax: (41) 22 908 20 01
Competition with international entries; seminars and conferences; retrospectives; special programmes; installations; Swiss art school programme

Vevey International Comedy Film Festival
October
La Grenette
CP 421
1800 Vevey
Tel: (41) 21 922 20 27
Fax: (41) 21 922 20 24
Competitive for medium and short films, hommage and retrospective

TUNISIA

Carthage Film Festival
October/November
The JCC Managing Committee
5 Avenue Ali Belahouane
2070 La Marsa
Tel: (216) 1 745 355
Fax: (216) 1 745 564
Official competition open to Arab and African short and feature films. Entries must have been made within two years prior to the festival, and

not have been awarded first prize at any previous international festival in an African or Arab country. Also has an information section, an international film market (MIPAC) and a workshop

TURKEY

International Istanbul Film Festival
April
Istanbul Foundation for Culture and Arts
Istiklal Cad Luvr
Apt No: 146
80070 Beyoglu
Istanbul
Tel: (90) 1 212 293 3133
Fax: (90) 1 212 249 7771
Two competitive sections, international and national. The International Competition for feature films on art (literature, theatre, cinema, music, dance and plastic arts) is judged by an international jury and the 'Golden Tulip Award' is presented as the Grand Prix. Entry by invitation

UNITED KINGDOM

Birmingham International Film and TV Festival
November
9 Margaret Street
Birmingham, B3 3BS
Tel: 0121 212 0777
Fax: 0121 212 0666
Non-competitive for features and shorts, plus retrospective and tribute programmes. The Festival hosts conferences debating topical issues in film and television production

Bite the Mango
September 1999
National Museum of Photography, Film & Television
Pictureville, Bradford BD1 1NQ
Tel: 01274 203320/203300
Fax: 01274 770217
Europe's only annual festival for South Asian and Black film and television. Entries accepted from South Asian and Black film and video-makers. Deadline for entries 13 August 1998

Black Sunday - The British Genre Film Festival
51 Thatch Leach Lane
Whitefield
Manchester M25 6EN
Tel: 0161 766 2566
Fax: 0161 766 2566

Non-competitive for horror, thriller, fantasy, science-fiction and film noir genres produced in the previous year. First choice festival for UK premieres of many of the above genre films. Special guests and retrospective programmes

Blackpool Film Festival
June/July
20 Glen Eldon Road
St Anne's-on-Sea
Lancashire FY8 2AU
Tel: 0253 721800
Fax: 0253 721800
Peter Stamford

Bradford Animation Festival
March
National Museum of Photography Film and Television
Pictureville
Bradford BD1 1NQ
Tel: 01274 203320/203300
Competitive festival for animated shorts in four categories: under 16s, non-professional, professional, experimental. Features closing awards night, interviews with animators, and international animation

Bradford Film Festival
September
National Museum of Photography, Film & Television
Pictureville
Bradford BD1 1NQ
Tel: 01274 770000/ 01274 732277/ 203320
Fax: 01274 770217
Non-competitive for feature films. Strands include widescreen with world's only Cinerama Screen and IMAX. Focus on national cinema of selected European countries

Brief Encounters - Bristol Short Film Festivals
PO Box 576
Bristol BS99 2BD
Tel: 117 9224636
Fax: 117 9222906
email: brief.encounters@dial.pipex
Competitive for short film in all categories (up to 40 mins). Thematic and specialised programmes and special events. Prizes awarded include Best British and International Productions. Closing date for entries is early June

British Animation Awards
March
c/o 219 Archway Rd
London N6 5BN
Tel: 0181 340 4563
The next animation awards are planned for 2000. A bi-annual event

Cambridge Film Festival
July
Arts Cinema
8 Market Passage
Cambridge CB2 3PF
Tel: 01223 504444
Fax: 01223 578956
email: festival@cambarts.co.uk
Non-competitive; new world cinema selected from international festivals. Also featuring director retrospectives, short film programmes, thematic seasons and revived classics. Conference for independent exhibitors and distributors. Public debates and post-screening discussions

Chichester Film Festival
19 August - Sept 1999
New Park Film Centre
New Park Road
Chichester
West Sussex PO19 1XN
Tel: 01243 784881/01243
Fax: 01243 539853
Contact: Roger Gibson

Cinemagic - International Film Festival for Young People
December
4th floor, 38 Dublin Road
Belfast BT2 7HN
Tel: 01232 311900
Fax: 01232 319709
Contact: Frances Cassidy
Competitive for international short and feature films aimed at 4-18 year-olds. The next festival will include the usual charity premieres, educational workpacks, practical workshops, directors talks and masterclasses with industry professionals. The Belfast event is to be held at Virgin Cinemas, Dublin Road

CineWomen
Cinema City
St.Andrew's Street
Norwich NR2 4AD
Tel: 01603 632366
Fax: 01603 7678238
email: j.h.morgan@uea.ac.uk
Jayne Hathor Morgan, Festival Director

Edinburgh International Film Festival
August
Filmhouse
88 Lothian Road
Edinburgh EH3 9BZ
Tel: 0131 228 4051
Fax: 0131 229 5501
email: info@edfilmfest.org.uk
Oldest continually running film festival in the world. Patron: Sean Connery. Programme sections: Gala

(World, European or British premieres); Rosebud (world premieres of drama and documentary films); Scene by Scene (illustrated lectures by film-makers); Retrospectives (film-makers and themes); Documentary (newly devised section devoted to non-fiction films); Mirrorball (music videos and music related projects); NBX (market place for UK productions); Also animation, outdoor cinema events and short films. Prizes for Best British Feature, Best Animation, Best New Film etc. Submissions must be received by Mid-May

Edinburgh Fringe Film and Video Festival

February
29 Albany Street
Edinburgh EH1 3QN
Tel: 0131 556 2044
Fax: 0131 557 4400
Competitive for low-budget/independent/innovative works from Britain and abroad. All submissions welcome

Edinburgh International Television Festival

August
2nd Floor
24 Neal Street
London WC2H 9PS
Tel: 0171 370 4519
Fax: 0171 836 0702
email: EITF@festival.demon.co.uk
A unique forum for discussion between programme makers and broadcasters, comprising around 30 workshops, lectures and debates

European Shot Film Festival

November
11 Holbein House
Holbein Place
London SW1 8NH
Tel: 0171 460 3901
Fax: 0171 259 9278
email: pearl@mail.bogo.co.uk
Website: http://www.bogo.co.uk/kohle/festival.html
Festival organiser: Fritz Kohle
Any full-time student may submit their work to the audience-centred event. There are no limitations to the chosen subject or genre of any of the films submitted. The audience is encouraged to participate in this event and awards the 'Best Film Award.' Recommendations are made by a panel of jurors in the following areas: direction, production, screenplay, camera, sound, post-production, acting and art-direction. The Festival is now also open to

non-students. Film makers on a low budget are invited to send in their films

Festival of Fantastic Films

September
33 Barrington Road
Altrincham
Cheshire WA14 1H2
Tel: 0161 929 1423
Fax: 0161 929 1067
email: 101341.3352@compuserve.com
Festival celebrates science fiction and fantasy film. Features guests of honour, interviews, signing panels, dealers, talks and over 30 film screenings

Foyle Film Festival

April
The Nerve Centre
2nd Floor
Northern Counties Building
8 Custom House Street
Derry BT48 6AE
Tel: 01504 267432
Fax: 01504 371738
email: shona@inerve-centre.org.uk
Shona McCarthy (Festival Director)
Northern Ireland's major annual film event celebrated its 10th year in 1996. The central venue is the Orchard cinema in the heart of Derry city centre

French Film Festival

November
13 Randolph Crescent
Edinburgh EH3 7TX
Tel: 0131 225 5366
Fax: 0131 220 0648
The UK's only festival devoted solely to French cinema including feature films and shorts. Section on first or second films qualifies for the Hennessy Audience Award. Retrospectives, panorama of new productions, and debates. Based at venues in three cities: Filmhouse, Edinburgh, Glasgow Film Theatre and Aberdeen Belmont Centre

Green Screen (London's International Environmental Film Festival)

November
45 Shelton Street
London WC2H 9JH
Tel: 0171 379 7390
Fax: 0171 379 7197
Non-competitive selection of international environmental films, question and answer sessions following every film showing with well known environmentalists, film-makers, media personalities and celebrities with environmental clout

IVCA Film and Video Festival

October
IVCA
Bolsover House
5-6 Clipstone Street
London W1P 8LD
Tel: 0171 580 0962
Fax: 0171 436 2606
email: 100434.1005@compuserve.com
Competitive for non-broadcast industrial/training films and videos, covering all aspects of the manufacturing and commercial world, plus categories for educational, business, leisure and communications subjects. Programme, Special and Production (Craft) Awards, and industry award for effective communication. Closing date for entries December

International Animation Festival, Cardiff

June
18 Broadwick Street
London W1V 1FG
Tel: 0171 494 0506
Fax: 0171 494 0807

International Celtic Film and Television Festival

March
The Library
Farraline Park
Inverness IV1 1LS
Tel: 01463 226 189
Fax: 01463 716 368
Competition for films whose subject matter has particular relevance to the Celtic nations

Italian Film Festival

April
82 Nicolson Street
Edinburgh EH8 9EW
Tel: 0131 668 2232
Fax: 0131 668 2777
A unique UK event throwing an exclusive spotlight on il cinema italiano over ten days in Edinburgh (Filmhouse) Glasgow (Film Theatre); and London (Riverside). Visiting guests and directors, debates, first and second films, plus a broad range of current releases and special focuses on particular actors or directors

KinoFilm

October/November
Manchester International Short Film and Video Festival
48 Princess Street
Manchester M1 6HR
Tel: 0161 288 2494
Fax: 0161 237 3423
John Wojowski

Kinofilm is dedicated to short films and videos from every corner of the world. Emphasis is placed on short innovative, unusual and off-beat productions. Films on any subject or theme can be submitted providing they are no longer than 30 minutes and were produced within the last two years and have not been previously submitted. All sections of film/video making community are eligible. Particularly welcome are applications from young film-makers and all members of the community who have never had work shown at festivals. Special categories include: Gay and Lesbian, Black Cinema, New Irish Cinema, New American Underground, Eastern European Work, Super 8 Film. Closing date for submissions for 1998 is August 98. Please telephone or fax for an application form. Entry fee £2.50 National, £5.00 International

Kino Festival of New Irish Cinema
March
48 Princess Street
Manchester M1 6HR
Tel: 0161 288 2494
Fax: 0161 237 3423
John.kino@good.co.uk
John Wojowski
Irish Film Festival devoted solely to new work. The festival features both short films and features. Low budget, innovative features especially welcome. Short films of all categories (inc student work) can be submitted. Special categories of new work by expatriate Irish Film makers. Work can be submitted any time before the deadline date. Deadline for submissions 25 January (late submissions only by arrangement). Work must have been completed within the last 18 months. Six awards (funding permitting)

Leeds International Film Festival
October
Town Hall
The Headrow
Leeds LS1 3AD
Tel: 0113 247 8389
Fax: 0113 247 8397
Non-competitive for feature films, documentaries and shorts, plus thematic retrospective programme. Lectures, seminars and exhibitions

 London Film Festival
November
National Film Theatre
South Bank
London SE1 8XT
Tel: 0171 815 1323/1324

Fax: 0171 633 0786
Non-competitive for feature films, shorts and video, by invitation only, which have not previously been screened in Great Britain. Films are selected from other festivals, plus original choices **(see p8)**

 London Jewish Film Festival
June
National Film Theatre
South Bank
London SE1 8XT
Tel: 0171 815 1323/1324
Fax: 0171 633 0786
Non-competitive for film and video made by Jewish directors and/or concerned with issues relating to Jewish identity and other issues. Some entries travel to regional film theatres as part of a national tour **(see p8)**

London Latin American Film Festival
September
Metro Pictures
79 Wardour Street
London W1V 3TH
Tel: 0171 434 3357
Fax: 0171 287 2112
Non-competitive, bringing to London a line up of contemporary films from Latin America and surveying current trends

 London Lesbian and Gay Film Festival
March
Festivals Office
National Film Theatre
South Bank
London SE1 8XT
Tel: 0171 815 1323/1324
Fax: 0171 633 0786
Non-competitive for film and videos of special interest to lesbian and gay audiences. Some entries travel to regional film theatres as part of a national tour from April to June **(see p8)**

London Pan-Asian Film Festival
The Asian Film Library
Suite 19, 2 Lansdowne Row
Berkeley Square
London W1X 8HL
Tel: 07970 506326
Fax: 0171 493 4935
Stephen Cremin, Festival Co-director

Raindance Film Showcase
October
81 Berwick Street
London W1V 3PF
Tel: 0171 287 3833
Fax: 0171 439 2243
Britain's only film market for

independently produced features, shorts and documentaries. Deadline 1 September

Sheffield International Documentary Festival
October
The Workstation
15 Paternoster Row
Sheffield S1 2BX
Tel: 0114 276 5141
Fax: 0114 272 1849
email: shefdoc@fdgroup.co.uk
Website: www.fdgroup.co.uk/neo/sidf
The only UK festival dedicated to excellence in documentary film and television. The week long event is both a public film festival and an industry gathering with session, screenings and discussions on all the new developments in documentary. The festival is non-competitive

Shots in the Dark -Crime, Mystery and Thriller Festival
June
Broadway Media Centre
14 Broad Street
Nottingham NG1 3AL
Tel: 0115 952 6600
Fax: 0115 952 6622
Non-competitive for all types of mysteries and thrillers. Includes previews of new movies, retrospectives, television events, special guests. Honorary Patron: Quentin Tarantino

 TV99: A Big Festival of the Small Screen
Festivals Office
National Film Theatre
South Bank
London SE1 8XT
Tel: 0171 815 1322/1323
Fax: 0171 633 0786
Festival of television where members of the public get the chance to see previews of new productions, meet programme-makers **(see p8)**

Television and Young People (TVYP)
August
24 Neal Street
London WC2H 9PS
Tel: 0171 379 4519
Fax: 0171 836 0702
email: TVYP@festival.demon.co.uk
Susanne Curran
Television and Young People is the UK's leading television event for young people, using top TV practitioners in a unique programme of masterclasses, workshops and debates. The educational arm of Edinburgh International Film Festival,

TVYP offers 150 free places to 17-21 year olds from the UK who are passionate about television

Video Positive
September/October
International Biennale of Video and Electronic Media Art
Foundation for Art and Creative Technology (FACT)
Bluecoat Chambers
School Lane
Liverpool L1 3BX
Tel: 0151 709 2663
Fax: 0151 707 2150
Non-competitive for video and electronic media art produced worldwide in the two years preceding the festival. Includes community and education programmes, screenings, workshops and seminars. Some commissions available

Welsh International Film Festival, Aberystwyth
November
c/o Premiere Cymru Wales Cyf
Unit 6G, Cefn Llan
Aberystwyth
Dyfed SY23 3AH
Tel: 01970 617995
Fax: 01970 617942
email: wff995@aber.ac.uk
Non-competitive for international feature films and shorts, together with films from Wales in Welsh and English. Also short retrospectives, workshops and seminars. D M Davies Award (£25,000) presented to the best short film submitted by a young film maker from Wales

Wildscreen
October 2000
PO Box 366, Deanery Road
College Green, Bristol BS99 2HD
Tel: 0117 909 6300
Fax: 0117 909 5000
email: wildscreen@cableinet.co.uk
International Festival of moving images from the natural world. Competitive: pandaawards include Conservation, Revelation, Newcomer, Children's, Outstanding Achievement, Photography, Craft. Eligible productions completed after 1 January 1998 and not entered in Wildscreen 98. Festival includes screenings, discussions, video kiosks, masterclasses and workshops

URUGUAY

International Children's Film Festival
March/April
Cinemateca Uruguaya

Carnelli 1311, Casilla de Correo 1170
11200 Montevideo
Tel: (598) 2 408 24 60
Fax: (598) 2 409 45 72
Competitive for fiction international films, documentaries and animation for children

International Film Festival of Uruguay
April
Cinemateca Uruguaya
Carnelli 1311, Casilla de Correo 1170
11200 Montevideo
Tel: (598) 2 48 24 60
Fax: (598) 2 49 45 72
Competitive for fiction and Latin American videos

USA

AFI Los Angeles International Film Festival
October
2021 N Western Avenue
Los Angeles
CA 90027
Tel: (1) 213 856 7707
Fax: (1) 213 462 4049
email: afifest@afionline.org
Competitive, FIAPF accredited. Features, documentaries, shorts by invitation

AFI National Video Festival
February
2021 N Western Avenue
Los Angeles
CA 90027
Tel: (1) 213 856 7707
Fax: (1) 213 462 4049
Non-competitive. Screenings in Los Angeles, Washington. Accepts: ¾" U-Matic, NTSC/PAL/SECAM (No Beta, no 1"

AFM (American Film Market)
9th Floor, 10850 Wiltshire Blvd,
Los Angeles
CA 90024
Tel: (1) 310 446 1000
Fax: (1) 310 446 1600
Annual market for film, television and video

Asian American International Film Festival
July
c/o Asian CineVision
32 East Broadway, 4th Floor
New York, NY 10002
Tel: (1) 212 925 8685
Fax: (1) 212 925 8157
Non-competitive, all categories and lengths. No video-to-film transfers

accepted as entries. Films must be produced, directed and/or written by artists of Asian heritage

Charleston International Film Festival - Worldfest Charleston
November
PO Box 56566
Houston, Texas
77256
Tel: (1) 713 965 9955
Fax: (1) 713 965 9960
Competitive for features and shorts. Production seminar and workshop series of three seminars. Cash awards, plus Worldfest Discovery programme. Winners introduced to organisers of top 200 international festivals. Screenplay category: winning screenplays submitted to top three US creative agencies

Chicago International Children's Film Festival
October
Facets Multimedia
1517 West Fullerton Avenue
Chicago IL 60614
Tel: (1) 773 281 9075
Fax: (1) 773 929 5437
email: kidsfest@facets.org
Competitive for entertainment films, videotapes and television programmes for children. Deadline for entries 29 May 1998

35th Chicago International Film Festival
7-21 October 1999
32 West Randolph Street
Suite: 600
Chicago
Illinois 60601 USA
Tel: (1) 312 425 9400
Fax: (1) 312 425 0944
email: filmfest@wwa.com
Website: http://www/chicago.ddbn.com/filmfest/
Contact: Michael Kutza
Founder & Artistic director
Competitive for feature films, documentaries, shorts, animation, student and First and Second Features

Cleveland International Film Festival
March
1621 Euclid Avenue, Suite 428
Cleveland
OH 44115
Tel: (1) 216 623 0400
Fax: (1) 216 623 0103
Non-competitive for feature, narrative, documentary, animation and experimental films. Competitive for shorts, with $2,500 prize money

Columbus International Film and Video Festival (a.k.a. The Chris Awards)

October
5701 High Street
Suite 200
Worthington
Ohio 43085
Fax: (1) 614 841 1666
email: chrisawd@infinet.com
Website:www.infinet.com/-chrisawd
The Chris Awards is one of the longest-running competitions of its kind in North America, specialising in honouring documentary, education, business and information films and videos, as well as categories for the arts and entertainment. Entrants compete within categories for the first place Chris statuette, second place Bronze plaque and third place Certificate of Honorable Mention Expanded public screenings. Entry deadline 1 July

Denver International Film Festival

October
1430 Larimer Square, Suite 201
Denver
CO 80202
Tel: (1) 303 595 3456
Fax: (1) 303 595 0956
email: Denver/film@csn.net
Non-competitive. New international features, tributes to film artists, independent features, documentaries, shorts, animation, experimental works, videos and children's films

Florida Film Festival

June
1300 South Orlando Avenue
Maitland
FL 32751
Tel: (1) 407 629 1088
Fax: (1) 407 629 6870
A 10 day event involving over 100 films, several seminars and social events. It includes an American Independent Film Competition with three categories: dramatic, documentary and short films

Fort Lauderdale International Film Festival

November
1402 East Solas Blvd Box 007
Fort Lauderdale
FL 33301
Tel: (1) 954 563 0500
Fax: (1) 954 564 1206
The festival typically features 40 - 50 full length features, plus documentaries, an art on film series, short subjects, as well as animation. Awards are presented for Best Film,

Best Foreign Language Film, Documentary, Short, Director, Actor, Actress and an Audience Award. The Festival also features an international student film competition with cash prizes totalling over 5,000 with an additional 5,000 in product grants from Kodak

Hawaii International Film Festival

November
1001 Bishop Street, Paacific Tower
Suite 745
Honolulu HI 96813
Tel: (1) 808 528 3456
Fax: (1) 808 528 1410
email: hiffinfo@hiff.org
Website: http://www.hiff.org
Entries must be produced in Asia, North America or the Pacific, or concern those areas and relate to the Festival's cross-cultural emphasis. Any genres welcomed. Awards given include best feature film promoting cultural understanding and best documentary film and audience award.Deadline 3 July 1998

Houston International Film and Video Festival - Worldfest Houston (+ Market)

April
PO Box 56566
Houston
TX 77256-6566
Tel: (1) 713 965 9955
Fax: (1) 713 965 9960
Competitive for features, shorts, documentary, television production and television commercials. Independent and major studios, experimental and video with inclusive film maker. Worldfest Discovery Programme where winners are introduced to organisers of top 200 international festivals. Screenplay category. Winning screenplays submitted to top three US creative agencies

Independent Feature Film Market

September
12th Floor
104 West 29th Street
New York
NY 1001-5310
Tel: (1) 212 465 8200
Fax: (1) 212 465 8525
cmail: IFPNY@ifp.org
Web site: www.ifp.org
The Independent Feature Film Market is the only market devoted to new, emerging American independent film

talent seeking domestic and foreign distribution. It is the market for discovering projects in development, outstanding documentaries, and startling works of fiction. Domestic and foreign filmmakers, distributors, and feature films. Sales agents, producers, festival representatives and casting directors attend to acquire and evaluate both completed films and projects in development

Miami Film Festival

January/February
Film Society of Miami
444 Brickell Avenue, Suite 229
Miami FL 33131
Tel: (1) 305 377 3456
Fax: (1) 303 577 9768
Non-competitive; screenings of 25-30 international films; all cate-gories considered, 35mm film only. Entry deadline 1 November

Mobius Advertising Awards Competition

11 February 1999
841 North Addison Avenue
Elmhurst, IL 60126-1291
Tel: (1) 630 834 7773
Fax: (1) 630 834 5565
email:Mobiusawards@.com
Website: www.mobiusawards.com
International awards competition for television and radio commercials produced or released in the 12 months preceding the annual 1 October entry deadline. Founded in 1971

National Educational Media Network (formerly National Educational Film & Video Festival)

November
655 Thirteenth Street
Oakland
CA 94612
Tel: (1) 510 465 6885
Fax: (1) 510 465 2835
email: nemn@nemn.org
Website: http://www.nemn.org
National Educational Media Network is the only US media organisation dedicated to recognising and supporting excellence in educational media, ranging from documentaries to moving image media designed especially for classroom and training programs. NEMN's internationally acclaimed annual Apple Awards competition is the largest in the US, with over 1,000 entrants yearly. The competition recognises excellence and innovation in educational film, video, television and multimedia works intended for national and

international distribution. Awards are given to media programs demonstrating technical and artistic skill that educate, inform and empower the end-user. MEM's annual media market each spring is the nation's primary gathering where distributors can view and acquire newly released educational productions. NEM's biannual Conference offers producers and distributors the latest on industry trends in educational and interactive media through workshops, panel discussions, and exhibits of media hardware, software and services

New York Film Festival

September/October
Film Society of Lincoln Center
70 Lincoln Center Plaza, 4th Floor
New York NY 10023
Tel: (1) 212 875 5610
Fax: (1) 212 875 5636
Non-competitive for feature films, shorts, including drama, documentary, animation and experimental films. Films must have been produced one year prior to the Festival and must be New York premieres

Nortel Palm Springs International Short Film Festival

August
PO Box 2230
Palm Springs
California
CA 92263
Tel: (1) 760 322 2930
Fax: (1) 760 322 4087
A showcase of international short films, animation and videos. Competition

Portland International Film Festival

February/March
Northwest Film Centre
1219 SW Park Avenue
Portland, OR 97205
Tel: (503) 221 1156
Fax: (503) 226 4842
email: info@nwfilm.org
www.nwfilm.org
Invitational survey of New World cinema. Includes over 100 features, documentary and short films from more than two dozen countries. Numerous visiting artists. Attendance for the 21st Festival is expected to be 25,000 drawn from throughout the North West of America

San Francisco International Film Festival

April/May
1521 Eddy Street

San Francisco
CA 94115-4102
Tel: (1) 415 929 5014
Fax: (1) 415 921 5032
Feature films, by invitation, shown non-competitively. Shorts, documentaries, animation, experimental works and television productions eligible for Golden Gate Awards competition section. Deadline for Golden Gate Awards entries early December

San Francisco International Lesbian & Gay Film Festival

June
Frameline
346 Ninth Street
San Francisco CA 94103
Tel: (1) 415 703 8650
Fax: (1) 415 861 1404
Largest lesbian/gay film festival in the world. Features, documentary, experimental, short film and video. Deadline for entries 15 February

Seattle International Film Festival

May/June
801 East Pine
Seattle
WA 98122
Tel: (1) 206 324 9996
Fax: (1) 206 324 9998
Jury prize for new director and American independent award. Other audience-voted awards. Submissions accepted 1 Jan to 15 March

Sundance Film Festival

21-31 January 1999
PO Box 16450
Salt Lake City
UT 84116
Tel: (1) 801 328 3456
Fax: (1) 801 575 5175
Competitive for American independent dramatic and documentary feature films. Also presents a number of international and American premieres and short films, as well as sidebars, special retrospectives and seminars

Telluride Film Festival

September
PO Box B-1156
53 South Main Street, Suite 212
Hanover NH 03755
Tel: (1) 603 643 1255
Fax: (1) 603 643 5938
email: tellufilm@aol.com
Website: http://telluridemm.com/filmfest.htm
Non-competitive. World premieres, archival films and tributes. Entry deadline 31 July

US International Film & Video Festival

June
841 North Addison Avenue
Elmhurst
IL 60126-1291
Tel: (1) 630 834 7773
Fax: (1) 630 834 5565
email:filmfestival
andmobiusaward@comcom
Website: www.filmfestawards.com
International awards competition for business, television, industrial and informational productions, produced or released in the 18 months preceding the annual 1 March entry deadline. Formerly the US Industrial Film and Video Festival, founded 1968

FILM SOCIETIES

Listed below are UK film societies which are open to the public (marked OP after the society name), those based in educational establishments (ST), private companies and organisations (CL), and some corporate societies (CP). Societies providing disabled access to their venues are marked (DA) - always contact the organiser beforehand to confirm details. Secretaries' addresses are grouped in broad geographical areas, along with the BFFS regional officers who can offer specific local information. Compiled by Tom Brownlie

BRITISH FEDERATION OF FILM SOCIETIES (BFFS) CONSTITUENT GROUPS

British Federation of Film Societies (BFFS)
The Secretary
BFFS
PO BOX 1DR
London W1A IDR
Tel: 0171 734 9300
Fax: 0171 734 9093

Eastern
Bedfordshire, Cambridgeshire, Essex, Hertfordshire, Lincolnshire, Norfolk, Suffolk

London
32 London Boroughs and the City of London

Midlands
Derbyshire (excluding High Peak District), Leicestershire, Northamptonshire, Nottinghamshire, Hereford and Worcester, Shropshire, Staffordshire, Warwickshire, Metropolitan districts of Birmingham, Coventry, Dudley, Sandwell, Solihull, Walsall, Wolverhampton

North West
Cheshire, High Peak district of Derbyshire, Lancashire, Metropolitan districts of Bolton, Bury, Knowsley, City of Liverpool, Manchester, Oldham, Rochdale, St Helens, Salford, Sefton, Stockport, Tameside, Trafford, Wigan, Wirral, Northern Ireland

Northern
Cleveland, Cumbria, Durham, Northumberland, Metropolitan districts of Gateshead, Newcastle, North Tyneside, South Tyneside, Sunderland

Scotland
Aberdeen, Angus, Ayrshire, Dumfries, Edinburgh, Faroe Islands, Fife, Glasgow, Isle of Lewis, Isle of Skye, Livingston, Perth, Shetland, Stirling, Strathblane, West Lothian

South East
Kent, Surrey, East Sussex, West Sussex

South West
Avon, Channel Islands, Cornwall, Devon, Dorset (except districts of Bournemouth, Christchurch and Poole), Gloucestershire, Somerset

Southern
Berkshire, Buckinghamshire, Hampshire, Isle of Wight, Oxfordshire, Wiltshire, Districts of Bournemouth, Christchurch, Poole

Wales
Dyfed, Cardiff, Cardigan, Gwent, Merthyr Tydfil, Mid-Glamorgan, Powys, Swansea

Yorkshire
Humberside, North Yorkshire, Metropolitan districts of Barnsley, Bradford, Calderdale, Doncaster, Kirklees, Leeds, Rotherham, Sheffield, Wakefield

EASTERN

Eastern Group BFFS
Gerry Dobson,
Kennel Cottage
Burton, Nr Lincoln,
Lincs LN1 2RD

Bedford Film Society *OP*
PJ Clark
33 The Ridgeway
Bedford MK41 8ES

Berkhamsted Film Society *OP*
DA
Dr Colin Davies

Seasons, Garden Field Lane
Berkhamsted
Herts HP4 2NN

Boston Film Society
Nigel Green
5 Grand Sluice Lane,
Boston Lincs, PE21 9HL

Bury St Edmunds Film Society
OP DA
Don Smith
Rectory Cottage, Drinkstone
Bury St Edmunds
Suffolk IP30 9SP

Chelmsford Film Club *OP DA*
Coline Hewitt
Sixth Ave
Chelmsford, Essex CM1 4ED

Epping Film Society
OP DA
Mrs A Berry
10 Bury Road
Epping, Essex CM16 5EU

Great Yarmouth Film Society *OP*
E C Hunt
21 Park Lane
Norwich, Norfolk NR2 3EE

Ipswich Film Society
Terry Cloke
4 Burlington Road
Ipswich
Suffolk IP1 2EU

Kings Lynn Centre for the Arts
OP
The Secretary
St Georges Guildhall Ltd
27 King Street
Kings Lynn
Norfolk PE30 1HA

Letchworth Film Society *OP DA*
Peter Griffiths
35 Broadwater Dale
Letchworth, Herts SG6 3HQ

Lincoln Film Society *OP*
Gerry Dobson
Kennel Cottage
Burton by Lincoln
Lincoln, Lincs LN1 2RD

Peterborough Film Society *OP*
Allan Bunch
196 Lincoln Road
Peterborough,
Cambs PE1 2NQ

St Georges Guildhall Ltd
Kings Lynn Centre for the Arts
27 King Street
King's Lynn
Norfolk PE30 1HA

Saint John's College Film Society *ST*
S Worthy
St John's College
Cambridge, Cambs CB2 1TP

Screen in the Barn
Brian Guthrie
Field House
Thrandeston, Diss
Norfolk 1P21 4BU

Stamford Schools' Film Society *ST*
J Dawson
Stamford School
St Paul's Street
Stamford, Lincs PE9 2BS

Trinity Arts Centre
Catherine Harrison
Trinity Street,
Gainsborough DN21 2AL

UEASU Film Society
ST DA
Di Anderson
Students Union, UEA
Norwich
Norfolk NR4 7TJ

University of Essex Film Society *ST*
Dee Mora
Students' Union Building
Wivenhoe Park
Colchester, Essex CO4 3SQ

Welwyn Garden City Film Society *OP DA*
Michael Massey
3 Walden Place
Welwyn Garden City
Herts AL8 7PG

LONDON

British Federation of Film Societies (BFFS)
The Secretary
PO BOX 1DR
London W1A 1DR
Tel: 0171 734 9300
Fax: 0171 734 9093

Avant-Garde Film Society *OP DA*
Chris White
9 Elmbridge Drive
Ruislip, Middx HA4 7XD

Barclays Bank Film Society *CL DA*
Liza Castellino
c/o IUKRB HR, 5th Floor
St Swithin's House
11-12 St Swithin's Lane
London EC4P 8AS

Bikkon Film Society
Mujibur Film Society
34 Broadhurst House
Joseph Street
London E3 4HY

Broomhill Film Society
Anthony Castro
55 Canadian Avenue
Catford
London SE6 3AX

Brunel Film Society
ST DA
The President
Brunel University
Uxbridge
Middx UB8 3PH

CD Film Society
Charles Drazin
12 Tolverne Road
London SW20 8RA

Chertsey Film Society
A Proctor
28 Wheatash Road
Addlestone
Surrey KT15 2ER

Chiswick Film Society *OP DA*
A V Downend
Hounslow Library
Treaty Centre
Hounslow TW3 1ES

Elgin Abseiling Film Club
c/o Richard Guard
BMC, 89.5 Worship Street
London EC2A 2BE

Gothique Film Society *OP DA*
R James
75 Burns Avenue
Feltham, Middx TW14 9LX

Hampden Community Association *CL DA*
Frances Holloway
Hampden Community Centre
150 Ossulston Street
London NW1 1EE

Holborn Film Society
OP DA
Noel Mcleod
48 Witley Court
Coram Street
London WC1N 1HD

Holloway Charm School
Jimmy Abatti
95 Seymour Road
London N8 OBH

Imperial College Union Film Society
ST DA
Ian Nicol
5 Sterling Place
South Ealing
London W5 4RA

Institut Français Film Society *OP*
The Secretary
(Cinema Dept)
17 Queensberry Place
London SW7 2DT

The Italian Cinema
Garwin Spencer-Davison
(Cinema Dept)
11 Hornton Street
London W8 7NP

John Lewis Partnership Film Society *CL DA*
Peter Allen
75 Burns Avenue, Feltham
Middlesex TW14 9LX

Kings College Hall
Baldwin Ho
Chanting Hill
London SE5 8AN

Lensbury Film Society *OP*
A Catto
Shell Centre, Room Y1085
York Road
Waterloo
London SE1 7NA

London Socialist Film Co-op
13 Foundling Road
Brunswick Centre
London WC1N 1QE

Lowe Howard - Spink

Sybil Dormer
Bowater House, 3rd Floor
63-114 Knightsbridge
London SW1X 7LT

National Physical Laboratory (NPL) Film Society *CL*

Roger Townsend
National Physical Laboratory
Queens Road
Teddington, Middx TW11 0LW

POSK Film Society

Zofia Binder, Membership Secretary
5 Rowan Road
London W6 7DT

Richmond Film Society

John Smith
16 Brackley Road
Chiswick
London W4 2HN

St Anne's Film Society

Erica Maran
17 Lancaster Park
Richmond
Surrey

Scandinavian Film Society *OP*

Françoise Cowie
14 Kimbell Gardens
London SW6 6QQ

South Indian Film Society

Ratnam Nithyananthan
179 Norval Road
North Wembley
Middlesex HA0 3SX

South London Film Society *OP*

Dr M Essex-Lopresti
14 Oakwood Park Road
Southgate
London N14 6QG

UCL Film Society *ST*

Simon Mesterten
University College London
25 Gordon Street
London WC1A 0AH

UCLU Film Society

Andrew Blackwell
25 Gordon Street
London WC1H OAH

University of Greenwich Film Society *ST*

William Clarke
University of Greenwich
Student's Union
Thomas Street, Woolwich
London SE18 6HU

Waltham Forest (Libs) Film Society *OP*

V Bates
William Morris Gallery
Lloyd Park
Forest Road
Walthamstow
London E17 4PP

Woolwich and District Co-op Film Society *OP*

P Graham
10 Harden Court
Tamar Street
Charlton
London SE7 8DQ

MIDLANDS

Midlands Group BFFS

Robert Johnson
The Villas, 86 School Lane,
Cookshill, Caverwall,
Stoke-on-Trent ST11 9EN

Aylesbury Vale Film Society

A Smart
Brooke Farm, Brooke End
Weston Turville
Bucks HP22 5RQ

Bablake School Film Society

J L Lawrence
Coventry School
Bablake
Coundon Road
Coventry CV1 4AU

Bishops Castle Film Society *OP*

J Parker
4 Lavender Bank
Bishops Castle
Shropshire SY9 5BD

Bromyard Conquest Film Society

OP DA
Barbara Koning
56 Old Road
Bromyard
Herefordshire HR7 4BQ

Hereford Film Club *OP*

Tony McQueen
12 Ainsley Close
Hereford HR1 1JH

Kinver Film Society

OP DA
P M Hassall
13 Bredon Avenue
Stourbridge
West Midlands DY9 7NR

Loughborough Students Union Film Society *ST*

Student Union Building
Ashby Road
Loughborough, Leics LE11 3TT

Malvern Film Society *OP*

Sylvia Finnemore
Ledbury Terr, 17 Wyche Road,
Malvern Worcs

New Kettering Film Society *OP*

Mr Meredith
Art Department
Tresham Institute
Rockingham Road
Kettering
Northants NN16 8JY

Nottingham Trent University Film Society

Laura Rowan
Byron House
Shakespeare Street
Nottingham

Open Film Society

David Reed
Faculty of Technology
The Open University
Walton Hall
Milton Keynes MK7 6AA

Peak Film Society

Paul Wooley
27 Hadfield Road
Hadfield

Pudleston Film Society

Robin Clarke
Keeders House, Brockmanton
Pudleston
Herefordshire HR6 OQU

Screen on the Hill Film Society

Ana Guimaraes
Media Development Manager
Stanley Hill
Amersham
Bucks HP7 9HN

Shrewsbury Film Society

B Mason
Pulley Lodge, Lwr Pulley Lane
Bayston Hill
Shrewsbury SY3 0DW

Solihull Film Society *OP DA*

Steve Wharam
2 Coppice Road
Solihull
West Midlands B2 9JY

Stafford Film Society *OP DA*

Mike Loveless

Coton Croft, Coton
Gnosall, Staffs ST20 0EQ

Stourbridge Film Society OP DA
M J Keightley
1a Pargeter Street
Stourbridge
West Midlands DY8 1AU

Three Rivers Film Society
Jon Taylor
222b New Road, Croxley Green
Rickmansworth, Herts, WD3 3HH

Tile Hill College Film Society
Mick Wheeler
Communications Unit
Tile Hill College of FE
Tile Hill Lane
Tile Hill, Coventry CV4 9SU

UCESU Film Society ST/OP
Jonathan King
University of Central England
Students Union
Perry Barr
Birmingham B3 3HQ

University of Keele Film Society ST
Nova Dudley
c/o Media & Communications
Keele University,
Staffs ST5 5BG

University of Warwick Film Society ST DA
The Secretary
Students Union
University of Warwick
Coventry CV4 7AL

Vale of Catmose Film Society
Peter Green
Vale of Catmose College
Oakham
Rutland LE15 6NJ

Weston Coyney & Caverswall Film Society
Robert Johnson
86 School Lane
Caverswall
Stoke-on-Trent ST11 9EN

NORTH WEST

North West Group BFFS
Chris Coffey
BFFS North West Group
64 Dale Crescent,
St Helens,
Merseyside WA9 4YE

Aldham Robarts Learning Resources Centre
Liverpool John Moores University
Mount Pleasant, Liverpool L3 5UZ

Birkenhead Library Film Society
Susan Boote
Central Library, Borough Road
Birkenhead
Wirral L41 2XB

Bolton Institute Film Society
Laurette Evans
Arts Office
Chadwick Campus
Bolton BL2 1JW

Bolton School Arts and Conferences
Duncan Kyle
Chorley New Road
Bolton BL1 4PA

Chester Film Society OP
Tony Slater
6 The Beaches
Pas Newton Lane, Upton
Cheshire CH1 5AG

Chorley Film Society
OP
John Leonard
55 Oak Croft
Clayton le Woods
Chorley, Lancs

Crewe & Alsager Film Society
Mark Nisbett
32 Gatefield Street
Crewe
Cheshire

Deeside Film Society OP DA
Peter Saunders
44 Albion Street
Merseyside L45 9UG

Ellesmere Port Library Film Society OP DA
Graham Fisher
Ellesmere Port Library
Civic Way
Ellesmere Port
Chester L65 0BG

Frodsham Film Society
Michael Donovan
58 The Willows
Frodsham
Cheshire WA6 7QS

Halton Film Society
Tim Leather
c/o Queens Hall, Victoria Road
Widnes
Cheshire WA8 7RS

Heswall Film Society
OP DA
Alvin Sant
18 Fairlawn Court, Bidston Road
Birkenhead, L43 6UX

Hulme Hall Film Society
David Butler
Houldsworth 51
Hulme Hall, Oxford Place
Manchester M14 5RR

Kino Film Club CL
John Wojonski
13 Mersey Crescent
West Didsbury
Manchester M20 2ZJ

Lancaster University Film Society
Alison Evans
Lonsdale College
University of Lancashire
Bailrigg
Lancs LA1 4GT

Lytham St Annes Film Society
OP DA
Alan Payne
18 Cecil Street
Lytham St Annes
Lancs FY8 5NN

Manchester and Salford Film Society OP DA
Tom Ainsworth
64 Egerton Road, Fallowfield
Manchester M14 6RA

Manchester University Film Society ST DA
H Fleming
Union Building
Oxford Road
Manchester M13 9PL

Preston Film Society OP DA
Michael Lockwood
14 Croftgate
Highgate Park, Fulwood
Preston
Lancs PR2 8LS

St Helen's Film Society CP
Chris Coffey
64 Dale Crescent
St Helens
Merseyside WA9 4YE

Society of Fantastic Films
Harry Nadler
5 South Mesnefield Road
Salford M7 0QP

Southport Film Guild OP
Irene Gunn

5 Chandley Close, Ainsdale
Southport
Merseyside PR8 2SJ

NORTHERN

Northern Group BFFS
c/o Scottish Group
28 Thornyflat Road,
Ayr KA8 0LX

Bede Film Society
Christian Errington
College of St Hild & St Bede
University of Durham
Durham DH1 1SZ

Centre Film Club *OP DA*
Mary Saunders
17 Benton Road
Middlesborough
Cleveland TS5 7PQ

Cleveland Film Group
Steven Moses
45 Oxford Road, Linthorpe
Middlesborough
Cleveland TS5 5DY

Darlington Arts Centre
Stef Hynd
Vane Terrace
Darlington
Co Durham DL3 7AX

Durham University Film Unit *ST DA*
Heidi Moore
Durham Student's Union
Dunelm House
New Elvet
Durham, DH1 3AN

Film Club at the Roxy *OP*
Phil Evans
55 Hibbert Road
Barrow-in-Furness
Cumbria LA14 5AF

Hartlepool Film Society *OP*
A Gowing
6 Warkworth Drive
Hartlepool, Cleveland TS26 0EW

Northumberland County Libraries *CP DA*
D Peacock
The Willows
Morpeth
Northumberland NE61 1TA

Sunderland University Film and Video Society
John Davison
The Gallery

Ashburne House
Ryhope Road
Tyne & Wear S22 7EF

Whitby Film Society
James Liddle
Lisvane
Grosment, nr Whitby,
N, Yorks YO22 5PE

SCOTLAND

Scottish Group BFFS
Ronald Currie
28 Thornyflat Road,
Ayr KA8 0LX

Aberdeen University Cinema Society
Garath Took
50/52 College Bounds
Old Aberdeen AB2 3DS

Academy Films 95/96
Andrew Morris
Tannoch, Moor Road
Strathblane G63 9EX

Auchtermuchty Film Society
Jim Scott
13 High Street, Fife KY14 7AP

Ayr & Craigie Film Society *OP DA*
Eleanor Danks
12 Lindston Place
Ayr
Ayrshire

Barony Film Society
Maureen Maciver
1 Lower Granton Road
Edinburgh EH5 2RX

Callander Film Society *OP*
David Hinton
25 Bridgend
Callander, Perthshire FK17 8AG

Carnoustie Cinema Club
Alison Groves
11 William Street
Carnoustie
Angus DD7 6DG

Cinema Fantastique
Allan Foster
Gamekeepers Cottage
Sprouston
Kelso
Roxburghshire TD5 8HN

Cove & Kilgreggan Film Society
Sally Garwood
Upper Flat, Cragowlet West,
Shore Road, Cove G84 OLS

Crieff Film Society *OP*
Charles Lacaille
5 Gilfillan Court
Commercial Lane
Comrie
Perthshire PH6 2DP

Eastwood Film Society
J Marchant
11 Beechlands Drive
Clarkston
Glasgow
Strathclyde G76 7XA

Edinburgh University Film Society *ST DA*
Societies' Centre
60 The Pleasance
Edinburgh EH8 9TJ

Electric Shadows - Glasgow University Film Club
Leah Panos
Flat 1/3, 61 Cecil Street
Glasgow G12 8RW

Filmsfelagid
Birgir Kruse
Shoga 2118
FR 165 Argir
Faroe Islands

Fleapit Film Club
E Taylor
PO Box 4711,
Glasgow G12 8YF

Glasgow Early & Silent Cinema Club
Aimara Reques
c/o GFVN, 3rd Floor, 34 Albion Street, Glasgow

Isleburgh Youth Film Club
Kaye Sandison
Creekhaven
Houl Road
Scalloway
Shetland ZE1 OXA

James Young High School Film Society
Jeremy Scott
Computing Studies Department
The James Young High School
Quentin Rise
Livingston EH54 6NS

Kelso High School
Stephanie Whitehead
Bowmont Street,
Kelso TD5 7EG

Lewis Film Society
Brendan O'Hanrahan
1 Marvig

South Lochs, Isle of Lewis

Linlithgow Film Society *OP*
Jenny Gilford
81 Belsyde Court
Linlithgow
West Lothian EH49 7RL

Mitchell Library
Periodicals Librarian
North Street, Glasgow G3 7DN

Portree Film Society
James Cryer
1 Camastianavaig
Braes by Portree
Isle of Skye IV51 9LQ

Queen Margaret College Film Society *OP DA*
Andrew Meneghini
Queen Margaret College
Grainger Stewart A6
Clerwood Terrace
Edinburgh EH12 8TS

Robert Burns Centre - Film Theatre *OP DA*
Kenneth Eggo
Dumfries Museum
The Observatory
Dumfries DG2 7SW

St Andrews University Film Society
ST DA
Belinda Cook
c/o General Office
Students Union
St Mary's Place, St Andrews
Fife KY16 9UZ

St Brides Film Society
George Williamson
St Brides, 10 Orwell Terrace,
Edinburgh EH11 2DY

Shetland Film Club
Stuart Hubbard
Nethaburn
Wester Quarff
Shetland ZE2 9EZ

Standard Life Film Society
Helen Wilson
Lyne Cottage
Elsrickle
Biggar ML12 6QZ

Stirling University Film Club
The President
SU3A, Stirling University
Stirling FK9 4LA

Strathallan School Film Society
Christopher Mayes
Strathallan School, Forgandenny
Perth PH2 9EG

Ten Day Weekend
Unit 14, Firhill Business Centre,
76 Firhill Road,
Glasgow G20 7BA

Tweedale Film Club *OP DA*
Jeanette Carlyle
Top Floor, 23 Marchmont Road
Edinburgh EH9 1HY

West Kilbride Film Club *OP DA*
Toni Cunningham
17 Alton Street
West Kilbride, Ayrshire KA23 9JN

SOUTH EAST

South East Group
c/o Southern Group, 1 Vanstone
Cottages, Bagshot Road,
Englefield Green, Egham,
Surrey TW20 ORS

Bradbourne School Film Society
Midge Adams, Media Department
Bradbourne Vale Road
Sevenoaks
Kent TN13 3LE

Brighton Film Society
Sylvia Alexander-Vine
West Hill Hall, Compton Ave,
Brighton BN1 3PS

Buckingham and Winslow Film Society
J Childs
22A Aylesbury Road
Wing
Leighton Buzzard

Chichester City Film Society *OP DA*
Roger Gibson
Westlands, Main Road
Hunston, Chichester
West Sussex PO20 6AL

Cranbrook Film Society *OP*
Vanessa Nicholson
Horserace House
Sissinghurst
Kent TN17 2AT

Dawson UK Ltd
PO Box 225
Attn: Subs Dept Al
Folkestone, Kent CT19 5EE

Deal Movie Club
Justin Linnane
19 Ranelagh Road
Deal, Kent CT13 0YF

Dover Film Society *OP*
C Eagleton
26 Church Street
Folkestone, Kent CT20 1SE

Eastbourne Film Society
Barbara Wilson
2 Chalk Farm Close
Willingdon
Eastbourne
East Sussex BN20 9HY

Faversham Film Society *OP*
Fliss Carlton
11 Dark Hill
Faversham, Kent ME13 7SP

Frame 25
Martin Gooch
40 Ivy Lane
Canterbury
Kent

Hastings College Film Society
Helen Dessent
Department of Art and Design
Hastings College
Archery Road
St Leonards on Sea TN38 0HX

HIADS Ltd
H Dyer
Station Theatre, Station Road
Hayling Island, Hants PO11 OEH

Hook Norton Film Society
Dr ME White
Wisteria House, High Street
Hook Norton, Oxon OX15 5NF

Kingston University Film Society
Alison Hardy
72 Norbiton Hall
Birkenhead Avenue
Kingston Upon Thames
Surrey KT 6RR

Lewes Film Guild *OP DA*
Mary Burke
6 Friars Walk
Lewes, East Sussex BN7 2LE

Lowestoft Film Society
Ken Jarmin
c/o Marina Theatre
Lowestoft
Suffolk NR32 1HH

Maidstone Film Society
Bruce Rylands
Overacre, Ulcombe Road
Langley Heath, Maidstone ME17
3JE

Medway Film Society *OP*
Caroline Reed
55 Maidstone Road

Rochester
Kent ME1 1RL

Platform Kino
Helen Greenfield
Platform Theatre, Burrell Road
Haywards Heath
West Sussex RH16 1TN

Rye Film Club
G Boudreau
The Old Grammar School
High Street
Rye, East Sussex TN31 7JF

Seaford Film Society
E Holland
25 Stafford Road
Seaford
East Sussex BN25 1UE

Stables Film Society
Fred Nash
c/o The Stables Theatre
32 Vale Road, Silverhill
St Leonards on Sea
East Sussex TN37 6PS

Stafford Film Theatre
Gerry McPherson
Orchard View, Mill Lane,
Acton Gate, Stafford ST14 ORA

Steyning Film Society
J Rampton
Woodbine Cottage, 124 High Street
Steyning , W Sussex BN44 3RD

Studio Film Society
Chris Archer
28 Beckenham Road
Beckenham BR3 4LS

Thanet Film Society
Camille Sutton
1 Vale Road, Broadstairs
Kent C10 2JE

Tunbridge Wells Film Society
Peter Warner
Flat 3, 44 Lime Hill Road
Tunbridge Wells
Kent TN1 1LL

Walton & Weybridge Film Society
Joan Westbrook
28 Eastwick Road
Walton on Thames
Surrey KT12 5AD

West Kent College of Further Education
D Davies
Brook Street
Tonbridge
Kent TN9 2PN

SOUTH WEST

South West Group BFFS
Brian Clay
The Garden Flat,
71 Springfield Road, Cotham
Bristol BS6 5SW

Bath Film Festival
Chris Baker
7 Terrace Walk
Bath
NE Somerset BA1 1LN

Bath Film Society
Karen Betts
Alexandra Cottage, Devizes Road
Box, Corsham, Wilts SN13 8DY

Bath Schools' Film Society *ST*
Jenny Wheals
47 Bobbin Lane
Westwood
Bradford on Avon
Wilts BA15 2DL

Blandford Forum Film Society *OP DA*
B Winkle
6A Bayfran Way
Blandford Forum
Dorset DT11 7RZ

Blundells 6th Form Film Club *ST*
Steve Goodwin
8 Alstone Road
Tiverton
Devon EX16 4JL

Bournemouth and Poole Film Society *OP*
Chris Stevenson
116 Wessex Oval
Wareham
Dorset BH20 4BS

Bridport Film Society
Mary Wood
9 Bowhayes
Greenways
Bridport
Dorset DT6 4EB

Cheltenham Film Society *OP DA*
A Thompson
29 Maidenhall, Highnam
Glos GL2 8DJ

Cheltenham Ladies College Film Society
Judi Bond
Bayshill Road
Cheltenham, Glos GL50 3EP

Cinema at the Warehouse *OP*
Rob Rainbow
Elwell House
Stocklinch
Ilminster
Somerset TA19 0JF

Cinematheque Yeovil *OP DA*
Nina Hatch
Peveril
42 Bowden Road
Templecombe
Somerset BA8 OLF

Cinsoc Exeter University *ST*
The President
Devonshire House
Stocker Road
Exeter, Devon EX4 4PZ

Dartmouth Film Society
Susie Ward
c/o 11 Victoria Road
Dartmouth Devon TQ6

Dorchester Film Society *OP DA*
Ann Evans
62 Casterbridge Road
Dorchester, Dorset DT1 2AG

ENTICKNAP - Film Society *ST*
The Secretary
College of St Mark & St John
Student Union, Derriford Road
Plymouth
Devon PL6 8BH

Exeter Film Society *OP*
H James
16 Pavilion Place
Exeter, Devon EX2 4HR

Falmouth Film Society *OP DA*
Antonio Villalon
Falmouth Arts Centre
Church Street, Falmouth
Cornwall TR11 3EG

Falmouth School of Art and Design *ST DA*
W Flint
Student Union Film Club
Woodlane
Falmouth
Cornwall TR11 4RA

Gloucester Film Society *OP DA*
C Toomey
8 Garden Way
Longlevens
Gloucester GL2 9UL

Holsworthy Film Society

Tony Langham
The Old Rectory
Pyworthy
Holsworthy
Devon EX22 6LA

Jersey Film Society *OP DA*

Paula Thelwell
Boulivot de Bas
Le Boulivot
Grouville, Jersey
Channel Islands JE3 9UH

Kingsbridge Theatre and Cinema

CP DA
William Stanton
138 Church Street
Kingsbridge, Devon TQ7 1DB

Lyme Regis Film Society *OP DA*

Suzanne Case
Little Park, Haye Lane
Lyme Regis
Dorset DT7 3NH

Mid Cornwall Film

Phill Webb
Restormel Arts
14 High Cross St
St Amstell
Cornwall PL25 4AN

The Octagon Film Society *OP*

Clifford Edwards
Okehampton College
Mill Road
Okehampton, Devon EX20 1PW

Reels On Wheels

Community Resource Centre
Milford Park, Milford Road
Yeovil, BA21 4QD

Regal Film Society

Victoria Thomas
Wagside Cottage, Cutcombe
Wheddon Cross
Somerset TA24 7AP

Screen Arts (Bridgewater Arts Centre) *OP DA*

Philippa Williams
11/13 Castle Street
Bridgewater
Somerset TA6 3DD

Chaftocbury Arts Centre Film Society *OP DA*

Paul Schilling
5 Well Lane
Enmore Green
Shaftesbury, Dorset SP7 8LP

Sherborne School Film Society

ST DA
Andrew Swift
Abbey Road
Sherborne, Dorset DT9 3AP

South Molton Film Society

Phillip Norman
20 Brook Meadow
South Molton
Devon

Stroud and District Film Society

OP DA
Tim Mugford
Manor Farm, Besbury Common
Minchinhampton
Stroud
Glos GL6 9ES

Totnes Film Society

Derek Bowerman
Top Flat, 78A High Street
Totness
Devon TQ9 55N

UPSU Cinematique

Nic Whelon
Rolle College, Douglas Avenue
Exmouth
Devon EX8 2AT

Wellesley Film Club

Louise Baker
11 Mantle Street
Wellington
Somerset TA21 8AR

SOUTHERN

Southern Group BFFS

Dudley Smithers
1 Vanstone Cottages, Bagshot
Road, Englefield Green, Egham,
Surrey TW20 ORS

Abingdon College & District Film Society *ST DA*

Mike Bloom
Abingdon College of FE
Northcourt Road
Abingdon, Oxon 0X14 1NN

Amersham & Chesham Film Society *OP*

A W Burrows, Hon Treasurer
73 Lye Green Road
Chesham, Bucks HP5 3NB

Ashcroft Arts Centre Film Society

OP DA
Steve Rowley
Osborn Road
Fareham, Hants PO16 7DX

Barton Peveril College 6th Form Film Society

T C Meaker
Cedar Road
Eastleigh
Hants SO5 5ZA

Bracknell Film Society

Ian Neall
Creedy Cottage, 30 Addiscombe
Road,
Crowthorn, Berks RG45 7JX

Bradford on Avon Film Society

John Holmes
22 Sladesbrook
Bradford on Avon
Wiltshire BA15 1SH

Charterhouse Film Society

Christopher O'Neill
Brooke Hall, Charterhouse
Godalming
Surrey GU7 2DX

Chiltern Film Club

Derek Goddard
9 Hospital Hill
Chesham
Bucks

The Dever Valley Film Society *OP*

Russell Alexander Smart
Gardener's Cottage
61 Church Street
Micheldever
Hants S021 3DB

Eton College Film Society *ST*

G.J.Savage
Eton College
Windsor, Berks SL4 6DW

Farnham Film Society

Pamela Woodroffe
c/o The Maltings, Bridge Square
Farnham
Surrey GU9 7QR

Forest Arts

Nadine Fry
Old Milton Road
New Milton, BH25 6DS

Harwell Film Society *CL DA*

Lorraine Watling
Building 156
Harwell Laboratory
Didcot, Oxon OX11 0RA

Havant College Film Society *ST*

P Turner
New Road
Havant, Hants PO9 1QL

Havant Film Society *OP DA*

M Short

14 South Street
Havant, Hants PO9 1DA

Henley-on-Thames Film Society
M Whittaker
10 St Andrews Rd
Henley-on-Thames
Oxon RG9 IHP

Horsham Film Society
Norman Chapman
Farthings
King James Lane
Henfield
Sussex BN5 9ER

Intimate Cinema Film Society
A Henk
10 Aston Way
Espom
Surrey KT18 5LZ

Marlborough College Senior Film Society *ST DA*
Ross Birkbeck
Marlborough College
Marlborough, Wilts SN8 1PA

Newbury Film Society *OP DA*
Stuart R Durrant
41 Jubilee Road
Newbury
Berks RG14 7NN

OBSU Film Society *ST*
Helen Critchley
Oxford Brookes Student Union
Oxford Brookes University
Gipsy Lane
Oxford OX3 0BT

Open Film Society *ST DA*
David Reed
Faculty of Technology
The Open University
Walton Hall
Milton Keynes
Bucks MK7 6AA

Oscar Film Unit
The Secretary
University of Surrey
Guildford
Surrey GU2 5XH

Oxford University Film Foundation
Joe Perkins
Balliol College
Oxford, OX1 3BJ

Pegasus Film Society *OP DA*
Deana Rankin
Pegasus Theatre
Magdalen Road
Oxford OX4 1RF

Radley College Film Society *ST*
C R Barker
Radley College
Abingdon, Oxon OX14 2HR

The Regal Film Society
Victoria Thomas
Wagside Cottage , Cutcombe
Wheddon Cross
Somerset TA24 7AP

Rewley House Film Theatre Club
OP DA
R T Rowley
Director
1 Wellington Square
Oxford OX1 2JA

St Anne's Film Society *ST DA*
Katherine Day
St Anne's College
Oxford OX2 6HS

Shere Film Society
Hugh Peacock
32 Wolsey Drive
Kingston
Surrey KT2 5ON

Slough Co-Operative Film Society
Jo Hughes
58 St Judes Road
Englefield Green
Egham
Surrey TW20 0BT

Southampton Film Theatre - The Phoenix *OP DA*
Dr Peter Street
24 The Parkway
Bassett, Southampton SO2 3PQ

Southampton University
The Secretary Union Films
Student's Union
Highfield
Southampton SO17 1BJ

Surbiton Film Club
Hugh Peacock
9 Bockhampton Road,
Kingston Upon Thames
Surrey KT2 5JU

Swindon Film Society *OP*
Julia Edwards
9 Westlecot Road
Old Twon, Swindon
Wilts SN1 4EZ

Talking Heads *OP*
Todd Mitchell
320 Portswood Road
Southampton
Hants SO17 2TD

Trowbridge College Film Society
Antoinette Midgley
Dept of General & Social Studies
Trowbridge College
College Road
Trowbridge, Wiltshire BA14 0ES

Union Films *ST DA*
Southampton University
Students' Union
Highfield
Southampton
Hants SO17 1BJ

University of Sussex Film Society
Film Society Pigeon Hole
Students Union, Falmer House
Sussex University
Falmer, Brighton BN1 9QF

Winchester College Film Society
ST
Dr J Webster
Winchester College
Winchester
Hants SO23 9NA

Winchester Film Society *OP DA*
Judith Altshul
17 Bramshaw Close
Winchester
Hants SO22 6LT

Windsor Arts Centre Film Society
OP
C Brooker
St Leonards Road
Windsor
Berks SL4 3DB

Woking's New Cinema Club
Barbara Millington
184 Alexandra Gardens
Knaphill
Woking
Surrey GU21 2DW

WALES

Welsh Group BFFS
Rupert Clarke
Felin Llwyngwair, Newport,
Pembs SA42 OLX

Abergavenny Film Society *OP DA*
Carol Phillips
Tybryn, Tal-y-coed
Monmouth
Gwent NP5 4HP

Brecon Film Society *OP DA*
Charlotee Davies
Cilgarrnyod, Pont Faen, Brecon

Cambria Cinema Club
M Wilson
Dewina , Llanddewi-Brefi
Tregaron
Ceredigion

Congress Film Club
David Giles
4 Fields Road, Oakfield
Cwmbran
Gwent NP44 3EF

Eirias High School Film Society
Jody Bigg
The Close
Penrhyn Bay
Llandudno LL30 3HZ

Fishguard Film Society *OP*
Frances Chivers
Iscoed-Pont Cilrhedyn
Abergwaun
Pembs SA65 9SB

Flicks *OP*
Sian Swann
10 St Catherine's Street
Carmarthen, Dyfed
SA31 1RE

Haverfordwest Film Society *OP*
DA
Kay Green
Summerhill
16A Haverfordwest Road
Letterston
Dyfed SA62 5UA

Maindy Film Club
Neal Hammond
Workmens Hall
Curch Road
Ton Pentre
Mid Glamorgan CF41 7EH

The Movie Club *OP*
Willie Jack
Crossroads
Aberhafesp
Newtown, Powys SY16 3LR

Newton Movie Club
Angus Eickoff
The Forge, Whitehouse Bridge,
Welshpool, Powys SY21

North East Wales Film Society
G. Jones
11 Cedar Gardens, Deeside,
Flintshire, Clwyd CH5 1XJ

Narbeth Film Society
Richard Swingler
Red Gables
Llanteg
Narbeth

Dyfed SA67 8PU

Phoenix Film Club *OP DA*
Neil Hammond
c/o Phoenix Centre, Church Road,
Ton Pentre, Rhondda, CF41 7EH

Presteigne Film Society *OP DA*
Heywood Hill
Harpton
New Radnor
Powys LD8 2RE

Sand Palace Film Society
OP DA
Charlotte Cortazzi
1 Home Cottage
Pen-Y-Wen
Stackpole
Pembroke, Dyfed SA17 5DG

Swansea Film Society *OP*
Daphne Evans
8 Ffordd Tlfan
Garden Village, Gorseinon
Swansea SA4 4HN

Theatr Mwldan Film Society *OP*
DA
Mary Champion
Pengarn Fawr
Cippyn
Cardigan, Dyfed SA43 3LT

UWCC Film Society
Steven Griffiths
220 Corporation Road
Grangetown
Cardiff

Valley Pictures
Dallas Stone
Centre House
Cefn Community Centre
Cefn Coed
Merthyr Tydfil CF48 2NA

The Windswept Film Society
Bel Crewe
Tan-y-Cefn, Nantmel, Nr
Rhayader,
Powys LD6 5PD

YORKSHIRE

Yorkshire Group BFFS
Richard Fort
8 Bradley Grove
Silsden
Keighley
West Yorks BD20 9LX

Ampleforth Film Society *ST DA*

S P Wright
St Dunstan's House
Ampleforth College
York YO6 4ER

Bradford Student Cinema
Tim Anglish
Students Union
University of Bradford
Richmond Road
Bradford BD7 1DP

Halifax Playhouse Film Club *OP*
DA
Sylvia Drake
7 North Road
Barkisland
Halifax HX4 OAH

Harrogate Film Society *OP*
James Lamb
19A Lynton Gardens
Harrogate
North Yorks

Hebden Bridge Film Society *OP*
Diane Stead
Mytholm House, Mytholm
Hebden Bridge
West Yorks HX7 6DS

Ilkley Film Society
OP DA
Richard Fort
8 Bradley Grove
Silsden
Keighley
West Yorks BD20 9LX

Kirklees Film Society
Alick Wilson
12 Manor Street
Huddersfield HD4 6NS

Leeds Film Society
David Griffiths
John Sowerby Community Theatre
Ralph Thoresby High School
Holt Park Village

The Old Meeting House Trust *OP*
DA
Martin Weyer
Knipes Hall
Helmsley
York YO6 5AE

Ryedale Film Society
Karin Doose
25 West End
Kirkby Moorside YO6 6AD

Saddleworth Film Society *OP*
Philip Adams

11 Huddersfield Road
Delph, Oldham

Scarborough Film Society *OP*

Tony Davison
29 Peasholm Drive
Scarborough
North Yorks YO12 7NA

Sheffield University SU Film Unit

ST DA

The Secretary
Student's Union
Sheffield University
Western Bank
Sheffield S10 2TG

University of Huddersfield Student Union Film Society *ST*

Karen McDonald
Student Union
Queensgate
Huddersfield, W Yorks HD1 3DH

Wakefield Film Society

Jean Durham
247 Dewsbury Road
Wakefield WF2 9BZ

York Student Cinema *ST DA*

Neil Potter
Student Union Corridor
Goodriche College
University of York
Heslington
York YO1 5DD

STUDENT GROUP

BFFS Student Group

Sharon Armour
23 Heathdene Road, Highfield,
Southampton, SO17 1PA

OVERSEAS

British Council Film Club & Video

Angelica Oserwalder
Rothenbaumchausse 34
2000 Hamburg 13
Germany

British Embassy Cinema Club

E Miskey
PO Box 393
Jeddah
Saudia Arabia

Cine Club of Calcutta

Sudhin Banerjee
2 Jawaharlal
Nehru Road
Calcutta 700 013
India

Irish Federation of Film Societies

Brenda Gannon
The Irish Film Centre
6 Eustace Street
Dublin 2

W Australian Federation of Film Societies

Barry King
P O Box 90
Subiaco 600B
Western Australia,
Australia

FUNDING

The list below is representative of the schemes available, but not definitive, since the volatile nature of funding, both public and private, means that schemes emerge and fall by the wayside unpredictably. These funding schemes cover development, distribution, exhibition etc as well as production. The funds available fall into three main categories: direct grants, production finance and reimbursable loans. In almost all cases these schemes are not open to students. **The Low Budget Funding Guide 98/99,** published by the BFI, provides useful additional information about funding

ADAPT (Access for Disabled People to Arts Premises Today)
The ADAPT Trust
8 Hampton Terrace
Edinburgh EH12 5JD
Tel: 01381 346 1999
Fax: 0131 346 1991
Development Manager: Stewart Coulter
Charitable trust providing advice and challenge funding to arts venues - cinemas, concert halls, libraries, heritage and historic houses, museums and galleries - throughout Great Britain. ADAPT also provides a consultancy service and undertakes access audits and assessments. Grants and Awards for 1999 advertised as available

ADAPT (Northern Ireland)
185 Stranmillis Road
Belfast
BT9 5DU
Tel: 01232 683463
Fax: 01232 661715

Arts Council of England
Visual Arts Department
14 Great Peter Street
London SW1P 3NQ
0171 312 0100
0171 973 6581
David Curtis, Gary Thomas
The Visual Arts Department supports the production of film and video work by artists, and related activities, including festivals and publications, through the following schemes:
The Artists' Film and Video National Fund: Production
Grants are for the production of single-screen film and moving-image electronic art for exhibition in galleries, cinemas and other spaces. The Fund seeks to support a range of works which are innovative in method and formal approach and which engage with and extend current debates in film and electronic media theory and practice, and which will achieve distribution

and exhibition in Britain and abroad. Students are not eligible. The annual deadline is in September
The Artists' Film and Video National Fund: Exhibiton and Initiatives
Supports activities that widen access to, and broaden the appreciation of film and video art. Grants are available to galleries and other organisations to commission and tour new work (installation or single-screen); to tour existing work; and for national festivals of work by artists. Deadlines in October and February
For guidelines, send an SAE marked 'Production' or 'Exhibiton' to Gary Thomas at the Arts Council.
Other funds may be available for Animate!, a co-commissioning scheme with Channel 4 for experimental animation, for publications and for multimedia projects; please check with the Visual Arts and Combined Arts Departments
NATIONAL LOTTERY DIVISION
Jeremy Newton, National Lottery Director
Carolyn Lambert: Director of Lottery Film
Film Production for Cinema
Funding is available for film production companies (usually based in England) proposing to produce a film which qualifies as British under the terms of the 1985 Films Act and which is intended for cinema release in the UK. ACE will not be the sole financier and will expect a substantial portion of production funding to have been raised already. In most cases the Arts Council Lottery's contribution will be within the range of 10 per cent to 50 per cent of total costs, although in exceptional circumstances applications will be considered for 75 per cent Lottery funding for projects with a total budget of under £750,000. Successful applicants need to provide a clear plan for completing the production on schedule and within budget. All

sections of the film making community, apart from full time students and individual film makers, are eligible for support.
Artists' Film and Video
Funding is available to producers, exhibitors and other commissioners of new work by artists. All moving image projects by artists of any nationality intended for exhibition in galleries, cinemas or other public spaces in England are eligible.
Greenlight Fund
The Greenlight Fund is managed by British Screen Finance, and is intended for part-financing higher budget films with directors of international repute. Amounts of up to £2 million may be awarded to suitable projects. The final decision on whether any particular project should be financed from the Fund rests with ACE.
Film Production Franchises
Im May 1997 ACE annouced three feature film production franchises to: **DNA Films, The Film Consortium**, and **Pathé Pictures.**

Arts Council of Northern Ireland
Lottery Unit
185 Stranmillis Road
Belfast BT9 5DU
Tel: 01232 667000
Fax: 01232 664766
Tanya Greenfield, Lottery Officer, Head of Unit

Arts Council of Wales
Lottery Unit, Holst House
Museum Place
Cardiff CF1 3NX
Tel: 01222 388288
Fax: 01222 395284
Jo Weston, Lottery Director

Arts for Everyone
A4E Interim Co-ordination Unit
Arts Council of England
14 Great Peter Street
London SW1P 3NQ
Tel: 0171 973 6582
Fax: 0171 973 6590
Sally Stote

BBC Bristol Television Features

Whiteladies Road
Bristol BS8 2LR
Tel: 0117 974 6746
Fax: 0117 974 7452
10x10 BBC2's scheme for New
Directors
Series Producer: Jeremy Howe
Produces 10 ten-minute films per
series, documentary, fiction and
anything in between. It is an
initiative to encourage and develop
new and innovative film making
talent in all genres, through the
provision of modest production
finance combined with practical
guidance. Now in its 11th series, the
scheme is open to any director with
no commissioned broadcast UK
Network directing credit. All
applications must include a showreel
with their proposals - which should be
treatments for documentary, scripts for
drama. Deadline for submissions for
series - to be confirmed

BFI Production

29 Rathbone Street
London W1P 1AG
Tel: 0171 636 5587
Fax: 0171 580 9456
Roger Shannon
New Directors Scheme
For film and video makers who are in
the early stages of their careers or for
people changing careers. Aims to
produce five or more short films a year
with budget ceilings of £35,000, all
productions to be made under Equity
and BECTU agreements. Advertised
annually in November; submissions
accepted in January with final
selection made by May **(see p8)**

British Council

11 Portland Place
London W1N 4EJ
Tel: 0171 389 3065
Fax: 0171 389 3041
Does not invest in production but
can assist in the co-ordination and
shipping of films to festivals, and in
some cases can provide funds for the
film-maker to attend when invited.
A limited amount of fundraising is
available for UK filmmakers to
attend European seminars/workshops
such as Arista, Eave, Sources etc

British Screen Finance

14-17 Wells Mews
London W1P 3FL
Tel: 0171 323 9080
Fax: 0171 323 9092
email: BS@cd-online.co.uk
Invests in British feature films
including films made under co-
production treaties with other

countries. Scripts should be
submitted with full background
information. All scripts are read.
Scripts submitted by producers with
a fully developed production
package are given priority, and
projects must have commercial
potential in the theatrical market.
British Screen's contribution rarely
exceeds £500,000, and is never more
than 30 per cent of a film's budget

Channel 4/MOMI Animators

Professional Residencies
Museum of the Moving Image
South Bank
London SE1 8XT
Tel: 0171 815 1376
Four Professional Residencies are
awarded to young or first time
animators. A fee of £2,850 plus a
budget of up to £1,600 towards
materials. At the end of residency at
MOMI, project will be considered
for commission by Channel 4

Croydon Film & Video Awards

Croydon Clocktower
Katharine Street
Croydon CR9 1ET
Tel: 0181 760 5400 ext 1048
Co-ordinator: Mark Wilcox
The Croydon Film & Video Awards
are an ongoing production initiative
co-funded by the LFVDA and the
London Borough of Croydon. The
£10,000 scheme was established in
1997 to support local film and video
makers to make fictional, documentary or experimental shorts

The Glasgow Film Fund

74 Victoria Crescent Road
Glasgow G12 9JN
Tel: 0141 337 2526
Fax: 0141 337 2562
Director: Eddie Dick
Marketing and Promotions:
Ela Zych-Watson
The Glasgow Film Fund is adminis-
tered on behalf of its funding bodies
by the Scottish Film Production
Fund. Set up in 1993, the Glasgow
Film Fund provides production
funding for feature films shooting in
the Glasgow area or produced by
Glasgow-based companies.
Applications are accepted for films
intended for theatrical release, with a
budget of at least £500,000. The
maximum investment made by the
Glasgow Film Fund in any one
project will be £150,000 or 20 per
cent of the production budget. The
board of the Scottish Film Production Fund meets regularly to
consider Glasgow Film Fund
applications: forms, dates of
meetings, deadlines and further

details are available from the GFF
office. Production credits include;
*Shallow Grave, Small Faces, The
Near Room, Carla's Song and The
Slab Boys*

Kraszna-Krausz Foundation

122 Fawnbrake Avenue
London SE24 0BZ
Tel: 0171 738 6701
Fax: 0171 738 6701
email: K-K@dial.pipex.com
Andrea Livingstone
Check this entry with the originals
Annual awards, with prizes for
books on the moving image (film,
television, video and related media),
alternating with those for books on
still photography. Books, to have
been published in previous two
years, can be submitted from
publishers in any language. Prize
money around £20,000, with awards
in two categories. 1997 winners:
Culture & History category: The
World According to Hollywood,
1918-1939 by Ruth Vasey (published by University of Exeter Press)
£10,000 prize;
Special commendation: David Lean
by Kevin Brownlow (published by
Richard Cohen Books) £1,000
Business, Techniques & Technology
catergory: The Encyclopaedia of
Animation Techniques by Richard
Taylor (published by Focal Press,
UK) £10,000 prize;
Special commendation: Advanced
Television Systems: Brave New TV
by Joan Van Tassel's (Focal Press,
USA) £10,000 prize. Grants are
offered for works which could not
otherwise be finished or published
and which are based in the UK.

London Production Fund

114 Whitfield Street
London W1
Tel: 0171 383 7766
Fax: 0171 383 7745
Co-ordinator: Maggie Ellis
The London Production Fund aims
to support and develop film, video
and television projects by independ-
ent film-makers living or working in
the London region. It is run by the
London Film and Video Develop-
ment Agency and receives financial
support from Carlton Television and
Channel 4. It has an annual budget
of approximately £200,000
Development Awards
Support of up to £3,000 each to assist in
the development of scripts, storyboards,
project packages, pilots etc
Project and Completion Awards
Offers support up to £15,000 each

for production or part-production costs. Awards will be made on the basis of written proposals and applicants' previous work. The Fund is interested in supporting as diverse a range of films and videos as possible

Nicholl Fellowships in Screenwriting

Academy of Motion Picture Arts and Sciences
8949 Wilshire Boulevard
Beverly Hills, CA 90211
USA
Tel: (1) 310 247 3059
Annual Screenwriting Fellowship Awards
Up to five fellowships of US$25,000 each to new screenwriters. Eligible are writers in English who have not earned money writing for commercial film or television. Collaborations and adaptations are not eligible. A completed entry includes a feature film screenplay approx 100-130 pages long, an application form and a US$30 entry fee. Send self-addressed envelope for rules and application form

Northern Ireland Film Commission

21 Ormeau Avenue
Belfast BT2 8HD
Tel: 01232 232444
Fax: 01232 239918

The Prince's Trust

18 Park Square East
London NW1 4LH
Tel: 0171 543 1269/1360
Fax: 0171 543 1315
Go and See Grants
Awards (£500 max) to help young people in difficulty explore collaborative projects with partners in European countries. Applicants must be under 26 and out of full-time education
Richard Mills Travel Fellowship
In association with the Gulbenkian Foundation and the Peter S Cadbury Trust, offers three grants of £1,000 for people working in community arts, in the areas of housing, minority arts, special needs, or arts for young people, especially the unemployed. The Fellowships are applicable to people under 35

Production Fund for Wales

Sgrin: The Media Agency for Wales
The Bank, 10 Mount Stuart Square
Cardiff Bay
Cardiff CF5 2PU
Tel: 01222 333300
Fax: 01222 333320
email: sgrin@sgrinwales.demon.co.uk
Website: http://www.sgrinwales.demon.co
Production coordinator: Gaynor Messer Price
Details on request

Scottish Arts Council

Lottery Department
12 Manor Place
Edinburgh EH3 7DD
Tel: 0131 226 6051
Fax: 0131 477 7240
David Bonnar

Scottish Screen

Dowanhill
74 Victoria Crescent Road
Glasgow G12 9JN
Tel: 0141 302 1700
Fax: 0141 302 1711
Chief Executive: John Archer
The Government body Scottish Screen was established in April 1997 to stimulate the film and television industry through production and development finance, training, education, marketing and the film Commission.
Production. Development assistance, advisor to the Scottish Arts Council Lottery fund on film production, principally fiction and short film. Short film schemes include financing of First Reels (£1,000 - £4,000) Prime Cuts (£23,000), Geur Ghearr (£45,000) Tartan Shorts (£45,000) and access to the Shiach Script library.
Development
Advice and Finance for the development of feature films. (Film script Development up to £15,000)
Media
At Scottish Screen the Media Antenna Scotland can help you access MEDIA II and other European support programmes

REGIONAL ARTS BOARDS

East Midlands Arts Board

Mountfields House
Epinal Way
Loughborough
Leics LE11 0QE
Tel: 01509 218292
Fax: 01509 262214
email: mckay.andy.ema@artsfs.org.uk
Head of Media Publishing and Visual Arts: Andy McKay
Film, Video and Broadcasting Officer: Caroline Pick
email: pick.caroline.ema@artsfs.org.uk
A limited number of small-scale, low budget Script Development Bursaries; Completion Grants; Materials Bursaries. A limited number (up to five) of awards for short films, including First Cut and Co-Production challenge schemes up to £10,000. Limited number of small awards for individual training

Eastern Arts Board

Cherry Hinton Hall
Cherry Hinton Road
Cambridge CB1 4DW
Tel: 01223 215355
Fax: 01223 248075
email: cinema@eastern-arts.co.uk
Website: www.arts.org.uk/ea/index.html
Cinema and Broadcasting Officer: Martin Ayres
Offers funding to people resident in Bedfordshire, Cambridgeshire, Essex, Hertfordshire, Lincolnshire, Norfolk and Suffolk
First Take
In conjunction with Eastern Arts Board and Anglia Television - commissions innovative work and co-finances projects by new producers, directors and writers for screening by Anglia. It also undertakes training and distribution initiatives
Media Arts Production Fund
Development up - to £500 for research, script development etc. Production award - up to £5,000 for sole or co-funding of production, post-production and completion of innovative moving image and audio projects across a range of genres.
Visual and Media Arts Artists Development Fund and Festivals and Events Fund
Support for individual initiatives, projects and events
Write Lines
Script reading and advice service for aspiring film, TV and radio scriptwriters

Visual and Media Arts Slide and Video Database

A non-selective, publicl accessible collection of the work of the region's visual and media artists. Open to film and video makers wishing to promote themselves and their work to potential commissioners.

Mailing list

Enables film and video makers, scriptwriters and animators to receive information on events and funding opportunities.

Media Courses Database

Information on further education and short courses

English Regional Arts Board

5 City Road
Winchester
Hampshire SO23 8SD
Tel: 01962 851063
Fax: 01962 842033
email: info.erab@artsfb.org.uk

London Arts Board

Elme House
133 Long Acre
Covent Garden
London WC2E 9AF
0171 240 1313
0171 240 4578
Trevor Phillips
Chief Executive: Sue Robertson

London Film and Video Development Agency

114 Whitfield Street
London W1P 5RW
Tel: 0171 383 7755
Fax: 0171 383 7745
Chief Executive: Steve McIntyre
Training and Education Officer: Tricia Jenkins
Projects Officer: Andrea Corbett
Provides revenue funding to a range of production, training facilities and exhibition organisations, deemed to be of crucial significance in maintaining the cultural infrastructure of the Capital, including the London Production Fund (qv). Also:

Project Funding

Support for training courses, film and video exhibition and festivals in London

Capital Grants

Grants of up to £10,000 available for capital funding, to be matched by at least double from other sources

The LFVDA is the assessing body for London film and video applications to the National Lottery. The LFVDA runs the London Production Fund (q.v.)

North West Arts Board

Manchester House
22 Bridge Street
Manchester M3 3AB

Tel: 0161 834 6644
Fax: 0161 834 6969
Media Officer: Film & Video, Howard Rifkin
Media Administrator: Louise Atherton
NWAB offers a range of funding schemes covering Production, Exhibition, Training and Media Education for those resident in the NWAB region. For details of these and other schemes please contact Louise Atherton on 0161 834 6644

Northern Arts Board

9-10 Osborne Terrace
Jesmond
Newcastle upon Tyne NE2 1NZ
Tel: 0191 281 6334
Fax: 0191 281 3276
email: jcl@norab.demon.co.uk
Head of Published & Broadcast Arts: Janice Campbell
Offers funding to people resident in County Durham, Cumbria, Northumberland, Teesside and Tyne & Wear

Northern Production Fund (NPF)

email: pmy@norab.demon.co.uk
Production Adviser: Paul Moody
The aim of NPF is to support the production of short and long form drama, for film, television and radio, animation, creative documentaries, and all forms of experimental filmmaking, including work for gallery exhibition. The foremost concern of NPF is for the quality of the production. We seek to support productions which are imaginative, innovative, thoughtful, courageous and powerful. NPF normally holds three meetings per year to consider applications under the small scale production, development and feature film development headings.

Company Support applications will normally be considered once a year - the application deadline is 12th June

Production

Support of up to £15,000 for production or part-production costs or completion costs.

Development

Support of up to £3,000 to assist in the development of scripts, storyboards, full treatments, pilot production, etc. This includes research and development for feature films, short drama for film or radio, animation, documentary projects and innovative television drama.

Feature Film Developments

A maximum of two £8,000 awards for feature film development will be available for projects each year. These awards will normally be made to production companies, working

with a Northern-based writer, who are able to demonstrate their ability to match the Northern Arts contribution. Matching funding may include the cost of feature film development expertise and/or the contribution of another funding partner.

Company Support

Support for companies is available to assist in the development of a programme of work. The level of support for any one company is not fixed and will normally be an appropriate proportion of the funding available under this heading. Company support will normally be awarded to support several projects rather than production costs.

Broadcaster Partnership Schemes

The Northern Production Fund also works in partnership with broadcastcrs to offer production schemes for short drama and documentary production. Previous partners include TTTV, Border TV, Channel 4 Independent Film & Video and BBC 10 X 10. These schemes have their own application procedures and deadlines which are available from Northern Arts on request.

New Voices

In partnership with Tyne Tees TV (in collaboration with Yorkshire and Humberside Arts, North West Arts, Yorkshire TV and Granada TV), provides funding for the writing, development and production of half hour screenplays by writers new to television. Six screenplays are fully developed through workshop sessions and three are then chosen to be made by independent producers for broadcast.

Writing On the Edge (pilot)

Border Television, Northern Arts, the Scottish Film Council, Cumbria and Northumberland County Councils and Dumfries and Galloway Regional Council joined forces for a new initiative aimed at nurturing and developing new television drama writing talent. Aimed at people who have not had scripts screened on TV, writers were invited to submit their ideas for original 10-15 minute long dramas with a clear potential for television production, of an original quality, with strong appeal to the Border TV region and capable of being filmed within the region.

Writers of scripts selected were invited to attend writing workshops with professionals from the film and TV industry to give expert tuition to help writers to develop scripts. The best scripts from these workshops will form the basis of three 10-15 minute long drama shorts to be filmed and transmitted by Border TV.

Hot Dox (pilot)
The aime of this new documentary initiative from Channel 4's Independent Film & Video Department is to offer first break opportunities to new directors based outside London to make their first documentary for a national network and to deliver new and innovative short documentaries to Channel 4's schedule. Channel 4 has contracted the Northern Production Fund and Sheffield Independent Film to provide two 10 minute documentaries from production companies in the Yorkshire & Northern Arts Regions. There are seven other similar schemes across Britain. All these films will be commissioned and scheduled for specific Independent Film & Video's late-night zones during 1997/98. The budget for each film is £25K and there will be no production fee

South East Arts
Union House
Eridge Road
Tunbridge Wells
Kent TN4 8HF
Tel: 01892 507200
Fax: 01892 549383
Exhibition Subsidy
Grants of up to £2,000 are awarded to venues in the region whose policy is educative. Full details on application
Production Grants
Grants for beginners of up to £1,000 awarded. Grants for those with film and video to show in support, maximum of £10,000. Deadline 31 May 96. Full details on application

South West Arts
Bradninch Place
Gandy Street
Exeter EX4 3LS
Tel: 01392 218188
Fax: 01392 413554
David Brieley
Chief Executive: Graham Long

South West Media Development Agency
59 Prince Street
Bristol BS1 4QH
Tel: 0117 927 3226
Fax: 0117 927 0210
Website:http://www.ex.ac.uk/brad/SWMDA
Judith Higginbottom (Director)
The South West Media Development Agency is the funding and development body for film, video and television in the south west. It provides advice and financial support for: low budget and independent production, animation, script development, artists' film and video commissioning, exhibition of independent art-house, historic and experimental cinema. Applicants for financial support must be resident in the South West Media Development region. For details of available funding, please contact Sarah-Jane Meredith

Southern Arts Board
13 St Clement Street
Winchester
Hants SO23 9DQ
Tel: 01962 855099
Fax: 01962 861186
Film and Video production grants available in two categories: production and completion. There are no fixed maximum limits for funding. However, total funding available is limited and any application for more than £8,000 is unlikely to be successful. Co-production funding is strongly encouraged. Artists' new media project grants are available to artists working with digital media, including time-based, site specific, CD-Rom and Internet projects. The David Altshul Award is a competitive award for creative achievement in film and video production available to those who live or work in the Southern region including students. Annual prize money of £1,000. Film and video production training grants to support Black and Asian practitioners and people with disabilities; production training up to £500; workshop production up to £2,500. Exhibition Development Fund to support programming, marketing, training and research. Media Education Development Fund for strategic development of regional media education. Full details on all the above on application

West Midlands Arts Board
82 Granville Street
Birmingham B1 2HJ
Tel: 0121 631 3121
Fax: 0121 643 7239
email: west.midarts@midnet.com
Media Officer: Film and Video: Laurie Hayward
email.
laurie.hayward.wma@artsfb.org.uk
"First Academy" Film and Video Production Scheme
A broadcast initiative supported by West Midlands Arts, Birmingham City Council, the Media Development Agency for the West Midlands (MDAWM), Central Broadcasting, BBC Resources Midlands and East, and the Midland Media Training Consortium. The aim is to produce a range of diverse programmes for regional television. Recipients of the award work with a Production Co-ordinator based at (MDAWM) to develop a project through training, production support and access to the broadcast industry. Awards are made to reflect two levels of experience. Level 1 is aimed at the first time film-maker. Five productions will be supported to a maximum budget of £4,000 plus production facilities from the broadcast industry. Level 2 seeks to encourage new and established film-makers who will make a continuing contribution to film and television culture in the region. Three productions will be supported up to a maximum of £13,000 plus production facilities from the broadcast industry. The scheme results in the Central Television 'First Cut' programme which will be broadcast in the Autumn of 1999. Application deadline January 1999.
New Work and Commissions
Offer artist film and video makers an opportunity to produce new work in film, video and new technology. The scheme seeks proposals which demonstrate innovation and experimentation. Awards are made for pieces up to ten minutes with a maximum budget of £5,000. The scheme favours work for screening in conventional, sites specific and other contexts. Application deadlines February/September.
Research and Development Awards
Enables makers to develop their proposals for future productions. Research and Development Awards are expected to range between £200 to £1000. Deadline February/September

Yorkshire and Humberside Arts Board
21 Bond Street
Dewsbury
West Yorks WF13 1AX
Tel: 01924 455555
Fax: 01924 466522
email: tony.dixon.yha@artsfb.org.uk
Short Film and Video Production Awards 1999
Awards of up to £10,000 are available for film or video productions of up to ten minutes. Yorkshire and Humberside Arts wishes to encourage original creative work in a variety of forms: fiction, documen-

tary, animation or experimental. Productions can be for theatrical distribution, broadcast television, or installation/display.
Deadline: March 1999 (to be confirmed). Guidelines available from Tony Dixon.

Development Awards 1999
Awards of £500 are available for projects to be developed to a stage where applications can be made to YHA for short film production funding. An award can be used in any way which advances the project. For example: scriptwriting, research, fees. Applications will be accepted from writers, producers, directors etc. Deadline late 1999 (to be confirmed - contact Tony Dixon for further information).

EUROPEAN AND PAN-EUROPEAN SOURCES

Eurimages
Council of Europe
Palais de l'Europe
67075 Strasbourg Cédex
France
Tel: (33) 3 88 41 26 40
Fax: (33) 3 88 41 27 60
Provides financial support for feature-length fiction films, documentaries, distribution and exhibition. The largest levels of funding are available for feature films (up to £500,000 per film) made by at least three member country partners

European Co-production Association
c/o France 2
22 Avenue Montaigne
75387 Paris
Cédex 08
Tel: (33) 1 4421 4126
Fax: (49) 1 4421 5179
Secretariat: Claire Heinrich
A consortium of European public service TV networks for the co-production of TV fiction series. Can offer complete finance. Development funding is also possible. Proposals should consist of full treatment, financial plan and details of proposed co-production partners. Projects are proposed directly to Secretariat or to member national broadcasters (Channel 4 in UK)

European Co-production Fund
c/o British Screen Finance
14-17 Wells Mews
London W1P 3FL
Tel: 0171 323 9080
Fax: 0171 323 0092
email: sara@britscrn.demon.co.uk
The Fund's aim is to enable UK producers to collaborate in the making of films which the European market demonstrably wishes to see made but which could not be made without the Fund's involvement. The ECF offers commercial loans, up to 30 per cent of the total budget and rarely more than £500,000, for full length feature films intended for theatrical release. The film must be a co-production involving at least two production companies, with no link of common ownership established in separate EU states

FilmFörderung Hamburg
Friedensallee 14-16
22765 Hamburg
Germany

Tel: (49) 40 398 370
Fax: (49) 40 398 3710
Eva Hubert
Producers of cinema films can apply for a subsidy amounting to at most 50 per cent of the overall production costs of the finished film. Foreign producers can also apply for this support. We recommend to co-produce with a German partner. It is necessary to spend at least 150 per cent of the subsidy in Hamburg. Part of the film should be shot in Hamburg. Financial support provided by the FilmFörderung Hamburg can be used in combination with other private or public funding, including that of TV networks

THE MEDIA II PROGRAMME

The MEDIA II Programme is an initiative of the European Union, managed by the European Commission in Brussels, Media II, which follows on from Media I, started in 1996 and will conclude in the year 2000.

European Commission

**Directorate General X:
Information, Communication,
Culture, Audio-visual
rue de la Loi, 200
1040 Brussels, Belgium**
Fax: (32) 2 299 92 14
Head of Programme: Jacques Delmoly

Who is Eligible?
All member states of the European Union and countries belonging to the European Economic Area are eligible for MEDIA II. The Programme is also open to Cyprus, Malta, Central and Eastern European countries subject to special agreements with the Commission.
Objectives
The Commission publishes in the **Official Journal of the European Commission** calls for projects and deadlines for submission for the following areas of support: Training, Development and Distribution.
MEDIA Antennae
At the time of publication MEDIA Antennae Glasgow had taken over the responsibilities from the UK MEDIA Desk as the information centre for European audiovisual funds policy and the MEDIA II Programme. It publishes newsletters, organises information seminars and events to promote increased understanding of the MEDIA II Programme and provide a consultation service to UK professionals about opportunities in Europe.

MEDIA Antenna Glasgow
**c/o Scottish Screen
74 Victoria Crescent Road
Glasgow G12 9JN**
Tel: 0141 334 4445
Fax: 0141 357 2345
email: louise.scott@dial.pipex.com
Contact: Louise Scott

MEDIA Antenna Cardiff
**c/o Sgrîn: The Media Agency for Wales
Llantrisant Road
Llandaff
Cardiff CF5 2PU**
Tel: 01222 578 370
Fax: 01222 578 654
email:antenna@scrwales.demon.co.uk
Contact: Gethin While

Training Support
The training encourages cooperation and exchange of know-how between partners who want to organise courses in initial and continuing training for graduates and professionals already working in the field. Support is in the form of grants.

Development Support
Development support provides loans to get financial and technical assistance for the development of fiction for cinema and television, creative documentaries, animated films, productions using new technologies and productions enhancing the European audiovisual heritage.

Industrial Platforms
Two initiatives were selected in 1996 by the Commission to run the Industrial Platforms for the areas of animation, new technologies and audiovisual heritage. These are respectively:

CARTOON (European Association of Animation Film)
**418 Boulevard Lambermont
1030 Brussels, Belgium**
Tel: (32) 2 245 12 00
Fax: (32) 2 245 46 89
Website: http://www.cartoonmedia.be
Contact: Corinne Jenart, Marc Vandeweyer
Provides financial assistance for graphic research, script writing or adaptation, and pilot film in the animation sector. Successful applicants are invited to become a member of the CARTOON Platform. Guidelines and application forms can be obtained from the UK MEDIA Desk and Antennae

Multimedia Investissements: 2MI
**Avenue de l'Europe, 4
94366 Bry-sur-Marne Cedex
France**

Tel: (33) 1 49 83 28 63
Fax: (33) 1 49 83 26 26
Contact: Jean-Bernard Tellio
email: mmi@2mi.fr
Website: http://www.2mi.com
This platform consists of 25 major European enterprises involved in multimedia and new audiovisual technologies. Its objectives are to co-finance projects, facilitate access to the market and establish networks within the sector to stimulate the production and distribution of European multimedia titles. Guidelines and application forms can be obtained from the MEDIA Antennae Glasgow.

Distribution Support
Distribution scheme aims at improving the transnational distribution of European films (in cinema and on all video formats and television programmes, as well as the promotion of audiovisual works at markets and festivals.

Exhibition Networks
Europa Cinemas
**54 rue Beaubourg
75003 Paris, France**
Tel: (33) 1 42 71 53 70
Fax: (33) 1 42 71 47 55
email: europacinema@magic.fr
Website: http://www.europa-cinemas.org
Claude-Eric Poiroux
Fatima Djoumer
This project encourages screenings and promotion of European films in a network of cinemas in European cities. It offers a financial support according to a certain percentage ie number of screens and the national market share of European films, for promotional activities and special events for European films.

MEDIA Salles
**Via Soperga, 2
20127 Milano, Italy**
Tel: (39) 02 669 4405
Fax: (39) 02 669 1574
Website: http://www.mediasalles.it/mediasalles/
Elisabetta Brunella
MEDIA Salles with Euro Kids Network proposes an initiative aimed at consolidating the availability of 'cinema at the cinema' for children and young people in Europe and at raising the visibility of European products with a younger audience. For further information please call the office in Italy.

INTERNATIONAL SALES

These companies acquire the rights to audiovisual products for sale to foreign distributors in all media
(see also Distributors (Non-Theatrical) and (Theatrical))

Action Time
Wrendal House
2 Whitworth Street West
Manchester M1 5WX
Tel: 0161 236 8999
Fax: 0161 236 8845
Keri Lewis Brown
Specialises in international format
sales of game shows and light
entertainment.

All American Fremantle International
57 Jamestown Road
London NW1 7DB
Tel: 0171 284 0880
Fax: 0171 916 5511
David Champtaloup, Doug Gluck,
Dinah Gray, Jennifer Chrein, Monica
Galer
London arm of NY-based Fremantle
Int. Produces and distributes game
shows and light entertainment
programmes

Allied Vision
The Glassworks
3-4 Ashland Place
London W1M 3JH
Tel: 0171 224 1992
Fax: 0171 224 0111
Peter McRae

Arts Council of England
14 Great Peter Street
London SW1P 3NQ
Tel: 0171 973 6454
Fax: 0171 973 6581
email: richard.gooderick.ace.@
artsfb.org.uk
Richard Gooderick
Arts Council Films is the in-house
distributor of the Arts Council
financed programmes. It also
specialises in the development
funding, and distribution of
innovative, high quality broadcast
programmes by working closely with
independent programme makers and
international broadcasters

August Entertainment
83 Marylebone High Street
London W1M 3DE
Tel: 0171 935 9498
Fax: 0181 935 9486

Eleanor Powell, Justine Griffiths
International sales agent and
packager for independent producers.
Films include: *Ivanhoe; Big Brass
Ring; Sin; Bicycle*

Australian Film Commission
2nd Floor, Victory House
99-101 Regent Street
London W1R 7HB
Tel: 0171 734 9383
Fax: 0171 434 0170
Pressanna Vasudevan
Australian government-funded body
set up to assist in development,
production and promotion of
Australian film, television and video
product

Jane Balfour Films
Burghley House
35 Fortess Road
London NW5 1AQ
Tel: 0171 267 5392
Fax: 0171 267 4241
email: Jbf@janebalfourfilms.co.uk
Jane Balfour, Mary Barlow, Sarah
Banbery, Méabh O'Donovan
International sales agent for
independent producers and some of
Channel 4, output handling drama,
documentaries, and specialised
feature films

BBC Worldwide Television/BBC Worldwide Publishing Woodlands
80 Wood Lane
London W12 0TT
Tel: 0181 576 2000
Fax: 0181 749 0538
Dr John Thomas, Hugh Williams,
Juliet Grimm
BBC Worldwide Television, a key
division of BBC Worldwide Ltd, was
created by the merger of the TV
activities of BBC Enterprises and
the channel businesses of the BBC
World Service Television. In
addition to a range of activity across
emerging technology such as video
on demand and interactive television, the division has two major
areas of operation: Programme, Sales
and Marketing - the sales and

licensing of BBC programmes and
international broadcasters, and the
generation of co-production business;
Channel Marketing - the development
of new cable and satellite delivered
television channels around the world

Beyond Films
3rd Floor
22 Newman Street
London W1V 3HB
Tel: 0171 636 9613
Fax: 0171 636 9614
Dee Emerson
Films: *Strictly Ballroom, Love &
Other Catastrophes, Love Serenade,
Children of the Revolution, Kiss or
Kill, Heaven's Burning, The Last
Bus Home, SLC Punk, Orphans*

The Box Office
3 Market Mews
London W1Y 7HH
Tel: 0171 499 3968
Fax: 0171 491 0008
email: paul@box-
office.demon.co.uk
International film and television
consultancy

bfi BFI Films
21 Stephen Steet
London W1P 2LN
Tel: 0171 957 8927
Fax: 0171 580 5830
Sales and distribution of mainly BFI
Production features and short films
including *Stella Does Tricks; Sixth
Happiness; Love is a Devil; 3 Steps
to Heaven; Made in Heaven;
Robinson in Space* **(see p8)**

British Home Entertainment
5 Broadwater Road
Walton-on-Thames
Surrey KT12 5DB
Tel: 01932 228832
Fax: 01932 247759
email: clive@bhe.prestel.co.uk
Clive Williamson
Video distribution/TV marketing. *An
Evening with the Royal Ballet,
Othello, The Mikado, The Soldier's
Tale, Uncle Vanya, Gulliver's
Travels, King and Country, The
Hollow Crown*

BRITE (British Independent Television Enterprises)

**The London Television Centre
Upper Ground
London SE1 9LT**
Tel: 0171 737 8603
Fax: 0171 928 8476
Nadine Nohr
International programme sales and distribution for Granada Television, London Weekend Television and Yorkshire-Tyne Tees Television. Leading titles include *Cracker, Prime Suspect, Agatha Christie's Poirot and Sherlock Holmes*

Carlton Film Distributors

**127 Wardour Street
London W1V 4AD**
Tel: 0171 437 9020
Fax: 0171 434 3689
Nicole Mackey
A library of 600 feature films plus TV series. Recent product includes *Circle of Friends*

Carlton International Media Limited

**35-38 Portman Square
London W1H 0NU**
Tel: 0171 224 3339
Fax: 0171 486 1707
Rupert Dilnott-Cooper, Louise Sexton, Philip Jones, Anthony Utley
International TV programme and film sales agent, now representing Carlton Television, Central Television, HTV, ITN Productions and Meridian Broadcasting as well as a growing number of independent production companies

Castle Target International

**Colet Court
100 Hammersmith Road
London W6 7JP**
Tel: 0181 974 1021
Fax: 0181 974 2674
Brian Leafe
Buddy's Song, The Monk, That Summer of White Roses, Conspiracy

CBC International Sales

**43-51 Great Titchfield Street
London W1P 8DD**
Tel: 0171 412 9200
Fax: 0171 323 5658
Susan Hewitt, Michelle Payne, Janice Russell
The programme sales division of Canadian Broadcasting Corporation and Société Radio Canada

CBS Broadcast International

**10 Dover Street
London W1X 3PH**
Tel: 0171 355 4422
Fax: 0171 355 4429

Sonja Mendes, Anne Hirsch
Wide range of US TV product

Channel 4 International

**124 Horseferry Road
London SW1P 2TX**
Tel: 0171 396 4444
Fax: 0171 306 8363
email: smowbray@channel4.co.uk
MD, Channel 4 International/Film Four International: Colin Leventhal; Director of Sales, Channel 4 International/Film Four International: Bill Stephens; Programme Sales Manager: Stephen Mowbray
Division of Channel 4 International Ltd dealing with the sales of programmes commissioned by Channel 4 and related activities such as the licensing of books, records and videos

Chatsworth Television Distributors

**97-99 Dean Street
London W1V 5RA**
Tel: 0171 734 4302
Fax: 0171 437 3301
Halina Stratton, Leigh Collins
Extensive library of documentary and special interest films. Also Chatsworth-produced light entertainment, drama and adventure series (*The Crystal Maze* and *Treasure Hunt*)

CiBy Sales

**10 Stephen Mews
London W1P 1PP**
Tel: 0171 333 8877
Fax: 0171 333 8878
Wendy Palmer, Fiona Mitchell, Francois Thos
Established in 1992, CiBy Sales is responsible for the international multi-media exploitation of films produced by French production company Ciby 2000 and other independent producers. Titles include: *Muriel's Wedding, The Piano, Secrets and Lies*

Cine Electra

**National House
60-66 Wardour Street
London W1V 3HP**
Tel: 0171 287 1123
Fax: 0171 722 4251
Julia Kennedy
Established in 1991, Cine Electra has continued to diversify its activities within international sales and production. From its acquisition, the company now represents a library of 150 titles including such directors as Peter Greenaway, Andrzej Wajda, Jacques Rivette and Idrissa Oucdraogo, as well as a comprehensive selection of shorts and documentaries

Circle Communications PLC

**45-49 Mortimer Street
London W1N 7TD**
Tel: 0171 636 9421
Fax: 0171 436 7426
Circle is an international television rights group. Based in the UK, and trading in all the major territories of the world, Circle provides a range of services for producers and broadcasters. Circle comprises distinct businesses principally engaged in the creation, acquisition, marketing and licensing of visual entertainment rights. The companies within Circle Communications are:
Carnival (Films & Theatre)
Pavilion International
Delta Ventures
Production Finance & Management
Independent Wildlife
Harlequin Films & Television
Oxford Scientific Films
La Plante International

Columbia TriStar International Television

**Sony Pictures, Europe House
25 Golden Square
London W1R 6LU**
Tel: 0171 533 1000
Fax: 0171 533 1246
Exec. V.P, European operations: John McMahon
Leslie Tobin Bacon, Senior V.P, General Manager, London
Lauren Cole, Senior V.P, International Network
European TV production and network operations and international distribution of Columbia TriStar's feature films and TV product

CTVC

**Hillside Studios
Merry Hill Road
Bushey
Watford WD2 1DR**
Tel: 0181 950 4426
Fax: 0181 950 6694
Ann Harvey
International programme sales and co-productions in documentary, music, children's, drama and arts programmes

The Walt Disney Television International

**3 Queen Caroline Street
Hammersmith
London W6 9PA**
Tel: 0181 222 1000
Fax: 0181 222 2795
MD: Etienne de Villiers
VP, Sales & Marketing: Keith Legoy
International television arm of a major US production company

DLT Entertainment UK Ltd
10 Bedford Square
London WC1B 3RA
Tel: 0171 631 1184
Fax: 0171 636 4571
John Reynolds; Martin Booth
Specialising in entertainment
programming. Recent titles include:
As Time Goes By, series seven for
BBC Television; *Bloomin' Marvel-
lous*, eight-part comedy series for
BBC Television

Documedia International Films Ltd
19 Widegate Street
London E1 7HP
Tel: 0171 625 6200
Fax: 0171 625 7887
Distributors of innovative and award
winning drama specials, drama
shorts, serials, tele-movies and
feature films; documentary specials
and series; for worldwide sales and
co-production. Library includes:
Drama serials - *Short Breaks* (first
and second series of 6 10 minutes),
Playhouse Pictures (first and second
series of 6 half hours), *Baker Street
Boys* (8 half hours and 4 hours), *Box
of Delights* (6 half hours and 3
hours), *Edge of Night* (13 half
hours); Drama specials - *Soulscapes*;
Telemovies/Features - *Deva's Forest,
Leaves and Thorns*; International
short drama - *Pile of Clothes,
JoyRidden, The Cage, Nazdrovia,
Thin Lines, Late Fred Morse, The
Extinguisher, The Summer Tree,
Arch Enemy, Tea and Bullets,
Isabelle, Peregrine, Beyond Reach,
Trauma, Edge of Night, Something
Wonderful*; Documentary series -
Horses of the World (8 hours), *Room
of Dreams* (4 half hours), *Open
Door on Photography* (13 half
hours), *History of European Art* (13
half hours), *In the Wild* (4 hours and
7 hours); Documentary specials -
*Positive Story, Caste in Half,
Strength and the Wager, Poison
People, Rock N Roll Juvenile, Over
the Wall in China, Graham Knuttel,
Dave Shepherd*. Represented
worldwide for sales to all media,
including internet/video on demand
and all television

EVA Entertainment
7a Langley Street
Covent Garden
London WC2H 9JA
Tel: 0171 836 3000
Fax: 0171 836 3300
email: info@eva.co.uk

Film Consultancy Services
19 Penfields House

Market Road
London N7 9PZ
Tel: 0171 700 7055
Fax: 0171 700 7055
Brian Sammes

Film Four International
124 Horseferry Road
London SW1P 2TX
Tel: 0171 306 8602
Fax: 0171 306 8361
Director of Sales, Channel 4
International/Film Four Interna-
tional: Bill Stephens; Heather
Playford-Denman
Film sales arm of Channel 4
International Ltd, selling feature
films which it finances or part-
finances. Recent titles include:
*Nothing Personal, Bandit Queen,
Carla's Song, Institute Benjamenta,
Shallow Grave, Sister My Sister,
Trainspotting* and *Beautiful Thing*

Goldcrest Films and Television
65/66 Dean Street
London W1V 5HD
Tel: 0171 437 8696
Fax: 0171 437 4448
Thierry Wase-Bailey
Major feature film production, sales
and finance company. Recent films
include *All Dogs Go to Heaven,
Black Rainbow, Rock-A-Doodle, The
Harvest, Me* and *Veronica, Painted
Heart*

The Samuel Goldwyn Company
St George's House
14-17 Wells Street
London W1P 3FP
Tel: 0171 436 5105
Fax: 0171 580 6520
Betsy Spanbock, Katerina Mattingley
Acquisition, development, sales,
distribution and marketing of films
and television product worldwide

Grampian Television
Queen's Cross
Aberdeen AB15 4XJ
Tel: 01224 846846
Fax: 01224 846800
Alistair Gracie (Controller) Hilary I.
Buchan (Head of Public Relations)
North Scotland ITV station
producing a wide range of program-
ming including documentaries,
sport, children's, religion and
extensive daily news and current
affairs to serve ITV's largest region.
Also making a mark in successful
network programmes such as The
National TV Awards. Regional
highlights include North Tonight,
the daily news magazine, the
environmental programme *Nature's
Prize*, nostalgia show *The Way It
Was*, travel series *Walking Back to*

Happiness, the interactive courtroom
debate *We the Jury* and a range of
Gaelic programming

HIT Entertainment
The Pump House
13-16 Jacob's Well Mews
London W1H 5PD
Tel: 0171 224 1717
Fax: 0171 224 1719
Peter Orton - Managing Director
Charlie Caminada - Sales Director

HIT Entertainment USA inc.
218 North Canon Drive
Suite A
Beverly Hills, Ca 90210
Dorian Langdon - Vice President
Distributors of children's, family and
natural history programming
including *Dennis and Gnasher,
Barney and Friends, The Wind in
the Willows and the Willows in
Winter, The World of Peter Rabbit
and Friends, Shakespeare - The
Animated Tales, TVNZ Library -
SPP, and Wild Media - TVNZ
Natural History, Partridge Films,
Scandinature Films*

Hemdale Communications
10 Avenue Studio
Sydney Close
Chelsea
London SW3 6HW
Tel: 0171 581 9734
Fax: 0171 581 9735
John Smallcombe
Titles include *Little Nemo -
Adventures in Slumberland, Quest of
the Delta Knights, The Legend of
Wolf Mountain, Savage Land* and
The Magic Voyage

Hollywood Classics
8 Cleveland Gardens
London W2 6HA
Tel: 0171 262 4646
Fax: 0171 262 3242
email:HollywoodClassicUK@
compusrve.com
Melanie Tebb
Hollywood Classics has offices in
London and Los Angeles and sells
back catalogue titles from major
Hollywood studios for theatrical
release in all territories outside
North America. Also represents an
increasing library of European and
independent American titles and has
all rights to catalogues from various
independent producers

Icon Entertainment International
37 Soho Square
London W1V 5DG
Tel: 0171 543 4300
Fax: 0171 543 4301
Ralph Kamp: Chief Executive

Jamie Carmichael: Head of Sales
Michaela Piper: Marketing Manager

ITC Entertainment Group

33 Foley Street
London W1P 7LB
Tel: 0171 255 3000
Fax: 0171 306 7800
Distributors of *Royce, Second
Chances, The Last Seduction,
Trouble Bound, When Love Kills,
Thunderbirds, Captain Scarlet,
Randall and Hopkirk (Deceased)*

ITEL

48 Leicester Square
London WC2H 7FB
Tel: 0171 491 1441
Fax: 0171 493 7677
CEO: Andrew Macbean; Director of
Programming: Paul Sowerbutts;
Director of Sales: Joel Denton
International television sales
company owned by MAI and Home
Box Office Inc. Represents Court
TV and other independent producers

J & M Entertainment

2 Dorset Square
London NW1 6PU
Tel: 0171 723 6544
Fax: 0171 724 7541
Julia Palau, Michael Ryan,
Anthony Miller
Specialise in sales of all media,
distribution and marketing of
independent feature films. Recent
films include: *Mute Witness, Night
Watch, The Road to Wellville,
Princess Caraboo, Sugar Hill,
T Rex, What's Eating Gilbert Grape?*

Kushner Locke International

83 Marylebone High Street
London W1M 3DE
Tel: 0171 935 9498
Fax: 0171 935 9486
Company lead by Donald Kushner
and Peter Locke has expanded its
existing successful television
production and distribution
activities into international theatrical
feature films, taking on Gregory
Cascante and Eleanor Powell,
partners of August Entertainment.
Products include: *Pinocchio*
(directed by Steve Barrow); *Basil;
Double Tap; Hell's Kitchen; Whole
Wide World*

Link Entertainment

7 Baron's Gate
33-35 Rothschild Road
Chiswick
London W4 5HT
Tel: 0181 996 4800
Fax: 0181 747 9452
email: info@linklic.demon.co.uk
Claire Derry, David Hamilton, Jo

Kavanagh-Payne
Specialists in children's programmes
for worldwide distribution and
character licensing. New properties
include: *Chatterhappy Ponies; The
Forgotten Toys Series; Pirates
Series III; Caribou Kitchen Series
III; The First Snow of Winter*

London Films

35 Davies Street
London W1Y 1FN
Tel: 0171 499 7800
Fax: 0171 499 7994
Andrew Luff
Founded in 1932 by Alexander
Korda, London Films is renowned
for the production of classics. Co-
productions with the BBC include
Poldark and I Claudius. More recent
series include *Lady Chatterley*
directed by Ken Russell and *Resort
to Murder*

London Television Service

Hercules House
Hercules Road
London SE1 7DU
Tel: 0171 261 8592
Fax: 0171 928 5037
Jackie Huxley
LTS is a specialist production and
distribution organisation that
handles the promotion and marketing of British documentary and
magazine programmes worldwide to
television, cable, satellite and non-
broadcast outlets. The flagship
science and technology series
Perspective has sold to television in
over 100 countries

Majestic Films & Television International

PO Box 13
Gloucester Mansions
Cambridge Circus
London WC2H 8XD
Tel: 0171 836 8630
Fax: 0171 836 5819
Guy East
Organises finance, sales, distribution
and marketing of feature films and
television productions throughout
the world. Recent titles include
*Immortal Beloved, Into the West,
Jane Eyre, The Man Without a Face,
A Man of No Importance*

MCA TV

1 Hamilton Mews
London W1V 9FF
Tel: 0171 491 4666
Fax: 0171 493 4702
Roger Cordjohn, Penny Craig
UK operation for the major US
corporation which owns Universal
Pictures

National Film Board of Canada

1 Grosvenor Square
London W1X 0AB
Tel: 0171 258 6484
Fax: 0171 258 6532
Jane Taylor
European sales office for documen-
tary, drama and animation produc-
tions from Canada's National Film
Board

NBD Television

Unit 2, Royalty Studios
105 Lancaster Road
London W11 1QF
Tel: 0171 243 3646
Fax: 0171 243 3656
Nicky Davies Williams, Charlotte
Felia, Carolyne Waters
Company specialising in music and
light entertainment. Clients include
Channel 4, Warner Bros Records,
BMG, PolyGram, Channel X, MPL,
Island Visual Arts, Celador, Talk
Back, Tiger Aspects and Fujisankei
Communications Inc.

Orbit Media Ltd

7-11 Kensington High Street
London W8 5NP
Tel: 0171 221 5548
Fax: 0171 727 0515
Chris Ranger, Jordan Reynolds
Specialises in vintage product from
the first decade of American TV:
The Golden Years of Television and
65 x 30 mins Series NoireTV series

Paramount Television

49 Charles Street
London W1X 8LU
Tel: 0171 629 1150
Fax: 0171 491 2086
Patrick Stambough

Pearson Television International

1 Stephen Street
London W1P 1PJ
Tel: 0171 691 6000
Fax: 0171 691 6060
Managing Director: Brian Harris
Represents worldwide sales efforts
for the companies under the Pearson
Television International marque,
including, ACI, Grundy, All
American and Thames, as well as
WizEnd and Alomo programmes.
Catalogue includes: (ACI) Volume 8
films, including, *First Do No Harm*,
starring Meryl Streep; *Love Kills*.
(Grundy) *Neighbours; Shortland
Street*. (Thames) *Homicide: Life on
the Street; Men Behaving Badly* (US
and UK versions), *Cull Red*.
(WizEnd, Alomo programmes) *Pie
in the Sky; Goodnight Sweetheart*

Photoplay Productions

21 Princess Road

London NW1 8JR
Tel: 0171 722 2500
Fax: 0171 722 6662
Kevin Brownlow, David Gill, Patrick
Stanbury
European dealer for the Blackhawk
16mm library of silent and early
sound films

Picture Music International
EMI House
43 Brook Green
London W16 7EF
Tel: 0171 605 5000
Fax: 0171 605 5050
Dawn Stevenson, Caroline Dare
Music concert and documentary
production and distribution for
television and video - rock, pop,
classical, entertainment, jazz and blues

PolyGram Film International
Oxford House
76 Oxford Street
London W1N 0H9
Tel: 0171 307 1300
Fax: 0171 307 1301
President: Aline Perry; Vice
President Distribution: Jam
Verheyen
Recent titles include: *Dead Man
Walking, Fargo, Sleepers, Portrait
of a Lady, The Game, Bean, The
Borrowers*

Portman Entertainment Ltd
167 Wardour Street
London W1V 3TA
Tel: 0171 468 3443
Fax: 0171 468 3469
Jane Baker
International feature film sales
division of the Portman Entertain-
ment Group (incorporating activities
previously handled by both Pinnacle
Pictures and Global Television).
Handles all media sales of both
Portman's own productions as well
as pre-financing and acquiring
feature films and television
programming from independent
producers for worldwide sales.
Recent titles include *Spanish Fly,
Wrestling with Alligators, Via
Satellite, Mayday, China Dream* and
Coming Home

Primetime Television Associates
Seymour Mews House
Seymour Mews
Wigmore Street
London W1H 9PE
Tel: 0171 935 9000
Fax: 0171 935 1992
Simon Willock, Richard Leworthy
Production and distribution.
Production includes: *Nicholas
Nickleby, Porgy and Bess, Great
Expectations, Othello*. Distribution

includes: *Home and Away, 99-1, In
the Wild, Finney, Our Friends in the
North, Bodyguards, Famous Five,
Eyes of the World*

Red Rooster Film & Television Entertainment
29 Floral Street
London WC2E 9DP
Tel: 0171 379 7727
Fax: 0171 379 5756
Linda James
Feature film producers and
distributors of quality television
fiction: *Wycliffe, Smokescreen, Body
& Soul, The Life & Times of Henry
Pratt, The Gift, Kersplat!, Coming
Up Roses, Just Ask for Diamond,
Crocodile Shoes, Samson Superslug.*
Titles include: *Wilderness, The
Sculptress, Body & Soul, The Gift,
Samson Superslug, Coming Up
Roses, Just Ast For Diamond*

Reuters Television
85 Fleet Street
London EC4P 4AJ
Tel: 0171 250 1122
Fax: 0171 542 4995
Distribution of international TV
news and sports material to
broadcasters around the world

RM Associates
46 Great Marlborough Street
London W1V 1DB
Tel: 0171 439 2637
Fax: 0171 439 2316
Sally Fairhead
In addition to handling the exclusive
distribution of programmes
produced/co-produced by RM Arts,
RM Associates works closely with
numerous broadcasters and
independent producers to bring
together a comprehensive catalogue
of music and arts programming

S4C
Parc Ty Glas
Llanishen
Cardiff CF4 5DU
Tel: 01222 747444
Fax: 01222 754444
Chairman: Prys Edwards
Chief Executive: Huw Jones
Director of Production: Huw Eirug
Director of Commercial Affairs:
Wyn Innes
For International Sales and Distribu-
tion please contact S4C International
Properties
Testament - The Bible in Animation
9x30'
Co-production between S4C, BBC
and Christmas Films. Produced by
Cartwn Cymru, Right Angle and
Christmas Films
Animated stories from the Bible

*Saints and Sinners - The History of
the Popes* 6x50'
Co-production with RTE and La5
Production company: Opus 30
An insight into the power of the
Papacy, of Popes past and present
Ancient Egypt 5x60'
Co-production with Discovery
Communications and La 5
Production company: John Gwyn
An indepth journey through a
dynasty that lasted 6,000 years
The Making of Maps 1x99'
Co-production with BBC + British
Screen
Production company: Gaucho
Film set in Wales during the Cuban
Missile Crisis of 1962 featuring the
loss of childhood innocence
Famous Fred 1x30'
A co-production between S4C,
Channel 4 and TVC
Production company -TVC
The fabulous furry adventures of
Fred - the home-loving cat who turns
into a rock star glamourpuss at night
Cameleon 1x120'
Production Company - Elidir Films
An idealistic, wayward young man
joins the army in search of a more
exciting life but goes absent without
leave after it fails to meet his
expectations
The Jesus Story 1 x 90' or 4x30'
Co-production with BBC +
Christmas Films + British Screen
Animated full length feature film
from the makers of *'Testament',
Shakespeare The Animated Tales
and 'Operavox The Animated
Operas'*
The Heather Mountain 1x80'
Production Compay - Llun y Felin
Productions
A feature film version of a classic
Welsh story follows the story of a
young girl growing up in North
Wales at the turn of the Century
Wild Islands 24x30' or 8x50' + 50'
Raptor special
Co-production with STE + RTE
Production Company - Performance
Films, Telesgop and Éamon de
Buitléar
A unique insight into the wildlife
and natural habitat of the national
regions which make up Britain and
Ireland

S4C International
Parc Ty Glas
Llanishen
Cardiff CF4 5DU
Tel: 01222 747444
Fax: 01222 754444
Rhianydd Darwin/Teleri Roberts/
Helen Howells
Distribute programmes plus co-

productions commissioned by S4C from independent producers - animation, drama, documentaries

Safir Films Ltd
49 Littleton Rd
Harrow
Middx HA1 3SY
Tel: 0181 423 0763
Fax: 0181 423 7963
email: Isafia@ibm.net
Lawrence Safir
Hold rights to numerous Australian, US and UK pictures, including Sam Spiegel's *Betrayal*

The Sales Company
62 Shaftesbury Avenue
London W1V 7DE
Tel: 0171 434 9061
Fax: 0171 494 3293
Alison Thompson, Rebecca Kearey and Joy Wong. The Sales Company is owned by British Screen, BBC Worldwide and Zenith Productions and handles international sales for their films, for all rights. Recent films include: *The Snapper, Priest, Butterfly Kiss, Antonia's Line, Stonewall, Land and Freedom, Small Faces, The Van, Mojo and I Went Down*. Also occasionally handles product from the international arena including *Safe, La Seconda Volta, Jerusalem* and *Private Confessions*

Scottish Television International
Cowcaddens
Glasgow G2 3PR
Tel: 0141 300 3000
Fax: 0141 300 3256
Ian Jones, Director
Anita Cox, Teleri Roberts
Sales and Marketing of STV, Grampian and Third Party Programming Worldwide, including: *McCallum, Taggart, Blobs, Hot Rod Dogs*

Screen Ventures
49 Goodge Street
London W1P 1FB
Tel: 0171 580 7448
Fax: 0171 631 1265
email: screenventures@easynet. co.uk
Christopher Mould, Mike Evans
Specialise in international film, TV and video licensing of music, drama and arts. Worldwide television sales agents for international record companies and independent producers. Screen Ventures is also an independent producer of television documentaries and music programming

Smart Egg Pictures
11&12 Barnard Mews

Barnard Road
London SW11 1QU
Tel: 0171 924 6284
Fax: 0171 924 5650
Tom Sjoberg, Judy Phang
Independent foreign sales company. Titles include *Spaced Invaders, Dinosaurs, Montenegro, The Coca-Cola Kid, Rave Dancing to a Different Beat, Phoenix and the Magic Carpet* and *Evil Ed*

Stranger Than Fiction Ltd
Suite 217
Golden House
29 Great Pulteney Street
London W1R 3DD
Tel: 0171 734 5489
Fax: 0171 734 5490
email: Gracarley@aol.com
Grace Carley
Financing and sales of low-budget cult-type movies, including *Deadline* (Anders Palm), *Meet the Feebles* (1989), Brain Dead (1992) and *Heavenly Creatures* (1993), all from Peter Jackson. Boutique-Style Sales Agency Dealing Primarily in Arthouse Features from the UK, US and Ireland

TCB Releasing
Stone House, Rudge
Frome
Somerset BA11 2QQ
Tel: 01373 830769
Fax: 01373 831028
Angus Trowbridge
Sales of jazz and blues music programmes to broadcast television and the home-video media

Trans World International
TWI House
23 Eyot Gardens
London W6 9TR
Tel: 0181 233 5000
Fax: 0181 233 5401
Eric Drossart, Bill Sinrich, Buzz Hornett
The world's largest independent producer and distributor of sports programmes, TWI is owned by Mark McCormack's IMG Group and specialises in sports and arts programming. Titles include: *Trans World Sport, Futbol Mundial, PGA European Tour productions, ATP Tour highlights, West Indies Test Cricket, Oddballs, A-Z of Sport, Goal!, The Olympic Series, Century* and *The Whitbread Round The World Race*

Turner International Television Licensing
CNN House
19 Rathbone Place
London W1P 1DF

Tel: 0171 637 6900
Fax: 0171 637 6925
Ross Portugeis
US production and distribution company of films and programmes from Hanna-Barbera (animation), Castle Rock, New Line, Turner Pictures Worldwide, World Championship Wrestling, Turner Original Productions (non-fiction), plus a library of over 2,500 films, 1,500 hours of television programmes and 1,000 cartoons from the MGM (pre-1986) and Warner Bros (pre-1950) studios

Twentieth Century Fox Television
31-32 Soho Square
London W1V 6AP
Tel: 0171 437 7766
Fax: 0171 439 1806
Stephen Cornish, Vice President
Randall Broman, Director of Sales
TV sales and distribution. A News Corporation company

UGC/UK
167-169 Wardour Street
London W1V 3TA
Tel: 0171 413 0838
Fax: 0171 734 1509
MD: Ralph Kamp; Television: Alison Trumpy; Theatrical: Jamie Carmichael
International sales of library, video and feature films. Also partake in feature film production: *Fresh, Six Days Six Nights, Somebody to Love, The City of Lost Children, Haunted, Leaving Las Vegas, The Rickshaw Boy*

VCI Programme Sales
VCI
76 Dean Street
London W1V 5HA
Tel: 0171 396 8888
Fax: 0171 396 8890
Paul Hembury
A wholly owned subsidiary of VCI PLC, responsible for all overseas activities. Distributes a wide variety of product including music, sport, children's, fitness, documentary, educational, special interest and features

Victor Film Company Ltd
2b Chandos Street
London W1M 9DG
Tel: 0171 636 6620
Fax: 0171 636 6511
Alasdair Waddell, Vic Bateman, Carol Philbin, Alexandra Roper, Eliana Celiberti
International sales agent for independent producers of commercial films. Titles include: *Clockwork Mice, Killing Time, Darklands, Aberration, Preaching to the Perverted*

Vine International Pictures

21 Great Chapel Street
London W1V 3AQ
Tel: 0171 437 1181
Fax: 0171 494 0634
Marie Vine, Barry Gill
Sale of feature films such as
*Rainbow, The Pillow Book, The Ox
and the Eye, Younger and Younger,
The Prince of Jutland, Erik the
Viking, Let Him Have It, Trouble in
Mind*

Warner Bros International Television

135 Wardour Street
London W1V 4AP
Tel: 0171 494 3710
Fax: 0171 287 9086
Richard Milnes, Donna Brett, Tim
Horan, Ian Giles
TV sales, marketing and distribu-
tion. A division of Warner Bros
Distributors Ltd, A Time Warner
Entertainment Company, LP

Worldwide Television News Corporation (WTN)

The Interchange
Oval Road, Camden Lock
London NW1 7EP
Tel: 0171 410 5200
Fax: 0171 413 8327 (Library)
Gerry O'Reilly, David Simmons
International TV news, features,
sport, entertainment, documentary
programmes and archive resources.
Camera crews in major global
locations, plus in-house broadcasting
and production facilities

Yorkshire-Tyne Tees Enterprises

15 Bloomsbury Square
London WC1A 2LJ
Tel: 0171 312 3700
Fax: 0171 312 3777
Sarah Doole, Susan Crawley, Ann
Gillham
International sales division of
Yorkshire-Tyne Tees TV

LABORATORIES

Bucks Laboratories Ltd
714 Banbury Avenue
Slough
Berks SL1 4LR
Tel: + 44 (0) 1753 576611
Fax: + 44 (0) 1753 691762
Contact: Darren Fagg, Harry F. Rushton
Comprehensive lab services in Super 35mm and 35mm, Super 16mm and 16mm, starting Sunday night. West End rushes pick up unit 10.30 pm. Also day bath. Chromakopy: 35mm low-cost overnight colour reversal dubbing prints. Photogard: European coating centre for negative and print treatment. Chromascan: 35mm and 16mm video to film transfer

Colour Film Services Group
10 Wadsworth Road
Perivale
Middx UB6 7JX
Tel: 0181 998 2731
Fax: 0181 997 8738
Film Laboratory: full 16mm and 35mm colour processing laboratory, with Super 16mm to 35mm blow up a speciality. Video Facility: broadcastt standard wet gate telecines and full digital edit suite. Video duplication, CD mastering and archiving to various formats. Superscan: unique tape to film transfer system in both Standard Resolution and High Resolution. Sounds Studios: analogue and digital dubbing, track laying, synching, voice overs and optical transfer bay

Colour-Technique
Cinematograph Film Laboratories
Finch Cottage, Finch Lane

Knotty Green
Beaconsfield HP9 2TL
Tel: 01494 672757
Specialists in 8mm, Super 8mm and 9.5mm blown up to 16mm with wet gate printing. Stretch printing 16 and 18 Fps to 24, 32 and 48 Fps. 16mm to 16mm optical copies with wet gate and stretch printing. World leader for archival film copying for 8mm, Super 8mm, 9.5mm and 16mm with wet gate printing from old shrunk films, B/w dupe negs and colour internegs. Also Super 8mm blown up to Super 16mm wet gate printing and stretch printing. 16mm to Super 16mm wet gate and stretch printing. Colour internegs and B&W dupe negatives. Super 8mm blown to 35 mm

East Anglian Film Archive
University of East Anglia
Norwich NR4 7TJ
Tel: 01603 592664
Fax: 01603 458553
Specialises in blow-up printing of Std 8mm, Super 8mm and 9.5 mm, b/w or colour, onto 16mm. Freeze frame available

Film and Photo Ltd
13 Colville Road
South Acton Industrial Estate
London W3 8BL
Tel: 0181 992 0037
Fax: 0181 993 2409
email: film-photo@demon.co.uk
Tony Scott
Post production motion picture laboratory. 35/16mm Colour & B/W reversal dupes. Tape to film transfers - Optical effects. Nitrate restoration/

preservation

The Film Clinic (Incorporating Les Latimer Optical Printing)
8-14 Meard Street
London W1V 3HR
Tel: 0171 734 9235
Fax: 0171 734 9471
Ray Slater, John Sears, Peter Davidson
Nitrate film handling (licensed nitrate vaults), examination and repair work, full optical printing services, specialised wet gate printing of shrunken and damaged film, scratch elimination treatment, ultrasonic cleaning, Vacuumate protection treatment, re-dimension treatment

Film Lab North Ltd
Croydon House
Croydon Road
Leeds LS11 9RT
Tel: 0113 243 4842
Fax: 0113 2434323
Full service in 16mm colour Negative Processing, 16mm colour printing, 35mm colour printing video transfer. Super 16mm a speciality - Plus 35mm colour grading and printing

The Film Treatment Centre (at The Machine Room)
54-58 Wardour Street
London W1V 3HN
Tel: 0171 734 3433
Fax: 0171 287 3773
David Atkinson, Paul Robinson
Scratch treatment, ultrasonic film cleaning, 35-16-Super 16-8 - Super

8, film examination and repair, perforation repair, nitrate film handling, nitrate storage vaults, 35mm/16mm mag track cleaning/renovation, FACT accredited

Filmatic Laboratories/Filmatic Television
16 Colville Road
London W11 2BS
Tel: 0171 221 6081
Fax: 0171 221 2718
Complete Super 16 and 16mm film processing laboratory and sound transfer service with full video post-production facility including Digital Wet Gate Telecines, D3, Digital Betacam, 1", Betacam SP and other video formats. On-line editing, duplication and standards conversion. Sync sound and A+B roll negative to tape transfer, Electronic Film Conforming (EFC), the system that produces the highest quality video masters from any original source, with frame accurate editing from film, non-linear disc or off-line video edit

London Filmmakers' Co-op
The Lux Centre
2-4 Hoxton Square
London N1 6NU
Tel: 0171 684 0202
Fax: 0171 684 2222
Paul Murray
16mm b/w printing and processing

The Machine Room Ltd
54-58 Wardour Street
London W1V 3HN
Tel: 0171 734 3433
Fax: 0171 287 3773
Contact: David Atkinson
Rank Cintel dry/wetgate telecine suites for transfer of 35mm, 16mm, Super 16 and Super 8. To all analogue and digital video tape formats. Digital online edit suite. Broadcast Standards conversion via alhemist or vistek vmc. Video duplication. Full film cleaning and treatments (see the Film Treatment Centre) Sponsor Members focal, FACT accredited, (Nitrate handling)

Metrocolor London
91-95 Gillespie Road
Highbury
London N5 1LS
Tel: 0171 226 4422
Fax: 0171 359 2353
Len Brown, Terry Lansbury,
Alan Douglas
Offers complete service for features, commercials, television productions and pop promos for 16mm, Super 16mm, 35mm and Super 35mm. Day

and night processing and printing colour, b/w and vnf. Overnight rushes and sound transfer. Overnight 'best-light' and 'gamma' Telecine rushes transfer and sync sound. Computerised logging and negative matching. Sound transfer to optical negative - Dolby stereo, Dolby SRD Digital stereo and DTS Timecode. Specialist Super 16mm services include: 35mm fully graded blow-up prints; 35mm fully graded blow-up immediates; fully graded prints re-formatted to standard 16mm retaining 1.66:1 aspect ratio

Rank Film Laboratories Group
North Orbital Road
Denham, Uxbridge
Middx UB9 5HQ
Tel: 01895 832323
Fax: 01895 832446
David Dowler
Part of the Rank Film Laboratories Group which includes Deluxe Hollywood and Deluxe in Toronto. Comprehensive worldwide laboratory services to the motion picture, commercials and television industries. Denham and Deluxe laboratories also include video transfer suites. Deluxe in Toronto includes complete sound mixing and dubbing suites. The Leeds Laboratory offers a full 16mm service. Also part of the group is well known special effects and optical house General Screen Enterprises

Soho Images
8-14 Meard Street
London W1V 3HR
Tel: 0171 437 0831
Fax: 0171 734 9471
Soho Laboratories offer day and night printing and processing of 16mm (including Super 16mm) and 35mm colour or b/w film

Technicolor Film Services
Technicolor Ltd
Bath Road
West Drayton
Middx UB7 0DB
Tel: 0181 759 5432
Fax: 0181 759 6270
West End pick-up and delivery point:
Goldcrest Ltd
1 Lexington Street
London W1R 3HP
Tel: 0171 439 4177
A 'Technicolor' logo in the end credits has always been synonymous with high quality film processing. For almost 60 years Technicolor has been at the forefront of film handling technology. A 24 hours-a-day service in all film formats; Europe's leading 65/70m laboratory facility (with specialist support for

large 'space theatre' formats) and a comprehensive sound transfer service, are highlights of Technicolor's broad based package. The laboratory is fully equipped to make SRD, SDDS and DTS prints too

LEGISLATION

This section of the Handbook has a twofold purpose, first to provide a brief history of the legislation relating to the film and television in the United Kingdom, and second to provide a short summary of the current principal instruments of legislation relating to film, television and video industries in the United Kingdom and in the European Community. Current legislation is separated into four categories: cinema and broadcasting; finance; copyright and European Union legislation. This section was compiled by Michael Henry of solicitors Henry Hepworth whose continued support we gratefully acknowledge

LEGISLATIVE HISTORY

Cinema

Legislation for the cinema industry in the United Kingdom goes back to 1909, when the Cinematograph Act was passed providing for the licensing of exhibition premises, and safety of audiences. The emphasis on safety has been maintained through the years in other enactments such as the Celluloid and Cinematograph Film Act 1922, Cinematograph Act 1952 and the Fire Precautions Act 1971, the two latter having been consolidated in the Cinemas Act 1985.

The Cinematograph Films (Animals Act) 1937 was passed to prevent the exhibition and distribution of films in which suffering may have been caused to animals. The Cinematograph (Amendment) Act 1982 applied certain licensing requirements to pornographic cinema clubs. Excluded from licensing were the activities of bona fide film societies and 'demonstrations' such as those used in shops, as well as exhibitions intended to provide information, education or instruction. Requirements for licensing were consolidated in the Cinemas Act 1985.

The Sunday Entertainments Act 1932 as amended by the Sunday Cinema Act 1972 and the Cinemas Act 1985 regulated the opening and use of cinema premises on Sundays.

The Sunday Entertainments Act 1932 also established a Sunday Cinematograph Fund for 'encouraging the use and development of cinematograph as a means of entertainment and instruction'. This was how the British Film Institute was originally funded.

Statutory controls were imposed by the Cinematograph Films Act 1927

in other areas of the film industry, such as the booking of films, quotas for the distribution and renting of British films and the registration of films exhibited to the public. This Act was modified by the Cinematograph Films Acts of 1938 and 1948 and the Film Acts 1960, 1966, 1970 and 1980 which were repealed by the Films Act 1985.

The financing of the British film industry has long been the subject of specific legislation. The National Film Finance Corporation was established by the Cinematograph Film Production (Special Loans) Act 1949. The Cinematograph Film Production (Special Loans) Act 1952 gave the National Film Finance Corporation the power to borrow from sources other than the Board of Trade. Other legislation dealing with film finance were the Cinematograph Film Production (Special Loans) Act 1954 and the Films Acts 1970 and 1980. The Cinematograph Films Council was established by the Cinematograph Films Act 1948, but like the National Film Finance Corporation, the Council was abolished by the Films Act 1985.

The Cinematograph Films Act 1957 established the British Film Fund Agency and put on a statutory footing the formerly voluntary levy on exhibitors known as the 'Eady levy'. Eady money was to be paid to the British Film Fund Agency, which in turn was responsible for making payments to British film-makers, the Children's Film Foundation, the National Film Finance Corporation, the British Film Institute and towards training film-makers. The Film Levy Finance Act 1981 consolidated the provisions relating to the Agency and the exhibitors' levy. The Agency was wound up in 1988 pursuant to a statutory order made under the Films Act 1985.

The British Film Institute used to obtain its funding from grants made by the Privy Council out of the Cinematograph Fund established under the Sunday Entertainments Act 1932 and also from the proceeds of subscriptions, sales and rentals of films. The British Film Institute Act 1949 allows for grants of money from Parliament to be made to the British Film Institute as the Lord President of the Privy Council thinks fit.

Broadcasting

The BBC first started as the British Broadcasting Company (representing the interests of some radio manufacturers) and was licensed in 1923 by the Postmaster General under the Wireless Telegraphy Act 1904 before being established by Royal Charter. The company was involved in television development from 1929 and in 1935 was licensed to provide a public television service.

The Independent Television Authority was established under the Television Act 1954 to provide additional television broadcasting services. Its existence was continued under the Television Act 1964 and under the Independent Broadcasting Act 1973, although its name had been changed to the Independent Broadcasting Authority by the Sound Broadcasting Act 1972 (which also permitted it to provide local sound broadcasting services).

The Broadcasting Act 1981 amended and consolidated certain provisions contained in previous legislation including the removal of the prohibition on certain specified people from broadcasting opinions expressed in proceedings of Parliament or local authorities, the extension of the IBA's functions to the provision of programmes for Channel 4 and the establishment of the Broadcasting Complaints Commission.

Cable programme services and satellite broadcasts were the subject of the Cable and Broadcasting Act 1984. This Act and the Broadcasting Act 1981 were repealed and consolidated by the Broadcasting Act 1990 which implemented proposals in the Government's White Paper *Broadcasting in the 1990's: Competition Choice and Quality* (Cm 517, November 1988). Earlier recommendations on the reform of the broadcasting industry had been made in the Report of the Committee on Financing the BBC (the Peacock Report) (Cmnd 9824, July 1986) and the Third Report of the Home Affairs Committee's inquiry into the Future of Broadcasting (HC Paper 262, Session 1987-88, June 1988).

CURRENT UK/EU LEGISLATION

BROADCASTING AND CINEMAS

Broadcasting Act 1996

The Broadcasting Act 1996 makes provision for digital terrestrial television broadcasting and contains provisions relating to the award of multiplex licences. It also provides for the introduction of radio multiplex services and regulates digital terrestrial sound broadcasting. In addition, the Act amends a number of provisions contained in the Broadcasting Act 1990 relating to the funding of Channel Four Television Corporation, the funding of Sianel Pedwar Cymru, and the operation of the Comataidh Craolidgh Gaialig (the Gaelic Broadcasting Committee). The Act also dissolves the Broadcasting Complaints Commission and Broadcasting Standards Council and replaces these with the Broadcasting Standards Commission. The Act also contains other provisions relating to the transmission network of the BBC and television coverage of listed events.

Broadcasting Act 1990

The Broadcasting Act 1990 established a new framework for the regulation of independent television and radio services, and for satellite television and cable television. Under the Act, the Independent Broadcasting Authority (IBA) and the Cable Authority were dissolved and replaced by the Independent Television Commission. The Radio Authority was established in respect of independent radio services. The Broadcasting Standards Council was made a statutory body and the Act also contains provisions relating to the Broadcasting Complaints Commission. Besides reorganising independent broadcasting, the Act provided for the formation of a separate company with responsibility for effecting the technical arrangements relating to independent television broadcasting – National Transcommunications Limited – as a first step towards the privatisation of the former IBA's transmission functions.

The Broadcasting Act 1990 repealed the Broadcasting Act 1981 and the Cable and Broadcasting Act 1984, amended the Wireless Telegraphy Act 1949, the Wireless Telegraphy Act 1967, the Marine [&c] Broadcasting (Offences) Act 1967, and the Copyright, Designs and Patents Act 1988, and also implements legislative provisions required pursuant to Directive 89/552 - see below.

The Broadcasting Act 1990 requires the British Broadcasting Corporation, all Channel 3 Licensees, the Channel Four Television Corporation, S4C (the Welsh Fourth Channel Authority) and the future Channel 5 Licensee to procure that not less than 25 per cent of the total amount of time allocated by those services to broadcasting "qualifying programming" is allocated to the broadcasting of a range and diversity of "independent productions". The expressions "qualifying programming" and "independent productions" are defined in the Broadcasting (Independent Productions) Order 1991.

Cinemas Act 1985

The Cinemas Act 1985 consolidated the Cinematographic Acts 1909 to 1952, the Cinematographic (Amendment) Act 1982 and related enactments. The Act deals with the exhibition of films and contains provisions for the grant, renewal and transfer of licences for film exhibition. There are special provisions for Greater London.

The Cinemas Act specifies the conditions of Sunday opening, and provides for exempted exhibition in private dwelling houses, and for non-commercial shows in premises used only occasionally.

Video Recordings Act 1984

The Video Recordings Act 1984 controls the distribution of video recordings with the aim of restricting the depiction or simulation of human sexual activity, gross violence, human genital organs or urinary or excretory functions. A system of classification and labelling is prescribed. The supply of recordings without a classification certificate, or the supply of classified recordings to persons under a certain age or in certain premises or in breach of labelling regulations, is prohibited subject to certain exemptions.

Classification certificates are issued by the British Board of Film Classification. It is an offence to supply or offer to supply, or to have in possession for the purposes of supplying, an unclassified video recording. Supplying recordings in breach of classification, supplying certain classified recordings otherwise than in licensed sex shops, supplying recordings in breach of labelling requirements and supplying recordings with false indications as to classification, are all offences under the Act. The Video Recordings Act provides for powers of entry, search and seizure and for the forfeiture of video recordings by the court.

Telecommunications Act 1984

The Telecommunications Act 1984 prohibits the running of a telecommunications system within the United Kingdom subject to certain exceptions which include the running of a telecommunication system in certain circumstances by a broadcasting authority. A broadcasting authority means a person who is licensed under the Wireless Telegraphy Act 1949 (see below) to broadcast programmes for general reception. Telecommunications systems include, among other things, any system for the conveyance of speech, music, other sounds and visual images by electric, magnetic, electro-magnetic, electro-chemical or electro-mechanical energy.

Wireless Telegraphy Acts 1967 and 1949

The 1967 Act provides for the Secretary of State to obtain information as to the sale and hire of television receiving sets. The Act allows the Secretary of State to prohibit the manufacture or importation of certain wireless telegraphy apparatus and to control the installation of such apparatus in vehicles.

The 1949 Act provides for the licensing of wireless telegraphy and defines "wireless telegraphy" as the sending of electro-magnetic energy over paths not provided by a material substance constructed or arranged for that purpose. The requirements to hold a licence under the Wireless Telegraphy Act 1949 or the Telecommunications Act 1984 are separate from the television and radio broadcast licensing provisions and cable programme source licensing provisions contained in the Broadcasting Act 1990.

Marine [&c] Broadcasting (Offences) Act 1967

The making of broadcasts by

wireless telegraphy (as defined in the Wireless Telegraphy Act 1949) intended for general reception from ships, aircraft and certain marine structures is prohibited under this Act.

The Cinematograph Films (Animals) Act 1937

The Cinematograph Films (Animals) Act 1937 provides for the prevention of exhibiting or distributing films in which suffering may have been caused to animals.

Celluloid and Cinematograph Film Act 1922

This Act contains provisions which are aimed at the prevention of fire in premises where raw celluloid or cinematograph film is stored or used. Silver nitrate film which was in universal use until the 1950s and was still used in some parts of the world (notably the former USSR) until the 1970s, is highly inflammable and becomes unstable with age. The purpose of the legislation was to protect members of the public from fire risks.

FINANCE

Finance (No 2) Act 1997

Section 48 Finance (No 2) Act 1997 introduced new rules for writing-off production and acquisition expenditure of British qualifying films costing £15 million or less to make. The relief applies to expenditure incurred between 2 July 1997 and 1 July 2000. Section 48 allows 100 per cent write-off for production or acquisition costs when the film is completed.

A British qualifying film is one certified as such by the Department of Culture Media and Sport under the Films Act 1985. In order to be certified a number of criteria must be met. These include using United Kingdom studios for a high proportion of the film and ensuring that the film is made by a company registered, managed in the United Kingdom or another European Union state.

The Inland Revenue made an announcement on 25 March 1998 that the Government intends to extend the time limit for relief under section 48 from 3 years to 5 years in a future Finance Bill. The relief will then apply to expenditure incurred between 2 July 1997 and 1 July 2002. The Film Review Group issued a report on 25 March 1998

which sets out an action plan for delivery by April 1999.

The Finance Act 1990, Capital Allowances Act 1990 and Finance (No 2) Act 1992

Section 80 and Schedule 12 to the Finance Act 1990 deals with the tax issues relating to the reorganisation of independent broadcasting provided for in the Broadcasting Act 1990.

Section 68 of the Capital Allowances Act 1990 replaces Section 72 of the Finance Act 1982 providing for certain expenditure in the production of a film, tape or disc to be treated as expenditure of a revenue nature.

Sections 41-43 of the Finance (No 2) Act 1992 amend the tax regime to provide accelerated relief for pre-production costs incurred after 10 March 1992 and production expenditure on films completed after that date. Section 69 of the Act makes certain consequential amendments to Section 68 of the Capital Allowances Act 1990.

Films Act 1985

The Films Act 1985 dissolved the British Film Fund Agency, ending the Eady levy system established in 1951. The Act also abolished the Cinematograph Film Council and dissolved the National Film Finance Corporation, transferring its assets to British Screen Finance Limited. The Act repealed the Films Acts 1960 - 1980 and also repealed certain provisions of the Finance Acts 1982 and 1984 and substituted new provisions for determining whether or not a film was 'British' film eligible for capital allowances.

National Film Finance Corporation Act 1981

The National Film Finance Corporation Act 1981 repealed the Cinematograph Film Production (Special Loans) Acts of 1949 and 1954 and made provisions in relation to the National Film Finance Corporation which has since been dissolved by the Films Act 1985. The National Film Finance Corporation Act 1981 is, however, still on the statute book.

Film Levy Finance Act 1981

Although the British Film Fund Agency was dissolved by the British Film Fund Agency (Dissolution) Order 1988, SI 1988/37, the Film Levy Act itself is still in place.

COPYRIGHT

Copyright, Designs and Patents Act 1988

This Act is the primary piece of legislation relating to copyright in the United Kingdom. The Act provides copyright protection for original literary, dramatic, musical and artistic works, for films, sound recordings, broadcasts and cable programmes, and for typographical arrangements of published editions.

The Act repeals the Copyright Act 1956 which in turn repealed the Copyright Act 1911, but the transitional provisions of the Copyright, Designs and Patents Act 1988 apply certain provisions of the earlier legislation for the purpose of determining ownership of copyright, type of protection and certain other matters. Because the term of copyright for original literary, dramatic and/or musical works is the life of the author plus 50 years, the earlier legislation will continue to be relevant until well into the next century. The provisions of the Act have been amended by EU harmonisation provisions contained in Directive 93/98 extending the term of copyright protection in relation to literary, dramatic, musical and artistic works originating in countries within the European Economic Area or written by nationals of countries in the EEA, to the duration of the life of the author or last surviving co-author plus, 70 years calculated from 31 December in the relevant year of decrease.

The Act provides a period of copyright protection for films and sound recordings which expires 50 years from the end of the calendar year in which the film or sound recording is made, or if it is shown or played in public or broadcast or included in a cable programme service, 50 years from the end of the calendar year in which this occurred. The provisions of the Act have been amended by EU harmonisation provisions contained in Directive 93/98 extending the term of copyright protection for films, to a period equal to the duration to the lifetime of the last to die of the persons responsible for the making of the film, plus 70 years calculated from 31 December in the relevant year of decrease.

The Act introduced three new moral rights into United Kingdom legislation. In addition to the right not to have a work falsely attributed to him or her, an author (of a literary dramatic musical or artistic work) or director (of a film) has the right to be identified in relation to their work, and the right not to permit their work to suffer derogatory treatment. A derogatory treatment is any addition, deletion, alteration or adaptation of a work which amounts to a distortion or mutilation of the work, or is otherwise prejudicial to the honour or reputation of the author or director. A person who commissions films or photographs for private and domestic purposes enjoys a new right of privacy established by the Act.

Another new development is the creation of a statutory civil right for performers, giving them the right not to have recordings of their performances used without their consent. United Kingdom copyright legislation was amended following a decision in Rickless -v- United Artists Corporation – a case which was brought by the estate of Peter Sellars and involved *The Trail of the Pink Panther*. The legislation is retrospective and protects performances given 50 years ago, not just in the United Kingdom, but in any country if the performers were "qualifying persons" within the meaning of the relevant Act. The performances which are covered include not only dramatic and musical performances, but readings of literary works, variety programmes and even mime. Numerous other provisions are contained in the Copyright, Designs and Patents Act including sections which deal with the fraudulent reception of programmes, the manufacture and sale of devices designed to circumvent copy-protection, and patent and design law.

EUROPEAN COMMUNITY LEGISLATION

Directive 89/552 – on television without frontiers

The objective of the Directive is to eliminate the barriers which divide Europe with a view to permitting and assuring the transition from national programme markets to a common programme production and distribution market. It also aims to establish conditions of fair competition without prejudice to the public interest role which falls to be discharged by television broadcasting services in the EC.

The laws of all Member States relating to television broadcasting and cable operations contain disparities which may impede the free movement of broadcasts within the EC and may distort competition. All such restrictions are required to be abolished.

Member States are free to specify detailed criteria relating to language etc. Additionally, Member States are permitted to lay down different conditions relating to the insertion of advertising in programmes within the limits set out in the Directive. Member States are required to provide where practicable that broadcasters reserve a proportion of their transmission time to European works created by independent producers. The amount of advertising is not to exceed 15 per cent of daily transmission time and the support advertising within a given one hour period shall not exceed 20 per cent.

Directive 92/100 – on rental rights

Authors or performers have, pursuant to the Directive, an unwaiveable right to receive equitable remuneration. Member States are required to provide a right for performers in relation to the fixation of their performances, a right for phonogram and film producers in relation to their phonograms and first fixations of their films and a right for broadcasters in relation to the fixation of broadcasts and their broadcast and cable transmissions. Member States must also provide a 'reproduction right' giving performers, phonogram producers, film producers and broadcasting organisations the right to authorise or prohibit the direct or indirect reproduction of their copyright works. The Directive also requires Member States to provide for performers, film producers, phonogram producers and broadcasting organisations to have exclusive rights to make available their work by sale or otherwise – known as the 'distribution right'.

Directive 93/83 on Satellite Transmission and Cable Retransmission

This Directive is aimed at eliminating uncertainty and differences in national legislation governing when the act of communication of a programme takes place. It avoids the cumulative application of several national laws to one single act of broadcasting.

The Directive provides that communication by satellite occurs in the member state where the programming signals are introduced under the control of a broadcaster into an uninterrupted chain of communication, leading to the satellite and down towards earth. The Directive also examines protection for authors, performers and producers of phonograms and broadcasting organisations, and requires that copyright owners may grant or refuse authorisation for cable retransmissions of a broadcast only through a collecting society.

Directive 98/98 on harmonising the term of protection of copyright and certain related rights

This Directive is aimed at harmonising the periods of copyright throughout the European Union where different states provide different periods of protection. Although the minimum term established by the Berne Convention on Copyright is 50 years *post mortem auctoris,* a number of states have chosen to provide for longer periods. In Germany the period of literary dramatic musical and artistic works is 70 years *pma,* in Spain 60 years (or 80 years for copyrights protected under the Spanish law of 1879 until its reform in 1987). In France the period is 60 years *pma* or 70 years for musical compositions.

In addition to the differences in the term of rights *post mortem auctoris,* further discrepancies arise in protection accorded by different member states through wartime extensions. Belgium has provided a wartime extension of 10 years, Italy 12 years, France six and eight years respectively in relation to the First and Second World Wars. In France, a further period of 30 years is provided in the case of copyright works whose authors were killed in action - such as Antoine de Saint-Exupéry.

The Directive also provides that rights of performers shall run from 50 years from the date of performance or if later, from the point at which the fixation of the performance is lawfully made available to the public for the first time, or if this has not occurred from the first assimilation of the performance. The rights of producers of phonograms

run 50 years from first publication of the phonogram, but expire 50 years after the fixation was made if the phonogram, but expire 50 years after the fixation was made if the phonogram has not been published during that time. A similar provision applies to the rights of producers of the first fixations of cinematographic works and sequences of moving images, whether accompanied or not by sound. Rights of broadcasting organisations run from 50 years from the first transmission of the broadcast.

The Directive provides that the person who makes available to the public a previously unpublished work which is in the public domain, shall have the same rights of exploitation in relation to the work as would have fallen to the author for a term of 25 years from the time the work was first made available to the public. The Directive applies to all works which are protected by at least one member state on 1 July 1995 when the Directive came into effect. As a result of the differing terms in European states, many works which were treated as being in the 'public domain' in the United Kingdom will have their copyright revived. Works by Beatrix Potter, James Joyce and Rudyard Kipling are all works which will benefit from a revival of copyright. The provisions relating to the term of protection of cinematographic films are not required to be applied to films created before 1 July 1994. Each member state of the European Union's required to implement the Directive. The precise manner of implementation and the choice of transitional provisions, are matters which each state is free to determine.

Directive 93/98 was implemented in the United Kingdom by the Rights in Performances Regulations 1995/ 3297 which took effect from 1 January 1996. The term of copyright protection for literary dramatic musical or artistic works expires at the end of the period of 70 years from the last day of the calendar year in which the author dies. Copyright in a film expires 70 years from the end of the calendar year in which the death occurs of the last to die of the principal director, the author of the screenplay, the author of the dialogue or the composer of the music specially created for and used for the film. The period of copyright previously applying to films under the Copyright, Designs and Patents

Act 1988 ended 50 years from the first showing or playing in public of a film, and the effect of the implementation of Directive 93/98 is to create a significant extension of the period in which a film copyright owner has the exclusive economic right to exploit a film. If, as anticipated, the United States of America also extends the duration of the copyright period applying to films, the value of intellectual property rights in audiovisual productions may increase significantly.

LIBRARIES

This section provides a directory of libraries and archives which have collections of books, periodicals and papers covering film and television. It includes the libraries of colleges and universities with graduate and post-graduate degree courses in the media. Most of these collections are intended for student and teaching staff use: permission for access should always be sought from the Librarian in charge. Where possible a breakdown of types of resources available is provided

BFI National Library
21 Stephen Street
London W1P 2LN
Tel: 0171 255 1444
Fax: 0171 436 2338
　　 0171 436 0165 (Information)

The BFI's own library is extensive and hold's the world's largest collection of documentation on film and television. It includes both published and unpublished material ranging from books and periodicals to news cuttings, press releases, scripts, theses, and files of festival material **(see p8)**

Reading Room opening hours:
Monday, Friday 10.30am - 5.30pm
Tuesday, Thursday 10.30am - 8.00pm
Wednesday 1.00pm - 8.00pm
Library library pass: £33.00
NFT Members pass: £25.00
Discount passes (£20.00) are available to Senior Citizens, Registered Disabled and Unemployed upon proof of eligibility. Students may also apply for a discounted library pass.

Day passes are available (£6.00) to anyone and space may be reserved by giving 48 hours notice.

Enquiry Lines:
The Enquiry Line is available for short enquiries. Frequent callers subscribe to an information service. The line is open from 10.00am to 5.00pm Monday to Friday via the BFI switchboard (0171 255 1444)

Research Services:
For more detailed enquiries, users should contact the Information Service by fax or mail.

RESOURCES

A　Specialist sections
B　Film/TV journals
C　Film/TV/CD Roms
D　Video loan service
E　Internet access
F　Special collections

Aberdeen

Aberdeen University Library
Queen Mother Library, Meston Walk, Aberdeen
Grampian AB24 3UE
Tel: 01224 272579
Fax: 01224 487048
email: library@abdn.ac.uk
Contact: University Librarian

Bangor

Normal College
Education Library
Bangor
Gwynedd LL57 2P
Tel: 01248 370171
Fax: 01248 370461
Contact: Librarian

Barnet

Middlesex University Cat Hill Library
Cat Hill, Barnet
Herts EN4 8HT
Tel: 0181 362 5042
Fax: 0181 440 9541
Contact: Art and Design Librarian

Bath

Bath University Library
Claverton Down
Bath BA2 7AY
Tel: 01225 826084
Fax: 01225 826229
Contact: University Librarian

Belfast

Belfast Central Library
Royal Avenue
Belfast
Co. Antrim BT1 1EA
Tel: 01232 332819
Fax: 01232 312886
Contact: Chief Librarian

Northern Ireland Film Commission
21 Ormeau Avenue
Belfast BT2 8HD
Tel: 01232 232 444
Fax: 01232 239 918
Contact: Information Officer
Resources: B, D

Queen's Film Theatre
25 College Gardens
Belfast BT9 6BS
Tel: 01232 667687 ext. 33
Fax: 01232 663733
Contact: Administrator/Programmer

Birmingham

BBC Pebble Mill
Information Research Library
Pebble Mill Road
Birmingham B5 7QQ
Tel: 0121 414 8922
Contact: Information Research Librarian
Resources:B,C,E

Vivid - Birmingham Centre for Media Arts
Unit 311 The Big Peg
120 Vyse Street
Birmingham B18 6ND
Tel: 0121 233 4061
Fax: 0121 212 1784
email:centre@vivid.globalnet.co.uk
Website:
www.wavespace.waverider.co.uk
Contact: Head of Service: Niky Rathbone
Resources:A,B,D,E

Birmingham University Library
Edgbaston
Birmingham B15 2TT
Tel: 0121 414 5817
Fax: 0121 414 5815
Contact: Librarian, Arts and Humanities

Central Broadcasting Ltd
Broad Street
Birmingham B1 2JP
Tel: 0121 643 9898
Contact: Reference Librarian

University of Central England
Birmingham Institute of Art &
Design
Gosta Green
Birmingham B4 7DX
Tel: 0121 331 5860
Contact: Library staff

Information Services
Franchise Street
Perry Barr
Birmingham B42 2SU
Tel: 0121 331 5300
Fax: 0121 331 6543
Contact: Dean of Information
Services

Brighton

Sussex University Library
Falmer
Brighton
East Sussex BN1 9QL
Tel: 01273 678163
Fax: 01273 678441
Contact: Information Services

University of Brighton Faculty of Art, Design and Humanities
St Peter's House Library
16-18 Richmond Place
Brighton BN2 2NA
Tel: 01273 571820
Contact: Librarian

Bristol

Bristol City Council
Leisure Services
Central Library, Reference
Library, College Green
Bristol BS1 5TL
Tel: 0117 927 6121
Fax: 0117 922 6775
Contact: Head of Reference &
Information Services

University of Bristol
University Library
Tyndall Avenue
Bristol BS8 1TJ
Tel: 0117 928 9017
Fax: 0117 925 5334
Website: www.bris.ac.uk/Depts/
Library
Contact: Librarian
Resources: A,B,C,E

University of Bristol Theatre Collection
29 Park Road
Bristol BS1 5LT
Tel: 0117 930 3215
Fax: 0117 973 2657
Contact: Keeper

West of England University at Bristol
Library, Faculty of Art, Media &
Design
Bower Ashton Campus
Clanage Road
Bristol BS3 2JU
Tel: 0117 966 0222 x4750
Fax: 0117 976 3946
Contact: Steve Morgan, Campus/
Subject Librarian, Art, Media and
Design

Canterbury

Canterbury College Library
Kent Institute of Art & Design
New Dover Road, Canterbury
Kent CT1 3AN
Tel: 01227 769371
Fax: 01227 451320

Christ Church College Library
North Holmes Road
Canterbury
Kent CT1 1QU
Tel: 01227 767700
Fax: 01227 767530
Website: www.cant.ac.uk./depts/
services/library/library1.html
Contact: Director of Library Services
Resources: A,B,C,D,E

Kent Institute of Art & Design at Canterbury
New Dover Road
Canterbury
Kent CT1 3AN
Tel: 01227 769371
Fax: 01227 451320
Contact: Learning Resources
Manager

Templeman Library
University of Kent at Canterbury
Canterbury
Kent CT2 7NU
Tel: 01227 764000
Fax: 01227 459025
Contact: Librarian

Cardiff

Coleg Glan Hafren
Trowbridge Road
Rumney
Cardiff CF3 8XZ
Tel: 01222 250250
Fax: 01222 250339
Contact: Learning Resources
Development Manager

University of Wales College
Cardiff, Bute, Library
PO Box 430
Cardiff CF1 3XT
Tel: 01222 874000
Fax: 01222 874192
Website: http://www.cardiff.ac.uk
Contact: Librarian

Carlisle

Cumbria College of Art and Design Library
Brampton Road
Carlisle
Cumbria CA3 9AY
Tel: 01228 25333 x206
Contact: Librarian

Chislehurst

Ravensbourne College of Design and Communication Library
Walden Road, Chislehurst
Kent BR7 5SN
Tel: 0181 289 4900
Fax: 0181 325 8320
email: library@rave.ac.uk
Contact: Librarian

Colchester

University of Essex
The Albert Sloman Library
Wivenhoe Park
Colchester CO4 3SQ
Tel: 01206 873333
Contact: Librarian

Coleraine

University of Ulster
Library
Coleraine
Northern Ireland BT52 1SA
Tel: 01265 32 4345
Fax: 01265 32 4928
Contact: Pro-Librarian
Resources: A,B,C,D

Coventry

Coventry City Library
Smithford Way
Coventry CV1 1FY
Tel: 01203 832314
Fax: 01203 832440
email: covinfo@discover.co.uk
Contact: Librarian - Karen Berry

Coventry University, Art & Design Library
Priory Street
Coventry CV1 5FB
Tel: 01203 838546
Fax: 01203 838686

Website: http://www.coventry.ac.uk./library/
Contact: Sub-Librarian, Art & Design
Resources: A,B,C,D,E

Warwick University Library
Gibbet Hill Road
Coventry CV4 7AL
Tel: 01203 524103
Fax: 01203 524211
Contact: Librarian
Resources: A,B,C,D,E,F*
* Collection of German film programme from the 1930s

Derby

Derby University Library
Kedleston Rd
Derby DE3 1GB
Tel: 01332 622222 x 4061
Fax: 01332 622222 x 4059
Contact: Librarian

University of Derby
Library and Learning Resources
Derby DE1 1RX
Tel: 01332 622222 Ext 3001
Website: http://www.derby.ac.uk/library/homelib.html
Contact: Subject Adviser, Art & Design
Resources: A,B,C,D,E

Dorking

Surrey Performing Arts Library
Vaughan Williams House
West Street
Dorking, Surrey RH4 1BY
Tel: 01306 887509
Fax: 01306 875074
email: p.arts@dial.pipex.com
Website: http://surreycc.gov.uk/libraries/direct/perfarts.html
Senior Librarian: G.Muncy
Contact: Librarian
Resources: A,B,C,D,E,F*
* Scripts

Douglas

Douglas Corporation
Douglas Public Library
Ridgeway Street, Douglas
Isle of Man
Tel: 01624 623021
Fax: 01624 662792
Contact: Borough Librarian

Dudley

Wolverhampton University
Dudley Learning Centre

University of Wolverhampton
Castle View
Dudley
West Midlands DY1 3BQ
Tel: 01902 323 560
Fax: 01902 323 354
Website: http://www.wlr.ac.uk.lib
Learning Centre Manager
Resources: A,B,C,D,E

Dundee

Library Duncan of Jordanstone College
University of Dundee
13 Perth Road
Dundee DD1 4HT
Tel: 01382 345255
Fax: 01382 229283
Contact: College Librarian
Resources: A,B,C,D,E,F*
* Few scripts

Egham

Royal Holloway University of London Library
Egham Hill
Egham
Surrey TW20 OEX
Tel: 01784 443330
Fax: 01784 437520
www.rhbnc.ac.uk
Contact: Librarian
Resources: A,B,C,D,E

Exeter

Exeter University Library
Stocker Road
Exeter
Devon EX4 4PT
Tel: 01392 263869
Fax: 01392 263871
Website: http://www.exe.ac.uk/@JACrawle/lib.film.html
Contact: Librarian
Tel: 01392 263869
Fax: 01392 263871
Contact: Librarian
Resources: A,B,C,D,E,F*
* The Bill Douglas Centre for the History of Cinema and Popular Culture

Farnham

Surrey Institute of Art & Design
Falkner Road, The Hart
Farnham
Surrey GU9 7DS
Tel: 01252 722441
Fax: 01252 733869
Contact: College Librarian

Gateshead

Gateshead Libraries and Arts Department
Central Library
Prince Consort Road
Gateshead
Tyne and Wear NE8 4LN
Tel: 0191 477 3478
Fax: 0191 477 7454
Contact: The Librarian
Resources: B,D,E

Glasgow

Glasgow Caledonian University Library
Cowcaddens Road
Glasgow G4 0BA
Tel: 0141 331 3858
Fax: 0141 331 3005
http://www.gcal.ac.uk/library/index.html
Contact: Assistant Academic Liaison Librarian for Language and Media
Resources: A,B,C,D,E

Glasgow City Libraries
Mitchell Library
North Street
Glasgow G3 7DN
Tel: 0141 287 2933
Fax: 0141 287 2815
Contact: Departmental Librarian, Art Department

Glasgow School of Art Library
167 Renfrew Street
Glasgow G3 6RQ
Tel: 0141 353 4551
Fax: 0141 332 3506
Contact: Principal Librarian

Scottish Council for Educational Technology
Dowanhill
74 Victoria Crescent Road
Glasgow G12 9JN
Tel: 0141 337 5000
Fax: 0141 337 5050
Website: http://www.sect.com
Contact: Librarian
Resources: D

Scottish Screen
74 Victoria Crescent Road
Glasgow G12 9JN
Tel: 0141 302 1700
Fax: 0141 302 1711
Contact: Chief Executive: John Archer
The Government body Scottish Screen was established in April 1997 to stimulate the film & television industry through production & development finance, training,

education, marketing and the film Commission
Resources: D,F*
*Production/Information
Access to the Shiach Script library containing over 100 feature and short film scripts. Video and Publication resource. Internet site Archive. Home of the National Archive, collection of factual documentary material mapping Scotland's social and cultural history. Available to broadcasters, programme makers, educational users and researchers. Distribution. Short film distribution back catalogue

University of Glasgow
The Library
Hillhead Street
Glasgow G12 8QQ
Tel: 0141 330 6704/5
Fax: 0141 330 4952
Contact: Librarian

Gravesend

VLV - Voice of the Listener and Viewer
101 King's Drive
Gravesend
Kent DA12 5BQ
Tel: 01474 352835
Fax: 01474 351112
Contact: Information Officer
In addition to its own VLV lds archives of the former independent Broadcasting Research Unit (1980-1991) and the former British Action for Children's Television (BACTV) (1988-1994) and makes these available for a small fee together with its own archives and library. VLV represents the citizen and consumer voice in broadcasting
Resources: A,B, D

Huddersfield

Kirklees Cultural Services
Central Library
Princess Alexandra Walk
Huddersfield HD1 2SU
Tel: 01484 221967
Fax: 01484 221974
Contact: Reference Librarian
Resources: C,D,E

Hull

Hull University Brynmor Jones Library
Cottingham Road
Hull
North Humberside HU6 7RX

Tel: 01482 465440
Fax: 01482 466205
Contact: Librarian

Humberside University

School of Art, Architecture and Design Learning Support Centre
Guildhall Road
Hull HU1 1HJ
Tel: 01482 440550
Fax: 01482 449627
Contact: Centre Manager

Isleworth

Brunel University
Twickenham Campus
300 St Margarets Road
Twickenham TW1 1PT
Tel: 0181 891 0121
Fax: 0181 891 0240
Contact: Director of Library Services
Resources: A,B,C,E

Keele

Keele Information Services
Keele University
Keele
Staffs ST5 5BG
Tel: 01782 583239
Fax: 01782 711553
Contact: Visual Arts Department
Resources: B,C,D,E

Kingston upon Thames

Kingston Museum & Heritage Service
North Kingston Centre
Richmond Road
Kingston upon Thames
Surrey KT2 5PE
Tel: 0181 547 6738 or 6755
http://www.kingston.ac.uk/muytexto.htm
Contact: T. Everson, Local History Officer
Resources: E,F*
*Eadweard Muybridge Collection

Kingston University Library Services
Art and Design Library
Knights Park
Kingston Upon Thames
Surrey KT1 2QJ
Tel: 0181 547 2000 x 4031
Fax: 0181 547 7011
Contact: Senior Faculty Librarian (Design)

Kingston University Library Services
Library & Media Services
Penrhyn Road
Kingston Upon Thames
Surrey KT1 2EE
Tel: 0181 547 2000
Contact: Head of Media Services

Leeds

Leeds City Libraries
Central Library
Municipal Buildings
Calverley Street
Leeds, West Yorkshire LS1 3AB
Tel: 0113 247 8265
Fax: 0113 247 8268
Contact: Director of Library Services

Leeds Metropolitan University
City Campus Library
Calverley Street
Leeds, West Yorkshire LS1 3HE
Tel: 0113 283 2600 x3836
Fax: 0113 242 5733
Contact: Tutor Librarian, Art & Design

Trinity and All Saints College Library
Brownberrie Lane
Horsforth
Leeds, West Yorkshire LS18 5HD
Tel: 0113 283 7100
Fax: 0113 283 7200
Website: http://www.tasc.ac.uk
Contact: Librarian
Resources:A,B,D,E

Leicester

Centre For Mass Communication Research
104 Regent Road
Leicester LE1 7LT
Tel: 0116 2523863
Fax: 0116 2523874
Contact: Director

De Montfort University Library
Kimberlin Library
The Gateway
Leicester LE1 9BH
Tel: 0116 255 1551
Fax: 0116 255 0307
Contact: Senior Assistant Librarian (Art and Design)

Leicester University Library
PO Box 248
University Road
Leicester LE1 9QD
Tel: 0116 252 2042
Fax: 0116 252 2066
Website: http//www.le.ac.uk

Contact: Librarian
Resources: A,B,E

Leicestershire Central Lending Library
54 Belvoir Street
Leicester LE1 6QL
Tel: 0116 255 6699
Contact: Area Librarian

Liverpool

Aldham Robarts Learning Resource Centre
Liverpool John Moores University
Mount Pleasant
Liverpool L3 5UZ
Tel: 0151 231 2121 x 3104
Contact: Site Librarian

Liverpool City Libraries
William Brown Street
Liverpool L3 8EW
Tel: 0151 225 5429
Fax: 0151 207 1342
Contact: Librarian

Liverpool Hope University College
Hope Park
Liverpool L16 9LB
Tel: 0151 291 2000
Fax: 0151 291 2037
Website: www.livhope.ac.uk
Contact: Director of Learning Resources
Resources: A,B,C,D,E

London

Barbican Library
Barbican Centre
London EC2Y 8DS
Tel: 0171 638 0569
Fax: 0171 638 2249
Contact: Librarian

BKSTS - The Moving Image Society
63-71 Victoria House
Vernon Place
London WC1B 4DA
Tel: 0171 242 8400
Fax: 0171 405 3560
email: movimage@bksts.demon.co.uk
Contact: Anne Fenlon

British Universities Film & Video Council Library
77 Wells Street
London W1P 3RE
Tel: 0171 393 1500
Fax: 0171 393 1555
Contact: Head of Information
Resources: B,D,E,F*
* British Newsreel issue sheets

Camberwell College of Arts Library
London Institute
Peckham Road
London SE5 8UF
Tel: 0171 514 6349
Fax: 0171 514 6324
Contact: College Librarian
Resources: A,B,E

Camden Public Libraries
Swiss Cottage Library
88 Avenue Road
London NW3 3HA
Tel: 0171 413 6527
swisslib@camden.gov.uk
Contact: Librarian
Resources: A,B,D,E

Carlton Screen Advertising Ltd
127 Wardour Street
London W1V 4NL
Tel: 0171 439 9531
Fax: 0171 439 2395
Contact: Secretary

The College of North East London Learning Resource Centre
High Road
Tottenham
London N15 4RU
Tel: 0181 442 3013
Fax: 0181 442 3091
Contact: Head of Learning Resources

Guildhall University Library Services
Calcutta House
Old Castle Street
London E1 7NT
Tel: 0171 320 1000
Fax: 0171 320 1177
Contact: Head of Library Services

Independent Television Commission Library
33 Foley Street
London W1P 7LB
Tel: 0171 306 7763
Fax: 0171 306 7750
Contact: Librarian
Resources: A,B,C,E, F*
* Press cuttings

Institute of Education Library (London)
20 Bedford Way
London WC1H OAL
Tel: 0171 580 1122
Fax: 0171 612 6126
Contact: Librarian

International Institute of Communications
Library and Information Service

Tavistock House South
Tavistock Square
London WC1H 9LF
Tel: 0171 388 0671
Fax: 0171 380 0623
Contact: Information & Library Manager

London Borough of Barnet Libraries
Hendon Library, The Burroughs
Hendon
London NW4 4BQ
Tel: 0181 359 2628
Fax: 0181 359 2885
Contact: Librarian

London College of Printing & Distributive Trades
Media School
Backhill
Clerkenwell EC1R 5EN
Tel: 0171 514 6500
Fax: 0171 514 6848
Contact: Head of Learning Resources

Middlesex University Library
Bounds Green Road
London N11 2NQ
Tel: 0181 362 5240
Contact: University Librarian

Roehampton Institute Library
Roehampton Institute London
Roehampton Lane
London SW15 5PH
Tel: 0181 392 3254
Contact: Faculty Librarian

Royal College of Art
Kensington Gore
London SW7 2EU
Tel: 0171 590 4224 - Library Desk
Tel: 0171 590 4444 - College
Fax: 0171 590 4500
Contact: Library Manager

Royal Television Society, Library & Archive
Holborn Hall
100 Grays Inn Road
London WC1X 8AL
Tel: 0171 430 1000
Fax: 0171 430 0924
Contact: Archivist

Slade/Duveen Art Library
University College London
Gower Street
London WC1E 6BT
Tel: 0171 387 7050 x 2594
Fax: 0171 380 7373
email: r.dar@ucl.ac.uk
Contact: Art Librarian: Ruth Dar

Thames Valley University Library
Learning Resources Centre
Walpole House
18-22 Bond Street W5 5AA

Tel: 0181 231 2248
Fax: 0181 231 2631
Contact: Humanities Librarian

University of East London
Greengate House Library
School of Art & Design
89 Greengate Street
London E13 0BG
Tel: 0181 590 7000 x 3434
Fax: 0181 849 3692
Contact: Site Librarian

University of London: Goldsmiths' College Library
Lewisham Way
London SE14 6NW
Tel: 0171 919 7168
Fax: 0171 919 7165
email: lbslpm@gold.ac.uk
http://www.gold.ac.uk
Contact: Subject Librarian: Media & Communications
Resources: A,B,C,D

University of North London
The Learning Centre
236-250 Holloway Road
London N7 6PP
Tel: 0171 607 2789 x 2720
Fax: 0171 753 5079
Contact: Film Studies Librarian

University of Westminster
Information Resource Services
Watford Road
Northwick Park
Harrow HA1 3TP
Tel: 0171 911 5000
http://www.wmin.ac.uk
Contact: Sub-Librarian
Resources: A,B,C,D,E

Westminster Reference Library
35 St Martins Street
London WC2H 7HP
Tel: 0171 641 4636
Fax: 0171 641 4640
Contact: Library Manager
Resources: A,B,C,E

Loughborough

Loughborough University Pilkington Library
Loughborough University
Loughborough LE11 3TU
Tel: 01509 222360
Fax: 01509 234806
Contact: Assistant Librarian

Luton

University of Luton Library
Park Square
Luton LU1 3JU

Tel: 01582 734111
Contact: Faculty Information Officer

Maidstone

Kent Institute of Art & Design at Maidstone
Oakwood Park
Maidstone
Kent ME16 8AG
Tel: 01622 757286
Fax: 01622 692003
Contact: College Librarian

Manchester

John Rylands University Library
Oxford Road
Manchester M13 9PP
Tel: 0161 275 3751/3738
Fax: 0161 273 7488
Contact: Lending Services Librarian

Manchester Arts Library
Central Library
St Peters Square
Manchester M2 5PD
Tel: 0161 234 1974
Fax: 0161 234 1963
Contact: Arts Librarian
Resources: A,B,D,E

Manchester Metropolitan University Library
All Saints Building
Grosvenor Square
Oxford Road
Manchester M15 6BH
Tel: 0161 247 6104
Fax: 0161 247 6349
Contact: Senior Subject Librarian

North West Film Archive
Manchester Metropolitan University
Minshull House
47-49 Chorlton Street
Manchester M1 3EU
Tel: 0161 247 3097
Fax: 0161 247 3098
email: n.w.filmarchive@mmu.ac.uk
http://www.mmu.ac.uk/services/library/wst.htm
Director: Maryann Gomes
Enquiries: Rachael Holdsworth
Resources: D,E,F*
* Ephemera

Morpeth

Northumberland Central Library
The Willows
Morpeth
Northumberland NE61 1TA
Tel: 01670 511156

Fax: 01670 518012
website: http//amenities@northumberland.gov.uk
Contact: The Librarian
Resources: A,B,C,D,E

Newcastle Upon Tyne

Newcastle Upon Tyne University Robinson Library
Newcastle Upon Tyne NE2 4HQ
Morpeth
Northumberland NE61 1TA
Tel: 0191 222 7713
Fax: 0191 222 6235
website: http//www.ncl.ac.uk/library
Contact: The Librarian
Resources: A,B,C,D,E

University of Northumbria at Newcastle Library Building
Ellison Place
Newcastle Upon Tyne NE1 8ST
Tel: 0191 227 4132
Fax: 0191 227 4563
Website: http://unn.ac.uk
Contact: Jane Shaw, Senior Officer, Information Services Department

Newport

University of Wales College Newport
Caerleon
Newport NP6 1XJ
Tel: 01633 430088
Fax: 01633 432108
Contact: Art and Design Librarian

Norwich

East Anglian Film Archive
Centre for East Anglian Studies
University of East Anglia
Norwich NR4 7TJ
Tel: 01603 592664
Fax: 01603 458553
Contact: Assistant Archivist

University of East Anglia
University Library
Norwich NR4 7TJ
Tel: 01603 456161
Fax: 01603 259490
Contact: Film Studies Librarian

Nottingham

Nottingham Central Library
Angel Row
Nottingham NG1 6HP
Tel: 0115 941 2121
Fax: 0115 953 7001
Contact: Librarian

Nottingham Trent University Library
Library and Information Services
Dryden Street
Nottingham NG1 4FZ
Tel: 0115 941 8418
Fax: 0115 941 5380
Website: http://www.ntu.ac.uk
Contact: Faculty Librarian (Art & Design)
Resources: A,B,C,D,E

Nottingham University Library
Hallward Library
University Park
Nottingham NG7 2RD
Tel: 0115 951 4584
Fax: 0115 951 4558
Website: http//
www.nottingham.ac.uk/library
Contact: Humanities Librarian
Resources: A,B,C,E

Plymouth

College of St Mark and St John Library
Derriford Road
Plymouth
Devon PL6 8BH
Tel: 01752 777188
Fax: 01752 711620
Contact: Head of Learning Resources

Plymouth College of Art & Design Library
Tavistock Place
Plymouth
Devon PL4 8AT
Tel: 01752 203412
Fax: 01752 203444
Contact: Librarian
Resources: A,B,C,D,E

Pontypridd

University of Glamorgan
Learning Resources Centre
Pontypridd
Mid Glamorgan CF37 1DL
Tel: 01443 482625
Fax: 01443 482629
Contact: Head of Learning Resource

Poole

Bournemouth & Poole College of Art & Design
Fern Barrow,
off Wallisdown Road, Poole
Dorset BH12 5HH
Tel: 01202 533011
Fax: 01202 537729
Contact: University Librarian

Bournemouth University Library
Talbot Campus, Fern Barrow,
off Wallisdown Road
Poole
Dorset BH12 5BB
Tel: 01202 595011
Contact: Librarian

Portsmouth

Highbury College Library
Cosham
Portsmouth
Hants PO6 2SA
Tel: 01705 283213
Fax: 01705 325551
Contact: College Librarian
Resources: E

Portsmouth University Library
Frewen Library
Cambridge Road
Portsmouth
Hampshire PO1 2ST
Tel: 01705 843222
Fax: 01705 843233
Website:http://.libr.port.ac.uk
Contact: University Librarian
Resources: A,B,C,D,E

Preston

University of Central Lancashire Library
St Peter's Square
Preston
Lancashire PR1 2HE
Tel: 01772 201201 x 2266
Fax: 01772 892937
Contact: Senior Subject Librarian

Reading

Reading University Bulmershe Library
Woodlands Avenue
Reading RG6 1HY
Tel: 0118 987 5123 ext 4824
Fax: 0118 931 8651
Contact: Faculty Team Manager (Education & Community Studies)

Rochdale

Rochdale Metropolitan Borough Libraries
Wheatsheaf Library
Wheatsheaf Centre
Baillie Street, Rochdale
Lancashire OL16 1AQ
Tel: 01706 864914
Fax: 01706 864992
Contact: Librarian

Salford

University of Salford, Academic Information Services (Library)
Adelphi Campus
Peru Street
Salford
Greater Manchester M3 6EQ
Tel: 0161 295 6183/6185
Fax: 0161 295 6103
Website: http//www.salford.ac.uk/
ais/homepage.html
Contact: Sue Slade (Faculty co-ordinator)
Contact: Andy Callen (Information Officer, Music & Media Productions)
Resources: A,B,C,D,E,F*
* Scripts

Sheffield

Sheffield Hallam University Library
Psalter Lane Site
Sheffield
South Yorkshire S11 8UZ
Tel: 0114 225 2721
Fax: 0114 225 2717
Website: http://www.shu.ac.uk/
services/ic/
Contact: Librarian, School of Cultural Studies
Resources: A,B,C,D,E

Sheffield Libraries & Information Services
Arts and Social Sciences Section
Central Library, Surrey Street
Sheffield S1 1XZ
Tel: 0114 273 4747/8
Fax: 0114 273 5009
Contact: Librarian

Sheffield University Library
Main Library
University of Sheffield
Western Bank
Sheffield
South Yorkshire S10 2TN
Tel: 0114 222 7200/1
Fax: 0114 273 9826
Contact: Head of Reader Services

Solihull

Solihull College
Chelmsley Campus,
Partridge Close
Chelmsley Wood
Solihull B37 6UG
Tel: 0121 770 5651
Contact: Librarian

Southampton

LSU College of Higher Education
The Avenue
Southampton SO17 1BG
Tel: 01703 228761
Fax: 01703 230944
Contact: Librarian

Periodical Office, Hartley Library
University of Southampton
University Road, Highfield
Southampton
Hants SO17 1BJ
Tel: 01703 593521
Fax: 01703 593007
Contact: Assistant Librarian, Arts
Resources: B,D,E,F*
* Personal papers, pressbooks

Southampton Institute
Mountbatten Library
East Park Terrace
Southampton
Hampshire SO17 1BJ
Tel: 01703 319000
Fax: 01703 3576161
Website: http://www.solent.ac.uk/
library/
Contact: Assistant Librarian, Arts
Resources: A,B,C,D,E

Stirling

University of Stirling
Library
Stirling FK9 4LA
Tel: 01786 467 235
Fax: 01786 51335
Contact: Librarian

Stoke on Trent

Staffordshire University Library
and Information Service
College Road
Stoke-On-Trent
Staffordshire ST4 2DE
Tel: 01782 294770/294809
Fax: 01782 744035
Contact: Art & Design Librarian

Sunderland

City of Sunderland Education & Community Services
City Library and Art Centre
Fawcett Street
Sunderland SR1 1RE
Tel: 0191-514 1235
Fax: 0191-514 8444
Contact: Librarian
Resources: D,E

Sunderland University Library
Langham Tower
Ryhope Road
Sunderland SR2 7EE
Tel: 0191 515 2900
Fax: 0191 515 2423
Contact: Librarian

Sutton

Sutton Central Library
Music and Arts Department
St Nicholas Way
Sutton
Surrey SM1 1EA
Tel: 0181 770 4764/5
Fax: 0181 770 4777
Contact: Arts Librarian

Swansea

Swansea Institute of Higher Education Library
Townhill Road
Swansea, SA2 OUT
Tel: 01792 481000
Fax: 01792 298017
Contact: Librarian

Teddington

Cinema Theatre Association
44 Harrowdene Gardens
Teddington
Middlesex TW11 0DJ
Tel: 0181 977 2608
Contact: Secretary

Uxbridge

Brunel University Library
Uxbridge
Middlesex UB8 3PH
Tel: 01895 274000
Fax: 01895 232806
Contact: Librarian

Warrington

Warrington Collegiate Institute
University College Library
Padgate Campus
Crab Lane
Warrington WA2 ODB
Tel: 01925 494284
Website: http://www.warr.ac.uk
Contact: College Librarian

Wellingborough

Tresham Institute of Further and Higher Education Library
Church Street
Wellingborough
Northamptonshire NN8 4PD
Tel: 01933 224165
Fax: 01933 441832
Contact: Librarian

Winchester

King Alfred's College Library
Sparkford Road, Winchester
Hampshire SO22 4NR
Tel: 01962 827306
Fax: 01962 827443
Website: http://www.kc.wkac.ac.uk
Contact: Librarian
Resources: A,B,C,D

Winchester School of Art
Park Avenue, Winchester
Hampshire SO23 8DL
Tel: 01962 842500
Fax: 01962 842496
Website:www.soton.ac.uk/
Contact: Head of Learning Resources
Resources: B,E

Wolverhampton

Light House
Media Reference Library
The Chubb Buildings
Fryer Street
Wolverhampton WV1 1HT
Tel: 01902 716055
Fax: 01902 717143
email: lighthse@waverider.co.uk
Contact Library: Carole Bayley
Exhibitions/Cultural Events: Evelyn Wilson
Chief Executive: Frank Challenger
Resources: A,B,E,F*
* Scripts, pressbooks

Wolverhampton Libraries and Information Services
Central Library
Snow Hill
Wolverhampton WV1 3AX
Tel: 01902 312 025
Fax: 01902 714 579
Contact: Librarian
Resources: B,D

Wolverhampton University
Art and Law Library
54 Stafford Street
Wolverhampton WV1 3AX
Tel: 01902 321597
Fax: 01902 322668
Contact: Art and Design Librarian

York

University College of Ripon and York St John Library
Lord Mayors Walk
York YO3 7EX
Tel: 01904 616700
Fax: 01904 612512
Website: http//www.ucrysj.ac.uk/
services/library/index.htm
Contact: Librarian
Resources:B,C,D,E,F*
* Ripon campus of college houses -
Yorkshire Film Archive

ORGANISATIONS

Listed below are the main trade/government organisations and bodies relevant to the film and television industries. A separate list of the Regional Film Commissions concludes the section

ABC (Association of Business Communicators)
1 West Ruislip Station
Ruislip
Middx HA4 7DW
Tel: 01895 622 401
Fax: 01895 631 219
Roger Saunders
Trade association of professionals providing the highest standards of audiovisual/video equipment/services for use in corporate communication

ACCS (Association for Cultural and Communication Studies)
Dept of Literature & Languages
Nottingham Trent University
Clifton Site
Nottingham NG11 8NS
Tel: 0115 941 8418 x3289
Fax: 0115 948 6632
Georgia Stone
Provides a professional forum for teachers and researchers in Media, Film, Television and Cultural Studies in both further and higher education. Its Executive Committee is drawn from the college and university sectors. Organises an Annual Conference, seeks to facilitate the exchange of information among members and liaises with various national bodies including 'sister' organisations: AME, NAHEFV and BUFVC

AFMA Europe
49 Littleton Road
Harrow
Middx HA1 3SY
Tel: 0181 423 0763
Tel: 0181 423 7963
Chairman: Lawrence Safir
Hold rights to numerous Australian, US and UK pictures, including Sam Spiegel's Betrayal

AIM (All Industry Marketing for Cinema)
22 Golden Square
London W1R 3PA
Tel: 0171 437 4383
Fax: 0171 734 0912
John Mahony
Unites distribution, exhibition and cinema advertising in promoting cinema and cinema-going. Funds film education, holds Cinema Days

for regional journalists, markets cinema for sponsorship and promotional ventures and is a forum for cinema marketing ideas

AMCCS (Association for Media, Cultural and Communication Studies)
Media and Cultural Studies
Middlesex University
Trent Park
Bramley Road
London N14 4XS
Tel: 0181 362 5065
Fax: 0181 362 5791

AME (Association for Media Education)
Television and Young People
24 Neal Street
London WC2H 9PS
Tel: 0171 379 4519
Fax: 0171 836 0702
email: TVYP@festival.demon.co.uk
Claire Bennett
Aims to promote media education at all levels; stimulate links with and between existing media education networks and provide a forum for the dissemination of effective ideas and practice. AME is open to anyone involved or interested in media education. It seeks to involve teachers and lecturers across all age phases as well as media professionals and cultural workers

AMPAS (Academy of Motion Picture Arts & Sciences)
8949 Wilshire Boulevard
Beverly Hills
CA 90211
USA
Tel: (1) 310 247 3000
Fax: (1) 310 859 9619
Organisation of producers, actors and others which is responsible for widely promoting and supporting the film industry, as well as for awarding the annual Oscars

AMPS (Association of Motion Picture Sound)
28 Knox Street
London W1H 1FS
Tel: 0171 402 5429
Fax: 0171 402 5429
Brian Hickin

Promotes and encourages science, technology and creative application of all aspects of motion picture sound recording and reproduction, and seeks to promote and enhance the status of those therein engaged

APRS - The Professional Recording Association
2 Windsor Square
Silver Street
Reading RG1 2TH
Tel: 0118 975 6218
Fax: 0118 975 6216
Mark Broad - Chief Executive
Represents the interests of the professional sound recording industry, including radio, TV and audio studios and companies providing equipment and services in the field. The long-established APRS Show is now run concurrently with the 'Vision & Audio' Exhibition at Earls Court, London

ASIFA
International Animated Film
Association, 61 Railwayside
Barnes
London SW13 0PQ
Tel: 0181 675 8299
Fax: 0181 675 8499
Pat Raine Webb
A worldwide association of individuals who work in, or make a contribution to, the animation industry, including students. Activities include involvement in UK and international events and festivals, an Employment Databank, Animation Archive, sales of animation artwork, children's workshops. The UK group provides an information service to members and a news magazine

Advertising Association
Abford House
15 Wilton Road
London SW1V 1NJ
Tel: 0171 828 2771/828 4831
Fax: 0171 931 0376
Andrew Brown
A federation of 28 trade associations and professional bodies representing advertisers, agencies, the advertising media and support services. It is the central organisation for the UK advertising business, on British and

European legislative proposals and other issues of common concern, both at national and international levels, and as such campaigns actively to maintain the freedom to advertise and to improve public attitudes to advertising. It publishes UK and European statistics on advertising expenditure, instigates research on advertising issues and organises seminars and courses for people in the communications business. Its Information Centre is one of the country's leading sources for advertising and associated subjects

Advertising Film and Videotape Producers' Association (AFVPA)
26 Noel Street
London W1V 3RD
Tel: 0171 434 2651
Fax: 0171 434 9002
Cecilia Garnett
Represents most producers of TV commercials. It negotiates with recognised trade unions, with the advertisers and agencies and also supplies a range of member services

Advertising Standards Authority (ASA)
Brook House
Torrington Place
London WC1
Tel: 0171 580 5555
Fax: 0171 631 3051

Amalgamated Engineering and Electrical Union (AEEU)
Hayes Court, West Common Road, Bromley, Kent BR2 7AU
Tel: 0181 462 7755
Fax: 0181 462 4959
Trade union representing - among others -people employed in film and TV lighting/electrical/electronic work

Arts Council of England
Visual Arts Department
14 Great Peter Street
London SW1P 3NQ
Tel: 0171 973 6443
Fax: 0171 973 6581
David Curtis, Gary Thomas
The Visual Arts Department works with national agencies for artists' film and video and supports the production of film and video work by artists, and related activities, including festivals and publications, through the Artists' Film and Video National Funds: Production, Exhibition

Arts Council of Wales
9 Museum Place
Cardiff CF1 3NX
Tel: 01222 394711
Fax: 01222 221447
Chief Executive: Emyr Jenkins

Asian Film Academy
Labour Party Buildings
rear of 142 - 144 Hounslow Road
Hanworth
London TW13 6AA
Tel: 0181 893 8385
(0589 585 647 mobile)
Fax: 0181 893 8385
The Academy provides information and services in respect of Asian arts and films and keeps a register of UK Asian Artists. Runs a training school in acting, video and film workshops

Association of Professional Composers (APC)
The Penthouse
4 Brook Street
London W1Y 1AA
Tel: 0171 629 4828
Fax: 0171 629 0993
email: a.p.c.@dial.pipex.com
Rosemary Dixon
APC represents composers from all sides of the profession - concert music, film, television, radio, theatre, electronic media, library music, jazz and so on. Its aims are to further the collective interests of its members and to inform and advise them on professional and artistic matters

Audio-Visual Association
Herkomer House
156 High Street
Bushey
Herts WD2 3DD
Tel: 0181 950 5959
Fax: 0181 950 7560
Mike Simpson FBIPP, Terry Bowles, FBIPP
The Audio Visual Association is a Special Interest Group within the British Institute of Professional Photography. With the Institute's current thinking of lateral represen-tation within all categories of imaging and imaging technology, the AVA represents those individuals involved in the various disciplines of audiovisual

Australian Film Commission (AFC)
European Marketing Branch
2nd Floor, Victory House
99-101 Regent Street
London W1R 7HB
Tel: 0171 734 9383
Fax: 0171 434 0170
Pressanna Vasudevan
The AFC is a statutory authority established in 1975 to assist the development, production and distribution of Australian films. The European marketing branch services producers and buyers, advises on co-productions and financing, and promotes the industry at markets and through festivals

Authors' Licensing & Collecting Society
Marlborough Court
14-18 Holborn
London EC1
Tel: 0171 395 0600
Fax: 0171 395 0660
email: alcs@alcs.co.uk
Website: http://www.alcs.co.uk
The ALCS is the British collecting society for all writers. Its principal purpose is to ensure that hard-to-collect revenues due to authors are efficiently collected and speedily distributed. These include the simultaneous cable retransmission of the UK's terrestrial and various international channels, educational off-air recording, BBC World Service and BBC Prime television. Contact the ALCS office for more information

BAFTA (British Academy of Film and Television Arts)
195 Piccadilly
London W1V OLN
Tel: 0171 734 0022
Fax: 0171 734 1792
Ron Allison
Chief Executive: Jane Clarke
Corporate Events Manager: Julie Chadwell
Events and Education Officer: Claire Bennett
Director of Corporate Affairs: Ron Allison
BAFTA was formed in 1946 by Britain's most eminent filmmakers as a non-profit making company. It aims to advance the art and technique of film and television and encourage experiment and research. Membership is available to those who have worked, or have been working actively within the film and/or television industries for not less than three years. BAFTA has facilities for screenings, conferences, seminars and discussion meetings and makes representations to parliamentary committees when appropriate. Its Awards for Film and Television are annual televised events. There are also Awards for Children's films and programmes and for Interactive Entertainment. The Academy has branches in Liverpool, Manchester, Glasgow, Cardiff, Los Angeles and New York.

See also under Awards and Preview Theatres

BARB (Broadcasters' Audience Research Board)

5th Floor, North Wing
Glenthorne House
Hammersmith Grove
London W6 0ND
Tel: 0181 741 9110
Fax: 0181 741 1943
Succeeding the Joint Industries' Committee on Television Audience Research (JICTAR), BARB commissions audience research on behalf of the BBC and ITV

BECTU (Broadcasting Entertainment Cinematograph and Theatre Union)

111 Wardour Street
London W1V 4AY
Tel: 0171 437 8506
Fax: 0171 437 8268
General Secretary: Roger Bolton
Deputy General Secretary: Roy Lockett
BECTU is the UK trade union for workers in film, broadcasting and the arts. Formed in 1991 by the merger of the ACTT and BETA, the union is 32,000 strong and represents permanently employed and freelance staff in television, radio, film, cinema, theatre and entertainment. BECTU provides a comprehensive industrial relations service based on agreements with the BBC, ITV companies, Channel 4, PACT, AFVPA and MFVPA, Odeon, MGM, Apollo, Society of Film Distributors, National Screen Services, independent exhibitors and the BFI itself. Outside film and television, the union has agreements with the national producing theatres and with the Theatrical Management Association, the Society of West End Theatres and others

BKSTS - The Moving Image Society

63-71 Victoria House
Vernon Place
London WC1B 4DF
Tel: 0171 242 8400
Fax: 0171 405 3560
email: movimage@bksts.demon.co.uk
Executive Director: Anne Fenton
Formed in 1931, the BKSTS is the technical society for film, television and associated industries. A wide range of training courses and seminars are organised with special rates for members. The society produces many publications including a monthly journal Image Technology and a quarterly Cinema

Technology both free to members. Corporate members must have sufficient qualifications and experience, however student and associate grades are also available. Biennial conference has become a platform for new products and developments from all over the world. The BKSTs also has a college accreditation scheme and currently accredits 9 courses within the HE + FE sector

BREMA (British Radio & Electronic Equipment Manufacturers' Association)

Landseer House
19 Charing Cross Road
London WC2H 0ES
Tel: 0171 930 3206
Fax: 0171 839 4613
email: 100612.3251@compuserve.com
Trade association for British consumer electronics industry

BTDA (British Television Distributors' Association)

Channel 4 Television
124 Horseferry Road
London SW1P 2TX
Tel: 0171 306 8741
Fax: 0171 306 8364
Frances Berwick
Informal industry lobbying association representing the distribution arms of the major ITV companies, BBC Enterprises, Channel 4 International and the major independent distributors

BUFVC (British Universities Film and Video Council)

55 Greek Street
London W1V 5LR
Tel: 0171 734 3687
Fax: 0171 287 3914
email: bufvc@open.ac.uk
Website: http://www.bufvc.ac.uk
Murray Weston
An organisation, funded via the Open University, with members in many institutions of higher education. It provides a number of services to support the production and use of film, television and other audiovisual materials for teaching and research. It operates a comprehensive Information Service, produces a regular magazine Viewfinder, catalogues and other publications such as the Researchers' Guide to British Film and Television Collections and the BUFVC Handbook for Film and Television in Education, organises conferences and seminars and distributes specialised film and video material. It runs a preview and

editing facility for film (16mm) and video (Betacam and other formats). Researchers in history and film and programme researchers come to the Council's offices to use the Slade Film History Register, with its information on British newsreels. BUFVC's off-air recording back-up service records all television programmes from the four UK terrestrial channels between 10.00am and 2.10am each day. The recordings are held for at least two months allowing educational establishments to request copies if they have failed to record the material locally under ERA licence

BVA (British Video Association)

167 Great Portland Street
London W1 5FD
Tel: 0171 436 0041
Fax: 0171 436 0043
Represents, promotes and protects the collective rights of its members who produce and/or distribute video cassettes for rental and sale to the public

British Academy of Songwriters, Composers and Authors (BASCA)

The Penthouse
4 Brook Street
Mayfair
London W1Y 1AA
Tel: 0171 629 0992
Fax: 0171 629 0993
email: basca@basca.org.uk
Represents its members' interests within the music industry. It issues standard contracts between publisher and songwriter. Founded in 1947 and administers the annual Ivor Novello Award

British Actors Equity Association

Guild House
Upper St Martin's Lane
London WC2H 9EG
Tel: 0171 379 6000
Fax: 0171 379 7001
email: equity@easynet.co.uk
General Secretary: Ian McGarry
Equity was formed in 1930 by professional performers to achieve solutions to problems of casual employment and short-term engagements. Equity has 40,000 members, and represents performers (other than musicians), stage managers, stage directors, stage designers and choreographers in all spheres of work in the entertainment industry. It negotiates agreements on behalf of its members with producers' associations and other employers. In some fields of work only artists with previous professional

experience are normally eligible for work. Membership of Equity is treated as evidence of professional experience under these agreements. It publishes Equity Journal four times a year

British Amateur Television Club (BATC)

Grenehurst
Pinewood Road
High Wycombe
Bucks HP12 4DD
Tel: 01494 528899
email: memsec@batc.org.uk
Website: http://www.batc.org.uk
Non-profit making organisation run entirely by volunteers. BATC publish a quarterly technical publication CQTV which is only available via subscription at £9 per annum. CQTV is all about TV hardware brought down to a practical home construction level with circuits for all aspects of television and satellite television. The circuits are backed by an extensive PCB service. Examples are simple fade to black, electronic testcard, sync generators and TV production Switchers

British Board of Film Classification (BBFC)

3 Soho Square
London W1V 6HD
Tel: 0171 439 7961
Fax: 0171 287 0141
The 1909 Cinematograph Films Act required public cinemas to be licensed by their local authority. Originally this was a safety precaution against fire risk but was soon interpreted by the local authorities as a way of censoring cinema owners' choice of films. In 1912, the British Board of Film Classification was established by the film industry to seek to impose a conformity of view-point: films cannot be shown in public in Britain unless they have the BBFC's certificate or the relevant local authorisation. The Board finances itself by charging a fee for the films it views. When viewing a film, the Board attempts to judge whether a film is liable to break the law, for example by depraving and corrupting a significant proportion of its likely audience. It then assesses whether its material greatly and gratuitously offensive to a large number of people. The Board seeks to reflect contemporary public attitudes. There are no written rules but films are considered in the light of the above criteria, previous

decisions and the examiners' personal judgement. It is the policy of the Board not to censor anything on political grounds. Five film categories came into effect in 1982, with the introduction of a '12' category in August 1989:
U - Universal: Suitable for all
PG - Parental Guidance: Some scenes may be unsuitable for young children
12 - Passed only for persons of 12 years and over
15 - Passed only for persons of 15 years and over
18 - Passed only for persons of 18 years and over
R 18 - For Restricted Distribution only, through segregated premises to which no one under 18 is admitted. The final decision, however, still lies with the local authority. In 1986 the GLC ceased to be the licensing authority for London cinemas, and these powers devolved to the Borough Councils. Sometimes films are passed by the BBFC and then banned by local authorities (*Straw Dogs, Caligula*). Others may have their categories altered (*Monty Python's Life of Brian, 9½ Weeks, Mrs Doubtfire*). Current newsreels are exempt from censorship. In 1985 the BBFC was designated by the Home Secretary as the authority responsible for classifying video works under the Video Recordings Act 1984. The film categories listed above are also the basis for video classification. However there is an additional category on video Uc -Universal, particularly suitable for young children

British Broadcasting Corporation (BBC)

Broadcasting House
Portland Place
London W1A 1AA
Tel: 0171 580 4468
The BBC provides two national television networks, five national radio networks, as well as local radio and regional radio and television services. They are funded through the Licence Fee. The BBC is a public corporation, set up in 1927 by Royal Charter. The existing charter expired at the end of 1996. Government proposals for the future of the BBC were published in a White Paper in July 1994. The BBC also broadcasts overseas through World Service Radio and Worldwide Television, but these are not funded through the Licence Fee

British Copyright Council

29-33 Berners Street
London W1P 4AA
Tel: 0181 371 9993

Provides liaison between societies which represent the interest of those who own copyright in literature, music, drama and works of art, making representation to Government on behalf of its member societies

The British Council

Films and Television Department
11 Portland Place
London W1N 4EJ
Tel: 0171 389 3065
Fax: 0171 389 3041
The British Council is Britain's international network for education, culture and technology. It is an independent, non-political organisation with offices in over 100 countries. Films and Television Department acts as a clearing house for international festival screenings of British short films and videos, including animation and experimental work. Using its extensive 16mm library, and 35mm prints borrowed from industry sources, it also ensures British participation in a range of international feature film events. The department arranges seminars overseas on themes such as broadcasting freedom and the future of public service television. It publishes the International Directory of Film and Video Festivals (biennial) and the annual British Films Catalogue. A 15-seat Preview Theatre (16mm, 35mm, video) is available for daytime use by UK filmmakers

British Design & Art Directors

9 Graphite Square
Vauxhall Walk
London SE11 5EE
Tel: 0171 582 6487
Fax: 0171 582 7784
Marketing Manager: Marcelle Johnson
A professional association, registered as a charity, which publishes an annual of the best of British and international design, advertising, television commercials and videos, and organises travelling exhibitions. Professional awards, student awards, education programme, lectures. Membership details are available on request

British Federation of Film Societies (BFFS)

BFFS
PO BOX 1DR
London W1A IDR
Tel: 0171 734 9300
Fax: 0171 734 9093
The BFFS exists to promote the

work of some 300 film societies in the UK

British Film Commission
70 Baker Street
London W1M 1DJ
Tel: 0171 224 5000
Fax: 0171 224 1013
email: info@bfc.co.uk
Website: www.britfilmcom.co.uk
The British Film Commission is funded by central Government to promote the UK as an international production centre and to encourage the use of British locations, services, facilities and personnel. It provides a computerised information service at no charge to enquirers and is a permanent source of information on all matters relevant to overseas producers contemplating production in the UK. For more details on area and city film commissions in the UK, see end of section

British Film Designers Guild
9 Elgin Mews
London W9 1JZ
Tel: 0171 286 6716
Fax: 0171 286 6716
Promotes and encourages activities of all members of the art department. Full availability and information service open to all producers

British Film Institute
21 Stephen Street
London W1P 2LN
Tel: 0171 255 1444
Fax: 0171 436 7950
Website http//:www.bfi.org.uk
Founded in 1933, the BFI was incorporated by Royal Charter in 1983; it is the UK national agency with responsibility for encouraging the arts of film and television and conserving them in the national interest. Approximately half the BFI's funding comes from the Department for Culture, Media and Sport. The rest is raised from the subscriptions of its members, provision of services, sponsorship and donations (see p8)

British Institute of Professional Photography
Fox Talbot House
Amwell End
Ware
Herts SG12 9HN
Tel: 01920 464011
Fax: 01920 487056
Company Secretary: Alex Mair
The qualifying body for professional photography and photographic processing. Members represent specialisations in the fields of photography, both stills and moving images

British Interactive Multimedia Association Ltd
5/6 Clipstone Street
London W1P 7EB
Tel: 0171 436 8250
Fax: 0171 436 8251
email: enquiries@bima.co.uk
Website: http://www.bima.co.uk
Norma Hughes, General Secretary

British Phonographic Industry Ltd (BPI)
25 Savile Row
London W1X 1AA
Tel: 0171 287 4422
Fax: 0171 287 2252
email: general@bpi.co.uk
John Deacon, Director General
BPI is the industry association for record companies in the UK. It provides professional negotiating skills, legal advice, information and other services for its 200 members. It protects rights, fights piracy and promotes export opportunities. Organises the BRIT Awards. Information service available

British Screen Advisory Council (BSAC)
19 Cavendish Square
London W1M 9AB
Tel: 0171 499 4177
Fax: 0171 306 0329
email: BSACouncil@aol.com
Director: Fiona Clarke-Hackston,
Chairman: David Elstein
General Manager: Anthea Hillman
BSAC is an independent, advisory body to government and policy makers at national and European level. It is a source of information and research for the screen media industries. BSAC provides a unique forum for the audio-visual industry to discuss major issues which effect the industry. Its membership embraces senior management from all aspects of television, film, and video. BSAC regularly commissions and oversees research on the audio-visual industry and uses this research to underpin its policy documents. In addition to regular monthly meetings, BSAC organises conferences, seminars and an annual reception in Brussels. BSAC is industry funded

British Screen Development (BSD)
14-17 Wells Mews
London W1P 3FL
Tel: 0171 323 9080
Fax: 0171 323 0092

Head of Development: Emma Berkofsky
BSD makes loans for the development of British and European cinema feature films. Films such as Photographing Fairies; Wilde; Before the Rain; Antonia's Line; Land and Freedom; Rob Roy; House of America; The Tango Lesson; Jilting Joe. It has a two-tier loan system: screenplay loans for new writers; development loans for production companies to pay writers, and ancillary costs. BSD also part-finances, administers and oversees the production of a variety of short films around the country. In 1998 some 17 short films will be commissioned with budgets of around £25,000. In Europe, BSD supports and sponsors candidates through the SOURCES, ACE and ARISTA programmes

British Screen Finance
14-17 Wells Mews
London W1P 3FL
Tel: 0171 323 9080
Fax: 0171 323 0092
email: BS@cd-online.co.uk
Since January 1986, British Screen, a private company aided by Government grant, has taken over the role and the business of the National Film Finance Corporation which was dissolved following the Films Act 1985. The Department of National Heritage has committed support until March 1999. British Screen exists primarily to support new talent in commercially viable productions for the cinema which might find difficulty in attracting mainstream commercial funding. Between 1986 and 1995 it invested in more than 100 productions, and hopes to support a further 8-12 in 1997. Recent successful films include *The Crying Game, Orlando, Naked, Land and Freedom* and *Richard III*. Through British Screen Development it also runs programmes of short films made in Scotland, Northern Ireland and English regions

British Society of Cinematographers (BSC)
11 Croft Road
Chalfont St Peter
Gerrards Cross
Bucks SL9 9AE
Tel: 01753 888052
Fax: 01753 891486
email:
BritCinematographers@compuserve.com
Frances Russell
Promotes and encourages the pursuit

of the highest standards in the craft of motion picture photography. Publishes a Newsletter and the BSC Directory

British Tape Industry Association

Ambassador House
Brigstock Road
Thornton Heath CR7 7JG
Tel: 0181 665 5395
Fax: 0181 665 6447
email: bbia@admin.co.uk
Trade association for the manufacturers of blank audio and videotape

Broadcasting Press Guild

c/o Richard Last,
Tiverton, The Ridge
Woking
Surrey GU22 7EQ
Tel: 01483 764895
Fax: 01483 764895
An association of journalists who write about TV radio, and the media in the national, regional and trade press. Membership by invitation. Monthly lunches with leading industry figures as guests. Annual Television and Radio Awards voted for by members

Broadcasting Research Unit

VLV Librarian
101 King's Drive
Gravesend
Kent DA12 5BQ
Tel: 01474 352835
Fax: 01474 351112
The Broadcasting Research Unit was an independent Trust researching all aspects of broadcasting, development and technologies, which operated from 1980-1991. Its publications and research are now available from the above address

Broadcasting Standards Commission

7 The Sanctuary
London SW1P 3JS
Tel: 0171 233 0544
Fax: 0171 233 0397
Chair: Lady Howe
Deputy Chairs: Jane Leighton and Lord Dubs
Director: Stephen Whittle
On 1 April 1997 the Broadcasting Complaints Commission merged with the Broadcasting Council to form the Broadcasting Standards Commission. The Broadcasting Standards Commission is the statutory body for both standards and fairness in broadcasting. It is the only organisation within the regulatory framework of UK broadcasting to cover television and radio. This includes BBC and commercial broadcasters as well as text, cable, satellite and digital services. As an independent organisation representing the interests of the consumer, the Broadcasting Standards Commission considers the portrayal of violence, sexual conduct and matters of taste and decency. As an alternative to a court of law, it provides redress for people who believe they have been unfairly treated or subjected to unwarranted infringement of privacy. The Commission has three main tasks which are set out in the 1996 Broadcasting Act: - produces codes of practice relating to standards and fairness; considers and adjudicates on complaints; monitors, researches and reports on standards and fairness in broadcasting. The Commission does not have the power to preview or to censor broadcasting

CFL Vision

PO Box 35
Wetherby
Yorks LS23 7EX
Tel: 01937 541010
Fax: 01937 541083
email: euroview@compuserve.com
Website: http://www.euroview.co.uk
CFL Vision began in 1927 as part of the Imperial Institute and is reputedly the oldest non-theatrical film library in the world. It is part of the COI and is the UK distributor for their audio-visual productions as well as for a large number of programmes acquired from both public and private sectors. Over 300 titles, mostly on video, are available for loan or purchase by schools, film societies and by industry

Cable Communications Association

5th Floor
Artillery House
Artillery Row
London SW1P 1RT
Tel: 0171 222 2900
Fax: 0171 799 1471
Chief Executive: Bob Frost
Represents the interests of cable operators, installers, programme providers and equipment suppliers. For further information on cable, see under Cable and Satellite

Campaign for Press and Broadcasting Freedom

8 Cynthia Street
London N1 9JF
Tel: 0171 278 4430
Fax: 0171 837 8868
email: cpbf@arcitechs.com
Website: http//www.architechs.com/cpbf
A broad-based membership organisation campaigning for more diverse, accessible and accountable media in Britain, backed by the trade union movement. The CPBF was established in 1979. The mail order catalogue is regularly updated and includes books on all aspects of the media from broadcasting policy to sexism; its bi-monthly journal Free Press examines current ethical, industrial and political developments in media policy and practice. CPBF acts as a parliamentary lobby group on censorship and media reform.

Celtic Film and Television Association

1 Bowmont Gardens
Glasgow G12 9LR
Tel: 0141 342 4947
Fax: 0141 342 4948
email: mail@celticfilm.co.uk
Contact: Frances Hendron, Director
Organises an annual competitive festival/conference, itinerant Scotland, Ireland, Cornwall, Wales and Brittany in March/April. Supports the development of television and film in Celtic nations and indigenous languages

Central Office of Information (COI)

Films and Television Division
Hercules Road
London SE1 7DU
Tel: 0171 261 8500
Fax: 0171 928 5037
Ian Hamilton
COI Films and Television Division is responsible for government filmmaking on informational themes as well as the projection of Britain overseas. The COI organises the production of a wide range of documentary films, television programmes, video programmes and audiovisual presentations including video disc production. It uses staff producers, and draws on the film and video industry for production facilities. It provides help to visiting overseas television teams

Chart Information Network

No8, Montague Close
London Bridge
London SE1 9UR
Tel: 0171 334 7333
Fax: 0171 921 5942
Supplies BVA members with detailed sales information on the sell-through video market. Markets and licenses the Official Retail Video Charts for broadcasting and publishing around the world

Children's Film and Television Foundation (CFTF)

Elstree Film Studios
Borehamwood
Herts WD6 1JG
Tel: 0181 953 0844
Fax: 0181 207 0860
In 1944 Lord Rank founded the Children's Entertainment Film Division to make films specifically for children. In 1951 this resulted in the setting up of the Children's Film Foundation (now CFTF), a non-profit making organisation which, up to 1981, was funded by an annual grant from the BFFA (Eady money). The CFTF no longer makes films from its own resources but, for suitable children's/family cinema/ television projects, is prepared to consider financing script develop- ment for eventual production by commercial companies. Films from the Foundation's library are available for hiring at nominal charge in 35mm, 16mm and video format

Church of England Communications Unit

Church House
Great Smith Street
London SW1P 3NZ
Tel: 0171 222 9011 x356/7
Fax: 0171 222 6672
Rev Jonathan Jennings
(Out of office hours: 0171 222 9233)
Responsible for liaison between the Church of England and the broadcasting and film industries. Advises the C of E on all matters relating to broadcasting

Cinema Advertising Association (CAA)

127 Wardour Street
London W1V 4NL
Tel: 0171 439 9531
Fax: 0171 439 2395
Bruce Koster
The CAA is a trade association of cinema advertising contractors operating in the UK and Eire. First established as a separate organisa- tion in 1953 as the Screen Advertis- ing Association, its main purpose is to promote, monitor and maintain standards of cinema advertising exhibition including the pre-vetting of commercials. It also commissions and conducts research into cinema as an advertising medium, and is a prime sponsor of the CAVIAR annual surveys

Cinema & Television Benevolent Fund (CTBF)

22 Golden Square
London W1R 4AD
Tel: 0171 437 6567
Fax: 0171 437 7186
The CTBF is the trade fund operating in the UK for retired and serving employees (actors have their separate funds) who have worked for two or more years in any capacity, in the cinema, film or independent television industries'. The CTBF offers caring help, support and financial assistance, irrespective of age, and the Fund's home in Wokingham, Berkshire offers full residential and convalescent facilities

Cinema and Television Veterans

Elanda House
9 The Weald
Ashford
Kent TN24 8RA
Tel: 01233 639967
An association open to all persons employed in the United Kingdom or by United Kingdom companies in the cinema and/or broadcast television industries in any capacity other than as an artiste, for a total of at least thirty years

Cinema Exhibitors' Association (CEA)

22 Golden Square
London W1R 3PA
Tel: 0171 734 9551
Fax: 0171 734 6147
The first branch of the CEA in the industry was formed in 1912 and consisted of cinema owners. Following a merger with the Association of Independent Cinemas (AIC) it became the only association representing cinema exhibition. CEA members account for the vast majority of UK commercial cinemas, including independents, Regional Film Theatres and cinemas in local authority ownership. The CEA represents members' interests - within the industry and to local, national and European Government. It is closely involved with legislation (current and proposed) emanating from the UK Government and the European Commission which affects exhibition

Cinema Theatre Association

44 Harrowdene Gardens
Teddington
Middx TW11 0DJ
Tel: 0181 977 2608
Adam Unger
The Cinema Theatre Association was formed in 1967 to promote interest in Britain's cinema building legacy, in particular the magnificent movie palaces of the 1920s and 1930s. It is the only major organisa- tion committed to cinema preserva- tion in the UK. It campaigns for the protection of architecturally important cinemas and runs a comprehensive archive. The CTA publishes a bi-monthly bulletin and the magazine Picture House

Cinergy

Minema Cinema
45 Knightsbridge
London SW1X 7NL
Tel: 0171 235 4226
Fax: 0171 235 3426
Creative Director: Nick Walker
Co-ordinatior: Damian Spandley
Multimeidator: Nick Perry
Cinergy is a multimedia cabaret club with a special emphasis on short films. It is open to the public and welcomes submissions from filmmakers on all formats and with high, low or no budgets.

Comataidh Craolaidh Gaidhlig (Gaelic Broadcasting Committee)

4 Harbour View, Cromwell Street
Stornoway
Isle of Lewis HS1 2DF
Tel: 01851 705550
Fax: 01851 706432
The Gaelic Television Fund and Comataidh Telebhisein Gaidhlig was set up under the provisions of the Broadcasting Act 1990. Funds made available by the Government were to be paid to the ITC for the credit of the fund to be known as the Gaelic Television Fund. The Fund was to be managed by the body known as the Gaelic Television Committee. Under the Broadcasting Act 1996 the Gaelic Television Fund was redesignated as the Gaelic Broad- casting Fund and the Gaelic Television Committee became the Gaelic Broadcasting Committee

Commonwealth Broadcasting Association

CBA Secretariat
Broadcasting House
London W1A 1AA
Tel: 0171 765 5144
Fax: 0171 765 5152
email: cba@bbc.co.uk
Website: http://oneworld.org/cba
An association of 63 public service broadcasting organisations in 52 Commonwealth countries

Composers' Guild of Great Britain

The Penthouse
4 Brook Street
London W1Y 1AA

Tel: 0171 629 0886
Fax: 0171 629 0993
General Secretary: Naomi Moskovic
The Guild represents composers of serious music, covering the stylistic spectrum from jazz to electronics. Its main function is to safeguard and assist the professional interests of its members, provide information for those wishing to commission music and information on recommended commission fees for broadcast and live performance

Critics' Circle
Film Section
4 Alwyne Villas
London N1 2HQ
Tel: 0171 226 2726
Fax: 0171 354 2574
Chairman: Christopher Tookey
Vice-Chairman: John Marriott
Hon. Secretary: Tom Hutchinson
Hon. Treasurer: Peter Cargin
The film section of the Critics' Circle brings together leading national critics for meetings, functions and the presentation of annual awards

Deaf Broadcasting Council
70 Blacketts Wood Drive
Chorleywood, Rickmansworth
Herts WD3 5QQ
Text Tel: 01923 283127
Fax: 01923 283127
email: dmyers@cix.co.uk
An umbrella consumer group working as a link between hearing impaired people and TV broadcasters - aiming for increased high quality access to television and video

Defence Press and Broadcasting Advisory Committee
Room 2235
Ministry of Defence
Main Building
Whitehall
London SW1A 2HB
Tel: 0171 218 2206
Fax: 0171 218 5857
Secretary: Rear Admiral David Pulvertaft
The Committee is made up of senior officials from the Ministry of Defence, the Home Office and the Foreign & Commonwealth Office and representatives of the media. It issues guidance, in the form of DA Notices, on the publication of information which it regards as sensitive for reasons of national security

Department for Culture, Media and Sport (DCMS) - Media Division (Films)
2-4 Cockspur Street

London SW1Y 5DH
Tel: 0171 211 6000
Fax: 0171 211 6249
Contacts:
For BFI, British Screen Finance (BSF), European Co-production fund (ECPF): Aidan McDowell
Tel: 0171 211 6429
For Enquiries concerning film which might be made under UK Co-production Agreements: Diana Brown
Tel: 0171 211 6433
For MEDIA Programme, British Film Commission (BFC), National Film and Television School (NFTS) and Audiovisual Eureka (AVE): Peter Wright
Tel: 0171 211 6435
The Department for Culture Media and Sport is responsible for Government policy on film, relations with the film industry and Government funding for: the British Film Institute, British Screen Finance, the European Co-Production Fund (administered by British Screen Finance), the British Film Commission and the National Film and Television School. It is also responsible for Government policy on and contribution to, the EC Media Programme and Audiovisual Eureka. It also acts as the UK competent authority for administering the UK's seven bilateral co-production agreements and the European Co-production Convention

Department for Education
Sanctuary Buildings
Great Smith Street
London SW1P 3BT
Tel: 0171 925 5000
Fax: 0171 925 6000
The DFE is responsible for policies for education in England and the Government's relations with universities in England, Scotland and Wales

Director of Creative Artists
185 Stranmillis Road
Belfast BT9 5DU
See entry for Northern Ireland Film Council for Northern Irish activities in this area

The Directors' and Producers' Rights Society
15-19 Great Titchfield Street
London W1P 7FB
Tel: 0171 631 1077
Fax: 0171 631 1019
email: dprs@dial.pipex.com
Suzan Dormer
The Directors' and Producers' Rights Society is a collecting society

which administers authorials rights payments on behalf of British film and television

Directors' Guild of Great Britain
15-19 Great Titchfield Street
London W1P 7FB
Tel: 0171 436 8626
Fax: 0171 436 8646
email: guild@dggb.co.uk
Website: http://www.dggb.co.uk
Sarah Wain
Represents interests and concerns of directors in all media. Publishes regular magazine DIRECT

ELSPA
Suite 1, Haddonsacre, Station Road, Offenham (Nr Evesham), Worcs WR11 5LW
Tel: 01386 830642
Fax: 01386 833871
Website: www.elspa.com
ELSPA was founded in 1989 to create a voice for the leisure software industry. It has over 100 members including leading companies in the leisure software market

Educational Policy Services, BBC
BBC White City
201 Wood Lane
London W12 7TS
Tel: 0181 752 4204
Fax: 0181 752 4441
email: Lucia.jones@bbc.co.uk
EPS supports the work of BBC Education radio and television production departments. It services the Educational Broadcasting Council representing professional users

Educational Television & Media Association (ETMA)
37 Monkgate
York YO3 7PB
Tel: 01904 639212
Fax: 01904 639212
email: josie.key@etma.u-net.com
Website: http://www.etrc.ox.ac.uk/ETMA.html
A dynamic association comprising a wide variety of users of television and other electronic media in education. International competition for educational videos (including Broadcast category) - deadline November. Journal of Educational Media (free to members), Annual Conference and Exhibition in the Spring. Regional meetings/workshops/training programmes. Patrons: BBC Education, Canford Audio, Channel 4, JVC Professional, MarCom Systems, Panasonic Broadcast, Sony Broadcast, Strand Lighting. New members welcome - contact Administrator

FIAF (International Federation of Film Archives)

1 rue Defacqz
1000 Bruxelles, Belgium
Tel: (322) 538 30 65
Fax: (322) 534 47 74
Christian Dimitriu, Senior Administrator
Indexes, with the help of archive specialists, more than 200 of the world's most important film and television periodicals. Publishes an annual volume of the International Index of Film Periodicals. Recently introduced the International Film Archive CD-Rom a compilation of databases originating from the work of FIAF Commissions. The International Index to film/TV Periodicals is the core work, currently covering 1979 to present, soon to be extended back to 1972. Also included is information on 20,000 silent films held by FIAF archives, a directory of film and TV documentation collections, plus invaluable film-related bibliographies

FOCAL (Federation of Commercial Audio-Visual Libraries)

Pentax House
South Hill Avenue
Northolt Road
South Harrow
Middx HA2 ODU
Tel: 0181 423 5853
Fax: 0181 423 5853
email: anne@focalltd.demon.co.uk
Website: http://www.
focalltd.demon.co.uk
Administrator: Anne Johnson
An international, non-profit making professional trade association representing commercial film/audiovisual libraries and interested individuals. Among other activities, it organises regular meetings, maximises copyright information, and produces a directory of libraries and quarterly journal

Federation Against Copyright Theft

7 Victory Business Centre
Worton Road
Isleworth
Middx TW7 6DB
Tel: 0181 568 6646
Fax: 0181 560 6364
R Dixon, Director General
DNL Lowe, Company Secretary
FACT, Federation Against Copyright Theft, is an investigative organisation funded by its members to combat counterfeiting, piracy and misuse of their products. The

members of FACT are major companies in the British and American film, video and television industries. FACT is a non-profit making company limited by guarantee. FACT assists all statutory law enforcement authorities and will undertake private criminal prosecutions wherever possible

Federation of Entertainment Unions (FEU)

1 Highfield
Twyford
Nr Winchester
Hants SO21 1QR
Tel: 01962 713134
Fax: 01962 713288
email: harris@interalpha.co.uk
The FEU is a lobbying and campaigning group and meets regularly with statutory bodies and other pressure groups ranging from the BBC and ITC and the British Film Commission through to the Parliamentary All Party Media Committee and the Voice of the Listener and Viewer. The Federation comprises British Actors' Equity Association, Broadcasting Entertainment Cinematograph and Theatre Union, Musicians' Union, National Union of Journalists, Writers' Guild of Great Britain and Amalgamated Engineering & Electrical Union (Electricians Section). It has three standing committees covering Film and Electronic Media, European Affairs and Training

The Feminist Library

5 Westminster Bridge Road
London SE1 7XW
Tel: 0171 928 7789
The Feminist Library provides information about women's studies, courses, services and current events. It has a large collection of fiction and non-fiction including books, pamphlets, papers etc. It holds a wide selection of journals and newsletters from all over the world and produces its own quarterly newsletter. Social events are held and discussion groups meet every Tuesday. The library is run entirely by volunteers. Membership library. Open Tuesday (11.00am-8.00pm) and Saturday (2.00-5.00pm)

Film Artistes' Association (FAA)

111 Wardour Street
London W1V 4AY
0171 437 8506
0171 437 8268
Spencer MacDonald
The FAA represents extras, doubles, stand-ins and small part artistes.

Under an agreement with PACT, it supplies all background artistes in the major film studios and within a 40 mile radius of Charing Cross on all locations

Film Complaints Panel

22 Golden Square
London W1R 3PA
Chief Administrator: Annette Bradford

Film Education

Alhambra House
27-31 Charing Cross Road
London WC2H 0AU
Tel: 0171 976 2291
Fax: 0171 839 5052
email: post@film-ed.u-net.com
website:http://
www.filmeducation.org
Ian Wall
Film Education is a registered charity supported by the UK film industry. For over a decade it has been at the forefront of the development of Film and Media Studies in schools and colleges and now has more than 20,000 named primary and secondary teacher contacts on its unique database.
The main aims of Film Education are to develop the use of film in the school curriculum and to facilitate the use of cinemas by schools. To this end it publishes a variety of free teaching materials, produces BBC Learning Zone programmes, organises Inset and runs a range of workshops and school screenings. All Film Education resources are carefully researched and written by teachers for teachers

Film Unit Drivers Guild

Guild House
51 North Road
Wimbledon
London SW19 1AQ
Tel: 0181 544 1141
Fax: 0181 544 1151
Micky or Jan Grover
FUDG represents its freelance members in the Film and Television industry when they are not on a production. It supplies them with work, such as pick ups and drops to any destination the client wishes to travel. Guild members are made up of professional film unit drivers and will look after all transportation needs

First Film Foundation

9 Bourlet Close
London W1P 7PJ
Tel: 0171 580 2111
Fax: 0171 580 2116
Director: Jonathan Rawlinson
Development and training provider for new writing, producing and

directing talent. Schemes include pan-European screen writers programme North By Northwest and New Directions, which promotes new British directors to the New York and LA film industries. The Foundation offers a script feedback service for feature scripts

German Federal Film Board and German Film Export Union

4 Lowndes Court
Carnaby Street
London W1V 1PP
Tel: 0171 437 2047
Fax: 0171 439 2947
Iris Kehr
UK representative of the German Federal Film Board (Filmförderungsanstalt), the government industry organisation, and the German Film Export Union (Export Union des Deutschen Films), the official trade association for the promotion of German films abroad. For full details see entries under Organisations (Europe)

Glasgow Film Fund

74 Victoria Crescent Road
Glasgow G12 9JN
Tel: 0141 337 2526
Fax: 0141 337 2562
Contact: Judy Anderson
The Glasgow Film Fund provides production funding for companies making feature films in the Glasgow area or produced by Glasgow-based production companies. Applications are accepted for films intended for theatrical release, with a budget of at £500,000. The maximum investment made by the Glasgow Film Fund in any one project will be £150,000, however, where there is an exceptionally high level of local economic benefit the GFF may consider raising its maximum investment to £250,000. Glasgow Film Fund application forms, meeting dates, submission deadlines and further details are available from the GFF office. Production credits include: *Shallow Grave, The Near Room, Small Faces, Carla's Song, The Slab Boys, Regeneration, The Life of Stuff, Orphans, My Name is Joe, The Acid House* and *The Debt Collector*

Guild of British Animation

26 Noel Street
London W1V 3RD
Tel: 0171 434 2651

Fax: 0171 434 9002
Cecilia Garnett
Represents interests of producers of animated films. AFVPA acts as secretariat for this association

Guild of British Camera Technicians

5-11 Taunton Road
Metropolitan Centre
Greenford
Middx UB6 8UQ
Tel: 0181 578 9243
Fax: 0181 575 5972
Office manager: Maureen O'Grady
Magazine Editors, Eyepiece: Charles Hewitt and Kerry-Anne Burrows
The Guild exists to further the professional interests of technicians working with film or video motion picture cameras. Membership is restricted to those whose work brings them into direct contact with these cameras and who can demonstrate competence in their particular field of work. By setting certain minimum standards of skill for membership, the Guild seeks to encourage its members, especially newer entrants, to strive to improve their art. Through its publication, Eyepiece: disseminates information about both creative and technical developments, past and present, in the film and television industry

Guild of British Film Editors

Travair
Spurlands End Road
Great Kingshill
High Wycombe
Bucks HP15 6HY
Tel: 0149 4712313
Fax: 0149 4712313
To ensure that the true value of film and sound editing is recognised as an important part of the creative and artistic aspects of film production

Guild of Film Production Accountants and Financial Administrators

Pinewood Studios
Pinewood Road
Iver
Bucks SL0 0NH
Tel: 01753 651767
Fax: 01753 652803
email: secretary@gfpa.org.uk
Website: www.gfpa.org.uk
John Seargent - Honorary Secretary
To update members with information on methods, practice and legislation affecting the film industry. To promote and maintain a high

standard of film production accounting, cost and financial administration

Guild of Film Production Executives

Pinewood Studios
Iver
Bucks SL0 0NH
Tel: 01753 656428
Fax: 01753 656850
Contact: Ann Runeckles

The Guild of Regional Film Writers

Honeysuckle House
9 Vaughan Road
Dibden
Southampton So45 5UL
Darren Vaughan, Chair
Founded in 1987, this self-funding organisation aims to encourage, support and promote the work of regional film journalists. They work closely with distributors, exhibitors and other key industry bodies, and three times a year members are invited to attend 'Cinema Days', an A.I.M. initiative, where new movies are screened and press conferences held. Prospective members should supply three relevant cuttings/tapes for approval

Guild of Stunt and Action Co-ordinators

72 Pembroke Road
London N8 6NX
Tel: 0171 602 8319
Sally Fisher
To promote the highest standards of safety and professionalism in film and television stunt work

Guild of Television Cameramen

1 Churchill Road
Whitchurch,
Tavistock
Devon PL19 9BU
Tel: 01822 614405
Fax: 01822 614405
Sheila Lewis
The Guild was formed in 1972 'to ensure and preserve the professional status of the television cameramen and to establish, uphold and advance the standards of qualification and competence of cameramen'. The Guild is not a union and seeks to avoid political involvement

Guild of Vision Mixers

147 Ship Lane
Farnborough
Hants GU14 8BJ

Tel: 01252 514953
Fax: 01252 656756
Peter Turl
The Guild aims to represent the interests of vision mixers throughout the UK and Ireland, and seeks to maintain the highest professional standards in vision-mixing

Hollywood Foreign Press Association

292 S. La Cienega Blvd, #316
Beverly Hills
CA 90211
USA
Tel: (1) 310 657 1731
Fax: (1) 310 657 5576
Journalists reporting on the entertainment industry for non-US media. Annual event: Golden Globe Awards - awarding achievements in motion pictures and television

IAC (Institute of Amateur Cinematographers)

24c West Street
Epsom
Surrey KT18 7RJ
Tel: 01372 739672
Encouraging amateurs interested in the art of making moving pictures and supporting them with a variety of services

ITV Network Centre

200 Gray's Inn Road
London WC1X 8HF
Tel: 0171 843 8000
Fax: 0171 843 8158
Director ITV Association: Barry Cox
Network Director: Marcus Plantin
A body wholly owned by the ITV companies which independently undertakes the commissioning and scheduling of those television programmes which are shown across the ITV Network through the ITV Association. It also provides a range of services to the ITV companies where a common approach is required

IVCA (International Visual Communication Association)

Bolsover House
5-6 Clipstone Street
London W1P 8LD
Tel: 0171 580 0962
Fax: 0171 436 2606
email: info@ivca.org
Chief Executive: Wayne Drew
The IVCA is the largest European Association of its kind, representing a wide range of organisations and individuals working in the established

and developing technologies of visual communication. With roots in video, film and business events industries, the Association has also developed significant representation of the new and fast growing technologies, notably business television, multimedia, interactive software and the internet. It provides business services for its members: legal help, internet service, insurance, arbitration etc. and holds events/seminars for training, networking and for all industry related topics

Imperial War Museum Film and Video Archive

Lambeth Road
London SE1 6HZ
Tel: 0171 416 5000
Fax: 0171 416 5379
See entry under Archives and Film Libraries

Incorporated Society of British Advertisers (ISBA)

44 Hertford Street
London W1Y 8AE
Tel: 0171 499 7502
Fax: 0171 629 5355
Deborah Morris
The ISBA was founded in 1900 as an association for advertisers, both regional and national. Subscriptions are based on advertisers' expenditure and the main objective is the protection and advancement of the advertising interests of member firms. This involves organised representation, co-operation, action and exchange of information and experience, together with conferences, workshops and publications. ISBA offer a communications consultancy service for members on questions as varied as assessment of TV commercial production quotes to formulation of advertising agency agreements

Incorporated Society of Musicians (ISM)

10 Stratford Place
London W1N 9AE
Tel: 0171 629 4413
Fax: 0171 408 1538
Chief Executive: Neil Hoyle
Professional association for all musicians: teachers, performers and composers. The ISM produces various publications, including the monthly Music Journal, and gives advice to members on all professional issues

Independent Film Distributors' Association (IFDA)

10a Stephen Mews
London W1P 0AX
Tel: 0171 957 8957
Fax: 0171 957 8968
IFDA was formed in 1973, and its members are mainly specialised film distributors who deal in both 16mm and 35mm and every type of film from classic features to 'popular music.' Supply many users including universities, schools, hospitals, prisons, independent cinemas, hotels, film societies, ships etc

Independent Television Commission (ITC)

33 Foley Street
London W1P 7LB
Tel: 0171 255 3000
Fax: 0171 306 7800
email: Publicaffairs@itc.org.uk
The ITC is the public body responsible for licensing and regulating commercially funded television services. These include Channel 3 (ITV), Channel 4, Channel 5, public teletext and a range of cable, local delivery and satellite services

Institute of Practitioners in Advertising (IPA)

44 Belgrave Square
London SW1X 8QS
Tel: 0171 235 7020
Fax: 0171 245 9904
The representative body for UK advertising agencies. Represents the collective views of its member agencies in negotiations with Government departments, the media and industry and consumer organisations

International Arts Bureau

4 Baden Place
Crosby Row
London SE1 1YW
Tel: 0171 403 6454
Phone enquiry service 0171 403 7001
Fax: 0171 403 2009
Director: Rod Fisher
The International Arts Bureau was established to undertake for the Arts Council of England, the Regional Arts Boards and the arts constituency and funding system, services previously dealt with by the Council's in-house International Affairs Unit

International Association of Broadcasting Manufacturers (IABM)

PO Box 979

Slough SL2 3DL
Tel: 01753 645682
Fax: 01753 645682
email: iabm@technocam.com
Secretariat: Brenda White
IABM aims to foster the interests of manufacturers of broadcast equipment from all countries. Areas of membership include liaison with broadcasters, standardisation, other technical information, an annual product Award for design and innovation and exhibitions. All companies active in the field of broadcast equipment manufacturing are welcome to join

International Teleproduction Society
c/o PACT
45 Mortimer Street
London W1N 7TD
Tel: 0171 331 6000
Fax: 0171 331 6700
The ITS UK Chapter (International Teleproduction Society) is the trade association serving the professional community of businesses that provide creative and technical services in pictures and sound

International Federation of the Phonographic Industry (IFPI)
IFPI Secretariat
54 Regent Street
London W1R 5PJ
Tel: 0171 878 7900
Fax: 0171 878 7950
Director General: Nicholas Garnett
An international association of 1,300 members in 71 countries, representing the copyright interests of the sound recording and music video industries

International Institute of Communications
Tavistock House South
Tavistock Square
London WC1H 9LF
Tel: 0171 388 0671
Fax: 0171 380 0623
The IIC promotes the open debate of issues in the communications field worldwide. Its current interests cover legal and policy, economic and public interest issues. It does this via its: bi-monthly journal Intermedia; through its international communications library; annual conference; sponsored seminars and research forums

Kraszna-Krausz Foundation
122 Fawnbrake Avenue
London SE24 0BZ
Tel: 0171 738 6701
Fax: 0171 738 6701
email: kk@dial.pipex.com

Administrator: Andrea Livingstone
The Foundation offers small grants to assist in the development of new or unfinished projects, work or literature where the subject specifically relates to the art, history, practice or technology of photography or the moving image (defined as film, television, video and related screen media)

London Screenwriters' Workshop
Holborn Centre for the Performing Arts
Three Cups Yard
Sandland St
London WC1R 4PZ
Tel: 0171 242 2134
Promotes contact between screenwriters and producers, agents, development executives and other film and TV professionals through a wide range of seminars. Practical workshops provide training in all aspects of the screenwriting process. Membership is open to anyone interested in writing for film and TV, and to anyone in these and related media

MEDIA (Media: Economic Development and Investment Agency)
Stonehills, Shields Road
Gateshead
Tyne and Wear NE10 0HW
Tel: 0191 495 0007
Fax: 0191 495 2266
An initiative taken by North East Media Development Trust to develop the audiovisual industry in the North of England. The Trust has established an audiovisual training centre, a media park offering sheltered work space and access to equipment to media practitioners, and a facilities centre providing leading edge technology

Mechanical-Copyright Protection Society (MCPS)
29/33 Berner Street
London W1P 4AA
Tel: 0181 664 4400
Fax: 0181 769 8792
email: info@mcps.co.uk
Website: http://www.mcps.co.uk
Contact: Non-retail Licensing Department. MCPS is an organisation of music publishers and composers, which issues licences for the recording of its members' copyright musical works in all areas of television, film and video production. Free advice and further information is available on request

Mental Health Media Council
The Resource Centre
356 Holloway Road
London N7 6PA
Tel: 0171 700 0100
Fax: 0171 700 0099
An independent charity founded in 1965, MHMC provides information, advice and consultancy on film/video use and production relevant to health, mental health, physical disability, learning difficulties and most aspects of social welfare. Resource lists on audiovisual material, quarterly newsletter Mediawise. Producers of video and broadcast programmes

Metier
Glyde House, Glydegate
Bradford BD5 0BQ
Tel: 01274 738 800
Fax: 01274 391 566
Chief Exec: Duncan Sones
A National Training Organisation, developing National and Scottish Vocational Qualifications for occupations in performing and visual arts, arts administration, and arts entertainment and technical support functions in the arts and entertainment sector. It is responsible for strategic action to improve the quality, availability and effectiveness of vocational training within its industrial sector

The Museum of Television and Radio
25 West 52 Street
New York, NY 10019
USA
Tel: (1) 212 621 6600/6800
Fax: (1) 212 621 6715
The Museum (formerly The Museum of Broadcasting) collects and preserves television and radio programmes and advertising commercials, and makes them available to the public. The collection, which now includes nearly 60,000 programmes, covers 70 years of news, public affairs programmes, documentaries, performing arts, children's programming, sports, and comedy. The Museum organises exhibitions, and screening and listening series

Music Film and Video Producers' Association
(MFVPA)
26 Noel Street
London W1V 3RD
Tel: 0171 434 2651
Fax: 0171 434 9002
The MFVPA was formed in 1985 to represent the interests of pop/music promo production companies. It

Top names, Top productions, Top locations!
The London Borough of Greenwich
welcomes film makers large and small

Four Weddings and a Funeral
The Madness of King George
Richard III
Oasis: 'Wonderwall' video
The Bill

filming in Greenwich

a one stop service

from commercials
to television
and pop videos
and of course
feature films

ADMIT ONE

ADMIT ONE

for a brochure
or a chat call
on 0181 312 5662
or 07970 139145

negotiates agreements with bodies such as the BPI and BECTU on behalf of its members. Secretariat support is run through AFVPA

Music Publishers' Association

3rd Floor, Strandgate
18/20 York Buildings
London WC2N 6JU
Tel: 0171 839 7779
Fax: 0171 839 7776
The only trade association representing UK music publishers. List of members available at £8.00

Musicians' Union (MU)

60-62 Clapham Road
London SW9 0JJ
Tel: 0171 582 5566
Fax: 0171 793 9185
Howard Evans, Session Organiser
Represents the interests of performing musicians in all areas

National Association for Higher Education in Film and Video (NAHEFV)

c/o London International Film
School, 24 Shelton Street
London WC2H 9HP
Tel: 0171 836 9642/240 0168
Fax: 0171 497 3718
The Association's main aims are to act as a forum for debate on all aspects of film, video and TV education and to foster links with industry, the professions and Government bodies. It was established in 1983 to represent all courses in the UK which offer a major practical study in film, video or television at the higher educational level

National Campaign for the Arts

Francis House
Francis Street
London SW1P 1DE
Tel: 0171 828 4448
Fax: 0171 931 9959/0171 233 6564
email: nca@ecna.org
Website: http://www.ecna.org./nca/
Director: Jennifer Edwards
The NCA specialises in lobbying, campaigning research and information. It provides facts and figures for politicians, journalists and any other interested parties. It is independent and funded by its members - individuals and organisations

National Council for Educational Technology (NCET)

Milburn Hill Road
Science Park
University of Warwick
Coventry CV4 7JJ
Tel: 01203 416994

Fax: 01203 411418
Formerly the National Council for Educational Technology (NCET), the new remit will be to ensure that technology supports the DfEE's objectives to drive up standards, in particular to provide the professional expertise the DfEE needs to support the future development of the National Grid for Learning. BECTA will also have a role in the further education sector's developing use of ICT, in the identification of ICT opportunities for special educational needs, and in the evalution of new technologies as they come on stream

National Film and Television School

Beaconsfield Studios
Station Road
Beaconsfield
Bucks HP9 1LG
Tel: 01494 671234
Fax: 01494 674042
Director: Stephen Bayly
The National Film and Television School provides advanced training and retraining in all major disciplines to professional standards. Graduates are entitled to BECTU membership on gaining employment. It is an autonomous non-profit making organisation funded by the Department for Culture, Media and Sport and the film and television industries. See also under Courses

National Film Trustee Company (NFTC)

14-17 Wells Mews
London W1P 3FL
Tel: 0171 580 6799
Fax: 0171 636 6711
An independent revenue collection and disbursement service for producers and financiers. The NFTC has been in business since 1971. It is a subsidiary of British Screen Finance

National Museum of Photography Film & Television

Pictureville
Bradford BD1 1NQ
Tel: 01274 203300
Fax: 01274 723155
Bill Lawrence, Head of Cinema
The world's only museum devoted to still and moving pictures, their technology and history. Features Britain's first giant IMAX film system; the world's only public Cinerama; interactive galleries and 'TV Heaven', reference library of programmes and commercials

National Screen Service

15 Wadsworth Road

Greenford
Middx UB6 7JN
Tel: 0181 998 2851
Fax: 0181 997 0840
2 Wedgwood Mews
12-13 Greek Street
London W1V 6BH
Tel: 0171 437 4851
Fax: 0171 287 0328
John Mahony, Brian McIlmail, Norman Darkins
Formed in 1926 as a subsidiary of a US corporation and purchased by its present British owner/directors in 1986. It distributes trailers, posters and other publicity material to UK cinemas and carries out related printing activity

National Union of Journalists

314 Gray's Inn Road
London WC1X 8DP
Tel: 0171 278 7916
Fax: 0171 837 8143
National Broadcasting Organiser: John Fray
Direct line to Broadcasting Office: 0171 843 3726
Represents all journalists working in broadcasting in the areas of news, sport, current affairs and features. It has agreements with all the major broadcasting companies and the BBC. It also has agreements with the main broadcasting agencies, WTN, Reuters Television and PACT

National Viewers' & Listeners' Association (NVALA)

All Saints House, High Street
Colchester
Essex CO1 1UG
Tel: 01206 561155
Fax: 01206 766175
General Secretary: John C Beyer
Founder & President Emeritus: Mary Whitehouse CBE
Concerned with moral standards in the media

Network of Workshops (NoW)

c/o Video in Pilton
30 Ferry Road Avenue, West
Pilton
Edinburgh EH4 4BA
Tel: 0131 343 1151
Fax: 0131 343 2820
A membership organisation which is open to all independent collective film and video groups who are committed to the cultural aims stated in the BECTU workshop declaration

Networking

c/o Vera Media
30-38 Dock Street
Leeds LS10 1JF
Tel: 0113 2428646

Fax: 0113 2451238
email: networking@vera-media.demon.co.uk
Membership organisation for women working, seeking work or interested/involved in film, video or television, colleges, libraries, careers depts., production companies welcome to join. Members receive quarterly 20-page newsletter with events, production info., letters, reports, news and views; entry in NET-WORKING index; individual advice and help; campaigning voice. £15 pa (UK), £18 (abroad)

New Producers Alliance (NPA)
9 Bourlet Close
London W1P 7PJ
Tel: 0171 580 2480
Fax: 0171 580 2484
email: administrator@npa.org.uk
Established in 1993 by a group of young producers building upon their shared desire to make commercial films for an international audience, NPA has now grown to a membership of more than 1,300. NPA is an independent networking organisation providing members with access to contacts, information, free legal advice and general help regarding film production. NPA publishes a monthly newsletter and organises meetings, workshops and seminars. Membership is available to producers, affiliates (individuals other than producers) and corporate bodies

Office of Fair Trading
Field House
15-25 Bream's Buildings
London EC4A 1PR
Tel: 0171 242 2858
Fax: 0171 269 8800
The Director General of Fair Trading has an interest in the supply of films for exhibition in cinemas. Following a report by the Monopolies and Mergers Commission (MMC) in 1994, the Director General has taken action to ensure that the adverse public interest findings of the MMC are remedied. Under the Broadcasting Act 1990, he also has two specific roles in relation to the television industry. In his report published in December 1992 he assessed the Channel 3 networking arrangement and from 1 January 1993 he had to monitor the BBC's progress towards a statutory requirement to source 25 per cent of its qualifying programming from independent producers

PACT (Producers Alliance for Cinema and Television)
45 Mortimer Street

London W1N 7TD
Tel: 0171 331 6000
Fax: 0171 331 6700
email: enquiries@pact.co.uk
Web: http://www.pact.co.uk
Chief Executive:
Membership Officer: David Alan Mills
PACT exists to serve the feature film and independent television production sector. Currently representing 1,400 companies, PACT is the UK contact point for co-production, co-finance partners and distributors. Membership services include a dedicated industrial relations unit, legal documentation and back-up, a varied calendar of events, courses and business advice, representation at international film and television markets, a comprehensive research programme, publication of a monthly newsletter The PACT Magazine, an annual members' directory, a number of specialist guidebooks, affiliation with European and international producers' organisations, plus extensive information and production advice. PACT works for participants in the industry at every level and operates a members' regional network throughout the UK with a divisional office in Scotland. PACT lobbies actively with broadcasters, financiers and governments to ensure that the producer's voice is heard and understood in Britain and Europe

Performing Right Society (PRS)
29-33 Berners Street
London W1P 4AA
Tel: 0171 580 5544
Fax: 0171 306 4050
Website: http:\\prs.co.uk
PRS is a non-profit making association of composers, authors and publishers of musical works. It collects and distributes royalties for the use, in public performances, broadcasts and cable programmes, of its members' copyright music and has links with other performing right societies throughout the world

Phonographic Performance (PPL)
1 Upper James Street
London W1R 3HG
Tel: 0171 534 1000
Fax: 0171 534 1111
Head of External Affairs:
Colleen Hue
Controls public performance and broadcasting rights in sound recordings on behalf of approximately 2,000 record companies in the UK. The users of sound

recordings licensed by PPL range from BBC and independent TV and Radio, pan-European satellite services, night clubs and juke boxes, to pubs, shops, hotels etc

Production Managers Association (PMA)
Ealing Studios
Ealing Green, Ealing
London W5 5EP
Tel: 0181 758 8699
Fax: 0181 758 8658
Represents over 140 broadcast production managers who all have at least three years experience and six broadcast credits. Provides a network of like-minded individuals

Radio, Electrical and Television Retailers' Association (RETRA)
Retra House, St John's Terrace
1 Ampthill Street
Bedford MK42 9EY
Tel: 01234 269110
Fax: 01234 269609
Fred Round
Founded in 1942, RETRA represents the interests of electrical retailers to all those who make decisions likely to affect the selling and servicing of electrical and electronic products

Reel Women
57 Holmewood Gardens
London SW2 3NB
Tel: 0181 678 7404
A networking organisation for all women in film, video and television. It places particular emphasis on the creative interaction between women from the broadcast, non-broadcast and independent sectors and higher education, aiming to provide a forum for debate around issues affecting women in all areas of production and training, as well as around broader concerns about the representation and position of women in the industry and on screen. Seminars, screenings and workshops are held as well as regular 'nights out'

The Royal Photographic Society
Milsom Street
Bath, Avon BA1 1DN
Tel: 01225 462841
Fax: 01225 448688
email: rps@rps.org
Website: http://www.rps.org
A learned society founded for the promotion and enjoyment of all aspects of photography. Contains a specialist Film and Video Group, secretary Tony Briselden, with a regular journal, meetings and the opportunity to submit productions

for the George Sewell Trophy and the Hugh Baddeley Trophy; and an Audiovisual group, secretary Brian Jenkins, LRPS, offering an extensive programme of events, seminars and demonstrations, and the bi-monthly magazine AV News. Membership open to both amateur and professional photographers

Royal Television Society
Holborn Hall
100 Grays Inn Road
London WC1X 8AL
Tel: 0171 430 1000
Fax: 0171 430 0924
Exec. Director: Michael Bunce
Dep. Exec. Director: Claire Price
The RTS, founded in 1927, has over 4,000 members in the UK and overseas, which are serviced by the Society's 17 regional centres. The Society aims to bring together all the disciplines of television by providing a forum for debate on the technical, cultural and social implications of the medium. This is achieved through the many lectures, conferences, symposia and workshops and master classes organised each year. The RTS does not run formal training courses. The RTS publishes a journal eight times a year Television. The RTS Television Journalism and Sports Awards are presented every year in February and the Programme Awards in May. There are also Craft & Design, Educational Television Awards and student video awards

Scottish Arts Council
12 Manor Place
Edinburgh EH3 7DD
Tel: 0131 226 6051
Fax: 0131 225 9833
Director: Seona Reid

Scottish Film Production Fund
74 Victoria Crescent Road
Glasgow G12 9JN
Tel: 0141 337 2526
Fax: 0141 337 2562
Director: Eddie Dick
Marketing and Promotions: Ela Zych-Watson
The Fund encourages and stimulates film production in Scotland through the provision of financial assistance for narrative fiction films. Its principle priorities are the development of feature films - funding of up to £15,000 is available for each project - and the production of short films through the Tartan Shorts, Geur Ghearr and Prime Cuts initiatives. The SFPF board meets regularly to consider applications. Total resources for script and production

support is £750,000. SFPF also administers the Glasgow Film Fund

Scottish Screen
Dowanhill
74 Victoria Crescent Road
Glasgow G12 9JN
Tel: 0141 334 4445
Fax: 0141 334 8132
Director: Maxine Baker
The national body for the promotion of all aspects of film and television culture including production, exhibition, training, media education and international co-operation. Home of the Scottish Film Archive. Supports initiatives such as First Reels and Movie Makers and now publishes an annual Screen Data Digest

Sgrin (Media Agency for Wales)
Screen Centre
Llantrieant Road
Cardiff CF5 2PU
Tel: 01222 578370
Fax: 01222 578654
email: media@scrwales.demon.co.uk
Website: http://
www.screenwales.org.uk/
Chief Executive: J. Berwyn Rowlands
Incorporating the Wales Film and Television Archive, Media Antenna Wales and the Production Fund. Formed in 1997 by the merger of the Wales Film Council and Screen Wales. Sgrin offers a coordinated vision for all aspects, cultural and industrial, of film television and related media in Wales

The Shape Network
c/o Ithaca
Unit 1, St John Fisher School
Sandy Lane West
Blackbird Leys
Oxford OX4 5LD
Tel: 01865 714652
Fax: 01865 714822
Chair: Anna Thornhill
A federation of independent arts organisations working to increase access to the arts for many groups of people who are usually excluded. The role of the Network is to share information, advice and support and to lobby on arts and disability issues nationally

The Short Film Bureau
68 Middle Street
Brighton
East Sussex BN1 1AL
Tel: 01273 235524
Fax: 01273 235528
email: matt@shortfilmbureau.com
Website; http://www.
shortfilmbureau.com

SKILLSET
The National Training Organisation for Broadcast, Film, Video and Multimedia
91-101 Oxford Street
London W1R 1RA
Tel: 0171 534 5300
Fax: 0171 534 5333
email: info@skillset.org
Website: www.skillset.org
Chief Executive: Dinah Caine
Director of Standards and Qualifications: Kate O'Connor
Communications Manager: Flora Teh-Morris
Founded and managed by the key employers and unions within the industry, SKILLSET operates at a strategic level providing relevant labour market and training information, encouraging higher levels of investment in training and developing and implementing occupational standards and the National and Scottish Vocation Qualifications based upon them. It seeks to influence national and international education and training policies to the industry's best advantage, strives to create greater and equal access to training opportunities and career development and assists in developing a healthier and safer workforce. SKILLSET is a UK-wide organisation

Society for the Study of Popular British Cinema
Department of English, Media and Cultural Studies
School of Humanities
Gateway House
De Montfort University
Leicester LE1 9BH
Fax: 0116 2577199
Contact: Alan Burton, Secretary
Society which produces a newsletter and encourages an interest in British films

Society of Authors' Broadcasting Group
84 Drayton Gardens
London SW10 9SB
Tel: 0171 373 6642
Fax: 0171 373 5768
Specialities: Radio, television and film scriptwriters

Society of Cable Telecommunication Engineers (SCTE)
Fulton House Business Centre
Fulton Road, Wembley Park
Middlesex HA9 0TF
Tel: 0181 902 8998
Fax: 0181 903 8719
Mrs Beverley K Allgood FSAE

Aims to raise the standard of cable telecommunication engineering to the highest technical level, and to elevate and improve the status and efficiency of those engaged in cable telecommunication engineering

Society of Film Distributors (SFD)
22 Golden Square
London W1R 3PA
Tel: 0171 437 4383
Fax: 0171 734 0912
General Secretary: D C Hunt
SFD was founded in 1915 and membership includes all the major distribution companies and several independent companies. It promotes and protects its members' interests and co-operates with all other film organisations and Government agencies where distribution interests are involved

Society of Television Lighting Directors
4 The Orchard, Aberthin
Cowbridge
South Glamorgan CF7 7HU
The Society provides a forum for the exchange of ideas in all aspects of the TV profession including techniques and equipment. Meetings are organised throughout the UK and abroad. Technical information and news of members' activities are published in the Society's magazine

Sovexportfilm
11b Paveley Drive
Morgans Walk
London SW11 3TP
Tel: 0171 358 1226
Fax: 0171 358 1226
Exports Russian films to different countries and imports films to Russia. Provides facilities to foreign companies wishing to film in Russia. Co-production information for producers

TAC (Welsh Independent Producers)
Gronant
Caernarfon
Gwynedd LL55 1NS
Tel: 01286 671123
Fax: 01286 678890
TAC is the trade association representing the 95 production companies working for Welsh broadcasters. It offers a full IR service and conducts negotiations on standard terms of trade with the broadcasters

Telefilm Canada/Europe
5 rue de Constantine
Paris 75007
Director: Sheila de La Varende

Canadian government organisation financing film and television productions. European office provides link between Canada, UK and other European countries

VLV - Voice of the Listener and Viewer
101 King's Drive
Gravesend
Kent DA12 5BQ
Tel: 01474 352835
Fax: 01474 351112
An independent non-profit making society which represents the citizen's voice in broadcasting and which supports the principle of public service in broadcasting. Founded in 1983, by Jocelyn Hay, VLV is the only consumer body speaking for listeners and viewers on the full range of broadcasting issues. VLV has over 2,000 members, more than 20 corporate members (most of which are registered charities) and over 50 colleges and university departments in academic membership. VLV is funded by its members and free from any sectarian, commercial or political links. Holds public lectures, conferences and seminars and arranges exclusive visits for its members to broadcasting centres in different parts of the country. Publishes a quarterly newsletter and briefings on broadcasting developments. Has responded to all parliamentary and public inquiries on broadcasting since 1984 and to all consultations by the ITC, Radio Authority, BBC and Broadcasting Standards Council since 1990. Is in frequent touch with MPs, civil servants, the BBC and independent broadcasters, regulators, academics and relevant consumer bodies. Holds the archive of the former independent Broadcasting Research Unit and of the former British Action for Children's Television (BACTV) and makes these available for a small fee together with its own archives and library. Set up the VLV Forum for Children's Broadcasting in 1994

Variety Club of Great Britain
St Martin's House
139 Tottenham Court Road
London W1P 9LN
Tel: 0171 387 3311
Fax: 0171 387 3322
Charity dedicated to helping disabled and disadvantaged children throughout Great Britain

Videola (UK)
Paramount House
162/170 Wardour Street
London W1V 3AT
Tel: 0171 437 2136
Fax: 0171 437 5413

Women in Film and Television (UK)
Garden Studios
11-15 Betterton Street
London WC2H 9BP
Tel: 0171 379 0344
Fax: 0171 379 1625
Director: Kate Norrish
Administrator: Donna Coyle
A membership organisation for women working in the film and television industries. WFTV aims to provide information and career support through a monthly programme of events that are free to members. In addition WFTV safeguards the interests of the members through its lobbying and campaigning. Annual Awards Ceremony at the Dorchester remains the only industry event that solely celebrates the achievements of women. WFTV is part of an international network of organisations - currently 37 chapters world-wide

Writers' Guild of Great Britain
430 Edgware Road
London W2 1EH
Tel: 0171 723 8074
Fax: 0171 706 2413
Website: http//www.writers.org.uk/guild
The Writers' Guild is the recognised TUC-affiliated trade union for writers working in film, television, radio, theatre and publishing

REGIONAL FILM COMMISSIONS

Bath Film Office
Abbey Chambers, Abbey Churchyard
Bath BA1 1LY
Tel: 01225 477711
Fax: 01225 477221
email: Richard_Angell@Bathne.gov.uk
Under the umbrella of the British Film
Commission, the Bath Film Office
offers a free service to the TV, film
and commercials industry. It maintains
a database of over 800 locations in
and around the city of Bath

Eastern Screen
Anglia TV
Norwich NR1 3JG
Tel: 01603 767077
Fax: 01603 767191
The film commission for the East of
England offering free help and
advice on locations, resources,
services and skills to anyone
intending to film within the region

Edinburgh and Lothian Screen Industries Office
Castlecliff
25 Johnston Terrace
Edinburgh EH1 2NH
Tel: 0131 622 7337
Fax: 0131 622 7338
George Carlaw, Ros Davis
The Film Commission for the City of
Edinburgh and the coastline,
countryside and counties of
Lothian. Advice on locations,
crews and facilities and local
authority liaison. A free service
provided by the City of Edinburgh
and West, East, and Midlothian
Councils, to encourage film, video
and television production in the
area

Film & Television Commission
North West England
Pioneer Buildings
65-67 Dale Street
Liverpool L2 2NS
Tel: 0151 330 6666
Fax: 0151 330 6611
FTC north west is the official film
commission for the north west of
England, working in partnership
with the film offices of Liverpool,
Manchester, Lancashire, Isle of Man
and CheshireFilm & Television
Commission
North West England
Pioneer Buildings
65-67 Dale Street
Liverpool L2 2NS
Tel: 0151 330 6666
Fax: 0151 330 6611
FTC north west is the official film
commission for the north west of
England, working in partnership
with the film offices of Liverpool,
Manchester, Lancashire, Isle of Man
and Cheshire

Isle of Man Film Commission
Illiam Dhone House
2 Circular Road
Douglas
Isle of Man 1M1 1PJ
Tel: 01624 685864
Fax: 01624 685454

Liverpool Film Office
Pioneer Buildings
67 Dale St
Liverpool L2 2NS
Tel: 0151 291 9191
Fax: 0151 291 9199
Film Commissioner: Brigid Marray
Locations Officer: Lynn Saunders
Provides a free film liaison service,
and assistance to all productions
intending to use locations, resources,
services and skills in the Merseyside
area. Undertakes research and
location scouting, liaises with local
agencies and the community. Offers
access to the best range of locations
in the UK through its extensive
locations library

London Film Commission
20 Euston Centre
Regent's Place
London NW1 3JH
Tel: 0171 387 8787
Fax: 0171 387 8788
email: lfc@london-film.co.uk
Christabel Albery, Film Commissioner
The London Film Commission
encourages and assists film and
television production in London and
holds databases of locations,
personnel and facilities. Funded by
Government, the film industry and
other private sector sponsors, it
works to promote London as a first
choice destination for overseas film-
makers. It collaborates with the
Local Authorities, the police and
other services to create a film
friendly atmosphere in the capital

Media Development Agency for the West Midlands incorporating Central Screen Commission

Unit 5 Holliday Wharf
Holliday Street
Birmingham B1 ITJ
Tel: 0121 643 9309
Fax: 0121 643 9064
email: info@mda-wm.org.uk
Website: http://www.mda-wm.org.uk
Kim Langford, Cathie Peloe
Regional Media Development Agency and Screen Commission, Directory & Database of local production facilities, crews and talents; location finding and liaison: low budget production funding including 1st Academy; information resources and counselling service; legal surgeries; seminars and masterclasses; media business development support; copyright registration scheme

The North Wales Film Commission

Council Offices
Shire Hall Street
Caernarfen
Gwynedd LL55 1SH
Tel: 01286 679685
Fax: 01286 673324
email: fil@gwynedd.gov.uk
Hugh Edwin Jones, Peter Lowther
Area film liaison office for information on filming in the county of Gwynedd and Anglesey. Information provided on locations, facilities and crew

Northern Ireland Film Commission

21 Ormeau Avenue
Belfast BT2 8HD
Tel: 01232 232 444
Fax: 01232 239 918
The Northern Ireland Film Commission promotes the growth of film and television culture and the industry in Northern Ireland

Northern Screen Commission (NSC)

Great North House
Sandyford Road
Newcastle Upon Tyne NE1 8ND
Tel: 0191 204 2311
Fax: 0191 204 2209
email:paul@filmhelp.demon.co.uk
Paul Mingard
Seeking to attract film, video and television production to the North of England, NSC can provide a full liaison service backed by a network of local authority contacts and public organisations. Available at no cost is a locations library, a database on local facilities and services as well as a full list of local crew or talent

Scottish Screen

74 Victoria Crescent Road
Glasgow G12 9JN
Tel: 0141 302 1700
Fax: 0141 302 1711
email:
info@scottishscreen.demon.co.uk
Website: http://
www.scottishscreen.demon.co.uk
Chief Executive: John Archer
The Government body Scottish Screen was established in April 1997 to stimulate the film & television industry through production & development finance, training, education, marketing and the film Commission
Production
Development assistance, advisor to the Scottish Arts Council Lottery fund on film production, principally fiction and short film. Short film schemes include financing of First Reels (£1,000-£4,000) Prime Cuts (£23,000), Geur Ghearr (£45,000) Tartan Shorts (£45,00) and access to the Shiach Script library
Development
Advice & Finance for the script development for feature films. (film script Development up to £15,000)
Media
At Scottish Screen the Media Antenna Scotland can help access to MEDIA II & other European support programmes
Exhibition
Support to Regional Film Theatres and active support of local & foreign film festivals. Development of cinema audiences
Film Commission
Offers advice on locations, script assessment, liaison contacts and production set up
Training
Producer support schemes, New entrants scheme, short courses for industry professionals and education of media in schools
Archive
Home of the National Archive, collection of factual documentary material mapping Scotland's social and cultural history. Available to broadcasters, programme makers educational users and researchers
Distribution
Short film distribution and promotion
Information
Library resource, videos, publications

South Wales Film Commission

The Media Centre
Culverhouse Cross
Cardiff Cf5 6XJ
Tel: (01222) 590240
Fax: (01222) 590511

email:
106276.223@compuserve.com
A member of the British Film Commission and AFCI, providing location, research and information on facilities and services across South Wales for film and television production

Scottish Highlands and Islands Film Commission

Comisean Fiolm na
Gaidhealtachd's nad Eilean Alba
PO Box 5549
Inverness 1V3 5YQ
Tel: 01463 702000
Fax: 01463 710848
The Scottish Highlands and Islands Film Commission provides a free, comprehensive liaison service to the film and television industry, including information and advice on locations, permissions, crew and services etc. We cover Argyll and Bute, Highland, Moray, Orkney, Shetland and the Western Isles, and have a network of local film liaison officers able to provide quick and expert local help, whatever your project

South West Film Commission

18 Belle Vue Road, Saltash,
Cornwall PL12 E5
Tel: 01752 841199
Fax: 01752 841254
email: swfilm@eurobell.co.uk
Film Commissioner: Sue Dalziel
Offers professional assistance to productions shooting in Devon, Cornwall, Somerset, Dorset, Bristol City and Gloucestershire

South West Scotland Screen Commission

Gracefield Arts Centre
28 Edinburgh Road
Dumfries DG1 1NW
Tel: 01387 263666
Fax: 01387 263666
An unrivalled variety and wealth of locations to suit any style of shoot or budget, plus a free location finding and film liaison service for South West Scotland

Yorkshire Screen Commission

The Workstation
15 Paternoster Row
Sheffield S1 2BX
Tel: 0114 2799115/2766511
Fax: 0114 2796522/2798593
Screen Commissioners: Sue Lathan, Information Co-ordinator: Shuna Frood, Crew Manager: Penny Finerty
Assists film & TV production companies in finding locations, facilities and crew in the Yorkshire region

ORGANISATIONS (EUROPE)

The following is a list of some of the main pan-European film and television organisations, the various MEDIA II projects instigated by the European Commission, and entries for countries of the European Union

ACE (Ateliers du Cinéma Européen/European Film Studio)

rue de Rivoli 68
75004 Paris
France
Tel: (33) 1 44 61 88 30
Fax: (33) 1 44 61 88 40
email: ace@i-t.fr
Director: Claudie Cheval
ACE was established as a joint project of the Media Business School and the European Producers Club, to work with producers during the development stages of their projects, guiding them in a way that maximises their chances of reaching the largest possible target audience

AGICOA (Association de Gestion Internationale Collective des Oeuvres Audio-Visuelles)

rue de St-Jean 26
1203 Geneva
Switzerland
Tel: (41) 22 340 32 00
Fax: (41) 22 340 34 32
Rodolphe Egli, Luigi Cattaneo
AGICOA ensures the protection of the rights of producers worldwide when their works are retransmitted by cable. By entering their works in the AGICOA Registers, producers can claim royalties collected for them

Audio-Visual EUREKA

Permanent Secretariat
rue de la Bonté 5-7
1050 Brussels, Belgium
Tel: (32) 2 538 04 55
Fax: (32) 2 538 04 39
Director: Karl-Gunnar Lidström
A programme which aims to stimulate the European audiovisual market by favouring the establishment of a network of partners around concrete projects which concern all spheres of the audio-visual sector. The promotion and support of labelled projects is a priority, as well as providing an active network of services, relations, information and exchanges

Bureau de Liaison Européen du Cinéma

c/o Fédération Internationale des Associations de Distributeurs de Films (FIAD)
boulevard Malesherbes 43
75008 Paris, France
Tel: (33) 1 42 66 05 32
Fax: (33) 1 42 66 96 92
Gilbert Grégoire
Umbrella grouping of cinema trade organisations in order to promote the cinema industry, including CICCE, FEITIS, FIAD, FIAPF, FIPFI and UNIC

Centre for Cultural Research

Am Hofgarten 17
53113 Bonn, Germany
Tel: (49) 228 211058
Fax: (49) 228 217493
Scharfschwerdtstr. 10
16540 Hohen Neuendorf
c/o IKM, Karlsplatz 2
1010 Vienna, Austria
Prof Andreas Johannes Wiesand
Research, documentation, and advisory tasks in all fields of the arts and media, especially with 'European' perspectives. Participation in arts and media management courses at university level. Produces publications and is founding seat of the European Institute for Comparative Cultural Policy and the Arts (ERIC Arts) with members in 21 European countries

Culturelink/IMO

Institute for International Relations
Ul. Lj. Farkasa Vukotinovica 2
10000 Zagreb
Croatia PO Box 303
Tel: (385) 1 45 54 522
Fax: (385) 1 48 28 361
Zrinjka Perusko Culek
Culturelink is a worldwide network for research and co-operation in cultural development. Areas of research/activities: cultural development, cultural policies, cultural identities, cultural co-operation, mass media and communications (media policies in Europe and Croatia, audio-visual co-operation between Eastern and Western Europe, cultural impact of media policies). Documentation, database, bulletin (quarterly)

EUTELSAT (European Telecommunications Satellite Organisation)

Tour Maine-Montparnasse
avenue du Maine 33
75755 Paris Cédex 15, France
Tel: (33) 1 45 38 47 47
Fax: (33) 1 45 38 37 00
Vanessa O'Connor
EUTELSAT operates a satellite system for intra-European communications of all kinds. Traffic carried includes Television and Radio channels, programme exchanges, satellite newsgathering, telephony and business communications

EURIMAGES

Council of Europe
Palais de l'Europe
avenue de l'Europe
67075 Strasbourg Cédex, France
Tel: (33) 3 03 88 41 26 40
Fax: (33) 3 0 3 88 41 27 60
Contact: Executive Secretary
Founded in 1988 by a group of Council of Europe member states. Its objective is to stimulate film and audio-visual production by partly financing the co-production, distribution and exhibition of European cinematographic and audio-visual works. Eurimages now includes 24 member states

Eurocréation Media

rue Debelleyme 3
75003 Paris, France
Tel: (33) 1 44 59 27 01
Fax: (33) 1 40 29 92 46
Jean-Pierre Niederhauser, Anne-Marie Autissier (Consultant)
Eurocréation Media develops consultation and expertise in the field of European audio-visual and cinema (research, support for the organisation and conception of European events, training activities)

European Academy for Film & Television

rue Verte 69
1210 Brussels, Belgium
Tel: (32) 2 218 66 07
Fax: (32) 2 217 55 72
Permanent Secretary: Dimitri Balachoff
The purpose of the Academy, a non-profit making association, is the research, development and disclosure of all matters relating to cinema and television chiefly in the European continent, and also in other continents, taking into account artistic, commercial, cultural, economic, financial, historical, institutional, pedagogical, trade union and technical aspects. Quarterly newsletter, ACANEWS. Other activities: Blue Angel Award, presented every year at the Berlin International Film Festival for the Best European Film in Competition

European Audio-visual Observatory

76 allée de la Robertsau
67000 Strasbourg, France
Tel: (33) 3 88 144400
Fax: (33)3 88 144419
Webstie: http:/www.obs.c_strasbourg.fr
Executive Director: Nils Klevjer
AAS. A Pan-European institution working in the legal framework of the Council of Europe. The Observatory is a public service centre providing information on the European television, film and video industries, aimed at the audio-visual industry, and available in English, French and German. It provides legal, economic and market, and film and television funding related information and counselling, and is working with a network of partner organisations on the developing harmonisation of data covering the whole of Europe. The Observatory also publishes a monthly newsletter (IRIS) on legal development in all of its 33 member States, as well as an annual Statistical Yearbook on Film, Television, Video and New Media. The Observatory provides information through an individualised document delivery service in the legal information area

European Broadcasting Union (EBU)

Ancienne Route 17a
1218 Grand-Saconnex
Geneva, Switzerland
Tel: (41) 22 717 2111
Fax: (41) 22 717 2200
Jean-Pierre Julien

The EBU is a professional association of national broadcasters with 117 members in 79 countries. Principal activities: daily exchange of news, sports and cultural programmes for television (Eurovision) and radio (Euroradio); Tv coproductions; technical studies and legal action in the international broadcasting sphere

European Co-production Association

c/o France 2
22 avenue Montaigne
75387 Paris Cedex 08
France
Tel: (33) 1 4421 4126
Fax: (33) 1 4421 5179
A consortium of, at present, six European public service television networks for the co-production of television programmes. Can offer complete finance. Proposals should consist of full treatment, financial plan and details of proposed co-production partners. Projects are proposed to the ECA Secretariat or to member national broadcasters

European Co-production Fund (ECF)

c/o British Screen Finance
14-17 Wells Mews
London W1P 3FL
Tel: 0171 323 9080
Fax: 0171 323 0092
In November 1996, the Department of National Heritage (DNH) (as it was - now the DCMS) announced that it would continue funding ECF at the current level of £2 million per year until March 2000. The Fund is available for investment in feature films made by European co-producers and for investment in European film development work. The ECF administers the fund (it is a wholly owned subsidy of BSF) for disbursing the Fund to producers in the form of loans on commercial terms. The Fund aims to improve the opportunity for UK producers to co-produce with EC partners. Successful recent films include: *Orlando, Before the Rain, Antonia's Line*

European Film Academy (EFA)

Segitzdamm 2, 6th Floor
10969 Berlin
Germany
Tel: (49) 30 615 30 91
Fax: (49) 30 614 31 31
Chairman: Nik Powell,
Director: Marion Döring
Promotes European cinema worldwide to strengthen its commercial and artistic position, to improve the knowledge and awareness of European cinema and to pass on the substantial experience of the Academy members to the younger generation of film professionals. The European Film Academy presents the annual European Film Awards

European Institute for the Media (EIM)

Kaistrasse 13
40221 Düsseldorf, Germany
Tel: (49) 211 90 10 40
Fax: (49) 211 90 10 456
Head of Research: Runar Woldt
Head of East-West: Dusoun Rejic
Acting Head of Library, Documentation and Statistics Centre: Helga Schmid
A forum for research and documentation in the field of media in Europe. Its activities include: research into the media in Europe with a political, economic and juridicial orientation; the organisation of conferences and seminars such as the annual European Television and Film Forum; East-West Co-operation Programme; the development of an advanced studies programme for students and media managers. Publication of the Bulletin in English/French/German, quarterly on European media development, and of the Ukrainian and Russian Bulletin as well as research reports. Officers in Kiev and Moscow. Organises the European Media Summer School, an annual course on media development for advanced students and professionals, and facilitates an information request service

FIAD (Fédération Internationale des Associations de Distributeurs de Films)

boulevard Malesherbes 43
75008 Paris, France
Tel: (33) 1 42 66 05 32
Fax: (33) 1 42 66 96 92
President: Gilbert Grégoire
President d'honneur: Luc Hemelaer
Vice President: Stephan Hutter
Secretaire General. Antoine Virenque
Represents the interests of film distributors

FIAF (International Federation of Film Archives)

rue Franz Merjay 190
1180 Brussels, Belgium
Tel: (32) 2 343 06 91
Fax: (32) 2 343 76 22

For further information about FIAF, see under Archives and Film Libraries

FIAPF (Fédération Internationale des Associations de Producteurs de Films)

avenue des Champs-Elysées 33
75008 Paris, France
Tel: (33) 1 42 25 62 14
Fax: (33) 1 42 56 16 52
An international level gathering of national associations of film producers (23 member countries). It represents the general interests of film producers in worldwide forums (WIPO, UNESCO, WCO, GATT) and with European authorities (EC, Council of Europe, Audio-visual EUREKA), it lobbies for better international legal protection for film and audio-visual producers

FIAT (International Federation of Television Archives)

National Film and Television Archive
21 Stephen Street
London W1P 2LN
Tel: 0171 957 8940
Fax: 0171 580 7503
FIAT membership is mainly made up of the archive services of broadcasting organisations. However it also encompasses national archives and other television-related bodies. It meets annually and publishes its proceedings and other recommendations concerning television archiving

Fédération Européenne des Industries Techniques de l'Image et du Son (FEITIS)

avenue Marceau 50
75008 Paris, France
Tel: (33) 1 47 23 75 76
Fax: (33) 1 47 23 70 47
A federation of European professional organisations representing those working in film and video services and facilities in all audio-visual and cinematographic markets

Fédération Internationale de la Presse Cinémaographique (FIPESCCI)

Schleissheimer Str 83
Munich
Tel: (49) 89 18 23 03
Fax: (49) 89 18 47 66
Klaus Eder, General Secretary

Fédération Internationale des Producteurs de Films Indépendants (FIPFI)

avenue Marceau 50
75008 Paris, France
Tel: (33) 1 47 23 70 30
Fax: (33) 1 47 20 78 17
Federation of independent film producers, currently with members in 21 countries. It is open to all independent producers, either individual or groups, provided they are legally registered as such. FIPFI aims to promote the distribution of independent films, to increase possibilities for co-production, to share information between member countries and seeks to defend freedom of expression

IDATE (Institut de l'audio-visuel et de télécommunications en Europe)

BP 4167
34092 Montpellier Cédex 5
France
Tel: (33) 67 14 44 44
Fax: (33) 67 14 44 00
Director: Yves Gassot
European study and research centre, a leader in the field of socio-economic analysis of the information and communication industries. Carries out strategic monitoring, market analysis, evaluation and consultancy, and research, and produces publications, conferences and training seminars

ISETU/FISTAV (International Secretariat for Arts, Mass Media and Entertainment Trade Unions/ International Federation of Audio-Visual Workers)

IPC, boulevard Charlemagne 1
PO Box 5
1040 Brussels, Belgium
Tel: (32) 2 238 09 51
Fax: (32) 2 230 00 76
General Secretary: Jim Wilson
Caters to the special concerns of unions and similar associations whose members are engaged in mass media, entertainment and the arts. It is a clearing house for information regarding multi-national productions or movement of employees across national borders, and acts to exchange information about collective agreements, legal standards and practices at an international level. It organises conferences, has opened a campaign in support of public service broadcasting, and has begun initiatives ranging from defending screen writers to focusing on the concerns of special groups

Institut de Formation et d'Enseignement pour les Métiers de l'Image det du Son (FEMIS)

rue Francoeur 6
75018 Paris
France
Tel: (33) 1 42 62 20 00
Fax: (33) 1 42 62 21 00
High level technical training in the audio-visual field for French applicants and those from outside France with a working knowledge of French. Organises regular student exchanges with other European film schools

Institut de Journalisme Robert Schuman - European Media Studies

rue de l'Association 32-34
1000 Brussels
Belgium
Tel: (32) 2 217 2355
Fax: (32) 2 219 5764
Anne de Boeck
Postgraduate training in journalism. Drawing students from all over Europe, it offers nine months intensive training in journalism for press, radio and television

International Cable Communications Council

boulevard Anspach 1, Box 34
1000 Brussels
Belgium
Tel: (32) 2 211 94 49
Fax: (32) 2 211 99 07
International body gathering European, Canadian, North American and Latin American cable television organisations

International Federation of Actors (FIA)

Guild House
Upper St Martin's Lane
London WC2H 9EG
Tel: 0171 379 0900
Fax: 0171 379 8260
Trade union federation founded in 1952 and embracing 60 performers' trade unions in 44 countries. It organises solidarity action when member unions are in dispute, researches and analyses problems affecting the rights and working conditions of film, television and theatre actors as well as singers, dancers, variety and circus artistes. It represents members in the international arena on issues such as cultural policy and copyright and publishes twice yearly newsheet FOCUS

The Prince's Trust Partners in Europe

8 Bedford Row
London WC1R 4BA
Tel: 0171 405 5799
Contact: Anne Engel
Offer 'Go and See' grants (max £500) towards partnership projects in Europe to people under 26 out of full time education

TV France International

5 rue Cernuschi
75017 Paris
Tel: (33) 1 40 53 23 00
Fax: (33) 1 40 53 23 01
email: tvfi@imaginet.fr
Website: http://www.tvfi.com
Olivier-René Veillon
TV France International was created in 1994 to support promotion and world sales of French television programmes. Members are producers, distributors and broadcasters. TV France International is also a key access to the French market, providing information about production and distribution companies, programmes and availability

UK EUREKA Unit

Department of Trade and Industry,
3rd Floor, Green Core
151 Buckingham Palace Road
London SW1W 9SS
Tel: 0171 215 1618
Fax: 0171 215 1700
For Advanced Broadcasting Technology: Brian Aldous
Tel: 0171 215 1737
A pan-European initiative to encourage industry-led, market-driven collaborative projects aimed at producing advanced technology products, processes and services

UNIC (Union Internationale des Cinémas)

15 Rue de Berri
75008 Paris
France
Tel: (33) 1 53 93 76 76
Fax: (33) 1 45 53 29 76
Defends the interests of cinema exhibitors worldwide, particularly in matters of law and economics. It publishes UNIC News and a Bulletin. Also provides statistical information and special studies concerning the exhibition sector, to members and others

URTI (Université Radiophonique et Télévisuelle Internationale)

General Secretariat

116, avenue du Président Kennedy
75786 Paris Cedex 16
France
Tel: (33) 1 42 30 39 98
Fax: (33) 1 40 50 89 99
President: Roland Faure
A non-governmental organisation recognised by UNESCO and founded in 1949, URTI is an association of professionals in the audio-visual field from all over the world. Promotes cultural programmes and organisation of projects including the International Grand Prix for Creative Documentaries, the Young Television Prize at the Monte Carlo International Television Festival, the Grand Prix for Radio (since 1989)

AUSTRIA

Animation Studio for Experimental Animated Films

University of Applied Arts Vienna
A - 1010 Wien Salzgries 14
Tel: (43) 1 7120392/71133-521
Fax: (43) 1 7120392
Hubert Sielecki

Association of Audio-visual and Film Industry

Wiedner Haupstrasse 63
1045 Wien
PO Box 327
Tel: (43) 1 50105/3010
Fax: (43) 1 50206/276
email: faf@wk.or.at
Dr Elmar Peterlunger

Austrian Film Commission

Stiftgasse 6
1070 Vienna
Tel: (43) 1 526 33 23/200
Fax: (43) 1 526 68 01
email: afilco@magnet.at
The Austrian Film Commission is a central information and promotion agency. The organization, financed by public funds, offers a wide variety of services for Austrian producers and creative artists, it acts as consultant whenever its productions are presented in international festivals, and it provides members of the profession in all sectors with comprehensive information as to current activity in the Austrian film industry. It is the aim of all activities to enhance the perception of Austrian film-making both at home and abroad. In addition to the major festivals in Berlin, Cannes and Venice, the Austrian Film Commission currently provides support for 70 international film festivals and fairs. The catalogue Austrian Films published annually, offers an overview, divided in sections, of current Austrian film-making. The Austrian Film Guide provides a concise source of addresses for Austrian producers, distributors, institutions associated with film, funding organisations etc. Also published is Austrian Film News, written about and for the Austrian film industry

Austrian Film Institute

Spittelberggasse 3
A-1070 Wien
Tel: (43) 1 526 97 30
Fax: (43) 1 526 97 30/440
email: oefi@filminstitut.or.at

Website: http://www.filminstitut.or.at
Film funding, Eurimages and
MEDIA II

Filmakademie Wien, National Film School, Vienna Hochschule für Musik und darstellende Kunst

Metternichgasse 12
A-1030 Wien
Tel: (43) 1 713 52 12 0
Fax: (43) 1 713 52 12 23

Wiener Filmfinanzierungsfonds (WFFO)

Stiftgasse 6 /2/3
1070 Vienna
Tel: (43) 1 526 50 88
Fax: (43) 1 526 50 88 - 20
The WFF concedes interest free
loans in order to support full-length
feature films and creative
documentaries for cinema and
television. Besides production, the
WFF assists scriptwriting, project
development, exploitation/
distribution and the improvement of
Viennese film industry infrastructure

BELGIUM

IDEM

227 Chaussee D'ixelles
1050 Brussels
Tel: (32) 2 640 77 31
Fax: (32) 2 640 98 56
Trade association for television
producers

Cinémathèque Royale de Belgique/Royal Film Archive

Rue Ravenstein 23
1000 Brussels
Tel: (32) 2 507 83 70
Fax: (32) 2 513 12 72
Gabrielle Claes
Film preservation. The collection
can be consulted on the Archive's
premises for research purposes.
Edits the Belgian film annual

Commission de Sélection de Films

Ministère de la Culture et des
Affaires Sociales
Direction de l'Audio-visuel
Boulevard Léopold II 40
1080 Brussels
Tel: (32) 2 413 22 39
Fax: (32) 2 413 22 42
Christiane Dano, Serge Meurant
Assistance given to the production
of short and long features, as well as
other audio-visual production by
independent producers

Commission du Film

Ministère de la Culture et des
Affaires Sociales
Direction de l'Audiovisuel
Boulevard Léopold II 44
1080 Brussels
Tel: (32) 2 413 22 21
Fax: (32) 2 413 20 68
Gives official recognition to Belgian
films; decides whether a film has
sufficient Belgian input to qualify as
Belgian

Film Museum Jacques Ledoux

Rue Baron Horta 9
1000 Brussels
Tel: (32) 2 507 83 70
Fax: (32) 2 513 12 72
Gabrielle Claes
Permanent exhibition of the
prehistory of cinema. Five
screenings per day - three sound, two
silent. Organises one mini festival a
year: L'Age d'Or Prize and prizes for
the distribution of quality films in
Belgium

Radio-Télévision Belge de la Communauté Française (RTBF)

Blvd Auguste Reyers 52
1044 Brussels
Tel: (32) 2 737 21 11
Fax: (32) 2 737 25 56
Administrateur Général: Jean-Louis
Stalport
Public broadcaster responsible for
French language services

VRT

Auguste Reyerslaan 52
1043 Brussels
Tel: (32) 2 741 3111
Fax: (32) 2 734 9351
Managing Director: Bert De Graeve
Television: Piet Van Roe
Radio: Chris Cleeren
Public television and radio station
serving Dutch speaking Flemish
community in Belgium

DENMARK

Danmarks Radio (DR)

Morkhojvej 170
2860 Soborg
Tel: (45) 35 20 30 40
Fax: (45) 35 20 26 44
Public service television and radio
network

DFI (Danish Film Institute)

Vognmagergade 10
DK - 1120 Københaun K
Tel: (45) 33 74 34 30
Fax: (45) 33 74 34 55
An autonomous self-governing body

under the auspices of the Ministry of
Culture, financed through the state
budget. Provides funding for the
production of Danish feature films,
shorts and documentaries, and also
supports distribution and exhibition.
Promotes Danish films abroad and
finances two community access
workshops

Det Danske Filminstitut / Filmmuseum

Filmhouse Denmark
Vognmagergade 10
Tel: (45) 3374 3400/3374 3575
Fax: (45) 3374 3599
The Film Museum, founded in 1941,
is one of the world's oldest film
archives. It has a collection of
25,000 titles from almost every
genre and country, and has daily
screenings. There is also an
extensive library of books and
pamphlets, periodicals, clippings,
posters and stills

Film-OG TV Arbejderforeningen

Danish Film and TV Workers
Union
Kongens Nytorv 21
Baghuset 3.sal
1050 Copenhagen K
Tel: (45) 33 14 33 55
Fax: (45) 33 14 33 03
Trade union which organises film,
video and television workers, and
maintains the professional, social,
economic and artistic interests of its
members. Negotiates collective
agreements for feature films,
documentaries, commercials,
negotiating contracts, copyright and
authors' rights. Also protection of
Danish film production

Producenterne

Kronprinsensgade 9B
1114 Copenhagen K
Tel: (45) 33 14 03 11
Fax: (45) 33 14 03 65
The Danish Producers' Association
of Film, Television, Video and AV

Statens Filmcentral

Vognmagergade 10. III
1120 Copenhagen K
Tel: (45) 33 74 35 00
Fax: (45) 33 74 35 65
Statens Filmcentral is the National
Film Board of Denmark, created in
1939. It is regulated by the Ministry
of Culture and produces, purchases
and rents out shorts and
documentaries on 16mm and video
to educational institutions/libraries
and private persons

FINLAND

AVEK - The Promotion Centre for Audio-visual Culture in Finland

Hietaniemenkatu 2
FIN - 00100 Helsinki
Tel: (358) 9 43152350
Fax: (358) 9 43152388
email: avek@avek.kopiosto.fi
Website: http:www.kopiostofi/avek
AVEK was established in 1987 to promote cinemas, video and television culture. It is responsible for the management of funds arising from authors' copyright entitlements and is used for authors' common purposes (the blank tape levy). AVEK's support activities cover the entire field of audio-visual culture, emphasis being on the production support of short films, documentaries and media art. The other two activity sections are training of the professionals working in the audio-visual field and audiovisual culture in general

Finnish Film Archive/Suomen Elokuva-arkisto

Pursimiehenkatu 29-31 A
PO Box 177
FIN-00151
Helsinki
Tel: (358) 9 615 400
Fax: (358) 0 615 40 242
email: sea@sea.fi
Website: htpp:www.sea.fi
Matti Lukkarila
Stock: 10,000 feature film titles; 30,000 shorts and spots; 18,000 video cassettes; 18,000 books and scripts; 330,000 different stills, 110,000 posters; and 40,000 documentation files. The archive arranges regular screenings in Helsinki and other cities. Documentation, database, publications (Finnish national filmography). Publications

Finnish Film Foundation

Kanavakatu 12
Fin-Helsinki
Tel: (358) 0 6220 300
Fax: (358) 0 6220 3050
Film funding for script, development and production of feature film and documentaries Audio post production and auditorio services. Distribution and screening support. International activities (cultural export and promotion of Finnish Film)

FRANCE

Les Archives du Film du Centre National de la Cinématographie

7 bis rue Alexandre Turpault
78390 Bois d'Arcy
Tel: (33) 1 30 14 80 00
Fax: (33) 1 34 60 52 25
Michelle Aubert
The film collection includes some 131,000 titles, mostly French features, documentaries and shorts from 1865 to date through the new legal deposit for films which includes all categories of films shown in cinemas including foreign releases. Since 1991, a special pluriannual programme for copying early films, including nitrate film, has been set up. So far, some 8,000 titles have been restored including the whole of the Lumière brothers film production from 1895 to 1905 which covers 1,400 short titles. A detailed catalogue of the Lumiére production is available in print and CD-Rom. Enquiries and viewing facilities for film are available on demand

Bibliothéque du Film (BIFI)

100 rue du Faubourg Saint-Antoine
75012 Paris
Tel: 01 53 02 22 30
Fax: 01 53 02 22 39
Website: http://www.bifi.fr
Contact: Laurent Billia
Centre d'Information et de Documentation
Contact: Fortunée Sellam
Director of the Mediathèque
Contact: Marc Vernet

Centre National de la Cinématographie (CNC)

rue de Lübeck 12
75016 Paris
Tel: (33) 1 45 05 1440
Fax: (33) 1 47 55 04 91
Director-General: Dominique Wallon, Press, Public & Internal Relations: Patrick Ciercoles
A government institution, under the auspices of the Ministry of Culture. Its areas of concern are: the economics of cinema and the audio-visual industries; film regulation; the promotion of the cinema industries and the protection of cinema heritage. Offers financial assistance in all aspects of French cinema (production, exhibition, distribution etc). In 1986, the CNC was made responsible for the system of aid offered to the production of films

made for television. These include fiction films, animated films and documentaries. The aim here corresponds to one of the principal objectives of public sector funding, where support is given to the French television industry while the development of a high standard of television is encouraged

Cinémathèque Française - Musée du Cinéma

7 avenue Albert de Mun
75016 Paris
Tel: (33) 1 45 53 21 86
Fax: (33) 1 42 56 08 55/47 04 79 34
Marianne de Fleury
Founded in 1936 by Henri Langlois, Georges Franju and Jean Mitry to save, conserve and show films. Now houses a cinema museum, screening theatres, library and stills and posters library

Fédération de la Production Cinématographique Française

rue du Cirque 5
75008 Paris
Tel: (33) 1 42 25 70 63
Fax: (33) 1 42 25 94 27
Alain Poiré, Pascal Rogard
National federation of French cinema production

Fédération Nationale des Distributeurs de Films

boulevard Malesherbes 43
75008 Paris
Tel: (33) 1 42 66 05 32
Fax: (33) 1 42 66 96 92
President: Fabienne Vonier, Délégué général: Antoine Virenque
National federation of film distributors

Fédération Nationale des Industries Techniques du Cinéma et de l'Audio-visuel (FITCA)

avenue Marceau 50
75008 Paris
Tel: (33) 1 47 23 75 76
Fax: (33) 1 47 23 70 47
A federation of technical trade associations which acts as intermediary between its members and their market. Maintains a database on all technical aspects of production, and helps French and European companies find suitable partners for research and development or commercial ventures

France 2

avenue Montaigne 22
75008 Paris
Tel: (33) 1 44 21 42 42

Fax: (33) 1 44 21 51 45
France's main public service
terrestrial television channel

Institut National de l'Audiovisuel (INA)
Avenue de l'Europe 4
94366 Bry-sur-Marne Cédex
Tel: (33) 1 49 83 21 12
Fax: (33) 1 49 83 31 95
Television and radio archive;
research into new technology;
research and publications about
broadcasting; production of over 130
first works for television and 15
major series and collections. INA
initiates major documentaries and
cultural series involving partners from
Europe and the rest of the world

TF1
1 Quai du Point du Jour
92656 Boulogne, Cédex
Tel: (33) 1 41 41 12 34
Fax: (33) 1 41 41 29 10
Privatised national television channel

GERMANY

ARD (Arbeitsgemeinschaft der öffentlichrectlichen Rundfunkanstalten der Bundesrepublik Deutschland)
Programme Directorate of
Deutsches Fernsehen
Arnulfstrasse 42
Postfach 20 06 22
80335 Munich
Tel: (49) 89 5900 01
Fax: (49) 89 5900 32 49
One of the two public service
broadcasters in Germany, consisting
of 13 independent broadcasting
corporations

BVDFP (Bundesverband Deutscher Fernseh - produzenten)
Widenmayerstrasse 32
80538 Munich
Tel: (49) 89 21 21 47 10
Fax: (49) 89 228 55 62
Trade association for independent
television producers

Bundesministerium des Innern (Federal Ministry of the Interior)
Postfach 170290
53108 Bonn
Tel: (49) 228 681 5566/7
Fax: (49) 228 681 5504
Detlef Flotho, Fabricle Beelitz
Awards prizes, grants funds for the
production and distribution of

German feature films, short films,
films for children and young people
and documentaries. Promotes film
institutes, festivals and specific
events. Supervisory body of the
Federal Archive for national film
production

Deutsches Filmmuseum
Schaumainkai 41
60596 Frankfurt/Main
Tel: (49) 69 21 23 33 69
Fax: (49) 69 21 23 78 81
email:filmmuseum@stadt-frankfurt.de
Permanent and temporary
exhibitions, incorporates the
Cinema, the municipally
administered cinémathéque. Film
archive and collections of
equipment, documentation, stills,
posters and designs, music and
sound. Library and videothéque

Deutsches Institut für Filmkunde
Schaumainkai 41
60596 Frankfurt/Main
Tel: (49) 69 9612200
Fax: (49) 69 620 060
The German Institute for Film
Studies is a non-profit making
organisation, and its remit includes
amassing culturally significant films
and publications and documents
about film; to catalogue them and
make them available for study and
research. It also supports and puts on
screenings of scientific, cultural and
art films

Export-Union des Deutschen Films (EXU)
Türkenstrasse 93
80799 München
Tel: (49) 89-390095
Fax: (49) 89-395223
Board of Directors: Antonio
Exacoustos, Benno Nowotny,
Jochem Strate
Managing Director: Christian Dorsch
PR Manager: Susanne Reinker
The Export-Union des Deutschen
Films (EXU) is the official trade
association for the promotion of the
export of German films, with
overseas offices located in London,
Paris, Rome, Madrid, Buenos Aires,
Tokyo and Los Angeles. The EXU
maintains a presence at all major
film and TV festivals (ie Berlin,
Cannes, Montreal, Toronto, Locarno,
Venice, MIP-TV, MIPCOM and
MIFED). It has a switchboard
function for German film companies
working abroad as well as for
foreign companies and buyers
looking for media outlets and
coproduction facilities in Germany

FFA
(Filmförderungsanstalt)
Budapester Strasse 41
10787 Berlin
Tel: (49) 30 254090-0
Fax: (49) 30 254090-57
Rolf Bahr, Dr. Karl Guhlke -
Directors General
The German Federal Film Board
(FFA), incorporated under public law,
is the biggest film funding institution
in the country. Its mandate is the all-
round raising of standards of quality in
German film and cinema and the
improvement of the economic
structure of the film industry. The
annual budget of about 60 million
Deutschmarks is granted by a levy
raised from all major German cinemas
and video providers. The
administrative council of 29 members
is a reprsentative cross section of the
German film industry including
members of the government's upper
and lower house as well as public and
private TV stations. Funding is offered
in the following areas: full-length
features, shorts, screenplays,
marketing, exhibition, additional
prints and professional training. The
Export-Union des Deutschen Films
e.V. largely represents the FFA's
interests abroad

FSK (Freiwillige Selbstkontrolle der Filmwirtschaft)
Kreuzberger Ring 56
65205 Wiesbaden
Tel: (49) 611 77 891 0
Fax: (49) 611 77 891 39
Film industry voluntary self-
regulatory body. Activities are: to
examine together with official
competent representatives which
films can be shown to minors under
18 year olds and under; to discuss
the examination of films with youth
groups; to organise seminars on the
study of film, videos and new media.
Adult films (age group from 18) are
approved only by delegates of films
and video industry

Film Förderung
Hamburg GmbH
Friedensalle14-16
22765 Hamburg
Tel: (49) 40 39837-0
Fax: (49) 40 39837-10
Managing director: Eva Hubert
Subsidies available for: script
development; pre-production; co-
production and distribution

Kunsthochschule für Medien Köln (Academy of Media Arts)
Peter-Welter-Platz 2

50676
Cologne
Tel: (49) 221 201890
Fax: (49) 221 2018917
The first academy of Arts in
Germany to embrace all the audio-
visual media. It offers an Audio-
visual Media graduate programme
concentrating on the areas of
Television/Film, Media Art, Media
Design and Art and Media Science

Stiftung Deutsche Kinemathek

Heerstr 18-20
14052 Berlin
Tel: (49) 30 3009030
Fax: (49) 30 30090313
Hans Helmut Prinzler
German Film Archive with collection
of German and foreign films, cine-
historical documents and equipment
(approx. 10,000 films, over a million
photographs, around 20,000 posters,
15,000 set-design and costume
sketches, projectors, camera and
accessories from the early days of
cinema to the 80s). Member of FIAF

ZDF (Zweites Deutsches Fernsehen)

ZDF-Strasse
PO Box 4040
55100 Mainz
Tel: (49) 6131 702060
Fax: (49) 6131 702052
A major public service broadcaster
in Germany

GREECE

ERT SA (Hellenic Broadcasting Corporation)

Messoghion 402
15342 Aghia Paraskevi
Athens
Tel: (30) 1 639 0772
Fax: (30) 1 639 0652
National public television and radio
broadcaster, for information,
education and entertainment

Greek Film Centre

10 Panepistimiou Avenue
10671 Athens
Tel: (30) 1 361 7633/363 4586
Fax: (30) 1 361 4336
Governmental organisation under the
auspices of the Ministry of Culture.
Grants subsidies for production,
promotion and distribution

Ministry of Culture

Cinema Department
Boulinas Street 20
10682 Athens
Tel: (30) 1 322 4737

IRELAND

An Chomhairle Ealaíon/The Arts Council

70 Merrion Square
Dublin 2
Tel: (353) 1 6180200
Fax: (353) 1 6761302
The Arts Council/An Chomhairle
Ealaion is the principal channel of
Government funding for the arts in
Ireland. In the area of film the
Council focuses its support on the
development of film as an art form
and on the individual film-maker as
artist. With a budget for film of
£975,000 in 1998 the Council
supports a national film centre and
archive, four film festivals and a
number of film resource
organisations. It administers an
awards scheme for the production of
short dramas, experimental films and
community video. It also co-operates
with the Irish Film Board and RTE
Television in Frameworks, an
animation awards scheme

Bord Scannán na hÉireann/Irish Film Board

Rockfort House
St. Augustine Street
Galway
Tel: (353) 91 561398
Fax: (353) 91 561405
email: film@iol.ie
Website: http//www.iol.ie/filmboard
Chief Executive: Rod Stoneman
Business Manager: Leslie Kelly
Applications Officer: Paddy Hayes
Information Co-ordinator: Cynthis
O'Murchu
Bord Scannán na hÉireann promotes
the creative and commercial
elements of Irish film-making and
film culture for a home and
international audience. Each year it
supports a number of film projects
by providing development and
production loans. Normally three
submission deadlines annually.
Dates and application procedures
available from the office

Film Censor's Office

16 Harcourt Terrace
Dublin 2
Tel: (353) 1 676 1985
Fax: (353) 1 676 1898
Sheamus Smith
The Official Film Censor is appointed
by the Irish Government to consider
and classify all feature films and
videos distributed in Ireland

Film Institute of Ireland

Irish Film Centre
6 Eustace Street, Temple Bar
Dublin 2
Tel: (353) 1 679 5744/677 8788
Fax: (353) 1 677 8755
email: info@ifc.ie
The Film Institute promotes film
culture through a wide range of
activities in film exhibition and
distribution, film/media education,
various training programmes and the
Irish Film Archive. Its premises, the
Irish Film Centre in Temple Bar, are
also home to Film Base, MEDIA
Desk, The Junior Dublin Film
Festival, The Federation of Irish
Film Societies, and Hubbard
Casting. The Building has
conference facilities, a bar cafe and
a shop as well as 2 cinemas seating
260 and 115

RTE (Radio Telefis Eireann)

Donnybrook
Dublin 4
Tel: (353) 1 208 3111
Fax: (353) 1 208 3080
Public service national broadcaster

ITALY

ANICA (Associazione Nazionale Industrie Cinematografiche e Audiovisive)

Viale Regina Margherita 286
00198 Rome
Tel: (39) 6 442 31 480
Fax: (39) 6 442 31 296/6 440 41 28
Gino de Dominicis
Trade association for television and
movie producers and distributors,
representing technical industries
(post-production companies/
dubbing/studios/labs); home video
producers and distributors;
television and radio broadcasters

Centro Sperimentale di Cinematografia Cineteca Nazionale

Via Tuscolana 1524
00173 Rome
Tel: (39) 6 722 941
Fax: (39) 6 721 1619

Fondazione Cineteca Italiana

Via Palestro 16
20121 Milan
Tel: (39) 2 799224
Fax: (39) 2 798289
Film Museum
Palazzo Dugnani
Via D Manin 2/b
Milan

Tel: (39) 2 6554977
Gianni Comencini
Film archive, film museum. Set up
to promote the preservation of film
as art and historical document, and
to promote the development of
cinema art and culture

Fininvest Television
Viale Europa 48
20093 Cologno Monzese, Milan
Tel: (39) 2 251 41
Fax: (39) 2 251 47031
Adriano Galliani
Major competitor to RAI, running
television channels Canale 5, Italia
Uno and Rete Quattro

Istituto Luce S.p.A
Via Tuscolana 1055
00173 Rome
Tel: (39) 6 722931/729921
Fax: (39) 6 7222493/7221127
Presiolente e Administratore
Delegato: Angelo Guglieluni
Diretore Ufficio Stampa e
Pubblicità: Patrizia de Cesari
Diretiore Commerciale: Leonardo
Tiberi
Created to spread culture and
education through cinema. It invests
in film, distributes films of cultural
interest and holds Italy's largest
archive

Museo Nazionale del Cinema
Via Montebello 15
10124 Turin
Tel: (39) 11 8154230
Fax: (39) 11 8122503
Giuliano Soria, Paolo Bertetto,
Sergio Toffetti, Donata Pesenti
Campagnoni, Luciana Spina. The
museum represents photography,
pre-cinema and cinema history. Its
collections include films, books and
periodicals, posters, photographs
and cinema ephemera

RAI (Radiotelevisione Italiana)
Viale Mazzini 14
00195 Rome
Tel: (39) 6 361 3608
Fax: (39) 6 323 1010
Italian state broadcaster

LUXEMBOURG

Cinémathèque Municipale - Ville de Luxembourg
rue Eugène Ruppert 10
2453 Luxembourg
Tel: (352) 4796 2644
Fax: (352) 40 75 19
Official Luxembourg film archive,
preserving international film

heritage. Daily screenings every year
'Live Cinema' performances - silent
films with music. Member of FIAF,
(13,000 prints/35mm, 16mm, 70mm)

CLT Multi Media
Blvd Pierre Frieden 45
1543 Luxembourg
Tel: (352) 42 1 42 2170
Fax: (352) 42 1 42 2756
Director of Corporate
Communications: Karin Schintgen
Radio, television; co-production/
distribution; press; rights aquisitions

THE NETHERLANDS

Ministry of Education, Culture and Science (OCW)
Film Department
PO Box 25.000
2700LZ Zoetermeer
Tel: (31) 79-3234368
Fax: (31) 79-3234959
Rob Docter, Séamus Cassidy
The film department of the Ministry
is responsible for the development
and maintenance of Dutch film
policy. Various different
organisations for production,
distribution, promotion and
conservation of film are subsidised
by this department

Nederlandse Omroep Stichting (NOS)
Postbus 26444
1202 JJ Hilversum
Tel: (31) 35 779 222
Fax: (31) 35 773 586
Louis Heinsman
Public corporation co-ordinating
three-channel public television

Netherlands Filmmuseum
Vondelpark 3
1071 AA Amsterdam
Tel: (31) 20 589 1400
Fax: (31) 20 683 3401
Film museum with three public
screenings each day, permanent and
temporary exhibitions, library, film
café and film distribution

Vereniging van Onafhankelijke Televisie Producenten (OTP)
Sumatralaan 45
PO Box 27900
1202 KV Hilversum
Tel: (31) 35 6231166
Fax: (31) 6280051
Director: Andries M. Overste
Trade association for independent
television producers (currently
14members)

PORTUGAL

Cinemateca Portuguesa - Museum do Cinema
Rua Barata Salgueiro 39
1200 Lisbon
Tel: (351) 1 54 62 79
Fax: (351) 1 352 31 80
President: João Bénard da Costa,
vice President: José Manuel Costa
National film museum and archive,
preserving, restoring and showing
films. Includes a public
documentation centre, a stills and
posters archive

Instituto Português da Arte Cinematográfica e Audiovisual (IPACA)
Rua S Pedro de Alcântara 45-1o
1250 Lisbon
Tel: (351) 1 346 66 34
Fax: (351) 1 347 27 77
President: Zita Seabra, Vice-
Presidents: Paulo Moreira, Salvato
Telles de Menezes
Assists with subsidies, improvement,
regulation and promotion of the
television and film industry

RTP (Radiotelevisão Portuguesa)
Avenida 5 de Outubro 197
1094 Lisbon Cedex
Tel: (351) 1 793 1774
Fax: (351) 1 793 1758
Maria Manuela Furtado
Public service television with two
channels: RTP1 - general, TV2 -
cultural and sports. One satellite
programme, RTP International,
covering Europe, USA, Africa, Macau

SPAIN

Academia de las Artes y de las Ciencias Cinematográficas de España
General Oraá 68
28006 Madrid
Tel: (34) 1 563 33 41
Fax: (34) 1 563 26 93

Filmoteca Española
Carretera de la Dehesa de la Villa
s/n, 28040 Madrid
Tel: (34) 1 549 00 11
Fax: (34) 1 549 73 48
Director: José Maria Prado; Deputy
Director: Catherine Gautier;
Documentation: Dolores Devesa
National Film Archive, member of
FIAF since 1958. Preserves 26,000
film titles including a large

collection of newsreels. Provides access to researchers on its premises. The library and stills departments are open to the public. Publishes and co-produces various books on film every year. Five daily public screenings with simultaneous translation or electronic subtitles are held at the restored Cine Doré, C/ Santa Isabel 3, in the city centre, where facilities include a bookshop and cafeteria

ICAA (Instituto de la Cinematografia y de las Artes Audio-visuales)
Ministerio de Cultura
Plaza del Rey No1
28071 Madrid
Tel: (34) 1 532 74 39
Fax: (34) 1 531 92 12
Enrique Balmaseda Arias-Dávila
The promotion, protection and diffusion of cinema and audiovisual activities in production, distribution and exhibition. Gives financial support in these areas to Spanish companies. Also involved in the promotion of Spanish cinema and audio-visual arts, and their influence on the different communities within Spain

RTVE (Radiotelevision Española)
Edificio Prado del Rey - 3a planta
Centro RTVE, Prado Del Rey,
22224 Madrid
Tel: (34) 1 5 81 70 00
Fax: (34) 1 5 81 77 57
Head of International Sales RTVE: Teresa Moreno
National public service broadcaster, film producer and distributor

SWEDEN

Oberoende Filmares Förbund (OFF)/Independent Film Producers Association
Box 27 121
102 52 Stockholm
Tel: (46) 8 665 12 21
Fax: (46) 8 663 66 55
email: off.se
OFF is a non-profit organisation, founded 1984, with some 300 members. OFF promotes the special interests of filmmakers and independent Swedish producers of documentaries, short and feature films. Our purpose is twofold: to raise the quality of Swedish audiovisual production and to increase the quantity of domestic production. OFF works on many

levels. The organisation partakes in public debate, organises seminars, publishes a quarterly newsletter, does lobby-work on a national level besides nordic and international networking. OFF aids its producers with legal counsel as well as copyright, economic and insurance policy advisement

Statens biografbyrå
Box 7728
103 95 Stockholm
Tel: (46) 8 24 34 25
Fax: (46) 8 21 01 78
The Swedish National Board of Film Classification (Statens biografbyrå) was founded in 1911. Films and videos must be approved and classified by the Board prior to showing at a public gathering or entertainment. For videos intended for sale or hire, there is a voluntary system of advance examination

Svenska Filminstitutet (Swedish Film Institute)
Box 27 126
Filmhuset
Borgvägen 1-5
S-10252 Stockholm
Tel: (46) 8 665 11 00
Fax: (46) 8 661 18 20
email: janerik.billinger@sfi.se
Jan-Erik Billinger: Head of the Information Department
The Swedish Film Institute is the central organisation for Swedish cinema. Its activities are to: support the production of Swedish films of high merit; promote the distribution and exhibition of quality films; preserve films and materials of interest to cinematic and cultural history and promote Swedish cinematic culture internationally

Sveriges Biografägareförbund
Box 1147
S 171 23 Solna
Tel: (946) 8 735 97 80
Fax: (946) 8 730 25 60
The Swedish Exhibitors Association is a joint association for Swedish cinema owners

Sveriges Filmuthyrareförening upa
Box 23021
S-10435 Stockholm
Tel: (946) 8 441 55 70
Fax: (946) 8 34 38 10
Kay Wall
The Swedish Film Distributors Association is a joint association for film distributors

Swedish Women's Film Association
Po Box 27182
S-10251 Stockholm
Visitors address: Filmhuset,
Borgvägen 5
Tel: (46) 8 665 1100/1293
Fax: (46) 8 666 3748
Anna Hallberg
Workshops, seminars, festivals and international exchange programme

THE MEDIA II PROGRAMME

The MEDIA II Programme is an initiative of the European Union, managed by the European Commission in Brussels. MEDIA II, which follows on from MEDIA I, started in 1996 and will conclude in the year 2000.

European Commission

Directorate General X:
Information, Communication,
Culture, Audio-visual
rue de la Loi, 200
1040 Brussels, Belgium
Fax: (32) 2 299 92 14
Head of Programme: Jacques Delmoly

MEDIA Antennae
At the time of publication MEDIA Antennae Glasgow had taken over the responsibilities from the UK MEDIA Desk as the information centre for European audiovisual funds policy and the MEDIA II Programme. It publishes newsletters, organises information seminars and events to promote increased understanding of the MEDIA II Programme and provide a consultation service to UK professionals about opportunities in Europe.

MEDIA Antenna Glasgow

c/o Scottish Screen
74 Victoria Crescent Road
Glasgow G12 9JN
Tel: 0141 334 4445
Fax: 0141 357 2345
email: louise.scott@dial.pipex.com
Contact: Louise Scott

MEDIA Antenna Cymru Wales

c/o Sgrîn: The Media Agency for Wales
Screen Centre
Llantrisant Road
Cardiff CF5 2PU
Tel: 01222 578 370
Fax: 01222 578 654
email: antenna@scrwales.demon.co.uk
Website: http://www.screenwales.org.uk/
Contact: Gethin While

The IO's

The MEDIA II Programme contracted three companies to act as Intermediary Organisations (IOs) for the five year period of MEDIA II to assist Brussels in administering and processing applications in each of the three areas of support (Training, Development and Distribution) and dealing with payments.

Intermediary Organisations

Training

Media Research and Consultancy Spain

(MRC) Madrid
Serrano 63, Esc 3 -6 izda
28006 Madrid, Spain
Tel: (34) 1 577 97 04
Fax: (34) 1 577 71 99
Head of office: Fernando Labrada

Development

European Media Development Agency

(EMDA) London
39c Highbury Place
London N5 1QP
Tel: (44) 171 226 9903
Fax: (44) 171 354 2706
email:101613.1775@compuserve.com@pmdf
Head of office: David Kavanagh

Distribution

D&S Media Service

Brussels, Munich and Dublin
Av- de Tevuren 35
B - 1040 Brussels
Tel: (322) 743 22 30
Fax: (322) 743 2245
Head of office: John Dick

Exhibition Networks

Europa Cinemas

54 rue Beaubourg
75003 Paris, France
Tel: (33) 1 42 71 53 70
Fax: (33) 1 42 71 47 55
email: europacinema@magic.fr
Website: http://www.europa-cinemas.org
Claude-Eric Poiroux
Fatima Djoumer
This project encourages screenings and promotion of European films in a network of cinemas in European cities. It offers a financial support according to a certain percentage ie number of screens and the national market share of European films, for promotional activities and special events for European films.

MEDIA Salles

Via Soperga, 2
20127 Milano, Italy
Tel: (39) 02 669 4405
Fax: (39) 02 669 1574
Website: http://www.mediasalles.it/mediasalles/
Elisabetta Brunella
MEDIA Salles with Euro Kids

Network proposes an initiative aimed at consolidating the availability of 'cinema at the cinema' for children and young people in Europe and at raising the visibility of European products with a younger audience. For further information please call the office in Italy.

Industrial Platforms

Two initiatives were selected in 1996 by the Commission to run the Industrial Platforms for the areas of animation, new technologies and audiovisual heritage. These are respectively:

CARTOON (European Association of Animation Film)

418 Boulevard Lambermont
1030 Brussels, Belgium
Tel: (32) 2 245 12 00
Fax: (32) 2 245 46 89
Website: http://www.cartoon-media.be
Contact: Corinne Jenart, Marc Vandeweyer
Provides financial assistance for graphic research, script writing or adaptation, and pilot film in the animation sector. Successful applicants are invited to become a member of the CARTOON Platform. Guidelines and application forms can be obtained from the Media Antennae

Multimedia Investissements: 2MI

Avenue de l'Europe, 4
94366 Bry-sur-Marne Cedex
France
Tel: (33) 1 49 83 28 63
Fax: (33) 1 49 83 26 26
Contact: Jean-Bernard Tellio
email: mmi@2mi.fr
Website: http://www.2mi.com
This platform consists of 25 major European enterprises involved in multimedia and new audiovisual technologies. Its objectives are to co-finance projects, facilitate access to the market and establish networks within the sector to stimulate the production and distribution of European multimedia titles. Guidelines and application forms can be obtained from the Media Antennae

Workshops

PILOTS

Pau Claris, 115, 5è. 4a.
E-08009 Barcelona, Spain
Tel: (34 -3) 487 37 73
Fax: (34-3) 487 39 52
Roger Gregory
PILOTS workshops on TV drama and comedy provide the opportunity to learn the best of American and European writing methods

PR COMPANIES

These are companies which handle all aspects of promotion and publicity for film and video production companies and/or individual productions

The Associates
34 Clerkenwell Close
London EC1R OAU
Tel: 0171 608 2204
Fax: 0171 250 1756
Catherine Flynn, Alison Marsh

Berlin Film Festival, PR Co-ordinator and British Representative
67 Parkway Drive
Queens Park
Bournemouth BH8 9JS
Tel: 01202 393033
Fax: 01202 301516
Soren Fischer

Tony Brainsby PR
16b Edith Grove
London SW10 0NL
Tel: 0171 834 8341
Fax: 0171 352 9451
Tony Brainsby

Byron Advertising, Marketing and PR
Byron House
Wallingford Road
Uxbridge
Middx UB8 2RW
Tel: 01895 252131
Fax: 01895 252137
Les Barnes

CJP Public Relations Ltd
29a Chippenham Mews
London W9 2AN
Tel: 0171 266 0167
Fax: 0171 266 0165
email: cjp@globalnet.co.uk
Carolyn Jardine

Cantorwise
109 Crouch Hill
Crouch End
London N8 9RD
Tel: 0181 347 7770
Fax: 0181 347 5550
Melanie Cantor

Jacquie Capri Enterprises
3rd Floor
46/47 Chancery Lane
London WC21 1JB
Tel: 0171 831 4545
Fax: 0171 831 2557

Emma Chapman Publicity
25 Frith Street

London W1V 5TR
Tel: 0171 734 9619
Fax: 0171 494 3884
Contact: Emma Chapman

Max Clifford Associates
109 New Bond Street
London W1Y 9AA
Tel: 0171 408 2350
Fax: 0171 409 2294
Max Clifford

Corbett and Keene
122 Wardour Street
London W1V 3LA
Tel: 0171 494 3478
Fax: 0171 734 2024
Ginger Corbett, Sara Keene

Warren Cowan/Phil Symes Associates
35 Soho Square
London W1V 6AX
Tel: 0171 439 3535
Fax: 0171 439 3737
Phil Symes, Warren Cowan

Dennis Davidson Associates (DDA)
Royalty House
72-74 Dean Street
London W1V 5HB
Tel: 0171 439 6391
Fax: 0171 437 6358
email: info@ddapr.com
Dennis Davidson, Stacy Wood, Chris Paton

Clifford Elson (Publicity)
223 Regent Street
London W1R 7DB
Tel: 0171 495 4012
Fax: 0171 495 4175
Clifford Elson, Patricia Lake-Smith

Entertainment Promotions
305 Gray's Inn Road
London WC1X 8QF
Tel: 0171 713 1234
Fax: 0171 713 1741/2
Murray Harkin

FEREF Associates
14-17 Wells Mews
London W1A 1ET
Tel: 0171 580 6546
Fax: 0171 631 3156
Peter Andrews, Ken Paul, Robin Behling, David Kemp, Brian Bysouth, Gareth Shepherd

Foresight Promotions
4 Albion Court, Galena Road
London W6 0QT
Tel: 0181 748 3550
Fax: 0181 741 8461
Tim Smith

Lynne Franks PR
327-329 Harrow Road
London W9 3RB
Tel: 0171 724 6777
Fax: 0171 724 8484
Julian Henry

Freud Communications Ltd
19-21 Mortimer Street
London W1N 8DX
Tel: 0171 580 2626
Fax: 0171 637 2626

GCI Group
1 Chelsea Manor Gardens
London SW3 5PN
Tel: 0171 349 5075
Fax: 0171 352 6244
Nick Oldham

Ray Hodges Communications
18 London End, Beaconsfield
Bucks HP9 2JH
Tel: 01494 672492
Fax: 01494 671644
Ms Ray Hodges

Sue Hyman Associates
70 Chalk Farm Road
London NW1 8AN
Tel: 0171 485 8489/5842
Fax: 0171 267 4715
email: sue.hyman.@btinternet.com
Sue Hyman

JAC Publicity
36 Great Queen Street
Covent Garden
London WC2B 5AA
Tel: 0171 430 0211
Fax: 0171 430 0222
Claire Forbes

Richard Laver Publicity
3 Troy Court
Kensington High Street
London W8 7RA
Tel: 0171 937 7322
Fax: 0171 937 8670
Richard Laver

Limelight Public Relations
9 Coptic Street

London WC1A 1NH
Tel: 0171 436 6949
Fax: 0171 323 6791
Fiona Lindsay, Linda Shanks

Mathieu Thomas
8 Westminster Palace Gardens
Artillery Row
London SW1P 1RL
Tel: 0171 222 0833
Fax: 0171 222 5784
Paul Mathieu, Amanda Slayton

McDonald and Rutter
14-18 Ham Yard
Gt. Windmill Street
London W1P 7PD
Tel: 0171 734 9009
Fax: 0171 734 1151
email: mcdonaldrutter@btinternet.com
Charles McDonald, Jonathan Rutter

Optimum Communications
Ludgate House
107-111 Fleet Street
London EC4A 2AB
Tel: 0171 583 1020
Fax: 0171 583 1116
Nigel Passingham

Orlando Kimber PR
POBox 5600
Newbury
Berkshire RG20 8YU
Tel: 01488 608888
Fax: 01488 608811
email: 100067.3216@compuserve.com

Porter Frith Publicity & Marketing
26 Danbury Street
London N1 8JU
Tel: 0171 359 3734
Fax: 0171 226 5897
Sue Porter, Liz Frith

Premier Relations
1 Meadway
Leatherhead
Surrey KT22 0LZ
Tel: 01372 842446
Fax: 01372 843819
Victoria Franklin

Riley Associates
6 Sherlock Mews
London W1
Tel: 0171 224 1525
Fax: 0171 224 0540
Tony Riley

S.S.A. Public Relations
Suite 323/324
The Linen Hall
162-168 Regent Street
London W1R 5TB
Tel: 0171 494 2755
Fax: 0171 494 2833
Céline Brook, Andrew O'Driscoll

S.S.A Public Relations is a full service public relations firm that provides trade and consumer publicity for a wide range of corporate and entertainment clients. The company specialises in key areas, representing television and theatrical film production and distribution companies

Judy Tarlo Associates
85 Ashworth Mansions
Grantully Road
London W9 ILN
Tel: 0171 286 6025
Fax: 0171 289 8969
Judy Tarlo, Louise Hanson

Peter Thompson Associates
134 Great Portland Street
London W1N 5PH
Tel: 0171 436 5991/2
Fax: 0171 436 0509
Peter Thompson, Amanda Malpass

Town House Publicity
45 Islington Park Street
London N1 1QB
Tel: 0171 226 7450
Fax: 0171 359 6026
email: townhouse@lineone.net
Mary Fulton

Stella Wilson Publicity
130 Calabria Road
London N5 1HT
Tel: 0171 354 5672
Fax: 0171 354 2242
Stella Wilson

PRESS CONTACTS

These are magazines and newspapers which cover issues relating to film, television and video. Circulation figures may have altered since going to press. Also listed are the news and photo agencies which handle media news syndication, and television and radio programmes concerned with the visual media

---

Arena (Bi-monthly)
Third Floor, Block A
Exmouth House
Pine Street
London EC1R 0JL
Tel: 0171 837 7270
Fax: 0171 837 3906
Film/TV editors: Mark Hooper
Magazine for men covering general interest, film, literature, music and fashion
Lead time: 6-8 weeks
Circulation: 100,000

Ariel (Weekly, Tues)
Room 123, Henry Wood House
386 Langham Place
London W1A 1AA
Tel: 0171 765 3623
Fax: 0171 765 3646
Editor: Robin Reynolds
BBC staff magazine
Lead time: Tuesday before publication
Circulation: 35,000

Art Monthly
Suite 17
26 Charing Cross Road
London WC2H 0DG
Tel: 0171 240 0389
Fax: 0171 497 0726
Editor: Patricia Bickers
Aimed at artists, art dealers, teachers, students, collectors, arts administrators, and all those interested in contemporary visual art
Lead time: 4 weeks
Circulation: 4,000 plus

Asian Times (Weekly, Tues)
138-148 Cambridge Heath Road
London E1 5QJ
Tel: 0171 702 8012
Fax: 0171 702 7937
Editor: Sewar Ahmed
National, weekly newspaper for Britain's English-speaking, Asian community
Press day: Thurs
Circulation: 30,000

The Big Issue (Weekly, Mon)
Fleet House
57-61 Clerkenwell Road
London EC1M 5NP
Tel: 0171 418 0418
Fax: 0171 418 0427
Editor: Becky Gardiner

Arts: Tina Jackson
Film editor: Xan Brooks
General interest magazine, with emphasis on homelessness. Sold by the homeless
Lead time: Tues, 3 weeks before
Circulation: ABC figure 142,937

The Box (Bi-Monthly)
38-42 Hampton Road
Teddington
Middx TW11 0JE
Tel: 0181 943 5096
Fax: 0181 943 5015
email: thebox@dial.pipex.com
Website: http://
www.thebox.haynet.com
Editor:Paul Simpson

Boz (incorporating Jazz Express)
Kettners Building
29 Romilly Street
London W1V 6HP
Tel: 0171 439 2120
Fax: 0171 434 1214
email: boz@pbpublications.co.uk
TV Editor: Khalil Hafiz Khairallah
General entertainment with a slant towards jazz and cabaret
Lead time: 4 weeks
Circulation: 10,000

British Film (Quarterly)
Arts and Entertainment
Publishing Ltd
24 Sandyford Place
Glasgow G3 7NG
Tel: 0141 221 4241
Fax: 0141 221 4247
Editor: Robert McColl
British Film covers Film making and broadcasting within the UK. Primarily a trade magazine distributed to all production companies and facility houses within the UK
Circulation: 20,000

British Film Facilities Journal
Compass Rose Publishing
12 Eton Street
Richmond upon Thames
Surrey TW9 1EE
Tel: 0181 332 1644
Fax: 0181 332 1755
email: info@compass-rose.co.uk
Website: http://www.compas-rose.co.uk

Editor: Colin Lenthall
Journal for those working in British film, TV and video industry

Broadcast (Weekly, Fri)
EMAP Media
33-39 Bowling Green Lane
London EC1R 0DA
Tel: 0171 505 8014
Fax: 0171 505 8050
Publisher/Editor: Jon Baker
Broadcasting industry news magazine with coverage of TV, radio, cable and satellite, corporate production and international programming and distribution
Press day: Wed. Lead time: 2 weeks
Circulation: 13,556

The Business of Film (Monthly)
Suite 3
2a New Cavendish Street
London W1M 7RP
Tel: 0171 486 1996
Fax: 0171 486 1969
Publisher/executive editor: Elspeth Tavares
Aimed at film industry professionals - producers, distributors, exhibitors, investors, financiers
Lead time: 2 weeks

Cable and Satellite Communications International (Monthly)
104 City View
463 Bethnal Green Road
London E2 9QY
Tel: 0171 613 5553
Fax: 0171 729 7723
email:de81@dial.pipex.com
Editor: Joss Armitage
Business magazine for professionals in the cable and satellite television industry
Circulation: 4,029

Capital Gay (Weekly, Thur)
1 Tavistock Chambers
Bloomsbury Way
London WC1A 2SE
Tel: 0171 242 2750
Fax: 0171 242 3334
Film editor: Pas Paschal
TV editor: Michael Mason
Newspaper for lesbians and gay men in the South East combining news,

features, arts and entertainment, what's on guide
Lead time: 1 week (Mon)
Circulation: 22,000

Caribbean Times
incorporating African Times (Weekly, Mon)
138-148 Cambridge Heath Road
London E1 5QJ
Tel: 0171 702 8012
Fax: 0171 702 7937
Editor: Clive Morgan
Tabloid dealing with issues pertinent to community it serves
Press day: Fri
Circulation: 25,000

City Life (Fortnightly)
164 Deansgate
Manchester M60 2RD
Tel: 0161 839 1416
Fax: 0161 839 1488
Editor: Chris Sharratt
Film editor: Melanie Dakin
What's on in and around Greater Manchester
Circulation: 20,000

COIL (journal of the moving image)
PO Box 14649
London EC2A 3RD
Tel: 0171 613 4946
Fax: 0171 613 4052
email: coil@backspace.org
Frequency: two issues per year
Editor: Giles Lane
Project Manager: Joan Johnston
Designer: Damian Jaques

Company (Monthly)
National Magazine House
72 Broadwick Street
London W1V 2BP
Tel: 0171 439 5000
Fax: 0171 439 5117
Editor: Fiona Mclutosh
Glossy magazine for women aged 18-30
Lead time: 10 weeks
Circulation: 272,160

Cosmopolitan (Monthly)
National Magazine House
72 Broadwick Street
London W1V 2BP
Tel: 0171 439 5000
Fax: 0171 439 5101
Editor: John Naughton
Arts/General: Sarah Kennedy
For women aged 18-35
Lead time: 12 weeks
Circulation: 461,080

Creation (Monthly)
MDI Ltd
30/31 Islington Green

London N1 8DU
Tel: 0171 226 8585
Fax: 0171 226 8586
Editor: Michael Taylor
Film, television, new media publication
Circulation: 8,000

Creative Review (Monthly)
St. Giles House
50 Poland Street
London W1V 4AX
Tel: 0171 439 4222
Fax: 0171 734 6748
Editor: Lewis Blackwell
Publisher: Morag Arman-Addey
Trade paper for creative people covering film, advertising and design. Film reviews, profiles and technical features
Lead time: 4 weeks
Circulation: 15,206

Daily Express
Ludgate House
245 Blackfriars Road
London SE1 9UX
Tel: 0171 928 8000
Fax: 0171 620 1654
Showbusiness editor: Annie Leask
Film: Jason Solomons
TV/Theatre critic: Robert Goe-Langton
Showbusiness Correspondent: David Wigg
National daily newspaper
Circulation: 1,227,971

Daily Mail
Northcliffe House
2 Derry Street
London W8 5TT
Tel: 0171 938 6000
Fax: 0171 937 4463
Chief showbusiness writer: Baz Bamigboye
Film: Christopher Tookey
TV: Peter Paterson
National daily newspaper
Circulation: 2,163,676

The Daily Star
Ludgate House
245 Blackfriars Road
London SE1 9UX
Tel: 0171 928 8000
Fax: 0171 922 7962
Film: Sandro Monetti
TV: Pat Codd
Video: Sandro Monetti and Pat Codd
National daily newspaper
Circulation: 654,866

Daily Telegraph
1 Canada Square
Canary Wharf
London E14 5DT
Tel: 0171 538 5000
Fax: 0171 538 6242

Arts Editor: Sarah Crompton
Film critic: Quentin Curtis
TV: Marsha Dunstan
National daily newspaper
Lead time: 1 week
Circulation: 1,117,439

Diva (Bi-monthly)
Ground Floor
Worldwide House
116-134 Bayham Street
London NW1 0BA
Tel: 0171 482 2576
Fax: 0171 284 0329
Editor: Gillian Rodgerson
Lesbian lifestyle magazine
Lead times: 4-6 weeks
Circulation: 15-20,000

Eclipse (Monthly)
Phoenix Magazines Limited
PO Box 33, Liskeard
Cornwall PL14 4YX
Tel: 01579 344313
Fax: (01579) 344313
email: phoenixmgs@aol.com
Editor: Simon Clarke
Magazine covering the entire spectrum of science fiction in books, cinema, television and comics, along with role playing and computer games. News, reviews, interviews, competitions, features, profiles, etc.
Lead time: six weeks
Circulation: 15,000

The Economist (Weekly)
25 St James's Street
London SW1A 1HG
Tel: 0171 830 7000
Fax: 0171 839 2968
Film/video/television (cultural): Tony Thomas;
(business): Frances Cairncross
International coverage of major political, social and business developments with arts section
Press day: Wed
Circulation: 327,689

Elle (Monthly)
Endeavour House
189 Shaftesbury Avenue
London WC2H 8JD
Tel: 0171 437 9011
Fax: 0171 434 0656
Editor: Marie O'Riordan
Arts Ed: Charlotte Moore
Glossy magazine aimed at 18-35 year old working women
Lead time: 3 months
Circulation: 205,623

Empire (Monthly)
Mappin House
4 Winsley Street
London W1N 4AR
Tel: 0171 436 1515
Fax: 0171 312 8249

email: empire@delphi.com
Editor: Ian Nathan
Quality film monthly incorporating features, interviews and movie news as well as reviews of all new movies and videos
Lead time: 3 weeks
Circulation: 161,503

The European
(Weekly, Thurs)
200 Gray's Inn Road
London WC1X 8NE
Tel: 0171 418 7777
Fax: 0171 713 1840/1870
Arts Editor: Andrew Harvey
Editor: Andrew Neil
In-depth coverage of European news, politics and culture
Press day: Thurs
Circulation: 160,511

Evening Standard (Mon-Fri)
Northcliffe House
2 Derry Street
London W8 5EE
Tel: 0171 938 2648
Fax: 0171 937 3193
Film: Alexander Walker, Neil Norman
Media editor: Victor Sebestyen
London weekday evening paper
Circulation: 438,136

Everywoman (Monthly)
9 St Alban's Place
London N1 0NX
Tel: 0171 704 8440
Fax: 0171 226 9448
Arts editor: Nina Rapi
Feminist magazine covering mainstream issues
Lead time: 6 weeks
Circulation: 15,000

The Express on Sunday
Ludgate House
245 Blackfriars Road
London SE1 9UX
Tel: 0171 928 8000
Fax: 0171 620 1656
Film: Chris Peachment
TV: Nigel Billen
National Sunday newspaper
Circulation: 1,159,759

FHM (Monthly)
Mappin House
London W1N 7AR
Tel: 0171 436 1515
Fax: 0171 312 8191
Editor: Ed Needham
Gavin Newsham
Anthony Noguera
Men's lifestyle magazine
Lead time: 6 weeks
Circulation: 644,110

The Face (Monthly)
Second Floor, Block A

Exmouth House
Pine Street
London EC1R 0JL
Tel: 0171 689 9999
Fax: 0171 689 0300
Film: Charles Gant, Adam Higginbotham
Visual-orientated youth culture magazine: emphasis on music, fashion and films
Lead time: 4 weeks
Circulation: 100,744

Film (Quarterly)
Suite 210
29 Great Pulteney Street
London W1R 3DD
Tel: 0171 734 9300
Fax: 0171 734 9093
Editor: Tom Brownlie
Thematically-based journal with information for Film Societies and other film exhibitors
Lead time: 2 weeks
Circulation: 2,000

Film Guide
(Monthly - Free)
Film Guide Ltd
30 North End Road
London W14 0SH
Tel: 0171 602 9790
Fax: 0171 602 2063
Editor: Alan Jones
Film news, features and interviews
Circulation: 125,000

Film Review
(Monthly + 4 specials)
Visual Imagination
9 Blades Court, Deodar Road
London SW15 2NU
Tel: 0181 875 1520
Fax: 0181 875 1588
Editor: Neil Corry
Reviews of films on cinema screen and video; star interviews and profiles; book and CD reviews
Lead time: 1 month
Circulation: 50,000

Financial Times
1 Southwark Bridge
London SE1 9HL
Tel: 0171 873 3000
Fax: 0171 873 3076
Arts: Annalena McAfee
Film: Nigel Andrews
TV: Christopher Dunkley
National daily newspaper
Circulation: 316,578

Flicks (Monthly)
25 The Coda Centre
189 Munster Road
London SW6 6AW
Tel: 0171 381 8811
Fax: 0171 381 1811
Managing Director: Val Lyon

Creative Director: Mike Rewney
Editor: Nick Thomas
Magazine of the film industry, free to cinema-goers throughout the country, or by subscription
Lead time: 6 weeks
Circulation: 400,000 A4 Flicks magazine, 600,000 A5 Flicks trailers

Gay Times (Monthly)
Ground Floor
Worldwide House
116-134 Bayham Street
London NW1 0BA
Tel: 0171 482 2576
Fax: 0171 284 0329
email: edit@gaytimes.co.uk
Arts editor: James Cary Parkes
Britain's leading lesbian and gay magazine. Extensive film, television and arts coverage. Round Britain guide
Lead time: 6-8 weeks
Circulation: 60,000

The Guardian
119 Farringdon Road
London EC1R 3ER
Tel: 0171 278 2332
Fax: 0171 837 2114
Film: Derek Malcolm, Johnathan Romney
TV critic: Nancy Banks-Smith
Media editor: John Mulholland
Arts editor: Claire Armitstead
Head of Press, PR & Corporate Affairs: Camilla Nicholls
Weekend editor: Deborah Orr
National daily newspaper
Circulation: 407,870

Harpers & Queen (Monthly)
National Magazine House
72 Broadwick Street
London W1V 2BP
Tel: 0171 439 5000
Fax: 0171 439 5506
Arts & Films: Anthony Quinn
Glossy magazine for women
Lead time: 12 weeks
Circulation: 93,186

The Herald
195 Albion Street
Glasgow G1 1QP
Grays Inn House
127 Clerkenwell Road
London EC1R 5DB
Tel: 0171 405 2121
Fax: 0171 405 1888
Film critic: William Russell (London address)
TV editor: Ken Wright
Scottish daily newspaper
Circulation: 107,527

The Hollywood Reporter (daily; weekly international, Tues)
32-34 Broadwick Street

London W1A 2HG
Tel: 0171 331 1950
Fax: 0171 331 1951
email:Hollywood_Reporter@VNU.co.uk
European bureau chief: Jeff Kaye
Showbusiness trade paper
Circulation: 39,000

i-D Magazine (Monthly)
Universal House
251 Tottenham Court Road
London W1P 0AE
Tel: 0171 813 6170
Fax: 0171 813 6179
Film & TV: David Sandhu
Youth/fashion magazine with film
features
Lead time: 8 weeks
Circulation: 45,000

Illustrated London News (2 pa)
20 Upper Ground
London SE1 9PF
Tel: 0171 805 5555
Fax: 0171 805 5911
Editor: Alison Booth
News, pictorial record and commen-
tary, and a guide to coming events
Lead time: 8-10 weeks
Circulation: 30,000

In Camera (Quarterly)
Professional Motion Imaging
PO Box 66, Hemel Hempstead
Herts HP1 1JU
Tel: 01442 844875
Fax: 01442 844987
Editor: Josephine Ober
Business editor: Giosi Gallotli
Journal for motion picture industry,
primarily for cinematographers, but
also for other technicians and
anyone in the industry
Lead time: 4 weeks
Circulation: 45,000

The Independent
1 Canada Square
Canary Wharf
London E14 5DL
Tel: 0171 293 2000
Fax: 0171 293 2047
Film: Sam Taylor
TV: Tom Sutcliffe, Gerard Gilbert
Media: Rob Brown
National daily newspaper
Circulation: 257,594

The Independent on Sunday
1 Canada Square
Canary Wharf
London E14 5DL
Tel: 0171 293 2000
Fax: 0171 293 2027
Film critic: Kevin Jackson
TV: Robin Boss
National Sunday newspaper
Lead time: 2 weeks
Circulation: 275,000

International Connection
1st Floor, 23 South Quay
Gt Yarmouth
Norfolk NR30 2RG
Tel: 07071 224091
Fax: 01493 330565
Film and TV industry business
magazine

Interzone (Monthly)
217 Preston Drove
Brighton BN1 6FL
Tel: 01273 504710
Editor: David Pringle
Film: Nick Lowe
Science-fiction magazine
Lead time: 8 weeks
Circulation: 10,000

Jewish Chronicle (Weekly, Friday)
25 Furnival Street
London EC4A 1JT
Tel: 0171 415 1616
Fax: 0171 415 9040
Editor: Edward J Temko
Film critic: Alan Montague
TV critic: Helen Jacobus
Lead time: 2 days
Press day: Wed
Circulation: 47,273

The List (Fortnightly, Thur)
14 High Street
Edinburgh EH1 1TE
Tel: 0131 558 1191
Fax: 0131 557 8500
email: editor@List.co.uk
Editor: Kathleen Morgan
Film editor: Alan Morrison
TV: Peter Ross
Glasgow/Edinburgh events guide
Lead time: 1 week
Circulation: 15,000

Mail on Sunday
Northcliffe House
2 Derry Street
London W8 5TS
Tel: 0171 938 6000
Fax: 0171 937 3829
Film: Jane Preston
TV critic: Brian Viner
National Sunday newspaper
Press day: Fri/Sat
Circulation: 2,137,872

Marie Claire (Monthly)
2 Hatfields
London SE1 9PG
Tel: 0171 261 5240
Fax: 0171 261 5277
Film: Anthony Quinn
Arts: Louise Clark
Women's magazine
Lead time: 3 months
Circulation: 457,034

Media Week (Weekly, Thur)
Quantum House
19 Scarbrook Road
Croydon CR9 ILX
Tel: 0181 565 4317
Fax: 0181 565 4394
email: mweeked@qpp-co.uk
Editor: Susannah Richmond
News magazine aimed at the
advertising and media industries
Press day: Wed
Circulation: 13,209 ABC July '96 -
June '97

Melody Maker (Weekly, Tues)
2nd Floor
King's Reach Tower
Stamford Street
London SE1 9LS
Tel: 0171 261 6229
Fax: 0171 261 6706
Editor: Mark Sutherland
Film: Robin Bresnark
Pop/rock music newspaper
Press day: Fri
Circulation: 46,895

Midweek (Weekly, Thur/West End, Mon/City)
7-9 Rathbone Street
London W1P 1AF
Tel: 0171 636 6651
Fax: 0171 255 2352
Editor: Bill Williamson
Film editor: Derek Malcolm
General interest male/female London
living and arts oriented. 18-35 target
age readership
Lead time: 2 weeks
Circulation: 100,000

The Mirror
1 Canada Square, Canary Wharf
London E14 5DP
Tel: 0171 293 3000
Fax: 0171 293 3409
Film: Simon Rose
TV : Tony Purnell
National daily newspaper with daily/
weekly film and television column
Circulation: 2,355,285
incorporating The Daily Record
(Scottish daily newspaper)

Morning Star
1-3 Ardleigh Road
London N1 4HS
Tel: 0171 254 0033
Fax: 0171 254 5950
Film/TV: Jeff Sawtell
The only national daily owned by its
readers as a co-operative. Weekly
film and TV reviews
Circulation: 9,000

Movie Plus (Monthly)
Inside Publications

16 Brand Street
Hitchin, Herts SG5 IJE
Tel: 01462 436785
Fax: 01462 436806
Editor: Carole Childs

Moving Pictures (Monthly)
151-153 Wardour Street
London W1V 3TB
Tel: 0171 287 0070
Fax: 0171 287 9637
Editor: Christian de Schutter
Worldwide coverage of television,
film, video and new media
Circulation: 8,500

Ms London (Weekly, Mon)
7-9 Rathbone Street
London W1P 1AF
Tel: 0171 636 6651
Fax: 0171 255 2352
Films: Dee Pilgrim
Free magazine with drama, video,
film and general arts section
Lead time: 2 weeks
Press day: Thurs
Circulation: 94,100

Neon (Monthly)
EMAP Metro
Mappin House
4 Winsley Street
London W1N 7AR
Tel: 0171 312 8775
Fax: 0171 580 6495
email:neon@dial.pipex.com
Editor: Adam Higginbotham
Magazine covering film and music
scene aimed at 15-25 year-olds.
Circulation: 100,000

19 (Monthly)
IPC Magazines
King's Reach Tower
Stamford Street
London SE1 9LS
Tel: 0171 261 6410
Fax: 0171 261 7634
Film: Corrine Barraclough
Magazine for young women
Lead time: 8 weeks
Circulation: 187,740

New Musical Express
(Weekly, Wed)
25th Floor
King's Reach Tower
Stamford Street
London SE1 9LS
Tel: 0171 261 5730
Fax: 0171 261 5185
Film/TV editor: Gavin Martin
Rock music newspaper
Lead time: Mon, 1 week before press
day
Circulation: 121,001

New Scientist
(Weekly, Sat avail Thur)

King's Reach Tower
Stamford Street
London SE1 9LS
Tel: 0171 261 7301
Fax: 0171 261 5134
Editor: Alun Anderson
Contains articles and reports on the
progress of science and technology
in terms which the non-specialist
can understand
Press day: Mon
Circulation: 120,744

New Statesman and Society
(Weekly, Fri)
7th Floor,
Victoria Station House
191 Victoria Street
London SW1E 5NE
Tel: 0171 828 1232
Fax: 0171 828 1881
Editor: Ian Hargreaves
Arts films: Laura Cumming
Independent radical journal of
investigation, revelation, politics and
comment
Press day: Mon
Circulation: 26,000

News of the World
News International
1 Virginia Street
London E1 9XR
Tel: 0171 782 4000
Fax: 0171 583 9504
Editor: Phil Hall
Films: Johnathon Ross
TV critic: Charles Catchpole
National Sunday newspaper
Press day: Sat
Circulation: 4,434,856

Nine to Five (Weekly, Mon)
7-9 Rathbone Street
London W1P 1AF
Tel: 0171 636 6651
Fax: 0171 255 2352
Film: Bill Williamson
Free London magazine
Press day: Wed
Circulation: 160,000

The Observer (Weekly, Sun)
119 Farringdon Road
London EC1R 3ER
Tel: 0171 278 2332
Fax: 0171 713 4250
Arts editor: Jane Ferguson
Film critic: Philip French
TV: Mike Bradley
National Sunday newspaper
Lead time: 1 week
Press day: Fri
Circulation: 450,831

Observer Life Magazine
(Weekly, Sun)
119 Farringdon Road

London EC1R 3ER
Tel: 0171 278 2332
Fax: 0171 239 9837
Supplement to The Observer

Options (Monthly)
King's Reach Tower
Stamford Street
London SE1 9LS
Tel: 0171 261 5000
Fax: 0171 261 7344
Film: Susy Feag
TV: Stuart Husband
Women's glossy magazine
Lead time: 3 months
Circulation: 146,692

The PACT Magazine
Producers Alliance for Cinema
and Television
published by MDI Ltd
30/31 Islington Green
London N1 8DU
Tel: 0171 226 8585
Fax: 0171 226 8586
Editor: Michael Taylor
PACT members' monthly
Circulation:2,000

The People (Weekly, Sun)
1 Canada Square
Canary Wharf
London E14 5AP
Tel: 0171 510 3000
Fax: 0171 293 3810
Films: Jane Simon
TV: Rachel Lloyd
National Sunday newspaper
Press day: Sat
Circulation: 1,932,237

Picture House (Annual)
Cinema Theatre Association
13 Tennyson Court, Paddockhall
Road, Haywards Heath,
West Sussex RH16 1EZ
Tel: 01444 455763
Documents the past and present
history of cinema buildings
Lead time: 8 weeks
Circulation: 2,000

The Pink Paper (Weekly, Thur)
Cedar House
72 Holloway Road
London N7 8NZ
Tel: 0171 296 6210
Fax: 0171 296 0026
Editor: Paul Clements
Film/TV: Mark O'Flaherty
Britain's national lesbian and gay
newspaper
Lead time: 14 days
Circulation: 52,525

PIX
21 Stephen Street
London W1P 2LN
0171 255 1444

0171 436 7950
Ilona Haberstadt
A counterpoint of images and critical texts, PIX brings together experimental, independent and commercial cinema from all over the world and explores its relation to other arts

Premiere (Monthly)
37-39 Millharbour
London E14 9TZ
Tel: 0171 972 6791
Fax: 0171 972 6791
Editor: Matt Mueller
A 16-page UK film supplement in issues of American Premiere sold in the UK, containing personality profiles, on the set reports, news and reviews
Lead time: 3 months
Circulation: 40,000

Press Gazette
33-39 Bowling Green Lane
London EC1R 0DA
Tel: 0171 505 8205
Fax:0171 505 8220
Editor: Roy Farndon
Weekly magazine covering all aspects of the media industry: journalism; advertising; broadcast; freelance
Press day: Thurs
Circulation: 8,500

Q (Monthly)
1st Floor
Mappin House
4 Winsley Street
London W1N 7AR
Tel: 0171 436 1515
Fax: 0171 312 8247
Editor: David Davies
Specialist music magazine for 18-45 year olds. Includes reviews of new albums, films and books
Lead time: 14 days
Circulation: 212,607

Radio Times (Weekly, Tues)
Woodlands
80 Wood Lane
London W12 0TT
Tel: 0181 576 3999
Fax: 0181 576 3160
Editor: Sue Robinson
Films: Barry Norman
Features: Kim Newson
Listings: Caroline Meyer
Weekly guide to UK television, radio and satellite programmes
Lead time: 14 days
Circulation: 1,406,152

Regional Film & Video
Flagship Publishing
164-165 North Street
Belfast BT1 IGF

Tel: 01232 319008
Fax: 01232 319101
Editor: Raymond Watson
Film and Video Trade Newspaper
Circulation: 12,000

Scotland on Sunday
20 North Bridge
Edinburgh EH1 1YT
Tel: 0131 556 3454
Fax: 0131 556 3454
email: editorial@scotsun.com
Film: Richard Mowe and Allan Hunter
TV: Tom Lappin
Scottish Sunday newspaper
Lead time: 10 days
Circulation: 110,000

Scottish Film (Quarterly)
Arts and Entertainment
Publishing Ltd
24 Sandyford Place
Glasgow G3 7NG
Tel: 0141 221 4241
Fax: 0141 221 4247
Editor: Robert McColl
Filmmaking and broadcasting within Scotland. Scottish Film is distributed throughout Scotland to all the production companies, facility houses and broadcasters
Circulation: 20,000

The Scotsman
20 North Bridge
Edinburgh EH1 1YT
Tel: 0131 225 2468
Fax: 0131 243 3686
Arts Editor: Robert Dowden Scott
Film critic: Trevor Johnston
Scottish daily newspaper
Circulation: 77,057

Screen (Quarterly)
The John Logie Baird Centre
University of Glasgow
Glasgow G12 8QQ
Tel: 0141 330 5035
Fax: 0141 330 8010
email: screen@arts.gla.ac.uk
Website: http://www.arts.gla.ac.uk/tfts/screen.html
Journal of essays, reports, debates and reviews on film and television studies. Organises the annual Screen Studies Conference
Circulation: 1,400

Screen Digest (Monthly)
Lyme House Studios
38 Georgiana Street
London NW1 0EB
Tel: 0171 482 5842
Fax: 0171 580 0060
email: screendigest@compuserve.com
Managing director: Allan Hardy
Editor: David Fisher
Executive editor: Ben Keen

Deputy editor: Mark Smith
International industry news digest and research report covering film, television, cable, satellite, video and other multimedia information. Has a centre page reference system every month on subjects like law, statistics or sales. Now also available on a computer data base via fax at 0171 580 0060 under the name Screenfax (see entry under Screenfax)

Screen Finance (Fortnightly)
FT Newsletters
30-31 Great Sutton Street
London EC1V 0DX
Tel: 0171 454 1185
Fax: 0171 490 1686
email: x 25@compuserve.com
Editor: Neil McCartney
Detailed analysis and news coverage of the film and television industries in the UK and Europe
Lead time: 1-3 days

Screen International (Weekly, Thur)
EMAP Media
33-39 Bowling Green Lane
London EC1R 0DA
Tel: 0171 505 8056/8080
Fax: 0171 505 8117
Editor: Boyd Farrow
Features: Mike Goodrich
International trade magazine for the film, television, video, cable and satellite industries. Regular news, features, production information from around the world
Press day: Wed
Features lead time: 3 weeks
Circulation: 8,360

Screenfax (Database)
Screen Digest
37 Gower Street
London WC1E 6HH
Fax: 0171 580 0060
Available on-line via Dialog, Profile, Data-Star, MAID and most other on-line databases, or by fax: 0171 580 0060. Provides customised print-outs on all screen media subjects with summaries of news developments, market research. See entry under Screen Digest

Shivers (Monthly)
Visual Imagination
9 Blades Court
Deodar Road
London SW15 2NU
Tel: 0181 875 1520
Fax: 0181 875 1588
Editor: David Miller
Horror film reviews and features
Lead time: 1 month
Circulation: 30,000

Sight and Sound (Monthly)
British Film Institute
21 Stephen Street
London W1P 2LN
Tel: 0171 255 1444
Fax: 0171 436 2327
Editor: Nick James
Includes regular columns, feature articles, a book review section and review/synopsis/credits of every feature film theatrically released, plus a brief listing of every video
Copy date: 4th of each month
Circulation: 26,000 **(see p8)**

South Wales Argus
Cardiff Road
Newport
Gwent NP9 1QW
Tel: 01633 810000
Fax: 01633 462202
Film & TV editor: Lesley Williams
Regional evening newspaper
Lead time: 2 weeks
Circulation: 32,569

The Spectator (Weekly, Thur)
56 Doughty Street
London WC1N 2LL
Tel: 0171 405 1706
Fax: 0171 242 0603
Arts editor: Elizabeth Anderson
Film: Mark Steyn
TV: James Delingpole
Independent review of politics, current affairs, literature and the arts
Press day: Wed
Circulation: 56,313

The Stage (incorporating Television Today) (Weekly, Thurs)
Stage House
47 Bermondsey Street
London SE1 3XT
Tel: 0171 403 1818
Fax: 0171 357 9287
Editor: Brian Attwood
Weekly trade paper covering all aspects of entertainment
Circulation: 37,512

Stage Screen & Radio (10 issues a year)
111 Wardour Street
London W1V 4AY
Tel: 0171 437 8506
Fax: 0171 437 8268
Editor: Janice Turner
Journal of the film, broadcasting, theatre and entertainment union BECTU. Reporting and analysis of these industries and the union's activities plus coverage of technological developments
Lead time: 4 weeks
Circulation: 34,600

Starburst (Monthly + 4 Specials + German language version)
Visual Imagination, 9 Blades Court, Deodar Road
London SW15 2NU
Tel: 0181 875 1520
Fax: 0181 875 1588
email: Star@cix.compulink.co.uk
Editor: David Richardson
Science fiction, fantasy and horror films, television and video
Lead time: 1 month
Circulation: 45,000

Subway Magazine
The Attic
62 Kelvingrove Street
Glasgow G3 7SA
Tel: 0141 332 9088
Fax: 0141 331 1477
Editor: Gill Mill

The Sun
PO Box 481
1 Virginia Street
London E1 9XP
Tel: 0171 782 4000
Fax: 0171 488 3253
Films: Nick Fisher
Showbiz editor: Dominic Mohan
TV editor: Danny Buckland
TV News: Sarah Crosbie
National daily newspaper
Circulation: 3,875,329

Sunday Express Magazine
Ludgate House
245 Blackfriars Road
London SE1 9UX
Tel: 0171 922 7150
Fax: 0171 922 7599
Editor: Katy Bravery
Supplement to The Express on Sunday
Lead time: 6 weeks

Sunday Magazine
1 Virginia Street
London E1 9BD
Tel: 0171 782 7000
Fax: 0171 782 7474
Editor: Judy McGuire
Deputy Editor: Jonathan Worsnop
Supplement to News of the World
Lead time: 6 weeks
Circulation: 4,701,879

Sunday Mirror
1 Canada Square
Canary Wharf
London E14 5AP
Tel: 0171 293 3000
Fax: 0171 293 3939
Film critic: Quentin Falk
TV: David Rowe, Pam Francis
National Sunday newspaper
Circulation: 2,268,263

Sunday Telegraph
1 Canada Square
Canary Wharf
London E14 5DT
Tel: 0171 538 5000
Fax: 0171 513 2504
Arts: John Preston
Film: Anne Billson
TV: Judy Rumbold
National Sunday newspaper
Circulation: 886,377

Sunday Times
1 Virginia Street
London E1 9BD
Tel: 0171 782 5000
Fax: 0171 782 5731
Film: Tom Shone
TV reviews: A A Gill
Video: George Perry
National Sunday newspaper
Press day: Wed
Circulation: 1,314,576

Sunday Times Magazine
Admiral House
66-68 East Smithfield
London E11 9XW
Tel: 0171 782 7000
Fax: 0171 867 0410
Editor: Robin Morgan
Supplement to Sunday Times
Lead time: 4 weeks
Circulation: 1,314,576

TV Quick (Weekly, Mon)
25-27 Camden Road
London NW1 9LL
Tel: 0171 284 0909
Fax: 0171 284 0593
Editor: Jon Gower
Mass market television magazine
Lead time: 3 weeks
Circulation: 799,000

TV Times (Weekly, Tues)
10th Floor
King's Reach Tower
Stamford Street
London SE1 9LS
Tel: 0171 261 7000
Fax: 0171 261 7777
Editor: Liz Murphy
Film editor: David Quinlan
Weekly magazine of listings and features serving viewers of independent TV, BBC TV, satellite and radio
Lead time: 6 weeks
Circulation: 981,311

TV Zone (Monthly + 4 specials)
Visual Imagination Limited
9 Blades Court
Deodar Road
London SW15 2NU
Tel: 0181 875 1520
Fax: 0181 875 1588

email: star@cix.compulink.co.uk
Editor: Jan Vincent-Rudzki
Magazine of cult television, past, present and future, with emphasis on science fiction and fantasy
Lead time: 1 month
Circulation: 45,000

Talking Pictures (Quarterly)
34 Darwin Crescent
Laira
Plymouth PL3 6DX
Tel: 01752 661044
Fax: 0171 737 4720
email: stntpublishingltd@btinternet.com
website: www.filmcentre.co.uk
Editor: Nigel Watson
Devoted to a serious look at film, computer entertainment, television and video
Lead time: 2 months
Circulation: 500

Tatler (Monthly)
Vogue House
1 Hanover Square
London W1R 0AD
Tel: 0171 499 9080
Fax: 0171 409 0451
Editor: Jane Procter
Arts: Celia Lyttleton
Smart society magazine favouring profiles, fashion and the arts
Lead time: 3 months
Circulation: 88,235

The Teacher (8 p.a.)
National Union of Teachers
Hamilton House
Mabledon Place
London WC1H 9BD
Tel: 0171 380 4708
Fax: 0171 387 8458
Editor: Mitch Howard
Circulation: 250,000 mailed direct to all NUT members and to educational institutions

Telegraph Magazine
1 Canada Square
Canary Wharf
London E14 5AU
Tel: 0171 538 5000
Fax: 0171 513 2500
TV films: Jessamy Calkin
Supplement to Saturday edition of the Daily Telegraph
Lead time: 6 weeks
Circulation: 1,300,000

Television (8 p.a.)
Royal Television Society
Holborn Hill
100 Gray's Inn Road
London WC1X 8AL
Tel: 0171 430 1000
Fax: 0171 430 0924
email: royaltvsociety@btinternet.com
Website: www.rts.org.uk

Editor: Peter Fiddick
Television trade magazine
Lead time: 2 weeks
Circulation: 5,000

Televisual (Monthly)
St. Giles House
50 Poland Street
London W1V 4AX
Tel: 0171 439 4222
Fax: 0171 287 0768
Editor: Mundy Ellis
Monthly business magazine for production professionals in the business of moving pictures
News lead time: 1 month
Features lead time: 2 months
Circulation: 8,040

Time Out (Weekly, Tues)
Universal House
251 Tottenham Court Road
London W1P 0AB
Tel: 0171 813 3000
Fax: 0171 813 6028
Film: Geoff Andrew
Video: Derek Adams
TV: Alkarim Jivani
London listings magazine with cinema and television sections
Listings lead time: 8 days
Features lead time: 1 week
Circulation: 110,000 plus

The Times
1 Virginia Street
London E1 9XN
Tel: 0171 782 5000
Fax: 0171 488 3242
Film/video critic: Geoff Brown
Film writer: David Robinson
TV: Matthew Bond
National daily newspaper
Circulation: 747,054

The Times Educational Supplement (Weekly, Fri)
Admiral House
66-68 East Smithfield
London E1 9XY
Tel: 0171 782 3000
Fax: 0171 782 3200
Editor: Caroline St John-Brooks
Film/TV editor: Janette Wolf
Press day: Tuesday
Lead time for reviews: copy 14-21 days
Circulation: 150,000

The Times Educational Supplement Scotland (Weekly, Fri)
37 George Street
Edinburgh EH2 2HN
Tel: 0131 220 1100
Fax: 0131 220 1616
Editor: Willis Pickard
Press day: Wed
Circulation: 8,000

The Times Higher Educational Supplement (Weekly, Fri)
Admiral House
66-68 East Smithfield
London E1 9XY
Tel: 0171 782 3000
Fax: 0171 782 3300
Film/TV editor: Sean Coughlan
Press day: Wed
Lead time for reviews: copy 10 days before publication
Circulation: 26,666

The Times Literary Supplement (Weekly, Fri)
Admiral House
66-68 East Smithfield
London E1 9XY
Tel: 0171 782 3000
Fax: 0171 782 3100
Arts editor: Will Eaves
Press day: Tues
Lead time: 2 weeks
Circulation: 34,044

Top Review
Julco House
5th Floor
26-28 Great Portland Street
London W1N 6AS
Tel: 07071 224091
Fax: 01493 330565
Film, video, car, computer and book reviews
Circulation: 60,000

Total Film (Monthly)
Future Publishing
30 Monmouth Street
Bath BA1 2BW
Tel: 01225 442244
Fax: 01225 732378
email: mops@futurenet.co.uk
Website: http//www.futurenet.co.uk
Editor: Emma Cochrane

Tribune (Weekly, Fri)
308 Gray's Inn Road
London WC1X 8DY
Tel: 0171 278 0911
Fax: 0171 833 0385
Review editor: Caroline Rees
Political and cultural weekly
Lead time: 14 days
Circulation: 10,000

Variety (Weekly, Mon)
First Floor,
151-153 Wardour Street,
London W1V 3TB
Tel: 0171 411 2828
Fax: 0171 411 2809
European editor: Adam Dawtrey
International showbusiness newspaper
Press day: Thurs
Circulation: 36,000

Video Home Entertainment
(Weekly, Fri)
Strandgate
18-20 York Buildings
London WC2 6JU
Tel: 0171 839 7774
Fax: 0171 839 4393
Editor: John Ferguson
Video trade publication for rental
and retail
Lead time: Monday before publication
Circulation: 7,613

View
Oakwood House
422 Hackney Road
London E2 7SY
Tel: 0171 729 6881
Fax: 0171 729 0988
Editor: Branwell Johnson
A weekly trade magazine for the
video industry covering news
relevant to the business from a retail
to distributor level. It carries a
complete listing of the month's
rental releases and a highlighted sell
through list. Regular features include
coverage from the US and interviews
with leading industry figures
Circulation: 8,000

Viewfinder (3 p.a.)
BUFVC
77 Wells Street
London W1P 3RE
Tel: 0171 393 1511
Fax: 0171 393 1555
Editor: Roland Glover
Periodical for people in higher
education and research, includes
articles on the production, study and
use of film, television and related
media. Deadlines: 10th Jan, 1st Apr,
1st Oct
Lead time: 6 weeks
Circulation: 5,800

Vogue (Monthly)
Vogue House
Hanover Square
London W1R 0AD
Tel: 0171 408 0559
Fax: 0171 493 1345
Editor: Alexandra Shulman
Films: Susie Forbes
Glossy magazine for women
Lead time: 12 weeks
Circulation: 201,187

The Voice (Weekly, Monday)
370 Coldharbour Lane
London SW9 8PL
Tel: 0171 737 7377
Fax: 0171 274 8994
Editor: Annie Stewart
Arts: Omega Douglas
Britain's leading black newspaper
with mainly 18-35 age group

readership. Regular film, television
and video coverage
Press day: Friday
Circulation: 52,000

The Web
Media House, Adlington Park
Macclesfield SK10 4NP
Tel: 01625 878888
Fax: 01625 879967
email: web@idg.co.uk
Editor: Mike Cowley
Focusing on lifestyle and culture on
the Net, film and television is
extensively covered with features,
leaders and listing
Lead time: 2 weeks

Western Mail
Thomson House
Cardiff CF1 1WR
Tel: 01222 223333
Fax: 01222 583652
Film: Carolyn Hitt
Daily newspaper
Circulation: 60,251

What's On In London
(Weekly, Tues)
180 Pentonville Road
London N1 9LB
Tel: 0171 278 4393
Fax: 0171 837 5838
Editor: Michael Darvell
Films & Video: David Clark
London based weekly covering
cinema, theatre, music, arts, books,
entertainment and video
Press day: Mon
Lead time: 10 days
Circulation: 42,000

What's On TV (Weekly, Tues)
King's Reach Tower
Stamford Street
London SE1 9LS
Tel: 0171 261 7769
Fax: 0171 261 7739
Editor: Mike Hollingsworth
TV listings magazine
Lead time: 3 weeks
Circulation: 1,676,000

Yorkshire Post
Wellington Street
Leeds
West Yorkshire LS1 1RF
Tel: 0113 238 8536
Fax: 0113 244 3430
TV editor: Angela Barnes
Regional daily morning newspaper
Deadline: 10.00 pm
Circulation: 100,126

NEWS AND PHOTO AGENCIES

Associated Press
12 Norwich Street
London EC4A 1BP
Tel: 0171 353 1515
Fax: 0171 583 0196

Bridge News
78 Fleet Street
London EC4Y 1HY
Tel: 0171 842 4000
Fax: 0171 583 5032
Business Information Service

Central Office of Information
Hercules Road
London SE1 7DU
Tel: 0171 928 2345
Fax: 0171 928 5037

Central Press Features
20 Spectrum House
32-34 Gordon House Road
London NW5 1LP
Tel: 0171 284 1433
Fax: 0171 284 4494
Film/TV: Chris King

Fleet Street News Agency
68 Exmouth Market
London EC1R 4RA
Tel: 0171 278 5661
Fax: 0171 278 8480

London News Service
68 Exmouth Market
London EC1R 4RA
Tel: 0171 278 5661
Fax: 0171 278 8480

Press Association
292 Vauxhall Bridge Road
London Sw1V 1AE
Tel: 0171 963 7244
Fax: 0171 963 7245

Reuters Ltd
85 Fleet Street
London EC4P 4AJ
Tel: 0171 250 1122
Fax: 0171 542 7921
Media: Mary Ellen-Barker

United Press International
408 The Strand
London WC2R 0NE
Tel: 0171 333 0990
Fax: 0171 333 1690

BBC TELEVISION

BBC
Television Centre
Wood Lane
London W12 7RJ
Tel: 0181 743 8000

BBC1 Omnibus
Television Centre
Wood Lane
London W12 7RJ
Tel: 0181 743 8000
Fax: 0181 895 6553

BBC2 Arena
Television Centre
Wood Lane
London W12 7RJ
Tel: 0181 895 6766
Fax: 0181 895 6974

Anglia Television
Anglia House
Norwich NR1 3JG
Tel: 01603 615151
Fax: 01603 615032

INDEPENDENT TELEVISION

Border Television
Television Centre
Carlisle CA1 3NT
Tel: 01228 25101
Fax: 01228 525101

Carlton Television
35-38 Portman Square
London W1H 0NU
Tel: 0171 486 6688
Fax: 0171 486 1132

Central Independent Television (East)
Carlton Studios
Lenton Lane
Nottingham NG7 2NA
Tel: 0115 986 3322
Fax: 0115 964 5018

Central Independent Television (South)
9 Windrush Court
Abingdon Business Park
Abingdon
Oxon OX14 1SA
Tel: 01235 554123
Fax: 01235 524024

Channel Five Broadcasting
22 Long Acre
London WC2E 9LY
Tel: 0171 550 5555
Fax: 0171 550 5554

Channel Four Television
124 Horseferry Road
London SW1P 2TX
Tel: 0171 396 4444
Fax: 0171 306 8353

Channel Television
Television House
Bulwer Avenue
St Sampsons
Guernsey GY2 4LA
Tel: 01481 41888
Fax: 01481 41889
The Television Centre
La Pouquelaye, St Helier
Jersey JE1 3ZD
Tel: 01534 816816
Fax: 01534 816689

GMTV
London Television Centre
Upper Ground
London SE1 9TT
Tel: 0171 827 7000
Fax: 0171 827 7249

Grampian Television
Queen's Cross
Aberdeen AB15 4XJ

Tel: 01224 846846
Fax: 01224 846802
North Tonight; Crossfire; Telefios;
Walking Back to Happiness; Top
Club; We the Jury; The Art Sutter
Show

Granada Television
Quay Street
Manchester M60 9EA
Tel: 0161 832 7211
Fax: 0161 827 2324
Albert Dock
Liverpool L3 4BA
Tel: 0151 709 9393
White Cross
Lancaster LA1 4XQ
Tel: 01524 606688
36 Golden Square
London W1R 4AH
Tel: 0171 734 8080
Bridgegate House
5 Bridge Place
Lower Bridge Street
Chester CH1 1SA
Tel: 01244 313966

HTV
Television Centre
Bath Road
Bristol BS4 3HG
Tel: 0117 977 8366
Fax: 0117 972 3122
HTV News; The West This Week,
West Eye View

HTV Wales
Television Centre
Culverhouse Cross
Cardiff CF5 6XJ
Tel: 01222 590590
Fax: 01222 590759

Independent Television News (ITN)
200 Gray's Inn Road
London WC1X 8XZ
Tel: 0171 833 3000

Meridian Broadcasting
TV Centre
Northam Road
Southampton SO14 0PZ
Tel: 01703 222555
Fax: 01703 335050
TV Weekly

S4C
Parc Ty Glas
Llanishen
Cardiff CF4 5DU
Tel: 01222 747444
Fax: 01222 754444
Head of Press and Public Relations:
David Meredith

Scottish Television
Cowcaddens
Glasgow G2 3PR

Tel: 0141 300 3000
Fax: 0141 332 9274

Tyne Tees Television
The Television Centre
City Road
Newcastle upon Tyne NE1 2AL
Tel: 0191 261 0181
Fax: 0191 232 7017

Ulster Television
Havelock House
Ormeau Road
Belfast BT7 1EB
Tel: 01232 328122
Fax: 01232 246695

Westcountry Television
Western Wood Way
Language Science Park
Plymouth PL7 5BQ
Tel: 01752 333333
Fax: 01752 333033

Yorkshire Television
The Television Centre
Kirkstall Road
Leeds LS3 1JS
Tel: 0113 243 8283
Fax: 0113 243 3655

BBC RADIO

BBC
Broadcasting House
Portland Place
London W1A 1AA
Tel: 0171 580 4468
Fax: 0171 637 1630

BBC Radio Bristol
Broadcasting House
Whiteladies Road
Bristol BS8 2LR
Tel: 0117 974 1111
Fax: 0117 923 8323

BBC CWR (Coventry & Warwickshire)
25 Warwick Road
Coventry CV1 2WR
Tel: 01203 559911
Fax: 01203 520080

BBC Radio Cambridgeshire
Broadway Court
Broadway
Peterborough PE1 1RP
Tel: 01733 312832
Fax: 01733 343768

BBC Radio Cleveland
PO Box 95FM
Broadcasting House
Newport Road
Middlesbrough TS1 5DG
Tel: 01642 225211
Fax: 01642 211356

BBC Radio Cornwall
Phoenix Wharf
Truro TR1 1UA
Tel: 01872 275421
Fax: 01872 240679

BBC Radio Cumbria
Hartington Street
Barrow-in-Furness
Cumbria LA14 5SC
Tel: 01228 835252
Fax: 01228 870008

BBC Radio Derby
PO Box 269
Derby DE1 3HL
Tel: 01332 361111
Fax: 01332 290794

BBC Radio Devon
PO Box 5
Broadcasting House
Seymour Road
Mannamead
Plymouth PL3 5YQ
Tel: 01752 260323
Fax: 01752 234599

BBC Essex
198 New London Road

Chelmsford
Essex CM2 9XB
Tel: 01245 262393
Fax: 01245 492983

BBC Radio Foyle
8 Northland Road
Londonderry BT48 7JD
Tel: 01504 378 600
Fax: 01504 378666

GLR
35c Marylebone High Street
London W1A 4LG
Tel: 0171 224 2424
Fax: 0171 487 2908

BBC GMR Talk
PO Box 951
Oxford Road
Manchester M60 1SD
Tel: 0161 200 2000
Fax: 0161 228 6110

BBC Radio Guernsey
Commerce House
Les Banques, St Peter Port
Guernsey GY1 2HS
Tel: 01481 728977
Fax: 01481 713557

BBC Hereford & Worcester
Hylton Road
Worcester WR2 5WW
Tel: 01905 748485
Fax: 01905 748006

BBC Radio Humberside
9 Chapel Street
Hull HU1 3NU
Tel: 01482 323232
Fax: 01482 226409

BBC Radio Jersey
18 Parade Road
St Helier
Jersey JE2 3PL
Tel: 01534 87000
Fax: 01534 32569

BBC Radio Lancashire
Darwen Street
Blackburn
Lancs BB2 2EA
Tel: 01254 262411
Fax: 01254 680821

BBC Radio Leeds
Broadcasting House
Woodhouse Lane
Leeds LS2 9PN
Tel: 0113 244 2131
Fax: 0113 242 0652

BBC Radio Leicester
Epic House
Charles Street
Leicester LE1 3SH
Tel: 0116 251 6688
Fax: 0116 251 1463

BBC Radio Lincolnshire
PO Box 219
Newport
Lincoln LN1 3XY
Tel: 01522 511411
Fax: 01522 511058

BBC Radio Merseyside
55 Paradise Street
Liverpool L1 3BP
Tel: 0151 708 5500
Fax: 0151 794 0909
Film and video reviewer: Ramsey
Campbell

BBC Radio Newcastle
Broadcasting Centre
Fenham
Newcastle Upon Tyne NE99 1RN
Tel: 0191 232 4141
Fax: 0191 232 5082

BBC Radio Norfolk
Norfolk Tower
Surrey Street
Norwich NR1 3PA
Tel: 01603 617411
Fax: 01603 633692

BBC Radio Northampton
Broadcasting House
Abington Street
Northampton NN1 2BH
Tel: 01604 239100
Fax: 01604 230709

BBC Radio Nottingham
PO York House
Mansfield Road
Nottingham NG1 3JB
Tel: 0115 955 0500
Fax: 0115 955 0501

BBC Radio Oxford
269 Banbury Road
Oxford OX2 7DW
Tel: 01865 311444
Fax: 01865 311996

BBC Radio Sheffield
Ashdell Grove
60 Westbourne Road
Sheffield S10 2QU
Tel: 0114 268 6185
Fax: 0114 266 4375

BBC Radio Solent
PO Box 900
Dorchester DT1 1TP
Tel: 01305 269654
Fax: 01305 250910

BBC Somerset Sound
14-16 Paul Street
Taunton TA1 3PF
Somerset
Tel: 01823 251641
Fax: 01823 332539

BBC Southern Counties
Broadcasting Centre, Guildford
GU2 5AP
Tel: 01483 306306
Fax: 01483 304952

BBC Radio Stoke
Cheapside
Hanley
Stoke-on-Trent ST1 1JJ
Tel: 01782 208080
Fax: 01782 289115

BBC Radio Sussex & Surrey
Broadcasting House
Guildford
Surrey GU2 5AP
Tel: 01483 306306
Fax: 01483 304952

BBC Three Counties Radio
PO Box 3CR , Hastings Street
Luton
Bedfordshire LU1 5XL
Tel: 01582 441000
Fax: 01582 401467

BBC Radio WM
PO Box 206
Birmingham B5 7SD
Tel: 0121 414 8484
Fax: 0121 414 8817

BBC Wiltshire Sound
Broadcasting House
Prospect Place
Swindon SN1 3RN
Tel: 01793 513626
Fax: 01793 513650

BBC World Service
Bush House
Strand
London WC2B 4PH
Tel: 0171 257 2171
Fax: 0171 240 3938

INDEPENDENT NATIONAL RADIO

Classic FM
Academic House
24-28 Oval Road
London NW1 7DQ
Tel: 0171 284 3000
Fax: 0171 713 2630

Longwave Radio
Atlantic 252
74 Newman Street
London W1P 3LA
Tel: 0171 436 4012
Fax: 0171 436 4015
Trium, Co Meath, Ireland
Tel/Fax: 00353 463655

Virgin 1215 AM
1 Golden Square
London W1R 4DJ
Tel: 0171 434 1215
Fax: 0171 434 1197

PREVIEW THEATRES

BAFTA
195 Piccadilly
London W1V 0LN
Tel: 0171 465 0277
Fax: 0171 734 1009
Formats: Twin 35mm all aspect
ratios. Dolby A, SR, SRD, DTS
sound. 35 Double head mono, twin/
triple track stereo plus Dolby Matrix.
Twin 16mm and super 16mm, 16
double head stereo plus Dolby
Matrix. BARCO 9200 Data Video
Projector VHS, Lo Band/Hi Band U-
matic, Beta, Beta SP, Digi Beta.
Interfaces for most PC outputs,
SVGA, MAC etc. 35mm slides
single, twin and disolve multi-wau
control, Audio, RGB Video Tie
Lines in Theatre. ISDN 2. Catering
by Roux Fine Dining. Seats:
Princess Anee Theatre, 213 Run Run
Shaw Theatre, 30 (not all formats
available), Function Room, up to
200

BUFVC
77 Wells Street
London W1P 3RE
Tel: 0171 393 1500
Fax: 0171 393 1555
email: bufvc@open.ac.uk
Formats: Viewing rooms equipped
with 16mm double-head, Betacam,
SVHS, VHS, lo-band and hi-band
U-Matic, Betamax, Phillips 1500
Seats: 15-20 max

British Film Institute
21 Stephen Street
London W1P 2LN
Tel: 0171 957 8976
Fax: 0171 580 5830
email: derek.young@bfi.org.uk
Formats: 35mm Dolby Opt/Mag
Stereo A/SR, Std 16mm Opt, Super
16 Mag Stereo A/SR, Large Screen
Video Projection PAL VHS, SVHS,
U-MATIC hi/lo band Triple
Standard, BETA SP, LASER DISC.
Disabled Access
Seats: 1: 36, 2: 36 (see p8)

Carlton Preview Theatre
127 Wardour Street
London W1V 4AD
Tel: 0171 437 9020 x257
Fax: 0171 434 3689
Formats: U-Matic, 16mm, 35mm,
double-head, Dolby SR Stereo, VHS,
U-Matic, slides. Lift to theatre
Seats: 58

Century Preview Theatres
31-32 Soho Square
London W1V 6AP
Tel: 0171 437 7766
Fax: 0171 434 2170
Formats: Century Theatre: 35mm
Dolby optical and magnetic stereo,
Dolby A & SR noise reduction,
2,000 double-head capacity,
spotlights and microphone for
conventions, DTS + SRD; Executive
Theatre: Dolby A + SR stereo optical
and magnetic 2,000 double-head
capacity
Seats: Century: 61, Executive: 38

Chapter Cinema
Market Road
Canton
Cardiff CF5 1QE
Tel: 01222 396061
Fax: 01222 225901
email: enquire@chapter.org
Formats: 35mm optical, 16mm
optical/sep mag, high quality video
projection, U-Matic/VHS - all
standards. Beta SP PAL2 Channel
infra-red audio amplification/
simultaneous translation system in
both screens. Reception space, bars
and restaurant
Seats: 1:194, 2:68

Columbia TriStar Films UK
Sony Pictures Europe House
25 Golden Square
London W1R 6LU
Tel: 0171 533 1095
Fax: 0171 533 1105
Formats: 35mm optical (SDDS,
Dolby "SR" + "A" type)/double
head, SVA Mag, 16mm optical
(Mono), Super 16 and Super 35.
BETA SP, BVU/U-Matic, VHS,
High Definition Video. Large
reception area. Seats: 80

The Curzon Minema
45 Knightsbridge
London SW1X 7NL
Tel: 0171 235 4226
Fax: 0171 235 3426
email:info@minema.com
Website: http://www.minema.com
Formats: 35mm and 16mm, video
and AV presentations

De Lane Lea Sound Centre
75 Dean Street
London W1V 5HA
Tel: 0171 439 1721

Fax: 0171 437 0913
Formats: 35mm and 16mm, Dolby
stereo A + SR with double-head
capacity. ¾ hi and lo-band video and
VHS. Bar and catering available.
Disabled access. Seats: 30

Edinburgh Film & TV Studios
Nine Mile Burn
Penicuik EH26 9LT
Tel: 01968 672131
Fax: 01968 672685
Formats: 16mm and 35mm double-
head stereo, U-Matic, VHS
Seats: 100

ICA
12 Carlton House Terrace
London SW1Y 5AH
Tel: 0171 930 0493
Fax: 0171 873 0051
Formats: Cinema: 35mm com-opt,
Dolby CP, 16mm com-opt, + Sep
mog; video projection Super 16.
Cinematheque: 16mm com-opt,
Super 8, large-screen video
projection. VHS, NTSC, lo-band U-
Matic. All video formats in both
cinema + cinematheque. Nash
function rooms available for
receptions: up to 250 capacity. Level
access to cinema and cinematheque.
Cafe Bar available exclusively till
noon. Seats: Cinema: 185,
Cinematheque: 45

Imperial War Museum
(Corporate Hospitality)
Lambeth Road
London SE1 6HZ
Tel: 0171 416 5394
Fax: 0171 416 5392/0171 416 5374
Formats: 35mm and 16mm;
Betacam, U-Matic, SVHS and VHS.
Catering by arrangement. Large
Exhibit Hall, capacity: 1,000
Disabled access
Seats: Cinema: 200

King's Lynn Arts Centre
27/29 King Street
King's Lynn
Norfolk PE30 1HA
Tel: 01553 765565
Fax: 01553 762141
Formats: 16mm, 35mm
Seats: 349

The Metro
11 Rupert Street
London W1V 7FS
Tel: 0171 287 3515

Fax: 0171 287 2112
Formats: 16mm and 35mm. Two
screens available from 10am-2pm
Seats: 1:195, 2:85

Mr Young's

14 D'Arblay Street
London W1V 3FP
Tel: 0171 437 1771
Fax: 0171 734 4520
Contact: Reuben/Andy/Derry
Formats: 16mm, Super 16mm,
35mm, Super 35mm, U-Matic, VHS,
Betacam SP, Dolby stereo double-
head optical and magnetic Dolby
SR. Large screen video projection.
Bar area, catering by request. Both
theatres non-smoking
Seats: 1: 42, 2: 25, 3:45

Pinewood Studios

Pinewood Road
Iver
Bucks SL0 0NH
Tel: 01753 656296
Fax: 01753 656014
email: helen_wells@rank.com
Contact: Helen Wells
Formats: 35mm, 70mm, U-Matic,
Dolby SR D, DTS, SDDS,Disabled
access. Lounge available
Seats: 115 seats

Planet Hollywood

13 Coventry Street
London W1
Tel: 0171 437 7827
Fax: 0171 439 7827
Formats: 35mm, 70mm, SVHS/
VHS, U-Matic, Laser Disc,
Lucasfilm Ltd THX Sound Sytem,
Dolby CP200 + SRD/DTS digital
stereo. Super 35mm with separate
magnetic tracks and remote volume
control. Microphone facilities. Lifts
for the disabled available
Seats: Cinema: 75, Dining area: 85,
120 (standing)

Prominent Facilities THX

68a Delancey Street
London NW1 7RY
Tel: 0171 284 1020
Fax: 0171 284 1202
Formats: 35mm Dolby optical and
magnetic, 2,000' double-head, rock
'n' roll. All aspect ratios, and Super
35, 24-25 30fps, triple-track,
interlock, Dolby A + SR stereo,
16mm double-head married. Fully
air conditioned, kitchen and
reception area. Wheelchair access.
Seats: 26

RSA

8 John Adam Street
London WC2N 6EZ
Tel: 0171 839 5049
Fax: 0171 321 0271

email: Conference@
fRSAUK.demon.co.uk
The Great Room
Video Formats: SVHS, Beta SP.
Other formats by arrangement.
Barcographics 8100 Projector for
Video and Data Projection. Loop
system for hard of hearing, disabled
access to all rooms. Full catering
available: Seats: 202
Durham House Street Auditorium
Video Formats: SVHS, Low band U-
matic. Other formats by arrangement. Sony 1252 Projector for Video
and Data Projection. Loop system
for hard of hearing, disabled access
to all rooms. Full catering available.
Seats: 60

Screen West

136-142 Brarnley Road
London W10 6SR
Tel: 0171 437 6292
State of the art Preview Theatre with
luxury seating for 74 people. Cutting
edge technology and full catering
facilities in the adjoining function room

Shepperton Studios

Studios Road
Shepperton
Middx TW17 0QD
Tel: 01932 562611/572350
Fax: 01932 568989
Formats: 35mm double-head and
married, Dolby A + SR, Video U-
Matic, NTSC, PAL, SECAM, VHS.
Seats: (35mm) 17

Twickenham Film Studios

St Margaret's
Twickenham
Middx TW1 2AW
Tel: 0181 607 8888
Fax: 0181 607 8889
Formats: 16mm, 35mm.
Seats: 31

Warner Bros

135 Wardour Street
London W1V 4AP
Tel: 0171 734 8400
Fax: 0171 437 5521
Formats: 35mm double-head, Dolby
Digital. VHS/U-Matic Beta video
projection. Disabled access. Seats: 40

Watershed Media Centre

1 Canons Road
Bristol BS1 5TX
Tel: 0117 9276444
Fax: 0117 9213958
email: watershed@online.redirect.co.uk
Formats: Super 8mm, 16mm double-
head, 35mm, VHS U-Matic lo-band,
Betacam SP, Dolby A + SR. Lift access,
for wheelchair spaces each theatre
(prior notification for C2 required)
Seats: 1: 200. 2: 55

PRODUCTION COMPANIES

Listed below are UK companies currently active in financing and/or making audio visual product for UK and international media markets. Film and video workshops are also active in this area. Not generally listed are the numerous companies making television commercials, educational and other non-broadcast material

LOTTERY FILM PRODUCTION FRANCHISES

On 15 May 1997 The Arts Council announced three National Lottery-funded commercial feature film production franchises. Each franchise will extend over a six year period. All franchise funds offered will serve as funding pre-allocations over that period and will be drawn down conditional upon Arts Council approval of individual film proposals.

Pathé Pictures
Kent House
Market Place
London W1N 8AR
Tel: 0171 323 5151
Fax: 0171 636 7594
Head of Production: Andrea Calderwood
Lottery award: £33 million
Number of films: 35
Pathé Productions is formed by six producers in association with Pathé Pictures. **Thin Man Films** and **Imagine Films**; **Allied Filmmakers** and **Allied Films Ltd**; **NFH**; **Pandora Productions**; **Sarah Radclyffe Productions**;**Fragile Films**; **MW Entertainment**

The Film Consortium
6 Flitcroft Street
London WC2 8DJ
Tel: 0171 691 4440
Fax: 0171 691 4445
Head of Production and Development: Colin Vanes
Lottery award: £30.25 million
Number of films: 39
The Film Consortium is formed by four production companies **Greenpoint Films, Parallax Pictures, Scala Productions**; **Skreba** in association with Virgin Cinemas. It intends to make four to five features a year with budgets in the range of £1.5-£6 million. The

Film Consortium (TFC) has a commitment to encourage the development of new writers, producers and directors

DNA Films
2-4 Noel Street
London W1 3RB
Tel: 0171 287 3209
Fax: 0171 287 3503
Lottery award: £29 million
Number of films: 16
Contact: Grace Hodge
Tel: 0171 485 4411
DNA Film Ltd is formed by **Toledo Pictures; Figment Films.** It intends to make three films per year. Each film's budget will be up to £4 million and will be fully funded by DNA Films.

A19 Film and Video
21 Foyle Street
Sunderland SR1 1LE
Tel: 0191 565 5709
Fax: 0191 565 6288
Documentary programmes for television. Education/training material for distribution. Low budget fiction work. Production support offered to local and regionally based film-makers, schools, community groups etc

ABTV
Agran Barton Television
The Yacht Club
Chelsea Harbour
London SW10 0XA
Tel: 0171 351 7070
Fax: 0171 352 3528/3645
Linda Agran, Nick Barton
The former Paravision (UK). Recent productions: *The Vanishing Man* (ITV); *Byzantium - The Lost Empire* (TLC & C4). Previous productions: *Seven Wonders of The World* (TLC & C4); *Moving Story* (Carlton); *The Wimbledon Poisoner* (BBC). In production: *The Vanishing Man Series 1* (ITV); *I'm Just a Girl* (C4)

Aardman Animations
Gas Ferry Road
Bristol BS1 6UN
Tel: 0117 984 8485
Fax: 0117 984 8486
Character led model animation studio producing films, commercials and television series. Aardman's first theatrical feature film, *Chicken Run*, is currently in production. Recent productions: *Rex the Runt*, Dir Richard Goleszowski 1997; *Stagefright*, Dir Steve Box 1997; *HumDrum*: Peter Peake 1997; *Wat's Pig*, Dir Peter Lord 1996; *Pop*, Dir Sam Fell 1996; *A Close Shave*, Dir Nick Park 1995 (1995 Academy Award Winner). Various commercials

Acacia Productions
80 Weston Park
London N8 9TB
Tel: 0181 341 9392
Fax: 0181 341 4879

email: acacia@dial.pipex.com
Recent productions: *Seeds of Hope for Rwanda* (25 mins); *A Future for Forests* (25 mins); *The Wokabout Somil* (25 mins); *Spirit of Trees* (8 x 30 min, C4). In production: *Wealth or Desolation?* (4 x 50 mins) international series of programmes about the global conservation of plant genetic resources

Action Time
Wrendal House
2 Whitworth Street
West Manchester M15WX
Tel: 0161 236 8999
Fax: 0161 236 8845
Entertainment programme devisors and producers in UK and Europe. Recent productions: *Body Heat, Catchphrase*

Adventure Pictures
6 Blackbird Yard
Ravenscroft Street
London E2 7RP
Tel: 0171 613 2233
Fax: 0171 256 0842
Produced Sally Potter's *Orlando* and *The Tango Lesson* with other features in development. Television documentaries include: *Death of a Runaway* (RTS award nomination 1992); *Child's Eye* (RTS award nomination 1995); *Looking for Billy; Let Me See My Children; Our House; Searching for Susan; Child's Eye; Home Alone; Stepfamilies; The Test; Men Who Pay For Sex; Footballer's Wives; The End is Nigh*

After Image Ltd
32 Acre Lane
London SW2 5SG
Tel: 0171 737 7300
Fax: 0171 326 1850
email: jane@arc.co.uk
Currently developing dramas. Recent Productions include *Television Songs of Seduction*, a music drama; *Pull*, a sculptural dance. 2 documentaries shot in Africa. 2 television operas called *Camera and The Empress* (C4) plus *The Score* (BBC2) a classical music magazine series

Agenda
Castell Close
Enterprise Park
Swansea SA7 9FH
Tel: 01792 410510
Fax: 01792 775469
Wales' largest independent production company which produces nightly magazine programme, Heno, for S4C. Entertainment, drama, features for S4C, C4, BBC, corporate sector

Alive Productions
37 Harwood Road
London SW6 4QP
Tel: 0171 384 2243
Fax: 0171 384 2026
TV programme production company including *Star Test* and *Star Chamber* (both for C4)

Allied Filmmakers/Allied Vision/ Allied Film Productions/Allied Troma Ltd
(See Pathé Pictures - Lottery Film Production Franchises)
The Glassworks
3/4 Ashland Place
London W1M 3JH
Tel: 0171 224 1992
Fax: 0171 224 0111
Recent productions: *Nostradamus, Midnight in Moscow, Beijing Express*. In production: *The Lawnmower Man II Beyond Cyberspace*

Amy International Productions
PO Box 17
Towcester
Northants NN12 8YJ
Tel: 01295 760256
Fax: 01295 760889
Such a Long Journey (in post production); *Dragon Under the Hill; The Liaison; Operation Farrier; Herculine Barbin*

Anglia TV Entertainment
48 Leicester Square
London WC2H 7FB
Tel: 0171 389 8651
Fax: 0171 930 8499
Touching Evil; Where the Heart is; Agatha Christie's Pale Horse

Anglo/Fortunato Films
170 Popes Lane
London W5 4NJ
Tel: 0181 840 4196
Fax: 0181 840 0279
Luciano Celentino
Feature film production company. 1994 Directed *Callan*; 1996 Produced, wrote directed *The Pinch*

The Animation Station
Leisure and Tourism Department
Cherwell District Council
Bodicote House
Bodicote, Banbury
Oxon OX15 4AA
Tel: 01295 252535
Fax: 01295 263155
Dex Mugan
A specialist arts education producer, distributor and trainer. Works in collaboration with innovative artists and performers from across the world, selecting and commissioning a broad range of high quality work

Animha Productions
121 Roman Road
Linthorpe
Middlesbrough TS5 5QB
Tel: 01642 813 137
Fax: 01642 813 137
email: animha@awn.com
Website: www.awn.com/animha

Animus Entertainments
67/71 Goswell Road
London EC1V 7EN
Tel: 0171 490 8234
Fax: 0171 490 8235
Ruth Beni

Antelope
29B Montague Street
London WC1B 5BH
Tel: 0171 209 0099
Fax: 0171 209 0098
email: antelope@antelope.co.uk
Mick Csaky, Annie Stogdale, Krishan Atcora, Simon Passmore
Dramas and documentaries for broadcast TV in UK, USA, Europe and Japan. Recent productions: *Flanders and Swann*; *Plácido Domingo - A Musical Life*; *Hiroshima; The Pier* (series I, II and III). *Cyberspace* (series)

Arena Films
2 Pelham Road
London SW19
Tel: 0181 543 3990
Fax: 0181 540 3992
Specialising in European co-production

Argo Productions
5 South Villas
Camden Square
London NW1 9BS
Tel: 0171 485 9189
Fax: 0171 485 6808
Broadcast documentaries. Recent productions: *Flying Squad* (1989 ITV, 8 x 30 mins); *Gracewell* (1991 C4 *True Stories*, 1 x 75 min); *Murder Squad* (1992 Thames TV, 1 x 60 min, 6 x 30 min); *Scotland Yard* (1994 ITV 6 x 30 min); *Murder Squad* (2nd Series 1996 2 x 60 mins + 1997, 3 x 60 mins); *Scotland Yard* (2nd series, 1995 ITV, 6 x 30 mins)

Ariel Productions
Goldcrest House
30-44 Brewer Street
London W1V 3HP
Tel: 0171 437 7972
Fax: 0171 437 6411

Arlington Productions Ltd
Pinewood Studios
Iver Heath
Bucks SL0 ONH
Tel: 01723 651700
Fax: 01723 656050

Film and TV filmmaker, formerly through Tyburn Film: *Masks of Death; A One-Way Ticket to Hollywood; Murder Elite; Legend of the Werewolf; Courier*

Artisan Films Ltd
Twickenham Film Studios
Saint Margarets, Twickenham
Middlesex TW1 2AW
Tel: 0181 607 8888
Fax: 0181 607 8889
email: Artisanfilms@msn.com
Recent productions: *Princess Caraboo*, Dir Michael Austin. In development: *Slow Train To Milan; Dorking Cocks; By Grand Central Station I Sat Down and Wept.* In production: *The Revengers' Comedies*

Richard Attenborough Productions
Twickenham Studios
St Margaret's
Twickenham TW1 2AW
Tel: 0181 607 8873
Fax: 0181 744 2766
Recent productions: *Grey Owl; In Love and War*

Aurum Productions
PO Box 14703
London SE1 9WQ
Tel: 0171 401 2700
Fax: 0171 401 2702
Nicholas Burgess-Jones, Director/Producer
Film and Video Production, Feature Films-Pop Promos, documentaries

Available Light Productions
12 Great George Street
Bristol BS1 5RS
Tel: 0117 929 1311
Fax: 0117 929 9039
Documentary series, singles and features. Recent productions: *Home Movies* (HTV); *Secrets of the Moors* (HTV); *The Levels* (HTV); *Running Bear* (C4); *Middle Ages* (HTV, Westcountry, C4); *Entertaining Mr Wedlock* (HTV); *Tracy Worcester* (HTV); *Hannah More* (HTV)

bfi BFI Production
29 Rathbone Street
London W1P 1AG
Tel: 0171 636 5587
Fax: 0171 580 9456
BFI Production fosters new talent and develops, produces and invests in a wide range of features, shorts, and documentaries. It has a small fund for commissioning feature scripts and finances two to three low budget feature films a year. Three feature films were completed in conjuction with Channel 4 during 1996 *Gallivant* Dir Andrew Kötting;

Stella Does Tricks Dir Coky Giedroyc; *A Bit of Scarlet*, Dir Andrea Weiss; *Under the Skin* Dir Carine Adler: *Sixth Happiness* Dir Waris Hussein and *Love is the Devil* Dir John Maybury (**see p8**)

Bandung
Block H
Carkers Lane
53-79 Highgate Road
London NW5 1TL
Tel: 0171 482 5045
Fax: 0171 284 0930
email: bandung@gn.apc.org
Recent productions: *Big Women*; a four-part Fay Weldon drama for Channel 4 *Letters from America* with Christopher Hitchens; *Dead West: the War on the American Desert*; *An Open Letter to India: Save the Taj Mahal; The Hanged Man: Nigeria's Shame;* Derek Jarman's *Wittgenstein*

Barraclough Carey Productions
Cambridge House
Cambridge Grove
London W6 0LE
Tel: 0181 741 4777
Fax: 0181 741 7674
Jenny Barraclough, George Carey, Chrissie Smith
Recent productions: *Europe* (BBC2); *Knife to the Heart* (BBC1); *Gunpower* (Discovery); *Rescue* (C4); *Moving Pictures* (BBC2); *Biteback* (BBC1); *Africa Express* (C4); *The Protection Racket* (BBC2); *The Reality Trip* (BBC2); *The New Jerusalem* (BBC2); *Empire of the Censors* (BBC2); *Fall of Saigon* (BBC1); *Visions of Heaven and Hell* (C4); *Messengers from Moscow* (BBC2); *Call of the Bagpipes* (BBC2); *Jackie Onassis* (C4); *Lost Civilisations* (NBC); *People's Parliament* (BCP North for C4)

Basilisk Communications
31 Percy Street
London W1P 9FG
Tel: 0171 580 7222
Fax: 0171 631 0572
Productions include *Temenos*, Dir Nina Danino; *The Gay Man's Guide to Safer Sex*, Dir David Lewis (Terrence Higgins Trust); *Blue*, Dir Derek Jarman; *Projections* Pet Shop Boys; *Daybreak* Dir Bernard Rudden

Peter Ratty Productions
Claremont House
Renfrew Road, Kingston
Surrey KT2 7NT
Tel: 0181 942 6304
Fax: 0181 336 1661
Recent C4 productions: *Swastika Over British Soil; A Time for*

Remembrance; The Divided Union; Fonteyn and Nureyev; The Algerian War; Swindle; Il Poverello.
Independent productions: *The Story of Wine; Battle for Warsaw; Battle for Dien Bien Phu; Birth of the Bomb; Search for the Super; Battle for Cassino; Operation Barbarossa; Farouk: Last of the Pharaohs*

Beambright
Debnershe
The Street
Shalford
Surrey GU4 8BT
Tel: 01483 539343
Fax: 01483 539343
Recent productions. Feature film: *Century*; television: *Frontiers*, a six-part drama series for Carlton

Beaver Films Ltd
Beaver Lodge, Richmond Green
Richmond-Upon-Thames
Surrey TW9 1NQ
Tel: 0181 940 7234
In production *Séance on a Wet Afternoon* (Remake)

Bedford Productions
2nd Floor
58-60 Berners Street
London W1P 3AE
Tel: 0171 436 7766
Fax: 0171 436 8786
Mike Dineen, Francis Megahy, Richard Mervyn
Television, documentary, drama production, and business to business programming

Bevanfield Films
2a Duke Street
Manchester Square
London W1M 5AA
Tel: 0171 487 4920
Fax: 0171 487 5472
Producers of animated and live-action programmes for television, video and cinematic release

The Big Bird Film Co
20c High Park Road
Kew
Richmond TW9 4BH
Alexander Moody
In pre-production: *Flight of the Swallow*

The Big Group
22 Stephenson Way
London NW1 2HD
Tel: 0171 383 2335
Fax: 0171 383 0357
Producing for television and film

Black Coral Productions
PO Box 333, Woodford Green
Essex IG9 6DB

Tel: 0941 102 820
Fax: 0181 504 3338
email: black.coral@virgin.net
Black Coral is actively involved in documentary and drama productions

Black Coral Training

130, Lea Valley Techno Park
Ashley Road
London N17 9LN
Tel: 0181 880 4861
Fax: 0181 880 4113
Black Coral Training is a non-profit making organisation specialising in 1-4 day foundation and intermediate level courses in: Production Management; Broadcast journalism; Scriptwriting for TV/Radio Drama; Composing Music for film and TV. Supported by Skillset & LFVDA and Middlesex University. Screenwriting Development aimed at readers, writers and script editors

Black Media Training Trust

Workshop 6,
Ace Business Centre
120 The Wicker, Sheffield,
South Yorks S3 8JD
Tel: 0114 275 6815
Fax: 0114 276 8292
Carl Baker
Ethnic minority, voluntary sector film and video-making projects offering training, multimedia and film and video production

Blue Dolphin Film & Video

40 Langham Street
London W1N 5RG
Tel: 0171 255 2494
Fax: 0171 580 7670
Production to date: *Blonde Fist*, Exec produced *The Big Freeze*
In development: *Beyond the Meadow, Nanatopia, Silk Estate*

Blue Heaven Productions Ltd

45 Leather Lane
London EC1N 7TJ
Tel: 0171 404 4222
Fax: 0171 404 4266
Producer of *The Ruth Rendell Mysteries* for Meridian/ITV Network. Developing *Greed*, 13 hour filmed drama for CBC Canada (in co-production with Telescene Montreal); developing *The Crocodile Bird* feature film

The Bridge

Sony Pictures Europe House
25 Golden Square
London W1R 6LU
Tel: 0171 533 1111
Fax: 0171 533 1105
Joint British company with Canal Plus in production: *Virtual Sexuality*

British Lion Film Corporation

Pinewood Studios
Iver Heath
Bucks SLO 0NH
Tel: 01753 651700
Fax: 01753 656391
Productions include: *The Wicker Man; Don't Look Now; Lady Jane* for Paramount Pictures and Turtle Diary, in association with United British Artists. *A Man for All Seasons* and *Treasure Island* for Turner Network Television (TNT). *A Prayer for the Dying* for Samuel Goldwyn Company; *The Crucifer of Blood* for TNT; *Death Train* for USA Network and Yorkshire TV

The Britt Allcroft Company PLC

3 Grosvenor Square
Southampton SO15 2BE
Tel: 01703 331661
Fax: 01703 332206
Recent productions: *Thomas the Tank Engine and Friends*, 104 stories filmed in live-action animation; *Shining Time Station*, 65 x 30 min live-action television shows with 5 x 1 hour specials; Britt Allcroft's *Magic Adventures of Mumfie*, 130 minute story filmed in classical animation with a score of 16 songs plus 1 x 22 minute Christmas special

Broadcast Communications

14 King Street
London WC2E 8HN
Tel: 0171 240 6941
Fax: 0171 379 5808
Through its wholly-owned subsidiaries Initial Film & TV, Hawkshead, and Bazal Productions, productions include: *The Brit Awards; Delia Smith's Winter Collection; The White Room; Ready, Steady, Cook!; 700 hrs a Year; Food & Drink*

Brook Lapping Productions

21-24 Bruges Place
Randolph Street
London NW1 0TF
Tel: 0171 482 3100
Fax: 0171 284 0626
Anne Lapping
Phillip Whitehead, Udi Eichler
In April 1997 Brian Lapping Associates merged with Brook Associates to form Brook Lapping Productions

Buena Vista Productions

Centre West, 3 Queen Caroline Street, Hammersmith
London W4 9PE
Tel: 0171 605 2400
International television production arm of The Walt Disney Studios

John Burder Films

7 Saltcoats Road
London W4 1AR
Tel: 0181 995 0547
Fax: 0181 995 3376
Corporate and broadcast worldwide, productions for Thorn EMI and many other sponsors. Including *The Common Sense Guides*, and *ABC of Guides*

Buxton Raven Productions Ltd

159-173 St. John Street
London EC1V 4QJ
Tel: 0171 296 0012
Fax: 0171 296 0014
email: buxtonraven@compuserve.com
Contacts: Jette Bonnevie, Jens Ravn
Recent feature productions include feature films *Robert Rylands Last Journey* (Spain/UK) and *Danish Language, Mimi & The Movers* and arts documentary, *Winter's Tale*. Projects in pre-production include the documentary, *The Working of Utopia* and features *Vinnie Got Blown Away* and *The Past is But a Dream* (Scandinavia). Projects in development include *Under My Skin* based on Sarah Dunant thriller, *Hell For Half A Crown* by Raoul Morris and a new original screenplay by Fay Weldon

Cabochon Film Productions - Celestino Coronado

16a Brechin Place
London SW7 4QA
Tel: 0171 373 6453
Fax: 0171 720 1302
Films include: *The Lindsay Kemp Circus; Miroirs; Hamlet; A Midsummer Night's Dream; Smoking Mirror*. Projects include: *Life is a Dream, The Incredible Orlando;* and *A Poet in New York*

Carlton Select

45 Fouberts Place
London W1V 2DN
Tel: 0171 434 3060
Fax: 0171 494 1421
Owned by Carlton Communications. Includes: *Soldier Soldier; Peak Practice; Birds of a Feather, Goodnight Sweetheart; The Good Sex Guide; The Camomile Lawn* and *Sharpe*. Carlton Food Network Europe's only dedicated food channel including shows by chefs Anthony Worrall Thompson, Brian Turner, Nanette Newman, Sophie Grigson, Tony Tobin, Aldo Zilli and Nancy Lam

Carnival (Films and Theatre)

12 Raddington Road
Ladbroke Grove

London W10 5TG
Tel: 0181 968 1818
Fax: 0181 968 0155
Recent productions: Firelight; *Up on the Roof; The Mill on the Floss; Shadowlands; Under Suspicion; Wilt; Bugs* (BBC); *Crime Traveller* (BBC); *The Infiltrator* (HBO); *The Fragile Heart* (C4); *Agatha Christie's Poirot* (LWT); *Jeeves & Wooster* (Granada)

Cartwn Cymru
Screen Centre
Llantrisant Road
Cardiff CF5 2PU
Tel: 01222 575999
Fax: 01222 575919
Animation production. Recent productions: *Turandot: Operavox* (BBC2/S4C Animated Operas); *Testament: The Bible in Animation;* (S4C/BBC2). Currently in production: *The Jesus Story*; animated feature film of the gospel

Catalyst Television
Brook Green Studios
186 Shepherd's Bush Road
London W6 7LL
Tel: 0171 603 7030
Fax: 0171 603 9519
Gardeners World (BBC); *Absolute Beginners* (C5); *Gardening From Scratch* (BBC)

Celador Productions
39 Long Acre
London WC2E 9JT
Tel: 0171 240 8101
Fax: 0171 836 1117
Television: primarily entertainment programming for all broadcast channels. Includes game shows, variety, with selected situation comedy, drama and factual output

Celtic Films
1-2 Bromley Place
London W1P 5HB
Tel: 0171 637 7651
Fax: 0171 436 5387
Sharpe (Carlton UK Productions), *Red Fox* (LWT), *Riszko* (ITV)

Chain Production
2 Clanricarde Gardens
London W2 4NA
Tel: 0171 229 4277
Fax: 0171 229 0861
Garwin Davison, Roberta Licurgo
Development and Co-Production
Feature Films, Previous Production with India, Italy, USA, Partners

Channel X
22 Stephenson Way
London NW1 2HD
Tel: 0171 387 3874

Fax: 0171 387 0738
XYZ; Jo Brand Through the Cakehole; The Smell of Reeves and Mortimer; The Unpleasant World of Penn and Teller; Funny Business, Phil Kay Feels..., Food Fight

Charisma Films
14-15 Vernon Street
London W14 0RJ
Tel: 0171 603 1164
Fax: 0171 603 1175

Chatsworth Television
97-99 Dean Street
London W1V 5RA
Tel: 0171 734 4302
Fax: 0171 437 3301
Sister company to Chatsworth distribution and merchandising companies. Producers of light entertainment and drama. Best known for the long running T*reasure Hunt* and *The Crystal Maze* (C4), *Mortimer's Law* (BBC). In development: *Wine Hunt* (C4)

Cheerleader Productions
60 Charlotte Street
London W1P 2AX
Tel: 0181 995 7778
Fax: 0181 995 7779
Specialists in live and post-produced sports programmes, documentaries, leisure, entertainment. Productions include *Sumo Wrestling, NHL Ice Hockey, Equestrian Sport, Motor Sport, Tennis, Water Sport, Basketball, Darts* and *Speedway*

The Children's Film Unit
Unit 8, Princeton Court
55 Felsham Road
London SW15 1AZ
Tel: 0181 785 0350
Fax: 0181 785 0351
A registered Educational Charity, the CFU makes low-budget films for television and PR on subjects of concern to children and young people. Crews and actors are trained at regular weekly workshops in Putney. Work is in 16mm and video, and membership is open to children from 8-18. Latest films for C4: *The Gingerbread House; Awayday;*
Administrator: Carol Rennie

Childsplay Productions
8 Lonsdale Road
London NW6 6RD
Tel: 0171 328 1429
Fax: 0171 328 1416
Television producers specialising in children's and family programming. Recent productions: Third series of *Pirates* - 12 episode comedy series (BBC Children's Department). *Children of the New Forest* - six part

serial for BBC Sunday evening transmission 1998

Chrysalis Visual Entertainment
The Chrysalis Building
13 Bramley Road
London W10 6SP
Tel: 0171 465 6259
Fax: 0171 465 6159
The following are all part of Chrysalis Visual Entertainment: Assembly Film and Television; Bentley Productions; Chrysalis TV; Chrysalis Sport; Lucky Dog Ltd; Red Rooster Film and Television Entertainment (qv); Stand and Deliver Productions; Watchmaker Productions. Also Chrysalis Distribution; Cactus TV; Chrysalis Films; IDTV

Cine Electra
National House
60-66 Wardour Street
London W1V 3HP
Tel: 0171 287 1123
Fax: 0171 722 4251
Specialises in European co-productions. Productions include: *The Baby of Mâcon* by Peter Greenaway; *The Ring with the Crowned Eagle* by Andrzej Wajda; *À Propos de Nice* (la suite) by Abbas Kiarostami

Cinema Verity Productions Ltd
The Mill House
Millers Way
1a Shepherds Bush Road
London W6 7NA
Tel: 0181 749 8485
Fax: 0181 743 5062
Class Act (Carlton TV); *She's Out* by Lynda La Plante (Carlton TV); *A Perfect State* (BBC)

Circus Films Ltd
Shepperton Studios
Shepperton
Middlesex TW17 OQD
Tel: 01932 572680/1
Fax: 01932 568989

Clark Television Production
Cavendish House
128-134 Cleveland Street
London W1P 5DN
Tel: 0171 388 7700
Fax: 0171 388 3366
Dispatches; The Black Bag; The Chrystal Rose Show; Class Action; Hard News; Chrystal's Style Guide; Frontline

Clio & Co
91 Mildmay Road
London N1 4PU
Tel: 0171 249 2551
Documentaries about women's history

The Comedy House

6 Bayley Street
London WC1B 3HB
Tel: 0171 304 0047
Fax: 0171 304 0048
Set up in 1990 to develop comedy films with British talent. Currently in development with six theatrical movies

The Comic Strip Ltd

Dean House
102 Dean Street
London W1V 5RA
Tel: 0171 734 1166
Fax: 0171 734 1105
Recent productions: *Four Men in a Car* - a one-off 30 minute comedy for C4. *The Comic Strip Presents* series 2 - 6 x 30 minute comedy films for BBC2; *Glam Metal Detectives* - 7 x 30 minute comedy drama series for BBC2

Compass Film Productions

1st Floor
175 Wardour Street
London W1V 3FB
Tel: 0171 734 8115
Fax: 0171 439 6456
Recent documentaries: *Loneliness Week* (BBC Wales), social action series; *Behind the Mask* (BBC 40 Minutes)

Connections Communications Centre Ltd

Palingswick House, 241 King Street, Hammersmith
London W6 9LP
Tel: 0181 741 1767
Fax: 0181 563 1934
email: @cccmedia.demon.co.uk
Jacqueline Davis
A registered charity producing promotional and educational videos for the voluntary sector. Currently in production Travelling Forward a 25 minute documentary commissioned by the Thalidomide Society

Contrast Films

311 Katherine Road
London E7 8PJ
Tel: 0181 472 5001
Fax: 0181 472 5001
Produce documentaries and feature films.Productions include: *Bangladesh 25: New Eastenders* (BBC Pebble Mill); *Rhythms* (C4); *Flame in my Heart* (C4)

Cosgrove Hall Films

8 Albany Road
Chorlton-cum-Hardy
Manchester M21 0AW
Tel: 0161 881 9211
Fax: 0161 881 1720
Award-winning animation subsidiary of Anglia Television. Producer of cartoon and model animation. Creators of: *Dangermouse; The Wind in the Willows; Count Duckula; The B.F.G.; Truckers; Noddy; Avenger Penguins, Wyrd Sisters*

Judy Couniham Films Ltd

1st Floor, 27/29 Berwick Street
London W1V 3RF
Tel: 0171 734 9870 ex219
Fax: 0171 734 9877
Previous productions include: *Antonia's Line, Before The Rain.* Productions in development include: *Crowd Pleaser, Janice Beard 45 wpm, Fish Out of Water.* Productions in pre-production: *Time to Love*

Countrywide Films Ltd

Production Office
Television Centre
Northam
Southampton SO14 0PZ
Tel: 01703 230286/712270
Cathedral (LWT); *Country Ways* (Meridian); *Country Faces* (BBC1); *Great House Cookery* (Meridian); *Pub People; Goodbye to All That* (TVS); *Land Girls* (Meridian); *Missa Luba* (Philips); *Country Ways* series 14 (in production); *Michael Barry's Undiscovered Cooks* (2 series)

DBA Television

7 Lower Crescent
Belfast BT7 1NR
Tel: 01232 231197
Fax: 01232 333302
Northern Ireland's leading production company

DLT Entertainment UK Ltd

10 Bedford Square
London WC1B 3RA
Tel: 0171 631 1184
Fax: 0171 636 4571
Specialising in entertainment programming. Titles include: Cinema Europe - *The Other Hollywood* (BBC); *As Time Goes By* (BBC). Also operate a joint venture with the Theatre of Comedy at the Shaftesbury Theatre. The parent company is a well known producer and distributor in New York, Los Angeles and Sydney

DNA Films

2-4 Noel Street
London W1 3RB
Tel: 0171 287 3209
Fax: 0171 287 3503
Contact: Grace Hodge
(See DNA Film Ltd - Lottery Film Production Franchises)

Dakota Films

12a Newburgh Street
London W1
Tel: 0171 287 4329
Fax: 0171 287 2303

Dan Films Ltd

37 Percy Street
London W1P 9FG
Tel: 0171 916 4771
Fax: 0171 916 4773
email: danfilmsltd@clara.net
Cilla Ware (Director)
Julie Baines (Producer); Sarah Daniel (Producer); Michael Daniels (Producer's Assistant); Jason Newmark (Head of Development); Dan Films produces feature films, television dramas and documentaries. Recent productions: *Silent Witness* (documentary); *Madagascar Skin* (35mm feature); *Butterfly Kiss* (35mm feature); *Everywhere* (16mm short). Projects in production/development include: *Los Angeles Without a Map* (35mm feature - pre-production); *Teeth* (35mm feature - in development); *The Art of Love* (drama series)

De Warrenne Pictures

2 Queen Victoria Terrace
Sovereign Court
London E1 9EU
Tel: 0171 481 8000
Fax: 0171 481 8624
Tom Waller
Feature film production company. Recent projects include *Monk Dawson* based on award winning novel by Piers Paul Read. In Production: *Butterfly Man*, by Kaprice Kea; In development: *Famine*, historical drama set in Ireland. *Sally*, based on the bestseller by Freya North

Direct Films

21B Brooksby Street
London N1
Tel: 0171 697 0747
Formerly Black Audio Films. Black independent films:
Three Songs on Pain, Light and Time; Mothership Connection; The Last Angel of History; Martin Luther King - Days of Hope; Who Needs A Heart; The Darker Side of Black

Dirty Hands

c/o Propaganda
2nd Floor
6-10 Lexington Street
London W1R 3HS
Tel: 0171 478 3207
Fax: 0171 734 7131

Distant Horizon

84-86 Regent Street
London W1R 5PF
Tel: 0171 734 8690
Fax: 0171 734 8691
Recent productions: *Sarafina!;
Chain of Desire; Cry the Beloved
Country; Captives; Dead Beat;
Yankee Zulu; The Mangler;
Scorpion Spring*

Diverse Productions

Gorleston Street
London W14 8XS
Tel: 0171 603 4567
Fax: 0171 603 2148
Established in 1982, Diverse is one
of Britain's leading independent
factual programme makers, and has
recently expanded into Interactive
media. Diverse has a tradition of
radical and innovative programming.
We are successful producers of
popular prime-time formats, strong
documentaries (one-offs and series),
investigative journalism, science,
business and history films, travel
series, arts and music, talk shows,
schools and education

Documedia International Films Ltd

Production Office
19 Widegate Street
London E1 7HP
Tel: 0171 625 6200
Fax: 0171 625 7887
Producers and distributors of
documentary and drama programming; corporate and Internet
adaptations. Recent productions
include *David Shepherd Biography*
(documentary special), *Short Breaks*
(drama series), *Playhouse Pictures*
(drama series), *Over the Wall in
China* (documentary special).
Programme library for worldwide
sales includes Drama serials - Short
Breaks (first and second series of 6,
10 minutes), *Playhouse Pictures*
(first and second series of 6 half
hours), *Baker Street Boys* (8 half
hours and 4 hours), *Box of Delights*
(6 half hours and 3 hours), *Edge of
Night* (13 half hours); Drama
specials - Soulscapes; Telemovies/
Features - *Deva's Forest, Leaves and
Thorns*; International short drama -
*Pile of Clothes, JoyRidden, The
Cage, Nazdrovia, Thin Lines, Late
Fred Morse, The Extinguisher, The
Summer Tree, Arch Enemy, Tea and
Bullets, Isabelle Peregrine, Beyond
Reach, Trauma, Edge of Night,
Something Wonderful,* Documentary
series - *Horses of the World* (8
hours), Room of Dreams (4 half
hours), *In the Wild* (4 hours and 7
hours); Documentary specials -
Positive Story, Caste in Half,

*Strength and the Wager, Poison
People, Rock n Roll Juvenile, Over
the Wall in China, Graham Knuttel.
David Shepherd.* Represented
worldwide for sales for all media,
including Internet/video on demand
and all television

Dogstar UK Ltd

5 Sherwood Street
London W1V 7RA
Tel: 0171 287 5944
Fax: 0171 287 1786
email: dogstar.co.uk
Liz Barron, Irena Brignull, Carolyn
Drebin, Alan Greenspan, Mike
Newell, David Parfitt

Domino Films

8 Stockwell Terrace
London SW9 0QD
Tel: 0171 582 0393
Fax: 0171 582 0437
Well-established company producing
wide range of factual programmes
which include: the award-winning
*Selling Murder; Secret World of Sex;
Lost Children of the Empire; Heil
Herbie.* Other productions include:
*Eve Strikes Back; Breadline Britain
1990s; Soviet Citizens; Take Three
Girls; Windows on the World*

Double Exposure

Unit 22-23
63 Clerkenwell Road
London EC1M 5PS
Tel: 0171 490 2499
Fax: 0171 490 2556
Production and distribution of
broadcast and educational docu-
mentaries in the UK and abroad

Downtown Pictures Ltd

4th Floor, Suite 2, St Georges
House, 14-17 Wells Street
London W1P 3FP
0171 323 6604
0171 636 8090
Martin McCabe, Alan McQueen,
Alan Latham, Anne Rigby

The Drama House

1 Hertford Place
London W1P 5RS
Tel: 0171 388 9140
Fax: 0171 388 3511
Joan Bakewell, Jack Emery
Film and television production.
Recent projects: *Breaking the Code*
(film for BBC1) with Derek Jacobi;
Witness Against Hitler (film for
BBC1) with James Wilby and Helen
McCrory; *Suffer the Little Children*
(BBC2) with Jane Horrocks; *A
Curse on the House of Windsor* (C4)
with Miriam Margolyes; *Reith to the
Nation* (C4) with Tim West; *Mister
Shaw's Mission Millions* (C4) with

Ian McKellen. In development: *The
Poet and the President* (film for
Channel 4); *Little White Lies*
(BBC1); *Terezin Requiem* (BBC2)

Dramatis Personae

19 Regency Street
London SW1P 4BY
Tel: 0171 834 9300
Nathan Silver, Nick Kent.
Consultant: Maria Aitken
Concerned primarily with self-
generated features on artistic skills
and human development having
broad cultural or social interest

Drumbeat Productions Ltd

17a Mercer Street
Covent Garden
London WC2H 9QJ
Tel: 0171 836 3710

Ecosse Films

12 Quayside Lodge
Watermeadow Lane
London SW6 2UZ
Tel: 0171 371 0290
Fax: 0171 736 3436
Douglas Rae, Emma Crawford,
Robert Bernstein
*Mrs. Brown, The Ambassador,
Unsuitable Job for a Woman*

Edinburgh Film & Video Productions

Edinburgh Film and TV Studios
Nine Mile Burn by Penicuik
Midlothian EH26 9LT
Tel: 01968 672131
Fax: 01968 672685
Major Scottish production company
established in 1961. Latest produc-
tion: *Sara*, an international
television co-production drama

EKco Television Ltd

PO Box 1552
126 Haberdasher Street
London N1 6EJ
Tel: 0171 490 1492/0121 449 3251

Elmgate Productions

Shepperton Studios, Studios Road
Shepperton
Middx TW17 0QD
Tel: 01932 562611
Fax: 01932 569918
Feature films, television films and series

Endboard Productions

114a Poplar Road
Bearwood
Birmingham B66 4AP
Tel: 0121 429 9779
Fax: 0121 429 9008
Producers of television programmes

English Film Company Ltd

250 Southlands Road

Bromley
Kent BR1 2EQ
Tel: 0181 460 7224
Fax: 0181 460 7224
Bachoo Sen

Enigma Productions
13 Queen's Gate Place Mews
London SW7 5BG
Tel: 0171 581 0238
Fax: 0171 584 1799
David Puttnam, Steven Norris
Film and television production
Recent productions: *Being Human;
War of the Buttons; The Burning
Season; Le Confessional*

Eon Productions
138 Piccadilly
London W1Z 9FH
Tel: 0171 493 7953

Equilibrium Films
28 Sheen Common Drive
Richmond TW10 5BN
Tel: 0181 898 0150/ 876 3637
Fax: 0181 898 0150/ 876 3637
Titles include: *The Tribe That Time
Forgot* - an Equilibrium Film
production in association with
WGBH Boston/Nova for PBS;
*Jaguar People;Yemen's Cultural
Drug: Dream or Nightmare;Yemen's
Jambiya Cult;Yemen's Lost Soul;
First Contact - Last Rites* - a Bare
Faced Production for BBC; *Egypt
Powerplays, Burma's Final
Solution, Conquering The Mountain
of Fire*

Faction Films
28-29 Great Sutton Street
London EC1V 0DU
Tel: 0171 608 0654/3
Fax: 0171 608 2157
Sylvia Stevens, Dave Fox, Peter Day,
Grant Keir
Group of independent film-makers.
Recent productions: *Three Kisses
and a Funeral; Prosecutor*

Festival Film and Television Ltd
Festival House
Tranquil Passage
Blackheath Village
London SE3 OBJ
Tel: 0181 297 9999
Fax: 0181 297 1155
The company concentrates mainly
on popular television dramas and
continues production of its
Catherine Cookson mini-series for
ITV. Recent completed productions
include: *The Tide of Life; The Girl;
The Wingless Bird; The Moth and
The Rag Nymph* (all three hour mini-
series for ITV). Also developing
feature films and children's drama

Figment Films Ltd
2 Noel Street
London W1
Tel: 0171 287 3209
Fax: 0171 287 3503
Productions include: *Shallow Grave;
Trainspotting, Twin Town, A Life
Less Ordinary*
(See DNA Film - Lottery Film
Production Franchises)

Film and General Productions
10 Pembridge Place
London W2 4XB
Tel: 0171 221 1141
Fax: 0171 792 1167
Recent productions: True Blue
(feature film); *SeeSaw* (co-
production with Scottish Television
for ITV); *The Queen's Nose* (BBC);
Sunny's Ear (Carlton for ITV)
Features in development: *Supersti-
tion; Constance and Carlotta; The
Bitter Sea; Boney and Betsy*

The Film Consortium
6 Flitcroft Street
London WC2 8DJ
Tel: 0171 691 4440
Fax: 0171 691 4445
Head of Production and Develop-
ment: Colin Vanes
Consists of: Greenpoint Films,
Parallax Pictures, Scala and Skreba
(See The Film Consortium - Lottery
Film Production Franchises)

Film Form Productions
64 Fitzjohn's Avenue
London NW3 5LT
Tel: 0171 794 6967
Fax: 0171 794 6967
Film/video production, drama and
documentary for television and
video distribution. Full crewing,
writers, producers and directors

Film Four International
124 Horseferry Road
London SW1P 2TX
Tel: 0171 396 4444
Fax: 0171 306 8361
International film sales and
distribution arm of C4, often credited
as a co-production partner for UK and
international productions. Decisions
on programming and finance relating
to these productions are initiated by
Channel Four Films, the film
programming strand of C4's drama
department. Recent productions:
*Trainspotting; Brassed Off; Fever
Pitch; Welcome to Sarajevo; Vigo -
Passion For Life, Martha - Meet
Frank, Daniel & Laurence; The Acid
House; Babymother*

FilmFair Animation
106 Gloucester Place

London W1H 3DB
Tel: 0171 935 1596
Fax: 0171 935 0229
Producers of model and cel
animation series, special effects and
commercials. Productions include:
*The Wombles; Paddington Bear;
Huxley Pig; Gingerbread Man;
Astro Farm ; The Dreamstone;
Brown Bear's Wedding; White
Bear's Secret; The Legend of
Treasure Island*

Filmworks
65 Brackenbury Road
Hammersmith
London W6 0BG
Tel: 0181 741 5631
Fax: 0181 748 3198
Recent productions: *On the Trail of
the Chinese Wildman; Struggle for
the Pole - In the Footsteps of Scott;
A Day in the Life of a Medical
Officer; Antarctic Challenge;
Anything's Possible. Current
animation series: Captain Star -
Inventing the Universe*

Fine Line Features
25-28 Old Burlington Street
London W1X 1LB
Tel: 0171 440 1000
Fax: 0171 439 6105
European film production
*Shine, Sweet Hereafter,
Deconstructing Harry*

The First Film Company
38 Great Windmill Street
London W1V 7PA
Tel: 0171 439 1640
Fax: 0171 437 2062
Feature film, television and
commercial production. *Dance with
a Stranger; Soursweet; The
Commitments; The Railway Station
Man; A Kind of Hush.* Among
projects in development: *Flying
Hero Class,* based on the novel by
Thomas Keneally; *Django
Reinhardt,* an original screenplay by
Shelagh Delaney; *No Man's Land,*
an original screenplay by John Forte

Flamingo Pictures
129a Newington Green Road
London N1 4RA
Project development

Flashback Productions
22 Kildare Terrace
London W2 5LX
Tel/Fax: 0171 727 9904
Tel/Fax: 0171 727 9904

Flashback Television Ltd
11 Bowling Green Lane
London EC1R OBD
Tel: 0171 490 8996

Fax: 0171 490 5610
Award-winning producers of a wide
range of factual programming
including lifestyle, history, natural
history and sport documentaries.
Recent productions include *Garden
Doctors: 3* series for Channel 4, *War
- The Inside Story*: a long running
series for Arts & Entertainment in
the US. Many projects in production
and development for a wide variety
of broadcasters in Britain, Europe
and the USA

Flashlight Films
15 Golden Square
London W1R 3AG
Tel: 0171 287 4252
Fax: 0171 287 4232
email: flashlightfilms@compuserve.com
Frank Mannion, Aaron Simpson

Focus Films
Rotunda Studios
Rear of 116-118 Finchley Road
London NW3 5HT
Tel: 0171 435 9004
Fax: 0171 431 3562
David Pupkewitz, Marsha Levin,
Lisa Disler, Malcolm Kohll
Film and television production. Past
productions: Janet Suzman's
critically acclaimed *Othello*; *Gad
Hollander's Diary of a Sane Man*,
Crimetime, psychological drama,
starring Stephen Baldwin and Pete
Postlethwaite. In development: *The
51st State; Johnny Riff, Spindrift,
The Complete History of the Breast;
Secret Society; Camden Girls*

Mark Forstater Productions
27 Lonsdale Road
London NW6 6RA
Tel: 0171 624 1123
Fax: 0171 624 1124
Recent productions: *Paper Mar-
riage; The Touch; La Cuisine
Polonaise; Grushko*, BBC drama
series; *Transcontinental;
Prowokator; Between the Devil and
the Deep Blue Sea*

Fragile Films
95-97 Dean Street
London W1N 3XX
Tel: 0171 734 5655
Fax: 0171 734 5553
(See Pathé Pictures - Lottery Film
Production Franchises)

Freedom Pictures
Ealing Studios
Ealing Green
London W5 5EP
Tel: 0181 567 6655
Fax: 0181 758 8698
In development: *Black Death, Face
Value, Mushroom Soup*

Freeway Films
67 George Street
Edinburgh EH2 2JG
Tel: 0131 225 3200
Fax: 0131 225 3667
33a Pembroke Square
London W8 6PD
Tel: 0171 937 9114
Fax: 0171 937 3938
Drama for television and cinema.
Recent productions: *Half the Picture*
(TV film for BBC2); *Màiri Mhór*
(TV film for BBC2); *Carrington*
(feature film for PolyGram). In
development: *The Silver Darlings*
(feature film); *A Pale View of Hills*
(feature); *A Storm From Paradise*
(feature); *Easy* (feature); *Foreign
Exchange* (feature); *Fritz's Jazz
Band* (feature)

Friday Productions
23a St Leonards Terrace
London SW3 4QG
Tel: 0171 730 0608
Fax: 0171 730 0608
Recent productions: *Goggle Eyes* (4
x 40 min), family comedy/drama
series for BBC, winner of Writers'
Guild award; *Harnessing Peacocks*
(1 x 120 min). In development:
From A View To A Death, ESF-
funded period black comedy scripted
by Andrew Davies, to be directed by
Hugh Laurie; *Acid Casuals*, to be
written and directed by Mike
Hodges; *Lux The Poet*, which is
being backed by British Screen; and
She Died a Lady, period detective
series starring Robert Hardy, being
funded by Carlton

Front Page Films
23 West Smithfield
London EC1A 9HY
Tel: 0171 329 6866
Fax: 0171 329 6844
Owners of the Richmond Film-
house. Past productions: *The Mini
Sagas*, six theatrical shorts which
were released by UIP alongside *A
Fish Called Wanda, Parenthood* and
The Naked Gun; Get Back, a feature
with Paul McCartney directed by
Richard Lester. In development: a
number of feature films

Fugitive Group
44 Bloomsbury Steet
London WC1B 3QJ
Tel: 0171 637 3300
Fax: 0171 637 4400
Productions: *Death Machine; The
Passion of Darkly Noon; The Krays;
The Reflecting Skin; Spooks of
Bottle Bay; La Passione; Operation
Good Guys*

Fulcrum Productions Limited
254 Goswell Road
London EC1V 7EB
Tel: 0171 253 0353
Fax: 0171 490 0206
email: 100070.1517@compuserve.com
Award winning company making a
range of documentaries, current
affairs and arts programmes for TV.
1997 includes: Channel 4 -
Dispatches: A Question of Sleaze;
Dispatches: *Murder in St James II;
Health Alert: Don't Swallow Your
Toothpaste!; The Third Party.* BBC2
- *The Terror & The Truth.* World
Bank - *The Capable State.* 1997 in
production: Discovery Channel -
Amazing Earth; Channel 5 - *First
on Five.* Channel 4 - *Seven Deadly
Sins: Sloth; Citizen's Arrest: Trike.*
BBC2 - The International Red Cross

Gainsborough (Film & TV) Productions
The Groom Cottage
Pinewood Studios
Pinewood Lane
Iver Heath, Iver
Bucks SL0 0NH
Tel: 0171 409 1925
Fax: 0171 408 2042
*Made Hazard of Hearts, The Lady
and the Highwayman, A Ghost in
Monte Carlo and Duel of Hearts*

John Gau Productions
Burston House
1 Burston Road
London SW15 6AR
Tel: 0181 788 8811
Fax: 0181 789 0903
email: Johngau@aol.com
Television documentary production.
Titles include: *Explorers of the
Titanic*, for C4; *The Great Outdoors*,
C4 (2 series); *Lights, Camera,
Action! A Century of the Cinema*,
series for LWT; *The Team - A Season
with McLaren* for BBC Bristol
(series); *Westminster - Condition
Critical* for BBC; *Where's the Bill?*
for BBC; *The Triumph of The Nerds*
for C4, *The Sales of the Century*
BBC; *The Culture of Science* for C4

Noel Gay Television
1 Albion Court
Gelene Road
Hammersmith
London W6 0QT
Tel: 0181 600 5200
Fax: 0181 600 5222
TV Drama and TV entertainment.
Associate companies: Grant Naylor
Production. Noel Gay Motion Picture
Company, Noel Gay Scotland, Rose
Bay Film Productions, Sunbeam
Productions, Pepper Productions

General Entertainment Investments

Bray Film Studios
Down Place
Windsor Road, Windsor
Berkshire SL4 5UG
Tel: 01628 22111
Fax: 01628 770381
Feature film producers/financiers. Recent productions include: Tropic of Ice, Anglo-Finnish co-production; Soweto, African music feature; Olympus Force, Anglo-Greek co-production

Gimlet Productions

21 Knowsley Road
London SW11 5BN
Tel: 0171 350 2878
Fax: 0171 350 2878
Specialise in documentaries

Global Vision Network

Elstree Film Studios
Boreham Wood
Hertfordshire WD6 1JG
Tel: 0181 324 2333

Bob Godfrey Films

199 Kings Cross Road
London WC1X 9DB
Tel: 0171 278 5711
Fax: 0171 278 6809
Prominent studio, with recent productions: *Henry's Cat*, Cable Ace Award 1994; *Small Talk*, entertainment short, Oscar nomination 1993; *How Kevin Saved the World*, children's television series. Commercials: *Bassett's Jelly Babies*

Goldcrest Films International

65-66 Dean Street
London W1V 5HD
Tel: 0171 437 8696
Fax: 0171 437 4448
Major feature film production, sales and finance company. Recent productions: *No Way Home; Bring Me The Head of Mavis Davis; Clockwatchers; Annabelle's Wish*

The Grade Company

34 Grosvenor Street
London W1X 9FG
Tel: 0171 409 1925
Fax: 0171 408 2042
Lord Grade, Marcia Stanton, John Hough
Something To Believe In, director/producer John Hough. In development: *Arsenic and Old Lace*, with Propaganda Films USA; USA production of *Jackattack*, exec. producer Lord Grade, producer John Hough, screenplay Paul Highlander

Granada Film

The London Television Centre
Upper Ground
London SE1 9LT
Tel: 0171 737 8681
Fax: 0171 737 8682
Head of Film: Pippa Cross
Established in 1989 - a subsidiary of the Granada Media Group. Feature films: *The Heart; Up On the Roof; Some Kind of Life; August; Jack & Sarah; The Field; My Left Foot; The Fruit Machine*. In development: *Dancing Queen; Empty Cradles; Giraffe; Prime Suspect; Seeing Red; Fredie's Hairfood Factory; Vanity Fair; Rock 'N' Roll; Passionate Woman; Risky Kisses; Roof World; Shampoo Planet*

Grand National Film Distributors Ltd

14 Kingsley Place
London N6 5EA
Tel: 0181 340 2000
Fax: 0181 340 3762

Grand Slam Sports

Durham House
Durham House Street
London WC2N 6HF
Tel: 0171 839 4646
Fax: 0171 839 8392
Ron Allison, Rick Waumsley, John Watts, Nick Sharrard
Television sports production, over 400 hours a year and more than 55 sports covered since 1988

Grasshopper Enterprises

50 Peel Street
London W8 7PD
Tel: 0171 229 1181
Fax: 0171 229 2070
Productions to date: *Mouse and Mole* an animated series of 26 5-minute films based on the picture books by Joyce Dunbar and James Mayhew, with the voices of Alan Bennett and Richard Briers; *The Story of Arion and the Dolphin*, 25 min animation based on the picture book by Vikrall Seth and Jane Ray, with the voices of Vikrall Seth, Jemma Redgrave, John Hallam and Conrad Nelson; *The Mousehole Cat*, 25 min animation with the Red Green and Blue Company, based on the picture book by Antonia Barber and *Nicola Bayley*, narrated by Sîan Phillips; *Grasshopper Island; Emma and Grandpa; East of the Moon; The Angel and the Soldier Boy; On Christmas Eve; A Pattern of Roses*

Greenpoint Films

5a Noel Street
London W1V 3RB
Tel: 0171 437 6492
Fax: 0171 437 0644
A loose association of ten filmmakers:

Simon Relph, Christopher Morahan, Ann Scott, Richard Eyre, Stephen Frears, Patrick Cassavetti, John Mackenzie, Mike Newell, David Hare and Christopher Hampton
(See The Film Consortium - Lottery Film Production Franchises)

Griffin Productions

46 Museum Street
London WC1
Tel: 0171 404 0505
Fax: 0171 404 0516
Current productions: *Human Bomb* (Showtime), *Place of Lions* (Universal), *Pimpernel* (Showtime), *Prince of Mars* (Universal), *Week to Remember* (BBC), *Quintessial Verse* (BBC)

The Gruber Brothers

5th Floor
41-42 Foley Street
London W1P 7LD
Tel: 0171 436 3413
Fax: 0171 436 3402
Richard Holmes, Stefan Schwartz, Neil Peplow
Recent productions: *Walking Ned, Shooting Fish*. In development: *The Lake, The Martin Amis Project, Twentieth Century Boy*

Halas & Batchelor

The Halas & Barchelor Collection Ltd
35a Pond Street
London NW3 2PN
Tel: 0171 435 8674
Fax: 0171 431 6835
Animation films from 1940

The Half Way Production House

Units 1 & 2 Taylors Yard
67 Alderbrook Road
London SW12 8AD
Tel: 0181 673 7926
Fax: 0181 675 7612
Georgina Hart, Emma Stewart, Trevelyan Evans
Training organisation and production company. Recent productions include: *Mumbo Gumbo; Hands Off; A Chemical Solution*. In development: *Ruby Ruby, Danny and Viras*

Hammer Film Productions Ltd

John Maxwell Building
Elstree Film Studios
Boreham Wood
Hertfordshire WD6 1JG
Tel: 0181 324 2282
Fax: 0181 324 2286
Website: www.hammerfilms.com
Currently developing British and US films based on old Hammer film themes

HandMade Films

15 Golden Square

London W1R 3AG
Tel: 0171 434 3132
Fax: 0171 434 3143
Gareth Jones, Hilary Davis, Jan
Roldanus
Completed films include *Intimate
Relations; Sweet Angel Mine; The
Wrong Guy; The Assistant; The
James Gang* and *Dinner at Fred's.*
Paragon Entertainment Corporations
acquired HandMade Films catalogue
of 23 feature films in August 1994.
Handmade Films intends to produce
4-6 films annually

Harcourt Films
58 Camden Square
London NW1 9XE
Tel: 0171 267 0882
Fax: 0171 267 1064
Producer of documentaries and arts
programmes. Recent productions: 90
minute TV special *The Capeman* for
HBO, One hour music docs.
American Beauty and *Graceland* for
BBC1, VH1, ISIS Productions,
World Music CD-Rom with Channel
4. Real world and EU Media
investment club. *Inner Space* for
BBC2; *Ladyboys* and *Telling Lies*
for C4 *Witness* and music series *On
the Edge* for C4; Nature of music
and the 14 hour, award-winning
Beats of the Heart series

Hartswood Films
Twickenham Studios
The Barons, St Margarets
Twickenham
Middx TW1 2AW
Tel: 0181 607 8736
Fax: 0181 607 8744
Men Behaving Badly (6 series); *the
English Wife*, single film; *A Woman's
Guide to Adultery* (3 x 1 hr mini-series)

Hat Trick Productions
10 Livonia Street
London W1V 3PH
Tel: 0171 434 2451
Fax: 0171 287 9791
Denise O'Donoghue
Jimmy Mulville, Mary Bell, Hilary
Strong
Specialising in comedy, light
entertainment and drama. Produc-
tions include: Father Ted; Drop the
Dead Donkey; Have I Got News For
You; Confessions; Whatever You
Want; Game On; The Peter Principle;
Clive Anderson All Talk; Room 101
and Whose Line is it Anyway? The
company's drama output includes: *A
Very Open Prison; Eleven Men
Against Eleven; Lord of Misrule;
Crossing the Floor; Gobble and
Underworld*

Hawkshead
48 Bedford Square
London WC1B 3DP
Tel: 0171 255 2551
Fax: 0171 580 8101
Tom Barnicoat, Frances Whitaker,
Angela Law, Nikki Cheetham
Recent productions: *The Sunday
Programme* (GMTV); *Grass Roots*
(Meridian); *John Berger* (BBC2/
Arte); *Dorset Detours* (Meridian/C4/
HTV West); *Delia Smith's Summer
Collection* (BBC2); *Delia Smith's
Winter Collection* (BBC2);
Metroland (Carlton); *Strings, Bows
and Bellows* (BBC 2); *3,000 Mile
Garden; Held in Trust* (ITV); *Mary
Berry's Ultimate Cakes* (BBC1)

Hemdale Communications
10 Avenue Studio
Sydney Close
Chelsea
London SW3 6HW
Tel: 0171 581 9734
Fax: 0171 581 9735
Past productions: *Terminator;
Return of the Living Dead; Body
Slam; River's Edge; Vampire's Kiss;
Shag; Staying Together;
Chattahoochee; Salvador; Platoon;
The Last Emperor*

Jim Henson Productions
30 Oval Road, Camden
London NW1 7DE
Tel: 0171 428 4000
Fax: 0171 428 4001
Producers of high quality children's/
family entertainment for television and
feature films, usually with a puppetry
or fantasy connection. Recent
productions: Muppet Treasure Island -
feature; *Gulliver's Travels* - mini
series; *Animal Show II* - children's
series; *Mr Willowby's Christmas* - 30
minute special; *The Wonderous World
of Dr Seuss* - children's series;
Muppets Tonight I & II - children's
series; *Aliens in the family* - children's
series; *Buddy* - feature

Hightimes Productions
5 Anglers Lane
Kentish Town
London NW5 3DG
Tel: 0171 482 5202
Fax: 0171 485 4254
The company specialises in
developing, packaging and producing light entertainment and comedy
ideas for television. It has expanded
its activities to include drama and
has a variety of projects at different
stages of development. Hightimes
packaged *Me and My Girl* (5 series)
for LWT and *The Zodiac Game* (2
series) for Anglia

Holmes Associates
38-42 Whitfield Street
London W1P 5RF
Tel: 0171 813 4333
Fax: 0171 637 9024
Long-established UK independent
production company for broadcast
television. Now also developing and
producing feature films including
Prometheus, the directorial debut of
poet Tony Harrison, supported by C4
and the National Lottery through
ACE. Other recent work includes
*Westminster Abbey With Alan
Bennett* (BBC2) and *The Cormo-
rant*, BBC film starring Ralph
Fiennes. Holmes has recently set up
a subsidiary DevCo, to develop and
package 'intellectual properties' for
TV and film, with a current slate of
12 projects

Horntvedt Television
London House
100 New Kings Road
London SW6 4LX
Tel: 0171 731 8199
Fax: 0171 731 8312
Corporate and promotion videos,
interactive CDs. Productions
include: *Return of Wind Power;
Racing Rock; Grosvenor Gardens
Mentor* for RICS

Hourglass Productions Ltd
4 The Heights
London SE7 8JH
Tel: 0181 858 6870/319 8949
Fax: 0181 858 6870
Founded by John Walsh, BBC
Young Film Maker of the Year 1985
and graduate of the London
International Film School, and
council member of the Directors
Guild of Great Britian; Hourglass
produces documentaries, dramas and
feature film productions. Productions include: Oscar-winning Ray
Harryhausen, HRH Elizabeth II, The
Comedy Store, actuality programming for Channel Five with Boyz &
Girlz and the internationally
successful feature film *Monarch*. For
1998/99 Hourglass has a production
slate for features and TV productions

Michael Hurll Television
8-12 Broadwick Street
London W1V 1FH
Tel: 0171 287 4314
Fax: 0171 287 4315
Recent productions: *Schofield's
Quest* (LWT)

ITM (Music) Productions
16 Raleigh Road
Southville
Bristol BS3 1QR
Tel: 0117 966 1116

Fax: 0117 966 1116
Specialist music, audio pre-and post-production company for film, television and video. Recent productions: full score for Czech Republic feature film *Nexus*; theme for *First Response* (HTV/ITV); theme for *Yndy Dwybble* (S4C)

Iambic Productions
89 Whiteladies Road
Bristol BS8 2NT
Tel: 0117 923 7 222
Fax: 0117 923 8343
All Iambic's documentaries from 1991 to 1998 have won or been nominated for major awards in the UK or US, including BAFTAs and an international Emmy. Recent productions include: *No Angel - A Life of Marlene Dietrich* (LWT/AMC/Amaya/Canal Plus); *Without Walls - In Search of the Holy Foreskin* (C4); *Aldous Huxley - Darkness and Light* (BBC); *Puccini The Bohemian* (BBC/NVC); *In Search of Humphrey Bogart* (LWT/RM Arts); *The Real Andre Previn Story* (BBC Omnibus/BBC Worldwide); *Pantoland* (4 part series for C4); *Bob Hope - The Other Side of the Road* (BBC/AMC/NVC)

Idealworld Productions Ltd
St George's Studios
93-97 St George's Road
Glasgow G3 6JA
Tel: 0141 353 3222
Fax: 0141 353 3221
Film and television production. 1998 Productions: *Deals On Wheels* (C4); *Tool Stories* (C4); *Equinox - Ekranoplan* (C4); *Beg To Differ* (C4); *Italian Cookbook* (C4); 1998 Transmissions: *Island Harvest* (BBC Scotland)

Illuminations Films/Koninck Studios
19-20 Rheidol Mews
Rheidol Terrace
London N1 8NU
Tel: 0171 226 0266
Fax: 0171 359 1151
email: griff@illumin.co.uk
Producers of fiction films for television and theatric release. Latest productions: *The Institute Benjamenta* by The Brothers Quay; *Robinson in Space* by Patrick Keiller; *Conspirators of Pleasure* by Jan Svankmajer; *Dance of the Wind* by Rajan Khosa: *The Falconer's Tale* by Chris Petit & Iain Sinclair; *Deadpan* by Steve McQueen

Illuminations Television
19-20 Rheidol Mews

Rheidol Terrace
London N1 8NU
Tel: 0171 226 0266
Fax: 0171 359 1151
email: illuminations@illumin.co.uk
Producers of cultural programmes for C4, BBC and others. Recent productions: *Richard II, Deborah Warner's* acclaimed production starring Fiona Shaw for BBC2 *Performance*; Tx 3 (BBC2), arts series; *The Net 4* (BBC2), magazine series about computers and the digital world; *The Turner Prize 1997* (C4); *Is Painting Dead?* arts discussion programme (C4); *Things to Come* (C4); and *Dope Sheet* (C4), series about animation

Illuminations Interactive
19-20 Rheidol Mews
Rheidol Terrace
London N1 8NU
Tel: 0171 226 0266
Fax: 0171 359 1151
email: terry@illumin.co.uk
Website: http://www.illumin.co.uk
Contact: Terry Braun
Partnered with Illuminations Television and Illuminations Films. Company developing and producing Interactive multi-media projects about history, culture and the visual and performing arts. Past interactive multimedia clients include: The Arts Council of England, the Museum of London, the Museum of Scotland, The Tate Gallery, The Horniman Museum and the European Community through its IMPACT initiative. Current interactive multimedia and World Wide Web clients include: The Tower of London, The Imperial War Museum, The Museum of the Moving Image, The Museum of London, The Royal Festival Hall, The Hayward Gallery, the British Film Institute, Granada Television and Channel 4 Television. Current Productions: CD-ROMs for British Museum, Channel 4 Television and English Nature

Illustra Television
13-14 Bateman Street
London W1V 6EB
Tel: 0171 437 9611
Fax: 0171 734 7143

Imagine Films Ltd
53 Greek Street
London W1V 5LR
Tel: 0171 287 4667
Fax: 0171 287 4668
(See Pathé Pictures - Lottery Film Production Franchises)
Producers: Simon Channing-

Williams; Stephanie Faugher; Finance Director: Eddie Kane

imaginary films
19 Ainsley St
London E2 ODL
Tel: 0171 613 5882
Fax: 0171 729 9280
email:brady@imagfilm.demon.co.uk
Website: www.imagfilm.co.uk
Recent productions: *Boy Meets Girl* (1994) Dir Ray Brady; *Little England* (1996) Dir Ray Brady, *Kiss Kiss Bang Bang* (1997) Dir Ray Brady

Impact Pictures Ltd
10/12 Carlisle Street (3rd floor)
London W1V 5RF
Registered address:
99 Kenton Road
Harrow
Middx HA3 OAN
Tel: 0171 734 9650
Fax: 0171 734 9652
James Roeber, Jeremy Bolt, Ariane Severin, Paul Anderson
Productions include: *Shopping*, Dir Paul Anderson and *Stiff Upper Lips*, Dir Gary Sinyor. In development: *Soldier* to be directed by Paul Anderson, with Gerry Weintraub. *Vigo* to be directed by Julien Temple with Nitrate Film/Channel 4

Indigo Productions
70 Greyhound Road
London N17 6XW
Tel: 0181 801 1445
Production company specialising in drama and experimental documentaries

Initial Film and Television
74 Black Lion Lane
Hammersmith
London W6 9BE
Tel: 0181 741 4500
Fax: 0181 741 9416

Insight Productions/Insight Pictures
Gidleigh Studio
Gidleigh, Chagford
Newton Abbot
Devon TQ13 8HP
Tel: 01647 432686
Fax: 01647 433141
Established in 1982, nearly 50 broadcast film credits; arts, entertainment, a 'Film on Four' *Playing Away*, environmental documentaries *Dartmoor the Threatened Wilderness, Taming the Flood* and *Camargue*. Recent productions: *New Forest*, environ-ment (50 min, C4 Encounters); *NatureLands: New Forest, North Kent Marshes, Abbotsbury, Reading and the Thames, landscape and*

environment (4 x 30 min, Meridian); *Flavio Titolo - Blind Sculptor,* portrait of blind artist (3 x 30 min, HTV/Westcountry). *Farm Fantasia -* a ballet for animals and humans (C4)

Interactive TV Productions

Mezzanine Floor
Mappin House
4 Winsley Street
London W1N 7AR
Tel: 0171 333 0444
Fax: 0171 333 0777
Recent productions and commissions include: *The Look*, a hybrid of *The Word* and *Passenger*, scheduled for release late 1998. *Dinner at 8*, a social comedy based around a restaurant in Fulham

International Broadcasting Trust (IBT)

2 Ferdinand Place
London NW1 8EE
Tel: 0171 482 2847
Fax: 0171 284 3374
An independent, non-profit television production company and educational charity, making programmes about development, environmental and human rights issues for UK and international broadcast. Recent productions include: *Guardians of Chaos* - for BBC2 with Michael Ignatieff follows the former UN Secretary General on a tour of Africa in trouble spots; *Too Close too Heaven* - a 3-part documentary series for C4 looking into the history of African/ American Black Gospel music; *The Battle For Rickety Bridge* a documentary for C4 about the involvement of older protesters in the environmental campaign against the Newbury Bypass; *Under the Blue Flag* - a BBC Schools TV, 4-part series looking at the role of the United Nations

J&M Entertainment Ltd

2 Dorset Square
London NW1 6PU
Tel: 0171 723 6544
Fax: 0181 724 7541
email: @jment.com
Recent productions: *American Werewolf in Paris; The Revenger's Comedies; All the Little Animals*

Kai Productions

1 Ravenslea Road
London SW12 8SA
Tel: 0181 673 4550
Fax: 0181 675 4760
Recent productions: *Homes on Wheels; L.A. Requiem; Cyberville; Walking on Water; Sounds of the West*

Kensington Films

60 Charlotte Street
London W1P 2AX
Tel: 0171 927 8458
Fax: 0171 927 8590
Developing a wide range of drama projects

Kilroy Television Company

Unit 1, Warstone Court
Warstone Road
Hockley
Birmingham B18 6JQ
Tel:0121 693 5515
Fax: 0121 693 5525
Produces over 100 hours of network television a year, specialises in current affairs, documentary and entertainment programming

Kinetic Pictures

Video and Broadcast Production
The Chubb Buildings
Fryer Street
Wolverhampton WV1 1HT
Tel: 01902 837777
Fax: 01902 717143
email: kinetic@waverider.co.uk
Contact: Gary J Crozier, Creative Director

King Rollo Films

Dolphin Court
High Street
Honiton
Devon EX14 8LS
Tel: 01404 45218
Fax: 01404 45328
Produce top quality animated entertainment for children. In autumn 1993 the company's highly acclaimed line of Spot films was launched under licence by Disney for the North American home video market and now world sales of these titles approach two million cassettes. Producers of the animated series: *Mr Benn; King Rollo; Victor and Maria; Towser; Watt the Devil; The Adventures of Spot; The Adventures of Ric; Anytime Tales; Art; Play It Again; It's Fun to Learn with Spot; Buddy and Pip, Spot's Magical Christmas, Little Mr Jakob; Philipp; Happy Birthday; Good Night, Sleep Tight; Spot and his Grandparents go to the Carnival*

Kohler

16 Marlborough Road
Richmond
Surrey TW10 6JR
Tel: 0181 940 3967
Cabiri, The Experiencer

Landseer Film & Television Productions

140 Royal College Street

London NW1 0TA
Tel: 0171 485 7333
Fax: 0171 485 7573
Documentary, drama, music and arts, children's and current affairs. Recent productions: *Kenneth MacMillan at 60* (BBC); *Biosphere II* (Central); *La Stupenda* (BBC Omnibus); *Winter Dreams* (BBC2); *Sunny Stories: Enid Blyton* (BBC Arena); *J.R.R.T.* (Tolkien Partnership); *Discovering Delius* (Delius Trust); *Should Accidentally Fall* (BBC/Arts Council); *Danny Kaye* (LWT South Bank Show); *Mister Abbott's Broadway* (BBC Omnibus)

Langham Productions (A division of the Man Alive Group)

Westpoint
33-34 Warple Way
London W3 0RG
Tel: 0181 743 7431
Fax: 0181 740 7454
Michael Latham, Michael Johnstone
Edward on Edward (ITV Network); *The Paranormal World of Paul McKenna Series II* (ITV Network); *Connections III* (Discovery Channel)

Helen Langridge Associates

75 Kenton Street
London WC1N 1NN
Tel: 0171 833 2955
Fax: 0171 837 2836
Feature development, television, commercials and music videos

Large Door Productions

2 Tunstall Road
London SW9 8BN
Tel: 0171 978 9500
Fax: 0171 978 9578
Founded in 1982 to specialise in documentaries about cinema and popular culture with an international emphasis. Recent productions: *French Cooking in Ten Minutes* (BBC2); *Dream Town* (BBC2). *Riding the Tiger* (Channel 4). Currently developing series on modern medicine, British cinema, broadcasting history and innovations in food

Laurol Productions

116-118 Grafton Road
London NW5 4BA
Tel: 0171 267 9399
Fax: 0171 267 8799
Documentary, current affairs, arts, environmental, drama-docs, and other factual programmes. Recent productions: *The Investigator* (C4); *The Health Farm* (C4); *Playing God* (BBC2); *Look Who's Talking, Babe* (BBC1); *Tinseltown* (C4)

Leda Serene

31 Holberton Gardens
London NW10 6AY
Tel: 0181 969 7094/0181 346 4482
Fax: 0181 964 3044
email: Is@ledaserene.demon.co.uk
Frances-Anne Solomon, Rene Mohandas
Productions: *I is a Long Memoried
Woman* (Arts Council/Yod Video,
Dance/drama); *Reunion* (BBC2
Drama/documentary); *Siren Spirits*
(BBC/BFI, 4x20mins Feature
Shorts); *What Mother Told Me* (C4,
Drama); *Speak Like A Child* (BBC/
BFI/ACE - Feature) In Development:
Cab War - Small-time Gangster
movie; *The Dinner Party* - Feature;
Homeland - South African
Docudrama; *Ego-Romeo* - Devised
Feature; *Exodus* - 5 x Radio Talking
Heads; *Valentines Day* by Kevin
Wong; *Conwoman* by Winsome
Pinnock

Little Bird Co

7 Lower James Street
London W1
Tel: 0171 434 1131
Fax: 0171 434 1803
James Mitchell, Jonathan Cavendish
Feature films: *Nothing Personal;
December Bride; Into the West; All
Our Fault; A Man of No Impor-
tance; My Mother's Courage.* TV:
The Hanging Gale; Divine Magic

Little Dancer

Avonway
Naseby Road
London SE19 3JJ
Tel: 0181 653 9343
Fax: 0181 653 9343
Recent productions in development:
Adios by Sue Townsend; *Wilderness
Years* by Sue Townsend. *Surfers* a
teen series; *Biff on TV*; digital Drama

Living Tape Productions

Ramillies House
1-2 Ramillies Street
London W1V 1DF
Tel: 0171 439 6301
Fax: 0171 437 0731
Producers of documentary and
educational programmes for
television and video distribution.
Recent productions: *Oceans of
Wealth*, major television series

Euan Lloyd Productions

Pinewood Studios
Iver Heath, Bucks SL0 0NH
Tel: 0171 937 9315
Fax: 0171 938 4024
Euan Lloyd
In development: Project No 9

London Films

35 Davies Street
London W1Y 1FN
Tel: 0171 499 7800
Fax: 0171 499 7994
Founded in 1932 by Alexander
Korda. Many co-productions with
the BBC, including *Scarlet
Pimpernel, Lady Chatterley, Resort
to Murder; I, Claudius, Poldark* and
Testament of Youth. Produced *The
Country Girls* for C4. In receipt of a
direct drama commission from a US
network for *Scarlet Pimpernel* and
Kim. Renowned for productions of
classics, most recently *Lady
Chatterley*, directed by Ken Russell

Dani Lukic Films Ltd

Pinewood Studios
Pinewood Road
Iver
Bucks SL0 0NH
Tel: 01753 656253
Fax: 01753 656844
Formed in November 1993 after
producing *Kleptophilia* (NFTS)
starring John Hannah

Lusia Films

7-9 Earlham Street
London WC2 9HL
Tel: 0171 240 2350
Fax: 0171 497 0446
Recent productions: *Murder and the
Feather Boa* (C4)

MRP Ltd (Marcel/Robertson Productions)

9 Gipsy Lane
London SW15
Tel: 0181 392 2725
Fax: 0181 392 1262
email: MRPltd@aol.com
Recent production: *Naked Flame.* In
development: *Hawk II*

MW Entertainments

48 Dean Street
Soho
London W1V 5HL
Tel: 0171 734 7707
Fax: 0171 734 7727
(See Pathé Pictures - Lottery Film
Production Franchises)

Malachite Productions

East Kirkby House
Spilsby
Lincolnshire PE23 4BX
Tel: 01790 763538
Fax: 01790 763409
Cambridge office: 01223 249018
Charles Mapleston, Nancy Thomas,
Nikki Crane
Producers of people-based documen-
tary programmes on music, design,
painting, photography, arts,
anthropology and environmental
issues for broadcast television. The
company also produces dramatised
documentary programmes, is
developing micro-budget fiction
films, and is experimenting with new
technologies to communicate in new
ways. Recent productions: *John
Clare's Journey; Clarke's Penny
Whistle; A Voyage with Nancy
Blackett; Small Silver Screens;
Sequins in my Dreams*

Malone Gill Productions Ltd

9-15 Neal Street
London WC2H 9PU
Tel: 0171 460 4683
Fax: 0171 460 4679
email: ikonic@compuserve.com
Recent productions: *The Face of
Russia* (PBS); *Vermeer* (ITV);
Highlanders (ITV); *Storm Chasers*
(C4/Arts and Entertainment
Network); *Nature Perfected* (C4/
NHK/ABC/Canal Plus/RTE); *The
Feast of Christmas* (C4/SBS);
Nomads (C4/ITEL)

Manhattan Films

217 Brompton Road
London SW3 2EJ
Tel: 0171 581 2408
Fax: 0171 591 0365
Recent productions: directed and
scripted *The Choice* in Switzerland.
In production: *Magic Days* to be
shot in London. In development:
Friendly Enemies in association with
US major. Pre-production: *Hidden
Memories* shooting in Ireland

Mike Mansfield TV

5th Floor
41-42 Berners Street
London W1P 3AA
Tel: 0171 580 2581
Fax: 0171 580 2582
5-7 Carnaby Street
London W1V 1PG
Tel: 0171 494 3061
Fax: 0171 494 3057
Recent productions: *Shirley Bassey
This Is My Life* (BBC1); *Helter
Skelter* 50 x 42 *Rock Shows* (LWT/
ITV); *Cue the Music* (185 x 1 hour,
ITV); *Whale On...!* (110 x 1 hour,
ITV); *Animal Country* (20x 30 min,
LWT); *Take That Concert of Hope
1994* (1 x 60 min, C4)

Jo Manuel Productions Ltd

11 Keslake Road
London NW6 6DG
Tel: 0181 930 0777
Fax: 0181 933 5475
Recent productions: *The Boy From
Mercury* directed by Martin Duffy
with Hugh O'Conor, Rita
Tushingham and Tom Courtenay.
Widow's Peak (Rank, Fineline,
British Screen), directed by John
Irvin with Mia Farrow, Joan

Plowright, Natasha Richardson. In pre-production: *Banjaxed* (to be directed by Declan Lowney and starring Mia Farrow); *The Secret Trials of Effie Grax* (to be directed by Anand Tucker)

Mass Productions Ltd
28 Poland Street
London W1V 3DB
Tel: 0171 734 1994
Fax: 0171 734 1916
Recent productions: *Before the Rain* (An Aim Production, 1993); *The Young Poisoner's Handbook* (1994); *Sweet Angel Mine* (1995)

Maya Vision
43 New Oxford Street
London WC1A 1BH
Tel: 0171 836 1113
Fax: 0171 836 5169
email: maya@mayavisn.demon.co.uk
John Cranmer
Recent productions: *In the footsteps of Alexander the Great* travel/history by Michael Wood, 4 x 1 hour for BBC/PBS; *Glasgow Kiss* docusoap by Dianne Barry, 4 x 25 mins for C4. *Just People* video diary by Ashley Irving, 1X1 hour for C4; *The Reunion* short by Jayne Parker, 1x10mins for Arts Council/BBC2 *Dance for the Camera*; *A Bit of Scarlet* feature documentary by Andrea Weiss, 75 mins for BFI / Channel 4; *3 Steps to Heaven* feature film by Constantine Giannaris, 90 mins for BFI/Channel 4. In production: *Random X* short drama by Clio Barnard, 15 mins for BFI New Directors; *Johnny Panic* short drama by Sandra Lahire 45 mins for Arts Council Lottery fund; *Hitler's Search for the Holy Grail* documentary, 1X1 hour for Channel 4; In development: 4x1 hour travel/history series by Michael Wood for BBC/PBS; 6 x 1 hour history series by John Triffitt for C4

Media Legal Origination
Media Legal, Burbank House
83 Clarendon Road
Sevenoaks
Kent TN13 1ET
Tel: 01732 460592
Production arm of Media Legal developing legal projects for film and TV

The Media Trust Ltd
3-6 Alfred Place
London WC1
Tel: 0171 637 4747
Fax: 0171 637 5757
email: mediatrust@theframe.com
The Media Trust helps other charities and voluntary organisations

to understand and access the media. Also produces the programme Voluntary Sector Television (BBC2)

Medialab
Unit 8 Chelsea Wharf
15 Lots Road
London SW10 0QH
Tel: 0171 351 5814
Fax: 0171 351 7898
Commercials, music videos, documentaries. Producers of MTV Live! concert series

Meditel Productions
172 Foundling Court
Brunswick Centre
London WC1N 1QE
Tel: 0171 833 4959
Fax: 0171 278 2603
Provides medical, science-based and factual documentaries for television. Past productions: AZT - Cause for *Concern*; *The AIDS Catch*; *AIDS and Africa* (C4 Dispatches); *Impotence - One in Ten Men* (C4); *HRT - Pause for Thought* (Thames TV This Week)

Mentorn Barraclough Carey
Mentorn House
140 Wardour Street
London W1V 4LJ
Tel: 0171 287 4545
Fax: 0171 287 3728
Entertainment, drama, entertainment news, documentaries, news and current affairs, children's and features. Recent productions: *Challenge Anneka; Exclusive!: The Ticket; VIP; Passport; Surprise Party; Capital Woman; The Bullion Boys; Scratchy & Co; Today's the Day; Space Precinct; Quisine; Robot Wars; Moving Pictures; The Street; Massive; First Edition; You Decide; The Poisoned Chalice; Africa Express; The People's Parliament; Focus on 5; Pound for Pound; The Plague; Fall of Saigon; Jackie O; The Unforgiving; Messengers from Moscow; Knife to the Heart; The Crimean War; Mad Cows and Englishmen; Cancer Wars; Science of the Impossible; Scare Stories; Tunnel Vision*

Merchant Ivory Productions
46 Lexington Street
London W1P 3LH
Tel: 0171 437 1200/439 4335
Fax: 0171 734 1579
Producer Ismail Merchant and director James Ivory together have made, among other films, for theatrical and television release: *Shakespeare Wallah; Heat and Dust; The Bostonians; A Room With a View; Maurice; Slaves of New*

York; Mr & Mrs Bridge; The Ballad of the Sad Cafe; Howards End; The Remains of the Day; In Custody; Jefferson in Paris; The Feast of July. Surviving Picasso; In development: *The Playmaker; Dan Leno and the Limehouse Golem; The Mystic Masseur; Man Eating Leopard; The Golden Bowl; Cott on Mary*

The Mersey Television Company
Campus Manor
Childwall Abbey Road
Liverpool L16 0JP
Tel: 0151 722 9122
Fax: 0151 722 1969
Independent production company responsible for C4 thrice-weekly drama series, *Brookside*

Mersham Productions
Newhouse, Mersham
Ashford
Kent TN25 6NQ
Tel: 01233 503636
Fax: 01233 502244
Lord Brabourne was a Governor of the BFI 1980-95. Amongst other films, he has produced in conjunction with Richard Goodwin four films based on stories by Agatha Christie, and *A Passage to India* directed by David Lean. He has also co-produced *Little Dorrit* and the TV series *Leontyne*

Metrodome Films
40 Crawford Street
London W1H 2BB
Tel: 0171 723 3494
Fax: 0171 724 6411
Paul Brooks/Alan Martin film production and finance. Recent productions: *Proteus; Killing Time; Darklands* and *Writer's Block*. UK distribution through Metrodome Distribution

Momentum Productions
63 Park Road
Teddington TW11 0AV
Tel: 0181 977 7333
Fax: 0181 977 6999
Specialists in on-screen marketing and promotion of feature films - film trailers, promos and commercials. Producers of corporate films

Morningside Productions Inc
8 Ilchester Place
London W14 8AA
Tel: 0171 602 2382
Fax: 0171 602 1047

Mosaic Films Ltd
2nd Floor
8-12 Broadwick Street
London W1V 1FH
Tel: 0171 437 6514

Fax: 0171 494 0595
email: 75337.1233@compuserve.com
Contact: Colin Luke
The Old Butcher's Shop
St Briavels
Glos. GL15 6TA
Tel: 01594 530708
Fax: 01594 530094
email: 100430.3545@compuserve.com
Contact: Adam Alexander
Recent Productions: *The Russian
Disease* for BBC Correspondent;
*Tunnels and Trees; Inside The
Fairmile Protest* a special for BBC2
and *United Kingdom!* a series for
BBC2

Moving Image Development Agency
109 Mount Pleasant
Liverpool L3 5TF
Tel: 0151 708 9858
Fax: 0151 708 9859

NVC Arts
The Forum
74-80 Camden Street
London NW1 0EG
Tel: 0171 388 3833
Fax: 0171 383 5332
Leading producer and distributor of
opera, dance, documentary, profiles,
rock and classical concerts

New Century Films Ltd
41b Hornsey Lane Gardens
Milton Park
Highgate London N6 1NY
Tel: 0181 348 7298
Fax: 0181 341 6334
In development: *The Wishing Tree*

New Realm Entertainments Ltd
2nd Floor, 25 Margaret Street
London W1
Tel: 0171 436 7800
Fax: 0171 436 0690

NewMedia
12 Oval Road
London NW1 7DH
Tel: 0171 916 9999
Fax: 0171 482 4957
Produced and published more than
20 titles for CD-Rom in Education
with the Department of Education
and international print publishers.
Title areas are: *Chemistry; Maths;
Learning to Read; Geography.* Also
producing discs for museums,
British National Space Agency and
the travel industry

North South Productions
Woburn Buildings
1 Woburn Walk
London WC1H 0JJ
Tel: 0171 388 0351
Fax: 0171 388 2398

Development, travel and other
international themes. Productions
include: BBC series *Only One
Earth*; C4 series *Wild India; Stolen
Childhood; How to Save the Earth;
Sex, Sin and Survival* (C4); *Earth
Tales II* (C4); *Worlds of Faith* (C4
schools); *Wild Islands* (C4). In
production: *Wild Australasia* (ITEL,
ABC, TVNZ)

Nova Productions
11a Winholme
Armthorpe
Doncaster DN3 3AF
Tel: (01302) 833422
Fax: (01302) 833422
Andrew White, Maurice White,
Gareth Atherton
Film and television production
company, specialising in documentary, entertainment, special event
and music promo production.
Producer of programmes released on
sell-through video on its own label
via subsidiary Nova Home Entertainment and on other labels. Game
show format development including
Bet To Win, Alphabeties and *The
Right Order.* Also training,
promotional and multi-camera OB
production for broadcast and non-
broadcast. Recent productions: *The
Doncaster Mansion House, More
Buses of the South Yorkshire PTE:
1974-1986, 50 Years of Sheffield's*
Transport - documentaries; *Beat Box*
- karaoke music quiz game show;
Remembrance Sunday 1997 - multi-
camera OB

Nub Television
Lower Ground Floor
6 Lyndhurst Gdns
London NW3 5NR
Tel: 0171 433 1484
Fax: 0171 435 8144
International documentaries located
in Spain, Nigeria, the West Indies
and UK. Recent production: bi-
monthly magazine series about
living with cancer

OG Films
Pinewood Studios
Pinewood Road
Iver Heath
Bucks SL0 0NH
Tel: 01753 651700
Fax: 01753 656844
Production, packaging, and
distribution company for feature
films and television

Observer Films
3rd Floor,
75 Farringdon Road
London EC1M 3JY
Tel: 0171 713 4343

Fax: 0171 404 1897
Television documentaries. Recent
productions: *Lords A-lobbying* (C4);
Correspondent (BBC) - *Beirut and
Haiti; The Yardies* (Network First);
Spy in the Camp (Dispatches);
*Frontline - Northern Ireland; The
Secrets of Porton Down* (Network
First); *Big Fish in China, profile of
Zhang Yimou; Anopthalmia; The
Battle of Twyford Down*

Orbit Media Ltd
7/11 Kensington High Street
London W8 5NP
Tel: 0171 221 5548
Fax: 0171 727 0515
Currently producing independent
shorts; comedy, music programmes for
television broadcast; documentaries,
community programmes. In production: *Saudi Dissidents* (documentary)

Orchid Productions
**Garden Studios, 11-15 Betterton
Street, Covent Garden**
London WC2H 9BP
Tel: 0171 379 0344
Fax: 0171 379 0801
Jo Kemp
**Sovereign House, Sovereign Street
Pendleton, Manchester M6 3LY**
Tel: 0161 737 2816
Fax: 0161 745 7248
Neil Molyneux

Orlando TV Productions
Up-the-Steps
Little Tew
Chipping Norton
Oxon OX7 4JB
Tel: 01608 683218
Fax: 01608 683364
email: 100564561@compuserve.com
Producers of TV documentaries:
Nova (WGBH-Boston); *Horizon*
(BBC); *QED* (BBC)

Oxford Film and Video Makers
The Stables
North Place, Headington
Oxford OX3 9HY
Tel: 01865 741682
Fax: 01865 742901
Oxford Film and Video Makers
supports a broad range of individual
productions - both broadcast and
non-broadcast - giving a voice to
people normally under-represented
in the media. Promoting experimental film and video art through
'Arteaters'. Supporting productions
with campaigning groups through
the CCD group. Production
company facility now also available

Oxford Film Company
6 Erskine Road
London NW3 3AJ

Tel: 0171 483 3637
Fax: 0171 483 3567
Released film *Restoration* adapted
by Rupert Walters and directed by
Michael Hoffman. In development:
Jackie, original screenplay Frank
Cottrell Boyce to be directed by
Anand Tucker

Oxford Scientific Films
Lower Road, Long Hanborough,
Oxford OX8 8LL
Tel: 01993 881881
Fax: 01993 882808
45-49 Mortimer Street
London W1N 7TD
Tel: 0171 323 0061
Fax: 0171 323 0161
Karen Goldie-Morrison
OSF, now part of Circle Communications, specialises in natural history
production and special effects
filming. Our work spans across the
media spectrum, from television
commercials to documentary films to
multimedia. Our extensive image
resources in the stock footage and
photo libraries add to our production
facilities and skills

PBF Motion Pictures
The Little Pickenhanger
Tuckey Grove
Ripley
Surrey GU23 6JG
Tel: 01483 224118/225179
Fax: 01483 224118

POW Productions
84 Wardour Street
London W1V 3LF
Tel: 0171 734 3938
Fax: 0171 437 0208

Pacesetter Productions Ltd
The Gardner's Lodge
Cloisters Business Centre
8 Battersea Park Road
London SW8 4BH
Tel: 0171 720 4545
Fax: 0171 720 4949

Pagoda Film & Television Corporation Ltd
Twentieth Century House
31-32 Soho Square
London W1V 6AP
Tel: 0171 534 3500
Fax: 0171 534 3501
In development: *Mary Stuart, That
Funny Old Thing, The Corsican Sisters*

Paladin Pictures
500 Chiswick High Road
London W4
Tel: 0181 956 2260
Fax: 0181 956 2262
Quality documentary, drama, music
and arts programming. Recent

productions include: *Plague Wars*
(BBC1 series/WGBH Frontline);
Purple Secret (C4-Secret History);
The Last Flight of Zulu Delta 576
(C4-Cutting Edge); *A Death In
Venice* (BBC2-The Works); *Brothers
& Sisters* (C4 series); *The Shearing
Touch* (ITV-South Bank Show).
Current Productions: *Georgiana -
Duchess of Devonshire* (C4
historical biography); *Blood Family
- a history of the Spencers* (C4
series); *Travels With My Tutu*, with
Deborah Bull (BBC2 series on
ballet); *The Secret Life of Daphne
du Maurier* (BBC); *The Assassin's
Wife* (2hr Movie of the Week)

Pandora Productions
2nd Floor
Kent House
Market Place
14-16 Grt Titchfield Street
London W1N 8AR
Tel: 0171 323 5151
Fax: 0171 462 4428
Lynda Myles
(See Pathé Pictures - Lottery Film
Production Franchises)

Panoptic Productions
35 Heddon Street
London W1R 7LL
Tel: 0171 287 3931
Fax: 0171 287 1518
Recent productions: *Out of Order,
The King and US, The Pope's
Divisions, The Ballad of Nicky Mouse*

paradogs Ltd
309 Panther House
38 Mount Pleasant
London WC1X OAP
Tel: 0171 833 1009
Steven Eastwood, Sophie Djian,
Duncan Western
Experimental, narrative, documen-
tary and pop promo

Parallax Pictures
7 Denmark Street
London WC2H 8LS
Tel: 0171 836 1478
Fax: 0171 497 8062
Recent productions: Ken Loach's
*Ladybird, Ladybird, Land and
Freedom; Carla's Song*; Les Blair's
*Jump the Gun; Bad Behaviour,
Bliss;* Philip Davis's *ID*; Christopher
Monger's *The Englishman Who
Went Up the Hill, But Came Down a
Mountain*
(See The Film Consortium - Lottery
Film Production Franchises)

Paramount-British Pictures Ltd
Paramount House
162-170 Wardour Street
London W1V

Tel: 0171 287 6767
Fax: 0171 734 0387
Sliding Doors, Event Horizon, Titanic

Parent Productions
7-9 Princelet Street
Spitafields
London E1 6QH
Tel: 0171 247 0511
Fax: 0171 247 0505
email: nephew@Parent.co.uk

Partridge Films
The Television Centre
Bath Road
Bristol BS4 3HG
Tel: 0117 972 3777
Fax: 0117 971 9340
Michael Rosenberg, Carol
O'Callaghan (Library)
Makers of wildlife documentaries
and videos for television and
educational distribution. Extensive
natural history stock shot library.
Recent productions: 6 wildlife docs
(BBC1); 2 x 30 min wildlife
programmes (BBC1); 6 wildlife docs
(C4); 3 wildlife specials (ABC
USA); 26 x 30 min syndicated
programmes (USA); 8 x 1 hour
wildlife programmes (PBS USA)

Pathé Pictures
Kent House
Market Place
London W1N 8AR
Tel: 0171 323 5151
Fax: 0171 636 7594
Head of Production: Andrea
Calderwood
Consists of: Thin Man Films and
Imagine Films, Allied Filmmakers
and Allied Films Ltd, NFH, Pandora
Productions, Sarah Radclyffe
Productions, Fragile Films and MW
Entertainment
(See Pathé Pictures - Lottery Film
Production Franchises)

Pelicula Films
7 Queen Margaret Road
Glasgow G20 6DP
Tel: 0141 945 3333
Fax: 0141 946 8345
Recent productions: feature film *As
An Eilean (From the Island)* in
Gaelic and English (C4/Grampian
TV/Gaelic Television Committee);
Back to Africa, documentary for
C4's Travellers' Tales series; *The
Land*, music documentary with
singer/song-writer Dougie McLean
(BBC TV); *The Transatlantic
Sessions* (BBC)

Pennies from Heaven
83 Eastbourne Mews
London W2 6LQ

Tel: 0171 402 0051/0181 576 1197
Kenith Trodd is a prolific producer of films for the BBC and others. Recent productions: *She's Been Away; Old Flames; They Never Slept; For The Greater Good; Common Pursuit; Femme Fatale; Bambino Mio*, all for the BBC; *Dreamchild; A Month in the Country*, for PFH and *Circle of Friends* for Savoy Pictures (1995); *My House in Umbria*. Much of his work has been with the late Dennis Potter, including *Karaoke* and *Cold Lazarus*

Penumbra Productions Ltd
80 Brondesbury Road
London NW6 6RX
Tel: 0171 328 4550
Fax: 0171 328 3844
Cinema and television productions, specialising in contemporary issues and in the relationship between 'North' and 'South'. Recent productions: *China Rocks; Repomen; Bombay and Jazz; Divided By Rape; Stories My Country Told Me*. In development: a cinema feature, Sold; several documentary series relating to history, the environment and the future

Persistent Vision Productions
299 Ivydale Road
London SE15 3DZ
Tel: 0171 639 5596
Fax: 0171 639 5596
email: calemon@dircon.co.uk
Carol Lemon, John Stewart
In distribution: *Crash; The Gaol;* In production, *The Break-In*. In development, a slate of five thriller, horror and sci-fi feature films

Photoplay Productions
21 Princess Road
London NW1 8JR
Tel: 0171 722 2500
Fax: 0171 722 6662
Kevin Brownlow, David Gill, Patrick Stanbury
Producers of documentaries and television versions of silent feature films. Specialist research and library services for early cinema. Production credits: *Hollywood; Unknown Chaplin; Keaton - A Hard Act to Follow; Harold Lloyd - The Third Genius; The Four Horsemen of the Apocalypse; Wings; D.W. Griffith - Father of Film; The Iron Horse; Cinema Europe - The Other Hollywood* (BBC2)

Picture Palace Films
19 Edis Street
London NW1 8LE

Tel: 0171 586 8763
Fax: 0171 586 9048
email: 100444.2737@compuserve.com
Website: www.picturepalace.com
Malcolm Craddock, Alex Usborne
Specialise in feature film and TV drama and documentary
In Production: *Kiszko* starring Olympia Dukakis (1 x 2hr film, co-production with Celtic Films); Irvine Welsh's *The Acid House* (feature). Recent Productions: *Sharpe* (5 series, 14 x 2 hours, Carlton), set in the Peninsular War *Little Napoleons* (4 x 1 hour, C4), *Tales From A Hard City* (90 mins, C4/Yorkshire TV/La Sept/ARTE). In development: *The Spire* (feature) by Roger Spottiswoode, *Rebellion* (TV drama BBC Northern Ireland, Irish Film Board, Irish Screen), *Kolymsky Heights* (feature) with Jon Amiel and *Sharpe's Tiger* (feature) with Celtic Films

Planet 24 Productions
The Planet Building, Thames Quay
195 Marsh Wall
London E14 9SG
Tel: 0171 345 2424/512 5000
Fax: 0171 345 9400
Recent productions: *The Big Breakfast* (C4); *Hotel Babylon* (ITV); *Gaytime TV* (BBC2); *Nothing But the Truth* (C4); *The Word* (C4); *Delicious* (ITV);*The Weekend Show* (BBC)

Polar Productions Ltd
3 Greek Street
London W1V 5LA
Tel: 0171 439 3011
Fax: 0171 439 3022

PolyGram Filmed Entertainment
8 St James's Square
London SW1Y 4JU
Tel: 0171 747 4000
Fax: 0171 747 4499

Popular Films Limited
77b Belsize Park Gardens
London NW3 4JP
Tel: 0171 419 0722
Fax: 0171 722 8154
Contact: Steven Kaye

Portman Productions
167 Wardour Street
London W1V 3TA
Tel: 0171 468 3400
Fax: 0171 468 3499
Major producer in primetime drama and feature films worldwide. Recent productions include: *Wrestling with Alligators* (feature); *Coming Home* (2x2); *Spanish Fly* (feature); *Rebecca* (2x2); *September* (2x2); *An Awfully*

Big Adventure (feature); *Famous Five*, series I & II; *Hostage* (feature); *Fall From Grace* (4 x1); *Friday on my Mind* (3x1); *Blackwater Trail* (feature); *Little White Lies* (feature); *Seventh Floor* (feature); *Crime Broker* (feature); *A Woman of Substance* (4x1); *Via Satellite* (feature)

Poseidon Productions
1st Floor
Hammer House
113 Wardour Street
London W1V 3TD
Tel: 0171 734 4441
Fax: 0171 437 0638
Recent productions: *Great Russian Writers* (C4), a mini-series; a documentary on dyslexia (C4); *Russian Composers* (BBC Wales); *Global Bears Rescue* (7 x 25 min, BBC Wales), animated series; *The Steal*, 35 mm feature film; *Rat-A-Tat-Tat* (C4 Schools), animated/live action series; animated series on the *Odyssey* (10 x 15 min), co-production with BBC. In production: *Technology* programme for Channel 4; *Night Witches*, feature film with BBC and *Screen Partners; Rat-A-Tat-Tat* (C4 Schools). In preparation: *Little Jerusalem*, a war drama set in Greece, a co-production with BBC/Fox/Poseidon

Praxis Films
PO Box 290
Market Rasen
Lincs LN3 6BB
Tel: 01472 398547
Fax: 01472 398683
email: 100625.232@compuserve.com
Film and video production of documentaries, current affairs and educational programmes. Recent productions: series for Yorkshire Television; films for C4's *Secret History, Dispatches, Cutting Edge, 3D* and *Secret Lives*. Extensive archive of sea, fishing, rural and regional material

Primetime Television
Seymour Mews House
Seymour Mews
Wigmore Street
London W1H 9PE
Tel: 0171 935 9000
Fax: 0171 935 1992
Richard Price, Ian Gordon, Victoria Hull
Independent television production/packaging company associated with distributors Primetime Television Associates (formerly RPTA). Specialise in international co-productions. Recent projects: *A Christmas Carol* (BBC/A&E);

Irving Berlin Story (Iambic/WNET NY); *Porgy & Bess* (BBC Homevale & Smith); *Let The Blood Run Free 2* (Media Arts/C4/RTL+/Canal+); *Beyond Belief!* (Nickelodeon/Avro/NOB); *Jose Carreras: A Life Story* (LWT/Iambic); *Re: Joyce* (BBC); *Ustinov on the Orient Express* (A&E/CBC/NOB/JMP); *Ethan Frome* (American Playhouse)

Prolific Films
90 Salusbury Road
London NW6 6PA
Tel: 0171 372 5495
Fax: 0171 372 5495
Julian Richards, Yvonne Michael
Feature film development and production. Produced three award winning short films for television *Bad Company* and *Pirates* for HTV and *Queen Sacrifice* for BBC Screenplay Firsts; (winner of the Thames Television Award for Best Film at the British Short Film Festival). Recent projects include *Calling All Monsters*, a screenplay developed for Amblin Entertainment and the lottery funded award winning British horror film *Darklands* for Metrodome Films. Current projects in development include: *Ravers* by James Handel, *The Wishbringer* by Chris Westwood, *The Monkey Farm* by Dave Mitchell and Warlord by Julian Richards

Prominent Features
68a Delancey Street
London NW1 7RY
Tel: 0171 284 0242
Fax: 0171 284 1004
Company formed by Steve Abbott, John Cleese, Terry Gilliam, Eric Idle, Anne James, Terry Jones and Michael Palin to produce in-house features. Productions include: *The Adventures of Baron Munchausen; Erik The Viking; A Fish Called Wanda; American Friends; Splitting Heirs, Brassed Off*

Prominent Television
68a Delancey Street
London NW1 7RY
Tel: 0171 284 0242
Fax: 0171 284 1004
Company formed by Steve Abbott, John Cleese, Terry Gilliam, Eric Idle, Anne James, Terry Jones and Michael Palin to produce in-house television programmes. Productions include: eight-part travel documentary series *Pole to Pole* (1992); four-part series on and around the Isle of Wight Palin's Column (1994): 10 part travel documentary series *Full Circle* with Michael Palin (1997) a journey around the Pacific rim

RM Arts
46 Great Marlborough Street
London W1V 1DB
Tel: 0171 439 2637
Fax: 0171 439 2316
RM Arts produces a broad range of music and arts programming and co-produces on an international basis with major broadcasters including BBC, LWT, C4, ARD and ZDF in Germany, NOS-TV in Holland, Danmarks Radio, SVT in Sweden and Thirteen/WNET in America

RSPB Film and Video Unit
The Lodge, Sandy
Beds SG19 2DL
Tel: 01767 680551
Fax: 01767 692365
Producers of *Osprey, Kingfisher, Where Eagles Fly, Barn Owl, The Year of the Stork* and *Flying for Gold*. Recent productions: The *Flamingo Triangle; Skydancer*. The unit also acts as an independent producer of environmental films and videos

Sarah Radclyffe Productions
5th Floor
83-84 Berwick Street
London W1V 3PJ
Tel: 0171 437 3128
Fax: 0171 437 3129
Sarah Radclyffe, Dixie Linder, Alison Jackson
Sarah Radclyffe previously founded and was co-owner of Working Title Films and was responsible for, amongst others, *My Beautiful Laundrette, Wish You Were Here,* and *A World Apart*. Sarah Radclyffe Productions was formed in 1993 and productions to date are: *Second Best,* Dir Chris Menges; *Sirens,* Dir. John Duigan; *Cousin Bette,* Dir. Des McAnuff; *Bent,* Dir. Sean Mathias (See Pathé Pictures - Lottery Film Production Franchises)

Ragdoll Productions
11 Chapel Street
Stratford Upon Avon
Warwickshire CV37 6EP
Tel: 01789 262772
Fax: 01789 262773
Specialist children's television producer of live action and animation, currently producing long-running series *Tots TV, Rosie and Jim* (for ITV); *Teletubbies; Brum* and *Open A Door* (for BBC)

Raw Charm
Ty Cefn
Rectory Road
Cardiff CF1 1QL
Tel: 01222 641511

Fax: 01222 668220
Pamela Hunt, Kate Jones Davies
Drama, documentaries, music, entertainment, reality. Recent productions: *Cop Swap* (HTV Wales); *Going Solo* (BBC2); *Kicking the Weed* (HTV West); *Murzikma* (C4); *Finishing School* (C4)

Rodney Read
45 Richmond Road
Twickenham
Middx TW1 3AW
Tel: 0181 891 2875
Fax: 0181 744 9603
email: Rodney_Read@Compuserve.com
Film and video production offering experience in factual and entertainment programming. Also provides a full range of back-up facilities for the feature and television industries, including 'making of' documentaries, promotional programme inserts, on air graphics and title sequences, sales promos, trailers and commercials. Active in production for UK cable and satellite

Recorded Picture Co
24 Hanway Street
London W1P 9DD
Tel: 0171 636 2251
Fax: 0171 636 2261
Managing Director: Jeremy Thomas
Films produced include: Nagisa Oshima's *Merry Christmas, Mr.Lawrence,* Stephen Frears' *The Hit,* Nicolas Roeg's *Insignificance, The Last Emperor* and *The Sheltering Sky* Dir Bernardo Bertolucci, and *The Naked Lunch, Crash* Dir David Cronenberg. Recent productions: *Stealing Beauty,* Dir Bernardo Bertolucci

Red Flannel Films
3 Cardiff Road
Taffs Well
Cardiff CF4 7RA
Tel: 01222 813086
Fax: 01222 813086
Clare Richardson, Caroline Stone, Carol White
A women's collective working in broadcast production. Recent productions for Channel 4: *Women Have No Country.* For HTV: *Valleys Live.* For BBC: *Sweet Dreams; The Time of Our Lives* (six part series); *All in the Game; Crackers Over Christmas*

Red Mullet Ltd
30 Percy Street
London W1
Mike Figgis

Red Rooster Film & Television Entertainment
29 Floral Street

London WC2E 9DP
Tel: 0171 379 7727
Fax: 0171 379 5756
Grainne Marmion, Sarah Williams,
Tim Vaughan, Julia Ouston
The Chrysalis Group has bought the
remaining 50 per cent holding in the
company, and continues to maintain
a £1m fund to develop television
programmes and feature films.
Recent productions: *Wilderness*
(ITV); *Beyond Fear* (Channel 5)

Redwave Films (UK) Ltd
26 Goodge Street
London W1P 1FG
Tel: 0171 436 2225
Fax: 0171 436 2772
Uberto Pasolini
Recent productions: *The Full Monty*

Redwing Film Company
1 Wedgwood Mews
12/13 Greek Street
London W1V 5LW
Tel: 0171 734 6642
Fax: 0171 734 9850
Film, television and commercial
productions. Recent productions: short
film for C4 *In the Time of Angels*. In
development: numerous commercials, film and television drama

Reeltime Pictures
70-72 Union Street
London SE1 0NW
Tel: 0171 620 3102
Fax: 0171 620 3104
Formed in 1984, Reeltime specialises in the production of drama and
documentaries connected with cult
film and television for theatrical and
non-theatrical release. The company
also purchases and distributes sell-
through video under its own label
and aims to promote independent
genre film and television production.
Titles include *Myth Makers, Doctor
Who's Return to Devils End* (doc);
Wartime, Downtime, PROBE and
The Stranger (drama)

Regent Productions
The Mews, 6 Putney Common
Putney
London SW15 1HL
Tel: 0181 789 5350
Fax: 0181 789 5332
Current productions: three new
series of the C4 quiz series *Fifteen-
to-One* (188 programmes). In
development: quiz shows; a seven-part
drama series; a current affairs project

Renegade Films
3rd Floor, Bolsover House
5/6 Clipstone Street
London W1P 8LD
Tel: 0171 637 0957

Fax: 0171 637 0959
email: renptism@dircom.co.uk
Robert Buckler, Amanda Mackenzie
Stuart, Ildiko Kemeny
Productions include: *Brothers in
Trouble, Pressure, The Last Place
on Earth, Facts of Life, Midnight
Expresso, The Star, The Sin Eater*
and (as Prisma Communications)
The Financial Times Business
Toolkit. Development projects
include: *The Go Kart, Room To
Rent, Hotel Sordide, The Thought
Gang*

Replay
36 Ritherdon Road
London SW17 8QF
Tel: 0181 672 0606
Production of broadcast and
corporate documentaries and drama

Richmond Light Horse Productions Ltd
3 Esmond Court
Thackeray Street
London W8
Tel: 0171 937 9315
Fax: 0171 938 4024

Ritefilms
20 Bouverie Road West
Folkestone
Kent CT20 2SZ
Tel: 01303 252335
Current affairs television news
gathering. Film production,
documentary and corporate

Riverfront Pictures
Dock Cottages, Peartree Lane
Glamis Road
Wapping
London E1 9SR
Tel: 0171 481 2939
Fax: 0171 480 5520
Specialise in music, arts and drama-
documentaries. Independent
productions for television. Recent
productions for C4 *Cutting Edge*
and BBC Arts

Roadshow Productions
c/o 6 Basil Mansions
Basil Street
London SW3 1AP
Tel: 0171 584 0542
Fax: 0171 584 1549

Roberts & Wykeham Films
7 Barb Mews
London W6 7PA
Tel: 0171 602 4897
Fax: 0171 602 3016
email: R&W@theframe.com
Sadie Wykeham, Joel Wykeham,
Gwynne Roberts
Recent productions: *Mandela's
Nuclear Nightmare* (C4 *Dispatches*);

The Road Back to Hell (BBC
Everyman); *Kurdistan: The Dream
Betrayed; Pocket Neutron; Trail of
Red Mercury* (C4 *Dispatches*)

Rogue Films
177 Wardour Street
London W1V 3FB
Tel: 0171 434 2222
Fax: 0171 494 7808
Commercials and music video
production - UK, Europe, USA

Roldvale Ltd
Birch Hall, Coppice Row
Theydon Bois
Essex CM16 7DR
Tel: 01992 815151
Fax: 01992 819514

Rose Bay Film Productions
1 Albion Court
Albion Place
London W6 0QT
Tel: 0181 600 5200
Fax: 0181 600 5222
TV and film drama; comedy;
features; entertainment

Samuelson Productions
23 West Smithfield
London EC1A 9HY
Tel: 0171 236 5532
Fax: 0171 236 5504
Arlington Road Dir Mark Pellington,
starring Tim Robbins, Jeff Bridges,
Joan Cusack and Hope Davis; *Wilde*
Dir Brian Gilbert, starring Stephen
Fry, Jude Law, Vanessa Redgrave
and Jennifer Ehle; *The Commis-
sioner* starring John Hurt and Armin
Mueller-Stahl; *Dog's Best Friend*
starring Richard Milligan and
Shirley Jones; Previously: *Tom and
Viv; Playmaker*; and documentaries
*Man, God and Africa; Vicars; The
Babe Business, Ultimate Frisbee*

Sands Films
119 Rotherhithe Street
London SE16 4NF
Tel: 0171 231 2209
Fax: 0171 231 2119
Recent productions: *The Butterfly
Effect; As You Like It. Amahl and
the Night Visitor*s. Shot May 97: *The
Nutcracker Story. The Kiss; The
Swan Princess*

Stefan Sargent Productions
5 Upper James Street
London W1R 3ED
Tel: 0171 287 2297
Fax: 0171 287 0296

Scala Productions
39-43 Brewer Street
London W1R 3FD
Tel: 0171 734 7060

Fax: 0171 437 3248
Nik Powell, Stephen Woolley, Elizabeth Karlsen, Amanda Posey Recent productions: Shane Meadows' *Twentyfourseven* starring Bob Hoskins. Mark Herman's *Little Voice* starring Michael Caine, Brenda Blethyn, Ewan McGregor, Jane Horrocks, Stephan Elliott's *Welcome To Woop Woop* starring Paul McGann and Susan Lynch; Michael Radford's *B.Monkey* starring Asia Argento, Jared Harris and Rupert Everett; David Caffrey's *Divorcing Jack* starring David Thewis and Rachel Griffiths; Chris Menges' *The Lost Son* starring Daniel Auteuil, Nastassja Kinski, Katrin Cartlidge, Ciaran Hinds; Set for 1998: *Jonathan Wild* to be directed by Richard Loncraine, *The Dwarves of Death* to be directed by Jean Stewart, *The Last September* to be directed by Deborah Warner; *Fanny and Elvis* to be directed by Kay Mellor

Scimitar Films
6-8 Sackville Street
London W1X 1DD
Tel: 0171 603 7272
Fax: 0171 602 9217
Michael Winner has produced and directed many films, including *Death Wish 3, Appointment with Death, A Chorus of Disapproval, Bullseye!* and *Dirty Weekend*

Scottish Television Enterprises
Cowcaddens
Glasgow G2 3PR
Tel: 0141 300 3000
Fax: 0141 300 3030
Alistair Moffat, Chief Executive Producers of: *Taggart*, detective series; *Hurricanes*, animation; *McCallum*, drama series; *Fun House*, children's series/game show; *Get Wet*, children's series; *Granne Bannanie*, children's animation

Screen Ventures
49 Goodge Street
London W1P 1FB
Tel: 0171 580 7448
Fax: 0171 631 1265
email: screenventures@easynet.co.uk
Christopher Mould, Jack Bond, Caroline Furness
Production and television sales company. Producing music, drama and documentaries. Recent projects: *Dani Dares* a five part documentary series for C4, *Mojo Working*, a 13-part music series for C4, *Genet*, a South Bank Show for LWT; *Vanessa Redgrave*, a South Bank Show for LWT. In development: *Falling*, a 90

minute feature backed by the European Script Fund

September Films Ltd
Silver House
35 Beak Street
London W1R 3LD
Tel: 0171 494 1884
Fax: 0171 439 1194
David Green, Elaine Day, Sara Langton
TV Includes: *Hollywood Lovers, Hollywood Sex, Hollywood Kids, Hollywood Pets, Hollywood Women* Feature Films: *House of America*

Seventh Art Productions
Brighton Media Centre
11 Jew Street
Brighton BN1 1UT
Tel: 01273 777678
Fax: 01273 323777
email: awilkie@fastnet.co.uk
Factual programming both for the national networks and international broadcasters. Established ten years ago by film-maker Phil Grabsky, with experience in all areas from finance to distribution. Currently in production on *I Caesar* 6x50' for BBC2/ A&E; *True..but Strange* series IV for Meridian. Recent titles include *24 Hours* for Meridian; *Garoghan's Guide*, inserts for BBC South's *Out & About*; Ancient Warriors 21 x 25 for the Discovery Network

Shooting Star Film Productions
Pinewood Studios, Iver Heath
Bucks SL0 0NH
Tel: 01753 651700
John Hough
Features, documentaries, television drama

Siren Film and Video Ltd
5 Charlotte Square
Newcastle-upon-Tyne NE1 4XF
Tel: 0191 232 7900
Fax: 0191 232 7900
email: sirenfilms@aol.com
Film and television production company specialising in work for and about children

Siriol Productions
Phoenix Buildings
3 Mount Stuart Square
Butetown
Cardiff CF1 6RW
Tel: 01222 488400
Fax: 01222 485962
email: siriol@baynet.co.uk
Formerly Siriol Animation.
Producers of high quality animation for television and the cinema.
Makers of: *SuperTed; The Princess and the Goblin; Under Milk Wood;*

Santa and the Tooth Fairies; Santa's First Christmas; Tales of the Tooth Fairies, The Hurricanes; Billy the Cat; The Blobs; Rowland the Reindeer

Skreba
5a Noel Street
London W1V 3RB
Tel: 0171 437 6492
Fax: 0171 437 0644
Ann Skinner, Simon Relph
Together produced *Return of the Soldier and Secret Places*, directed by Zelda Barron. Other projects include: *Bad Hats; A Profile of Arthur J Mason; Honour, Profit and Pleasure; The Gourmet*. Relph produced *Blue Juice, Comrades, Camilla*, and *The Slab Boys*, and co-produced Louis Malle's *Damage*. Skinner executive produced on *Heavenly Pursuits* and produced *The Kitchen Toto, A Very British Coup, One Man's War, God on the Rocks, A Pin for the Butterfly* and two television series of *Chandler & Co* (See The Film Consortium - Lottery Film Production Franchises)

Skyline Films
PO Box 821U
London W41 WH
Tel: 0836 275584
Fax: 0181 354 2219
Steve Clark-Hall, Mairi Bett
Recent productions: T*he Winter Guest* (with Ed Pressmann), *Love and Death on Long Island, Small Faces, Margaret's Museum, Still Crazy (for Margot Tandy)*

Soho Communications
8 Percy Street
London W1P 9FB
Tel: 0171 637 5825
Fax: 0171 436 9740
Recent productions: *Where Have All the Flowers Gone?*, 1 hour TV documentary; *Forbidden Planet*, 1 x 30 min TV documentary; Channel One for London, Sky Televison

Sony Pictures Europe UK Ltd
Sony Pictures Europe House
25 Golden Square
London W1R 6LU
Tel: 0171 533 1111
Fax: 0171 533 1105
Recent productions: *First Knight, Les Miserables, Spice World*. In production: *Virtual Sexuality*

Specific Films
25 Rathbone Street
London W1P 1AG
Tel: 0171 580 7476
Fax: 0171 494 2676
Michael Hamlyn, Christian Routh

Formed in 1991 with PolyGram to produce comedy feature films. Recent productions: *the Last Seduction 2*, Dir Terry Marcel; *PAWS* (exec. prod) Dir Carl Zwicky; *The Adventures of Priscilla, Queen of the Desert*, Dir Stephan Elliott; *Mr Reliable* Dir Nadia Tass. Developing a number of feature film projects

Speedy Films
8 Royalty Mews, Dean Street
London W1V 5AW
Tel: 0171 494 4043
Fax: 0171 434 0830
Animated shorts, commercials and titles. Recent productions: *Voles*, a TV pilot for Nickelodeon, 1994; *Abductees*, 11 minute short for C4, 1995

Spice Factory
81 The Promenade
Peacehaven
Brighton
East Sussex BN10 8LS
Tel: 01273 585275
Fax: 01273 585304
email: sfactory@fastnet.co.uk
Films, Games & Television Production Company. Films in Pre-Production: *New Blood* US$4M, *Pilgrim* US$6M. Films in development: *Black Tuesday* US$30M, Director attached Bill Duke; *Fry* US$6M Director attached Alberto Sciamma; *Crush Hour* US$15M Director attached Alberto Sciamma; *Somma* US$8M, *Breaking the Code* US$5M; *Il Principe* US$30M; *the Four Horse Men* US$30M; *Endgame* US$30M; *In Too Deep* US$5M; *Hotel Elephant*; *Kremlin Contract* US$15M, *King of Kansas* US$20M; *Rick 6* US$6M; Television Projects in development include: *James Herbert's Shrine* (with ITEL); *Deckies* (with ITEL); *Taken For Ride* (Meridian Entertainment Show); *Food For Love* (Meridian Entertainment Show)

Splice Partnership
Unit 21
44 Earlham Street
Covent Garden
London WC2H 9LA
Tel: 0171 240 2570
Fax: 0171 379 5210
Frazer Diamond/Mike Key Independent film, television and multimedia production

Stagescreen Productions
12 Upper St Martin's Lane
London WC2H 9DL
Tel: 0171 497 2510
Fax: 0171 497 2208
Film, television and theatre company

whose work includes: *A Handful of Dust; Death of a Son* (BBC TV); *Where Angels Fear to Tread; Foreign Affairs* (TNT)

Steel Bank Film Co-op
Brown Street
Sheffield S1 2BS
Tel: 0114 272 1235
Fax: 0114 279 5225
Jessica York, Simon Reynell, Noemie Mendelle, Chrissie Stansfield
Film and television production, including social documentaries, arts programmes, and drama. Programmes include: *Clocks of the Midnight Hours; Custom-Eyes; Crimestrike; For Your Own Good; Spinster; Take a Deep Breath; Temp'est; Turkish Delight; The Gift*

Stephens Kerr
8-12 Camden High Street
London NW1 0JH
Tel: 0171 916 2124
Fax: 0171 916 2125
Eleanor Stephens, Jean Kerr
In production: Food File 4 (C4). Recent programming for C4: *The Love Weekend; Food File 3; Nights; Men Talk; Sex Talk; Love Talk; Talking About Sex*

Robert Stigwood Organisation
Barton Manor
East Cowes
Isle of Wight
PO32 6LB
Tel: 01983 280 676
Fax: 01983 293 923
Robert Stigwood, David Land, David Herring
Theatre and film producer Stigwood involved in *Evita*

Swanlind Communication
The Wharf
Bridge Street
Birmingham B1 2JR
Tel: 0121 616 1701
Fax: 0121 616 1520
Swanlind Communication has been concentrating on developing its employee communication service even further. Offerings now includes: communication consultancy; strategic communication design and implementation; broadcast television production; conference design and presentation; print and publishing; multimedia programming design and production; professional television systems design; installation and maintenance

TKO Communications
PO Box 130, Hove
East Sussex BN3 6QU

Tel: 01273 550088
Fax: 01273 540969
A division of the Kruger Organisation, making music programmes for television, satellite and video release worldwide as well as co-producing various series and full length feature films. Recent co-production with Central TV music division: *21st Anniversary Tour of Glen Campbell*, The Hollywood Series, 39 x 30 min specials about people and events that shaped the historical development of Hollywood and the world of film plus *Country Superstars* such as a 13 episode series called *Country Music Comes to Europe* and featuring Kris Kristofferson, Nitty Gritty Dirt Band, Hoyt Axton, Johnny Cash, Daniel O'Donnell and others. *On the Road Again* starring Willie Nelson and King of the Country Hits with Conway Twitty. Latest productions Jerry Lee Lewis - A live concert presentation

TNTV
6e The Courtyard
44 Gloucester Avenue
London NW1 8JD
Tel: 0171 483 3526
Fax: 0171 483 3521
Directors: Ben Noakes, Mick Taylor
Specialists in human interest and lifestyle films for terrestrial and cable networks and in the production of electronic press kits and corporate films. Recent credits include: *Lloyd Grossman's World Wide Culinary Tours* and a series of cookery strands with Susan Brookes (both for Granada), an eight-part health and fitness series for BBC Daytime and four series of fashion films for BBC1

TV Cartoons
39 Grafton Way
London W1P 5LA
Tel: 0171 388 2222
Fax: 0171 383 4192
John Coates, Norman Kauffman
Productions include: *The Snowman*, Academy Award nominated film and the feature film *When the Wind Blows*, both adaptations from books by Raymond Briggs; *Granpa*, a half hour television special for C4 and TVS; half hour special of *Raymond Briggs' Father Christmas* (C4); *The World of Peter Rabbit & Friends* (9 x 30 min), based on the books by Beatrix Potter, *The Wind in the Willows*, a TVC production for Carlton UK Television and Willows in Winter, Famous Fred Academy Award Nominated 1998 (C4&S4C). In production: *The Bear* (C4)

TV Choice

22 Charing Cross Road
London WC2H OHR
Tel: 0171 379 0873
Fax: 0171 379 0263
Chris Barnard, Norman Thomas
Producer and distributor of dramas
and documentaries about business,
technology and finance. TV Choice
videos and learning packs are used
in education and training in the UK
and overseas. Co-producers of
feature film *Conspiracy* with new
features in development

TVF

375 City Road
London EC1V 1NA
Tel: 0171 837 3000
Fax: 0171 833 2185
David Pounds
Production company specialising in
factual output. Titles include
*Dispatches, Without Walls, Equinox,
Horizon, Incredible Evidence,
Fawlty Premises, High Interest,
Face Value, Expletives Deleted, The
Computer Triangle, The Money
Programme, Exhausted Nature, Sex
Games On the Line, For Queen and
Country*

Talent Television

2nd Floor Regent House
235 Regent Street
London W1R 7AG
Tel: 0171 434 1677
Fax: 0171 434 1577
John Kaye Cooper, Managing Director
Recent credits: Bill Bailey's *Cosmic
Jam* - PolyGram, Brian Conley's
Alive & Dangerous - ITV, *Buddy -
The Video* - VCI, *The Making of
Beauty & the Beast* - Buena Vista/
Carlton. Forthcoming Productions:
Miss UK - Sky TV, *Jim Davidson -
Video* - VVL, *Julian Clary Special* -
ITV, *BAFTA Awards* - ITV, *The Fat
Boys* (Hale and Pace) - BBC
Comedy Special, Hale & Pace series
co-production with BBC

Talisman Films Limited

5 Addison Place
London W11 4RJ
Tel: 0171 603 7474
Fax: 0171 602 7422
Richard Jackson
Alan Shallcross, Neil Dunn, Caroline
Oulton
Production of the whole range of
drama for television together with
theatric features. Particularly
committed to co-producing with
continental European and North
American partners. Recent produc-
tions: *Remember Me?* (Feature, C4)
starring Robert Lindsay, Rik Mayall,

Imelda Staunton; *Rob Roy* (Feature
for United Artists) starring Liam
Neeson, Jessica Lange, Tim Roth;
Just William (series I + II, BBC)
starring Oliver Rokison, Lindsay
Duncan, Miriam Margolyes; *The
Rector's Wife* (4 x 60 mins, C4),
drama serial starring Lindsay
Duncan. In production: *The Secret
Adventures of Jules Verne* (22 part tv
series); *The Blue Castle* (feature film)

TalkBack Productions

36 Percy Street
London W1P 0LN
Tel: 0171 323 9777
Fax: 0171 637 5105
Productions include: *Smith and Jones*
(+ 5 previous series, BBC1); *They
Think It's All Over* (+3 previous series,
BBC1); *Never Mind The Buzzcocks* (+
1 previous series, BBC2); *The Lying
Game* (BBC1); *In Search of Happi-
ness* (BBC1); *Brass Eye* (Channel 4);
*Knowing Me Knowing You... with
Alan Partridge* (BBC2); *The Day
Today* (BBC2); *Murder Most Horrid*
(+ 2 previous series, BBC2)

Tall Man Films

81 Berwick Street
London W1V 3PF
Tel: 0171 439 8113
Fax: 0171 494 2006
The production of feature films for
the world market. In post-produc-
tion: *Room 36*

David Taylor Associates

1 Lostock Avenue
Poynton
Nr Stockport
Cheshire SK12 1DR
Tel: 01625 850887
Fax: 01625 850887
Corporate; commercials; broadcast

Richard Taylor Cartoon Films

River View, Waterlood Drive
Clun, Craven Arms
Shropshire SY7 8JD
Tel: 01588 640 073
Fax: 01588 640 074
Production of all forms of drawn
animation

Team Video Productions

Canalot
222 Kensal Road
London W10 5BN
Tel: 0181 960 5536
Recent productions: *Why Civil
Liberties?; Count to Five* (Team);
Representing Your Members
(GPMU); *Energy Alternative*
(Oxfam); *Bangladesh* (Christian
Aid); *Why Poetry?* (Team)

Television History Workshop (THW)

27 Old Gloucester Street
Queen Square
London WC1N 3AF
Tel: 0171 405 6627
Fax: 0171 242 1426
Recent productions: From Butler to
Baker, a two-part documentary on
50 years of education; *Radio Wars*,
radio and revolution in Cuba; Art
Adventures, school's television;
Goodbye, Dear Friend, documentary
about pet loss; *Not the Vicar of
Dibley*, documentary about a female
priest in inner London; Good Health,
2 x schools sex education

Teliesyn

Helwick House
19 David Street
Cardiff CF1 2EH
Tel: 01222 667556
Fax: 01222 667546
Involved in feature film, television drama
and television documentary/feature. In
production: *Reel Truth*, docu/drama for
S4C and C4; *Cyber Wales* and
Answering Back for BBC Wales;
Dragon's Song for C4. In development:
Video Pirates (6 x 30 min drama) *Coron
yr Wythnons* plus a number of
feature film and drama series

Tempest Films

33 Brookfield
Highgate West Hill
London N6 6AT
Tel: 0181 340 0877
Fax: 0181 340 9309
In production: Bliss, pilot (90 mins) for
Carlton television; *Beck*, six part series
for BBC. In development: *The Devil's
Whore*, 10 part series for the BBC;
two low budget films and four series

Testimony Films

12 Great George Street,
Bristol BS1 5RS
Tel: 0117 925 8589
Fax: 0117 925 7668
Steve Humphries
Specialists in social history
documentaries. Recent productions:
*Labour of Love: Bringing up
Children in Britain 1900-1950* (6 x
40 min, BBC2); *Forbidden Britain:
Our Secret Past* (6 x 40 min,
BBC2); *A Man's World: The
Experience of Masculinity* (6 x 40
minutes, BBC2); *The Call of the
Sea: Memories of a Seafaring
Nation* (6 x 40 minutes, BBC2); *The
Roses of No Man's Land* (1 x 60
minutes C4); *Sex in a Cold Climate*
(1 x 60 minutes C4); *Hooked:
Britain in Pursuit of Pleasure* (6 x
30 minutes C4)

Thin Man Films
9 Greek Street
London W1V 5LE
Tel: 0171 734 7372
Fax: 0171 287 5228
Simon Channing-Williams, Mike Leigh
Recent productions: *Career Girls*;
Secrets & Lies
(See Pathé Pictures - Lottery Film
Production Franchises)

Third Eye Productions
Unit 101, Canalot Studios
222 Kensal Road
London W10 5BN
Tel: 0181 969 8211
Fax: 0181 960 8790
Television productions covering the
worlds of arts, music, ethnography
and developing world culture

Tiger Aspect Productions
5 Soho Square
London W1V 5DE
Tel: 0171 434 0672
Fax: 0171 287 1448
Harry Enfield and Chums (2 series
BBC); *The Vicar of Dibley* (BBC1)
The Thin Blue Line (BBC1) *The
Village* (7 series for Meridian)
Howard Goodall's Organ Works
(Ch4) *Hospital* (Ch5) *Deacon
Brodie* (Screen One for BBC 1)

Time and Light Productions
5 Darling Road
London SE4 1YQ
Tel: 0181 692 0145
Fax: 0181 692 0145
email: Rojatpen@aol.com
mobile/voicemail: 0958 317360
Specialises in the development and
production of film, television and
radio featuring the work of artists,
writers and musicians. Productions
include: *Sea, See, C* with Henry
Moore; *Time and Light*, with Jo
Spence, *When Shura met Hobie*, a
portrait of Shura Cherkassky, *A
Shadow into the Future* with Geoff
Dyer, *Will it be a Likeness?* with
John Berger; *To the Wedding* by
John Berger; *Fugitive Pieces* by
Anne Michaels for BBC Radio 3

Tiny Epic Video Co
37 Dean Street
London W1V 5AP
Tel: 0171 437 2854
Fax: 0171 434 0211

Toledo Pictures
30 Oval Road, Camden
London NW1 75E
Tel: 0171 485 4411
Fax: 0171 485 4422
Duncan Kenworthy
(See DNA - Lottery Film Production
Franchises)

Topaz Productions
Manchester House
46 Wormholt Road
London W12 0LS
Tel: 0181 749 2619
Fax: 0181 749 0358
In production: ongoing corporate
productions

Trans World International
TWI House
23 Eyot Gardens
London W6 9TR
Tel: 0181 233 5400
Fax: 0181 233 5401
Television and video sports
production and rights representation
branch of Mark McCormack's
International Management Group,
TWI produces over 2,500 hours of
broadcast programming and
represents the rights to many leading
sports events including Wimbledon,
British Open, US Open, and World
Matchplay golf. Productions include:
*Trans World Sport; Futbol Mundial;
PGA European Tour; ATP Tour
Highlights; West Indies, Indian and
Pakistan Test cricket; Oddballs; A-Z
of Sport; High 5; The American Big
Match and Blitz; The Olympic
Collection* and *The Whitbread
Round The World Race*

Transatlantic Films
Studio 1
3 Brackenbury Road
London W6 OBE
Tel: 0181 735 0505
Fax: 0181 735 0605
email: mail@transatlanticfilms.com
Revel Guest, Justin Albert
Recent Programming: *Horse Tales*
13x30 mins stories about the special
bond between people and horses, for
Animal Planet; *Amazing Animal
Adaptors* - 1x60 mins special for
Discovery Channel; *History's
Turning Point* - 26x30 mins about
decisive moments in history, for
Discovery Europe. Current Produc-
tion: *Trailblazers* - 13x60 mins,
travel and adventure series for
Discovery Europe, Travel Channel;
Three Gorges - 2x60 mins, the
building of the World's biggest dam
in China, for Discovery Channel/TLC

Paul Trijbits Productions
5A Noel Street
London W1V 3RB
Tel: 0171 439 4343
Fax: 0171 434 4447
Alexei Boltho

Triple Vision Ltd
Folly Lodge, Folly Lane
N Wootton

Shepton Mallet
Avon BA4 4EL
Tel: 01749 890 610
Film and television production
company. Documentary, drama, arts,
pop promos. Music clips for
PolyGram and Philips Classical label.
Features in development: *Tryptich;
Heartland; Dragons Hill; Underworld*

Tribune Productions Ltd
22 Bentley Way
Stanmore
Middlesex HA7 3RP
Tel: 0181 420 7230
Fax: 0181 207 0860

Troma UK
Allied Troma Ltd
The Glassworks
B-4 Ashland Place
London W1M 3JH
Tel: 0171 224 1992
Fax: 0171 224 0111

Try Again Limited
Leigh Grove Farmhouse
Leigh Grove
Bradford on Avon
Wilts BA15 2RF
Tel: 01225 862 705
Fax: 01225 862 205
Michael Darlow, Rod Taylor, Chris
Frederick
Produces drama, music, arts,
documentary programmes. Recent
productions include: *War Cries -
Angels of Mercy?* C4 1 x 30 mins.
*Martin Parr and the Ladies of the
Valley* BBC 1 x 40 mins; 1:4; *The
Lost Child - The Works* BBC 1 x 30
mins. *Something of a Different Pace
- The Works* BBC 1 x 30 mins

Twentieth Century-Fox Productions Ltd
20th Century House
31-31 Soho Square
London W1V 6AP
Tel: 0171 437 7766
Fax: 0171 734 3187
*Recent productions: The Full Monty,
Braveheart, Stealing Beauty; Titanic*

Twentieth Century Vixen
13 Aubert Park
Highbury
London N5 1TL
Tel: 0171 359 7368
Fax: 0171 359 7368
Film/video production and distribu-
tion, mainly feature documentaries.
Recent projects: documentary for
BBC2 on the *New Marilyn*
women's nightclub in Shinjuku,
Tokyo, *Shinjuku Boys* and *Law
Stories* film about Shani'a Law
Courts in Iran for C4

Twenty Twenty Television
10 Stucley Place
London NW1 8NS
Tel: 0171 284 2020
Fax: 0171 284 1810
The company continues to produce programmes exclusively for broadcast television, specialising in worldwide investigative journalism, documentaries, productions, factually based drama, science and childrens programmes. Recent productions include: *The Big Story* (Carlton); *Secret Lives Walt Disney* (C4); *Un Blues* (C4) and *Cutting Edge* (C4)

Ty Gwyn Films
Y Ty Gwyn
Llanllyfni
Caernarfon
Gwynedd LL54 6DG
Tel: 01286 881235
Specialises in film drama. In production: 8 x 60 minute television series on film

UBA (United British Artists)
21 Alderville Road
London SW6 3RL
Tel: 01984 623619
Fax: 01984 623733
Production company for cinema and TV projects. Past productions include: *Keep the Aspidistra Flying* for OFE; *Sweeney Todd* for Showtime/Hallmark; *Champions* for Embassy; *Ghost Hunter* for Granada; Wind-prints for MCEG Virgin Vision; *Taffin* for MGM; *Castaway* for Cannon; *The Lonely Passion of Judith Hearne* for HandMade Films; *Turtle Diary* for the Samuel Goldwyn Company

Uden Associates
Chelsea Wharf
Lots Road
London SW10 0QJ
Tel: 0171 351 1255
Fax: 0171 376 3937
Film and television production company for broadcast through C4, BBC and corporate clients. Recent productions: *Classic Trucks*, 6 x 30 min for C4 and various projects for Equinox, *Cutting Edge (Nurses, A is for Accident)* and *Short Stories*, *Classic Ships*, 6 x 30 min for C4, *Autoerotic II*, 3 x 30 min for C4, *Secret Lives: The Young Freud*

Unicorn Organisation
Pottery Lane Studios
34a Pottery Lane, Holland Park
London W11 4LZ
Tel: 0171 229 5131
Fax: 0171 229 4999

Union Pictures
36 Marshall Street
London W1V 1LL
Tel: 0171 287 5110
Fax: 0171 287 3770
Recent productions include: *The Crow Road; Deadly Voyage; Masterchef; Junior Masterchef; The Roswell Incident*

United Artists Corporation, Ltd (MGM/United Artists)
Pinewood Studios
Iver Heath
Buckinghamshire SL0 ONH
Tel: 01723 651700
Fax: 01723 656844
Recent productions: *Tomorrow Never Dies, Man in the Iron Mask*

Universal Pictures Ltd
1 Hamilton Mews
London W1V 9FF
Tel: 0171 491 4666
Fax: 0171 493 4702
Recent productions: *The Jackal, DragonHeart, Fierce Creatures*

V.U.E
387B King Street
London W6 9NH
Tel: 0181 741 1144
Fax: 0181 748 3597
Peter Greenaway

Vendetta Films
Albery Theatre
St Martins Lane
London WC2
Tel: 0171 240 3000
Fax: 0171 240 3401
Recent productions include: *Decadence*. In pre-production: Film *Stars Don't Die* in Liverpool

Vera Productions Ltd
3rd Floor
66/68 Margaret Street
London W1N 7FL
Tel: 0171 436 6116
Fax: 0171 436 6117/6016
Contact: Elaine Morris

Video Visuals
37 Harwood Road
London SW6 4QP
Tel: 0171 384 2243
Fax: 0171 384 2027
Currently produces *The Chart Show* for ITV

Videotel Productions
Ramillies House
1/2 Ramillies Street
London W1V 1DF
Tel: 0171 439 6301
Fax: 0171 437 0731
Producers of educational and

training packages for television and video distribution

Viz
4 Bank Street
Inverkeithing
Fife KY11 1LR
Tel: 01383 412811
Fax: 01383 418103
Founded 1971. 1996 productions: *Carlo Scarpa* directed by Murray Grigor for C4 and ACE. Opened the Montreal Festival of Art Films

Wall To Wall Television
8-9 Spring Place
Kentish Town
London NW5 3ER
Tel: 0171 485 7424
Fax: 0171 267 5292
Alex Graham, Jane Root
Producers of quality innovative programming including, drama: *Statement of Affairs; You Me and It; Plotlands*. Natural history: *Baby It's You*. Leisure: *Eat Your Greens; Sophie's Meat Course; For Love or Money*. Entertainment: *Big City; Heartland; The Big Country*. Arts and culture: *Fantasy by Gaslight; Rwandan Stories*

Warner Bros. Productions Ltd
135 Wardour Street
London W1V 4AP
Tel: 0171 494 3710
Fax: 0171 287 9086
Recent productions: *Eyes Wide Shut, The Avengers, Surviving Picasso*

Warner Sisters Film & TV Ltd, Cine Sisters Ltd
Canalot Studios
222 Kensal Road
London W10 5BN
Tel: 0181 960 3550
Fax: 0181 960 3880
Directors: Lavinia Warner, Jane Wellesley, Anne-Marie Casey and Dorothy Viljoen
Founded 1984. Drama, Comedy. TV and Feature Films. Output includes *A Village Affair; Dangerous Lady; Dressing for Breakfast; The Spy Who Caught a Cold; Capital Sins; The Bite; The Jump* - and feature film *Jilting Joe*. Developing a wide range of TV and feature projects

Watershed Television
11 Regent Street
Clifton
Bristol BS8 4HW
Tel: 0117 973 3833
Fax: 0117 973 3722
Film and video production. Broadcast, corporate, commercials

White City Films
79 Sutton Court Road
London W4 3EQ
Tel: 0181 994 6795
Fax: 0181 995 9379
Current affairs and documentary
productions

David Wickes Productions
169 Queen's Gate
London SW7 5HE
Tel: 0171 225 1382
Fax: 0171 589 8847
David Wickes, Heide Wilsher

Woodfilm Productions
59 Campden Street
London W8 7EL
Tel: 0171 243 8600
Producers of arts, features and
television drama. Recent produc-
tions: *The Pantomime Dame* (The
Arts Council); *The Future of Things
Past; Stairs; Go For It; Sophie; Say
Hello to the Real Dr Snide* (C4); *Off
the Streets* (Carlton); *The Care
Takers* (People First, C4)

Dennis Woolf Productions
Silver House
31-35 Beak Street
London W1R 3LD
Tel: 0171 494 4060
Fax: 0171 287 6366
Specialising in current affairs:
Dispatches; documentaries: *Cutting
Edge;* music: Epitaph: Charles
Mingus; and studio reconstructions
of contemporary trials: *The Court
Report* series (C4); *The Trial of
Klaus Barbie* (BBC)

Workhouse Television
Granville House
St Peter Street
Winchester
Hants SO23 9AF
Tel: 01962 863449
Fax: 01962 841026

Working Title Films
Oxford House
76 Oxford Street
London W1N 9FD
Tel: 0171 307 3000
Fax: 0171 307 3001/2/3
Tim Bevan, Eric Fellner
Recent film productions: *The
Hudsucker Proxy; Four Weddings
and a Funeral; French Kiss; Loch
Ness; Moonlight and Valentino;
Fargo; Dead Man Walking*

Working Title Television
77 Shaftesbury Avenue
London W1V 8HQ
Tel: 0171 494 4001
Fax: 0171 255 8600
Simon Wright, Analisa Barreto,

Hazel Finch
Recent productions: *Edward II;
Amnesty - The Big 30; Further Tales
of the Riverbank; TV Squash; The
Borrowers* (series 1 & 2); *Tales of
the City; The Baldy Man*

World Film Services
17 Golden Square
London W1R 4BB
Tel: 0171 734 3536
Fax: 0171 734 3585

World Wide Group
21-25 St Anne's Court
London W1V 3AW
Tel: 0171 434 1121
Fax: 0171 734 0169
Ray Townsend, Chris Courtenay Taylor
Catherine Cookson film series
(ITV); *Finders Keepers* (ITV);
Oliver's Travels (BBC Drama);
*Placido Domingo's Tales at the
Opera* (BBC Music and Arts):
Geography of Russia (C4 Schools);
*The Last Supper and The Last Soviet
Citizen* (both for BBC Arena)

The Worldmark Production Company Ltd
7 Cornwall Crescent
London W11 1PH
Tel: 0171 792 9800
Fax: 0181 792 9801
Drummond Challis, David Wooster
Current productions: *Gary's Golden
Boots* 8x30min for BBC1; *Living
with Lions* 6x30min. Recent
productions: *England 2006, Japan
2002 - FIFA World Cup Bid
Promotions. The Delaunax Cup - A
History of the European Football
Championships*

Year 2000 Film and Television Productions
3 Benson Road
Blackpool FY3 7HP
Tel: 01253 395403
Fax: 01253 395403
Michael Hammond
Film and television producers

X-Dream International
Stones, Wickham St Pauls
Halstead
Essex CO9 2PS
Tel: 01787 269089
Fax: 01787 269029
Specialises in the production,
distribution and representation of
sports TV, children's TV and film

Yorkshire Film Co
Capital House
Sheepscar Court
Meanwood Road
Leeds LS7 2BB

Tel: 0113 244 1224
Fax: 0113 244 1220
Producers of satellite/broadcast
sports documentaries, news
coverage, corporate and commercials
in film and video

Zenith North
11th Floor
Cale Cross House
156 Pilgrim Street
Newcastle upon Tyne NE1 6SU
Tel: 0191 261 0077
Fax: 0191 222 0271
email: zenithnorth@dial.pipex.com
Ivan Rendall, Peter Mitchell
(Managing Director), John Coffey
Productions include: *Byker Grove*
(BBC1); *Blues and Twos* (Carlton/
ITV); *The Famous Five* (ITV);
Animal Ark (HTV); *Dear Nobody*
(BBC); *Pass the Buck* (BBC);
Network First (Carlton ITV); music
specials for S4C; variety of regional
productions for Tyne Tees TV

Zenith Productions
43-45 Dorset Street
London W1H 4AB
Tel: 0171 224 2440
Fax: 0171 224 3194
email: zenith@zenith.tv.co.uk
Film and television production company.
Recent feature films: Todd Haynes'
Velvet Goldmine; Nicole Holofcener's
Walking and Talking. Recent television
drama: *The Uninvited* (ITV);
Hamish Macbeth (3 series, BBC
Scotland); *Rhodes* (BBC1)

Zephyr Films
24 Colville Road
London W11 2BJ
Tel: 0171 221 8318
Fax: 0171 221 9289

Zooid Pictures
66 Alexander Road
London N19 5PQ
Tel: 0171 281 2407
Fax: 0171 281 2404
email: pictures@zooid.sonnet.co.uk
Website: http://www.sonnet.co.uk/
zooid.co.uk
Producers of experimental and
television documentaries, various
shorts; documentaries

PRODUCTION STARTS

These are feature-length films intended for theatrical release with a significant British involvement (whether creative, financial or UK-based) which went into production between January and December 1997. Single television dramas in production for the same period are indicated with ❏

Categories

Category A
Films where the cultural and financial impetus is from the UK and the majority of personnel are British

Category B
Majority UK Co-Productions. Films in which, although there are foreign partners, there is a UK cultural content and a significant amount of British finance and personnel

Category C
Minority UK Co-Productions. Foreign (non-US) films in which there is a small UK involvement in finance or personnel

Category D
American films with a UK creative and/or minor financial involvement

Category D¹
American financed or part financed films made in the UK. Most films have a British cultural content

Category D²
American films with some UK financial involvement

1997 UK PRODUCTION STARTS

TV Films in italics. UK Film Category in bold brackets.

1997
Anorak of Fire
Coming Home
The General
The Governess
Heart
Shadow Run
Speak Like a Child

JANUARY
King Lear
Martha - Meet Frank, Daniel and Lawrence
Spice World
2 Vigo: A Passion for Life
2 Buskers
24 The Croupier
27 The Big Lebowski

FEBRUARY
The Red Violin
Resurrection Man
Up 'n' Under
3 Appetite
10 The Land Girls
10 A Price Above Rubies
16 The Commissioner
16 The Life of Stuff
24 Basil
24 Dad Savage

MARCH
Jilting Joe
Jinnah-Quaid-E-Azam
3 Hard Edge
3 *Our Boy*
3 Time Enough
3 24 7 TwentyFourSeven
3 Velvet Goldmine
4 Letters from a Killer
13 Keep the Aspidistra Flying

APRIL
Lost in Space
Sex and Chocolate
1 *Perfect Blue*
1 Sliding Doors
1 Tomorrow Never Dies
4 Lock, Stock and Two Smoking

Barrels
10 *Prince of Hearts*
12 Talisman
14 Orphans
21 *Cold Enough for Snow*
21 A System Devoured
29 Sunset Heights

MAY
The Avengers
Nutcracker Prince
The Woman in White
4 Crossmaheart
5 Bumping the Odds
7 Girls' Night
19 I Want You
25 Hooligans

JUNE
Los Angeles Without a Map
Love Is the Devil
Saving Private Ryan
5 Us Begins With You
6 The Misadventures of Margaret
16 All the Little Animals
16 Day Release
16 Table 5
16 Theory of Flight
22 The Tichborne Claimant
23 A Kind of Hush
23 St Ives
26 Star Wars Prequel 1

JULY
3 Among Giants
5 Prometheus
5 The Real Howard Spitz
14 The Parent Trap
20 Lucia
21 If Only

AUGUST
4 Treasure Island
6 Vent de Colère
8 Underground
11 *Bravo Two Zero*
11 Waking Ned Devine
12 What Rats Won't Do
13 Beach Boys
18 Dancing at Lughnasa
18 Devil's Snow
18 Get Real
25 Divorcing Jack
28 Parting Shots

SEPTEMBER
1 Elizabeth
1 *Mothertime*
1 My Name Is Joe
7 Dangerous Obsession
8 Bedrooms & Hallways
8 Captain Jack
9 An Urban Ghost Story
10 Wolves of Kromer
11 Tom's Midnight Garden
15 Plunkett and Macleane
15 The Secret Laughter of Women
17 Simple Gifts
20 Laid Up
21 Babymother
21 The Hi-Lo Country
21 Titanic Town
22 Himalaya - A Chief's Childhood
22 Parting Shots
29 Dance Me Outside
29 Fast Food

OCTOBER
  Mistress of the Craft
3 Dirty British Boys
4 Alien Blood
6 The Death and Loss of Sexual
   Innocence
7 A Soldier's Daughter Never Cries
10 Little Voice
14 Neville's Island
17 Comic Act
20 The Dance
21 Hideous Kinky
26 The Last Seduction 2
27 The Wisdom of Crocodiles

NOVEMBER
  Rogue Trader
  This Could Be the Last Time
10 Understanding Jane
12 Woundings
15 The Jolly Boys' Last Stand
16 Getting Hurt
18 Sugar Sugar

DECEMBER
1 Spoonface Steinberg
8 Hilary and Jackie
8 The Final Cut

Alien Blood
4 October
Production Companies: West Coast Films
Locations/Studio: Lake District
Exec Prod: Roberta Moore
Producers: Jon Sorensen
Director: Jon Sorensen
Screenplay: Jon Sorensen
Director of Photography: Peter Rowe
Cast: Francesca Manning, Rebecca Sterling, Glyn Whiteside, Vanessa Stevens, Catherine Whitaker
Category A

All the Little Animals
16 June
Production Companies: Recorded Picture Co
Locations/Studio: Cornwall, London, Isle of Man
Exec Prod: Chris Auty
Producers: Jeremy Thomas
Director: Jeremy Thomas
Screenplay: Eski Thomas
Film Editor: John Victor Smith
Director of Photography: Mike Molloy
Cast: John Hurt, Daniel Benzali
Category A

Among Giants
3 July
Production Companies: A Kudos Film production with the participation of British Screen, the Arts Council of England and BBC Films in Association with Capitol Films and the Yorkshire Media Production Fund
Locations/Studio: Yorkshire
Exec Prod: Jana Edelbaum
Producers: Stephen Garrett
Director: Sam Miller
Screenplay: Simon Beaufoy
Film Editor: Paul Green
Director of Photography: Witold Stok
Cast: Pete Postlethwaite, Rachel Griffiths, James Thornton, Rob Jarvis, Lenny James
Category A

❑ Anorak of Fire
Production Companies: BBC Television
Locations/Studio: North Yorkshire
Exec Prod: David M. Thompson
Producers: Tatiana Kennedy
Director: Elijah Moshinsky
Screenplay: Stephen Dinsdale
Film Editor: Eva Lind
Director of Photography: Patrick Duval
Cast: Kenny Doughty, Laura Crossley, William Ash, Stuart Callaghan, Kay Wragg

Appetite
3 February

Production Companies: Alternative Cinema Company/ Loudmouse Productions/ 101 Films
Locations/Studio: Douglas, Isle of Man
Exec Prod: Matthew Payne, Mark Vennis
Producers: Simon Johnson
Director: George Milton
Screenplay: George Milton
Film Editor: Rod Edge
Director of Photography: Peter Thwaites
Cast: Ute Lemper, Trevor Eve, Christien Anholt, Edward Hardwicke
Category B

The Avengers
May
Production Companies: Jerry Weintraub Productions (in association with Warner Bros)
Locations/Studio: London and Pinewood Studios
Exec Prod: Susan Ekins
Producers: Jerry Weintraub
Director: Jeremiah Chechik
Screenplay: Don MacPherson
Film Editor: Mick Audsley
Director of Photography: Roger Pratt
Cast: Uma Thurman, Ralph Fiennes, Sean Connery
Category D[1]

Babymother
21 September
Production Companies: Channel Four Films, Formation Films
Locations/Studio:
Exec Prod: Margaret Matheson
Producers: Parminder Vir
Director: Julian Henriques
Screenplay: Julian Henriques
Film Editor: Brad Fuller
Director of Photography: Ron Fortunato
Cast: Anjela Lauren Smith, Wil Johnson, Caroline Chikezie, Jocelyn Eisen, Don Warrington
Category A

Basil
24 February
Production Companies: Kushner-Locke
Locations/Studio: Tenby, South Wales
Exec Prod: Donald Kushner, Peter Locke
Producers: Radha Bharadwaj
Director: Radha Bharadwaj
Screenplay: Radha Bharadwaj
Director of Photography: David Johnson
Cast: Christian Slater, Jared Leto, Derek Jacobi, Claire Forlani, Rachel Pickup
Category D[1]

Beach Boys
13 August
Production Companies: Ragged Productions

Locations/Studio: Brighton, Birmingham
Producers: Justin Edgar
Director: Richard Gossage
Screenplay: Richard Gossage
Director of Photography: Nick Osborne
Cast: Simon Lowe, Steven Emeries, Polly Sands, Colin Higgins
Category A

Bedrooms & Hallways
8 September
Production Companies: Berwin & Dempsey
Locations/Studio: London
Producers: Dorothy Berwin, Ceci Dempsey
Director: Rose Troche
Screenplay: Robert Farrar
Film Editor: Chris Blunden
Director of Photography: Daf Hobson
Cast: Simon Callow, Jennifer Ehle, Christopher Fulford, Julie Graham, Tom Hollander, Kevin McKidd
Category A

The Big Lebowski
27 January
Production Companies: Working Title Films
Locations/Studio: Los Angeles
Exec Prod: Eric Fellner, Tim Bevan
Producers: Ethan Coen
Director: Joel Coen
Screenplay: Joel Coen, Ethan Coen
Film Editor: Roderick Jaynes
Director of Photography: Roger Deakins
Cast: Jeff Bridges, John Goodman, Steve Buscemi, John Turturro
Category D²

❏ Bravo Two Zero
Production Companies: BBC.Distant Horizon
Locations/Studio: South Africa
Exec Prod: David M Thompson
Producers: Ruth Caleb
Director: Tom Clegg
Screenplay: Andy McNab, Troy Kennedy Martin
Cast: Sean Bean

❏ Bumping the Odds
5 May
Production Companies: Wall to Wall Television in association with Halcyon Productions for BBC Scotland
Locations/Studio: Glasgow and Loch Lomond
Exec Prod: Jo Willett, Andrea Calderwood
Producers: Ian Madden
Director: Rob Rohrer
Screenplay: Rona Munro
Film Editor: Sue Wyatt
Director of Photography: Barry Ackroyd

Cast: Joseph McFadden, Sharon Small, Shirley Henderson, Gary Sweeney

Buskers
2 January
Production Companies: Lyndania Films
Locations/Studio: London
Producers: Jeymes
Director: Jeymes
Screenplay: Jeymes
Film Editor: Paul Murray
Director of Photography: Paul Murray
Cast: Jeymes, Clive Manu, Tanya Samuel
Category A

Captain Jack
8 September
Production Companies: Viva Films for Granada Films, Baltic Media, Winchester Films, the Arts Council of England
Locations/Studio: Scotland, Whitby, Shepperton
Exec Prod: Pippa Cross, Chris Craib, William Sargent
Producers: John Goldschmidt
Director: Robert Young
Screenplay: Jack Rosenthal
Film Editor: Eddie Mansell
Director of Photography: John McGlashan
Cast: Bob Hoskins, Sadie Frost, Anna Massey, Gemma Jones, Peter Macdonald, Maureen Lipman, Patrick Malahide
Category A

❏ Cold Enough for Snow
21 April
Production Companies: BBC Films
Locations/Studio: London, Exeter, Luton, and Amsterdam
Exec Prod: Tessa Ross
Producers: Ann Scott
Director: Piers Haggard
Screenplay: Jack Rosenthal
Film Editor: Michael Parker
Director of Photography: Nina Kellgren
Cast: Maureen Lipman, David Ross, Tom Wilkinson, Anna Carteret, Nicholas Le Prevost

Comic Act
17 October
Production Companies: 3pider Pictures
Locations/Studio: London
Producers: Paulo Branco
Director: Jack Hazan
Screenplay: Jack Hazan
Director of Photography: Richard Branczik
Cast: Suki Webster, Stephen Moyer, David Schneider
Category A

❏ Coming Home
Production Companies: Portman Productions, Tele-Muenchen
Exec Prod: Tim Buxton, Rikolt von Gagern
Producers: David Cunliffe
Director: Giles Foster
Screenplay: John Goldsmith
Film Editor: Colin Green
Director of Photography: Simon Archer
Cast: Peter O'Toole, Joanna Lumley, Emily Mortimer, Katie Ryder-Richardson, Patrick Ryecart, Penelope Keith, Carol Drinkwater

The Commissioner
16 February
Production Companies: Metropolis Filmproduktion (Germany), New Era Vision Ltd (UK), Saga Film Production (Belgium).
Locations/Studio: Cologne, Brussels, London
Exec Prod: George Reinhart
Producers: Christina Kallas, Luciano Gloor
Director: Danny Hiller
Screenplay: David Ambrose, Adrian Hodges
Film Editor: Denise Vindvogel
Director of Photography: Witold Stok
Cast: John Hurt, Armin Mueller-Stahl, Rosana Pastor, Alice Krige, Johan Leysen
Catergory C

Crossmaheart
4 May
Production Companies: Lexington Films
Locations/Studio: Northern Ireland
Producers: Don Boyd
Director: Henry Herbert
Screenplay: Colin Bateman
Film Editor: Adam Ross
Director of Photography: Peter Butler
Cast: Gerard Rooney, Maria Lennon
Category A

The Croupier
24 January
Production Companies: Little Bird/Tatfilm Production, Channel Four Films, Filmstiftung NRW and WDR
Locations/Studio: Germany and UK
Exec Prod: James Mitchell
Producers: Jonathan Cavendish
Director: Mike Hodges
Screenplay: Paul Mayersberg
Film Editor: Les Healey
Director of Photography: Mike Garfarth
Cast: Clive Owen, Gina McKee, Alex Kingston, Kate Hardie, Alexander Morton
Category B

Dad Savage

24 February
Production Companies: PolyGram
Filmed International, Sweet Child
Films
Locations/Studio: Lincolnshire,
Twickenham Studios
Producers: Gwynneth Lloyd, Robert
Jones
Director: Betsan Morris-Evans
Screenplay: Steven Williams
Film Editor: Guy Bensley
Director of Photography: Gavin Finney
Cast: Patrick Stewart, Kevin
McKidd, Helen McCrory, Joseph
McFadden
Category A

The Dance

20 October
Production Companies: Isfilm/
Oxford Film, Nordisk Film
Productions, Hamburger Kino
Kompanie
Producers: Agust Guomundsson
Director: Agust Guomundsson
Director of Photography: Ernest
Vincze
Cast: Gunnar Helgason, Baldur
Trausti Hreinsson, Palina Jonsdottir,
Dofri Hermannsson
Category C

Dancing at Lughnasa

18 August
Production Companies: Ferndale
Films/Pandora/Samson Film
Locations/Studio: Irleand
Exec Prod: Jane Barclay, Sharon Harel
Producers: Noel Pearson
Director: Pat O'Connor
Screenplay: Frank McGuinness
Film Editor: Humphrey Dixon
Director of Photography: Kenneth
MacMillan
Cast: Meryl Streep, Michael
Gambon, Catherine McCormack,
Kathy Burke, Brid Brennan
Category C

Dangerous Obsession

7 September
Production Companies: Working
Title Films, Film Development Corp
and Alberto Ardissone in association
with Baltic Media/Isle of Man
Commission
Locations/Studio: Isle of Man
Producers: Alan Latham, Clifford
Haydn-Tovey
Director: Gerry Lively
Screenplay: John Howlett
Film Editor: David Spiers
Director of Photography: Adam
Santelli
Cast: Sherilyn Fenn, Ray Winstone,
Tim Dutton, Oliver Tobias
Category A

Day Release

16 June
Production Companies: Bolt-On
Media/Liscombe Holdings
Locations/Studio: Licombe Park,
Buckinghamshire
Exec Prod: Nicholas Bonsor, Martin
Darvill
Producer: Tim Purcell
Director: Charlie Gauvain
Film Editor: Kevin Waters
Cast: Simon Green, Manouk van der
Meulen, Elaine Hallam
Category A

The Death and the Loss of Sexual Innocence

6 October
Production Companies: New Line
Cinema
Locations/Studio: UK, Italy, Africa
Producers: Annie Stewart, Mike
Figgis
Director: Mike Figgis
Cast: Saffron Burrows, Julian Sands,
Johanna Torrel, Hanne Klintoe,
Stefano Dionisi
Category D[1]

Devil's Snow

18 August
Production Company: Eye-Cue
Productions
Location: Huddersfield
Producer: Benjamin Johns
Director: Richard Hellawell
Cast: David Smith, Gabriel
Swartland, Sian Foulkes, Paul
Zarins, Sohail Khan
Category A

Dirty British Boys

3 October
Production Companies: Firestorm
Pictures
Locations/Studio: the Midlands
Producers: Fraz Hussein
Director: Assad Raja
Screenplay: Assad Raja
Director of Photography: Ken Koh
Cast: Shashi Kapoor, Assad Raja,
Paul Usher, George Christopher
Category A

Divorcing Jack

25 August
Production Companies: BBC Films,
Winchester Films and Scala
Productions in association with the
Arts Council of England and the
Arts Council of Northern Ireland, a
Scala Production in association with
IMA Films
Locations/Studio: Belfast
Exec Prod: Nik Powell, Stephen
Woolley, David Thompson
Producers: Robert Cooper
Director: David Caffrey
Screenplay: Colin Bateman
Film Editor: Nick Moore
Director of Photography: James
Welland
Cast: David Thewlis, Rachel
Griffiths, Richard Grant, Laura
Fraser
Category B

Elizabeth

1 September
Production Companies: A PolyGram
Filmed Entertainment presentation
of a Working Title Films
Locations/Studio: London
Producers: Alison Owen, Tim
Bevan, Eric Fellner
Director: Shekhar Kapur
Screenplay: Michael Hirst
Film Editor: Jill Bilcock
Director of Photography: Remi
Adefarasin
Cast: Cate Blanchett, Geoffrey Rush,
Richard Attenborough, Eric Cantona
Category A

Fast Food

29 September
Production Companies: Fast Food
Films
Locations/Studio: London
Exec Prod: Tessa Gibbs
Producers: Phil Hunt
Director: Stewart Sugg
Screenplay: Stewart Sugg
Film Editor: Jeremy Gibb
Director of Photography: Simon
Reeves
Cast: Emily Woof, Douglas
Henshall, Stephen Lord
Category A

The Final Cut

Production Companies: Fugitive
Films
Locations/Studio: London
Exec Prod: Jim Beach
Producers: Dominic Anciano, Rau
Burdis
Director: Dominic Anciano, Rau
Burdis
Screenplay: Dominic Anciano, Rau
Burdis
Director of Photography: John Ward
Cast: Ray Winstone, Sadie Frost,
Jude Law
Category A

The General

Production Companies: Nattore
Limited, Merlin Films, J&M
Entertainment,
Location: Ireland
Exec Prod: Kieran Corrigan
Producers: John Boorman
Director: John Boorman
Screenplay: John Boorman
Film Editor: Ron Davis
Director of Photography: Seamus Deasy

Cast: Brendan Gleeson, Adrian Dunbar, Sean McGinley, Maria Doyle Kennedy, Angeline Ball
Category C

Get Real
Production Companies: Graphite Films
Locations/Studio: Basingstoke
Exec Prod: Anat Singh, Helena Spring
Producers: Stephen Taylor
Director: Simon Shore
Screenplay: Patrick Wilde
Film Editor: Barrie Vince
Director of Photography: Allan Almond
Cast: Ben Silverstone, Brad Gorton, Charlotte Brittain,
Category A

☐ Getting Hurt
16 November
Production Companies: BBC Films, Mayfair Entertainment
Locations/Studio: London, Birmingham
Exec Prod: David Thompson, Robert Cooper
Producers: Gareth Neame
Director: Ben Holt
Screenplay: Andrew Davies
Film Editor: Jerry Leon
Director of Photography: Brian Tufano
Cast: Ciaran Hinds, Amanda Ooms, David Hayman, Ingrid Lacey

☐ The Gift
Production Companies: Tetra Films for the BBC
Locations/Studio: Cambridge, Slough, Middlesex and Bray Studios
Exec Prod: Tessa Ross
Producers: Alam Horrox
Director: Danny Hiller
Screenplay: Lucy Gannon
Film Editor: Jeremy Strachan
Director of Photography: Steve Saunderson
Cast: Amanda Burton, Neil Dudgeon, Trevor Peacock, Crispin Bonham Carter, Philip Whitchurch

Girls' Night
7 May
Production Companies: Granada Film
Locations/Studio; Manchester and Las Vegas
Exec Prod: Pippa Cross
Producers: Bill Boyes
Director: Nick Hurran
Screenplay: Kay Mellor
Film Editor: John Richards
Director of Photography: David Odd
Cast: Brenda Blethyn, Julie Walters, Kris Kristofferson, George Costigan, Philip Jackson
Category A

The Governess
Production Companies: British Screen, Arts Council of England, BBC Films, Parallax Pictures
Locations/Studio: Scotland, London
Exec Prod: Sally Hibbin
Producers: Sarah Curtis
Director: Sandra Goldbacher
Screenplay: Sandra Goldbacher
Film Editor: Isabel Lorente
Director of Photography: Ashley Rowe
Cast: Minnie Driver, Tom Wilkinson, Johnathan Rhys Meyers, Harriet Walter, Florence Hoath, Burce Myers
Category A

Hard Edge
3 March
Production Companies: DMS Films
Locations/Studio: London, Reading, Salsibury
Exec Prod: Daniel M. San
Producers: Daniel M. San, Caleb Lindsay
Director: Caleb Lindsay
Screenplay: Caleb Lindsay
Film Editor: Anthony B. Sloman
Director of Photography:
Cast: Luke Shaw, Simon Bateso, Bryan Marshall, David O'Kelly, Matt Lane
Category A

Heart
Production Companies: Granada Film
Locations/Studio: Liverpool and Manchester
Exec Prod: Gub Neal, Pippa Cross
Producers: Nicola Shindler
Director: Charles McDougall
Screenplay: Jimmy McGovern
Film Editor: Edward Mansell
Director of Photography: Julian Court
Cast: Christopher Eccleston, Saskia Reeves, Kate Hardie, Rhys Ifans, Bill Paterson
Category A

Hideous Kinky
21 October
Production Companies: L Films, Greenpoint Films
Locations/Studio: Marrakech, Morocco
Exec Prod: Mark Shivas, Simon Relph
Producers: Ann Scott
Director: Gillies MacKinnon
Screenplay: Billy MacKinnon
Film Editor: Pia Di Ciaula
Director of Photography: John de Borman
Cast: Kate Winslet, Said Taghmaoui,
Category B

The Hi-Lo Country
21 September
Production Companies: Working Title Films (US), PolyGram Pictures
Locations/Studio: New Mexico
Producers: Martin Scorsese, Barbara De Fina, Tim Bevan, Eric Fellner, Rudd Simmons
Director: Stephen Frears
Screenplay: Walon Green
Director of Photography: Oliver Stapleton
Cast: Woody Harrelson, Billy Crudup, Patricia Arquetta, Penelope Cruz, Sam Elliot
Category D^2

Hilary and Jackie
8 December
Production Companies: Oxford Films
Locations/Studio: Shepperton
Producers: Anand Paterson, Nicholas Kent
Director: Andy Tucker
Screenplay: Frank Cottrell-Boyce
Director of Photography: David Johnson
Cast: Emily Watson, Rachel Griffiths, James Frain
Category A

Himalaya - A Chief's Childhood
22 September
Production Companies: Antelope
Locations/Studio: Nepal, Himalaya
Exec Prod: Jaques Perrin, Christophe Barratier
Producers: Jacques Perrin, Mick Csasky
Director: Eric Valli
Screenplay: Eric Valli
Film Editor: Marie-Josephe Yoyette
Director of Photography: Jean-Paul Meurisse
Cast: Tilen Labrang, Tashie Dhamtso, Gurgon Kyiap, Aggal Lama, Karma Tensing Nyima
Category C

Hooligans
25 May
Production Companies: Liquid Films, Bord Scannán na hÉireann
Locations/Studio: Dublin
Exec Prod: Terry Ollinwood, Denis Wigman
Producers: Nicholas O'Neil, Kees Kasander
Director: Paul Tickell
Screenplay: James Mathers
Cast: Darren Healy, Jeff O'Toole, Vivianna Verveen
Category C

I Want You
19 May
Production Company: Revolution Films

Locations/Studio: Hastings and Dungeness
Exec Prod: Stewart Till
Producers: Andrew Eaton
Director: Michael Winterbottom
Screenplay: Eoin McNamee
Film Editor: Trevor Waite
Director of Photography: Slawomir Idziak
Cast: Rachel Weisz, Allessandro Nivola, Luka Petrusic, Labina Mitevska, Carmen Ejogo, Ben Daniels, Graham Crowden, Geraldine O'Rawe
Category A

If Only
21 July
Production Companies: An ESICMA production for Paragon Corp (Can), HandMade Films (UK) in association with CLT-Ufa International (Fr), Mandarin Films (Fr) and Wild Rose Productions (US).
Locations/Studio: Ealing Studios, London
Exec Prod: Jon Slan, Gareth Jones,
Producers: Juan Gordon,
Director: Maria Ripoll
Screenplay: Rafa Russo
Film Editor: Nacho Ruiz-Capillas
Director of Photography: Javier G Salmones
Cast: Douglas Henshall, Lena Headey, Mark Strong, Neil Stuke, Charlotte Coleman, Elizabeth McGovern, Penelope Cruz
Category C

Jilting Joe
March
Production Companies: Warner Sisters Film & Television
Locations/Studio: London
Exec Prod: Jane Wellesley, Andrea Calderwood
Producers: Anne-Marie Casey
Director: Dan Zeff
Screenplay: Rosamund Orde-Powlett
Film Editor: Victoria Bodell
Director of Photography: Julian Court
Cast: James Purefoy, Geraldine Somerville, Benjamin Whitrow, Matilda Ziegler, Angus Wright, Oriana Bonet
Category A

Jinnah -Quaid-E-Azam
March
Production Companies: Petra Films
Locations/Studio: Pakistan and London
Exec Prod: Prof Akbar Ahmed
Producers: Jamil Dehlavi
Director: Jamil Dehlavi
Screenplay: Jamil Dehlavi, Prof

Akbar Ahmed
Film Editor: Budge Tremlett
Director of Photography: Nic Knowland
Cast: Christopher Lee, Shashi Kapoor, James Fox, Maria Aitken, Richard Lintern
Category C

The Jolly Boys' Last Stand
15 November
Production Companies: Jolly Productions, The Bigger Picture Co, Function Films
Locations/Studio: London and Humberside
Exec Prod: Tom McCabe,
Producers: Craig Woodrow
Director: Christopher Payne
Screenplay: Christopher Payne
Film Editor: Tullio Brunt
Director of Photography: Robin Cox
Cast: Andy Serkis, Milo Twomey, Rebecca Craig, Matt Wilkinson, Anton Saunders
Category A

Keep the Aspidistra Flying
13 March
Production Companies: Aspidistra Productions Ltd
Locations/Studio: Ealing Studios, London
Exec Prod: John Wolstenholme, Robert Bierman
Producers: Peter Shaw
Director: Robert Bierman
Screenplay: Alan Pater
Film Editor: Bill Wright
Director of Photography: Giles Nuttgens
Cast: Richard E. Grant, Helena Bonham Carter
Category A

A Kind of Hush
23 June
Production Companies: First Film Co
Locations/Studio: London
Producers: Roger Randall-Cutler
Director: Brian Stirner
Screenplay: Brian Stirner
Film Editor: David Martin
Director of Photography: Jacek Petrycki
Cast: Harley Smith, Marcella Plunkett, Ben Roberts, Paul Williams, Nathan Constance, Mike Phibbins, Peter Saunders, Roy Hudd, Jeanie Drynan
Category A

❑ King Lear
January
Production Companies: Chesterhead Productions, BBC Films WGBH Boston
Exec Prod: Simon Curtis, Rebecca Eaton
Producers: Sue Birtwistle

Director: Richard Eyre
Director of Photography: Roger Pratt
Cast: Ian Holm, Timothy West, Barbara Flynn, Amanda Redman, Michael Bryant

Laid Up
20 September
Production Companies: Rented Films
Locations/Studio: London
Producers: Rob Gunns, Andrew Stephen Powell
Director: Rob Gunns
Director of Photography: Edward Wright
Cast: Selina Giles, Danny Edwards, Helena Clavert, Robert Reina, Ricky Dearman
Category A

The Land Girls
10 February
Production Companies: Greenpoint Films, West Eleven Films
Locations/Studio: Somerset, London
Exec Prod: Ruth Jackson
Producers: Simon Relph
Director: David Leland
Screenplay: David Leland, Keith Dewhurst
Film Editor: Nick Moore
Director of Photography: Henry Braham
Cast: Catherine McCormack, Rachel Weisz, Anna Friel, Steven Mackintosh
Category B

The Last Seduction 2
26 October
Production Companies: Specific Films/Ty Cefn
Locations/Studio: Barcelona, Cardiff
Exec Prod: Michael Hamlyn
Producers: David Ball, Clare Wise
Director: Terry Marcel
Screenplay: David Cummings
Film Editor: Belinda Cottrell
Director of Photography: Geza Sinkovics
Cast: Joan Severance, Con O'Neill, Beth Goddard, Dean Williamson, Rocky Taylor
Category A

The Life of Stuff
16 February
Production Companies: Prairie Pictures, BBC Films, The Scottish Arts Council Lottery Fund, The Glasgow Film Fund
Locations/Studio: Glasgow
Exec Prod: Mark Shivas, Eddie Dick
Producers: Lynda Myles
Director: Simon Donald
Screenplay: Simon Donald
Film Editor: Justin Krish
Director of Photography: Brian Tufano

Cast: Ewen Bremner, Liam Cunningham, Jason Flemyng, Ciaran Hinds, Gina McKee
Category A

Little Voice
10 October
Production Companies: Scala Proudctions, Miramax films
Locations/Studio:
Exec Prod: Nik Powell, Stephen Woolley
Producers: Elizabeth Karlsen
Director: Mark Herman
Script Supervisor: Angela Wharton
Film Editor: Michael Ellis
Director of Photography: Andy Collins
Cast: Jane Horrocks, Michael Caine, Brenda Blethyn, Ewan McGregor, Jim Broadbent
Category D[1]

Lock, Stock and Two Smoking Barrels
4 April
Production Companies: Ska Productions, Steve Tisch Co [US]
Locations/Studio:
Exec Prod: Steve Tisch, Peter Morton, Stephen Marks, Angard Paul, Trudie Styler, Gareth Jones
Producers: Matthew Vaughn
Director: Guy Ritchie
Screenplay: Guy Ritchie
Film Editor: Jeremy Gibbs, Niven Howie
Director of Photography: Tim Maurice-Jones
Cast: Nick Moran, Jason Flemyng, Steve Mackintosh, PH Moriarty, Dexter Fletcher, Vinnie Jones
Category A

Los Angeles without a Map
June
Production Companies: Dan Films, Arts Council of England, Baltic Media, Euro American Films SA, European co-Production Fund/BSR, Finnish Film Foundation, Marianna Films, Yorkshire Media Production Agency
Locations/Studio: Los Angeles, Las Vegas, Bradford
Producers: Julie Baines, Sarah Daniel
Director: Mika Kaurismäki
Screenplay: Richard Rayner, Mika Kaurismäki
Director of Photography: Michel Amathieu
Cast: David Tennant, Vinessa Shaw, Julie Delpy, Vincent Gallo
Category B

Lost in Space
April
Production Companies: New Line Cinema, Prelude Pictures

Locations/Studio: Shepperton
Exec Prod: Mace Neufeld, Robert Rehme
Producers: Mark Koch, Akiva Goldsman, Stephen Hopkins
Director: Stephen Hopkins
Screenplay: Akiva Goldsman
Film Editor: Peter Levy
Director of Photography: Ray Lovejoy
Cast: Gary Oldman, William Hurt, Matt LeBlanc
Category D[1]

Love Is the Devil
Production Companies: BBC Films, and the British Film Institute in association with Premiere Heure
Locations/Studio: London, Paris
Exec Prod: Ben Gibson, Frances-Anne Solomon
Producers: Chiara Menage
Director: John Maybury
Screenplay: John Maybury
Film Editor: Danile Goddard
Director of Photography: John Mathieson
Cast: Derek Jacobi, Daniel Craig, Tilda Swinton, Adrian Scarborough, Karl Johnson
Category A

Lucia
20 July
Production Companies: Lexington Films
Locations/Studio: East Lothian
Exec Prod: Henry Herbert
Producers: Stephanie Mills, Alison Kerr
Director: Don Boyd
Screenplay: Don Boyd
Director of Photography: Dewald Aukema
Cast: Richard Coxon, Amanda Boyd, Andrew Greenan, Mark Holland, John Daszak
Category A

Martha - Meet Frank, Daniel and Laurence
January
Production Companies: A Banshee production for Channel Four Films
Locations/Studio: London, Minneapolis, Twickenham
Producers: Grainne Marmion
Director: Nick Hamm
Screenplay: Peter Morgan
Film Editor: Michael Bradsell
Director of Photography: David Johnson
Cast: Monica Potter, Rufus Sewell, Tom Hillander, Joseph Fiennes
Category A

The Misadventures of Margaret
6 June
Production Companies: A Lunatics

and Lovers production for Mandarin (France) TFI France and Granada Film
Locations/Studio: London, Paris, New York
Exec Prod: Andy Harries, Pippa Cross, Dominique Green
Producers: Brian Skeet
Director: Brian Skeet
Screenplay: Brian Skeet
Film Editor: Clare Douglas
Director of Photography: Romain Shields
Cast: Parker Posey, Jeremy Northam, Brooke Shields, Elizabeth McGovern
Category B

Mistress of the Craft
October
Production Companies: Armadillo Films, Vista Street Entertainment
Locations/Studio: London, Elstree
Exec Prod: Jerry Feifer
Producers: Jonathan Blay, Elisar Cabrera
Director: Elisar Cabrera
Screenplay: Elisar Cabrera
Director of Photography: Alvin Leong
Cast: Wendy Cooper, Stephanie Beaton, Kerry Knowlton, Eileen Daly, Sean Harry
Category A

❏ Mothertime
1 September
Production Companies: BBC
Locations/Studio: London
Exec Prod: David Thompson
Producers: Josh Golding
Director: Matthew Jacobs
Screenplay: Matthew Jacobs
Director of Photography: Peter Hannan
Cast: Kate Maberley, Gina McKee, Anthony Andrews, Imogen Stubbs, Megan de Wolf

My Name Is Joe
1 September
Production Companies: Parallax Pictures, Road Movies Vierte Produktionen
Locations/Studio: Glasgow
Exec Prod: Ulrich Felsburg
Producers: Rebecca O'Brien
Director: Ken Loach
Screenplay: Paul Laverty
Film Editor: Jonathan Morris
Director of Photography: Barry Ackroyd
Cast: Peter Mullan
Category B

❏ Neville's Island
14 October
Production Companies: PrimeTime Production

Locations/Studio: Lake District, Surrey, London
Exec Prod: Andre Ptaszynski, Richard Price
Producers: Judy Craymer
Director: Terry Johnson
Screenplay: Tim Firth
Film Editor: Martin Sharpe
Director of Photography: Paul Wheeler
Cast: Jeff Rawle, Martin Clunes, David Bamber, Timothy Spall, Sylvia Syms

The Nutcracker Prince
May
Production Companies: IMAX Corp, Sands Film Ltd
Locations/Studio: London
Exec Prod: Andrew Gellis
Producers: Olivia Stockman, Lorne Orleans
Director: Christine Edzard
Screenplay: Christine Edzard
Director of Photography: Noel Archampauly
Cast: Miriam Margolyes, Heathcote Williams, Patrick Pearson, Harriet Thorpe
Category B

Orphans
14 April
Production Companies: Channel Four Films presents in association with The Scottish Arts Council and The Glasgow Film Fund, An Antoine Greenbridge Production
Exec Prod: Paddy Higson
Producers: Frances Higson
Director: Peter Mullan
Screenplay: Peter Mullan
Film Editor: Colin Monie
Director of Photography: Grant Scott Cameron
Cast: Douglas Henshll, Gary Lewis, Stephen Cole, Rosemarie Stevenson, Frank Gallagher
Category A

❏ Our Boy
3 March
Production Companies: Wall to Wall Productions for BBC Films
Locations/Studio: London
Exec Prod: Alex Graham, Tessa Ross
Producers: Jo Willett
Director: David Evans
Screenplay: Tony Grounds
Film Editor: Chris Ridsdale
Director of Photography: Oliver Curtis
Cast: Ray Winstone, Pauline Quirke, Neil Dudgeon, Philip Jackson, Perry Fenwick

Owd Bob
15 June
Production Companies:

Kingsborough Greenlight Pictures
Exec Prod: Harry Alan Towers
Producers: Pieter Kroonenburgh
Director: Rodney Gibbons
Screenplay: Sharon Buckingham, Harry Alan Towers
Director of Photography: Keith Young
Cast: James Cromwell, Colm Meaney
Catergory A

The Parent Trap
14 July
Production Companies: Walt Disney Pictures
Locations/Studio: California, London, Shepperton
Producers: Charles Shyer
Director: Nancy Myers
Screenplay: Nancy Myers, Charles Shyer, David Swift
Film Editor: Stephen A. Rotter
Director of Photography: Dean Cundey
Cast: Dennis Quaid, Natasha Richardson
Category D[1]

Parting Shots
28 September
Production Companies: Scimitar Films Produciton
Locations/Studio: London
Exec Prod: Michael Winner
Producers: Michael Winner
Director: Michael Winner
Screenplay: Michael Winner
Film Editor: Arnold Crust
Director of Photography: Ousama Rawi
Cast: Chris Rea, Felicity Kendal, John Cleese, Bob Hoskins, Ben Kingsley
Category A

❏ Perfect Blue
1 April
Production Companies: BBC Films
Locations/Studio: London
Exec Prod: David Thompson
Producers: Elinor Day
Director: Kieron Walsh
Film Editor: Kant Pan
Director of Photography: Zoran Djordevic
Cast: Inday Ba, Michclle Austin, Philip Glenister, Ruth Gemmell, Katie Carr

Plunkett and Macleane
15 September
Production Companies: Working Title Films
Locations/Studio: Prague, Thame House, Shepperton, Barrandov Studios
Producers: Eric Fellner, Tim Bevan, Rupert Harvey

Director: Jake Scott
Screenplay: Selwyn Roberts
Film Editor: Oral Ottey
Director of Photography: John Mathieson
Cast: Robert Carlyle, Liv Tyler, Jonny Lee Miller, Michael Gambon, Ken Stott
Category A

A Price Above Rubies
10 February
Production Companies: Channel Four Films/Miramax Films
Locations/Studio: New York
Exec Prod: Bob Weinstein, Harvey Weinstein
Producers: Lawrence Bender, John Penotti
Director: Boaz Yakin
Screenplay: Boaz Yakin
Film Editor: Arthur Coburn
Director of Photography: Adam Holender
Cast: Renee Zellweger, Christopher Eccleston
Category D[2]

❏ The Prince of Hearts
10 April
Production Companies: BBC Screen One
Locations/Studio: Cambridge, London
Producers: Mike Darbon
Director: Simon Curtis
Screenplay: Lee Hall
Director of Photography: John Daly
Cast: Robson Green, Rupert Penry-Jones

Prometheus
5 July
Production Companies: A Holmes Associates Production in association with the Arts Council of England for Channel Four Films
Exec Prod: Michael Kustow
Producers: Andrew Holmes
Director: Tony Harrison
Screenplay: Tony Harrison
Film Editor: Luke Dunkley
Director of Photography: Alistair Cameron
Cast: Michael Feast, Jonathan Waintridge, Fern Smith, Steve Huison, Walter Sparrow
Category A

The Real Howard Spitz
5 July
Production Companies: Writer's Block Ltd
Locations/Studio: Halifax, Nova Scotia; Los Angeles; CineSite, Canada
Exec Prod: Alan Martin
Producers: Paul Brooks, Christopher Zimmer
Director: Vadim Jean
Screenplay: Jurgen Wolff

Film Editor: Pia Di Ciaula
Director of Photography: Glen MacPherson
Cast: Kelsey Grammer, Amanda Donohoe, Joseph Rutten, Patrick McKenna

Red Mercury
Production Companies: Red Mercury/Theatre East
Producers: Jonathan Reason, Stuart Croll
Director: Steve Rogers
Director of Photography: Gene Talvin
Cast: Suzanne Cave, Dean Pidoux, Daniel Pilpott, Andy Lucas , Antonia Mirto
Category A

The Red Violin
February
Production Companies: New Line Cinema/Channel Four (UK)/Mikado (Italy)/Rhombus Media (Can)
Locations/Studio: Montreal, Italy, China, Oxford
Producers: Niv Fichman
Director: François Girard
Screenplay: François Girard, Don McKellar
Film Editor: Gaetan Huot
Director of Photography: Alain Dostie
Cast: Samuel L Jackson, Greta Scacchi
Category C

Resurrection Man
February
Production Companies: Revolution Films, PolyGram Filmed Entertainment
Locations/Studio: Manchester, North of England
Exec Prod: Michael Winterbottom
Producers: Andrew Eaton
Director: Marc Evans
Screenplay: Eoin McNamee
Film Editor: John Wilson
Director of Photography: Pierre Aim
Cast: Stuart Townsend, James Nesbitt, Geraldine O'Rawe, Brenda Fricker, John Hannah
Category A

Rogue Trader
November
Production Companies: Granada Films
Locations/Studio:
Exec Prod: Pippa Cross, David Frost
Producers: Janette Day, Paul Raphael, Clare Chapman
Director: James Dearden
Screenplay: James Dearden
Film Editor: Catherine Creed
Director of Photography: Jean-François Robin

Cast: Ewan McGregor, Anna Friel, Tom Wu, Nigel Lindsay
Category A

Saving Private Ryan
June
Production Companies: Dream Works SKG/Paramount Pictures
Locations/Studio: Wexford, Ireland, Hatfield, Hertfordshire
Producers: Steven Spielberg, Ian Bryce, Mark Gordon, Gary Levinsohn
Director: Steven Spielberg
Screenplay: Robert Rodat, Frank Darabont
Film Editor: Michael Kahn
Director of Photography: Janusz Kaminski
Cast: Tom Hanks, Tom Sizemore, Edward Burns, Matt Damon, Jeremy Davies
Category D[1]

The Sea Change
September
Production Companies: Winchester Films
Locations/Studio: UK, Spain
Exec Prod: Gary Smith, Chris Craib
Producers: Billy Hurman
Director: Michael Bray
Screenplay: Michael Bray
Film Editor: Bryan Oates
Director of Photography: Joseph M Civit
Cast: Maryam D'Abo, Sean Chapman, Andree Bernard, Ray Winstone
Category B

The Secret Laughter of Women
15 September
Production Companies: Paragon entertainment Corp, HandMade Films, ELBA Films, European Co-production Fund, The Arts Council of England
Locations/Studio: London, Southern France
Exec Prod: Gareth Jones
Producers: O. O. Sagay
Director: Peter Schwabach
Screenplay: O. O. Sagay
Director of Photography: Martin Fuhrer, Jacques Renoir
Cast: Colin Firth, Nia Long, Fissy Roberts, Joke Jacobs, Bella Enaiioro
Category A

❑ Sex and Chocolate
April
Production Companies: BBC Drama
Locations/Studio: London and Paris
Exec Prod: Ruth Caleb, George Faber
Producers: Sophie Clarke-Jervoise
Director: Gavin Millar
Screenplay: Tracey Schofield
Film Editor: Angus Newton

Director of Photography: John Else
Cast: Dawn French, Phil Daniels, Michael Maloney, Julia Carling, Jan Alphonse

Shadow Run
Production Companies: Geoff Reeve Film
Producers: Geoff Reeve
Director: Geoff Reeve
Screenplay: Desmond Lowden
Film Editor: Robert Morgan
Director of Photography: Eddy Van Der Enden
Cast: Michael Caine, James Fox, Christopher Cazenove, Nigel Havers, Ken Colley
Category A

Sliding Doors
1 April
Production Companies: Mirage Enterprises
Locations/Studio: London, Shepperton
Exec Prod: Sydney Pollack
Producers: Bill Horberg, Philippa Braithwaite
Director: Peter Howitt
Screenplay: Peter Howitt
Film Editor: John Smith
Director of Photography: Remi Adefarasin
Cast: Gwyneth Paltrow, John Hannah, John Lynch, Jeanne Tripplehorn, Zara Turner, Douglas McFerran, Paul Brightwell, Nina Young, Phyllida Law, Kevin McNally
Category D[1]

A Soldier's Daughter Never Cries
7 October
Production Companies: Merchant Ivory Productions
Locations/Studio: Paris, Long Island, New York
Exec Prod: Jane Barclay, Sharon Harel
Producers: Ismail Merchant
Director: James Ivory
Screenplay: Ruth Prawer Jhabvala, James Ivory
Film Editor: Noëlle Boisson
Director of Photography: Jean-Marc Fabre
Cast: Kris Kristofferson, Barbara Hershey, LeeLee Sobieski, Jane Birkin
Category A

🎬 Speak Like a Child
Production Companies: BBC Films and British Film Institute in association with the Arts Council of England present a Leda Serene Production).
Locations/Studio: London and Northumberland
Exec Prod: Frances Anne Solomon, David Thompson, Ben Gibson

Producers: Fiona Morham, Lazell Daley
Director: John Akomfrah
Screenplay: Danny Padmore
Film Editor: Annabel Ware
Director of Photography: Jonathan Collinson
Cast: Fraser Aures, Daniel Newman, Alison Mac, Carla Henry, Gavin Green
Category A

Spice World
January
Production Companies: Spice Girls presents in association with PolyGram Filmed Entertainment and Icon Entertainment International, A Fragile Films production)
Locations/Studio: Twickenham
Exec Prod: Simon Fuller
Producers: Barnaby Thompson, Uri Fruchtmann
Director: Bob Spiers
Screenplay: Kim Fuller
Film Editor: Andrea McCarthur
Director of Photography: Clive Tickner
Cast: Emma Bunton, Geri Halliwell, Melanie Brown, Melanie Chisholm, Victoria Adams, Richard E Grant, Claire Rushbrook, Richard Briers, Michael Barrymore, Alan Cummings, Stephen Fry, Hugh Laurie, Frank Bruno, Barry Humphries, Jools Holland
Category A

❏ Spoonface Steinberg
1 December
Production Companies: BBC Films
Locations/Studio: London
Producers: Suzan Harrison, Simon Curtis
Director: Betsan Morris Evans
Screenplay: Lee Hall
Film Editor: Pamela Power
Director of Photography: Dafydd Hobson
Cast: Ella Jones, Helen McCrory, Mark Strong, Linda Bassett, Becky Simpson

St Ives
23 June
Production Companies: Little Bird, TATfilm, Compaigne des Phares et Balises
Locations/Studio: N.Ireland, France, Germany, Ireland
Exec Prod: James Mitchell
Producers: Jonathan Cavendish
Director: Harry Hook
Screenplay: Allan Cubitt
Film Editor: John MacDonald
Director of Photography: Robert Alazraki
Cast: Jean-Marc Barr, Richard E.Grant, Anna Friel
Category C

Star Wars Prequel 1
26 June
Production Companies: JAK Prods
Locations/Studio: Italy, Tunisia, Leavesden Studios
Exec Prod: George Lucas
Producers: Rick McCallum
Director: George Lucas
Screenplay: George Lucas
Film Editor: Paul Martin
Director of Photography: David Tattersall
Cast: Ewan McGregor, Natalie Portman, Liam Neeson, Jake Lloyd, Terence Stamp
Category D[1]

Sugar Sugar
18 November
Production Companies: Sweet Tooth Films
Locations/Studio: London
Exec Prod: Sarah Davies, Bradley Souber
Director: Bradley Souber
Screenplay: Bradley Souber
Director of Photography: Lincoln Ascott
Cast: Sarah Manners, Jason Traynor
Category A

Sunset Heights
29 April
Production Companies: Northlands Film Production, Scorpio Productions
Locations/Studio: Northern Ireland, Ireland
Exec Prod: James Flynn
Producers: Denis Bradley, Daniel Figuero, Zugi Kamasa
Director: Colm Villa
Screenplay: Colm Villa
Film Editor: Jeanine Hurley
Director of Photography: Roger Bonnici
Cast: Toby Stephens, Jim Norton, James Cosmo, Patrick O'Kane, Joe Rea
Category B

Table 5
6 June
Production Company: Raw Talent Productions
Director: Elliot Grove
Screenplay: Elliot Grove
Director of Photography: James Solan
Cast: Markus Napier, Jackie Sawris, Aron Paramour, Alex McSweeney
Category A

Talisman
12 April
Production Companies: Ealing Touch Film Productions
Locations/Studio: London, Ealing
Producers: Mark Collins-Cope
Director: Xian Vassie

Screenplay: Xian Vassie
Director of Photography: Roger Eaton
Cast: Manu Monade, Antona Mirto, William Neenan, Ellie Fairman, Sean Brosnan
Category A

The Theory of Flight
16 June
Production Companies: Distant Horizon, BBC Films
Locations/Studio: Wales, UK
Exec Prod: David M. Thompson
Producers: Anat Singh, Ruth Caleb, Helena Spring
Director: Paul Greengrass
Screenplay: Richard Hawkins
Film Editor: Mark Day
Director of Photography: Ivan Strasburg
Cast: Helena Bonham Carter, Kenneth Branagh, Gemma Jones, Sue Jones Davies, Holly Aird
Category A

❏ This Could Be the Last Time
November
Production Companies: BBC Films
Locations/Studio: London, Disneyland, Paris
Producers: Colin Ludlow,
Director: Gavin Millar
Screenplay: Geoffrey Case
Film Editor: Angus Newton
Director of Photography: Nigel Walters
Cast: Joan Plowright, Nicholas Scellier, Penelpe Wilton, Dorothy Tutin

The Tichborne Claimant
22 June
Production Companies: The Bigger Picture Co
Locations/Studio: Isle of Man
Producers: Tom McCabe
Director: David Yates
Screenplay: Joe Fisher
Film Editor: Jamie Trevill
Director of Photography: Peter Thwaites
Cast: John Kani, Robert Pugh, Robert Hardy
Category A

Time Enough
3 March
Production Companies: Time Films Ltd
Locations/Studio: London and Sussex
Producers: Jonathan Weissler
Director: Foster Marks
Screenplay: Steve Lunnon
Film Editor: Maria Walker
Director of Photography: Ross Fall
Cast: Alex Wheeler, Foster Marks, Dexter Fletcher, Carmon Squire
Category A

Titanic Town
21 September
Production Companies: Company Pictures, BBC Films, British Screen, Arts Council of Northern Ireland
Locations/Studio: Belfast, London
Exec Prod: David Thompson, Robert Cooper, Rainer Mockert
Producers: George Faber, Charles Pattinson
Director: Roger Michell
Screenplay: Anne Devlin
Film Editor: Kate Evans
Director of Photography: John Daly
Cast: Julie Walters, Ciaran Hinds, Ciaran McMenamin
Category A

Tomorrow Never Dies
1 April
Production Companies: Eon Productions, United Artists (USA)
Locations/Studio: France, Germany, Souh East Asia, Mexico, UK
Producers: Michael G Wilson, Barbara Broccoli
Director: Roger Spottiswoode
Screenplay: Bruce Feirstein
Film Editors: Dominique Fortin, Michel Arcand
Director of Photography: Robert Elswit
Cast: Pierce Brosnan, Jonathan Pryce, Michelle Yeoh, Judi Dench, Desmond Llewelyn, Samantha Bond, Ricky Jay, Gotz Otto
Category D[1]

Tom's Midnight Garden
11 September
Production Companies: Hyperion Studio Jadein
Locations/Studio: Pinewood Studios, Isle of Man
Exec Prod: Marie Vine
Producers: Charles Salmon, Thomas. L.Wilhite, Adam Shapiro
Director: Willard Carroll
Screenplay: Willard Carroll
Director of Photography: Gavin Finney
Cast: Greta Scacchi, James Wilby, Joan Plowright
Category A

Treasure Island
4 August
Production Companies: Kingsborough Greenlight Pictures
Locations/Studio: UK
Exec Prod: John Buchanan
Producers: Pieter Kroonenburgh
Director: Peter Rowe
Screenplay: Peter Rowe
Film Editor: Ion Webster
Director of Photography: Marc Charlebois
Cast: Kevin Zegers, Jack Palance, David Robb, Christopher Benjamin
Category C

24 7 TwentyFourSeven
3 March
Production Companies: A Scala production for BBC Films
Locations/Studio: Nottingham and the Peak District
Exec Prod: Stephen Woolley, Nik Powell, George Faber
Producers: Imogen West
Director: Shane Meadows
Screenplay: Robyn Slovo
Film Editor: Bill Diver
Director of Photography: Ashley Rowe
Cast: Bob Hoskins, Mat Hand, Sun Hand, Sarah Thom, Frank Harper
Category A

Underground
8 August
Production Companies: Creative Film Services
Locations/Studio: South London
Producers: Chris Leeson
Director: Paul Spurrier
Screenplay: Paul Spurrier
Director of Photography: Paul Spurrier
Cast: Billy Smith, Nick Sutton, Zoe Small, Joel Beckett, Nick Frost
Category A

Understanding Jane
10 November
Production Companies: DMS Films, Flash Point Pictures
Locations/Studio: London
Producers: Daniel M San
Director: Caleb Lindsay
Screenplay: Jim Mummery
Director of Photography: Christian Koerner
Cast: Kevin McKidd, Amelia Curtis, Louisa Milwood-Haigh, John Simm
Category A

Up 'n' Under
February
Production Companies: Touchdown Films/Colour Features
Locations/Studio: Cardiff, Wales
Exec Prod: David Ball
Producers: Mark Thomas
Director: John Godber
Screenplay: John Godber
Film Editor: Chris Lawrence
Director of Photography: Alan Trow
Cast: Gary Olsen, Tony Slattery, Griff Rhys Jones, Neil Morrisey, Samantha Janus
Category A

An Urban Ghost Story
9 September
Production Companies: Living Spirit Pictures
Exec Prod: David Harwick
Producer: Chris Jones

Director: Genevieve Jolliffe
Screenplay: Chris Jones, Genevieve Jolliffe
Film Editor: Eddie Hamilton
Director of Photography: Jon Walker
Cast: Jason Connery, Stephanie Buttle, Nicola Stapleton, James Cosmo, Alan Owen
Category A

Us Begins with You
5 June
Production Companies: Bill Kenwright Films
Exec Prod: Anthony Edwards, Dante Di Loreto
Producer: Bill Kenwright
Director: Willi Patterson
Screenplay: Geoff Morrow
Film Editor: Peter Beston
Cast: Anthony Edwards, Jenny Seagrove, Charles Dance
Category A

Velvet Goldmine
3 March
Production Companies: Zenith Productions, Killer Films (USA)
Locations/Studio: London
Exec Prod: Scott Meek, Michael Stipe, Sandy Stern
Producers: Christine Vachon
Director: Todd Haynes
Screenplay: Todd Haynes
Film Editor: James Lyons
Director of Photography: Maryse Alberti
Cast: Ewan McGregor, Christian Bale, Jonathan Rhys Meyers, Toni Collette
Category B

Vent de Colère
6 August
Production Companies: Millennium Films (UK), CLC (Lyon)
Locations/Studio: Vercors, France
Exec Prod: Patrick Cassavetti
Producers: Emma Hayter, Daniel Charrier
Director: Michael Raeburn
Screenplay: Michael Raeburn
Film Editor: Stephanie Mahet
Director of Photography: Chris Seager
Cast: Patrick Bouchitey, Eric Boucher, Coralie Zohanero, Agoumi
Category C

Vigo: A Passion for Life
2 January
Production Companies: Impact Films, Nitrate Films, Channel Four Films, Little Magic Films, Road Movies, Tournesol, Canal Plus, European Script Fund
Locations/Studio: London, Paris
Exec Prod: Kiki Miyake
Producers: Amanda Temple, Jeremy Bolt

Director: Julien Temple
Screenplay: Peter Ettedgui, Anne Devlin
Film Editor: Marie Therese Boiche
Director of Photography: John Mathieson
Cast: Romane Bohringer, James Frain, Jim Carter, Paola Dionisotti, William Scott-Mason
Category B

Waking Ned Devine
11 August
Production Companies: Tomboy Films, The Gruber Bros in association with Mainstream (Fr), Bonaparte (Fr), the Isle of Man Film Commission and Overseas Film Group, with the participation of Canal Plus
Locations/Studio: Isle of Man
Exec Prod: Alexandre Heylen
Producers: Glynis Murray, Richard Holmes
Director: Kirk Jones
Screenplay: Kirk Jones
Film Editor: Alan Strachan
Director of Photography: Henry Braham
Cast: Ian Bannen, David Kelly, Fionnula Flanagan, Susan Lynch, James Nesbitt, Brenda F Dempsey
Category B

What Rats Won't Do
12 August
Production Companies: A PolyGram Filmed Entertainment presentation of a Working Title Films Production
Locations/Studio: London
Producers: Nicky Kentish Barns, Simon Wright, Tim Bevan, Eric Fellner
Director: Alastair Reid
Screenplay: Steve Coombes, Dave Robinson, William Osborne
Director of Photography: Brian Tufano
Cast: James Frain, Natascha McElhone
Category A

The Wisdom of Crocodiles
27 October
Production Companies: Zenith Productions
Locations/Studio: London
Exec Prod: Scott Meek, Dorothy Berwin
Producers: David Lascelles, Carolyn Choa
Director: Po Chih Leong
Screenplay: Paul Hoffmann
Director of Photography: Oliver Curtis
Cast: Jude Law, Elina Löwensohn, Kerry Fox, Timothy Spall, Jack Davenport
Category A

Wolves of Kromer
10 September
Production Companies: Discodog Productions
Locations/Studio:
Producers: Charles Lambert
Director: Will Gould
Screenplay: Charles Lambert
Film Editor: Carol Salter
Director of Photography: Laura Remacha
Cast: Lee William, James Layton
Category A

❑   The Woman in White
May
Production Companies: Woman in White Production for BBC Television and WGBH Boston
Locations/Studio: London, Northamptonshire, Lincolnshire, North Yorkshire, Oxfordshire
Exec Prod: Jonathan Powell, David Thompson
Producers: Gareth Neame
Director: Tim Fywell
Screenplay: David Pirie
Film Editor: Robin Sales
Director of Photography: Richard Greatrex
Cast: Tara Fitzgerald, James Wilby, Simon Callow, Ian Richardson, Corin Redgrave

Woundings
12 November
Production Companies: Muse Productions, Stone Canyon Entertainment, Isle of Man Film Commission
Locations/Studio: Isle Of Man
Exec Prod: Jordan Gertner
Producers: Chris Hanley, Brad Schlei, Keith Hayley, Lisa Collins
Director: Roberta Hanley
Screenplay: Roberta Hanley
Director of Photography: Alun Bollinger
Cast: Guy Pearce, Jonathan Schaech, Ray Winstone
Category D[1]

RELEASES

Listed here are feature-length films, both British and foreign, which had a theatrical release in the UK between January and December 1997. Entries quote the title, distributor, UK release date, certificate, country of origin, director/s, leading players, production company/ies, duration, gauge (other than 35mm), the **Sight and Sound** (S&S) reference. UK films or films with some UK involvement are followed by a Film Category reference. Re-releases are indicated by an asterisk and a **Monthly Film Bulletin** (MFB) reference

UK Film Categories

Category A
Films where the cultural and financial impetus is from the UK and the majority of personnel are British

Category B
Majority UK Co-Productions. Films in which, although there are foreign partners, there is a UK cultural content and a significant amount of British finance and personnel

Category C
Minority UK Co-Productions. Foreign (non-US) films in which there is a small UK involvement in finance or personnel

Category D
American films with a UK creative and/or minor financial involvement

Category D¹
American financed or part financed films made in the UK. Most films have a British cultural content

Category D²
American films with some UK financial involvement

1997 UK FILM RELEASES

UK Films (including co-productions and minority co-productions) in bold with category in brackets. * Indicates re-release.

JANUARY
3   **Shine**
3   Sleepers
10 The Mirror Has Two Faces
10 **Robinson in Space**
10 Some Mother's Son
10 The Starmaker/L'uomo delle stelle
17 The Ghost and the Darkness
17 Picture Bride
17 The Preacher's Wife
17 **Walking and Talking**
24 The Arrival
24 Flirting with Disaster
24 The Frighteners
24 Like Grains of Sand
24 Set It Off
24 That Thing You Do!
24 Welcome to the Dollhouse
24 Your Beating Heart/Un Coeur qui bat
31 **Carla's Song**
31 **Extreme Measures**
31 Looking For Richard
31 **Quadrophenia** *

FEBRUARY
2   Costa Brava (Family Album)
7   Fly Away Home
7   His Girl Friday *
7   **The Proprietor/La Propriétaire**
7   Ransom
7   Ridicule
7   White Man's Burden
14 **Conspirators of Pleasure/Spiklenci slasti**
14 **Fierce Creatures**
14 **Hamlet**
14 Powder
14 Trees Lounge
21 Flirt
21 Michael
14 Harriet the Spy
14 **In Love and War**
21 Grace of My Heart
21 The Phantom
28 Bound
28 The Crucible

28 Mars Attacks!
28 **Portrait of a Lady**

MARCH
7   **Blood and Wine**
7   Ghosts of Mississippi
7   Irma Vep
7   Jerry Maguire
7   Love Serenade
7   Normal Life
7   She's The One
7   Swann
14 The English Patient
14 Evening Star
14 Mother Night
14 Never Talk to Strangers
21 Space Jam
21 Star Wars Episode IV A New Hope *
21 **Trojan Eddie**
28 Basquiat
28 Dante's Peak
28 **Driftwood**
28 Larger Than Life
28 Love Lessons/Lust och fägring stor/Laererinden
28 **The Railway Children** *
28 William Shakespeare's Romeo + Juliet

APRIL
4   **Fever Pitch**
4   Hard Men
4   Korea
4   A Self-Made Hero/Un héro très discret
4   Tokyo Fist
11 Bits and Pieces/Il cielo è sempre più blu
11 Citizen Kane *
11 The Empire Strikes Back *
11 Mandela
11 **The Near Room**
11 The People vs. Larry Flynt
11 **Total Eclipse**
18 The Addiction
18 Box of Moonlight
18 Everyone Says I Love You
18 The Funeral
18 Metro
18 **The Saint**
18 **Twin Town**
25 **Cold Comfort Farm**
25 Eddie

25 Return of the Jedi *
25 The Spiral Staircase *
25 Vertigo *

MAY
2 **The Boy from Mercury**
2 Donnie Brasco
2 Female Perversions
2 It Takes Two
2 Kids Return
2 **Kolya**
2 Liar Liar
2 **Margaret's Museum**
2 Scream
2 Sebastian
9 Anaconda
9 Crying Freeman
9 Ghost From the Past
9 Moll Flanders
16 High School High
16 Dangerous Ground
16 Killer a Journal of Murder
16 Lush Life
16 Microcosmos/Microcosmos Le
   Peuple de l'herbe
16 **La Passione**
16 The Relic
16 See How They Fall/Regarde les
   hommes tomber
16 When We Were Kings
23 Anna Karenina
23 Beavis and Butt-Head Do
   America
23 Jungle2Jungle
23 Love and Other Catastrophes
23 Mon homme
23 No Way Home
23 Space Truckers
30 Absolute Power
30 Big Night
30 **Gridlock'd**
30 The Spitfire Grill
30 Turbulence

JUNE
6 Adrenalin Fear the Rush
6 **Alive and Kicking/Indian**
   Summer
6 Con Air
6 Crash
6 Drifting Clouds/Kauas pilvet
   karkaavat/Au loin s'en vont les
   nuages
6 Entertaining Angles The
   Dorothy Day Story
6 The Fifth Element/Le
   Cinquiéme Elément
6 The Informer *
6 Men, Women:Users Manual/
   Hommes femmes Mode
   d'emploi
13 The Associate
13 Johns
13 Shadow Conspiracy
13 The Square Circle/Daayraa
13 Trigger Happy
20 A Bit of Scarlet
20 The Chamber
20 The Devil's Own

20 **Intimate Relations**
20 **Kama Sutra A Tale of Love**
20 **Madame Butterfly**
20 Marvin's Room
20 Private Parts
27 Batman & Robin
27 Battle of Algiers *
27 **Frantz Fanon Black Skin**
   White Mask
27 love jones

JULY
4 City of Industry
4 One Fine Day
4 Pink Flamingos *
4 **Preaching to the Perverted**
4 The Quest
4 Rumble in the Bronx/Hongfan
   Qu
4 Unhook the Stars/Décroches les
   étoiles
5 Jour de Fête
11 Bang
11 Get on the Bus
11 Murder at 1600
11 Sélect Hotel
11 Someone Else's America/
   L'Amérique des autres
11 Stephen King's Thinner
11 Swingers
11 Unforgettable
18 **The Butterfly Effect/El efecto**
   mariposa
18 Lady and the Tramp *
18 The Lost World Jurassic Park
18 **Remember Me?**
25 Broken English
25 The Disappearance of Kevin
   Johnson
25 Idiot Box
25 Love! Valor! Compassion!
25 **Palookaville**
25 **Portraits Chinois**
25 Warriors of Virtue

AUGUST
1 Addicted to Love
1 BAPS
1 Men in Black
8 **Bean**
8 Grosse Pointe Blank
8 Ma vie sexualle (How I Got Into
   an Argument)/Comment je
   me suis disputé...("ma vie
   sexuelle")
8 **Roseanna's Grave**
8 Tierra
10 Hercules
15 Cloud-capped Star/Meghe
   Dharka Tara
15 **Heat And Dust** *
15 Speed 2 Cruise Control
22 **Event Horizon**
22 **Jump the Gun**
22 Albino Alligator
22 Keys to Tulsa
22 Lost Highway
22 Plein Soleil/Delitto in pieno
   sole/Blazing Sun/Purple Noon *

22 Romy and Michele's High
   School Reunion
29 Conspiracy Theory
29 **The Full Monty**
29 **The Slab Boys**

SEPTEMBER
5 L'Appartement
5 Austin Powers International
   Man Of Mystery
5 **Fetishes**
5 The Honeymoon Killers *
5 **Mrs. Brown**
5 Night Falls on Manhattan
5 That Old Feeling
12 **Air Force One**
12 **Hell Is a City** *
12 187
12 The Watermelon Woman
19 **Career Girls**
19 Deep Crimson
19 **Gallivant**
19 My Best Friend's Wedding
19 **Photographing Fairies**
20 Spawn
26 Contact
26 Dancehall Queen
26 **Face**
26 **The Leading Man**
26 Picture Perfect
26 The Sweet Hereafter

OCTOBER
3 Booty Call
3 First Strike/Jingcha gushi 4 Zhi
   Jiandan Renwu
3 **Head Above Water**
3 Latin Boys Go to Hell
3 Volcano
10 Father's Day
10 The Game
10 **House of America**
10 **Nil by Mouth**
10 Pusher
10 Temptress Moon/Fengyue
17 The Blue Angel *
17 Hard Eight
17 **Shooting Fish**
17 A Simple Wish
17 SubUrbia
17 Wilde
18 Free Willy 3 The Rescue
24 **Darklands**
24 **A Life Less Ordinary**
24 **Ma vie en rose**
24 The Peacemaker
31 **An American Werewolf in**
   Paris
31 Fools Rush In
31 L.A. Confidential
31 **Smalltime**
31 Smilla's Feeling for Snow/
   Fräulein Smillas Gespür für
   Schnee
31 **Snow White A Tale of Terror**

NOVEMBER
7 Face/Off
7 **The Gambler**

Absolute Power
**Rank-Castle Rock/Turner -
30 May**
(15) USA, Dir Clint Eastwood
with Clint Eastwood, Gene
Hackman, Ed Harris, Laura Linney,
Scott Glenn, Dennis Haysbert
**Castle Rock Entertainment
presents a Malpaso production**
121 minutes 5 seconds
S&S June 1997 p44

Addicted to Love
Warner Bros - 1 August
(15) USA, Dir Griffin Dunne
with Meg Ryan, Matthew
Broderick, Kelly Preston, Tcheky
Karyo, Maureen Stapleton, Nesbitt
Blaisdell
**Warner Bros An Outlaw
production in association with
Miramax Films**
100 minutes 20 seconds
S&S September 1997 p36

The Addiction
Guild - 18 April
(18) USA, Dir Abel Ferrara
with Lili Taylor, Christopher
Walken, Annabella Sciorra, Edie
Falco, Paul Calderon, Fredro Starr
Fast Films, Inc
82 minutes 19 seconds
S&S April 1997 p34

Adrenalin Fear the Rush
Columbia TriStar - 6 June
(18) USA, Dir Albert Pyun
with Christophe Lambert, Natasha
Henstridge, Norbert Weisser,
Elizabeth Barondes, Craig Davis,
Xavier Declie
**Largo Entertainment, Inc./Toga
Productions, Inc**
**A Largo Entertainment Inc in
association with Filmwerks
presentation**
76 minutes 7 seconds
S&S July 1997 p34

Air Force One
Buena Vista - 12 September
(15) USA, Dir Wolfgang Petersen
with Harrison Ford, Gary Oldman,
Wendy Crewson, Paul Guilfoyle,
William H.Macy, Liesel Matthews
**© Beacon Communications Corp.
Buena Vista International
presentation
In association with Beacon
Pictures A Radiant production**
124 minutes 25 seconds
S&S October 1997 p42

Albino Alligator
Electric Pictures - 22 August
(18) USA/France, Dir Kevin Spacey
with Matt Dillon, Faye Dunaway,

Gary Sinise, William Fichtner, Viggo
Mortensen, John Spencer
**Motion Picture Corporation of
America UGC D.A. International
presents in association with
Motion Picture Corporation of
America**
96 minutes 46 seconds
S&S August 1997 p36

Alien Resurrection
**20th Century Fox - 28th
November**
(18) USA, Dir Jean-Pierre Jeunet
with Sigourney Weaver, Winona
Ryder, Dominique Pinon, Ron
Perlman, Gary Dourdan, Michael
Wincott
**Twentieth Century Fox Film
Corporation
A Brandywine production**
108 minutes 45 seconds
S&S December 1997 p36

Alive and Kicking/Indian Summer
Film Four Distributors - 6 June
(15) UK, Dir Nancy Meckler
with Jason Flemyng, Antony Sher,
Dorothy Tutin, Anthony Higgins,
Bill Nighy, Phillip Voss
**Channel Four Television
Corporation Channel Four Films
present an M.P. production**
99 minutes 5 seconds
S&S July 1997 p35
Category A

An American Werewolf in Paris
Entertainment - 31 October
(15) USA/Luxembourg/France/UK,
Dir Anthony Waller
with Tom Everett Scott, Julie Delpy,
Vince Vieluf, Phil Buckman, Julie
Bowen
**© Stonewood Communications B.V./
Hollywood Pictures Company.
In co-operation with Avrora
Media/Delux Productions/
Président Films
In association with Propoganda
Films A J&M Entertainment
presentation in association with
Cometstone Pictures**
102 minutes 31 seconds
S&S December 1997 p38
Category C

Anaconda
Columbia TriStar - 9 May
(15) USA/Brazil, Dir Luis Llosa
with Jennifer Lopez, Ice Cube, Jon
Voight, Eric Stolz, Jonathan Hyde
**Columbia Pictures Industries, Inc
A CL Cinema Line Films
Corporation production Skylight
Cinema Foto Art Ltd**

89 minutes 14 seconds
S&S July 1997 p35

Anna Karenina
Warner Bros - 23 May
(15) USA, Dir Bernard Rose
with Sophie Marceau, Sean Bean,
Alfred Molina, Mia Kirshner, James
Fox, Fiona Shaw
**Icon Distribution, Inc, A Warner
Bros presentation**
108 minutes 1 second
S&S June 1997 p45

L'Appartement
Artificial Eye - 5 September
(15) France/Spain/Italy, Dir Gilles
Mimouni
with Romaine Bohringer, Vincent
Cassel, Jean-Phillipe Écoffey,
Monica Bellucci, Sandrine
Kiberlain, Olivier Granier
© **IMA Films/UGC Images/La
Sept Cinéma/M6 Films.
In association with Mate
Production Cecchi Gori Group
Tiger Cinematografica s.r.l.
Supported by funds from the
Fonds Eurimages of the Council of
Europe With the participation of
Centre National de la
Cinématographie/Canal+/Sofica
Sofinergie 3**
116 minutes 10 seconds (Subtitles)
S&S October 1997 p43

The Arrival
Entertainment - 24 January
(12) USA, Dir David Twohy
with Charlie Sheen, Lindsay Crouse,
Teri Polo, Richard Schiff, Leon
Rippy, Tony T.Johnson
**Live Film and Mediaworks Inc.
Live Entertainment presents a
Steelworks Film/Thomas G. Smith
production**
115 minutes 1 second
S&S March 1997 p40

The Associate
PolyGram - 13 June
(PG) USA, Dir Donald Petrie
with Whoopi Goldberg, Dianne
Wiest, Tim Daly, Bebe Neurwirth,
Eli Wallach, Austin Pendleton
**PolyGram Film Productions B.V.
PolyGram Filmed Entertainment
presents
An Interscope Communications in
association with Frédéric
Golchan/René Gainville**
113 minutes 19 seconds
S&S August 1997 p37

Austin Powers International Man of Mystery
Guild - 5 September

(15) USA, Dir Jay Roach
with Mike Myers, Elizabeth Hurley,
Michael York, Mimi Rogers, Robert
Wagner, Seth Green
**New Line Productions Inc New
Line Cinema presents in
association with Capella
International/KC Medien
A Moving Pictures/Eric's Boy
production**
94 minutes 47 seconds
S&S September 1997 p37

BAPS
Entertainment - 1 August
(15) USA, Dir Robert Townsend
with Halle Berry, Martin Landau, Ian
Richardson, Natalie Desselle, Troy
Beyer, Luigi Amodeo
**New Line Cinema Corporation
An Island Pictures production**
92 minutes 7 seconds
S&S August 1997 p38

Bang
PolyGram - 11 July
(18) USA, Dir Ash
with Darling Narita, Peter Greene,
Everlast, Michael Arturo, James
Sharpe, Luis Guizar
**A Renegade Film Inc presentation
Eagle Eye Films Inc/Asylum Films
Ltd**
99 minutes 51 seconds
S&S August 1997 p38
US release title: **The Big Bang
Theory**

Basquiat
Guild - 28 March
(15) USA, Dir Julian Schnabel
with David Bowie, Dennis Hopper,
Gary Oldman, Jeffrey Wright,
Benicio Del Toro, Claire Forlani
**Eleventh Street Productions
Miramax Films and John Kirk
present
A Peter Brant-Joseph Allen
production**
106 minutes 34 seconds
S&S April 1997 p35

Batman & Robin
Warner Bros - 27 June
(15) USA, Dir Joel Schumacher
with Arnold Schwarzenegger,
George Clooney, Chris O'Donnell,
Uma Thurman, Alicia Silverstone,
Michael Gough
Warner Bros
125 minutes
S&S August 1997 p39

Battle of Algiers/La Battaglia di Algeri *
BFI - 27 June
(18) Italy/Algeria 1966, Dir Gillo
Pontecorvo
with Brahim Haggiag, Jean Martin,
Yacef Saadi, Sami Kerbash
Igor Film, Casbah Film
135 minutes (b/w)
S&S March 1997 p69 (article)
MFB v.38 n.447 April 1971 p68

Bean
PolyGram - 8 August
(PG) UK, Dir Mel Smith
with Rowan Atkinson, Peter
MacNicol, Pamela Reed, Harris
Yulin, Burt Reynolds, Larry Drake
**PolyGram Film Production BV/
NV PolyGram SA
PolyGram Filmed Entertainment
presents A Working Title production
In association with Tiger Aspect**
89 minutes 58 seconds
S&S August 1997 p41
Category B

Beavis and Butt-head Do America
UIP - 23 May
(12) USA, Dir Mike Judge
with the voices of Mike Judge,
Robert Stack, Cloris Leachman,
Jacqueline Barba, Pamela Blari, Eric
Bogosian
**MTV Networks
A Paramount Pictures
presentation in association with
Geffen Pictures
An MTV production**
80 minutes 57 seconds
S&S June 1997 p46

Big Night
Electric Pictures - 30 May
(15) USA, Dir Stanley Tucci,
Campell Scott
with Minnie Driver, Ian Holm,
Isabella Rossellini, Tony Shaulhoub,
Stanley Tucci, Caroline Aaron
**Rysher Entertainment Inc
A Timpano production**
109 minutes 28 seconds
S&S June 1997 p47

A Bit of Scarlet
BFI - 20 June
(15) USA, Dir Andrea Weiss
Narrator Ian McKellen
**BFI and Channel Four
The British Film Institute and
Channel Four present a Maya
Vision production**
74 minutes 38 seconds
S&S August 1997 p42

Bits and Pieces/Il cielo è sempre più blu
Electric Pictures - 11 April
(15) Italy, Dir Antonello Grimaldi
with Asia Argento, Luca

Barbareschi, Margherita Buy,
Roberto Citran, Enrico Le Verso,
Ivan Marescotti
Fandango s.r.l/Colorado Film
Production C.F.P. s.r.l
109 minutes 3 seconds (subtitles)
S&S April 1997 p36

Blood and Wine

20th Century Fox - 7 March
(15) USA/UK, Dir Bob Rafelson
with Jack Nicholson, Stephen Dorff,
Jennifer Lopez, Judy Davis, Harold
Perrineau Jr, Michael Caine
Blood and Wine Productions, Inc
Made in association with Marmont
Productions, Inc/Recorded Picture
Company/Majestic Films
100 minutes 23 seconds
S&S March 1997 p41
Category D[2]

The Blue Angel/ Der blaue Engel *

BFI - 17 October
(Not submitted) Germany 1930 Dir
Josef von Sternberg
with Emil Jannings, Marlene
Dietrich, Kurt Gerron, Rosa Valetti
Ufa
106 minutes (b/w)
No S&S reference
MFB v.21 n.247 August 1954 p115

Booty Call

Columbia TriStar - 3 October
(18) USA, Dir Jeff Pollack
with Jamie Foxx, Tommy Davidson,
Vivica A.Fox, Tamala Jones, Scott
LaRose, Ric Young
© **Columbia Pictures Industries,**
Inc.
A Turman/Morrissey Company
production
79 minutes 15 seconds
S&S December 1997 p39

The Borrowers

PolyGram - 5 December
(U) UK, Dir Peter Hewitt
with John Goodman, Jim Broadbent,
Mark Williams, Celia Imrie, Hugh
Laurie, Ruby Wax
© **PolyGram Filmed Entertainment.**
A Working Title production
86 minutes 33 seconds
S&S December 1997 p40
Category B

Bound

Guild - 28 February
(18) USA, Dir Andy Wachowski,
Larry Wachowski
with Jennifer Tilly, Gina Gershon, Joe
Pantoliano, John P. Ryan, Christopher
Meloni, Richard C. Sarafian
Dino De Laurentiis Company In
association with Summit

Entertainment and Newmarket
Capital Group
108 minutes 33 seconds
S&S March 1997 p42

Box of Moon Light

First Independent - 18 April
(15) USA, Dir Tom DiCillo
with John Turturro, Sam Rockwell,
Catherine Keener, Lisa Blount,
Annie Corley, Rica Martens
Lakeshore Entertainment Corp.
Largo Entertainment in
association with JVC
Entertainment, Inc
A Lemon Sky production
111 minutes 51 seconds
S&S May 1997 p38

The Boy from Mercury

Blue Dolphin - 2 May
(PG) UK/France, Dir Martin Duffy
with Rita Tushingham, Tom
Courtenay, Hugh O'Conor, James
Hickey, Ian McElhinney, Joanne
Gerrard
Mercurian Productions Ltd/Blue
Dahlia Productions, SA./Blue Rose
Films/Joe Manuel Productions
Ltd/Le Studio Canal + SA
Produced with the assistance of
Bord Scánnan na hÉireann/Irish
Film Board
Made in association with Radio
Telefis Éireann
Produced with the support of
investment incentives for the Irish
Film Industry provided by the
Government of Ireland
Developed with the support of the
European Script Fund
87 minutes
S&S April 1997 p 37
Category C

Broken English

First Independent - 25 July
(18) New Zealand, Dir Gregor
Nicholas
with Rade Serbedzija, Aleksandra
Vujcic, Julian Arahanga, Marton
Csokas, Madeline McNamara
Communicado (Broken English)
Limited Village Roadshow
presents A Communicado
production In association with the
New Zealand Film Commission.
NZ on Air
92 minutes 15 seconds
S&S August 1997 p42

The Butterfly Effect/El efecto mariposa

Blue Dolphin - 18 July
(15) Spain/France/UK, Dir
Fernando Colomo
with Maria Barranco, Coque Malla,

Rosa Maria Sardá, James Fleet, Peter
Sullivan, Cécile Pallas
Fernando Colomo P.C./
Mainstream S.A./Portman
Productions Limited/Oceandeep
Ltd With the participation of TVE
Television Española/Canal+
España/Canal+ France/Eurimages
108 minutes 54 seconds
S&S August 1997 p43
Category C

Career Girls

Film Four Distributors - 19
September
(15) UK, Dir Mike Leigh
with Katrin Cartlidge, Lynda
Steadman, Mark Benton, Kate
Byers, Andy Serkis, Joe Tucker
Channel Four Television
Corporation/Thin Man Films
Limited/Matrix Film Partnership
87 minutes 3 seconds
S&S September 1997 p38
Category A

Carla's Song

PolyGram - 31 January
(15) UK/Germany/Spain, Dir Ken
Loach
with Robert Carlyle, Oyanka
Cabezas, Scott Glenn, Salvador
Espinoza, Louise Goodall, Richard
Loza
Channel Four Television
Corporation and The Glasgow
Film Fund A Parallax Picture in
co-production with Road Movies
Dritte Productionen and Tornasol
Films S.A.
With the support of The Institute
of Culture, Nicaragua and ARD/
DEGETO Film, Filmstiftung
NordrheinWestfalen, Television
Española and Alta Films, the
European Script Fund, and the
Scottish Film Production Fund
125 minutes 3 seconds
S&S February 1997 p38
Category B

The Chamber

UIP - 20 June
(12) USA, Dir James Foley
with Chris O'Donnell, Gene
Hackman, Faye Dunaway, Lela
Rochon, Robert Prosky, Raymond
Barry
Universal City Studios, Inc
A Universal Pictures and Imagine
Entertainment presentation
A Brian Grazer/Davis
Entertainment production
112 minutes 51 seconds
S&S July 1997 p36

Chasing Amy

Metrodome - 14 November

(18) USA, Dir Kevin Smith
with Ben Affleck, Joey Lauren
Adams, Jason Lee, Dwight Ewell,
Jason Mewes, Kevin Smith, Ethan
Suplee
**Miramax Films presents A View
Askew production**
113 minutes 12 seconds
S&S October 1997 p36

Citizen Kane *

Gala - 11 April
(U) USA 1941, Dir Orson Welles
with Orson Welles, Dorothy
Comingore, Everett Sloane, Joseph
Cotten
**RKO Radio Pictures, Mercury
Productions**
109 minutes (b/w)
No S&S reference
MFB v.8 n.96 December 1941 p164

City of Industry

PolyGram - 4 July
(15) USA, Dir John Irvin
with Harvey Keitel, Stephen Dorff,
Timothy Hutton, Famke Janssen,
Wade Dominguez, Michael Jai
White
**Largo Entertainment, Inc in
association with JVC
Entertainment Inc**
96 minutes 53 seconds
S&S August 1997 p44

 ### Close-Up/Namayeh Nazdik

BFI - 19 December
(Not submitted) Iran, Dir Abbas
Kiarostami
with Hossain Sabzian, Mohsen
Makhmalbaf, Abolfazl Ahankhah,
Mahrokh Ahankhah, Nayer Mohseni
Zonoozi
© **The Institute for the Intellectual
Development of Children and
Young Adults**
100 minutes (subtitles)
S&S December 1997 p 40

 ### Cloud-capped Star/Meghe Dhaka Tara

BFI - 15 August
(Not submitted) India, Dir Ritwik
Ghatak
with Supriya Choudhury, Anil
Chatterjee, Bijon Bhattacharya, Guita
De, Gita Ghatak, Dwiju Bhawal
Chitrakalpa
126 minutes 33 seconds (b/w)
(subtitles)
S&S September 1997 p39

Cold Comfort Farm

Feature Film Company - 25 April
(PG) UK, Dir John Schlesinger
with Eileen Atkins, Kate Beckinsale,
Sheila Burrell, Stephen Fry, Freddie

Jones, Joanna Lumley, Ian McKellen
**BBC A BBC Television/Thames
Television production**
103 minutes 33 seconds
S&S May 1997 p39
Category A

Con Air

Buena Vista - 6 June
(15) USA, Dir Simon West
with Nicolas Cage, John Cusack,
John Malkovich, Steve Buscemi,
Ving Rhames, Colm Meany
**Touchstone Pictures/Jerry
Bruckheimer, Inc**
115 minutes 13 seconds
S&S July 1997 p37

Conspiracy Theory

Warner Bros - 29 August
(15) USA, Dir Richard Donner
with Mel Gibson, Julia Roberts,
Patrick Stewart, Cylk Cozart,
Stephen Kahan, Terry Alexander
**Warner Bros A Silver Pictures
production in association with
Shuler Donner/Donner
Productions**
135 minutes 9 seconds
S&S September 1997 p39

Conspirators of Pleasure/ Spiklenci slasti

ICA - 14 February
(Not submitted) Czech Republic/
Switzerland/UK, Dir Jan Svankmajer
with Petr Meissel, Gabriela
Wilhelmová, Barbora Hrzánová, Anna
Weltlinská, Jiri Lábus, Pavel Novy
**Produced by Athanor in co-
prodcution with Pierre Assouline/
Delfilm S.A. Geneva and Keith
Griffiths/Koninck London. This
film was supported by The
Eurimages Fund of the Council of
Europe, The State Fund of the
Czech Republic for Czech
Cinematography and The Federal
Office of Culture (Switzerland)**
75 minutes
S&S February 1997 p39
Category C

Contact

Warner Bros - 26 September
(PG) USA, Dir Robert Zemeckis
with Jodie Foster, Matthew
McConaughey, Tom Skerritt, Angela
Bassett, John Hurt, David Morse,
Rob Lowe
© **Warner Bros.
A Southside Amusement Company
production**
149 minutes 37 seconds
S&S October 1997 p44

CopLand

Buena Vista - 5 December

(15) USA, Dir James Mangold
with Sylvester Stallone, Harvey
Keitel, Ray Liotta, Robert De Niro,
Peter Berg, Janeane Garofalo
**A Woods Entertainment
production Developed with the
assistance of the Sundance
Institute**
104 minutes 37 seconds
S&S December 1997 p41

Costa Brava (Family Album)

DTK - 2 February
(15) Spain, Dir Marta Balletbò-Coll
with Desi del Valle, Marta Ballebò-
Coll, Montserrat Gausachs, Josep
Maria Brugués, Ramon Mari, Sergi
Schaaff
Marta Balletbò-Coll
92 minutes (subtitles)
S&S March 1997 p43

Crash

Columbia TriStar - 6 June
(18) Canada, Dir David Cronenberg
with James Spader, Holly Hunter,
Deborah Kara Unger, Rosanna
Arquette, Elias Koteas, Peter
MacNeil
**Alliance Communications
Corporation in Trust A Jeremy
Thomas & Robert Lantos
presentation Produced with the
participation of Telefilm Canada/
TWN - The Movie Network**
100 minutes 11 seconds
S&S June 1997 p48

The Crucible

20th Century Fox - 28 February
(15) USA, Dir Nicholas Hytner
with Daniel Day-Lewis, Winona
Ryder, Paul Scofield, Joan Allen,
Bruce Davison, Rob Campbell
**Twentieth Century Fox Film
Corporation**
123 minutes 21 seconds
S&S March 1997 p44

Crying Freeman

Guild - 9 May
(18) France/Canada/USA/Japan, Dir
Christophe Gans
with Mark Dasascos, Julie Condra,
Rae Dawn Chong, Byron Mann,
Masaya Kato, Yoko Shimada
**Crying Freeman Productions Inc
Samuel Hadida presents in
association with Toei Video
Company/Fuji Television/
Tohokushinsha Film Corporation
A Davis Films/Ozla Pictures/
Yuzna Films production**
101 minutes 45 seconds
S&S May 1997 p40

Curdled

NFT/Miramax - 2 December

(Not submitted) USA Dir Reb Braddock with William Baldwin, Bruce Ramsey, Lois Chiles, Mel Gorham, Barry Corbin, Carmen Lopez
Tinderbox, Band Apart
88 mins
No S&S reference

Dancehall Queen

Island Jamaica - 26 September
(Not submitted) Jamaica, Dir Don Letts with Audrey Reid, Carl Davis, Paul Campbell, Pauline Stone-Myrie
Island Digital Media, Jamaica Films, Hawk's Nest Production
98 mins
No S&S reference

Dangerous Ground

Entertainment - 16 May
(18) South Africa/USA, Dir Darrell James Roodt
with Ice Cube, Elizabeth Hurley, Sechaba Morojele, Eric 'Waku' Miyeni, Ving Rhames, Thokozani Nkosi
Investec Merchant Bank A New Line Cinema presentation
95 minutes 29 seconds
S&S June 1997 p49

Dante's Peak

UIP - 28 March
(12) USA, Dir Roger Donaldson with Pierce Brosnan, Linda Hamilton, Jamie Renée Smith, Jeremy Foley, Elizabeth Hoffman, Charles Hallahan
Universal City Studios Inc A Pacific Western production
108 minutes 25 seconds
S&S April 1997 p38

Darklands

Metrodome - 24 October
(18)* UK, Dir Julian Richards with Craig Fairbass, Rowena King, Jon Finch, David Duffy, Roger Nott, Richard Lynch
© Darklands Ltd.
A Lluniau Lliw Cyf production Produced with the participation of The Arts Council of Wales (Euryn Ogwen Williams. Jo Weston, Lucy Jones)
89 minutes 42 seconds
S&S October 1997 p37
* Not submitted for theatrical release. Video certificate and time.
Category A

Deep Crimson/Profundo carmesi/Carmin profond

Metro/Tartan - 19 September
(15) Mexico/France/Spain Dir Arturo Ripstein

with Daniel Giménez Cacho, Regina Orozco, Marisa Paredes, Verónica Merchant, Julieta Egurrola, Patricia Reyes Spindola
Ivania Films/IMCINE/MK2 Productions/Wanda Films With Fondo de Fomento a la Calidad Cinematográfica, Gobierno del Estado de Sonora, Televisión Española, Les Productions Traversière With the collaboration of Canal+ Made with the support of FONCA
114 minutes 48 seconds (subtitles)
S&S September 1997 p40

Def Jam's How to Be a Player

PolyGram - 28 November
(18) USA, Dir Lionel C. Martin with Bill Bellamy, Natalie Desselle, Lark Voorhies, Mari Morrow, Pierre, Jermaine 'Big Hugg' Hopkins
**PolyGram Filmed Entertainment presents
An Island Pictures production in association with Outlaw Productions**
93 minutes 45 seconds
S&S January 1998 p37

The Devil's Own

Columbia TriStar - 20 June
(15) USA, Dir Alan J. Pakula with Harrison Ford, Brad Pitt, Margaret Colin, Rubén Blades, Treat Williams, George Hearn
Columbia Pictures Industries Inc A Lawrence Gordon presentation
111 minutes 9 seconds
S&S June 1997 p50

The Disappearance of Kevin Johnson

The Bedford Communications Group - 25 July
(15) USA, Dir Francis Megahy with Pierce Brosnan, James Coburn, Dudley Moore, Alexander Folk, Bridget Bass, Carl Sundstrom
Makani Kai Productions Inc & Wobblyscope Inc
105 minutes 44 seconds
S&S September 1997 p41

Donnie Brasco

Entertainment - 2 May
(18) USA, Dir Mike Newell with Al Pacino, Johnny Depp, Michael Madsen, Bruno Kirby, James Russo, Anne Heche
Mandalay Entertainment A Baltimore Pictures/Mark Johnson production
126 minutes 23 seconds
S&S May 1997 p40

Drifting Clouds/Kauas pilvet karkaavat/Au loin s'en vont les nuages

Metro/Tartan - 6 June
(PG)Finland/Germany/France, Dir Aki Kaurismäki
with Kati Outinen, Kari Väänänen, Elina Salo, Sakari Kuosmanen, Markku Peltola, Matti Onnismaa, Matti Pellonpää
Sputnik Oy In association with YLE TV-1 (Eila Werning)/ Pandora Film (Reinhard Brunig)/ Pyramide Productions S.A. (Fabienne Vonier) Production supported by the Finnish Film Foundation/The Nordic Film and Television Fund
96 minutes 56 seconds (subtitles)
S&S June 1997 p51

Driftwood

Blue Dolphin Film/Goldcrest - 28 March
(18) Ireland/UK, Dir Ronan O'Leary with James Spader, Anne Brochet, Barry McGovern, Anna Massey, Aiden McHugh, John Cleere
Deadwood Limited Goldcrest Films International in association with Mary Breen-Farrelly presents A Setanta Film Produced with the support of investment incentives for the Irish Film Industry provided by the Government of Ireland
100 minutes 19 seconds
S&S April 1997 p39
Category C

Eddie

First Independent - 25 April
(15) USA, Dir Steve Rash with Whoopi Goldberg, Frank Langella, Dennis Farina, Richard Jenkins, Lisa Ann Walter, John Benjamin Hickey
Hollywood Pictures Company In association with PolyGram Filmed Entertainment/Island Pictures
100 minutes 23 seconds
S&S May 1997 p41

8 Heads in a Duffel Bag

Carlton - 28 November
(15) USA/UK, Dir Tom Schulman with Joe Pesci, Andy Comeau, Kristy Swanson, Todd Louiso, George Hamilton, Dyan Cannon
Rank Film Distributors presents in association with Orion Pictures A Brad Krevoy & Steve Stabler production
94 minutes 50 seconds
S&S December 1997 p42
Category D²

The Empire Strikes Back *

Twentieth Century Fox - 11 April
(U) USA 1980/1997, Dir Irvin
Kershner
with Mark Hamil, Harrison Ford,
Carrie Fisher, Billy Dee Williams
Lucas Film
Twentieth Century Fox
Corporation
124 mins
No S&S reference
MFB v.47 n.558 July 1980 p129

The English Patient

Buena Vista - 14 March
(15) USA, Dir Anthony Minghella
with Ralph Fiennes, Juliette
Binoche, Willem Dafoe, Kristin
Scott Thomas, Naveen Andrews,
Colin Firth
Tiger Moth Productions Inc,
Miramax Films
161 minutes 31 seconds
S&S March 1997 p45

Entertaining Angels The Dorothy Day Story

Warner Bros - 6 June
(12) USA, Dir Michael Ray Rhodes
with Moira Kelly, Heather Graham,
Melinda Dillon, Lenny Von Dohlen,
Martin Sheen, Paul Lieber
Paulist Pictures
111 minutes 23 seconds
S&S July 1997 p38

Evening Star

Entertainment - 14 March
(15) USA, Dir Robert Harling
with Shirley MacLaine, Bill Paxton,
Juliette Lewis, Miranda Richardson,
Ben Johnson, Scott Wolf
Rysher Entertainment, Inc
A Paramount Pictures/Rysher
Entertainment presentation
128 minutes 52 seconds
S&S May 1997 p42

Event Horizon

UIP - 22 August
(15) USA, Dir Paul Anderson
with Laurence Fishburne, Sam Neill,
Kathleen Quinlan, Joely Richardson,
Richard T. Jones, Jack Noseworthy
© Paramount Pictures.
A Golar production in association
with Impact Pictures
95 minutes 46 seconds
S&S October 1997 p46
Category D[1]

Everyone Says I Love You

Buena Vista - 18 April
(12) USA, Dir Woody Allen
with Alan Alda, Woody Allen, Drew
Barrymore, Lukas Haas, Goldie
Hawn, Gaby Hoffmann

Magnolia Productions, Inc/
Sweetland Films B.V.
100 minutes 53 seconds
S&S April 1997 p40

Excess Baggage

Columbia TriStar - 21 November
(12) USA, Dir Marco Brambilla
with Alicia Silverstone, Benico Del
Toro, Christopher Walken, Jack
Thompson, Nicholas Turturro,
Michael Bowen
© Columbia Pictures Industries,
Inc.
A First Kiss production
101 minutes 3 seconds
S&S October 1997 p37

Extreme Measures

Rank-Castle Rock/Turner - 31
January
(15) USA/UK, Dir Michael Apted
with Hugh Grant, Gene Hackman,
Sarah Jessica Parker, David Morse,
Bill Nunn
Columbia/Castle Rock
Entertainment presents a Simian
Films production
117 minutes 45 seconds
S&S October 1997 p47
Category D[2]

Face

UIP - 26 September
(18) UK, Dir Antonia Bird
with Robert Carlyle, Ray Winstone,
Steven Waddington, Philip Davis,
Damon Albarn, Lena Headey
© BBC/Distant Horizon Limited.
Produced by Daigoro Face
Productions Ltd
With the participation of British
Screen
105 minutes 11 seconds
S&S October 1997 p47
Category A

Face/Off

Buena Vista - 7 November
(18) USA, Dir John Woo
with John Travolta, Nicolas Cage,
Joan Allen, Alessandro Nivola, Gina
Gershon, Dominique Swain
© Paramount Pictures
Corporation and Touchstone
Pictures.
A Douglas/Reuther production
A WCG Entertainment production
138 minutes 42 seconds
S&S October 1997 p38

Fathers' Day

Warner Bros - 10 October
(12) USA, Dir Ivan Reitman
with Robin Williams, Billy Crystal,
Julia Louis-Dreyfus, Natassja Kinski,
Charles Hofheimer, Bruce Greenwood
© Warner Bros.

A Silver Pictures production in
association with Northern Lights
Entertainment
99 minutes 4 seconds
S&S October 1997 p48

Female Perversions

Feature Film Company - 2 May
(18) USA/Germany, Dir Susan
Streifeld
with Tilda Swinton, Amy Madigan,
Karen Sillas, Frances Fisher, Laila
Robins, Clancy Brown
MAP Films, Inc. Transatlantic
Entertainment presents a Mindy
Affrime production Co-production:
Kinowelt Filmproduktion/Degeto
Film for ARD
113 minutes 6 seconds
S&S May 1997 p43

Fetishes

ICA - 5 September
(Not submitted) USA/UK Dir
Nicholas Broomfield
Lafayette Films
84 minutes
No S&S reference
Category D[1]

Fever Pitch

Film Four Distributors - 4 April
(15) UK, Dir David Evans
with Colin Firth, Ruth Gemmell,
Neil Pearson, Lorraine Ashbourne,
Mark Strong, Holly Aird
Film Four Distributors, Ltd,
Channel Four Films present a
Wildgaze Films production
102 minutes 20 seconds
S&S April 1997 p41
Category A

Fierce Creatures

UIP - 14 February
(PG) USA/UK, Dir Robert Young,
Fred Schepisi
with John Cleese, Jamie Lee Curtis,
Kevin Kline, Michael Palin, Ronnie
Corbett, Carey Lowell, Robert Lindsay
Universal City Studios Inc
A Fish Productions/Jersey Films
production
93 minutes 5 seconds
S&S March 1997 p46
Category D[1]

The Fifth Element/Le Cinquième Elément

Guild - 6 June
(15) France, Dir Luc Besson
with Bruce Willis, Gary Oldman, Ian
Holm, Milla Jovovich, Chris Tucker,
Luke Perry
Gaumont
126 minutes 18 seconds
S&S July 1997 p39

First Strike/Jingcha gushi 4 Zhi Jiandan Renwu

Entertainment - 3 October
(15) Hong Kong/USA, Dir Stanley Tong
with Jackie Chan, Chen Chun Wu, Jackson Lou, Bill Tung, Jouri Petrov, Grishajeva Nonna
© **Paragon Films Ltd/New Line Productions Inc.**
A Raymond Chow/Golden Harvest production
83 minutes 41 seconds
S&S October 1997 p49

Flirt

Artificial Eye - 21 February
(15) USA/Germany/Japan, Dir Hal Hartley
with Bill Sage, Martin Donovan, Parker Posey, Hannah, Sullivan
True Fiction Pictures
In association with Pandora Films/ NDF - Nippon Development & Finance With the support of Filmboard Berlin-Brandenburg GmbH
83 minutes 50 seconds
S&S March 1997 p47

Flirting with Disaster

Buena Vista - 24 January
(15) USA, Dir David O.Russell
with Ben Stiller, Patricia Arquette, Téa Leoni, Mary Tyler Moore, George Segal, Alan Alda, Lily Tomlin
Miramax International
92 minutes 11seconds
S&S February 1997 p42

Fly Away Home

Columbia TriStar- 7 February
(U) USA, Dir Carroll Ballard
with Jeff Daniels, Anna Paquin, Dana Delany, Terry Kinney, Holter Graham, Jeremy Ratchford
Columbia Pictures Industries, Inc
A Sandollar Production
106 minutes 50 seconds
S&S February 1997 p43

Fools Rush In

Columbia TriStar - 31 October
(12) USA, Dir Andy Tennant
with Matthew Perry, Salma Hayek, Jon Tenney, Carlos Gomez, Tomas Milian
© **Columbia Pictures Industries, Inc.**
109 minutes 1 seconds
S&S October 1997 p50

Frantz Fanon Black Skin White Mask

BFI - 27 June
(Not submitted) UK/France, Dir Isaac Julien

with Colin Salmon, Al Nedjari, John Wilson, Ana Remalho, Norin Ni Dubhgail
Normal Films Ltd In association with the BBC, the Arts Council of England and L'Institut National de l'Audiovisuel Supported by the National Lottery through the Arts Council of England and by a grant from the Ford Foundation
71 minutes (partly subtitled)
S&S September 1997 p42
Category B

Free Willy 3 The Rescue

Warner Bros - 18 October
(15) USA, Dir Sam Pillsbury
with Jason James Richter, August Schellenberg, Annie Corley, Vincent Berry, Patrick Kilpatrick, Tasha Simms
© **Warner Bros.**
Production Limited/ Monarchy Enterprises B.V./Regency Entertainment (USA) Inc
A Shuler Donner/Donner production
85 minutes 33 seconds
S&S October 1997 p51

The Frighteners

UIP - 24 January
(15) New Zealand/USA, Dir Peter Jackson
with Michael J. Fox, Trini Alvarado, Peter Dobson, John Astin, Jeffrey Combs, Dee Wallace Stone
Wingnut Films
Universal Pictures
109 minutes
S&S February 1997 p43

Frisk

Dangerous to Know - 21 November
(Not submitted) USA, Dir Todd Verow
with Michael Gunther, Craig Chester, Parker Posey, James Lyons, Alexis Arquette, Raoul O'Connell, Jaie Laplante
Strand Releasing presents an Industrial Eye production This project was developed in part by Cinemart, Rotterdam Film Festival
83 minutes
S&S January 1998 p40

Full Contact/Xia Dao Gao Fei

Made in Hong Kong - 14 November
(18) Hong Kong, Dir Ringo Lam
with Chow Yun-Fat, Simon Yam, Ann Bridgewater, Anthony Wong, Bonnie Fu, Lee Kin-Sang
Golden Princess Film Production Limited
96 minutes 43 seconds (subtitles)
S&S January 1998 p41

The Full Monty

20th Century Fox - 29 August
(15) USA/UK, Dir Peter Cattaneo
with Robert Carlyle, Tom Wilkinson, Mark Addy, Lesley Sharp, Emily Woof, Steve Huison, Paul Barbar
Twentieth Century Fox Film Corporation Fox Searchlight Pictures presents A Redwave Films production Developed by Channel Four Television Corporation
91 minutes 27 seconds
S&S September 1997 p43
Category B

The Funeral

Guild - 18 April
(18) USA, Dir Abel Ferrara
with Christopher Walken, Chris Penn, Vincent Gallo, Benicio Del Toro, Annabella Sciorra, Isabella Rossellini
October Film presents in association with MDP Worldwide A C*P production
98 minutes 59 seconds
S&S April 1997 p42

A Further Gesture

Film Four Distributors - 12 December
(15) UK/Germany/Japan/Ireland, Dir Robert Dornhelm
with Stephen Rea, Alfred Molina, Rosana Pastor, Brendan Gleeson, Pruitt Taylor Vince, Maria Doyle Kennedy
Channel Four Films in association with NDF International Ltd. and Pony Canyon Inc presents A Zephyr Films Samson Films/Road Movies Dritte Production Produced with the support of investment incentives for the Irish Film Industry including production finance from Board Scannán na hÉireann/The Irish Film Board Supported by Eurimages/Filmstiftung Nordrhein-Westfalen Developed with the support of the European Script Fund
100 minutes 50 seconds
S&S January 1998 p42
Category B

G.I. Jane

First Independent - 14 November
(15) USA/UK, Dir Ridley Scott
with Demi Moore, Viggo Mortensen, Anne Bancroft, Jason Beghe, Scott Wilson, Lucinda Jenney
© **Trap-Two Zero Productions Inc and Hollywood Pictures Company. Presented in association with Scott**

Free and Largo Entertainment A
Roger Birnbaum/Scott Free/
Moving Pictures production
Produced in association with First
Independent Films Ltd
125 minutes 14 seconds
S&S November 1997 p42
Category D[2]

 Gallivant
Electric Pictures - 19
September
(15) UK, Dir Andrew Kötting
with Gladys Morris, Eden Kötting,
Andrew Kötting, Douglas
Templeton, Gary Parker
**A Tall Stories production in
association with The British Film
Institute and Channel 4 Television
Supported from the proceeds of
the National Lottery through the
Arts Council of England and with
the assistance from the European
Union's 16:9 Action Plan**
103 minutes 31 seconds
S&S September 1997 p44
Category A

The Gambler

**Film Four Distributors - 7
November**
(15) UK/Netherlands/Hungary, Dir
Károly Makk
with Michael Gambon, Jodhi May,
Polly Walker, Dominic West, John
Wood, Johan Leysen
© Trendraise Company Ltd/
Channel Four Television
Corporation.
**A Channel Four Films and UGC
DA International presentation in
association with Hungry Eye
Pictures/KRO Drama A Gambler
Productions (UK) production in
association with Hungry Eye
Lowland (The Netherlands) and
Objektiv Filmstudio (Hungary)
Supported by Nederlands Fonds
voor de Film, Coproductiefonds
Binnenlandse/Omroep and
Eurimages Developed with the
support of the European Script
Fund**
97 minutes 15 seconds
S&S November 1997 p40
Category C

The Game

PolyGram - 10 October
(15) USA, Dir David Fincher
with Michael Douglas, Sean Penn,
James Rebhorn, Deborah Kara
Unger, Peter Donat, Carroll Baker
© PolyGram Filmed
Entertainment, Inc.
A Propaganda Films production
128 minutes 1 seconds
S&S November 1997 p41

George of the Jungle

Buena Vista - 19 December
(U) USA, Dir Sam Weisman
with Brendan Fraser, Leslie Mann,
Thomas Haden Church, Richard
Rowntree, Greg Cruttwell
**Walt Disney Pictures presents A
Mandeville Films Avnet/Kerner
production**
91 minutes 38 seconds
S&S January 1998 p43

Get on the Bus

Columbia TriStar - 11 July
(15) USA, Dir Spike Lee
with Richard Belzer, DeAundre
Bonds, Andre Braugher, Thomas
Jefferson Byrd, Gabriel Casseus,
Albert Hall
**Columbia Pictures Industries, Inc
40 Acres & a Mule A Spike Lee
Joint This film was completely
funded by 15 African American
men: Larkin Arnold, Jheryl
Busby, Reggie Rock Blytheewood,
Reuben Cannon, Johnnie L.
Cochran Jr, Lemuel Daniels,
Danny Glover, Calvin Grogsby,
Robert Guillaume, Robert
Johnson, Olden Lee, Spike Lee,
Charles D. Smith, Will Smith,
Wesley Snipes**
121 minutes 50 seconds
S&S July 1997 p40

The Ghost and the Darkness

UIP - 17 January
(15) USA, Dir Stephen Hopkins
with Michael Douglas, Val Kilmer,
Bernard Hill, John Kani, Tom
Wilkinson
**Paramount Pictrues. Constellation
Films presents a Douglas/Reuther
production**
109 minutes 51 seconds
S&S February 1997 p45

Ghosts from the Past

Rank - 9 May
(15) USA, Dir Rob Reiner
with Alec Baldwin, Whoopi
Goldberg, James Woods, Craig T.
Nelson, Susanna Thompson, Lucas
Black
Castle Rock Entertainment
130 minutes 24 seconds
S&S May 1997 p44
US title: **Ghosts of Mississippi**

Grace of My Heart

UIP - 21 February
(15) USA, Dir Allison Anders
with Illeana Douglas, Matt Dillon,
Eric Stolz, Bruce Davison, Patsy
Kensit, Jennifer Leigh Warren
**Universal City Studios, Inc
Gramercy Pictures presents a**

Cappa production
115 minutes 27 seconds
S&S March 1997 p48

Gridlock'd

PolyGram - 30 May
(18) USA/UK, Dir Vondie Curtis-
Hall
with Tim Roth, Tupac Shakur,
Thandie Newton, Charles Fleishcher,
Howard Hesseman, James Pickens Jr
**PolyGram Film Productions B.V.
An Interscope Communications
produced in association with DEF
Pictures/Webster and Dragon
Pictures**
91 minutes 9 seconds
S&S June 1997 p52
Category D[1]

Grosse Pointe Blank

Buena Vista - 8 August
(15) USA, Dir George Armitage
with John Cusack, Minnie Driver,
Alan Arkin, Dan Aykroyd, Joan
Cusack, Jeremy Piven
**Hollywood Pictures Company In
association with Caravan Pictures/
New Crime production**
107 minutes 25 seconds
S&S August 1997 p45

Hamlet

**Rank-Castle Rock/Turner - 14
February**
(PG) USA, Dir Kenneth Branagh
with Kenneth Branagh, Julie
Christie, Billy Crystal, Gérard
Depardieu, Charlton Heston, Derek
Jacobi, Jack Lemmon, Robin
Williams, Richard Attenborough,
Richard Briers, Judi Dench, Reece
Dinsdale, Ken Dodd, John Gielgud,
John Mills, Timothy Spall, Kate
Winslett
Castle Rock Entertainment
242 minutes 16 seconds
S&S February 1997 p46
Category D[1]

Hard Eight

Entertainment - 17 October
(18) USA, Dir Paul Thomas
Anderson
with Philip Baker Hall, John C.
Reilly, Gwyneth Paltrow, Samuel L.
Jackson, F.William Parker, Philip
Seymour Hoffman
**Rysher Entertainment presents a
Green Parrot production in
association with Trinity Developed
with the assistance of the
Sundance Institute**
101 minutes 25 seconds
S&S January 1998 p44

Hard Men

Entertainment - 4 April

(15) USA, Dir J.K. Amalou
with Vincent Regan, Ross Boatman,
Lee Ross 'Mad' Frankie Fraser, Ken
Campbell, Mirella D'Angelo
Venture Movies/DaciaFilms
86 minutes 53 seconds
S&S February 1997 p47

Harriet the Spy
UIP - 14 February
(PG) USA, Dir Bronwen Hughes
with Michelle Trachtenberg, Vanessa
Lee Chester, Gregory Smith, Rosie
O'Donnell, J.Smith-Cameron,
Robert Joy
Paramount Pictures
In association with Nickelodeon
Movies
A Rastar production
101 minutes 29 seconds
S&S March 1997 p49

Head Above Water
Warner Bros - 3 October
(15) USA/UK, Dir Jim Wilson
with Harvey Keitel, Cameron Diaz,
Craig Sheffer, Billy Zane, Shay
Duffin
Head Above Water Productions
Inc/Firmjewel Limited InterMedia
Films presents in association with
Fine Line Features
A Tig Productions/Majestic Films
production
91 minutes 49 seconds
S&S September 1997 p44
Category D[1]

Heat And Dust *
First Independent - 15 August
(15) UK, Dir James Ivory
with Christopher Cazenove, Greta
Scacchi, Julian Glover, Susan
Fleetwood, Patrick Godfrey
Merchant Ivory Productions
Limited
130 minutes
No S&S reference
MFB v.50 p.580 January 1983 p14

Hell Is a City *
Barbican - 12 September
(15) UK 1959, Dir Val Guest
with Stanley Baker, John Crawford,
Donald Pleasence, Maxine Audley,
Billie Whitelaw
Hammer Film Productions,
Associated British Picture
Corporation
98 minutes
No S&S reference
MFB v.27 n.316 May 1960 p64

Hercules
Buena Vista - 10 October
(U) USA, Dir John Musker, Ron
Clements
with the voices Tate Donovan,

Joshua Keaton, Roger Bart, Danny
DeVito, James Woods, Susan Egan,
Rip Torn, Samantha Eggar
© **Disney Enterprises, Inc.**
A Walt Disney Pictures presentation
92 minutes 42 seconds
S&S October 1997 p52

High School High
Columbia TriStar - 16 May
(15) USA, Dir Hart Bochner
with Jon Lovitz, Tia Carrere, Mekhi
Phifer, Guillermo Diaz, John
Neville, Malinda Williams
TriStar Pictures Inc A David
Zucker production
99 minutes 14 seconds
S&S July 1997 p42

His Girl Friday *
BFI - 7 February
(U) USA 1939, Dir Howard Hawks
with Cary Grant, Rosalind Russell,
Ralph Bellamy, Gene Lockhart,
Porter Hall, Ernest Truex
Columbia Pictures Corporation
92 minutes (b/w)
S&S March 1997 p64

Home Alone 3
20th Century Fox - 19 December
(PG) USA, Dir Raja Gosnell
with Alex D. Linz, Olek Krupa, Rya
Kihlstedt, Lenny Von Dohlen, David
Thornton, Haviland Morris
Twentieth Century Fox presents a
John Hughes production
102 minutes 24 seconds
S&S January 1998 p44

The Honeymoon Killers *
BFI - 5 September
(Not submitted) USA 1969, Dir
Leonard Kastle
with Shirley Stoller, Tony Lo
Bianco, Mary Jane Higby
Roxanne Productions
106 mins (b/w)
No S&S reference
MFB v.37 n.436 May 1970 p98

House of America
First Independent - 10 October
(15) UK/Netherlands, Dir Marc
Evans
with Siân Philips, Steven Mackin-
tosh, Lisa Palfrey, Matthew Rhys,
Pascal Laurent, Richard Harrington
September Films Ltd/Stichting
Bergen With the participation of
British Screen In Association with
the Arts Council of Wales/HTV/
The Dutch Film Fund Supported
by the National Lottery through
the Arts Council of Wales
96 minutes 20 seconds
S&S September 1997 p45
Category B

I Know What You Did Last Summer
Entertainment - 12 December
(18) USA, Dir Jim Gillespie
with Jennifer Love Hewitt, Sarah
Michelle Gellar, Ryan Phillippe,
Freddie Prinze Jr, Johnny Galecki,
Bridgette Wilson
Mandalay Entertainment presents
a Neal H. Moritz production
101 minutes 17seconds
S&S January 1998 p45

Idiot Box
Beyond Films/NFT - 25 July
(Not submitted) Australia, Dir David
Caesar
with Ben Mendelsohn, Jeremy Sims
Central Park Films
84 minutes
No S&S reference

In Love and War
Entertainment - 14 February
(15) USA, Dir Richard
Attenborough
with Sandra Bullock, Chris
O'Donnell, Mackenzie Astin, Emilio
Bonucci, Ingrid Lacey, Margot
Steinberg
New Line Productions, Inc
In association with Dimitri Villard
Productions
114 minutes 29 seconds
S&S March 1997 p50
Category D[1]

Incognito
Warner Bros - 14 November
(15) USA, Dir John Badham
with Jason Patric, Irène Jacob,
Thomas Lockyer, Ian Richardson,
Simon Chandler, Pip Torrens
James G. Robinson presents a
Morgan Creek Production
107 minutes 19 seconds
S&S January 1998 p46

The Informer *
Gala - 6 June
(PG) USA 1935, Dir John Ford
with Victor McLaglen, Heather
Angel, Preston Foster
RKO Radio Pictures
91 minutes (b/w)
No S&S reference
MFB v.2 n.19 August 1935 p104

Intimate Relations
20th Century Fox - 20 June
(15) UK/Canada, Dir Philip
Goodhew
with Julie Walters, Rupert Graves,
Matthew Walker, Laura Sadler, Holly
Aird, Les Dennis
Intimate Relations Limited/
Chandlertown X1 Limited

Partnership (Paragon
Entertainment Corportation)
A HandMade Films presentation
A Boxer Films and Paragon
Entertainment Corporation
production
99 minutes 14 seconds
S&S July 1997 p42
Category B

Inventing the Abbotts

20th Century Fox - 21 November
(15) USA, Dir Pat O'Connor
with Joaquin Phoenix, Billy Crudup,
Will Patton, Kathy Baker, Jennifer
Connelly, Michael Sutton
© Twentieth Century Fox Film
Corporation.
An Imagine Entertainment
production
106 minutes 50 seconds
S&S November 1997 p43

Irma Vep

ICA - 7 March
(Not submitted) France, Dir Olivier
Assayas
with Maggie Cheung [Zhang
Manyu], Jean-Pierre Léaud, Nathalie
Richard, Antoine Basler, Nathalie
Boutefeu, Alex Bescas
Dacia Films
98 minutes (subtitles)
S&S March 1997 p51

It's a Wonderful Life *

Feature Film - 5 December
(U) USA 1946, Dir Frank Capra
with James Stewart, Henry Travers,
Donna Reed, Lionel Barrymore,
Thomas Mitchell
RKO Radio Pictures, Liberty Films
129 minutes (b/w)
No S&S reference
MFB v.14 n.160 April 1947 p50

It Takes Two

Entertainment - 2 May
(15) USA, Dir Andy Tennant
with Kirstie Alley, Steve Guttenberg,
Mary-Kate Olsen, Ashley Olsen,
Phillip Bosco, Jane Sibbett
Rysher Entertainment, Inc An Orr &
Cruickshank produced in association
with Dualstar Productions
97 minutes 22 seconds
S&S June 1997 p53

Jerry Maguire

Columbia TriStar - 7 March
(15) USA, Dir Cameron Crowe
with Tom Cruise, Cuba Gooding Jr,
Renee Zellweger, Kelly Preston,
Jerry O'Connell, Jay Mohr
TriStar Pictures, Inc
A Gracie Films production
138 minutes 54 seconds
S&S March 1997 p52

Johns

Metrodome - 13 June
(18) USA, Dir Scott Silver
with David Arquette, Lukas Haas,
Wilson Cruz, Keith David,
Christopher Gartin
Bandeira Entertainment
95 minutes 39 seconds
S&S June 1997 p53

Jour de Fête *

Gala - 5 July
(U) France 1949/1995, Dir Jacques
Tati
with Jacques Tati, Guy Decomble,
Paul Frankeur
Panoramic Film, Ministère de la
Culture, Centre National de la
Cinématograph
79 minutes
No S&S reference
MFB v.17 n.196 April-May 1950
p59

Jungle 2Jungle

Buena Vista - 23 May
(PG) USA, Dir John Pasquin
with Tim Allen, Sam Huntington,
JoBeth Williams, Lolia Davidovich,
Martin Short, Valerie Mahaffey
Disney Enterprises, Inc and TF1
International
104 41 minutes 41 seconds
S&S June 1997 p54

Jump the Gun

Film Four Distributors - 22
August
(15) UK/South Africa, Dir Les Blair
with Baby Cele, Lionel Newton,
Michele Burgers, Thulani Nyembe,
Rapulana Seiphemo, Danny Keogh
Channel Four Television
Corporation Channel Four Films
presents a Parallax/Xencat
Pictures production Developed
with the support of the European
Script Fund
124 minutes 5 seconds
S&S September 1997 p46
Category C

Kama Sutra A Tale of Love

Film Four Distributors - 20 June
(18) India/UK/Japan/Germany, Dir
Mira Nair
with Naveen Andrews, Sarita
Choudhury, Ramon Tikaram, Rekha,
Indira Varma, Pearl Padamsee
Rasa Film, Inc. and NDF
International Ltd A NDF
International Ltd. Pony Canyon
Inc. Pandora Film in association
with Channel Four Films
presentation A Mirbai Films
production
114 minutes 26 seconds

S&S July 1997 p44
Category B

Keep the Aspidistra Flying

First Independent - 28 November
(15) UK, Dir Robert Bierman
with Richard E. Grant, Helena Bonham
Carter, Julian Wadham, Jim Carter
© Bonaparte Films Ltd.
Overseas Filmgroup presents in
association with The Arts Council
of England and Bonaparte Films
Limited An UBA/Sentinel Films
production Developed in
association with 'Freewheel'
 101 minutes 10 seconds
S&S November 1997 p44
Category A

Keys to Tulsa

PolyGram - 22 August
(18) USA, Dir Leslie Greif
with Eric Stoltz, Cameron Diaz,
Randy Graff, Mary Tyler Moore,
James Coburn, Deborah Kara Unger
© A PolyGram Filmed
Entertainment presentation.
An ITC Entertainment Group/
Peyton/Greif production
113 minutes 39 seconds
S&S October 1997 p53

Kids Return

ICA Projects - 2 May
(Not submitted) Japan, Dir Takeshi
Kitano
with Masanobu Ando, Ken Kaneko,
Leo Morimoto, Hatsuo Yamaya,
Mitsuko Oka, Ryo Ishibashi
Office Kitano with support from
Ota Publishing An Office Kitano/
Bandai Visual presentation
107 minutes (subtitles)
S&S May 1997 p45

Killer a Journal of Murder

First Independent - 16 May
(18) USA, Dir Tim Metcalfe
with James Woods, Robert Sean
Leonard, Ellen Greene, Cara Buono,
Robert John Burke, Richard Riehle
Spelling films Inc Oliver Stone
and Spelling Films present An
Ixtlan production in association
with Breakheart Films
91 minutes 38 seconds
S&S June 1997 p55

Kiss Me, Guido

UIP - 19 December
(15) USA/UK, Dir Tony Vitae
with Nick Scotti, Anthony Barrile,
Anthony DeSando, Molly Pricee,
Craig Chester, David Deblinger
© Paramount Pictures
Corporation.
© Kiss Me Productions, Inc.
Capitol Films and Kardana/Swinsky

Films present A Redeemable
Features production
89 minutes 12 seconds
S&S December 1997 p44
Category D[1]

Kitchen/Wo Ai Chufang

Alliance Releasing - 26 December
(15) Japan, Dir Yim Ho
with Jordan Chan, Yasuko Tomita,
Law Kar-Ying, Karen Mok, Lau Siu-
Ming, Lo Koon-Lan
© Harvest Crown Limited (Hong
Kong)/Amuse Inc (Tokyo).
A Pineast Pictures Limited and
Harvest Crown Limited production
124 minutes (subtitles)
S&S January 1998, p48

Kolya/Kolja

Buena Vista - 2 May
(12) Czech Republic/UK/France, Dir
Jan Sverák
with Zdenek Sverák, Andrej
Chalimon, Libuse Safránková,
Ladislav Smoljak
Biograf Jan Sverák/Portobello
Pictures/Ceska Televize/Pandora
Cinema In co-production with
Ceska Televize/CinemArt
Supported by Centrum Ceskeho
Video/Eurimages/French Ministry
of Culture (Centre National de la
Cinématographie)/The Czech
Republic State Fund for Support
and Development of
Cinematography
105 minutes 1 second (subtitles)
S&S May 1997 p46
Category C

Korea

Barbican - 4 April
(Not submitted) Ireland, Dir Cathal
Black
with Donal Donnelly, Andrew Scott,
Fiona Molony, Vass Anderson
Cathal Black Production, Zweites
Deutsches Fernsehen, Association
rélative à la télévision Bord na
hÉireann, Black Star/Films,
Radio-Telefis Éireann, Nederlande
Omrop Stichting, Groupewest
European pour la Circular
80 minutes
No S&S reference

L.A. Confidential

Warner Bros - 31 October
(15) USA, Dir Curtis Hanson
with Kevin Spacey, Russell Crowe,
Guy Pearce, James Cromwell, Kim
Basinger, Danny DeVito
© Monarchy Enterprises B.V. and
Regency Entertainment (USA).
137 minutes 45 seconds
S&S November 1997 p45

The Lady and the Tramp *

Buena Vista - 18 July
(U) USA 1955, Dir Hamilton Luske
with the voices of Peggy Lee, Larry
Roberts, Bill Baucom, Verna Felton
Walt Disney Productions
75 minutes
No S&S reference
MFB v.22 n.261 October 1955 p149

Larger Than Life

Guild - 28 March
(PG) USA, Dir Howard Franklin
with Bill Murray, Janeane Garofalo,
Matthew McConaughey, Keith
David, Pat Hingle, Jeremy Piven
United Artists Pictures Inc
Trilogy/RCS/Majestic Pictures
presents In association with
United Artists Pictures A Trilogy
Entertainment Group production
TRM Pictures
92 minutes 59 seconds
S&S April 1997 p43

Last Summer in the Hamptons

Revere Releasing - 14 November
(15) USA, Dir Henry Jaglom
with Victoria Foyt, Viveca Lindfors,
Jon Robin Baitz, Savannah Bouchér,
Roscoe Lee Browne, André Gregory
107 minutes 52 seconds
© The Rainbow Film Company.
A Jagtoria Film production
S&S October 1997 p54

Latin Boys Go to Hell

Dangerous to Know - 3 October
(18) USA/Germany/Spain/Japan, Dir
Ela Troyano
with Irwin Ossa, John Bryant Davila,
Alexis Artiles, Mike ruiz, Jenifer Lee
Simard, Guinevere Turner
© Latin Boys Production.
A Jürgen Brüning/Fernando
Colomo P.C. production In
association with Strand Releasing/
Stance/GM Films/Pro Fun This
film has been partially funded by
The Jerome Foundation and The
New York State Council on the Arts
69 minutes 33 seconds
S&S December 1997 p45

Lawn Dogs

Carlton - 21 November
(15) UK, Dir John Duigan
with Sam Rockwell, Christopher
McDonald, Kathleen Quinlan, Bruce
McGill, Mischa Barton, David Barry
Gray
© Rank Film Distributors Limited
From Toledo Pictures.
100 minutes 44 seconds
S&S November 1997 p46
Category A

The Leading Man

Guild - 26 September
(15) UK, Dir John Duigan
with Jon Bon Jovi, Anna Galiena,
Lambert Wilson, Thandie Newton,
Barry Humphries
J&M Entertainment Ltd
99 minutes 48 seconds
S&S September 1997 p47

Liar Liar

UIP - 2 May
(15) USA, Dir Tom Shadyac
with Jim Carrey, Maura Tierney,
Justin Cooper, Cary Elwes, Anna
Haney, Jennifer Tilly, Amanda
Donohoe
Universal City Studios, Inc
Universal Pictures and Imagine
Entertainment present
86 minutes 35 seconds
S&S May 1997 p48

A Life Less Ordinary

PolyGram - 24 October
(15) UK, Dir Danny Boyle
with Ewan McGregor, Cameron
Diaz, Holly Hunter, Delroy Lindo,
Ian Holm, Dan Hedya
© LifeLess Limited.
PolyGram Filmed Entertainment
A Figment film Developed and
supported by Channel 4 Films
101 minutes 41 seconds
S&S November 1997 p47
Category D[2]

Like Grains of Sand/Nagisa no Sindbad

ICA - 24 January
(Not submitted) Japan, Dir Ryosuke
Hashiguchi
with Yashinori Okada, Kota Kusano,
Ayumi Hamazaki
Toho/Pia Corporation A YES
(Young Entertainment Square)
presentation
129 minutes (subtitles)
S&S February 1997, p47

Looking for Richard

20th Century - 31 January
(12) USA, Dir Al Pacino
with Al Pacino, Harris Yulin,
Penelope Allen, Alec Baldwin,
Kevin Spacey, Estelle Parsons,
Winona Ryder
Twentieth Century Fox Film
Corporation
A Michael Hadge/Al Pacino
production
109 minutes
S&S February 1997 p48

Lost Highway

PolyGram - 22 August
(18) USA, Dir David Lynch

with Bill Pullman, Patricia Arquette, Balthazar Getty, Robert Blake, Natasha Gregson Wagner, Richard Pryor
Lost Highway Productions Inc. A CiBy 2000/Asymmetrical production
134 minutes 4 seconds
S&S September 1997 p48

The Lost World Jurassic Park
UIP - 18 July
(PG) USA, Dir Steven Spielberg
with Jeff Goldblum, Julianne Moore, Pete Postlehwaite, Arliss Howard, Richard Attenborough, Vince Vaughan
Universal City Studios Inc/ Ambline Entertainment Inc A Universal Picture
134 minutes
S&S July 1997 p44

Love and Other Catastrophes
20th Century Fox - 23 May
(15) Australia, Dir Emma-Kate Croghan
with Frances O'Connor, Alice Garner, Radha Mitchell, Matthew Dyktynski, Matt Kay, Suzi Dougherty
Screwball Five Pty Ltd Beyond Films and Screwball Five Pty Ltd present Produced in association with the Australian Film Commission
78 minutes 34 seconds
S&S May 1997 p48

love jones
Entertainment - 27 June
(15) USA, Dir Theodore Witcher
with Larenz Tate, Nia Long, Isiah Washington, Lisa Nicole Carson, Khalil Kain, Leonard Roberts
New Line Productions Inc An Addis/Wechsler production
108 minutes 37 seconds
S&S July 1997 p46

Love Lessons/Lust och fägring stor/Laererinden
Gala - 28 March
(15) Sweden/Denmark, Dir Bo Widerberg
with Johan Widerberg, Marika Lagercrantz, Tomas von Brömssen, Karin Huldt, Björn Kjellman, Nina Gunke
Per Holst Film Produced in co-operation with The Danish Filminstitute, Hans Hansen/ Nordisk Film & TV Fond, Bengt Forslund/The Swedish Filminstiute, Per Lysander/ Egmont Film AB/TV2 Danmark/ Sveriges Television Drama Developed with the support of the

European Script Fund
130 minutes 10 seconds (subtitles)
S&S April 1997 p44

Love Serenade
Barbican/Australian Film Commission - 7 March
(Not submitted) Australia, Dir Shirley Barrett
with Miranda Otto, Rebecca Frith, George Sherstov
Jan Chapman Productions
101 minutes
No S&S reference

Love! Valor! Compassion!
Entertainment - 25 July
(15) USA, Dir Joe Mantello
with Jason Alexander, Stephen Spinella, Stephen Bogardus, Randy Becker, John Benjamin Hickey
New Line Productions, Inc Fine Line Features presents A Krost/ Chapin production
114 minutes 3 seconds
S&S September 1997 p49

Lush Life
NFT - 16 May
(Not submitted) USA, Dir Michael Elias
with Jeff Goldblum, Forest Whitaker, Kathy Baker
Chanticleer Films
101 minutes
No S&S reference

Macbeth
CJP Public Relations -
(12) UK, Dir Jeremy Freeston
with Jason Connery, Helen Baxendale, Graham McTavish, Kenny Bryans, Kern Falconer, Hildegard Neil
Cromwell Productions In association with Lamancha Productions and Grampian Television
129 minutes 2 seconds
S&S June 1997 p56

Madame Butterfly
Blue Dolphin - 20 June
(15) France/Japan/Germany/UK, Dir Fédéric Mitterrand
with Ying Huang, Richard Troxell, Ning Liang, Richard Cowan, Jing Ma Fan, Christopheren Nomura
Erato Films/Idéale Audience/ France 3 Cinéma A co-production with Imalyre/VTCOM France Télécom/Sony Classical With the participation of Canal+/BBC Television/ZDF/S4C With the support of Fondation d'Enterprse/France Télécom With the participation of Centre National de la Cinématographie and of the European Script Fund

134 minutes 6 seconds
S&S July 1997 p47
Category C

Ma vie en rose
Buena Vista - 24 October
(15) France/Belgium/ UK/ Switzerland, Dir Alain Berliner
with Michèle Laroque, Jean-Philippe Écoffey, Hélène Vincent, Georges Du Fresne, Daniel Hanssens, Laurence Bibot
© Haut et Court/WFE/Freeway Films/ CAB/La Sept Cinéma/ RTBF/TF1 Films Production. With the support of Canal+/ Cofimage 8/Avance sur Recettes of Centre National de la Cinématographie, Centre du Cinéma et de l'Audiovisuel de la Communauté française de Belgique/Eurimages/Procirep/ Fondation Martini & Rossie/ European Co-production
89 minutes 1 second (subtitles)
S&S November 1997 p48
Category C

Ma vie sexuelle (How I Got Into an Argument)/Comment je me suis disputé...("ma vie sexuelle")
Pathé - 8 August
(15) France, Dir Arnaud Desplechin
with Mathieu Amalric, Emmanuelle Devos, Emmanuel Salinger, Marianne Denicourt, Thibault de Montalembert, Chiara Mastroianni
Why Not Productions/La Sept Cinéma/France 2 Cinéma With the participation of Canal + and Centre National de la Cinématographie With the support of Procirep
180 minutes 4 seconds (subtitles)
S&S August 1997 p46

 The Magnificent Ambersons *
BFI - 27 December
(U) USA 1942, Dir Orson Welles
with Joseph Cotten, Dolores Costello, Agnes Moorhead, Tim Holt, Anne Baxter
RKO Radio Pictures, Mercury
88 minutes (b/w)
No S&S reference
MFB v.9 p130

Mandela
Nubian Tales - 11 April
(PG) USA, Dir Jo Menell, Angus Gibson
with (interviewees) Nelson Mandela, Mabel Mandela, Mandla Mandela, Walter Sisulu, Anthony Sampson - (narrator) Patrick Shai

Island Pictures. A Clinica Esterico production
122 minutes 37 seconds
S&S April 1997 p45

Margaret's Museum
Metrodome - 2 May
(15) Canada/UK, Dir Mort Ransen
with Helena Bonham Carter, Clive
Russell, Craig Olejnik, Kate Nelligan,
Kenneth Welsh, Andrea Morris
Glace Bay Pictures Inc/Télé-
Action/Skyline Film and Television
A Ranfilm/Imagex/Télé-Action/
Skyline production Produced with
the participation of Téléfilm
Canada/British Screen Finance
Limited/The National Film Board
of Canada/Nova Scotia Film
Development Corporation/Société
de Développement des Enerprises
Culturelles du Québec
(Programme de crédits d'impôt)
99 minutes 38 seconds
S&S May 1997 p49
Category C

Marius et Jeannette
Porter Frith - 5 December
(15) France, Dir Robert Guédiguian
with Ariane Ascaride, Gérard
Meylan, Pascale Roberts, Jacques
Boudet, Frédérique Bonnal, Jean-
Pierre Darroussin
© La Sept Cinéma/Agat Films &
Cie.
With the participation of Canal+
With the participation of the city
of Aubagne
102 minutes 9 seconds (subtitles)
S&S December 1997 p46

Mars Attacks!
Warner Bros - 28 February
(12) USA, Dir Tim Burton
with Jack Nicholson, Glenn Close,
Annette Bening, Pierce Brosnan,
Danny DeVito, Martin Short
Warner Bros
105 minutes 41 seconds
S&S March 1997 p53

Marvin's Room
Buena Vista - 20 June
(12) USA, Dir Jerry Zaks
with Meryl Streep, Leonardo
DiCaprio, Diane Keaton, Robert De
Niro, Hume Cronyn, Gwen Verdon
A Miramax Film Corp.
A Scott Rudin/Tribeca production
98 minutes 13 seconds
S&S July 1997 p48

Maximum Risk
Columbia TriStar - 7 November
(15) USA, Dir Ringo Lam
with Jean-Claude Van Damme,
Natasha Henstridge, Jean-Hugues

Anglande, Zach Grenier, Stéphanie
Audran, Frank Senger
© Columbia Pictures Industries,
Inc.
100 minutes 49 seconds
S&S December 1997 p47

Men in Black
Columbia TriStar - 1 August
(PG) USA, Dir Barry Sonnenfeld
with Tommy Lee Jones, Will smith,
Linda Fiorentino, Vincent
D'Onofrio, Rip Torn, Tony Shalhoub
Columbia Pictures Industries, Inc
An Amblin Entertainment
production
98 minutes
S&S August 1997 p47

Men, Women: Users Manual/ Hommes femmes Mode d'emploi
Gala - 6 June
(12) France, Dir Claude Lelouch
with Fabrice Luchini, Bernard Tapie,
Alessandra Martines, Pierre Arditi,
Ticky Holgado, Ophélie Winter
Les Film13/TF1 Films Production
An U.G.C./UFS presentation In
association with Canal+
122 minutes 47 seconds (subtitles)
S&S June 1997 p57

Metro
Buena Vista - 18 April
(18) USA, Dir Thomas Carter
with Eddie Murphy, Michael
Rapaport, Kim Miyori, Art Evans,
James Carpenter, Donal Logue
Touchstone Pictures
In association with Caravan
Pictures
117 minutes 23 seconds
S&S May 1997 p50

Michael
Rank-Castle Rock/Turner - 21
February
(PG) USA, Dir Nora Ephron
with John Travolta, Andie
MacDowell, William Hurt, Bob
Hoskins, Robert Pastorelli, Jean
Stapleton
Turner Pictures Worldwide, Inc
A Turner Pictures presentation
An Alphaville production
104 minutes 58 seconds
S&S March 1997 p53

Microcosmos/Microcosmos Le Peuple de l'herbe
Guild - 16 May
(U) France/Switzerland/Italy, Dir
Claude Nuridsany, Marie Pérennou
with Ladybird with seven spots,
Swallow-tail butterfly, Climbing
caterpillar, Bee gathering pollen

from the sage flower, Long-tailed
blue caterpillar, Burgundy snails,
New-born caterpillar of the Jason
butterfly, Argiope Spider, Bombyle
(gathering fly) Processionary
caterpillars, Red ants, Gathering
ants, Polist wasps, Sacred scarab
beetle, Pheasant, Water spiders,
Notonects, Argyronet Spider (with
its diving bell) Young Agrion
dragon-flies, Eucera bee in love with
the Ophyrs orchid, Drosera
carnivorous plant, Rhinocertos
beetle, lule, Stag beetles, Bucephal
caterpillars, diablotin, Great peacock
moth, Cousin mosquito and its
metamorphosis
Galatée Films/France 2 Cinéma/
Bac Films A Jacques Perrin
presentation A co-production of
Bac Films/Delta Images/Les
Productions JMH (Switzerland)/
Urania Films SRL (Italy)/
Télévision Suisse romande with
the participation of Canal+/
Conseil Général de l'Aveyron/
Sivom des Monts et Lacs du
Lévezou/Crédit Agricole/Région
Midi Pyrénées/Centre National de
la Cinématographie/Ministère de
la Recherche/Enseignement
Supérieur/Agence Jules Verne/
Office Fédéral de la Culture
(Bern) with grant from Fondation
Gan pour le Cinéma and Procirep
Cinéma Supported by Fonds
Eurimages du Conseil de l'Europe
75 minutes 17 seconds
S&S June 1997 p58

The Mirror Has Two Faces
Columbia TriStar - 10 January
(15) USA, Dir Barbara Streisand
with Barbara Streisand, Jeff Bridges,
Pierce Brosnan, George Segal, Mimi
Rogers, Brenda Vaccaro, Lauren
Bacall
TriStar Pictures Inc
In association with Phoenix
Pictures An Arnon Milchan-
Barwood Films Production
126 minutes 25 seconds
S&S February 1997 p49

Moll Flanders
20th Century Fox - 9 May
(12) USA, Dir Pen Densham
with Robin Wright, Morgan
Freeman, Stockard Channing, John
Lynch, Brenda Fricker, Geraldine
James
Spelling Films Inc A Spelling
Films/Metro-Goldwy-Mayer
Pictures presentation a Trilogy
Entertainment Group production
Produced with the support of
investment incentives for the Irish
Film Industry

122 minutes 30 seconds
S&S June 1997 p59

Mon homme

Artificial Eye - 23 May
(18) France, Dir Bertrand Blier
with Anouk Grinberg, Gérard
Lanvin, Valéria Bruni-Tedeschi,
Olivier Martinez, Dominique
Valadié, Jacques François, Michel
Galabru
Les Films Alain Sarde/Plateau A
With Studio Images 2 and the
participation of Canal+
99 minutes 6 seconds (subtitles)
S&S June 1997 p59

Mother Night

Entertainment - 14 March
(15) USA, Dir Keith Gordon
with Nick Nolte, Sheryl Lee, Alan
Arkin, John Goodman, Kirsten
Dunst, Arye Gross
New Line Productions, Inc
Fine Line Features presents a
Whyaduck production
113 minutes 45 seconds
S&S March 1997 p56

Mrs Brown

Buena Vista - 5 September
(PG) UK/USA/Ireland, Dir John
Madden
with Judi Dench, Billy Connolly,
Anthony Sher, Geoffrey Palmer,
Richard Pasco, David Westhead
BBC Scotland An Ecosse Films
production for BBC Films and
WGBH/Boston in association with
Irish Screen Mobil Masterpiece
Theatre presentation
103 minutes 35 seconds
S&S September 1997 p50
Category B

Murder at 1600

Warner Bros - 11 July
(15) USA, Dir Dwight Little
with Wesley Snipes, Diane Lane,
Alan Alda, Daniel Benzali, Ronny
Cox, Dennis Miller
Warner Bros Productions Limited/
Monarchy Enterprises B.V. In
association with Regency
Enterprises
106 minutes 37 seconds
S&S July 1997 p49

My Best Friend's Wedding

Columbia TriStar - 19 September
(15) USA, Dir P.J.Hogan
with Julia Roberts, Dermot Mulroney,
Cameron Diaz, Rupert Everett, Philip
Bosco, M.Emmet Walsh
TriStar Pictures Inc A Jerry
Zucker/Predawn production
104 minutes 47 seconds
S&S September 1997 p50

The Myth of Fingerprints

Feature Film Company - 28
November
(15) USA, Dir Bart Freundlich
with Arija Bareikis, Blythe Danner,
Hope Davis, Laurel Holloman, Brian
Kerwin, James LeGros
© **GM - Good Machine.**
A Good Machine production in
association with Eureka Pictures
90 minutes 12 seconds
S&S December 1997 p48

The Near Room

Metrodome - 11 April
(18) UK, Dir David Hayman
with Adrian Dunbar, David O'Hara,
David Hayman, Julie Graham, Tom
Watson, James Ellis
Inverclyde Productions A Glasgow
Film Fund and British Screen
presentation
89 minutes 50 seconds
S&S April 1997 p46
Category A

Never Talk to Strangers

Columbia TriStar- 14 March
(18) Canada/USA, Dir Peter Hall
with Rebecca De Mornay, Antonio
Banderas, Dennis Miller, Len
Cariou, Harry Dean Stanton
Never Talk To Strangers
Productions, Inc TriStar Pictures
and Peter Hoffman present an
Alliance Production
85 minutes 54 seconds
S&S May 1997 p51

Night Falls on Manhattan

First Independent - 5 September
(15) USA, Dir Sidney Lumet
with Andy Garcia, Richard Dreyfuss,
Lena Olin, Ian Holm, James
Gandolfini, Colm Feore, Ron Leibman
Spelling Films Inc A Mount/
Kramer production
113 minutes 25 seconds
S&S September 1997 p51

Nil by Mouth

20th Century Fox - 10 October
(15) USA, Dir Gary Oldman
with Ray Winstone, Kathy Burke,
Charlie Creed-Miles, Laila Morse,
Edna Doré, Chrissie Cotterill
© **SE8 Group Limited.**
Luc Besson presents
128 minutes 20 seconds
S&S October 1997 p55
Category B

No Way Home

Blue Dolphin - 23 May
(18) USA, Dir Buddy Giovinazzo
with Tim Roth, James Russo,
Deborah Dara Unger, Joseph Ragno,
Catherine Kellner, Saul Stein

Back Alley Productions, Inc An
Orenda Films production
92 minutes 50 seconds
S&S April 1997 p46

Normal Life

First Independent - 7 March
(18) USA, Dir John McNaughton
with Luke Perry, Ashley Judd, Bruce
Young, Jim True, Dawn Maxey,
Scott Cummins
Normal Life Productions Inc
A Spelling Films International
presentation
In association with Fine Line
Features
102 minutes 27 seconds
S&S March 1997 p57

Nothing to Lose

Buena Vista - 21 November
(15) USA, Dir Steve Oedekerk
with Tim Robbins, Martin Lawrence,
John C. McGinley, Giancarlo
Esposito, Michael McKean, Susan
Barnes
© **Touchstone Pictures.**
97 minutes 49 seconds
S&S November 1997 p48

One Eight Seven

Warner Bros - 12 September
(15) USA, Dir Kevin Reynolds
with Samuel L. Jackson, John Heard,
Kelly Rowan, Clifton Gonzalez
Gonzalez, Tony Plana, Karina
Arroyave
Anna K. Production C.V. Icon
Entertainment International
presents An Icon production
118 minutes 53 seconds
S&S September 1997 p52

One Fine Day

20th Century Fox - 4 July
(PG) USA, Dir Michael Hoffman
with Michelle Pfeiffer, George
Clooney, Mae Whitman, Alex
D.Linz, Charles Durning, Jon Robin
Baitz
Twentieth Century Fox Film
Corporation A Fox 2000 Pictures
presentation
In association withVia Rosa
Productions
108 minutes 45 seconds
S&S March 1997 p57

One Night Stand

Entertainment - 28 November
(18) USA, Dir Mike Figgis
with Wesley Snipes, Nastassja
Kinski, Kyle MacLachian, Ming-na-
Wen, Robert Downey Jr
© **New Line Productions, Inc.**
A Red Mullet production
102 minutes 52 seconds
S&S December 1997 p49

Palookaville

Metrodome - 25 July
(15) USA, Dir Alan Taylor
with William Forsythe, Vincent
Gallo, Adam Trese, Gareth Williams,
Lisa Gay Hamilton, Bridgit Ryan
**Public Television Playhouse Inc
Playhouse International Pictures
presents in association with The
Samuel Goldwyn Company and
Redwave Films Produced in
association with American
Playhouse which is supported by
funds from Public Television
Stations, the Corporation for
Public Broadcastingand the
National Endowment for the Arts**
91 minutes 45 seconds
S&S August 1997 p48
Category D[2]

Paradise Road

20th Century Fox - 5 December
(15) USA, Dir Bruce Beresford
with Glenn Close, Frances McDormand,
Pauline Collins, Cate Blanchett,
Jennifer Ehle, Julianna Margulies
**© Twentieth Century Fox Film
Corporation/YTC Motion Picture
Investments and Village
Roadshow Pictures Pty Ltd.
Project developed with the
assistance of the Australian Film
Commission/Pacific Film and
Television Commission Produced
with the assistance of the
Queensland Government**
121 minutes 30 seconds
S&S December 1997 p50

La Passione

Warner Bros - 16 May
(15) UK, Dir John B.Hobbs
with Shirley Bassey, Sean Gallagher,
Thomas Orange, Paul Shane, Jan
Ravens, Carmen Silvera
**W.E.A. International Inc Warner
Vision International presents A
Fugitive Production**
91 minutes 3 seconds
S&S June 1997 p60
Category A

The Peacemaker

UIP - 24 October
(15) USA, Dir Mimi Leder
with George Clooney, Nicole
Kidman, Armin Mueller Stahl,
Marcel Iures, Alexander Baluev,
Rene Medvesek
© Dream Works LLC.
123 minutes 51 seconds
S&S November 1997 p49

The People vs. Larry Flynt

Columbia TriStar - 11 April
(15) USA, Dir Milos Forman
with Woody Harrelson, Courtney

Love, Edward Norton, James
Cromwell, Crispin Glover, James
Carville
**Columbia Pictures Industries, Inc
An Ixtlan Production In
association with Phoenix Pictures**
129 minutes 33 seconds
S&S March 1997 p58

Persons Unknown

Metro - 12 December
(18) USA, Dir George Hickenlooper
with Joe Mantegna, Kelly Lynch,
Naomi Watts, J.T. Walsh, Xander
Berkeley, Jon Favreau
**Promark Entertainment Group
and Spectacor Films in association
with Videal**
99 minutes 5 seconds
S&S January 1998 p50

The Phantom

UIP - 21 February
(12) USA/Australia, Dir Simon
Wincer
with Billy Zane, Kristy Swanson,
Treat Williams, Catherine Zeta
Jones, James Remar, Cary-Hirouyuki
Tagawa
**Paramount Pictures In association
with Robert Evans and The Ladd
Company A Village Roadshow
Pictures production**
100 minutes 29 seconds
S&S February 1997 p51

Photographing Fairies

Entertainment - 19 September
(15) UK, Dir Nick Willing
with Toby Stephens, Emily Woof,
Frances Barber, Phil Davis, Ben
Kingsley, Rachel Shelley
**The Starry Night Film Company
Ltd PolyGram Filmed
Entertainment presents with the
participation of British Screen in
association with The Arts Council
of England and The BBC In
association with Dogstar Films
Developed with the support of The
European Script Fund Developed
with the assistance of British
Screen Finance Ltd Supported by
The National Lottery through the
Arts Council of England**
106 minutes 18 seconds
S&S September 1997 p53
Category A

Picture Bride

Artificial Eye - 17 January
(12) USA, Dir Kayo Hatta
with Youki Kudoh, Akira Takayama,
Tamlyn Tomita
**Thousand Cranes Filmworks In
association with Miramax Films
Produced in association with
Cécile Co., Ltd**

94 minutes 51 seconds
S&S February 1997 p52

Picture Perfect

20th Century Fox - 26 September
(PG) USA, Dir Glenn Gordon Caron
with Jennifer Aniston, Jay Mohr,
Kevin Bacon, Olympia Dukakis,
Illeana Douglas, Kevin Dunn
**© Twentieth Century Fox Film
Corporation.
A 2 Arts production**
101 minutes 40 seconds
S&S December 1997 p51

Pink Flamingos *

Entertainment - 4 July
(18) USA 1972, Dir John Waters
with Divine, David Lochary, Mary
Vivian Pearle, Mink Stole
Dreamland Productions
95 minutes
No S&S reference
MFB v.45 n.528 January 1978 p11

Plein Soleil/Delitto in pieno sole/ Blazing Sun/ Purple Noon *

BFI - 22 August
(PG) France/Italy 1960, Dir René
Clément
with Alain Delon, Marie Laforêt,
Maurice Ronet, Erno Crisa, Frank
Latimore, Bill Kearns
**Paris Film Production Titanus
Rome**
118 minutes 47 seconds (subtitles)
S&S September 1997 p57

The Portrait of a Lady

PolyGram - 28 February
(12) UK/USA, Dir Jane Campion
with Nicole Kidman, John Malkovich,
Barbara Hershey, Mary Louise Parker
**PolyGram Film Productions BV
A Propoganda production**
144 minutes
S&S March 1997 p60
Category D[1]

Portraits Chinois

Film Four Distributors - 25 July
(15) France/UK, Dir Martine
Dugowson
with Helena Bonham Carter,
Romane Bohringer, Marie
Trintignant, Elsa Zylberstein, Yvan
Attal, Sergio Castellito
**IMA Films/UGC Images/France 2
Cinéma/Polar Productions With
the participation of Canal +/
Soficas Sofinergie 3/Sofinergie 4
and The European Co-production
Fund (UK)
In association with Channel Four
Films**
122 minutes 51 seconds

S&S August 1997 p49
Category C

Powder
Buena Vista - 14 February
(12) USA, Dir Victor Salva
Mary Steenburgen, Sean Patrick
Flanery, Lance Henriksen, Jeff
Goldblum
**Hollywood Pictures In association
with Caravan Pictures**
112 minutes
S&S May 1996 p57

The Preacher's Wife
Buena Vista - 17 January
(U) USA, Dir Penny Marshall
with Denzel Washington, Whitney
Houston, Courtney B. Vance,
Gregory Hines
**Touchstone Pictures/The Samuel
Goldwyn Company In association
with Parkway Productions and
Mundy Lane Entertainment**
123 minutes 43 seconds
S&S March 1997 p61

Preaching to the Perverted
Entertainment - 4 July
(15) UK, Dir Stuart Urban
with Guinevere Turner, Christien
Anholt, Tom Bell, Julie Graham,
Georgina Hale, Julian Wadham
**PTTP Films Ltd A Cyclops Vision
production Developed with the
support of the European Script
Fund**
99 minutes 30 seconds
S&S August 1997 p50
Category A

Prince Valiant/Prinz Eisenherz
Entertainment - 19 December
(PG) Germany/UK/Ireland/USA, Dir
Anthony Hickox
with Stephen Moyer, Katherine
Heigl, Thomas Kretschmann,
Edward Fox, Udo Kier, Warwick
Davis
**Paramount Pictures presents a
Constantin/Lakeshore production
in association with Hearst
Entertainment Inc/Legacy Film
Productions and Celtridge Ltd
Supported by FFA,
FilmFernsehFonds Bayern and
Filmboard Berlin-Brandenberg In
collaboration with Babelsberg
Film**
91 minutes 22 seconds
S&S January 1998 p52
Category D[1]

Private Parts
Entertainment - 20 June
(18) USA, Dir Betty Thomas
with Howard Stern, Robin Quivers,
Mary McCormack, Fred Norris, Paul

Giamatti, Gary Dell'Abate, Jackie
Martling
**Rysher Entertainment, Inc A
Paramount Pictures/Rysher
Entertainment presentation
Northern Lights Entertainment**
109 minutes 23 seconds
S&S July 1997 p49

The Proprietor/La Propriétaire
Warner Bros - 7 February
(12) UK/France/Turkey/USA, Dir
Ismail Merchant
with Jeanne Moreau, Sean Young,
Sam Waterson, Christopher
Cazenove, Neil Carter, Jean-Pierre
Aumont
**Merchant Ivory Productions Ltd
Ognon Pictures & Fez Production
Filmcilik
In association with Largo
Entertainment, Canal +, Channel
Four
With the support of The
Eurimages Fund of the Council of
Europe**
113 minutes 5 seconds
S&S February 1997 p53
Category B

Pusher
Metrodome - 10 October
(18) Denmark, Dir Nikolas Winding
Refn
with Kim Bodnia, Zlatko Buric,
Laura Drasbaek, Slavko Labovic
**© Balboa Enterprise ApS.
A Henrik Danstrup/Nikolas Winding
Refn production With the support
from the Danish Film Institute**
109 minutes 55 seconds (subtitles)
S&S October 1997 p56

Quadrophenia *
Feature Film - 31 January
(15) UK 1979, Dir Franc Roddam
with Phil Daniels, Leslie Ash, Philip
Davis, Mark Wingett, Sting,
Raymond Winstone
Who Films, Polytel Films
120 minutes
S&S p16 February 1997 (article)
MFB v.46. n.548 September 1979
p198

The Quest
UIP - 4 July
(15) USA, Dir Jean-Claude Van
Damme
with Jean-Claude Van Damme,
Roger Moore, James Remar, Janet
Gunn, Jack McGee, Aki Aleong
**Universal City Studios Inc A
Universal Pictures presentation
An MDP Worldwide presentation**
94 minutes 30 seconds
S&S August 1997 p51

The Railway Children *
BFI - 28 March
(U) UK 1970, Dir Lionel Jeffries
with Dinah Sheridan, Bernard
Cribbins, William Mervyn, Iain
Cuthbertson, Jenny Agutter, Sally
Thomsett
Associated British Production
108 mins
No S&S reference
MFB v.38 n.445 February 1971

Ransom
Buena Vista - 7 February
(15) USA, Dir Ron Howard
with Mel Gibson, Rene Russo,
Brawley Nolte, Gary Sinise, Delroy
Lindon, Lili Taylor
Touchstone Pictures
121 minutes 29 seconds
S&S February 1997 p54

Regeneration
Artificial Eye - 21 November
(15) UK/Canada, Dir Gillies
MacKinnon
with Jonathan Pryce, James Wilby,
Jonny Lee Miller, Stuart Bunce,
Tanya Allen, David Hayman
**© Rafford Films Limited/Norstar
Entertainment Inc in Trust.
A Rafford Films Limited/Norstar
Entertainment Inc/BBC Films/The
Scottish Arts Council Lottery
Fund presentation Produced with
the participation of Telefilm
Canada/The Scottish Film
Production Fund/The Glasgow
Film Fund**
113 minutes 51 seconds
S&S December 1997 p52
Category B

The Relic
PolyGram - 16 May
(15) USA, Dir Peter Hyams
with Penelope Ann Miller, Tom
Sizemore, Linda Hunt, James
Whitmore,Clayton Rohner, Chi
Muoi Lo, Thomas Ryan
**Paramount Pictures Corporation
A Cloud Nine Entertainment/
PolyGram Filmed Entertainment/
Toho-towa, Tele München/BBC
presentation In association with
paramount Pictures/Marubeni
A Pacific Western production**
109 minutes 34 seconds
S&S June 1997 p61

Remember Me?
Film Four Distributors - 18 July
(PG) UK, Dir Nick Hurran
with Robert Lindsay, Rik Mayall,
Imelda Staunton, Brenda Blethyn,
James Fleet, Haydn Gwynne
Channel Four Television

Corporation Channel Four Films presents a Talisman production
77 minutes 5 seconds
S&S August 1997 p52
Category A

Return of the Jedi *

20th Century Fox - 25 April
(U) USA 1983/1997, Dir Richard Marquand
with Mark Hamill, Harrison Ford, Carrie Fisher, Billy Dee Williams
Lucasfilm
132 minutes
No S&S reference
MFB v.50 n.594 July 1983 p181

Ridicule

Electric Pictures - 7 February
(15) France, Dir Patrice Leconte
with Fanny Ardant, Charles Berling, Bernard Giraudeau, Judith Godrèche, Jean Rochefort
Epithète, Cinéa, France 3 Cinéma With the participation of Canal+, Investimage 4, Polygram Audio Visuel, La Procirep, Gras Savoye, Centre National du Cinéma
102 minutes 15 seconds (subtitles)
S&S February 1997, p55

 ## Robinson in Space

BFI - 10 January
(PG) UK, Patrick Keiller
with the voice of Paul Scofield
BBC A Koninck production for BBC Films In association with The British Film Institute
81 minutes 59 seconds
S&S January 1997 p44
Category A

Romy and Michele's High School Reunion

Buena Vista - 22 August
(12) USA, Dir David Mirkin
with Mira Sorvino, Lisa Kudrow, Alan Cumming, Julia Campbell, Janeane Garofalo, Vincent Ventresca
Touchstone Pictures A Laurence Mark production in association with Bungalow 78 Productions
91 minutes
S&S September 1997 p54

Roseanna's Grave

PolyGram - 8 August
(12) USA/UK, Dir Paul Weiland
with Jean Réno, Mercedes Ruhl, Polly Walker, Mark Frankel, Luigi Diberti, Roberto Della Casa
Roseanna's Grave Fine Line/ Spelling Films presents in association with PolyGram Filmed Entertainment a Hungry Eye/ Trjbits & Worrell/Remote production in association with Remil & Associates

97 minutes 55 seconds
S&S August 1997 p53
US Title: For Roseanna
Category D²

Rumble in the Bronx /Hongfan Qu

Buena Vista - 4 July
(15) USA, Dir Stanley Tong
with Jackie Chan, Anita Mui, François Yip, Bill Tung, Marc Akerstream, Garvin Cross, Morgan Lam
[export version] Paragon Films Ltd New Line Productions, Inc A New Line Cinema presentation A Raymond Chow/Golden Harvest production
89 minutes 16 seconds
S&S July 1997 p51

The Saint

UIP - 18 April
(12) USA, Dir Phillip Noyce
with Val Kilmer, Elisabeth Shue, Rade Serbedzija, Valery Nikolaev, Henry Goodman, Alun Armstrong
Paramount Pictures In association with Rysher Entertainment A David Brown/Robert Evans production
116 minutes 3 seconds
S&S May 1997 p52
Category D¹

Scream

Buena Vista - 2 May
(18) USA, Dir Wes Craven
with David Arquette, Neve Campbell, Courteney Cox, Matthew Lillard, Rose McGowan, Skeet Ulrich, Drew Barrymore
Miramax Film Corporation Dimension Films presents a Woods Entertainment production
111 minutes 7 seconds
S&S May 1997 p53

Sebastian

Dangerous to Know - 2 May
(15) Norway/Sweden, Dir Svend Wam
with Hampus Björck, Nicolai Cleve Broch, Ewa Fröling, Helge Jordal, Rebecka Hemse, Lena Olander
Mefistofilm AS A Mefistofilm AS/ Miramar Films AB/Nordisk Film AS production With the assistance of Produktejonfondet for Kino & Fjernsynsfilm/Nordisk Film & TV-Fond (Svenska Filminstitutet Bengt Forslund/Peter Hald)
84 minutes 7 seconds (subtitles)
S&S July 1997 p52

See How They Fall/Regarde les hommes tomber

M.I.H.K. - 16 May

(18) France, Dir Jacques Audiard
with Jean-Louis Trintignant, Jean Yanne, Mathieu Kassovitz, Bulle Ogier, Christine Pascal, Yvon Back
Bloody Mary Productions/ France 3 Cinéma/ CEC Rhône-Alpes With the participation of la Région Rhône-Alpes/Centre National de la Cinématographie/Canal +/ Cofimage 5 With the support of Procirep Cinéma
9 minutes 53 seconds (subtitles)
S&S June 1997 p62

Sélect Hotel

M.I.H.K. - 11 July
(18) France, Dir Laurent Bouhnik
with Julie Gayet, Jean-Michael Fête, Serge Blumental, Marc Andreoni, Sabine Bail, Eric Aubrahn
Climax Productions With the support of Centre National de la Cinématographie and Procirep With the assistance of DAL and SIS Racisme
85 minutes 26 seconds (subtitles)
S&S July 1997 p52

A Self-Made Hero/Un héro très discret

Artificial Eye - 4 April
(15) France, Dir Jacques Audiard
with Mathieu Kassovitz, Anouk Grinberg, Sandrine Kiberlain, Jean-Louis Trintignant, Albert Dupontel, Nadia Barentin
Aliceléo/Lumière/France 3 Cinéma/M6 Films/Initial Groupe In association with Cofimage 7/ Studio Images 2/Le Studio Canal + With the participation of Canal +/ Centre de la National de la Cinématographie/Ministère de la Culture/La Sacem and the financial support of La Procirep
105 minutes 48 seconds (subtitles)
S&S April 1997 p47

Set It Off

Entertainment - 24 January
(18) USA, Dir F. Gary Gray
with Jada Pinkett, Queen Latifah, Vivica A. Fox, John C. McGinley, Kimberly Elise, Blair Underwood
New Line Productions Inc A Peak Production
122 minutes 27 seconds
S&S March 1997 p62

Seven Years in Tibet

Entertainment - 21 November
(PG) USA/UK, Dir Jean-Jacques Annaud
with Brad Pitt, David Thewlis, B.D.Wong, Mako, Danny Denzongpa, Victor Wong
© Mandalay Entertainment.

A Reperage and Vanguard Films/
Applecross production
135 minutes 46 seconds
S&S December 1997 p53

Shadow Conspiracy
Entertainment - 13 June
(15) USA, Dir George P.Cosmatos
with Charlie Sheen, Donald
Sutherland, Linda Hamilton,
Stephen Lang, Ben Gazzara,
Nicholas Turturro
Cinergi Pictures Enertainment
Inc/ Cinergi Productions N.V. An
Andrew G. Vajna presentation
102 minutes 35 seconds
S&S July 1997 p53

She's The One
20th Century Fox - 7 March
(15) USA, Dir Edward Burns
with Jennifer Aniston, Maxine
Bahns, Edward Burns, Cameron
Diaz, John Mahoney, Mike McGlone
Twentieth Century Fox Film
Corporation Fox Searchlight
Pictures present a Good Machine/
Marlboro Road Gang production
In association with South Fork
Pictures Developed with the
assistance of The Sundance
Institute
96 minutes
S&S February 1997 p56

Shine
Buena Vista - 3 January
(12) Australia/UK, Dir Scott Hicks
with Armin Mueller-Stahl, Noah
Taylor, Geoffrey Rush, Lynn
Redgrave, Googie Withers
Australian Film Finance
Corporation Limited/Momentum
Films Pty Limited/The South
Australian Film Corporation and
Film Victoria. In association with
Pandora Cinema/British
Broadcasting Corporation Script
developed with the assistance from
The Australian Film Commission
Developed with assistance from
Ronin Films, Colour & Movement
Films, Great Scott Productions,
Roadshow, Coote & Carroll Pty Ltd
and the NSW Film and TV Office
105 minutes 37 seconds
S&S January 1997 p44
Category C

Shooting Fish
Entertainment - 17 October
(12) UK, Dir Stefan Schwartz
with Dan Futterman, Stuart
Townsend, Kate Beckinsale,
Nickolas Grace, Claire Cox, Ralph
Ineson
© The Gruber Brothers
(Entrepreneurs) Ltd.

Winchester Films presents a
Gruber Brothers film in association
with the Arts Council of England
and Tomboy Films Supported by
the National Lottery through the
Arts Council of England
112 minutes 29 seconds
S&S October 1997 p56
Category A

A Simple Wish
UIP - 17 October
(U) USA, Dir Michael Ritchie
with Martin Short, Kathleen Turner,
Mara Wilson, Robert Pastorelli,
Amanda Plummer, Francis Capra
© Universal City Studios
Productions, Inc and The Bubble
Factory LLC.
A Sheinberg production
89 minutes 30 seconds
S&S November 1997 p51

The Slab Boys
Film Four Distributors - 29
August
(15) UK, Dir John Byrne
with Robin Laing, Russell Barr, Bill
Gardiner, Louise Berry, Julie Wilson
Nimmo
Channel Four Television
Corporation/Skreba Slab Boys Ltd
Channel Four Films presents in
association with The Scottish Arts
Council, The Arts Council of
England and The Glasgow Film
Fund A Skreba film in association
with Wanderlust Films Supported
by the National Lottery through
the Scottish Arts Council and
theArts Council of England
Developed with the support of the
European Script Fund
97 minutes 48 seconds
S&S September 1997 p55
Category A

Sleepers
PolyGram - 3 January
(15) USA, Dir Barry Levinson
with Kevin Bacon, Robert De Niro,
Dustin Hoffman, Bruno Kirby, Jason
Patric
Warner Bros PolyGram Film
Productions B.V. PolyGram Film
Production B.V. PolyGram Filmed
Entertainment presents a
Propoganda Films/Baltimore
Pictures production
146 minutes 44 seconds
S&S January 1997 p45

ⓑⓕⓘ Smalltime
BFI - 31 October
(18) UK, Dir Shane Meadows
with Mat Hand, Dena Smiles, Shane
Meadows, Gena Kawecka, Jimmy
Hynd

©A Big Arty production in
association with The British Film
Institute.
60 minutes 59 seconds
S&S November 1997 p52
Category A

Smilla's Feeling for Snow/ Fräulein Smillas Gespür für Schnee
20th Century Fox - 31 October
(15) Germany/Denmark/Sweden, Dir
Bille August
with Julia Ormond, Gabriel Byrne,
Richard Harris, Robert Loggia, Jim
Broadbent, Mario Adorf
© Constantin Film Produktion.
GmbH In co-operation with Smilla
Film A/S/Greenland Film
Production AB/Bavaria Film
GmbH This film was supported by
Eurimages/FFA, Film Fernseh
Fonds Bayern/Danish Film
Institute/Hans Hansen/Nordic
Film & TV Fund
121 minutes 7 seconds
S&S November 1997 p52 (subtitles)
US title: Smilla's Sense of Snow

Snow White A Tale of Terror
PolyGram - 31 October
(15) UK, Dir Michael Cohn
with Sigourney Weaver, Sam Neill,
Gil Bellows, Taryn Davis, David
Conrad, Brian Glover
© PolyGram Film Productions
B.V.
PolyGram Filmed Entertainment
presents An Interscope
Communications production
100 minutes 11 seconds
S&S November 1997 p53
Category A

Someone Else's America/ L'Amérique des autres
Film Four Distributors - 11 July
(15) France/United Kingdom/
Germany, Dir Goran Paskaljevic
with Tom Conti, Miki Manojlovic,
Maria Casares, Zorka Manojlovic,
Segej Trifunovic, Jose Ramo Rosario
An Antoine and Martine de
Clermont-Tonnerre, David Rose,
Karin Bamborough, Helga Bähr,
presentation A co-production
between Mact Productions/
Intrinsica Films/Lichtblick
Filmproduction/Stefi 2 In
association with Pandora Cinema
This film was supported by The
Eurimages Fund of the Council of
Europe/Film Fonds Hamburg/
Hamburger Filmbüro With the
participation of Canal +/The
European Co-production Fund
(UK)/Lightworks Edition Systems/

Guisana Ltd/Singidunum Film Productions/Iguana Films Developed with the support of the European Script Fund /Channel Four Television (UK)
91 minutes
S&S May 1997 p54

Some Mother's Son
Rank-Castle Rock/Turner - 10 January
(15) USA/Ireland, Dir Terry George with Helen Mirren, Fionnula Flanagan, Aidan Gillen, David O'Hara, John Lynch
Castle Rock presents
A Hell's Kitchen production
111 minutes 29 seconds
S&S January 1997 p46

Space Jam
Warner Bros - 21 March
(U) USA, Dir Joe Pytka with Michael Jordan, Wayne Knight, Theresa Randle, Manner 'Mooky' Washington, Eric Gordon
Warner Bros
87 minutes 12 seconds
S&S April 1997 p48

Space Truckers
Entertainment - 23 May
(12) USA/Ireland, Dir Stuart Gordon with Dennis Hopper, Stephen Dorff, Debi Mazar, George Wendt, Vernon Wells, Barbara Crampton, Shane Rimmer
Pachyderm Productions, A Peter Newman/Internal Production in association with Mary Breen-Farrelly Productions Produced with the support of Investment Incentives for the Irish Film Industry
96 minutes 17 seconds
S&S June 1997 p62

Spawn
Entertainment - 20 September
(12) USA, Dir Mark A.Z. Dippé with John Leguizamo, Michael Jai White, Martin Sheen, Theresa Randle, Melinda Clarke, Miko Hughes
© New Line Cinema Productions, Inc.
In association with Todd McFarlane Entertainment A Dippé/Goldman/Williams production
96 minutes 11 seconds
S&S October 1997 p57

Speed 2 Cruise Control
20th Century Fox - 15 August
(PG) USA, Dir Jan De Bont with Sandra Bullock, Jason Patric, Willem Dafoe, Temuera Morrison, Brian McCardie, Christine Firkins

Twentieth Century Fox Film Corporation Twentieth Century Fox presents a Blue Tulip production
124 minutes 59 seconds
S&S August 1997 p53

Spice World
PolyGram - 26 December
(PG) UK, Dir Bob Spiers with Posh/Victoria Adams, Baby/Emma Bunton, Sporty/Melanie Chisholm, Ginger/Geri Halliwell, Scary/Melanie Brown, Richard E. Grant
© Five Girls Limited.
The Spice Girls present in association with PolyGram Filmed Entertainment and Icon Entertainment International a Fragile Films production Developed in association with Brackmount Films
92 minutes 49 seconds
S&S February 1998 p49
Category A

 The Spiral Staircase *
BFI - 25 April
(PG) USA 1945, Dir Robert Siodmak with Dorothy McGuire, George Brent, Ethel Barrymore, Kent Smith
RKO Radio Pictures
83 mins (b/w)
No S&S reference
MFB v.13 n.146 February 1946 p20

The Spitfire Grill
Rank-Castle Rock/Turner - 30 May
(12) USA, Dir Lee David Zlotoff with Alison Elliott, Ellen Burstyn, Marcia Gay Harden, Will Patton, Kieran Mulroney, Gailard Sartain
Castle Rock Entertainment. A Gregory Production in association with The Mendocino Corporation
116 minutes 21 seconds
S&S June 1997 p63

The Square Circle/Daayraa
Blue Dolphin - 13 June
(15) India, Dir Amnol Palekar with Nirmal Pandey, Sonali Kulkarni, Faiyyaz, Rekha Sahay, Nina Kulkarni, Hyder Ali
Gateway Entertainment Pvt. Ltd
97 minutes 51 seconds (subtitles)
S&S June 1997 p64

The Starmaker/L'uomo delle stelle
20th Century Fox - 10 January
(18) Italy, Dir Giuseppe Tornatore with Sergio Castellitto, Tiziana Lodato, Leopoldo Trieste, Nicola Di Pinto, Franco Scaldati, Tony Sperandeo

Cecchi Gori Group Tiger Cinematografica Made by Francesco Tornatore/ Sciarlò s.r.l.
In association with Summit Entertainment N.V.
In collaboration with RAI Radiotelevisione Italiana A Mario & Vittorio Cecchi Gori presentation
106 minutes 34 seconds (subtitles)
S&S January 1997 p47

Star Wars
Episode IV A New Hope *
20th Century Fox - 21 March
(U) USA 1977/1997, Dir George Lucas with Mark Hamill, Harrison Ford, Carrie Fisher, Peter Cushing, Alec Guinness, Anthony Daniels, Kenny Baker, Peter Mayhew, David Prowse
Twentieth Century-Fox Film Corporation A Lucasfilm Ltd production
124 minutes 35 seconds
S&S April 1997 p50

Stephen King's Thinner
Warner Bros - 11 July
(15) USA, Dir Tom Holland with Robert John Burke, Joe Mantegna, Lucinda Jenney, Michael Constantine, Kari Wuhrer, Joy Lenz
Spelling Films Inc
92 minutes 17 seconds
S&S August 1997 p55

SubUrbia
Carlton - 17 October
(18) USA, Dir Richard Linklater with Jayce Bartok, Amie Carey, Nicky Katt, Ajay Naidu, Parker Posey, Giovanni Ribisi
Castle Rock Entertainment A Detour Film production
120 minutes 39 seconds
S&S October 1997 p59

Swann
Guild - 7 March
(15) Canada/United Kingdom, Dir Anna Benson Gyles with Miranda Richardson, Brenda Fricker, Michael Ontkean, David Cubitt, Sean McCann, John Neville
Shaftesbury (Swann) Films, Inc/ Greenpoint (Swann) Limited A Majestic Films in association with Norstar Entertainment presentation Produced with the participation of Telefilm Canada/ Ontario Film Development Corporation/Foundation to Unerwrite New Drama In association with The Canadian Broadcasting Corporation and CityTV

95 minutes 46 seconds
S&S March 1997 p63

The Sweet Hereafter
Electric Pictures - 26 September
(15) Canada, Dir Atom Egoyan
with Ian Holm, Maury Chaykin,
Gabrielle Rose, Peter Donaldson,
Bruce Greenwood, David Hemblen
© The Sweet Heareafter a division
of Speaking Parts Limited.
An Alliance Communication
presentation An Ego Film Arts
production Produced with the
participation of Telefilm Canada/
the Harold Greenberg Fund/TMN -
The Movie Network (an Astral
Communications Network) With the
assistance of The Government of
Canada, Canadian Film or Video
Production Tax Credit Program
112 minutes 5 seconds
S&S October 1997 p60

Swingers
Guild - 11 July
(15) USA, Dir Doug Liman
with Jon Favreau, Vince Vaughn,
Ron Livingston, Patrick Van Horn,
Alex Desert, Heather Graham
Doug Liman Productions A
Miramax Films presentation In
association with Independent
Pictures An Alfred Shay
production
96 minutes 22 seconds
S&S July 1997 p55

The Tango Lesson
Artificial Eye - 28 November
(PG) UK/France/Argentina/Japan/
Germany, Dir Sally Potter
with Sally Potter, Pablo Vernon,
Gustavo Naveira, Fabian Salas,
David Toole, Carolina Iotti
© Adventure Pictures (Tango)
Limited.
A co-production with OKCK Films
(Argentina)/PIE (France)/NDF
(Japan)/Imagica (Japan)/. Pandora
Film (Germany)/Cinema Projects
(Germany), Sigma Pictures
(Holland) With the participation of
The Arts Council of England. The
European Co-Production Fund
(UK)/The Sales Company/
Eurimages/Medien Filmgesellschaft
Baden-Württemberg/NPS TV/Cobo
Fund (Holland) Supported by the
National Lottery through the Arts
Council of England
101 minutes 44 seconds, (subtitles)
(b/w - some sequences in colour)
S&S December 1997 p54
Category B

Temptress Moon/Fengyue
Artifical Eye - 10 October

(15) Hong Kong/China, Dir Chen
Kaige
with Leslie Cheung [Zhang
Guorong], Gong Li, Kevin Li [Lin
Jianhua] He Saifei, Xie Tian, Xie
Tian, Zhang Shi
© Tomson (Hong Kong) Films Co.,
Ltd.
Shanghai Film Studios
116 minutes 1 second (subtitles)
S&S October 1997 p61

That Old Feeling
UIP - 5 September
(12) USA, Dir Carl Reiner
with Bette Midler, Dennis Farina,
Paula Marshall, Gail O'Grady, David
Rasche, Jamie Denton
Universal City Studios Inc/The
Bubble Factory LLC A Sheinberg
production in association with Boy
of the Year and All Girl
Productions
105 minutes 6 seconds
S&S September 1997 p56

That Thing You Do!
20th Century Fox - 24 January
(PG) USA, Dir Tom Hanks
with Tom Everett Scott, Liv Tyler,
Johnathon Schaech, Steve Zahn,
Etan Embry, Tom Hanks
Twentieth Century Fox Film
Corporation
A Clinica Estetico production
In association with Clavius Base
107 minutes 35 seconds
S&S February 1997 p57

This World, Then the Fireworks
First Independent - 5 December
(18) USA, Dir Michael Oblowitiz
with Billy Zane, Gina Gershon,
Sheryl Lee, Rue McClanahan,
Seymour Cassel, William Hootkins
© Largo Entertainment Inc.
A Muse, Balzac's Shirt, Wyman
production
100 minutes 3 seconds
S&S December 1997 p55

Tierra
Metro/Tartan - 8 August
Spain, Dir Julio Medem
with Carmelo Gómez, Emma Suárez,
Karra Elejalde, Silke Klein, Nancho
Novo, Txema Blasco
Sogetel, S.A. In collaboration with
Sogepaq. S.A. In association with
Canal+
125 minutes (subtitles)
S&S August 1997 p56

Tokyo Fist
Blue Dolphin - 4 April
(18) Japan, Dir Shinya Tsukamoto
with Kahori Fujii, Shinya
Tsukamoto, Naomasa Musaka,

Naoto Takenaka, Koichi Wajima,
Tomoroh Taguchi, Nobu Kanoka
Kaijyu Theater Co., Ltd
87 minutes 38 seconds (subtitles)
S&S April 1997 p51

Tomorrow Never Dies
UIP - 12 December
(15) USA, Dir Roger Spottiswoode
with Pierce Brosnan, Jonathan
Pryce, Michelle Yeoh, Teri Hatcher,
Joe Don Baker
© Danjaq LLC and United Artists
Corporation.
Albert R. Broccoli's Eon
Productions Limited presentation
Made by Eon Productions Ltd
119 minutes 8 seconds
S&S February 1998 p52
Category D[1]

Total Eclipse
Feature Film Company - 11 April
(18) France/UK/Belgium, Dir
Agnieszka Holland
with Leonardo DiCaprio, David
Thewlis, Romane Bohringer,
Dominique Blanc, Felicite Pasotti
Cabarbaye, Nita Klein
FI Production/Portman Production/
K2/S.F.P. Cinema With the
participation of the European
Coproduction Fund (UK) and Canal
Plus and Le Studio Canal Plus
111 minutes 11 seconds
S&S April 1997 p52
Category C

Trees Lounge
Electric Pictures - 14 February
(15) USA, Dir Steve Buscemi
with Steve Buscemi, Chloe Sevigny,
Anthony LaPaglia, Elizabeth Bracco,
Mark Boone Junior, Seymour Cassel
Live Film and Mediaworks Inc.
Live Entertainment presents an
Addis - Wechsler/Hanley/Wyman
production In association with
Seneca Falls Productions
94 minutes 53 seconds
S&S February 1997 p58

Trial And Error
Entertainment - 14 November
(12) USA, Dir Jonathan Lynn
with Michael Richards, Jeff Daniels,
Charlize Theron, Jessica Steen,
Austin Pendleton, Alexandra
Wentworth
New Line Cinema presentsa
Larger Than Life production
97 minutes 59 seconds
S&S January 1998 p54

Trigger Happy
First Independent - 13 June
(15) USA, Dir Larry Bishop
with Ellen Barkin, Gabriel Byrne,

Richard Dreyfuss, Jeff Goldblum,
Diane Lane, Larry Bishop, Gregory
Hines, Kyle MacLachlan
**Ring-A-Ding Productions, LCC
A Dreyfus/James Productions in
association with Skylight Films
and United Artists Pictures
presentation**
92 minutes 52 seconds
S&S June 1997 p65
US Title: **Mad Dog Time**

Trojan Eddie
Film Four Distributors - 21 March
(15) UK/Ireland, Dir Gillies
MacKinnon
with Richard Harris, Stephen Rea,
Brendan Gleeson, Sean McGinley,
Angeline Ball, Brid Brennan
**Channel Four Corporation An
Initial Films production with
Stratford Production in Ireland
Channel Four Films presents for
Bord Scannán na Éireann/Irish
Film Board, Irish Screen and
Channel 4 Produced with the
support of investment incentives for
the Irish Film Industry provided by
the Government of Ireland**
105 minutes 10 seconds
S&S April 1997 p53
Category C

Turbulence
Enertainment - 30 May
(18) USA, Dir Robert Butler
with Ray Liotta, Lauren Holly,
Brendan Gleeson, Ben Cross, Rachel
Ticotin, Jeffrey DeMunn, John Finn
Rysher Entertainment, Inc
99 minutes 55 seconds
S&S May 1997 p55

Twin Town
PolyGram - 18 April
(18) UK, Dir Kevin Allen
with Llyr Evans, Rhys Ifans, Dorien
Thomas, Dougray Scott, Diddug
Williams, Ronnie Williams
**PolyGram Films (UK) Limited
PolyGram Filmed Entertainment
presents A Figment Films
production in association with
Agenda Developed in association
with Aimimage Productions Ltd**
99 minutes 5 seconds
S&S April 1997 p54
Category A

 ## Under the Skin
BFI - 28 November
(18) UK, Dir Carine Adler
with Samantha Morton, Claire
Rushbrook, Rita Tushingham,
Christine Tremarco, Stuart
Townsend, Matthew Delamere
© **British Film Institute/Channel
Four.**

**A Strange Dog production for the
British Film Institute and Channel
Four In association with Rouge
Films and the Merseyside Film
Production Fund Screenplay
developed by the British Film
Institute and Channel Four**
82 minutes 59 seconds
S&S December 1997 p56
Category A

Unforgettable
20th Century Fox - 11 July
(15) USA, Dir John Dahl
with Ray Liotta, Linda Fiorentino,
Peter Coyote, Christopher
McDonald, Kim Cattrall, Kim Coates
**Metro-Goldyn-Mayer Pictures
Inc/Spelling Films Inc A Dino De
Laurentiis Company production**
116 minutes 43 seconds
S&S July 1997 p56

Unhook the Stars/Décroches les étoiles
Artificial Eye - 4 July
(15) France, Dir Nick Cassavetes
with Gena Rowlands, Marisa Tomei,
Gérard Depardieu, Jake Lloyd,
David Sherrill, David Thornton
**Hachette Première & Gérard
Depardieu presentation**
105 minutes 3 seconds
S&S July 1997 p57

Up on the Roof
**Rank-Castle Rock/Turner - 7
November**
(15) UK, Dir Simon Moore
with Billy Carter, Clare Cathcart,
Adrian Lester, Amy Robbins, Daniel
Ryan
© **Rank Film Distributors Limited/
Granada Film Limited.
In association withCarnival Films/
Production Line**
100 minutes 58 seconds
S&S November 1997 p55
Category A

Vertigo *
UIP - 25 April
(PG) USA 1958, Dir Alfred Hitchcock
with James Stewart, Kim Novak,
Barbara Bel Geddes, Tom Helmore
**Alfred J Hitchcock Productions,
Paramount Pictures Corporation**
128 mins
S&S April 1997 p14 (article)
MFB v.23 n.296 September 1958
p111

Volcano
20th Century Fox - 3 October
(12) USA, Dir Mick Jackson
with Tommy Lee Jones, Anne
Heche, Gaby Hoffmann, Don

Cheadle, Jacqueline Kim, Keith David
© **Twentieth Century Fox Film
Corporation.
Fox 2000 Pictures presents
A Shuler Donner/Donner and
Moritz Original production**
103 minutes 50 seconds
S&S October 1997 p62

Walking and Talking
Electric Pictures -17 January
(15) UK/USA, Dir Nicole
Nolofcener
with Catherine Keener, Anna Heche,
Todd Field, Liev Schreiber Randall
Batinkoff
**Zenith Productions Zenith
presents in association with
Channel Four films, TEAM,
Pandora, Miado and Electric A
Good Machine/Zenith Production
Developed with the assistance of
The Sundance Institute**
85 minutes 26 seconds
S&S January 1997 p54
Category D[2]

Warriors of Virtue
Entertainment - 25 July
(PG) USA/China, Dir Ronny Yu
with Angus MacFadyen, Mario
Yedidia, Marley Shelton, Jack Tate,
Doug Jones, Don W. Lewis
**IJL Creations, Inc/Law Brothers
Entertainment International Ltd
Four Brothers Productions**
102 minutes 41 seconds
S&S August 1997 p57

The Watermelon Woman
Dangerous to Know - 12 September
(Not submitted) USA, Dir Cheryl
Dunye
with Cheryl Dunye, Guinevere
Turner, Valarie Walker, Lisa Marie
Bronson, Irene Dunye, Brian Freeman
© **Cheryl Dunye & Dancing Girl
Productions Inc.**
81 minutes
S&S October 1997 p63

Welcome to Sarajevo
**Film Four Distributors - 21
November**
(15) UK/USA, Dir Michael
Winterbottom
with Stephen Dillane, Woody
Harrelson, Marisa Tomei, Emira
Nurevic, Kerry Fox, Goran Visnjic
© **Channel Four Television
Corporation/Miramax Film Corp.
A Dragon Pictures (Sarajevo) Ltd**
101 minutes 36 seconds
S&S November 1997 p56
Category D[1]

Welcome to the Dollhouse
Artificial Eye - 24 January

(15) USA, Dir Todd Solondz
with Heather Matarazzo, Victoria
Davis, Christina Brucato, Christine
Vidal, Siri Howard, Brendan Sexton Jr
Suburban Pictures
87 minutes 41 seconds
S&S February 1997 p59

When We Were Kings
PolyGram - 16 May
(PG) USA, Dir Leon Gast
with Muhammad Ali, George
Foreman, Don King, James Brown,
B.B.King,
**DAS Films Ltd PolyGram Filmed
Entertainment/DAS Films present
A David Soneberg production A
film by Leon Gast and Taylor
Hackford**
87 minutes 17 seconds
S&S May 1997 p56

White Man's Burden
20th Century Fox - 7 February
(15) USA, Dir Desmond Nakano
with John Travolta, Harry Belafonte,
Kelly Lynch, Margaret Avery, Tom
Bower, Andrew Lawrence
**UGC presents a Lawrence Bender/
Band Apart production**
89 minutes 5 seconds
S&S February 1997 p60

Wilde
PolyGram - 17 October
(15) USA, Dir Brian Gilbert
with Stephen Fry, Jude Law, Vanessa
Redgrave, Jennifer Ehle, Gemma
Jones, Judy Parfitt
**© Samuelson Entertainment Ltd/
NDF International Ltd.
A Samuelson production In
association with Dove
International INc/NDF
International Ltd/Pony Canyon
Inc/Pandora Film/Capitol Films
and BBC Films
With the participation of The
Greenlight Fund Produced in
association with Wall-to-Wall
Television Ltd Developed with the
assisstance of British Screen
Finance Ltd and the support of
the European Script Fund
Supported by the National Lottery
through the Arts Council of
England**
116 minutes 46 seconds
S&S October 1997 p65

Will it Snow for Christmas?/
Y'aura t'il de la neige à Noël
Artificial Eye - 7 November
(12) France, Dir Sandrine Veysset
with Dominique Reymond, Daniel
Duval, Jessica Martinez, Alexander
**© Ognon Pictures.
With the participation of Centre**

**National de la Cinématographie/
Canal+ With the support of
Fondation GAN pour le Cinéma**
90 minutes 52 seconds (subtitles)
S&S November 1997 p57

William Shakespeare's Romeo +
Juliet
20th Century Fox - 28 March
(12) USA, Dir Baz Luhrmann
with Leonardo DiCaprio, Claire
Danes, Brian Dennehy, John
Leguizamo, Pete Postlethwaite, Paul
Sorvino
**Twentieth Century Fox Film
Corporation a Bazmark
production**
120 minutes 3 seconds
S&S April 1997 p54

Your Beating Heart/Un Coeur qui
bat
Artificial Eye - 24 January
(15) USA, Dir François Dupeyron
with Dominique Faysse, Thierry
Fortineau, Jean-Marie Winling,
Steve Kalfa, Daniel Laloux,
Christophe Pichon
**Hachette Première et Cie/FR3
films Production/U.G.C./Avril SA
With the participation of Centre
National de la Cinématographie/
Canal+/Soficas Investimage2/
Investimage3
This film benefitted from the
financial support of Fondation
Gan pour le Cinéma**
99 minutes 15 seconds
S&S February 1997 p61

SPECIALISED GOODS & SERVICES

Agfa-Gevaert
Motion Picture Division
27 Great West Road
Brentford
Middx TW8 9AX
Tel: 0181 231 4310
Fax: 0181 231 4315
Major suppliers to the Motion
Picture and Television Industries of
Polyester based Colour Print Film
and Optical Sound Recording Film

Angels - The Costumiers
40 Camden Street
London NW1 OEN
Tel: 0171 387 0999
Fax: 0171 383 5603
email: angelscos.cos.uk
Chairman: Tim Angel
Contact: Richard Green - Production
Director
Contact: Jonathan Lipman -
Production Director
World's largest Costume Hire
Company. Extensive ranges covering
every historical period, including
contemporary clothing, civil and military
uniforms. Full in-house ladies and men's
making service, millinery department, jewellry, glasses and watch
hire. Branches also in Shaftesbury
Avenue and Paris. Additional services:-
experinced personal costumiers,
designers office space, reference
library and shipping department

Angels Wigs
40 Camden Street
London NW1 OEN
Tel: 0171 387 3999
Fax: 0171 383 5603
email: wigs@angelsacos.co.uk
Ben Stanton
All types of styles of wigs and
hairpieces in either human hair
bespoke or synthetic ready-to-wear.
Large stocks held, ready to dress, for
hire including legal wigs. In house
craftsmen to advise on style or period.
Facial hair made to order for sale

Any Effects
64 Weir Road
London SW19 8UG
Tel: 0181 944 0099
Fax: 0181 944 6989
Contact: Julianne Pellicci
Managing Director: Tom Harris
Mechanical (front of camera) special
effects. Pyrotechnics: simulated
explosions, bullet hits. Fine models
for close up camera work. Weather:
rain, snow, fog, wind. Breakaways:
shatterglass, windows, bottles,
glasses, collapsing furniture, walls,
floors. Specialised engineering rigs
and propmaking service

Barclays Bank
27 Soho Square
London W1A 4WA
Tel: 0171 445 5700
Fax: 0171 445 5784
Geoff Salmon
Large business centre providing a
comprehensive range of banking
services to all aspects of the film and
television industry

Charlie Bennett Underwater Productions
4 Richmond Way
Shepherds Bush
London W12 8LY
Tel: 0181 450 5421
Ifafa, Main Street
Ashby Parva
Leicestershire LE17 5HU
Tel: 01455 209 405
Mobile: 0402 263 952
Contact: Charlie Bennett
Underwater services to the film and
television industry, including
qualified diving personnel and
equipment; underwater video, stills
photography and scuba instruction.
Advice, logistics and support offered
on an international scale with
fluency in Spanish, Portuguese, and
French. Registered HSE Diving

Boulton-Hawker Films
Hadleigh
Ipswich
Suffolk IP7 5BG
Tel: 01473 822235
Fax: 01473 824519
Time-lapse, cinemicrography and
other specialised scientific filming
techniques

Dolly Brook Casting Agency
52 Sandford Road
East Ham

London E6 3QS
Tel: 0181 472 2561/470 1287
Fax: 0181 552 0733
Russell Brook
Specialises in walk-ons, supporting
artistes, extras and small parts for
films, television, commercials,
modelling, photographic, voice-
overs, pop videos

Bromley Casting (Film & TV Extras Agency)
77 Widmore Road
Bromley BR1 3AA
Tel: 0181 466 8239
Fax: 0181 466 8239
Providing quality background artisits
to the UK film and TV industry

Cabervans
Caberfeidh
Cloch Road
Gourock
Nr. Glasgow PA19 1BA
Tel: 01475 638775
Fax: 01475 638775
Make-up and wardrobe units, dining
coaches, motorhomes for use as
production office, green room, etc.
Experienced drivers

Calibre Films
187 Wardour Street
London W1V 3FA
Tel: 0171 437 1552
Fax: 0171 437 1558
email: info@hexgo.ftech.co.uk
Contacts: Adam Sutcliffe and Paul
Grindey
Offers experience in all aspects of
the negotiation and drafting of
agreements, film production, film
financing and international co-
productions, with an emphasis on
concise and effective documents,
and a practical 'business affairs'
approach to legal matters for all
those involved in the film-making
and distribution process

Central Casting Inc
13-14 Dean Street
London W1
Tel: 0171 437 4211

Connections Communications Centre Ltd
Palingswick House
241 King Street
Hammersmith
London W6 9LP
Tel: 0181 741 1767

Fax: 0181 563 1934
email: @cccmedia.demon.co.uk
Jacqueline Davis
A registered charity producing
promotional and educational videos
for the voluntary sector. Currently in
production Travelling Forward a 25
minute documentary commissioned
by the Thalidomide Society

Cool Million
Mortimer House
46 Sheen Lane
London SW14 8LP
Tel: 0181 878 7887
Fax: 0181 878 8687
Dot O'Rourke
Promotional merchandising, launch
parties and roadshows

De Wolfe Music
Shropshire House
2nd Floor East
11/20 Capper Street
London WC13 6JA
Tel: 0171 631 3600
Fax: 0171 631 3700
email: dewolfe_Music@
Compuserve.com
Warren De Wolfe, Alan Howe
World's largest production music
library. Represents 25 composers for
commissions, television and film
scores. Offices worldwide, sound FX
department, 3 x 24-track recording
studies all with music to picture
facilities, also digital editing

Deloitte & Touche
Hill House
1 Little New Street
London EC4A 3TR
Tel: 0171 936 3000
Fax: 0171 583 8517/1198
Gavin Hamilton-Deeley, Mark Attan,
Robert Reed
Advisors to film, television and
broadcasting organisations. Business
plans and financial models for
companies, tax planning and
business advice for individuals, and
information on legal and regulatory
developments affecting the sector

Diverse Design
Gorleston Street
London W14 8XS
Tel: 0171 603 4567
Fax: 0171 603 2148
email: design@diverse.co.uk
Steve Billinger, Daniel Creasey
Titles, series, format design, content
graphics. Recent work: *Berkeley
Square, The History of Tom Jones;
Real Woman; Full Circle with
Michael Palin; Cold War; Booked;
French Express; Escape; In the
Footsteps of Alexander the Great*

Downes Agency
96 Broadway
Bexleyheath
Kent DA6 7DE
Tel: 0181 304 0541
Fax: 0181 301 5591
Agents representing presenters and actors experienced in the fields of presentations, documentaries, commentaries, narrations, television dramas, feature films, industrial videos, training films, voice-overs, conferences and commercials

EOS Electronics AV
EOS House
Weston Square
Barry
South Glamorgan CF63 2YF
Tel: 01446 741212
Fax: 01446 746120
Specialist manufacturers of video animation, video time laspsing and video archiving equipment.
Products: Supertoon Low Cost School Animation System, AC 580 Lo-band Controller, BAC900 Broadcast Animation Controller, LCP3 Compact Disc, Listening Posts

ETH Screen Music
17 Pilrig Street
Edinburgh EH6 5AN
Tel: 0131 553 2721
Harald Tobermann
Producer and publisher of original music for moving images. Complete creative team - composers, arrangers, musicians

Eureka Location Management
51 Tonsley Hill
London SW18 1BW
Tel: 0181 870 4569
Fax: 0181 871 2158
Suzannah Holt
Finds and manages locations for film and television in Britain and abroad. Offices in London and Toronto

FTS Bonded
Aerodrome Way
Cranford Lane
Hounslow
Middx TW5 9QB
Tel: 0181 897 7973
Fax: 0181 897 7979
Inventory management, worldwide freight, courier services, technical facilities including film checking and tape duplication, storage and distribution

Faunus The Florists
Interior and Exterior Floral Design
69 Walmgate
York YO1 2TZ
Tel: 01904 613044

Contact: Robert Hale
Suppliers and designers of interior floral decoration

Film Finances
1-11 Hay Hill
Berkeley Square
London W1X 7LF
Tel: 0171 629 6557
Fax: 0171 491 7530
G J Easton, J Shirras, D Wilder, H Penallt Jones
Provide completion guarantees for the film and television industry

The Film Stock Centre Blanx
70 Wardour Street
London W1V 3HP
Tel: 0171 494 2244
Fax: 0171 287 2040
D John Ward
ndependent distributor of major manufacturers' motion picture film stock, professional video tape, Polaroid, audio tape, related products. Impartial advice, competitive prices, SOR, special deals for low-budgets. Weekdays 9.00am to 6.30pm

Harkness Hall Ltd
The Gate Studios
Station Road
Borehamwood
Herts WD6 1DQ
Tel: 0181 953 3611
Fax: 0181 207 3657
email: sales@harknesshall.com
Ian Sim, Robert Pickett
Projection screens and complete screen systems, fixed and portable, front or rear, flat, curved, flying, roller etc. Curtain tracks, festoons, cycloramas, raise and lower equipment, stage equipment, installation and maintenance

Henry Hepworth
Media Law Solicitors
5 John Street
London WC1N 2HH
Tel: 0171 242 7999
Fax: 0171 242 7988
Michael Henry
A new specialist media and intellectual property practice with a distinctive high quality client base which is active across the entire spectrum of the copyright and intellectual property industries

Hirearchy Classic and Contemporary Costume
45 Palmerston Road
Boscombe
Bournemouth
Dorset BH1 4HW
Tel: 01202 394465

Specialists in the hire of original 20th Century costume including accessories, militaria, luggag and textiles. Medieval, Tudor and Victorian costume also available

Hothouse Models & Effects
Studio 6
Sussex Road
Colchester
Essex CO3 3HQ
Tel: 01206 764434
Fax: 01206 764435
Jeremy Clarke
Large scale props and sets; working models; high detail close-ups, miniatures, creatures/puppets, advertisements, conceptual design

Kodak Limited
Professional Motion Imaging
PO Box 66
Hemel Hempstead
Herts HP1 1JU
Tel: 01442 61122
Fax: 01442 844458
A Kennedy
Suppliers of the full range of Kodak colour negative and print films, including the new family of Vision colour negative films

Lip Service Casting
4 Kingly Street
London W1R 5LF
Tel: 0171 734 3393
Fax: 0171 734 3373
Susan Mactavish
Voiceover agency for actors, and voiceover casting agency. Publishers of The Voice Analysis - a breakdown of actors' vocal profiles

MBS Underwater Video Specialists
1 Orchard Cottages
Coombe Barton
Shobrooke, Crediton
Devon
Tel: 01363 775 278
Fax: 01363 775 278
email: mbscm@mail.eclipse.co.uk
Website: http://www.eclipse.co.uk.mbs
Contact: Colin Munro
MBS provides underwater stills photography and videography services, specialising in underwater wildlife shots. We can provide full HSE registered dive teams for UK based work, and cover all aspects of diving safety and support, vessel servicing and specialist underwater equipment supply

Marcus Stone Casting
Georgian House
5 The Pavilions
Brighton BN2 1RA

Tel: 01273 670053
Fax: 01273 670053
Supplies television, film extras. Up to 1,000 extras available for crowd scenes

Media Education Agency
5A Queens Parade
Brownlow Road
London N11 2DN
Tel: 0181 888 4620
David Lusted
Consultancy, lectures and teacher in-service education (INSET) in film, television and media studies. Contacts include academics, educationists, broadcasters, writers and actors

Midland Fire Protection Services
National Fire and Rescue
The Fire Station
Courtaulds
256 Foleshill Road
Coventry CV6 5AB
Tel: 01203 685252/0836 651408 (mobile)
Fax: 01203 685252
Specialists in fire and rescue cover for location, studio and stage work. Special services, firefighters, action vehicles, fully equipped fire and rescue appliances, 5,000 gallons of water storage systems available, throughout the UK 24 hour service

Moving Image Touring Exhibition Service (MITES)
Foundation For Art & Creative Technology (FACT)
Bluecoat Chambers
Liverpool L1 3BX
Tel: 0151 707 2881
Fax: 0151 707 2150
Simon Bradshaw
Courses for artists, gallery curators, technicians and exhibitors concerned with the commissioning and presentation of moving image art works. Also development, advice, consultation services and an extensive exhibition equipment resource

Nicholson Graham & Jones
110 Cannon Street
London EC4N 6AR
Tel: 0171 648 9000
Fax: 0171 648 9001
Annmarie Pryor, Marketing Manager
A City law firm and founder member of the international GlobaLex network in the UK, USA, Europe and the Far East. The Intellectual Property Group handles film and television production, financing and distribution, cable, satellite and telecommunications work, book and

newspaper publishing, syndication, advertising, merchandising, sponsorship and sports law. Also advise on technology transfer, patent , trade mark, service mark, know-how arrangements and franchising as well as computer hardware and software agreements and all intellectual property copyright, moral and performers' right issues

Olswang
90 Long Acre
London WC2E 9TT
Tel: 0171 208 8888
Fax: 0171 208 8800
email: olsmail@olswang.co.uk
Website:http//www.olswang.co.uk
One of the UK's leading entertainment and media law firms. It provides specialist advice in all aspects of broadcasting, satellite, cable, multimedia, IT & telecommunications, media convergence and music law, to the European and US markets

Oxford Scientific Films (OSF)
Long Hanborough
Oxford OX8 8LL
Tel: 01993 881 881
Fax: 01993 882 808
45-49 Mortimer Street
London W1N 7TD
Tel: 0171 323 0061
Fax: 0171 323 0161
Experts in macro, micro, time-lapse, high-speed and other specialist photography for productions ranging from natural history documentaries to television commercials requiring special effects

Pirate Models and Effects
St Leonards Road
London NW10 6ST
Tel: 0181 930 5000
Fax: 0181 930 5001
email: help@pirate.co.uk
Web: www.pirate.co.uk
Michael Ganss
Established in 1988, specialising in the design and production of technical models and physical effects for film, commercials and TV station idents. We build the most technically sophisticated models and effects available - using extensive computing and traditional facilities and skilled senior staff, who also have extensive experience in related fields such as industrial design and product prototyping

Pirate Motion Control
St Leonards Road
London NW10 6ST
Tel: 0181 930 5000

Fax: 0181 930 5001
email: help@pirate.co.uk
Web: www.pirate.co.uk
Michael Ganss
Pirate Motion Control have the largest, most versatile video motion control studio facility in the UK. Located near AFM

ProDigital
3 George Street
West Bay
Dorset DT6 4EY
Tel: 01308 422 866
Sound equipment, service and maintenance. Specialises in location sound equipment for the film and television industry - particularly DAT recorders

Radcliffes Transport Services
3-9 Willow Lane
Willow Lane Industrial Estate
Mitcham
Surrey CR4 4NA
Tel: 0181 687 2344
Fax: 0181 687 0997
Ken Bull
Specialist transport specifically for the film and television industry, both nationally and internationally. Fleet ranges from transit vans to 40' air ride articulated vehicles with experienced staff

Richards Butler
Beaufort House
15 St Botolph Street
London EC3A 7EE
Tel: 0171 247 6555
Fax: 0171 247 5091
email: law@richards-butler.com
Richard Philipps, Barry Smith, Stephen Edwards, Martin Boulton
Richards Butler is an international law firm which has been associated with the media and entertainment industry for over 60 years

The Screen Company
182 High Street
Cottenham
Cambridge CB4 4RX
Tel: 01954 250139
Fax: 01954 252005
Manufacture, supply and installation of all types of front and rear projection screens for video, slide, film and OHP

Security Archives Ltd
1-8 Capitol Park
Capitol Way
London NW9 0EQ
Tel: 0181 205 5544
Fax: 0181 200 1130
Secure storage for film, video and audio tape in bomb-proof vaults with thermohydrographic controls

and Halon fire suppression. 24hr collection and delivery, computerised, bar-coded management and tracking of clients' material

Snow-Bound
37 Oakwood Drive
Heaton
Bolton BL1 5EE
Tel: 01204 841285
Fax: 01204 841285
Suppliers of artificial snow and the machinery to apply it for the creation of snow/winter scenes. The product is life-like (not poly beads or cotton wool) adheres to any surface and is fire-retardent, non-toxic and safe in use, and eco-friendly

Stanley Productions
36 Newman Street
London W1P 3PD
Tel: 0171 636 5770
Fax: 0171 636 5660
Richard Hennessy
Europe's largest distributor of video tape and equipment. Full demonstration facilities with independent advice on suitable equipment always available

Ten Tenths
106 Gifford Street
London N1 0DF
Tel: 0171 607 4887
Fax: 0171 609 8124
Props service specialising in vehicles (cars, bikes, boats and planes) ranging from 1901 to present day - veteran, vintage, classic, modern - with complementary wardrobe facilities

Wrap it up
116a Acton Lane,
Chiswick
London W4 5HH
Tel: 0181 995 3357 (Mobiles 0973 198154)
Fax: 0181 2348 3030
Wrap it up provides production services which include transcription, post production scripts, voice scripts and logging of rushes for production companies. Recent work: September Films - Teenagers British lifestyles - Transcription and Post Production Scripts. Horizon BBC - Transcription. Dennis and Gnasher, Tony Collingwood Productions - Voice Scripts

Zooid Pictures Limited
66 Alexander Road
London N19 5PQ
Tel: 0171 281 2407
Fax: 0171 281 2404
email: pictures@zooid.co.uk
Website: http://www.zooid.co.uk

Richard Philpott
For over 20 years, Zooid has been a one-stop media resources supplier and researcher for all copyright materials including film/video, stills, illustration, animation and sound, from archives, libraries, agencies, private collections and museums worldwide, for use in film, television, book publishing, CD-Rom, multimedia, presentations and on-line services. Zooid manage all aspects from first briefing through to licensing Zooid use advanced digital technologists and license their management system, Picture Desk, to leading international publishers

STUDIOS

BBC Television Centre Studios
Wood Lane
London W12 7RJ
Tel: 0181 576 7666
Fax: 0181 576 8806
8 full-facility television studios
TC1 10,250 sq ft
TC3 8,000 sq ft
TC4 and TC8 8,000 sq ft (digital and widescreen capable)
TC6 8,000 sq ft (digital)
TC2, TC5 and TC7 3,500 sq ft
Other studios include The Music Studio, The Virtual Studio and studios at Elstree

The Boilerhouse
8 Nursery Road
Brixton
London SW9 8BP
Tel: 0171 737 7777
Fax: 0171 737 5577
Clive Howard, Michael Giessler
100 sq metre studio, dry/wet stage, special effect facilities, variable tank systems, rain rigs. Productions: Adidas, Cadbury Chocolate ads

Bray Studios
Down Place
Water Oakley
Windsor Road
Windsor SL4 5UG
Tel: 01628 622111
Fax: 01628 770381
Studio manager: Beryl Earl
STAGES
1 (sound) 955 sq metres
2 (sound) 948 sq metres
3 (sound) 238 sq metres
4 (sound) 167 sq metres
TELEVISION
Pie in the Sky, SelecTV; *Our Friends in the North*, BBC; *Murder Most Horrid*, Talkback; *Midsummer Nights Dream*, Edenwood Productions; *Emma*, Mai Productions; *Mo Jo*, Portobello Pictures; *Velvet Goldmine*, Velvet Goldmine Productions Ltd

Capital FX
21A Kingly Court
London W1R 5LE
Tel: 0171 439 1982
Fax: 0171 734 0950
email: enquiries@capital.fx.co.uk
Website: www.capital.fx.co.uk
Graphic design and production, optical effects, film and laser subtitling

De Lane Lea Dean Street Studio
75 Dean Street
London W1V 5HA

Tel: 0171 439 1721/ 0171 432 3877 (direct line 24 hours)
Fax: 0171 437 0913
Studio manager: Dick Slade
STAGE
1 86 sq metres
40x23x18 SYNC
lighting rig, film and TV make-up rooms, one wardrobe, one production office, full fitted kitchen

Ealing Studios
Ealing Studios
Ealing Green
London W5 5EP
Tel: 0181 567 6655
Fax: 0181 758 8658
email: info@ealing-studios.co.uk
Vicki Harvey-Piper
STAGES
1 (new) - bluescreen/motion control = area 222 sqm
2 (sound) 873.92m2
3A (sound) 439m2
3B (sound) 453m2
3A/B (combined) 955m2
4 (modelstage silent) 430m2
5 (sound) 91m2
FILMS
Remember Me?
The Secret Agent
Jane Eyre
TELEVISION
Tales from the Crypt Series 7 (HBO), *My Night With Reg* (BBC), *The Demon Headmaster I & II* (BBC), *Broken Glass* (BBC), *The Fragile Heart* (Carnival)

Halliford Studios
Manygate Lane
Shepperton
Middx TW17 9EG
Tel: 01932 226341
Fax: 01932 246336
Charlotte Goddard
STAGES
A 334 sq metres
B 223 sq metres

Holborn Studios
49/50 Eagle Wharf Road
London N1 7ED
Tel: 0171 490 4099
Fax: 0171 253 8120
Studio manager: Ivan Merrell
STAGES
4 2,470 sq feet
6 2,940 sq feet
7 2,660 sq feet
18 roomsets 3,125 sq feet
Also eight fashion studios, set building, E6 lab, b/w labs, KJP in house, canal-side restaurant and bar. Productions; National Lottery Stills;

Advertisements for Scratch cards; Saatchis - photographer Dave Stewart

Isleworth Studios
Studio Parade
484 London Road
Isleworth
Middx TW7 4DE
Tel: 0181 568 3511
Fax: 0181 568 4863
STAGES
A 292 sq metres
B 152 sq metres
C 152 sq metres
D 152 sq metres

Jacob Street Studios
9-19 Mill Street
London SE1 2BA
Tel: 0171 232 1066
Fax: 0171 252 0118

Lamb Studio
Bell Media Group, Lamb House
Church Street, Chiswick Mall
London W4 2PD
Tel: 0181 996 9960
Fax: 0181 996 9966
email: paul@belmedia.demon.co.uk
Sound proofed, air-conditioned studio. Total floor area of 575 sq ft. Average ceiling height of 12ft. Free parking, production office, kitchen, make-up room. Easy, access from central London, M4, M3 and M25. Ideal for talking heads, interviews, small dramas, pack shots, motion control, training. Post-production facilities also available

Leavesden Studios
PO Box 3000
Leavesden
Herts WD2 7LT
Tel: 01923 685 060
Fax: 01923 685 061
Studio manager: Daniel Dark
STAGES
1A 32,076sq feet
1B 28,116 sq feet
1C 12,285 sq feet
1D 12,808 sq feet
1E 26,868 sq feet
Effects 15,367 sq feet
Back Lot 100 acres
180 degrees of clear and uninterrupted horizon
Further 200,000 sq.ft of covered space available
FILMS
GoldenEye, Mortal Kombat, Annihilation; Episode One of the Star Wars prequels

Leavesden is proud to have been the UK studio for...

STAR WARS
Episode 1

JAMES BOND'S GOLDENEYE

MORTAL KOMBAT: ANNIHILATION

LEAVESDEN
Studios

LEAVESDEN STUDIOS, P.O. BOX 3000, LEAVESDEN, HERTS. WD2 7LT, ENGLAND. TELEPHONE: (01923) 685 060 FACSIMILE: (01923) 685 061
STUDIO MANAGER: DANIEL DARK
A MEMBER OF THE DKH/GEORGE TOWN HOLDINGS GROUP

Millennium Studios

Elstree Way
Herts WD6 1SF
Tel: 0181 236 1400
Fax: 0181 236 1444
Contact: Kate Tufano
'X' Stage: 327 sq metres sound stage with flying grid and cyc. Camera room, construction workshop, wardrobe, dressing rooms, edit rooms, hospitality suite and production offices are also on site.
Recent productions: Carlton's *Bliss*; *Showcareer's Basil*

Pinewood Studios

Pinewood Road
Iver
Bucks SL0 0NH
Tel: 01753 651700
Fax: 01753 656844
Managing Director: Steve Jaggs
STAGES
A 1,685 sq metres
(Tank: 12.2m x 9.2m x 2.5m)
B 827 sq metres
C 827 sq metres
D 1,685 sq metres
(Tank: 12.2m x 9.2m x 2.5m)
E 1,685 sq metres
(Tank: 12.2m x 9.2m x 2.5m)
F 700 sq metres
(Tank: 6.1m x 6.1m x 2.5m)
G 247 sq metres
H 300 sq metres
J 825 sq metres
K 825 sq metres
L 880 sq metres
M 880 sq metres
N/P 768 sq metres
South Dock (silent)
1,548 sq metres
007 (silent) 4,225 sq metres (Tank: 90.5m x 22.3m x 2.7m Reservoir: 15.3m x 28.7m x 2.7m)
Large Process 454 sq metres
Small Process 166 sq metres
Exterior Lot 52 acres, comprising formal gardens and lake, woods, fields, concrete service roads and squares
Exterior Tank 67.4m narrowing to 32m x 60.4m x 1.06m deep.
Capacity 764,000 galls - Inner tank: 15.5m x 12.2m x 2.7m
Backing: 73.2m x 18.3m
Largest outdoor tank in Europe
FILMS
Interview with the Vampire; Mary Reilly; First Knight; Loch Ness; Jack and Sarah; Mission: Impossible; Fierce Creatures; The Fifth Element; The Saint; Fairytale - A True Story; Event Horizon; Incognito; The Jackal; Eyes Wide Shut; The Avengers; Tomorrow Never Dies
TELEVISION

Chandler & Co; Space Precinct; Scavengers; Class Act; Karaoke; Cold Lazarus; House of Cards - The Final Cut; You Bet!; Ken Russell's Treasure Island; Poldark; Mirad - A Boy From Bosnia; Ivanhoe; The Designated Mourner; The Preventers; JOnathan Creek; The Vanishing Man

Riverside Studios

Crisp Road
Hammersmith
London W6 9RL
Tel: 0181 237 1000
Fax: 0181 237 1011
Jon Fawcett
Studio One 529 sq metres
Studio Two 378 sq metres
Studio Three 112 sq metres
Plus preview cinema, various dressing rooms, offices
TELEVISION
T.F.I. Friday, 'Collins & McConies Movie Club', Channel 4 Sitcom Festival, 'This Morning with Richard Not Judy'

Rotherhithe Studios

119 Rotherhithe Street
London SE16 4NF
Tel: 0171 231 2209
Fax: 0171 231 2119
O Stockman, C Goodwin
STAGES
1 Rotherhithe 180 sq metres
Pre-production, construction, post-production facilities, costume making, props
FILMS
The Nutcracker Story (IMAX 3D)

Shepperton Studios

Studio Road
Shepperton
Middx TW17 0QD
Tel: 01932 562 611
Fax: 01932 568 989
Paul Olliver
STAGES
A 1,668 sq metres
B 1,115 sq metres
C 1,668 sq metres
D 1,115 sq metres
E 294 sq metres
F 294 sq metres
G 629 sq metres
H 2,660 sq metres
I 657 sq metres
J 1,394 sq metres
K 1,114 sq metres
L 604 sq metres
M 259 sq metres
T 261 sq metres
P 276 sq metres
R 948 sq metres
S 929 sq metres
FILMS
Lost in Space; Hamlet; 101

Dalmatians; Carrington; Evita; In Love and War;

Stonehills Studios

Shields Road
Gateshead
Tyne and Wear NE10 0HW
Tel: 0191 495 2244
Fax: 0191 495 2266
Studio Manager: Nick Walker
STAGES
1 1,433 sq feet
2   750 sq feet
The North's largest independent television facility comprising of Digital Betacam Edit Suite with the BVE 9100 Edit Controller, and Abekas ASWR 8100 mixer, A57 DVE and four machine editing, including two DVW 500s. Also three Avid off-line suites, 2D Matador and 3D Alias graphics and a Sound Studio comprising a Soundtracs 6800 24-track 32 channel desk and Soundscape 8-track digital editing machine
TELEVISION
Germ Genie, BBC 2; *The Spark, Border; Come Snow Come Blow,* Granada

Teddington Studios

Broom Road
Middlesex TW11 9NT
Tel: 0181 977 3252
Fax: 0181 943 4050
Ewart Needham
STAGES
1 653 sq metres
2 372 sq metres
3 120 sq metres
4 74 sq metres
TELEVISION
Kilroy; Birds of a Feather; Over the Rainbow; This is Your Life; Men Behaving Badly; Goodnight Sweetheart, Des O'Connor Tonight

Theed Street Studios

12 Theed Street
London SE1 8ST
Tel: 0171 928 1953
Fax: 0171 928 1952
Bill Collom
STAGE
A 151 sq metres
TELEVISION
Metropolis for BBC *Continuing Education; Reality on the Rocks* for C4; *Lost Civilisations* for Time Life

Three Mills Island Studios

Three Mill Lane
London E3 3DU
Tel: 0181 522 0849
Fax: 0181 522 0848
Edwin Shirley
STAGES
5A 93ft x 80ft

5B 30ft x 80ft
5C 150ft x 80ft
6A 120ft x 80ft
6B 210ft x 60ft
9B 160ft x 52ft

Twickenham Film Studios

St Margaret's
Twickenham
Middx TW1 2AW
Tel: 0181 607 8888
Fax: 0181 607 8889
Gerry Humphreys, Caroline Tipple
(Stages)
STAGES
1 702 sq metres
with tank 37 sq metres x 2.6m deep
2 186 sq metres
3 516 sq metres
2 x dubbing theatres; 1 x ADR/Foley
theatre; 40 x cutting rooms;
Lightworks, Avid 35/16mm

Wembley Studios

10 Northfield Industrial Estate
Beresford Avenue
Wembley
Middlesex HAO 1RT
Tel: 0181 903 4296
Fax: 0181 900 1353
STAGES

Studio 290 sq metres
Cyc 193 sq metres
Power: 900 amps 3 phase
Production offices, dressing rooms,
kitchen

Westway Studios

8 Olaf Street
London W11 4BE
Tel: 0171 221 9041
Fax: 0171 221 9399
Steve/Kathy
STAGES
1 502 sq metres (Sound Stage)
2 475 sq metres
3 169 sq metres
4 261 sq metres

TELEVISION COMPANIES

Below are listed all British television companies, with a selection of their key personnel and programmes. The titles listed are a cross-section of productions initiated (but not necessarily broadcast). For details of feature films made for television, see Production Starts

BBC TELEVISION

British Broadcasting Corporation
Television Centre
Wood Lane
London W12 7RJ
Tel: 0181 743 8000
Chairman: Sir Christopher Bland
Director-General: John Birt
Chief Executive BBC Worldwide:
Rupert Gavin
Chief Executive BBC Broadcast:
Will Wyatt
Chief Executive BBC Production:
Ronald Neil
Chief Executive BBC News: Tony
Hall
Chief Executive BBC Resources:
Rod Lynch
Director of Television: Alan Yentob

BBC North
New Broadcasting House
Oxford Road
Manchester M60 1SJ
Tel: 0161 200 2020

BBC Northern Ireland
Broadcasting House
Ormeau Avenue
Belfast BT2 8HQ
Tel: 01232 338000
Fax: 01232 338800

BBC Pebble Mill
Birmingham B5 7QA
Tel: 0121 414 8888
Fax: 0121 414 8634

BBC Scotland
Broadcasting House
Queen Margaret Drive
Glasgow G12 8DG
Tel: 0141 339 8844

BBC South
Broadcasting House
Whiteladies Road
Bristol BS8 2LR
Tel: 0117 973 2211

BBC Wales
Broadcasting House

Llandaff
Cardiff CF5 2YQ
Tel: 01222 564888

The Animation Unit
BBC Bristol
Features
Whiteladies Road
Bristol BS8 2LR
Tel: 0117 974 2483
Fax: 0117 923 9790
email: animation.unit@bbc.co.uk
Executive Producer: Colin Rose
Commissions mainly half-hour
specials and series of 30 and 10
minutes for adult, youth and family
audiences from experienced
animation directors. Editorially, the
emphasis is on entertainment,
innovation, story and script. Does
not commission children's programming. Occasionally produces longer
form films and buys work by new
directors

BBC TV Arts
BBC Television Centre
Wood Lane
London W12 7RJ
Tel: 0181 895 6500
Fax: 0181 895 6586
Head of Arts: Kim Evans
Managing Editor, Arts: Alex Graham
Arts Features and Regional
Programmes: Keith Alexander

Arena
Editor: Anthony Wall

The Works
Editor: Mike Poole

Omnibus
Editor: Gillian Greenwood

One Foot in the Past
Editor: Basil Comely

Home Front
Series Producer: Franny Moyle

Looking Good
Series Producer: Jeanine Josman

The Bookworm
Series Producer: Mary Sackville-West

Renaissance
Executive Producer: Nick Rossiter

Tx.
Editor: John Wyver

The Renaissance
Executive Producer: Nick Rossiter

BBC Broadcast Programme Acquisition
Centre House
56 Wood Lane
London W12 7RJ
Tel: 0181 743 8000
Fax: 0181 749 0893
Controller, Programme Acquisition:
Alan Howden
Head of Programme Acquisition:
Sophie Turner Laing
Selects and presents BBC TV's
output of feature films and series on
both channels
Business Unit: May Grainger
Contact for commissioned material
and acquisition of completed
programmes, film material and
sequences for all other programme
departments
Business Development Executive:
Paul Eggington
Contact for sub-licensing of material
acquired by (but not produced by)
the BBC

BBC TV Children's Programmes
Television Centre
Wood Lane
London W12 7RJ
Tel: 0181 743 8000
Head Of Children's Programmes:
Lorraine Heggessey
Head Of Children's Commissioning:
Roy Thompson

Blue Peter
Programme Editor: Oliver
Macfarlane
Presenters: Stuart Mills, Katy Hill,
Romana D'Annunzio, Konnie Huq
Continuing x 25 mins
Blue Peter began in October 1958.
The programme is named after the
blue and white flag which is raised
within 24 hours of a ship leaving

harbour: the idea is that the programme is like a ship setting out on a voyage, having new adventures and discovering new things

Byker Grove
Produced by Zenith North for BBC TV
Executive Producer: Matthew Robinson
Producer: Helen Gregory
20 x 25 minutes
The gritty drama with hard-hitting storylines continues at the Byker Grove youth centre

Dear Mr Barker
Senior Producer: Jane Tarleton
Presenter: Paul Hendy
15 x 15 minutes
Answering young children's curious questions in an innovative way

Record Breakers
Producer: Melissa Hardinge
13 x 25 minutes
Factual programme discovering record breakers from all over the world

The Really Wild Show
Executive Producer: Eric Rowan
Producer: David Wallace
13 x 25 minutes
Amazing wildlife stories from around the globe

Short Change
Senior Producer: Roy Milani
8 x 20 minutes
Consumer journalists take on important issues for children

Smart
Executive Producer: Christopher Pilkington
Producer: Christopher Tandy
12 x 25 minutes
A fun art programme encouraging viewers to 'have a go' with everyday objects and easy-to-find materials

Teletubbies
Produced by Ragdoll for BBC TV
Producer: Anne Wood
Entertaining and educational daily programme for pre-school viewers

Fully Booked
Produced by BBC Scotland
Series Producer: Ed Gray
Producers: Margaret Blythe
Presenters: Gail Porter, Chris Jarvis, Tim Vincent
22 x 2 hours
Live variety television - including celebrities, competitions and music - set in a hotel

Grange Hill
Producer: Diana Kyle
Script Editor: Jyoti Patel
Writers: Leigh Jackson, Chris Ellis,

Alison Fisher, Sarah Daniels, Diane Whitley, Jeff Povey, Annie Wood, Tim O'Mara
20 x 25 minutes
Fictional characters face true-to-life situations at a large comprehensive school

Live and Kicking
Editor: Chris Bellinger
Senior Producer: Angela Sharp
A winning formula of top celebrities, chart-topping bands, comedy, news, reviews and competitions

Newsround
Editor: Ian Prince
Senior Producer: Marshall Corwin
Presenters: Chris Rogers, Kate Sanderson, Lizo Mzimba
Continuing x 10 minutes
Daily news bulletin for children

BBC Classical Music
Broadcasting House
Langham Place
London W1A 1AA
Head of Classical Music: Roger Wright
BBC Classical Music, Television
East Tower
BBC Television Centre
Wood Lane
London W12 7RJ
Tel: 0181 895 6500
Fax: 0181 895 6586

BBC Classical Music, Television
East Tower
BBC Television Centre
Wood Lane
London W12 7RJ
Tel: 0181 895 6500
Fax: 0181 895 6586
Head of Classical Music, Television: Avril MacRory
Editor, Music Programmes: Peter Maniura
Dance and BBC Young Musicians
Executive Producer: Bob Lockyer

BBC Community Programme Unit
White City
201 Wood Lane
London W12 7TS
Tel: 0181 752 4705
Fax: 0181 752 4666
Head: Paul Hamann
Over the Edge; From the Edge;
The Day That Changed my Life; Video Diaries; Video Nation Shorts
Private Investigations; Prison Weekly

BBC TV Documentaries, History, Community and Disability Programmes
BBC White City
201 Wood Lane
London W12 7TS

Tel: 0181 752 6322
Head of Department: Paul Hamann
Chief Assistant: Margaret Magnusson

Inside Story
BBC1 - Editor: Olivia Lichtenstein

Rough Justice
BBC1 - Executive Producer: Elizabeth Clough

Modern Times
BBC2 - Series Editor: Stephen Lambert

Timewatch
BBC2 - Series Editor: Laurence Rees
Executive Producer: Clare Paterson

Reputations
BBC2 - Series Editor: Janice Hadlow
Executive Producer: Clare Paterson

BBC Drama Series & Serials
Television Centre
Wood Lane
London W12 7RJ
Tel: 0181 743 8000
Head of Serials: Michael Wearing
Head of Series: Jo Wright
Head of Drama in Scotland: Andrea Calderwood
Head of Drama in Northern Ireland: Robert Cooper
Head of Drama in Wales: Pedr James
Head of Drama in Birmingham: Tony Virgo
Returning productions for BBC1 include *Ballykissangel, Backup, Bugs, Casualty, Dalziel & Pascoe, Dangerfield, Hamish Macbeth, Hetty Wainthropp Investigates, Pie In the Sky, Preston Front, Silent Witness, This Life* (BBC2) and *Eastenders*
New series include:

Playing the Field
(6 x 50, BBC1)
Producer: Greg Brenman for Tiger Aspect
Directors: Paul Seed, Catherine Morshead,
Writer: Kay Mellor
Cast includes: Melanie Hill, Saira Todd, Lesley Sharp

The Ambassador
(6 x 50, BBC1)
Producer: Douglas Rae, Ecosse Films for BBC Northern Ireland
Writer: Russell Lewis
Cast includes: Pauline Collins

Invasion: Earth
(6 x 50, BBC1)
Producer: Chrissy Skinns for BBC Scotland/Sci-Fi Channel
Devised by Jed Mercurio

Berkeley Square
(10 x 50, BBC1)
Producer: Alison Davis
Director: Lesley Manning
Writer: Deborah Cook
Cast includes: Victoria Smurfit,
Clare Wilkie, Tabitha Wady

New Serials include:
Tom Jones
(1 x 90, 4 x 55, BBC1)
Producer: Suzan Harrison
Director: Metin Huseyin
Writer: Simon Burke, from Henry
Fielding's novel
Cast: Max Beesley, Samantha
Morton, Brian Blessed, Frances de la
Tour, Kathy Burke, James d'Arcy

The Lakes
(5 x 50, BBC1)
Producer: Charles Pattinson
Writer: Jimmy McGovern
Director: David Blair
Cast: John Simm, Emma Cunniffe,
Clare Holman, Robert Pugh, Mary
Jo Randle

Our Mutual Friend
(4 x 90, BBC2)
Producer: Catherine Wearing
Writer: Sandy Welch, from Charles
Dickens' novel
Cast: Anna Friel, Pam Ferris, Paul
McGann, Timothy Spall

Holding On
(8 x 55, BBC2)
Producer: David Snodin
Director: Adrian Shergold
Writer: Tony Marchant
Cast: Phil Daniels, David Morrissey,
Saira Todd

A Respectable Trade
(4 x 50, BBC1)
Producer: Ruth Baumgarten
Director: Suri Krishnamma
Writer: Phillipa Gregory, from her
novel
Cast includes: Warren Clarke, Emma
Fielding, Anna Massey, Ariyon
Bakare, Jenny Agutter, Simon
Williams

Looking After Jo-Jo
(4 x 50, BBC2)
Producer: Deirdre Keir for BBC
Scotland
Director: John Mackenzie
Writer: Frank Deasy
Cast includes: Robert Carlyle

The Beggar Bride
(2 x 75, BBC1)
Producer: Kate Harwood
Director: Diarmuid Lawrence
Writer: Lizzie Mickery from Gillian
White's novel
Cast includes: Keeley Hawkes, Joe
Duttine, Nicholas Jones

Bright Hair
(2 x 60, BBC1)
Producer: Eileen Quinn for
Monogram Productions
Director: Chris Menaul
Writer: Peter Ransley
Cast includes: Emilia Fox, James
Purefoy

Real Women
(3 x 50, BBC1)
Producer: Debbie Shewell
Director: Phil Davis
Cast includes: Pauline Quirke,
Michelle Collins, Frances Barber,
Lesley Manville, Gwyneth Strong

BBC Education for Adults
BBC White City
Wood Lane 201
London W12 7TS
Tel: 0181 752 5252
Head of Commissioning: Fiona
Chesterton

BBC TV Entertainment
Television Centre
Wood Lane
London W12 7RJ
0181 743 8000
Head of Comedy: Geoffrey Perkins
Head of Comedy Entertainment: Jon
Plowman
Head of Light Entertainment:
Michael Leggo
Head of Factual Entertainment: Tony
Moss
Head of Independent Commission-
ing: Bill Hilary

Situation Comedy 98/99 (Independents)
Game On
Hat Trick Productions
Producer: Geoffrey Perkins, Sioned
Wiliam
Director: John Stroud
Writer: Bernadette Davis
Cast: Samantha Janus, Neil Stuke,
Mathew Cottle

Unfinished Business
Alomo Productions
Producer: Sioned Wiliam
Director: tba
Writers: Laurence Marks, Maurice
Gran
Cast: Henry Goodman, Harriet Walter

Birds Of A Feather
Alomo Productions
Producer: Charlie Hanson
Director: Terry Kinane
Writers: Laurence Marks, Maurice
Gran
Cast: Pauline Quirke, Linda Robson,
Lesley Joseph

Goodnight Sweetheart
Alomo Productions

Executive Producer: Allan
McKeown
Producer: John Bartlett
Director: tba
Writers: Laurence Marks, Maurice
Gran
Cast: Nicholas Lyndhurst

Situation Comedy 98/99 (In-house Productions)
2point4 children
Producer/Director: tba
Writer: Andrew Marshall
Cast: Belinda Lange, Gary Olsen,
Clare Buckfield, John Pickard

Last Of The Summer Wine
Producer/Director: Alan J W Bell
Writer: Roy Clarke
Cast: Bill Owen, Peter Sallis, Brian
Wilde, Kathy Staff

Jonathan Creek
Executive Producer: Susan Belbin
Director: tba
Writer: David Renwick
Cast: Alan Davis, Caroline Quentin

Roger, Roger
Producer: Gareth Gwenlan
Director: Tony Dow
Writer: John Sullivan
Cast: Robert Daws, Lesley Vickerage

Light Entertainment
Big Break
Producer: Geoff Miles
Host: Jim Davidson
Referee: John Virgo

Jim Davidson's Generation Game
Producer: Jon Beazley

Noel's House Party
Executive Producer: Mike Leggo
Producer: Guy Freeman
Director: tbc
Script Associate: tbc

The National Lottery Live
Producer: Peter Estall
Director: Duncan Cooper

Auntie's Sporting Bloomers
Producer: Tom Webber
Host: Terry Wogan

The Other Half
Producer: Jon Beazley
Director: Richard Valentine
Host: Dale Winton

Dad
Producer: Marcus Mortimer
Writer: Andrew Marshall
Cast: George Cole, Kevin McNally,
Julia Hills, Toby Ross Bryant

BBC TV Feature & Events
BBC White City
201 Wood Lane

London W12 7TS
0181 752 5909
Head of Department: Anne Morrison
Deputy Head: Jane Lush

Watchdog
Series Editor: Helen O'Rahilly

Crimewatch UK
Series Producer: Seetha Kumar

Holiday
Series Editor: Jane Lush

Esther
BBC 1
Series producer: Patsy Newey
Presenter: Esther Rantzen

Sky At Night
Series Editor: Pieter Morpurgo

Have I Got News For You
(Hat Trick Production)
Executive Producer: Jane Lush

BBC Films and Single Drama
Head of BBC Films and Single
Drama: David M. Thompson
Development Executive: Tracey
Scoffield
BBC Films
Feature films include: Udayan
Prasad's *My Son the Fanatic*, written
by Hanif Kureishi; John Madden's
Mrs Brown, written by Jeremy
Brock; Richard Kwietniowski's *Love
and Death on Long Island*; Gillies
MacKinnon's *Regeneration*, written
by Allan Scott based on the novel by
Pat Barker; Antonia Bird's *Face* by
Ronan Bennett; Jez Butterworth's
Mojo; David Hare's *The Designated
Mourner*, written by Wallace Shawn;
Stephen Poliakoff's *The Tribe*; Mike
Barker's *The James Gang*, written
by Stuart Hepburn; Shane Meadows'
24:7; Malcolm Mowbray's *The
Revengers' Comedies*, adapted from
the plays by Alan Ayckbourn; John
Henderson's *Bring Me the Head of
Mavis Davis* by Craig Strachan;
Frances-Anne Solomon's *Peggy Su!*
by Kevin Wong; Waris Hussein's
Trying to Grow by Firdaus Kanga;
Marleen Gorris' *Mrs. Dalloway*,
adapted by Eileen Atkins from the
novel by Virginia Woolf; Phillip
Saville's *Metroland* by Adrian
Hodges; and Robert Bierman's *Keep
The Aspidistra Flying*, written by
Alan Plater from the novel by
George Orwell

Films for BBC1
Feature-length films for BBC1
include *Hostile Waters* by Troy
Kennedy Martin, directed by David
Drury; *The Fix*, written and directed
by Paul Greengrass, *The Gift* by
Lucy Gannon, directed by Danny

Hiller; *Sex & Chocolate* by Tony
Grounds, directed by Gavin Millar;
Our Boy, also by Tony Grounds,
directed by David Evans; *The
Woman in White*, written by David
Pirie from the novel by Wilkie
Collins, directed by Tim Fywell; *The
Prince of Hearts* by Lee Hall,
directed by Simon Curtis; and St.
Ives adapted by Allan Cubitt from
Robert Louis Stevenson, directed by
Harry Hook

Films for BBC2
Original and distinctive films for
BBC2 include *Eight Hours From
Paris*, written and directed by
Philippa Lowthorpe; *Flight by
Tanika Gupta*, directed by Alex
Pillai, *The Scar* by Amber Films;
Stone, Scissors, Paper by Richard
Cameron, directed by Stephen
Whittaker; Billy and Gillies
Mackinnon's *Small Faces*, directed
by Gillies MacKinnon; and *Bumping
the Odds* by Ross Munro, directed
by Rob Rohrer

Love Bites
The second season of Love Bites
included: *In Your Dreams* by Ol
Parker, directed by Simon Cellan
Jones; *The Perfect Blue* by Nick
Collins, directed by Kieron J. Walsh;
Anorak of Fire by Stephen Dinsdale,
directed by Elijah Moshinsky; and
Theory of Flight by Richard Hawkins,
directed by Paul Greenglass

Brief Encounters
Short 35mm films commissioned
jointly by the BBC and Channel 4;
Daphne And Apollo by Billie
Reynolds, directed by Clare Kilner;
Dual Balls, written and directed by
Dan Zeff; *I Love My Mum* written by
Smita Bhide and directed by Alrick
Riley; *Shark Hunt* written and
directed by Dominic Lees; Jo
Hodges' *Silent Film*, directed by
Malcolm Venville; *The Sin Eater*
written and directed by Terence
Gross; *We & Dry* written and
directed by John McKay; and *Closer*,
written by Bille Eltringham, directed
by Simon Beaufoy and Bille
Eltringham

Drama Serials
Head of Drama Serials: Michael
Wearing

The Tenant of Wildfell Hall
(BBC1)
Written by: David Noakes and Janet
Barron from the novel by Anne
Bronte
Producer: Suzan Harrison
Director: Mike Barker
Cast: Tara Fitzgerald, Rupert Graves,

Toby Stephens, Pam Ferris and
Kenneth Cranham
3 x 52 mins

Nostromo
BBC2
Adaptation of Joseph Conrad's novel
by John Hale
Producer: Fernando Ghia
Director: Alastair Reid
Cast: Claudio Amendola, Colin
Firth, Claudia Cardinale, Albert
Finney, Serena Scott Thomas,
Joaquim De Almeida, Lothaire
Bluteau, Brian Dennehym, Ruth
Gabriel and Fernando Hilbeck
4 x 80 mins

Holding On
BBC2
Writer: Tony Marchant
Producer: David Snodin
Director: Adrian Shergold
Cast: Phil Daniels, David Morrissey,
Annette Badland, Saira Todd and
Sean Gallagher
8 x 50 mins

Have Your Cake
BBC 1
Developed by Eileen Quinn from an
original script by Rob Heyland
Producer: Eileen Quinn/Dave Edwards
Director: Paul Seed
Cast: Sinead Cusack, Miles
Anderson, Holly Aird
4 x 50 mins

Common as Much II
BBC1
Written by William Ivory
Producer: Catherine Wearing
Director: Metin Huseyin
Cast: Edward Woodward, Roy Hudd,
Kathy Burke, Neil Dudgeon,
Michelle Holmes, June Whitfield,
Frank Finlay, Saee Jaffrey
6 x 55 mins

The Crow Road
BBC2
Written by Bryan Elsley from the
novel by Iain Banks
Producer: Bradley Adams
Director: Gavin Miller
Cast: Joseph McFadden, Bill
Paterson, Elizabeth Sinclair,
Dougram Scott, Valerie Edmund,
Peter Capaldi
1 x 60 mins

Harvest Moon
BBC1
Written by: Michael Chaplin
Producer: Ruth Caleb
Directors: Tristram Powell, Lesley
Manning
Cast: David Calder, Emma Fielding,
Andrew Howard, Geraldine James,
Robert Pugh, John Standing, Ray

Stevenson, Keith Barron, Liz Frazer, Robert Glenister, Anastasia Hille
6 x 50 mins

Ivanhoe
BBC1
Written by: Deborah Cook from the novel by Sir Walter Scott
Producer: Jeremy Gwilt
Director: Stuart Orme
Cast: Ralph Brown, Jimmy Chisholm, Trevor Cooper, James Cosmo, Rory Edwards, Aden Gillet, Ciaran Hinds, David Horovitch, Christopher Lee, Susan Lynch, Sian Phillips, Ronald Pickup, Victoria Smurfit, Steven Waddington
6 x 50 mins

The Missing Postman
BBC1
Written by: Mark Wallington, adapted from his novel
Producer: Gareth Neame
Director: Alan Dossor
Cast: James Bolam
2 x 75 mins

Harpur and Iles
BBC1
Written by: Dom Shaw from the novels by Bill James
Producer: Jane Dauncey
Director: Jim Hill
Cast: Hywel Bennett, Aneirin Hughes, Patrick Robinson, Jim Carter
2 x 50 mins

Plotlands
BBC1
Written by Jeremy Brock
Producer: Louis Marks
Director: John Strickland
Cast: Saskia Reeves, Richard Lintern, David Ryall, Ger Ryan, Rebecca Callard, Phillip Whitchurch, Richard Cordery, Petra Markham

The Locksmith
BBC1
Written by: Stephen Bill
Producer: Irving Teitelbaum
Directors: Alan Dossor, Lawrence Gordon Clark, Chris Bernard
Cast: Warren Clarke, Chris Gascoyne, Sheila Kelley, Sarah-Jane Potts, Polly Hemingway
6 x 50 mins

Music programmes
Songs Of Praise
Editor: Hugh Faupel
52 x 35 mins
A musical celebration of life and faith. Regular presenter Pam Rhodes, Sir Harry Secombe, Diane Louise Jordan and Kevin Woodford, travel the country and the world

Other programmes include:
Carols From King's (BBC2)
Holy Cow (BBC1)
Ancient Voices (BBC2)
The Temptation Game (BBC1)

Drama Series
Acting Head of Drama Series: Jo Wright

Ballykissangel
Series 2, BBC1
Producer: Chris Griffin
Executive Producer: Tony Garnett
Writers: Kieran Prendiville, John Forte, Rio Fanning, Niall Leonard, Jo O'Keefe
Principal cast: Stephen Tompkinson, Dervla Kirwin, Tony Doyle
8 x 50 mins
(World Production for BBC Northern Ireland)

Backup
Series 2, BBC1
Producer: Hilary Salmon
Writers: Roy Mitchell, Susan Wilkins, Avril E Russell, Simon Andrew Stirling, Peter Gibbs
Principal cast: Martin Troakes, Katrina Levon, Nick Miles, Christopher John Hall, Calum McPherson, William Tapley, James Gaddas, Matthew Rhys
6 x 50 mins
(BBC Midlands and East)

Beck
Series 1 BBC 1
Producer: Jacky Stoller, Paul Hines
Writer: Paul Hines
Principal cast: Amanda Redman, Caroline Loncq, David Hunt, David Herlihy
6 x 50 mins
(BBC Production in association with Tempest Films and TV)

The Brokers' Man
Series 1, BBC1
Producer: Adrian Bate
Writers: Al Hunter-Ashton, Tim O'Mara
Principal cast: Kevin Whately
2 x 50 mins
(BBC in association with Bentley Productions)

Bugs
Series 3, BBC1
Producer: Brian Eastman
Writers: Stephen Gallagher, Colin Brake, Frank de Palma, Terry Borst
Principal cast: Craig McLachlan, Jaye Griffiths, Jesse Birdsall
10 x 50 mins
(A Carnival Films Production)

Casualty
Series 11, BBC1
Producer: Rosalind Anderson

Principal cast: Derek Thompson, Clive Mantle, Sorcha Cusack, Ian Bleasdale, Julia Watson, Jonathan Kerrigan, Gray O'Brien, Ganiat Kasumo
24 x 50 mins

The Crime Traveller
Series 1, BBC 1
Producer: Brian Eastman
Writers: Anthony Horowitz, Colin Brake
Principal cast: Michael French
8 x 50 mins
(A Carnival Production)

Crocodile Shoes
Series 2, BBC 1
Producer: Peter Richardson
Writer: Jimmy Nail
Principal cast: Jimmy Nail, Melanie Hill
6 x 50 mins
(A Big Boy Production)

Dalziel and Pascoe
Series 2, BBC1
Producer: Paddy Higson
Writers: Alan Plater, Malcolm Bradbury
Principal cast: Warren Clarke, Colin Buchanan, Susannah Corbett
3 x 90 mins
(BBC Midlands & East in association with Portobello Pictures)

Dangerfield
Series 3, BBC1
Producer: Peter Wolfes
Writers: Peter J Hammond, Don Shaw, Barbara Cox, Suj Ahmed, Alick Rowe, Tony Etchells
Principal cast: Nigel Le Vaillant, Tim Vincent, Tamsin Malleson
10 x 50 mins
(BBC Midlands & East)

EastEnders
BBC1
Producer: Jane Harris
The continuing thrice-weekly serial set in Albert Square

Hamish Macbeth
Series 3, BBC1
Producer: Charles Salmon
Writers: Daniel Boyle, Stuart Hepburn, Dominic Minghella
Principal cast: Robert Carlyle
6 x 50 mins
(Zenith/Skyline Production for BBC Scotland)

The Hello Girls
Series 1, BBC1
Producers: Laurence Brown, Jacinta Peel
Writer: Ruth Carter
Principal cast: Letitia Dean, Amy Marston, Maggie McCarthy, Helen Shields, Stephanie Turner

Hetty Wainthropp Investigates
Series 2, BBC1
Producer: Carol Parks
Writers: David Cook, John Bowen,
Jeremy Paul
Principal cast: Patricia Routledge,
Derek Benfield, Dominic Monaghan
6 x 50 mins

Insiders
Series 1, BBC1
Producer: Nicholas Brown
Writer: Lucy Gannon
Principal cast: Julia Ford, Robert
Cavanah, Adrian Rawlins, Idris Elba
6 x 50 mins
(BBC Pebble Mill)

The Lab
BBC1
Producer: Peter Norris
Writer: Michael Stewart
Principal cast: to be announced

The Lakes
Series 1, BBC1
Producer: Charles Pattinson
Writer: Jimmy McGovern
Principal cast: to be announced
4 x 50 mins

Pie in the Sky
Series 4, BBC1
Producers: Chrissy Skinns
Writers: Andrew Payne, Richard
Maher, Robert Jones
Principal cast: Richard Griffiths,
Maggie Steed, Malcolm Sinclair,
Bella Enahoro, Samantha Janus
6 x 50 mins
(WitzEnd Production)

Plotlands
Series1, BBC1
Producer: Louis Marks
Writer: Jeremy Brock
Principal cast: Saskia Reeves,
Richard Lintern, David Ryall
6 x 50 mins
(Wall to Wall Productions)

Preston Front
Series 3, BBC1
Producer: Julian Murphy
Writer: Tim Firth
Principal cast: Colin Buchanan, Paul
Haigh, Susan Wooldridge, Caroline
Catz, Adrian Hood
7 x 40 mins
(BBC Pebble Mill)

Silent Witness
Series 2, BBC1
Producer: Alison Lumb
Writers: John Milne, Jacqueline
Holborough, Gillian Richmond,
Peter Lloyd
Principal cast: Amanda Burton,
Mick Ford
8 x 50 mins

This Life
Series 2, BBC2
Producer: Jane Fallon
Writer: Amy Jenkins
Principal cast: Jack Davenport,
Amita Dhiri, Jason Hughes, Andrew
Lincoln, Daniella Nardini
(A World Production)

BBC News
Television Centre
Wood Lane
London W12 7RJ
Tel: 0181 743 8000
Fax: 0181 743 7882
Chief Executive, BBC News: Tony
Hall
Controller, Programme Policy: Peter
Bell
Director, Continuous News: Jenny
Abramsky
Head of News Programmes: Richard
Clemmow
Head of Political Programmes: Mark
Damazer
Head of Current Affairs: vacant
Head of Newsgathering: Richard
Sambrook
Head of Business Programmes:
Helen Boaden
Main news programmes/Networks:
Breakfast News 7-9 am
One O'Clock News
Six O'Clock News
Nine O'Clock News
Newsnight 10.30 - 11.15 pm
BBC News 24 (24-hour news for
UK)
BBC World (24-hour news for
viewers outside UK)
Ceefax
Other programmes include:
Breakfast with Frost
Business Breakfast
Correspondent
Here & Now
The Money Programme
On The Record
Panorama
Pound For Pound
Question Time
Scrutiny
Westminster
The Record
Working Lunch

BBC Religion
New Broadcasting House
Oxford Road
Manchester M60 1SJ
Tel: 0161 200 2020
Head of BBC Religion: Rev. Ernest
Rea
Managing Editor, Religion: Helen
Alexander

Everyman
Editor: Richard Denton
50 mins

Reflective religious documentary
series

Heart of the Matter
Series Producer: Anne Revell
50 mins
Explores topical moral and religious
dilemmas through short film reports
and debates, chaired by Joan
Bakewell

First Light
Senior Producer: Hugh Faupel
30 mins
Sunday morning prayer and
reflection

BBC School Programmes
White City
201 Wood Lane
London W12 7TS
Tel: 0181 752 4344
Fax: 0181 752 5421
BBC School Programmes transmits
between 500 and 600 hours of
programming every year, approximately 100 hours new. It produces
programming across the age range
from pre-school to post 16 aimed at
supporting teachers and pupils in
schools across the UK. All programmes are supported by an
extensive range of resource materials
including, in some cases, CD-Roms

Head of Commissioning: Frank
Flynn
Head of Programmes for BBC
Learning: Liz Cleaver

BBC TV Science
201 Wood Lane
London W12 7TS
Tel: 0181 752 6178
Fax: 0181 752 6810
Head: Glenwyn Benson
Manager: Nigel Gamble

Horizon
BBC2
Editor: John Lynch
Single subject documentaries
presenting science to the general
public and analysing the implications of new discoveries

QED
BBC2
Editor: Michael Morley
The human stories behind the
science headlines

Tomorrow's World
BBC1
Editor: Saul Nassé
Popular weekly programme
investigating the latest in science,
technology and medicine

Animal Hospital
BBC1

Executive Producers: Phil Dolling and Sarah Hargreaves
Rolf Harris and his team report weekly from the RSPCA's Hospital

Trust Me, I'm a Doctor
BBC2
Executive Producer: Micheal Morley
A critical insiders guide to medicine, presented by GP and stand-up comic Phil Hammond, with reporter Donna Bernard

BBC TV Sport
BBC TV Centre
Wood Lane
London
Tel: 0181 743 8000
Fax: 0181 749 7886
Controller of Television Sport, Broadcast: Jonathan Martin, OBE
Head of Sport, Production: Bob Shennan
Head of Development, Sport: John Rowlinson
Head of Finance & Business: David Cole
Executive Editor, Football and Deputy Head of Sport, Production: Niall Sloane
Executive Editor, Magazines and Documentaries: Philip Bernie
Executive Editor: Television: Dave Gordon
Executive Editor, Radio: Gordon Turnbull
Executive Producers, Studios and Olympics: Martin Hopkins
Executive Producer, OBs: Malcolm Kemp

BBC TV Youth and Entertainment Features
New Broadcasting House
Oxford Road
Manchester M60 1SJ
Tel: 0161 200 2020
Head of Department: John Drury
Programmes include:
A Question of Sport; MasterChef; Junior MasterChef; Blockbusters The Sunday Show; The Travel Show Today's the Day; University Challenge; The Mrs Merton Show Rough Guide to the World; Jane Asher's Good Living; Wipe Out

INDEPENDENT TELEVISION COMPANIES

Anglia Television
Anglia House
Norwich NR1 3JG
Tel: 01603 615151
Fax: 01603 631032
Chairman: David McCall
Managing Director: Graham Creelman
Controller of Programmes and Production: Malcom Allsop

Network programmes:
The Time ... The Place ...
Vanessa
Morning Worship
Survival
There's Only One Barry Fry
Regional programmes:
A Century of Change
Anglia Gold
Backstage
Badminton
Band X
Champions of the Future
Clappers
Countrywide
Cover Story
Cover Story Crime Special
Crawshaw Points Constable Country
Crime Special Update
Crown and Country
First Take
Five A Side Football
Gardens Without Borders
Go Fishing
Hazel Soan Painting
Heirloom
Homemaker
Jungle on your Doorstep
Kick-Off Live!
Lizas Country
Magic and Mystery
Midweek Kick-Off
Nelson
Nurses
Out to Lunch With Brian Turner
Signs and Wonders
Silverstone Karting Finals
Street Wise
The Warehouse

Border Television
The Television Centre
Carlisle CA1 3NT
Tel: 01228 25101
Chairman: James Graham
Managing Director: Peter Brownlow
Controller of Programmes: Neil Robinson

Forum
Producer: various
10 x 60 mins
Local current affairs and community issues programme

Innovators
Producer: Ian Fisher
6 x 30 mins
A series looking at the latest developments in science and technology

Supplement
Producer: John Mapplebeck
10 x 30 mins
Interviewer Eric Robson talks to significant and interesting people from the region

Blessed Are They...
8 x 20 mins religious series based on The Beatitudes on the Sermon on the Mount

Carlton UK Television
101 St Martin's Lane
London WC2N 4AZ
Tel: 0171 240 4000
Fax: 0171 240 4171
Chairman: Nigel Walmsley
Chief Executive: Clive Jones
Director of Programmes: Andy Allan
Managing Director, Carlton Sales: Martin Bowley
Finance Director: Mike Green
Business Affairs Director: John Egan
Controller of Regional and Public Affairs: Hardeep Kalsi
Controller of Compliance and Legal Affairs: Don Christopher
ITV covers 15 regional licence holders including Carlton Television which is made up of three licensees - Carlton Broadcasting, Central Broadcasting and Westcountry.
Carlton Broadcasting is responsible for the ITV regional licence for London and the South East. Central Broadcasting is responsible for the ITV regional licence for East, West, and South Midlands. Westcountry is responsible for the ITV regional licence for the South West of England.

Carlton Productions produces Network and Regional programmes for both Carlton and Central Broadcasting and for other national and international markets. Carlton Sales sells airtime and sponsorship for all three broadcasters.

Carlton Television also operates two facilities operations - Carlton Studios in Nottingham supplying studios and related services and Carlton 021 the largest commercial operator of Outside Broadcast Services in Europe.

High fibre television

PRODUCTIONS
35-38 Portman Square
London W1H 9FH
Tel: 0171 486 6688
Fax: 0171 486 1132

Carlton Studios
Lenton Lane
Nottingham NG7 2NA
Tel: 0115 9863322
Fax: 0115 9645552
Director of Programmes: Andy Allan
Director of Drama and Co-production: Jonathan Powell
Controller of Entertainment and Comedy: John Bishop
Controller of Factual Programmes: Steve Clark
Controller of Children's & Young People's Programmes: Michael Forte
Controller of Network Affairs: Claire Lummis
Finance Director: Martin McCausland
Controller of Community Programmes Unit: Peter Lowe
Controller of Commissioning and Network Business Affairs: Tom Betts
Production Executive, Carlton Films: William Turner

BROADCASTING
London Television Centre
Upper Ground
London SE1 9LT
Tel: 0171 620 1620
Fax: 0171 827 7500
Central House
Broad Street
Birmingham B1 2JP
Tel: 0121 643 9898
Fax: 0121 643 4897
101 St Martin's Lane London
WC2N 4AZ
Tel: 0171 240 4000
Fax: 0171 240 4171

Controller of Broadcasting: Coleena Reid
Finance Director: Ian Hughes
Controller of Promotions: Jim Stokoe
Head of Acquisitions: George McGhee
Head of Presentation, Carlton Broadcasting: Wendy Chapman
Head of Presentation and Programme Planning, Central Broadcasting: David Burge

CARLTON BROADCASTING
101 St Martin's Lane
London WC2N 4AZ
Tel: 0171 240 4000
Fax: 0171 240 4171
Chairman: Nigel Walmsley
Managing Director: Colin Stanbridge

CENTRAL BROADCASTING
Central House
Broad Street
Birmingham B1 2JP
Tel: 0121 643 9898
Fax: 0121 643 4897
Carlton Studies
Lenton Lane
Nottingham NG7 2NA
Tel: 0115 986 3322
Fax: 0115 964 5552
Unit 9, Windrush Court
Abingdon Business Park
Abingdon OX1 1SA
Tel: 01235 554123
Fax: 01235 524024
Chairman: Nigel Walmsley
Managing Director: Ian Squires
Controller of News & Operations: Laurie Upshon
Controller of Sport: Gary Newbon
Editor: Central News West: John Boileau
Editor: Central News East: Dan Barton
Editor: Central News South: Phil Carrodus

WESTCOUNTRY
Langage Science Park
Western Wood Way
Banham
Plymouth PL7 5BG
Tel: 01752 333333
Fax: 01752 333444
Chairman: Clive Jones
Managing Director: Mark Haskell
Director of Programmes: Jane McClosky
Head of Presentation: Graham Stevens

CARLTON SALES
101 St Martin's Lane
London WC2N 4AZ
Tel: 0171 240 4000
Fax: 0171 240 4171
Elizabeth House, 3rd Floor
St Peter's Square
Manchester M2 3DF
Tel: 0161 237 1881
Fax: 0161 237 1970
Chairman: Clive Jones
Managing Director: Martin Bowley
Sales Director: Steve Platt
Agency Negotiation Director: Gary Digby
Sales Operations Director: Jeremy Hallsworth
Sales Administration Director: Ron Coomber
Marketing Director: Fran Cassidy
Head of Sponsorship: David Prosser
Head of Client Services: Caroline Hunt

CARLTON STUDIOS
Lenton Lane
Nottingham NG7 2NA
Tel: 0115 9863322

Fax: 0115 9645552
Managing Director: Ian Squires
Director of Operations: Paul Flanaghan
Production Controller: John Revill

CARLTON 021
12-13 Gravelly Hill Industrial
Estate, Gravelly Hill
Birmingham B24 8HZ
Tel: 0121 327 2021
Fax: 0121 327 7021
Managing Director: Ed Everest
Business Manager: Mike McGowan
Head of Operations: Rob Hollier
Chief Engineer: John Fisher

Central Independent Television
Central Court
Gas Street
Birmingham B1 2JT
Tel: 0121 643 9898
Fax: 0121 616 1531
Chairman: Leslie Hill
Managing Director: Ron Henwood

Channel 5 Broadcasting
22 Long Acre
London WC2E
Tel: 0171 550 5555
Fax: 0171 550 5554
Britain's fifth terrestrial channel launched at the end of March 1997
Chief Executive: David Elstein
Director of Programming: Dawn Airey

Channel Four Television
124 Horseferry Road
London SW1P 2TX
Tel: 0171 396 4444
Fax: 0171 306 8353
Website: www.channel4.com
Executive Members
Chief Executive and Director of Programmes: Michael Jackson
Managing Director: David Scott
Director of Advertising Sales and Marketing: Andy Barnes
Director of Business Affairs: Janet Walker
Director and General Manager: Frank McGettigan
Director of Strategy and Development: David Brook
Non-Executive Members
Chairman: Vanni Treves
Deputy Chairman: Bert Hardy
Murray Grigor
Sarah Radclyffe
Usha Prashar, CBE
Company Secretary: Andrew Yeates
Deputy Director of Programmes: Karen Brown
Head of Programmes Nations and Regions: Stuart Cosgrove
Head of Factual Programmes & Features: Steve Hewlett
Head of News, Current Affairs &

Business: David Lloyd
Head of Drama & Animation: Gub Neal
Head of Film: Paul Webster
Head of Entertainment: Kevin Lygo
Senior Commissioning Editor, Documentaries: Peter Moore
Commissioning Editor, Children & Young People: Andi Peters
Commissioning Editor: Daytime: Fiona Chesterton
Daytime Strategy Director: Julia Le Stage
Commissioning Editor, Schools: Paul Ashton
Commissioning Editor, Sport & The Big Breakfast: Mike Miller
Commissioning Editor, Arts: Janey Walker
Commissioning Editor, Education: Mark Galloway
Commissioning Editor, Independent Film & Video: Robin Gutch
Commissioning Editor, Multicultural Programmes: Yasmin Anwar
Commissioning Editor, Religion and Features: Peter Grimsdale
Commissioning Editor, Science and Talks: Sara Ramsden
Commissioning Editor, Current Affairs: Dorothy Byrne
Commissioning Editor, Entertainment: Graham K. Smith
Commissioning Editor, Entertainment: Caroline Leddy
Commissioning Editor, Animation: Clare Kitson
Head of Purchased Programmes: Mari Macdonald
Controller of Acquisition: June Dromgoole
Head of Corporate Affairs: Sue Robertson
Head of Marketing: Wendy Lanchin
Head of Research: Hugh Johnson
Head of Presentation: Steve White
Head of Rights/Corporation Secretary: Andrew Yeates
Head of Personnel and Administration: Gill Monk
Head of Programme Finance: Maureen Semple-Piggott
Head of Finance: Tony Moore
Chief Engineer: Peter Marchant
Head of Information Systems: Geoff Balls
Head of Business Affairs: Andrew Brann
Head of Business Affairs for Film: Andrew Hildebrand
Head of Programme Planning & Strategy: Rosemary Newell
Head of Legal Compliance: Jan Tomalin
Director of 124 Facilities: Catherine Houston
Chief Executive, Channel Four Learning: Davina Lloyd

Chief Press Office: Chris Griffin-Beale

Channel Television
Television Centre, La Pouquelaye
St Helier
Jersey JE1 3ZD
Tel: 01534 816816
Fax: 01534 816817
Television Centre, Bulwer Avenue
St Sampsons
Guernsey GY1 2BH
Tel: 01481 723451
Fax: 01481 710739
Chief Executive: John Henwood
Managing Director: Michael Lucas
Head of Programmes: Karen Rankine
Head of Resource and Transmission: Tim Ringsdore
Controller, Production and Development: Philippe Bassett

GMTV
London Television Centre
Upper Ground
London SE1 9TT
Tel: 0171 827 7000
Fax: 0171 827 7001
Chairman: Charles Allen
Managing Director: Christopher Stoddart
Director of Programmes: Peter McHugh
Managing Editor: John Scammell
Editor: Gerry Melling
Head of Press & PR: Sue Brealey
Presenters: Eamonn Holmes, Fiona Phillips, Lorraine Kelly, Penny Smith, Matthew Lorenzo
6am-9.25am 7 days a week

Grampian Television
Queen's Cross
Aberdeen AB15 4XJ
Tel: 01224 846846
Fax: 01224 846800
Chairman: Dr Calum A MacLeod CBE

Granada Television
Granada Television Centre
Quay Street
Manchester M60 9EA
Tel: 0161 832 7211
36 Golden Square
London W1R 4AH
Tel: 0171 734 8080
Granada News Centre
Albert Dock, Liverpool L3 4BA
Tel: 0151 709 9393
Fax: 0151 709 3389
Granada News Centre
Bridgegate House
5 Bridge Place
Lower Bridge Street
Chester CH1 1SA
Tel: 01244 313966
Fax: 01244 320599

Granada News Centre
White Cross, Lancaster LA1 4XQ
Tel: 01524 60688
Fax: 01524 67607
Granada News Centre
Daisyfield Business Centre
Appleby Street
Blackburn BB1 3BL
Tel: 01254 690099
Fax: 01254 699299
Chief Executive: Gerry Robinson
Joint Managing Directors: Andrea Wonfor, Jules Burns
Commercial Director: Katherine Stross
Director of Production and Resources: Brenda Smith
Director of Broadcasting: Julia Lamaison
Sales Director: Mick Desmond
Director of Production: Max Graesser
Director of Public Affairs: Chris Hopson
Technical Director: Roger Pickles
Controller of Drama: Sally Head
Controller of Entertainment and Comedy: Andy Harris
Controller of Factual Programmes: Dianne Nelmes
Controller of Programme Services and Personnel: David Fraser
Head of Film: Pippa Cross
Head of Technical Operations: Chris Hearn
Head of Entertainment: Bill Hilary
Head of Features: James Hunt
Head of Factual Drama: Ian McBride
Head of Regional Affairs: Rob McLoughlin
Head of Planning and Marketing: Colin Marsden
Head of Production Services: Jim Richardson
Head of Music: Iain Rousham
Head of Regional Programmes: Mike Spencer
Head of Design and Post Production: Mike Taylor
Head of Current Affairs and Documentaries: Charles Tremayne
Head of Transmission Operations: Peter Williams
Head of Comedy: Antony Wood
Coronation Street
Cracker
This Morning
World in Action
You've Been Framed
Revelations
What the Papers Say

HTV
The Television Centre
Culverhouse Cross
Cardiff CF5 6XJ
Tel: 01222 590590

Fax: 01222 597183
HTV West
The Television Centre
Bath Road
Bristol BS4 3HG
Tel: 0117 977 8366
Fax: 0117 972 2400
16 Old Bond Street
London W1X 3DB
Tel: 0171 499 7100
Fax: 0171 409 0564
99 Baker Street
London W1M 2AJ
Tel: 0171 486 3808
Fax: 0171 486 2593
Chairman: Louis Sherwood
Chief Executive: Chris Rowlands
Director of Broadcasting and Deputy
Chief Executive: Ted George
Deputy Director of Broadcasting and
Director of Programmes (Wales):
Menna Richards
Head of Factual Programmes
(Wales): Elis Owen
Head of Human Resources: Linda
Chamberlain-Jones
Director of Programmes (West):
Jeremy Payne
Head of News & Current Affairs
(West): Ken Rees

Independent Television News

200 Gray's Inn Road
London WC1X 8XZ
Tel: 0171 833 3000
Fax: 0171 430 4700
Chairman: Sir David English
Chief Executive: Stewart Purvis
Editor-in-Chief: Richard Tait
Editor of ITN programmes on ITV:
Nigel Dacre
Senior Press Officer: Anne Were
ITN is the news provider nominated
by the Independent Television
Commission to supply news
programme for the ITV network.
Subject to review, this licence is for
a ten year period from 1993. ITN
also provides news for Channel 4,
Channel 5 and for the Independent
Radio News (IRN) network. ITN is
recognised as one of the world's
leading news organisation whose
programmes and reports are seen in
every corner of the globe. In addition
to its base in London, ITN has
permanent bureaux in Washington,
Moscow, South Africa, the Middle
East, Hong Kong, and Brussels as
well as at Westminster and eight
other locations around the UK.
News At Ten
Channel 4 News
Early Evening News
Lunchtime News
Morning and Afternoon Bulletins
Night-time Bulletins
Weekend Programmes

Five News
Five News Early
5.30am Morning News
Radio News
Travel News
The Big Breakfast News on
Channel 4
ITN World News
House to House on Channel 4
First Edition on Channel 4
Special Programmes
News Archives
The Westminster Television
Centre
ITN Productions

LWT (London Weekend Television)

The London Television Centre
Upper Ground
London SE1 9LT
Tel: 0171 620 1620
Fax: 0171 261 1290
Chief Executive: Charles Allen
Managing Director: Steve Morrison
Controller of Arts: Melvyn Bragg
Controller of Drama: Sally Head
Controller of Entertainment: John
Kaye Cooper
Controller of Factual and Regional
Programmes: Simon Shaps
London's Burning
The South Bank Show
Gladiators
Blind Date

Meridian Broadcasting

TV Centre
Northam Road
Southampton SO14 0PZ
Tel: 01703 222555
Fax: 01703 335050
48 Leicester Square
London WC2H 7LY
Tel: 0171 839 2255
Fax: 0171 925 0665
West Point
New Hythe
Kent ME20 6XX
Tel: 01622 882244
Fax: 01622 714000
1-3 Brookway
Hambridge Lane
Newbury, Berks RG14 5UZ
Tel: 01635 522322
Fax: 01635 522620
Chairman: Clive Hollick
Chief Executive: Roger Laughton
Director of Public Affairs: Simon
Albury

S4C

Parc Ty Glas
Llanishen
Cardiff CF4 5DU
Tel: 01222 747444
Fax: 01222 754444
Chairman: Elan Closs Stephens

Chief Executive: Huw Jones
Director of Productions: Huw Eirug
Director of S4C International: Wyn
Innes
Gogs
A Mind to Kill
The Making of Maps
The Friendly Witches of
Somwhereland
Tales From a Piano
Testament - The Bible in
Animation

Scottish Television

Cowcaddens
Glasgow G2 3PR
Tel: 0141 300 3000
Fax: 0141 300 3030
20 Lincoln's Inn Field
London WC2A 3ED
Tel: 0171 446 7000
Fax: 0171 446 7010
Chairman: Gus Macdonald
Managing Director: Andrew
Flanagan
Chief Executive Scottish Television
Enterprises: Alistair Moffat
Director of Broadcasting/Head of
Regional Programmes: Blair Jenkins

Doctor Finlay
Executive Producer: Robert Love
Taggart
Executive Producer: Robert Love
Glasgow detective series

High Road
Producer: John Temple
Drama serial set on Loch Lomond

Wheel of Fortune
Executive Producer: Sandy Ross
Popular network game show

Other programmes include:
Chart Bite
The Home Show
Kirsty
NB
Scotland Today
Scotsport
Scottish Passport
Scottish Women
Win, Lose or Draw

Tyne Tees Television

The Television Centre
City Road
Newcastle Upon Tyne NE1 2AL
Tel: 0191 261 0181
Fax: 0191 261 2302
Chairman: Sir Ralph Carr-Ellison TD
Deputy Chairman: R H Dickinson
Managing Director: John Calvert
Director of Broadcasting: Peter
Moth
Group Head of Engineering: John
Nichol
Controller of News: Graeme
Thompson

Controller of Programme Administration and Planning: Peter MacArthur
Controller of Operations: Margaret Fay
Head of Current Affairs: Sheila Browne
Head of Light Entertainment: Christine Williams
Head of Young Peoples Programmes: Lesley Oakden
Head of Sports: Roger Tames

Chain Letters
Production Company: Tyne Tees Television
Producer: Christine Williams
Director: Ian Bolt
Presenter: Ted Robbins

Ulster Television
Havelock House
Ormeau Road
Belfast BT7 1EB
Tel: 01232 328122
Fax: 01232 246695
Chairman: J B McGuckian
Managing Director: J D Smyth
General Manager: J McCann
Controller of Programming: A Bremner
Head of Public Affairs: M McCann

UTV Live at Six
Five days a week hour-long news and features programme. Includes a wide range of strands - environment, health, home, entertainments, local communities, consumer affairs and sport

Westcountry Television
Western Wood Way
Langage Science Park
Plymouth PL7 5BG
Tel: 01752 333333
Fax: 01752 333444
Chairman: Sir John Banham
Deputy Chairman: Frank Copplestone
Chief Executive: Stephen Redfarn
Managing Director: John Prescott Thomas
Finance Director: Mark Haskell
Commercial Director: Caroline McDevitt
Press & Publicity: Mark Clare
Controller of News and Current Affairs: Richard Myers
Controller of Features: Jane McClosky
Controller of Operations & Engineering: Sim Harris

Yorkshire Television
The Television Centre
Leeds LS3 1JS
Tel: 0113 2 438283
Fax: 0113 2 445107

Charter Square
Sheffield S1 3EJ
Tel: 0114 2 723262
Fax: 0114 275 4134

23 Brook Street
The Prospect Centre
Hull HU2 8PN
Tel: 01482 24488
Fax: 01482 586028

88 Bailgate
Lincoln LN1 3AR
Tel: 01522 530738
Fax: 01522 514162

8 Coppergate
York YO1 1NR
Tel: 01904 610066
Fax: 01904 610067
Director of Broadcasting: Mike Best
Group Controller of Factual Programmes: Chris Bryer
Company Secretary: Simon Carlton
Controller of Commercial Affairs: Filip Cieslik
Head of Religion: Pauline Duffy
Head of Site Services: Peter Fox
Deputy Controller of Factual Programmes: Peter Gordon
Managing Director, Broadcasting: Richard Gregory
Group Head of Risk Management: John Hastings
Managing Director, Programme Production: David Holdgate
Deputy Financial Controller - North East: Nick Holmes
Head of News and Current Affairs: Clare Morrow
Controller of Drama, YTV: Carolyn Reynolds
Group Controller of Entertainment: David Reynolds
Group Controller of Drama: Keith Richardson
Group Head of Engineering: John Nichol
General Manager: Peter Rogers
Head of Personnel, YTTTV: Sue Slee
Head of Sales & Planning: John Surtees
Head of Children's and Education: Patrick Titley
Director of Programmes: John Whiston

Emmerdale
Executive Producer: Keith Richardson
Producer: Kieran Roberts
Cast: Clive Hornby, Richard Thorp, Stan Richards, Malandra Burrows, Christopher Chittell, Glenda McKay, Claire King, Leah Bracknell, Peter Amory
Three nights a week serial which was first transmitted in 1972

Heartbeat (series 7)
Executive Producer: Keith Richardson
Producer: Gerry Mill
Directors: Tim Dowd, Tom Cotter, Ken Horn, Garth Tucker, Brian Farnham
Writers: Veronica Henry, Ron Rose, Peter Barwood, Jane Hollowood, Garry Lyons, James Stevenson, Susan Wilkins, Peter Gibbs, Brian Finch, David Lane, Bill Lyons
Cast: Nick Berry, Juliette Gruber, Derek Fowlds, Bill Maynard, Jason Durr, Tricia Penrose, Kazia Pelka, William Simons
24 x 1 hr

A Touch of Frost
4 x 120 mins
Producer: Martyn Auty
Executive Producer: David Reynolds, Richard Briers, Philip Burley
In Association with Excelsior Productions Ltd
Directors: David Reynolds, Paul Seed, Graham Theakston, Sandy Johnson
Writers: Michael Russell, Sian Orrells, Malcolm Bradbury
Cast: David Jason

Bruce's Price is Right
YTV/Fremantle (UK) Productions Ltd
Production Executives: David Reynolds, David Champtaloup
Executive Producers: Arch Dyson, Keith Stewart
Producer: Howard Huntridge
Director: Bill Morton
Presenter: Bruce Forsyth

Other programmes include:
Life & Crimes of William Palmer
March In Windy City
The Big Bang
The Big Bag
The Scoop
3D

VIDEO LABELS

These companies acquire the UK rights to all forms of audio-visual product and arrange for its distribution on videodisc or cassette at a retail level (see also under Distributors). Listed is a selection of titles released on each label

Arrow Film Distributors
18 Watford Road
Radlett
Herts WD7 8LE
Tel: 01923 858306
Fax: 01923 869673
Neil Agran
La Bonne Annee
The Brood
Les Diaboliques
Europa Europa
Fritz the Cat
Ginger and Fred
Montenegro
La Retour de Martin Guerre
Scanners
Wages of Fear

Art House Productions
39-41 North Road
Islington
London N7 9DP
Tel: 0171 700 0068
Fax: 0171 609 2249
Richard Larcombe
Les Biches
Bicycle Thieves
Buffet Froid
Django
La Grande Bouffe
La Grande Illusion
The Harder They Come
Mephisto
Miranda
The Navigator
The Spirit of the Beehive
The Turning
Ultra

Artificial Eye Video
14 King Street
London WC2E 8HN
Tel: 0171 240 5353
Fax: 0171 240 5242
Robert Beeson, Roz Arratoon
Amateur
Beyond The Clouds
Clerks
The Confessional
D'Artagnan's Daughter
Exotica
The Bait
Heavy
The Horseman on the Roof
Mrs Parker and the Vicious Circle
A Self-Made Hero
A Summer's Tale
Three Colours: Red

Through the Olive Trees
Underground
Welcome to the Dollhouse
When the Cat's Away

BBC Video
Woodlands
80 Wood Lane
London W12 0TT
Tel: 0181 576 2236
Fax: 0181 743 0393
Dr Who
Match of the Day series
One Foot in the Grave
Pingu
Pole to Pole
Red Dwarf
Steptoe and Son

Blue Dolphin Film & Video
40 Langham Street
London W1N 5RG
Tel: 0171 255 2494
Fax: 0171 580 7670
Joseph D'Morais
Video releases to date:
A Great Day in Harlem
A Fistful of Fingers
Invaders from Mars
Destination Moon
Flight to Mars
Mister Frost
The Square Circle
Loaded

Braveworld
Multimedia House
Trading Estate Road
Park Royal
London NW10 7NA
Tel: 0181 961 0011
Fax: 0181 961 1413
Amadeus
Babette's Feast
One Flew Over the Cuckoo's Nest
Pelle the Conqueror
The Running Man

Buena Vista Home Video
Beaumont House
Kensington Village
Avonmore Road
London W14 8TS
Tel: 0171 605 2400
Fax: 0171 605 2795
Distribute and market Walt Disney, Touchstone, Hollywood Pictures and Henson product on video

Carlton Home Entertainment
The Waterfront
Elstree Road
Elstree
Herts WD6 3BS
Tel: 0181 207 6207
Fax: 0181 207 5789
Beatrix Potter on video x 7 titles
The Lover's Guide 1,2,3, and *4*
The Rank Classics Collection
The Ed Wood Collection
Elle MacPherson - The Body Workout
Cindy Crawford: Shape Your Body
Old Bear and Friends 1,2,3 and *4*
The Alexander Korda Film Collection
Soldier Soldier
Inspector Morse
Kavanagh QC
Tots TV
Rosie & Jim

CIC UK
Glenthorne House
5-17 Hammersmith Grove
London W6 0ND
Tel: 0181 846 9433
Fax: 0181 741 9773
A Universal/Paramount Company
Clear and Present Danger
The Flintstones
Forrest Gump
Jurassic Park
The Paper
River Wild
Schindler's List
The Shadow
True Lies

Columbia TriStar Home Video
Sony Pictures Europe House
25 Golden Square
London W1R 6LU
Tel: 0171 533 1200
Fax: 0171 533 1105
Devil in a Blue Dress
First Knight
Higher Learning
It Could Happen to You
Legends of the Fall
Little Women
Mary Shelley's Frankenstein
Only You
The Quick and the Dead
Street Fighter

 ## Connoisseur Video
10a Stephen Mews

London WIP 0AX
Tel: 0171 957 8957/8
Fax: 0171 957 8968
A joint venture between the British
Film Institute and Argos Film,
France, Connoisseur distributes over
200 titles covering five decades of
world cinema on its Connoisseur and
Academy lables **(see p8)**.
Recent Connoisseur releases:
The Mahabharata (Box Set)
Eyes Without a Face
Faraway So Close
L'Avventura
Sanjuro
Recent Academy releases:
Martin Scorsese Directs
*Bob's Birthday and the Best of
British Animation*
Pumping Iron 2 - The Women
Atomic Cafe
Winsor McCay - Animation Legend

Curzon Video
13 Soho Square
London W1V 5FB
Tel: 0171 437 2552
Fax: 0171 437 2992
Belle Epoque
Daens
Deadly Advice
Decadence
L'Enfer
Fausto
The Hour of the Pig
*How to be a Woman and Not Die in
the Attempt*
In Custody
Mina

Dangerous to Know
66 Offley Road
Kennington Oval
London SW9 0LS
Tel: 0171 735 8330
Fax: 0171 793 8488
email: dangeroustoknow@dtk.co.uk
For a Lost Soldier
I Like You, I Like You Very Much
Midnight Dancers
Only the Brave
Pink Narcissus
Postcards from America
Via Appia
The Satin Spider
Thin Ice
Two of us
Zero Patience

Electric Pictures Video (Alliance Releasing)
184-192 Drummond Street
London NW1 3HP
Tel: 0171 580 3380
Fax: 0171 636 1675
Angel Baby
Arizona Dream
The Baby of Macon

Before the Rain
Belle de Jour
Blood Simple
Butterfly Kiss
The Celluloid Closet
Cold Fever
*The Cook, The Thief, His Wife and
Her Lover*
Death and the Maiden
Delicatessen
La Dolce Vita
Drowning by Numbers
The Eighth Day
The Flower of my Secret
I Shot Andy Warhol
Kansas City
Kika
Ladybird, Ladybird
Love and Human Remains
Orlando
Priest
Prospero's Books
Raise the Red Lantern
Red Firecracker, Green Firecracker
Ridicule
The Runner
Shanghai Triad
The Story of Qiu Ju
Trees Lounge
The White Balloon
The Young Poisoner's Handbook
Walking and Talking

Entertainment in Video
27 Soho Square
London W1V 5FL
Tel: 0171 439 1979
Fax: 0171 734 2483
Kingpin
Last Man Standing
Leaving Las Vegas
Living in Oblivion
Nixon
Seven
Twelfth Night
Up Close and Personal

First Independent Films
99 Baker Street
London W1M 1FB
Tel: 0171 317 2500
Fax: 0171 317 2502/2503
Above the.Rim
Automatic
Dumb and Dumber
The Lawnmower Man II
Little Odessa
Mortal Kombat
Nostradamus
Rainbow
Sleep With Me

FoxGuild Home Entertainment
Twentieth Century House
31-32 Soho Square
London W1V 6AP
Tel: 0171 753 0015
Fax: 0171 434 1435

Airheads
Braveheart
Johnny Mnemonic
Judge Dredd
The Scout
The Shawshank Redemption
Stargate
Trapped in Paradise
Wes Craven's New Nightmare

Granada LWT International
London Television Centre
Upper Ground
London SE1 9LT
Tel: 0171 620 1620
Fax: 0171 928 8476
Brideshead Revisited
Cracker - series 1 & 2
Gladiators
Hale & Pace - Greatest Hits
Jeeves & Wooster - series 1 & 2
Jewell in the Crown
London's Burning - series 1 & 6
Nicholas and Alexandra
Rik Mayhall Presents... - series 1 & 2

Imperial Entertainment (UK)
Main Drive, GEC Estate
East Lane
Wembley
Middx HA9 7FF
Tel: 0181 904 0921
Fax: 0181 904 4306/908 6785
UK distributor of feature films
including Danielle Steele video titles

Jubilee Film and Video
Egret Mill
162 Old Street
Ashton-Under-Lyne
Manchester OL6 7ST
Tel: 0161 330 9555

Le Channel Ltd
10 Frederick Place
Weymouth
Dorset DT4 8HT
Tel: 01305 780446
Fax: 01305 780446
Art videos about famous paintings of
the Western World. Palettes is a
collection of very high standard
videos about famous paintings of the
Western World. Adapted from the
French, the films have been
researched by leading art historians
and curators. Each Palette narrates
the creation of a painting the story
of a painter, the progression of a
Palette: Claude; Leonardo; Monet
Poussin, Seurat, Vermeer

Lighthouse
Lumiere Video
167-169 Wardour Street
London W1V 3TA
Tel: 0171 413 0838
Fax: 0171 734 1509
A catalogue of French, Spanish and

Italian titles from the Initial Groupe, as well as the entire Weintraub library
Fort Saganne
La Marge
Plein Soleil

Lumiere Classics
167-169 Wardour Street
London W1V 3TA
Tel: 0171 413 0838
Fax: 0171 413 0838
Eating Roual
Eva

Lumiere Home Video
167-169 Wardour Street
London W1V 3TA
Tel: 0171 413 0838
Fax: 0171 734 1509
The Avengers
12 titles from Hammer
Hue and Cry
Laughter in Paradise
Bruce Lee
Mighty Max
Moby Dick
Mona Lisa (widescreen)

Mainline Pictures
37 Museum Street
London WC1A 1LP
Tel: 0171 242 5523
Fax: 0171 430 0170
Bandit Queen
The Diary of Lady M
A Flame in my Heart
Go Fish
Let's Get Lost
Luck, Trust & Ketchup
The Premonition
Ruby in Paradise
The Wedding Banquet
Lovers
Crazy Love
Metropolitan
Chain of Desire

Media Releasing Distributors
27 Soho Square
London W1V 5FL
Tel: 0171 437 2341
Fax: 0171 734 2483
Day of the Dead
Eddie and the Cruisers
Kentucky Fried Movie
Return of Captain Invincible
Distributed through Entertainment in Video (qv)

Medusa Communications & Marketing Ltd
Medusa Pictures Video Division
Regal Chambers, 51 Bancroft
Hitchin
Herts SG5 1LL
Tel: 01462 421818
Fax: 01462 420393
American Yakuza 2

The Babysitters
Crash
Evolver
The Final Cut
Freefall
F.T.W.
If These Walls Could Talk
Rent a Kid
When the Bough Breaks
A Woman Scorned 2

MGM Home Entertainment (Europe) Ltd
Glenthorne House
Hammersmith Grove
Tel: 0181 563 8383
Fax: 0181 563 2896

Nova Home Entertainment
11a Winholme
Armthorpe
Doncaster DN3 3AF
Tel: (01302) 833422
Fax: (01302) 833422
Contact: Andrew White, Maurice White, Gareth Atherton
Sell-through video distributor, a subsidiary of Nova Productions, with a catalogue based specialist & local interest documentaries and nostalgia programming

Odyssey Video
15 Dufours Place
London W1V 1FE
Tel: 0171 437 8251
Fax: 0171 734 6941
Ambush in Waco
Beyond Control
Burden of Proof
Honour Thy Father & Mother
Lady Boss
Lucky/Chances
Out of Darkness
A Place for Annie
Remember
War & Remembrance

Orbit Media Ltd
7/11 Kensington High Street
London W8 5NP
Tel: 0171 221 5548
Fax: 0171 727 0515
Website: http://www.btinternet.com/orbitmedia
Chris Ranger, Jordan Reynolds
A range of screen classics and literary classics

Out on a Limb
Battersea Studios
Television Centre
Thackeray Road
London SW8 3TW
Tel: 0171 498 9643
Fax: 0171 498 1494
Being at Home with Claude
Forbidden Love
My Father is Coming

No Skin Off My Ass
Seduction: The Cruel Woman
Virgin Machine

Picture Music International
20 Manchester Square
London W1A 1ES
Tel: 0171 486 4488
Fax: 0171 465 0748
Blur: Showtime
Cliff Richard: The Hit List
David Bowie: The Video Collection
Iron Maiden: The First Ten Years
Kate Bush: The Line, The Cross and The Curve
Peter Gabriel: Secret World Live
Pet Shop Boys: Videography
Pink Floyd: Pulse
Queen; Box of Flix
Tina Turner: Simply the Best

PolyGram Video
1 Sussex Place
Hammersmith
London W6 9XS
Tel: 0181 910 5000
Fax: 0181 910 5406
Video catalogue includes music videos, feature films, sport, comedy, special interest and children's videos. New titles available through retail/sell-through outlets include:
Trainspotting
Barney
Chubby Brown
Brookside

PolyGram Visual Programming
Oxford House
76 Oxford Street
London W1N 0HQ
Tel: + 44 (0) 171 307 7600
Fax: + 44 (0) 171 307 7639
Austin Shaw
A subsidiary of Polygram Filmed Entertainment producing and acquiring programmes for worldwide video distribution across all genres including music, children's, comedy, sport and general interest
Mr Bean
Bee Gees - 'Keppell Road'
Barney
The Thin Blue Line
E17 - Greatest Hits

Quadrant Video
37a High Street
Carshalton
Surrey SM5 3BB
Tel: 0181 669 1114
Fax: 0181 669 8831
Sports video cassettes

SIG Video Gems Ltd
The Coach House
The Old Vicarage
10 Church Street
Rickmansworth

Herts WD3 1BS
Tel: 01923 710599
Fax: 01923 710549
Black Beauty (TV series)
The Great Steam Trains
Minder
Moonlighting
Professionals
Return of the Incredible Hulk
Rumpole of the Bailey
Ruth Rendell
Sweeney
UK Gold Comdey Compilation
UK Gold Action/Drama Compilation

Screen Edge
28-30 The Square
St Annes-on-Sea
Lancashire FY8 1RF
Tel: 01253 712453
Fax: 01253 712362
email: king@visicom.demon.co.uk
Website: http//www.visionary.co.uk
Rhythm Thief
Der Todesking
Pervirella

Tartan Video
Metro Tartan House
79 Wardour Street
London W1V 3TH
Tel: 0171 494 1400
Fax: 0171 439 1922
Cinema Paradiso
Golden Balls
Jamón Jamón
Kwaidan
Man Bites Dog
Seventh Seal
The Umbrellas of Cherbourg
La Haine

Telstar Video Entertainment
The Studio
5 King Edward Mews
Byfeld Gardens
London SW13 9HP
Tel: 0181 846 9946
Fax: 0181 741 5584
A sell-through video distributor of
music, sport, special interest,
comedy, children and film pro-
grammes
The Best Kept Secret in Golf
Foster & Allen: By Request
Harry Secombe Sings
Hollywood Women
John Denver: A Portrait
*Michael Crawford: A Touch of
Music in the Night*

Thames Video Home Entertainment
Pearson Television International
1 Stephen Street
London W1P 1PJ
Tel: 0171 691 6000
Fax: 0171 691 6079

Mr Bean
Tommy Cooper
Wind in the Willows
World at War
Men Behaving Badly
The Bill
The Sweeney

THE (Total Home Entertainment)
National Distribution Centre
Rosevale Business Park
Newcastle under Lyme
Staffs ST5 7QT
Tel: 01782 566566
Fax: 01782 565400
Exclusive distributors for Paradox,
Pride, Kiseki, International
Licencing & Copyright Ltd,
Grosvenor, Empire and distributors
of over 3,000 other titles - Quantum
Leap, Green Umbrella

20:20 Vision Video UK
Horatio House
77-85 Fulham Palace Road
London W6 8JA
Tel: 0181 748 4034
Fax: 0181 748 4546
Little Women
Street Fighter
The Quick and the Dead

Visionary Communications
28-30 The Square
St Annes-on-Sea
Lancashire FY8 1RF
Tel: 01253 712453
Fax: 01253 712362
email: king@visicom.demon.co.uk
Website: http//www.visionary.co.uk
Scorpio Rising
The Pope of Utah
Three Films' Burroughs'/Gysin
Destroy All Rational Thought
Cyberpunk
Angelic Conversation
In the Shadow of the Sun
The Gun is Loaded
Alice
Freaks
Island of Lost Souls
Mystery of the Wax Museum

WEBSITES

The worldwide web is awash with websites connected with film and television. This section contains a small selection of some useful websites based on other information in this book

Archive and Film Libraries

France - La Vidéoteque de paris
http://www.vdp.fr/

National Film and Television Archive
http://www.bfi.org.uk

Awards

BAFTA
http://www.bafta.org/

Berlin
http://www.berlinale.de

Cannes
http://www.cannes.fest.com/1998/index/html

Emmys
http://www.emmys.org/

Karlovy Vary
http://www.iffkv.cz/

Oscars
http://www.oscar.com/

Books

BFI Publishing
http://www.bfi.org.uk

Oxford University Press
http://www.oup.co.uk/

Routledge
http://www.routledge.com/

Booksellers

Blackwell's
http://www.blackwell.co.uk/bookshops

Cinema Store
http://www.atlasdigital.com/cinemastore

NFT Bookshop
http://www.bfi.org.uk

Cable and Satellite

Atlantic Telecom Group
http://www.atlantictelecom.co.uk

BSkyB
http://www.sky.co.uk

Bravo
http://www.bravo.co.uk

Carlton Select
http://www.carltonselect.co.uk

Cartoon Network
http://www.cartoon-network.co.uk/toonet.app/

The Discovery Channel TV
http://www.discovery.com/digitnets/international/europe/europe.html

Eurosport
http://www.eurosport-tv.com

Granada Plus
http://www.gsb.co.uk/plus/home/html

MTV
http://www.mtv.co.uk

NTL
http://www.cabeltel.co.uk

QVC: The Shopping Channel
http://www.qvc.com

The Sci-Fi Channel Europe
http://www.scifi.com/sfeurope/index.html

Telewest Communications
http://www.telewest.co.uk

Careers

Film Education
http://www.filmeducation.org

Skillset
http://www.skillset.org

CD-Roms

International Film Index
http://www.bowker-saur.com/service/

The Knowledge
http://www.mfplc.com

Variety's Video Directory Plus
http://www.bowker-saur.co.uk/service/

Cinemas

ABC Cinemas
http://www.abccinemas.co.uk/

Cinemas in the UK
http://www.aber.ac.uk/~jwp/cinemas/

Cineworld
http://www.Cineworld.co.uk/

UCI (UK) Ltd
http://www.uci-cinemas/co.uk/

Courses

Bath University
http://www.bath.ac.uk

Birmingham University
http://www.bham.ac.uk/

Bradford Universtiy
http://www.brad.ac.uk

Bristol University
http://www.bristol.ac.uk/

Canterbury Christ Church College
http://www.chive.cant-col.ac.uk/

De Montfort University
http://www.dmu.ac.uk/

East Anglia University
http://www.cpca3.uea.ac.uk/welcome.html

Exeter University
http://www.ex.ac.uk/

Glasgow University
http://www.gla.ac.uk

Glasgow Caledonian University
http://www.gcal.ac.uk

Goldsmith's College, London
http://www.goldsmiths.ac.uk

King Alfred's College, Winchester
http://www.wkac.ac.uk

Liverpool Institute for Performing Arts
http://www.lipa.ac.uk/

Liverpool John Moores University
http://www.livjm.ac.uk/intro.html

London College of Printing and Distributive Trades
http://www.linst.ac.uk/lcp/index.html

London International Film School
http://www.tecc.co.uk/lifs/index.html

London Screenwriters Workshop
http://www.lsw.org.uk

Napier University, Edinburgh
http://www.napier.ac.uk/

University of Newcastle upon Tyne
http://www.ncl.ac.uk/\ncrif

Northern School of Film and
Television
http://www.lmu.ac.uk/

Portsmouth University
http://www.port.ac.uk

Queens University, Belfast
http://www.qub.ac.uk

Reading University
http://www.reading.ac.uk

Roehampton Institute, London
http://www.roehampton.ac.uk

Royal College of Art, London
http://www.rca.ac.uk/Design

Salford University
http://www.salford.ac.uk/

Sheffield University
http://www.2.shef.ac.uk/

Sheffield Hallam University
http://www.shu.ac.uk/

South Bank University, London
http://www.sbu.ac.uk/

Southampton Universtiy
http://www.soton.ac.uk/

Sussex University
http://www.susx.ac.uk

Trinty and All Saints College, Leeds
http://www.tasc.ac.uk

Ulster University
http://www.ulst.ac.uk

University College London
http://www.ucl.ac.uk/

University of Wales College,
Newport
http://www.newport.ac.uk

Warwick University
http://www.csv.warwick.ac.uk

University of Westminster
http://www.wmin.ac.uk/

Wolverhampton University
http://www.wlv.ac.uk/

Distributors (Non-Theatrical)

CFL Vision
http://www.euroview.co.uk

Cinenova
http://www.luton.ac.uk/cinenova

London Electronic Arts
http://www.lea.org.uk

Distributors (Theatrical)

Alliance Releasing
http://www.alliance.ca/
motionpictures/distribution.html

BFI Films
http://www.bfi.org.uk

Buena Vista
http://www.bvimovies.com/

Cinenova
http://www.luton.ac.uk/cinenova

Dangerous to Know
http://www.dtk.co.uk

Walt Disney
http://www.disney.co.uk

Twentieth Century Fox
http://www.foxmovies.com/

UIP (United International Pictures)
http://www.UIP.com/

Warner Bros
http://www.wb.com/
frame_moz3_day.html

Facilities

Edinburgh Film Workshop Trust
http://www.efwt.demon.co.uk

Hillside Studios
http://
www.ourworld.compuserve.com/
homepages/hillside_studios

Hull Time Based Arts
http://www.htba.demon.co.uk

Humphries Video Services
http://www.hvs.co.uk

Studio Pur
http://www.ace.mdx.ac.uk/
hyperhomes/houses/pur/
index.html

Videolondon Sound
http://www.ftech.net/~videolon

Festivals

Austria
Viennale - Vienna International Film
Festival
http://www.viennale.or.at

Belgium
Brussels International Film Festival
http://ffb.cinebel.com
Flanders International Film Festival -
Ghent
http://www.filmfestival.be

Brazil
Gramado International Film Festival
- Latin and Brazilian Cinema
http://www.viadigital.com. br/
gramado

São Paulo International Film Festival
http://www.mostra.org

Canada
The Atlantic Film Festival
http://www.atlanticfilm.com
Banff Television Festival
http://www.cochran.com/banfftv
Festival International du Film Sur
L'Art (International Festival of Films
on Art)
http://www.maniacom.com/
fifa.html
Montreal International Festival of
Cinema and New Media
http://www.fcmm.com
Vancouver International Film
Festival
http://viff.org

Croatia
World Festival of Animated Films -
Zagreb
http://animafest.hr

Denmark
International Odense Film Festival
http://www.filmfestival.dk

France
Festival International de Films de
Femmes
htttp://www.gdebussac.fr/filmfem

Germany
Berlin International Film Festival
http://www.berlinale.de
Feminale, 9th International Women's
Film Festival
http://www.dom.de./filmworks/
Feminale
Nordic Film Days Lübeck
http://www.luebeckk.de/filmtage
Oberhausen International Short Film
Festival
http://www.shortfilm.de

Italy
The Pordenone Film Fair
http://www.cinetec@del.frivli.org/
gcm
Pordenone Silent Film Festival (La
Giornate del Cinema Muto)
http://www. cinetec@delfrivli.org/
gcm/
Cinema Giovani - Torino Film
Festival
http://www.torinofilmfest.org
Venice Film Festival
http://www.labiennale.it

Japan
International Animation Festival
Hiroshima
http://www.city.hiroshima.jp

The Netherlands
Dutch Film Festival
http://www.nethlandfilm.nl

Portugal

Fantasporto - Oporto International
Film Festival
http://www.caleida.pt.fantasporto
Festival Internacional de Cinema da
Figueira da Foz
http:\\www.ficff.pt

Spain

International Short Film Contest
'Ciudad de Huesca'
**http://www.huesca-
filmfestival.com**
San Sebastian International Film
Festival
**http://www.ddnet.es/
san_sebastian_film_festival**
Valladolid International Film
Festival
http://www.seminic.com

Sweden

Gotebörg Film Festival
http://goteborg.filmfestival.org
Uppsala International Short Film
Festival
http://www.shortfilmfestival.com

United Kingdom

Edinburgh Festival
**http://www.ed.ac.uk/~eif/
home.html**
European Shot Film Festival
**http://www.bogo.co.uk/kohle/
festival.html**
Sheffield International Documentary
Festival
http://www.fdgroup.co.uk/neo/sidf

USA

35th Chicago International Film
Festival
**http://www/chicago.ddbn.com/
filmfest/**
Columbus International Film and
Video Festival (a.k.a. The Chris
Awards)
http://www.infinet.com/-chrisawd
Honolulu HI 96813
http://www.hiff.org
Independent Feature Film Market
http://www.ifp.org
Mobius Advertising Awards
Competition
http://www.mobiusawards.com
National Educational Media
Network (formerly National
Educational Film & Video Festival)
http://www.nemn.org
Portland International Film Festival
http://www.nwfilm.org
Telluride Film Festival
**http://telluridemm.com/
filmfest.htm**
US International Film & Video
Festival
http://www.filmfestawards.com

Film Societies

British Federation of Film Societies
**http://www.shef.ac.uk/uni/union/
susoc/fu/sg/bffs/index.html**

Funding

Arts Council of England
**http://www.artscouncil.org.uk/
intro.html**

BFI Production
http://www.bfi.org.uk

Production Fund for Wales
http://www.sgrinwales.demon.co

Regional Arts Board
http://www.arts.org.uk/

Scottish Arts Council
http://www.sac.org.uk

Scottish Screen
**http://
www.scottishscreen.demon.co.uk.**

International Sales

Arts Council of England
**http://www.artscouncil.org.uk/
intro.html**

BFI Films
http://www.bfi.org.uk

Walt Disney Company
http://www.wboc.org/cowdc.html

Pearson Television International
**http://www.pearson.com/
index.htm**

Warner Bros
**http://www.wb.com/
frame_moz3_day.html**

Libraries

BFI National Library
http://www.bfi.org.uk

Coventry University, Art & Design
Library
http://www.coventry.ac.uk/library/

Library Association
http://www.la-hq.org.uk/

Kingston Museum & Heritage
Service
**http://www.kingston.ac.uk/
muytexto.htm**

North West Film Archive
**http://www.mmu.ac.uk/services/
library/wst.hym**

Southampton Institute
http://www.solent.ac.uk/library/

Surrey Performing Arts Library
**http://surreycc.gov.uk/libraries/
direct/perfarts.html**

Vivid - Birmingham Centre for
Media Arts
**http://
www.wavespace.waverider.co.uk**

Winchester School of Art
http://www.soton.ac.uk/

Organisations

American Film Institute
http://www.afionline.org/

Arts Council of England
**http://www.artscouncil.org.uk/
intro.html**

BAFTA
http://www.bafta.org/

BUFVC (British Universities Film
and Video Council)
http://www.bufvc.ac.uk

BBC
http://www.bbc.co.uk

British Council
http://www.britfilms.co

British Film Commission
http://www.britfilmcom.co.uk

BFI
http://www.bfi.org.uk

BKSTS - The Moving Image Society
http://www.bksts.demon.co.uk

Department for Culture, Media and
Sport (DCMS)
http://www.culture.gov.uk/

Directors' Guild of Great Britain
http://www.dggb.co.uk

Film Education
http://www.filmeducation.org

Laurel and Hardy Museum,
Ulverston
**http://www.wwwebgides.com/
britain/cumbria/furness/laure.html**

National Museum of Photography:
Film and Television
http://www.nmsi.ac.uk/nmpft/

New Producers Alliance
http://www.npa.org.uk

PACT (Producers Alliance for
Cinema and Television)
http://www.pact.co.uk

Sgrîn (Media Agency for Wales)
http://www.sgrinwales.demon.co

Scottish Screen
**http://
www.scottishscreen.demon.co.uk**

Skillset
http://www.skillset.org

Writers' Guild of Great Britain
http://www.writers.org.uk/guild

Organisations (Europe)

European Audio-visual Observatory
http://www.ob.c_strasbourg.fr

European Film Academy
http://
www.europeanfilmacademy.org.

Belgium - The Flemish Film Institute
http://www.vfi-filminsituutbe

Denmark - Danish Film Institute
http://www.dfi.dk

Finland - AVEK - The Promotion
Centre for Audio-visual Culture in
Finland
http://www.kopiostofi/avek
Finnish Film Archive
http://www.sea.fi
The Finnish Film Foundation
http://www.ses.fi/ses

France - Bibliothéque du Film (BIFI)
http://www.bifi.fr
TV France International
http://www.tvfi.com

Germany - Filmförderungsanstalt
http://www.ffa.de

Iceland - Icelandic Film Fund
http://www.centrum.is/filmfund

Ireland - Bord Scannán na
hÉireann/Irish Film Board
http://www.iol.ie/filmboard
Film Institute of Ireland
http://www.iftn.ie/ifc

Poland - Polish Cinema Database
http://www.info.fuw.edu.pl/Filmy/

Portugal - Portuguese Film and
Audiovisual Institute
http://www.nfi.no/nfi.htm

Scotland - Scottish Screen
http://
www.scottishscreen.demon.co.uk

UK - British Film Institute
http://www.bfi.org.uk

Wales - Sgrîn, Media Agency for Wales
http://www.sgrinwales.demon.uk

Press Contacts

British Film Facilities Journal
http://www.compas-rose.co.uk

Documenter -
internet documentary film journal
http://www.documenter.com

Empire
http://www.empireonline.co.uk

Filmwaves
http://www.real.co.uk

Flicks
http://www.flicks.co.uk

Guardian online
http://www.guardian.co.uk/
guardian//

Premiere
http://www.premieremag.com/

Radio Times
http://www.radiotimes.beeb.com

Screen
http://www.arts.gla.ac.uk/tfts/
screen/html

Sunday Times
http://www.sunday-times.co.uk

Talking Pictures
http://www.filmcentre.co.uk

Television
http://www.rts.org.uk

Time Out
http://www.timeout.co.uk/

Total Film
http://www.futurenet.co.uk

Variety
http://www.variety.com

Preview Theatres

BAFTA
http://www.bafta.org/

BFI
http://www.bfi.org.uk

The Curzon Minema
http://www.minema.com

Production Companies

Illuminations
http://www.illumin.co.uk

imaginary films
http://www.imagfilm.co.uk

Mosaic Films
http://www.mosaicfilms.com

Praxis Films
http://www.icehouse.co.uk

Talkback Productions
http://www.talkback.co.uk

Third Eye Productions
http://www.thirdeye.demon.co.uk

Twentieth Century Fox
http://www.fox.co.uk

Zooid Pictures
http://www.sonnet.co.uk/
zooid.co.uk

Specialised Goods and Services

MBS Underwater Video Specialists
http://www.eclipse.co.uk.mbs

Olswang
http://www.olswang.co.uk

Pirate Models and Effects
http://www.pirate.co.uk

Zooid Pictures Limited
http://www.zooid.co.uk

Studios

Capital FX
http://www.capital.fx.co.uk

Elstree Film Studios
http://www.elstreefilmstudios.co.uk

Hillside Studios
http://www.ctvc.co.uk

Millenium Studios
http://www.elstree-online.co.uk

Television Companies

Anglia Television
http://www.anglia.tv.co.uk/

BBC
http://www.bbc.co.uk/

BBC Scotland
http://www.bbc.co.uk/aberdeen/
index/html

BBC Wales
http://www.bbc.wales.com/

Border Television
http://www.border-tv.com/

Carlton Television
http://www.carltontv.co.uk/

Channel Television
http://www.channeltv.co.uk

Channel 4
http://www.cityscape.co.uk/
channel4/

Granada Television
http://www.granada.co.uk/

HTV
http://www.htv.co.uk/

London Weekend Television (LWT)
http://www.lwt.co.uk/

Meridian Broadcasting Ltd
http://www.meridan.tv.co.uk/

S4C
http://www.s4c.co.uk/

Scottish Television
http://www.stv.co.uk/

Ulster Television
http://www.utvlive.com

Video Labels

BFI Films
http://www.bfi.org.uk

Blockbuster Entertainment
http://www.blockbuster.com

Buena Vista
http://www.bvimovies.com/

Walt Disney
http://www.disney.co.uk

Twentieth Century Fox
http://www.foxmovies.com/

UIP (United International Pictures)
http://www.UIP.com/

Warner Bros
**http://www.wb.com/
frame_moz3_day.html**

Websites

Cyber Film School
http://www.cyberfilmschool.com

Film Festival Server
http://www.filmfestivals.com/

Hammer House of Horror
**http://www.futurenet.co.uk/
netmag/hammer**

Hollywood Online
http://www.hollywood.com

Internet Movie Database
http://www.uk.imdb.com/

Workshops

City Eye
http://www.interalpha.net/cityeye

Edinburgh Film Workshop Trust
http://www.ukefwt.demon.co.uk

Hull Time Based Arts
http://www.htba.demon.co.uk

London Electronic Arts
http://www.lea.org.uk

Vivid
**http://
www.wavespace.waverider.co.uk/
'vivid'**

WORKSHOPS

The film and video workshops listed below are generally non-profit distributing and subsidised organisations. Some workshops are also active in making audio-visual products for UK and international media markets. Those workshops with an asterisk after their name are BECTU-franchised

Amber Side Workshop*
5 Side
Newcastle upon Tyne NE1 3JE
Tel: 0191 232 2000
Fax: 0191 230 3217
Murray Martin
Film/video production, distribution and exhibition. Most recent productions include: Letters to Katiya, 1 hour documentary; Eden Valley 90 minute feature film; The Scar 115 minute feature film. The Workshops National Archive is based at Amber. Large selection of workshop production on VHS, a substantial amount of written material and a database. Access by appointment

AVA (Audio Visual Arts)
1 Newcastle Chambers
Angel Row
Nottingham NG1 6HL
Tel: 0115 948 3684
Chris Ledger
Independent video production company which has stayed small, maintaining a reputation for quality and integrity. Formed by experienced arts workers, the company specialises in arts promos, documentaries and documentation, used widely overseas as well as throughout the UK. Clients include individual artists, art galleries, educational or community organisations, regional arts boards and local authorities

Belfast Film Workshop
37 Queen Street
Belfast BT1 6EA
Tel: 01232 648387
Fax: 01232 246657
Alastair Hrron, Kate McManus
Film co-operative offering film/video/animation production and exhibition. Offers both these facilities to others. Made *Acceptable Levels* (with Frontroom); *Thunder Without Rain: Available Light*; a series on six Northern Irish photographers, various youth animation pieces and a series of videos on traditional music

Black Media Training Trust (BMTT)
Workstation
15 Paternoster Row
Sheffield S12 BX
Tel: 01142 492207
Fax: 01142 492207
Contact: Carl Baker
Film and video training. Commercial media productions facility and training resource within and for all Asian, African and African Caribbean communities for community development purposes. Also various commercial media consultancy and project services and facilities hire. Funded by National Lottery Single Regeneration Budget and church urban fund

Caravel Media Centre
The Great Barn Studios
Cippenham Lane
Slough SL1 5AU
Tel: 01753 534828
Fax: 01753 571383
Denis Statham
Training, video production, distribution, exhibition and media education. Offers all these facilities to others. Runs national video courses for independent video-makers

The Children's Film Unit
Unit 8, Princeton Court
55 Felsham Road
London SW15 1AZ
Tel: 0181 785 0350
Fax: 0181 785 0351
A registered educational charity, the CFU makes low-budget films for television and PR on subjects of concern to children and young people. Crews and actors are trained at regular weekly workshops in Putney. Work is in 16mm and video and membership is open to children from 10 - 18. Latest films for Channel 4: *Emily's Ghost; The Higher Mortals; Willies War; Nightshade; The Gingerbread House; Awayday*. For the Samaritans: *Time to Talk*. For the Children's Film and Television Foundation: *How's Business*

Cinema Action
27 Winchester Road
London NW3 3NR
Tel: 0171 267 6878
35mm and 16mm film; video, photo, computer graphics, multimedia production, distribution and exhibition. Co-operative work. Productions include: *Rocking the Boat; So That You Can Live; The Miners' Film; People of Ireland; Film from the Clyde; Rocinante; Bearskin; Precious Lives*

City Eye
1st Floor, Northam Centre
Kent Street
Northam
Southampton SO14 5SP
Tel: 01703 634177
Fax: 01703 575717
email: cityeye@interalpha.co.uk
Website: www.interalpha.net/cityeye
Film and video equipment hire. Educational projects. Production and post-production services. Screenings. Community arts media development. Training courses all year in varied aspects of video, film, photography and radio. Committed to providing opportunities for the disadvantaged/under-represented groups. 50 per cent discount on all non-profit/educational work

Connections Communications Centre
Palingswick House
241 King Street
London W6 9LP
Tel: 0181 741 1767
Fax: 0181 563 9134
email:
connections@cccmedia.demon.co.uk
Training in technical operations, video editing, production management and video technology. Betacam SP and SVHS editing facilities and production equipment available for hire

Cornwall Media Resource Trust Ltd
Royal Circus Buildings
Back Lane West
Redruth
Cornwall TR15 2BT

Tel: 01209 218288
Chair: Andy Lancorte
Film/video production and multimedia facilities, training and arts/community based projects. Hi-band SP video production kit with three machine editing and time code non-linear editing facilities also available. Information resource. Open to individuals and groups. Undertakes commissions and cultural production work. Special work and hire. Special rates for unfunded and grant-aided work

Counter Image
3rd Floor, Fraser House
36 Charlotte Street
Manchester M1 4FD
Tel: 0161 228 3242
Fax: 0161 228 3242
Independent media charity. Offers production facilities and co-production support to independent film and video makers and photographers. Digital imaging facilities. Provides a range of short training courses, facilities and a database of film related services, training and practitioners in the North West region. Productions include: *Concrete Chariots; Take One Simple Test; Special Offer; Andrew's Story; Seen and Not Heard*

Creu Cof (Ceredigion Media Association)
Blwch Post 86
Aberystwyth
Dyfed SY23 1AA
Tel: 01970 624001
Fax: 01970 625119
Media education for all ages and interests with specific reference to Welsh speakers, and to rural themes and issues. Runs longer term projects

Cultural Partnerships
90 De Beauvoir Road
London N1 4EN
Tel: 0171 254 8217
Fax: 0171 254 7541
Heather McAdam, Karen Merkel, Lol Gellor, Inge Black
Arts, media and communications company. Offers various courses in digital sound training. Makes broadcast and non-broadcast films and videos and radio programmes. Production-based training forms a vital part of the work. Studio facilities for dry/wet hire: fully air-conditioned and purpose built, 8000 sq ft multi-purpose studio. Analogue and Digital audio studios. Live audio studio

Depot Studios
Bond Street

Coventry CV1 4AH
Tel: 01203 525074
Contact: Deborah Martin-Williams, Audrey Droisen, Mike Roberts
U-matic, Hi-8 and VHS video, 16-track digital sound recording studio. Provides facilities, training, information, production support and, as well, produces its own productions. Has a particular brief to develop work with 16-25 age group

Despite TV
113 Roman Road
London E2 0HU
Tel: 0181 983 4278
Video co-operative providing training through production for people living or working in Tower Hamlets, Hackney, Newham. Facilities for hire include hi-band SP editing and Amiga graphics. Hi8 hire and crew available. Produces both local topic and single-issue magazine tapes which are available for hire, as well as documentary for Channel 4 *Battle of Trafalgar*, plus sub-commission The Bailiff Cometh

Edinburgh Film Workshop Trust*
29 Albany Street
Edinburgh EH1 3QN
Tel: 0131 557 5242
Fax: 0131 557 3852
email: post@efwt.demon.co.uk
Website:
http:\\www.efwt.demon.co.uk
David Halliday, Cassandra McGrogan, Robin MacPherson, Edward O'Donnelly
Scotland's only franchised workshop. Broadcast, non-broadcast and community integrated production. Facilities include Betacam production; lo-band and hi-band U-Matic production; VHS, lo-band and hi-band edit suites; Nelson Hordell 16mm Rostrum camera; 8mm and 16mm cameras; film cutting room. Women's unit. Projects 1996 include: *The Butterfly Man* (C4), Short Stories: *Huskies* (C4)

Edinburgh Film Workshop Trust
(Animation Workshop)
29 Albany Street
Edinburgh EH1 3QN
Tel: 0131 557 5242
Fax: 0131 557 3852

Education on Screen Productions
64 All Saints Road
Kings Heath
Birmingham
Tel: 0121 444 3147
Fax: 0121 434 5154
Mike Kalemkerian, Neil Gammie
Independent company with extensive experience of drama/workshop based

projects offering unique video production service to organisations in the West Midlands. Workshop-based approach encourages, where required, close and active client involvement particularly in pre-shoot stages. Also offers training projects to schools, colleges and businesses in the West Midlands

Exeter & Devon Arts Centre, Media Centre
Bradninch Place
Gandy Street
Exeter EX4 3LS
Tel: 01392 219741
Fax: 01392 499929
email: video@ eurobell.co.uk
Video and multimedia training and facilities. On line non-linear editing, Betcam-SP

Film Work Group
Top Floor
Chelsea Reach
79-89 Lots Road
London SW10 0RN
Tel: 0171 352 0538
Loren Squires, Nigel Perkins
Video and film post-production facilities and graphic design. 3-machine hi-band SP with effects and video-graphics, 2-machine lo-band off-line, and 6-plate Steenbeck. Special rates for grant aided and non-profit projects

First Take Video
Merseyside Innovation Centre
131 Mount Pleasant
Liverpool L3 5TF
Tel: 0151 708 5767
Peter Goodwin, Mark Bareham, Lynne Harwood
Offers training and production services to the voluntary, arts, community and education sectors across the North West. Joint projects with other arts groups and special needs groups. Training courses, both basic and intermediate, for the public in video production and editing. Commissioned productions for local authorities

Fosse Arts Centre Audio, Video and Animation Studio
Fosse Building
Leicester City Council
Mantle Road
Leicester LE3 5HG
Tel: 0116 251 5577
Fax: 0116 251 5145
Alan Wilson, Brian McDowell, Liz Soden
An audio-visual resource providing training in all aspects of video production, computer graphics and

animation, video editing and 8-track and midi sound recording. Offers a regular programme of short video and audio courses and City & Guilds 770 course. Has a wide range of video and audio equipment for hire at a low cost including VHS/SVHS shooting kits. Video studio with full lighting rig. Animation studio and programme of animation courses, editing facilities - 2 or 3-machine VHS & SVHS and MII, 8-track recording studio with midi, sampling, and DAT facilities and two fully equipped music rehearsal studios

Four Corners Film Workshop
113 Roman Road
London E2 0QN
Tel: 0181 981 4243
Fax: 0181 983 4441
Holds film production courses in S8mm and 16mm and film theory classes. A full programme runs all year round. Provides subsidised film and video equipment for low budget independent film-makers. Has a 40 seat cinema, S8mm, 16mm, Hi8 and U-Matic production and post-production facilities

Fradharc Ur
11 Scotland Street
Stornoway
Isle of Lewis PA87
Tel: 01851 703255
The first Gaelic film and video workshop, offering VHS and hi-band editing and shooting facilities. Production and training in Gaelic for community groups. Productions include: *Under the Surface, Na Deilbh Bheo; The Weaver; A Wedding to Remember; As an Fhearran*

Glasgow Film and Video Workshop
3rd Floor
34 Albion Street
Glasgow G1 1LH
Tel: 0141 553 2620
Fax: 0141 553 2660
Ian Reid, Paul Cameron, Aimara Recques
GFVW is an open access training/ access resource for film and video makers. Offers equipment hire and training courses at subsidised rates. Facilities include; BetaSP, MiniDV, DVCPro, S-VHS and VHS, cameras and edit suites (including 2 AVIDs). Super 8 and 16mm production and support. Runs production and community based projects, distributes and promotes the exhibition of films and videos made through the workshop

Y Gweithdy Fideo/The Video Workshop*
Chapter Arts Centre
Market Road
Canton
Cardiff CF5 1QE
Tel: 01222 342755
Fax: 01222 644479
Terry Dimmick, Emyr Jenkins, Diana Bianchi
Film/video production, distribution and exhibition in English and Welsh, plus broadcast programmes. Offers production facilities to others. Working with community and arts organisations on social, political and cultural issues

HAFAD 1st Chance Project
(Hammersmith and Fulham Action for Disability)
Beaufort House
Lillie Road
London SW6 1VF
Tel: 0171 386 5616
Fax: 0171 386 7343
Contact: Richard Day
VHS and SVHS cameras and wheelchair-accessible edit suite for hire. Remote control adaptation for camera - can be operated by manual joystick or footplate
NB: Due to move late 1997

The Half Way Production House
Units 1&2, Taylors Yard
67 Alderbrook Road
London SW12 8AD
Tel: 0181 673 7926
Fax: 0181 675 7612
Georgina Hart, Emma Stewart
Production and training centre offers vocational courses in 16mm film production, Hi8 and non-linear editing for unemployed people aged 18-25. Industry links to help successful trainees gain work placements. Productions include: Hands Off, A Chemical Solution and a European training initiative in development

Hull Community Artworks
Northumberland Avenue
Hull HU2 0LN
Tel: 01482 226420
Film/video production, distribution and exhibition. Offers production and exhibition facilities to others. Holds regular training workshops

Hull Time Based Arts
8 Posterngate
Hull HU1 2JN
Tel: 01482 215050
Fax: 01482 589952
email: Ron@htba.demon.co.uk
Website: http://www.htba.demon.co.uk

Film/video production, exhibition and education. Also promotes, produces and commissions experimental film, video, performance and music. Equipment for hire includes video projectors, Avid non-linear editing suites [Media Composer 8000 and 400] with output to Betacam SP and DVC Pro, with full training provided. DVC Pro cameras and production facilities are also available

Intermedia Film and Video
19 Heathcoat Street
Nottingham NG1 3AF
Tel: 0115 950 5434
Fax: 0115 955 9956
Malcolm Leick, Graham Forde
Offers facilities and training in 16mm production, Betacam SP, Lightworks non-linear editing. Has new three camera component studio. Provides production support for programme makers new to broadcast. Plus advice, information and customised training

Ipswich Media Project
13 All Saints Road
Ipswich
Suffolk IP1 4DG
Tel: 01473 250685
Super 8 film, VHS, SVHS equipment, familiarisation training and media work

Island Arts
The Tiller Centre
Tiller Road
London E14 8PX
Tel: 0171 987 7925
Fax: 0171 538 3314
Equipment and edit suite hire for individuals and groups. Courses throughout the year, production and workshop facilities

Jackdaw Media Educational Trust
The Annexe
13-15 Hope Street
Liverpool L1 9BH
Tel: 0151 709 5858
Fax: 0171 707 2020
email: anim@jackdaw.u net.com
Laura Knight, Yvonne Eckersley,
The national animation resource based on the North West. Hands-on animation sessions available for schools (mainstream and special), colleges, community groups, galleries and museums. All ages, abilities and needs catered for. Television and other commercial production work accepted. EOS video animation system, 16mm rostrum camera and 6-plate Steenbeck available for hire

Jubilee Arts Co Ltd
84 High Street
West Bromwich
West Midlands B70 6JW
Tel: 0121 553 6862
Fax: 0121 525 0640
email: @jubart.demon.co.uk
Jubilee Arts is a unique multi-media community arts team, formed in 1974. Skills include photography, video, drama, audio visual, music/ sound, computers, training and graphic design. We work with communities, using the arts as a tool to create opportunities for positive ways for people to express themselves. Jubilee Arts works in partnership with a wide range of groups, agencies and voluntary and statutory bodies

Lambeth Video
Unit F7
245a Coldharbour Lane
London SW9 8RR
Tel: 0171 737 5903
Lambeth Video is a part-funded workshop. Runs production-based training. Runs South London Documentary Project which is open to women and black applicants. Offers production bursaries to four new directors each year and traineeships on large-scale productions. Write or telephone for application forms

Leeds Animation Workshop (A Women's Collective)*
45 Bayswater Row
Leeds LS8 5LF
Tel: 0113 248 4997
Fax: 0113 248 4997
Jane Bradshaw, Terry Wragg, Stephanie Munro, Janis Goodman, Milena Dragic
Production company making films on social issues. Distributing over 20 short films including - *A World of Difference, Did You Say Hairdressing? Waste Watchers, No Offence, Through the Glass Ceiling, Out to Lunch, Give us a Smile, All Stressed Up.* They also offer short training courses

Lighthouse
Brighton Media Centre
9-12 Middle Street
Brighton BN1 1AL
Tel: 01273 384222
Fax: 01273 384233
email: info@lighthouse.org.uk
Jane Finnis, Caroline Freeman
A training and production centre, providing courses, facilities and production advice for video and digital media. Avid off- and online edit suites. Apple Mac graphics and animation workstations. Digital video capture and manipulation. Output to/from Betacam SP. SVHS offline edit suite. Post Production and Digital Artists equipment bursaries offered three times a year

Line Out
Fosse Neighbourhood Centre
Mantle Road
Leicester LE3 5HG
Tel: 0116 262 1265
Fax: 0116 251 5145
Kofi Boafo, Sue Wallin
Professional video facilities. Hi-band SP shooting kit, 3-machine hi-band SP edit suite, Super 8 flat bed editor, all available for hire. Runs short courses, workshops and media education programmes accredited via the Open College network and currently developing courses for the NVQ route. A membership organisation providing advice, information and guidance to enter the professional media industry

London Deaf Access Project
1-3 Worship Street
London EC2A 2AB
Tel: 0171 588 3522 (voice) Tel: 0171 588 3528 (text)
Fax: 0171 588 3526
email: idap@adirect.co.uk
Translates information from English into British Sign Language (BSL) for Britain's deaf community, encourages others to do likewise and provides a consultancy/monitoring service for this purpose. Promotes the use of video amongst deaf people as an ideal medium for passing on information. Runs workshops and courses for deaf people in video production, taught by deaf people using BSL. Works with local authorities and government departments ensuring that public information is made accessible to sign language users. Titles include: School Leavers, Access to Women's Services, Health issues

London Electronic Arts
2-4 Hoxton Square
London N1 6NU
Tel: 0171 684 0101
Fax: 0171 684 111
email: info@lea.org.uk
Website: http:// www.lea.org.uk
London Electronic Arts (formerly London Video Access) is Britain's national centre for video and new media art. Offers a complete range of services including production based training, facility hire (production and post-production), distribution and exhibition of video and new media art and film

London Fields Film and Video
10 Martello Street
London E8 3PE
Tel: 0171 241 2997
Film and video production and distribution. Also provides production facilities and support for others

London Film Makers' Co-op
Lux Centre, 2-4 Hoxton Square,
London N1 6NU
Tel: 0171 684 0201
Fax: 0171 739 6366
Paul Sukhija, Patricia Diaz, Paul Murray
Film workshop, distribution library and cinema enables film-makers to control the production, distribution and exhibition of their films. Workshop runs regular practical and theoretical film courses. Distribution has 2,000 experimental films for hire, from a 20s to current work. Cinema screenings twice weekly. Work with cultural aesthetic/political aims which differ from the industry

London Screenwriters' Workshop
Holborn Centre for the Performing Arts, Three Cups Yard, Sandland Street, London WC1R 4PZ
Tel: 0171 242 2134
Alan Denman, Paul Gallagher
The LSW is an educational charity whose purpose is to help new and developing writers learn the craft of screenwriting for film and television. It offers writing workshops, industry seminars, a newsletter and a script reading service. It is open at very reasonable cost to everyone. Membership £25.00 p.a

Media Arts
Town Hall Studios
Regent Circus
Swindon SN1 1QF
Tel: 01793 463224
Fax: 01793 463223
Ann Cullis, Shahina Johnson
Film & video production and training centre. Offers short courses and longer term media projects. First stop scheme offers funding for first time film/video makers. Also offers media education services, equipment hire, screenwriting advice and undertakes production commissions. Organises screenings and discussions

The Media Workshop
Peterborough Arts Centre
Media Department
Orton Goldhay
Peterborough PE2 0JQ
Tel: 01733 237073

Fax: 01733 235462
email:peterborougharts@mailhost.
cityscape.co.uk
Video, multimedia and photography
production, workshops and
exhibitions. Offering DVCPRO,
SVHS production/edit facilities and
Media 100 non-linear editing. Also
full multimedia authoring and design

Mersey Film and Video

13-15 Hope Street
Liverpool L1 9BQ
Tel: 0151 708 5259
Fax: 0151 707 8595
email: mfv@hopestreet.u-net.com
Production facilities for: BETA SP,
DVC PRO, MINI DV, multi-media
stations, photoshop, Dolly and
track, jibarm, lights, mics, mixer,
DAT etc. Post Production on Avid,
MC1000+3DFX, SVHS, Hi8, FX &
music library. Guidance and help for
production, scripting, funding,
budgets

Migrant Media

90 De Beauvoir Road
London N1 4EN
Tel: 0171 254 9701
Fax: 0171 241 2387
Ken Fero, Ivan Ali Fawzi,
Soulyman Garcia
Media production training and
campaigning for migrants and
refugees. Focus on African and
Middle Eastern communities.
Networks internationally on media/
political issues. Broadcast credits
include: *After the Storm* (BBC);
Sweet France (C4), *Tasting Freedom*
(C4), *Justice For Joy* (C4)

Moving Image Touring Exhibition Services (MITES)

Moviola
Bluecoat Chambers
Liverpool L1 3BX
Tel: 0151 707 2881
Fax: 0151 707 2150
Courses for artists, gallery curators,
technicians and exhibitors. Starting
with Computers for Artists;
Multimedia for Artists; Design
and Project Management for
Multimedia; Making the Most of
Image and Design; Dealing with
Exhibition Technology and
Multimedia in the Gallery. Also
development, advice, consultation
services and an extensive exhibition
equipment resource

The Old Dairy Studios

156b Haxby Road
York YO3 7JN
Tel: 01904 641394
Fax: 01904 692052

Digital video production facilities
inc. Fast video system, 32 Track
digital recording studio, audio visual
facilities with Adobe Photoshop,
Radio Production and Midi
Composition Studios are available.
Courses in video production and
editing, sound engineering, radio
production, midi composition and
digital imaging. Working with
people with disabilities, unemployed
people, people aged between 12 and
25 as well as with members of the
community in general

Oxford Film and Video Makers

The Stables
North Place
Headington
Oxford OX3 7RF
Tel: 01865 741682 (01865 60074
course enquiries)
Fax: 01865 742901
email: ofvm@ox39hy.demon.co.uk
Offers courses covering the entire
production process in both film and
video and particularly specialised
courses to meet the needs of target
groups. Oxford Film and Video
Makers supports a broad range of
individual productions - both
broadcast and non-broadcast - giving
a voice to people normally under-
represented in the media. Promoting
experimental film and video art
through 'Arteaters'. Supporting
productions with campaigning
groups through the Video 'CCD'
group. Equipment hire includes a
Beta SP and digital non linear
editing system (FAST). Production
Company facility now also available

Panico Workshop

Panico Studio
No1 Falconberg Court
London W1V 5FG
Tel: 0171 734 5120
Fax: 0171 734 3398
Bob Doyle, Jacqui Wetherill
Prominent Studios is the only
professional feature film studios
offering training and workshop
facilities. Access to all that is required
for 35mm, 16mm or video production.
New members must either do the
Panico Foundation Course, or have
had previous experience

Passing Light Syndicate

Beam Ends
Tincleton,
nr Dorchester DT2 8QP
Tel: 01305 849019
Lee Berry, John Rampley
Independent broad based film, video,
mixed media creative outfit of
varying members, assistance with
fundraising, grants, advice. Broad-

based creative collaboration re-film
and video production, installation,
performance, Internet and othe site
specific spectatctulars etc. Always
open to new people, ideas and
projects. Independent productions
include: Polsow Hudol; Owth Ober
Yn

Paxvision

The Albany
Douglas Way
London SE8 4AG
Tel: 0181 692 6322
0973 416447 (mobile)
Fax: 0181 692 6322
Video project providing services,
video facilities and training
(beginners to advanced) mainly
targeted at the local community,
independent video and film makers
and disadvantaged groups or
individuals. Also collaborates with
local community groups and artists
in video production; undertakes
commissions on production and
music scoring. Facilities include
Betacam SP and SVHS production
equipment, VHS edit suite and
SVHS edit suite with MX50
Visionmixer, Amiga 2000 computer,
graphics and titling, Adobe Premiere
Non Linear-editing (PC). All
services available at community and
commerical rates

Picture This Independent Film & Video

Kingsland House, Gas Lane
St Philips
Bristol BS2 0QL
Tel: 0117 972 1002
Fax: 0117 972 1750
Film and video production and
training workshop. Specialises in
broadcast TV, independent and
artists' film and video community
projects and time-based media
installations. Offers short courses
and long term training projects.
Super 8mm, 16mm film, DV/DVC
Pro and Beta SP. Production and
post-production facilities available,
including Avid non-linear, 3-
machine tape editing and 16mm film
editing

The Place in the Park

Bellvue Road
Wrexham
LL13 7NH
Tel: 01978 358522
Video/Film production access centre,
offering subsidised facilities hire.
Equipment includes Beta SP and
SVHS shooting and editing kit. The
Place in The Park acts as a focal
point/contact centre for independent

film and video makers in the North Wales region and beyond

Platform Films and Video
3 Tankerton House
Tankerton Street
London WC1H 8HP
Tel: 0171 278 8394 Mobile: 0973 278 956
Fax: 0171 278 8394
Chris Reeves
Film/video production and distribution. Also equipment hire including complete Sony BVW400P shooting kit, Panasonic Hi-Fi sound VHS edit suite, Avid Media Composer 400, 9Gb, 20" monitors, Pro-Tools, title tool, Previs 2 fx, Sanyo 220 video projector. Titles include: *Green Party General and Local election broadcasts* 1998; *From The Edge* inserts for BBC2's *Disability Programme* Unit's magazine programme 1995-97; *Proud Arabs and Texan Oilmen* for C4's *Criticial Eye* 1993; *The People's Flag* series for C4 1987-88; *The Miners' Campaign Tapes* 1984-85; *The Cause Of Ireland* for C4 1983

Plymouth Arts Centre
38 Looe Street
Plymouth PL4 0EB
Tel: 01752 660060
Fax: 01752 250101

Real Time Video
The Arts and Media Centre
21 South Street
Reading RG1 4QU
Tel: 01734 585627
Fax: 01734 504911
Clive Robertson
Community access video workshop, video production, training, exhibition and consultancy. Runs workshops and projects using video as development and self advocacy tool. Organises screenings and offers training in production, post-production, computer graphics and community video practice. SVHS edit suite with video and audio-processing, and graphics available for hire (reduced rates for non-profit work)

Sankofa Film and Video*
Spectrum House
Unit K
32-34 Gordon House Road
London NW5 1LP
Tel: 0171 485 0848
Fax: 0171 485 2869
Maureen Blackwood, Johann Insanally,
Film production and 16mm editing facilities, training in film production and scriptwriting, screenings.

Productions include: *The Passion of Remembrance, Perfect Image; Dreaming Rivers; Looking for Langston; Young Soul Rebels; In between; A Family Called Abrew; Des'ree EPK; Home Away From Home; Father Sons; Unholy Ghosts; Is it the design on the Wrapper?* + *Vacuum*

Second Sight
Zair Works
111 Bishop Street
Birmingham B5 6JL
Tel: 0121 622 4223
Fax: 0121 622 5750
Second Sight provides high quality accessible practical training to women and men from novice to industry professionals. Video Production company specialising in arts and social issues. Provides consultancy, information and assessment to individuals and companies in training and production

Sheffield Independent Film
5 Brown Street
Sheffield S1 2BS
Tel: 0114 272 0304
Fax: 0114 279 5225
Colin Pons
A resource base for independent film and video-makers in the Sheffield region. Regular training workshops; access to a range of film and video equipment; technical and administrative backup; office space and rent-a-desk; regular screenings of independent film and video. Regular producers sessions to help producers keep abreast of developments in the industry. Administer Yorkshire Media Production Agency. A grant giving up to £80,000 or 25 per cent of a film's budget

(SHIFT) Sheffield Independent Film and Television
5 Brown Street
Sheffield S1 2BS
Tel: 0114 272 0304
Fax: 0114 279 5225
Linda Marshall
Registered charity which facilitates and provides access to training in film and video for those who are currently under-represented in the media industry

Signals, Essex Media Centre
21 St Peters Street
Colchester CO1 1EW
Tel: mini com 01206 560255
Fax: 01206 369086
email: carmel@signals-media.demon.co.uk
Film/video resource. Betacam production and post-production

equipment. Services in training, production, media education and equipment hire. Productions include: *Three Hours in High Heels is Heaven* (C4), *Coloured* (Anglia TV), *Cutting Up* (C4); *Garden of Eve* (Anglia TV) and *Fork in the Road*

Star Productions Studios
1 Cornthwaite Road,
Clapton
London E5 9QD
Tel: 0181 986 4470/5766
Fax: 0181 533 6597
Raj Patel, Carlos Homer,
Khalid Hussein
Film/video production company working from an Asian perspective. Multi-lingual productions. Offering studios for hire, video editing suite, 16mm cutting room. Production and exhibition facilities. Output includes community documentaries, video films of stage plays, feature films and corporate videos on health issues. Recent productions: *Tangled Web*, feature film; *Getting Back*, short drama

Swingbridge Video
Norden House
41 Stowell Street
Newcastle upon Tyne NE1 4YB
Tel/Fax: 0191 232 3762
email: Swingvid@aol.com
Contact: Hugh Kelly
A producer of both broadcast and non-broadcast programmes, including drama and documentary formats and specialising in socially purposeful and educational subjects. Offers training and consultancy services to public sector, community and voluntary organisations. Also provides a tape distribution service. Productions include: *White Lies; An English Estate; Happy Hour; Where Shall We Go?; Sparks; Set Your Free; Mean Streets* and many more

33 Video
Luton Community Arts Trust
33-35 Guildford Street
Luton
Beds LU1 2NQ
Tel: 01582 419584
Fax: 01582 459401
Production and training

Trade Films*
36 Bottle Bank
Gateshead
Tyne and Wear NE8 2AR
Tel: 0191 477 5532
Fax: 0191 478 3681
Film, television and video production - fiction, documentary and factual programmes. Hi-band U-Matic and 16mm post-production

facilities available. Recent productions include: *Border Crossing* (128 min drama for C4/ZDF/Tyne Tees Television); *The People's Canal* (2 x30 min for Granada); *Real Resources* (25 min factual programme for the European Trade Union College); *Hazard* (interactive multimedia CD Rom on health and safety)

Trilith

Corner Cottage, Brickyard Lane
Bourton, Gillingham
Dorset SP8 5PJ
Tel: 01747 840750/840727
Trevor Bailey, John Holman
Specialises in rural television and video on community action, rural issues and the outlook and experiences of country born people. Also works with organisations concerned with physical and mental disability and with youth issues. Produces own series of tapes, undertakes broadcast and tape commissions and gathers archive film in order to make it publicly available on video. Distributes own work nationally. Recent work includes broadcast feature and work with farmers and others whose lives revolve around a threatened livestock market, and a production scripted and acted by people with disabilities

Valley and Vale Community Arts Ltd

The Sardis Media Centre
Betws
Mid Glamorgan CF32 8SU
Tel: 01656 729246/871911
Fax: 01656 729185/870507
Video production, distribution and exhibition. Open access workshop offering training to community groups in VHS, lo-band U -Matic and Hi8 formats

Vera Productions

30-38 Dock Street
Leeds LS10 1JF
Tel: 0113 2428646
Fax: 0113 2451238
Alison Garthwaite,
Catherine Mitchell
Video production - documentary, education, arts equality, public sector, health. Training (ESF/other) for women/mixed. Co-ordinates production and exhibition. Information resource. Runs membership organisation (networking) for women in film, video and television

Video Access Centre

25a SW Thistle Street Lane
Edinburgh EH2 1EW
Tel: 0131 220 0220
Fax: 0131 220 0017
Audrey Hutchison, Stephen Flitton
A membership-based association which provides resources and training for individuals and community groups to work with video. Courses are short and at basic or specialist level. Has VHS & SVHS, Super 8 and 16mm production facilities, runs bi-monthly newsletter and information service

Video in Pilton*

30 Ferry Road Avenue
Edinburgh EH4 4BA
Tel: 0131 343 1151
Fax: 0131 343 2820
Hugh Farrell, Joel Venet,
Lorna Simpson
Training and production facilities in the local community; documentary and fiction for broadcast. Contact address for UK Network of Workshops (see Organisations)

Vivid

Birmingham Centre for Media Arts
Unit 311
The Big Peg
120 Vyse Street
Birmingham B18 9ND
Tel: 0121 233 4061
Fax: 0121 212 1784
email: vividd@waverider.co.uk
Website: www.wavespace.waverider.co.uk/'vivid/
Nicky Edmonds, Marian Hall, Andrew Robinson, Jayne Murray
Facilities include 16mm film production, Beta SX, DV Cam, Hi8 video production equipment, Avid, Video Machine and linear video editing, EOS animation, 5x4 medium format and 35mm photographic equipment and darkrooms, Power Macs and PCs running Photoshop, premiere, director etc

WAVES (Women's Audio-Visual Education Scheme)

London Women's Centre
Wesley House
4 Wild Court
London WC2B 4AU
Tel: 0171 430 1076
Fax: 0171 242 2765
WAVES is the nation centre of training for women in audio-visual media culture, promoting the ability of women both to make their own work and to gain industry employment. WAVES initiatives include: Director's Course; Multi-skilling; Foundation in Broadcasting; Recording Live Events, and inter-related modular short courses in film

theory, film language and production. WAVES supports women filmmakers by providing information, contacts, equipment and a unique cultural space within which to operate, whilst the educational and training programme begins to redress the inequalities of opportunities faced by specific groups of women, as well as ensuring that all courses are accessible, culturally and physically

WFA

Media and Cultural Centre
9 Lucy Street
Manchester M15 4BX
Tel: 0161 848 9785
Fax: 0161 848 9783
Main areas of work include media access and training, including City and Guilds 770 National Certificate, with a full range of production, post-production and exhibition equipment and facilities for community, semi-professional and professional standards. Video production unit (BECTU). Distribution and sale of 16mm films and videos, booking and advice service, video access library. Cultural work, mixed media events. Bookshop/outreach work

West Yorkshire Media Services

Leeds Metropolitan University
Brunswick Terrace
Leeds LS2 813U
Te: 0113 2831713
Fax: 0113 2831713
email: m.spadafora@lmu.ac.uk
18 month Certificate in Film and Video Production courses accredited by Leeds Metropolitan University. A free course that welcomes applications from women and people people from minorities. Other courses and projects as per programme offers a thourough grounding in all aspects of film and video production.

WITCH (Women's Independent Cinema House)*

Holmes Building
46 Wood Street
Liverpool L1 4AH
Tel: 0151 707 0539
The Black Women's Media Project is a vocational training programme, offering media based skills to young black women in between 16-25 years. Training is given in video production, photography and media skills, including: camera technique, sound recording, lighting, scripting and production and editing. With MOCF accreditation for a duration of 12 weeks. Additional workshops

in photography, scriptwriting and video. Exhibition Just the Job

Welfare State International (WSI)
The Ellers
Ulverston
Cumbria LA12 0AA
Tel: 01229 581127
Fax: 01229 581232
A consortium of artists, musicians, engineers and performers. Film/video production, hire and exhibition. Output includes community feature films *King Real and the Hoodlums* (script Adrian Mitchell) and work for television. Titles include: *Piranha Pond* (Border TV), RTS Special Creativity Award; *Ulverston Town Map*, community video; Community Celebration, Multinational Course leading to Lantern Procession (video) and Rites of Passage publications include: The Dead Good Funerals Book available from WSI

Western Eye Television
Easton Business Centre
Felix Road
Bristol BS5 0HE
Tel: 0117 941 5854
Fax: 0117 941 5851
Video production for business and voluntary sector. Health and technology specialists. Also produce sell-through programmes eg *The Complete Teach Yourself Juggling Kit*

The Wheel
Wild Court
Off Kingsway
London WC2B 4AU
Tel: 0171 831 6946
Fax: 0171 831 9059
Linda Eziquiel, Joy de Shong
A women's resource providing facilities for women and women's organisations to hire with a sliding scale of charges. Facilities include playback and screening space. All equipment for use on the premises only. Rooms for hire range from the theatre (capacity 150) to a variety of seminar/meeting and rehearsal spaces. The resource also houses a women's health and leisure suite with therapies, sauna, gym, courses and cafe/bar

Women's Media Resource Project
89a Kingsland High Street
London E8 2PB
Tel: 0171 254 6536
Training/workshops. Offers accredited training - City and Guilds in Sound Engineering. Workshops/ summer schools in Sound Engineering, Sound for Video Sound Technologies. Hire of equipment and studios, send SAE for leaflet. Recording/video packages for theatre, dance, artists, bands

ABBREVIATIONS

ABC Association of Business Communicators

ABSA Association of Business Sponsorship of the Arts

ACCS Association for Cultural and Communication Studies

ACE Arts Council of England/Ateliers du Cinéma Européen

ADAPT Access for Disabled People to Arts Premises today

AEEU Amalgamated Engineering and Electrical Union

AETC Arts and Entertainment Training Council

AFC Australian Film Commission

AFCI Association of Film Commissioners International

AFECT Advancement of Film Education Charitable Trust

AFI American Film Institute/ Australian Film Institute

AFM American Film Market

AFVPA Advertising Film and Videotape Producers' Association

AGICOA Association de Gestion Internationale Collective des Oeuvres Audiovisuelles

AIM All Industry Marketing for Cinema

AMCCS Association for Media, Cultural and Communications Studies

AME Association for Media Education

AMFIT Association for Media Film and Television Studies in Higher and Further Education

AMPAS Adcademy of Motion Picture Arts and Sciences (USA)

AMPS Association of Motion Picture Sound

APC Association of Professional Composers

APRS The Professional Recording Association

AVEK The Promotion Centre for Audio Visual Culture in Finland

BAFTA British Academy of Film and Television Arts

BARB Broadcasters' Audience Research Board

BASCA British Academy of Songwriters, Composers and Authors

BATC British Amateur Television Club

BBC British Broadcasting Corporation

BBFC British Board of Film Classification

BCS British Cable Services

BECTU Broadcasting Entertainment Cinematograph and Theatre Union

BFB Black Film Bulletin

BFC British Film Commission

BFFS British Federation of Film Societies

BFI British Film Institute

BIEM Bureau Internationale des Sociétés gérant les Droits d'Enregistrement

BIPP British Institute of Professional Photography

BKSTS British Kinematograph Sound and Television Society

BNFVC British National Film and Video Catalogue

BPI British Phonographic Industry

BREMA British Radio and Electronic Equipment Manufacturers' Association

BSAC British Screen Advisory Council

BSC British Society of Cinematiographers Broadcasting Standards Commission

BSD British Screen Development

BSkyB British Sky Broadcasting

BSS Broadcasting Support Services

BTDA British Television Distributors Association

BUFVC British Universities Film and Video Council

BVA British Video Association

CAA Cinema Advertising Association

CARTOON European Association of Animation Film

CAVIAR Cinema and Video Industry Audience Research

CD Compact Disc

CDI Compact Disc Interactive

CD ROM Compact Disc Read Only Memory

CEA Cinematograph Exhibitors' Association

CEPI Co-ordination Europénne des Producteurs Indépendants

CFTF Children's Film and Television Foundation

CFU Children's Film Unit

C4 Channel 4

CICCE Comité des Industries Cinématographiques et Audiovisuelles des Communautés Européenes et de l'Europe Extracommunautaire

CILECT Centre Internationale de Liaison des Ecoles de Cinéma et de Télévision

CNN Cable News Network

COI Central Office of Information

CPBF Campaign for Press and Broadcasting Freedom

CTA Cable Television Association/ Cinema Theatre Association

CTBF Cinema and Television Benevolent Fund

DAT Digital Audio Tape

DBC Deaf Broadcasting Council

DCMS Department for Culture Media and Sport

DBS Direct Broadcasting by Satellite

DFE Department for Education

DFI Danish Film Institute

DGGB Directors' Guild of Great Britain

DTI Department of Trade and Industry

DVI Digital Video Interactive

EAVE European Audiovisual Entrepreneurs

EBU European Broadcasting Union

ECF European Co-Production Fund

EDI Euopaïsches Dokumentarfilm Institut/Entertainment Data International

EFA European Film Academy

EFCOM European Film Commissioners

EFDO European Film Distribution Office

EGAKU European Committee of Trade Unions in Arts, Mass Media and Entertainment

EIM European Institute for the Media

EITF Edinburgh International Television Festival

EMG Euro Media Garanties

ENG Electronic News Gathering

EU European Union

EUTELSAT European Telecommunications Satellite Organisation

FAA Film Artistes' Association

FACT Federation Against Copyright Theft

FAME Film Archive Management and Entertainment

FBU Federation of Broadcasting Unions

FEITIS Fédération Européene des Industries Techniques de l'Image et du Son

FEMIS Institut de Formation et d'Enseignement pour les Métiers de l'Unage et du Son

FEPACI Fédération Pan-Africain des Cinéastes

FESPACO Festivale Pan-Africain des Cinémas de Ougadougou

FEU Federation of Entertainment Unions

FIA International Federation of Actors

FIAD Fédération Internationale des Associations de Distributeurs de Films

FIAF Fédération Internaionale des Archives du Film

FIAPF International Federation of Film Producers Associations

FIAT Fédération Internationale des Archives de Télévision

FICC Fédération Internationale des Ciné-Clubs

FIFREC International Film and Student Directors Festival
FIPFI Fédération Internationale des Producteurs de Films Indépendants
FIPRESCI Fédération Internationale de la Presse Cinématographique
FOCAL Federation of Commercial Audio Visual Libraries
FTVLCA Film and Television Lighting Contractors Association
FX Effects/special effects
HBO Home Box Office
HDTV High Definition Television
HTV Harlech Television
HVC Home Video Channel
IABM International Association of Broadcasting Manufacturers
IAC Institute of Amateur Cinematographers
ICA Institue of Contemporary Arts
IDATE Insitut de l'Audiovisuel et des Télécommunications en Europe
IFDA Independent Film Distributors' Association
IFFS International Federation of Film Societies (aka FICC)
IFPI International Federation of the Phonographic Industry
IFTA International Federation of Television Archives (aka FIAT)
IIC International Institute of Communications
ILR Independent Local Radio
INR Independent National Radio
IPA Institute of Practitioners in Advertising
ISBA Incorporated Society of British Advertising
ISETV International Secretariat for Arts, Mass Media and Entertainment Trade Unions
ISM Incorporated Society of Musicians
ITC Independent Television Commission
ITN Independent Television News
ITV Independent Television
ITVA Independent Television Association
IVCA International Vsual Communications Association
IVLA International Visual Literacy Association
JICTAR Joint Industries' Committee for Television Audience Research
LAB London Arts Board
LFF London Film Festival
LFMC London Film Makers' Co-op
LFVDA London Film and Video Development Agency
LSW London Screenwriters' Workshop
LVA London Video Access
LWT London Weekend Television
MBS Media Business School
MCPS Mechanical Copyright Protection Society
MEDIA Mesures pour Encourager le Développement de l'Industrie Audiovisuelle

MENU Media Education News Update
MFVPA Music, Film and Video Producers' Association
MGM Metro Goldwyn Mayer
MHMC Mental Health Media Council
MIDEM Marché International du Disque et de l'Edition Musicale
MIFED Mercato Internazionale del TV, film e del Documentario
MIPCOM Marché Internaional des Films et des Programmes pour la TV, la Vidéo, le Câble et le Satellite
MIP-TV Marché International de Programmes de Télévision
MOMI Museum of the Moving Image
MPA Motion Picture Association of America
MPEAA Motion Picture Export Association of American
MU Musicians' Union
NAHEFV National Association for Higher Education in Film and Video
NAVAL National Audio Visual Aids Library
NCA National Campaign for the Arts
NCC National Cinema Centre
NCET National Council for Educational Technology
NCVQ National Council for Vocational Qualifications
NFDF National Film Development Fund
NFT National Film Theatre
NFTC National Film Trustee Company
NFTS National Film and Television School
NFTVA National Film and Television Archive
NHMF National Heritage Memorial Fund
NIFC Northern Ireland Film Council
NMPFT National Museum of Photography, Film and Television
NoW Network of Workshops
NPA New Producers Alliance
NSC Northern Screen Commission
NTSC National Television Standards Committee
NUJ National Union of Journalists
NUT National Union of Teachers
NVALA National Viewers' and Listeners'Association
PACT Producers Alliance for Cinema and Television
PAL Programme Array Logic/ Phase Alternation Line
PPL Phonographic Performance
PRS Performing Right Society
RAB Regional Arts Board
RETRA Radio, Electrical and Television Retailers' Association
RFT Regional Film Theatre
RTBF Radio Television Belge de la

Communanté Française
RTS Royal Television Society
S4C Siandel Pedwar Cymru
S&S Sight and Sound
SAC Scottish Arts Council
SBFT Scottish Broadcast & Film Training
SCALE Small Countries Improve their Audio-visual Level in Europe
SCTE Society of Cable Television Engineers
SECAM Séquentiel couleur â mémoire
SFA Short Film Agency
SFC Scottish Film Council
SFD Society of Film Distributors
SFPF Scottish Film Production Fund
SFX Special Effects
SIFT Summary of Information on Film and Television
SMATV Satellite Mater Antenna Television
SOURCES Stimulating Outstanding Resources for Creative European Scriptwriting
TVRO Television receive-only
UA United Artists
UCI United Cinemas International
UIP United International Pictures
UNESCO United Nations Educational, Scientific and Cultural Organisation
UNIC Union International des Cinémas
URTI Université Radiophonique et Télévisuelle Internationale
VCPS Video Copyright Protection Society
VCR Video Cassette Recorder
VHS Video Home System
VLV Voice of the Listener and Viewer
WGGB Writers' Guild of Great Britain
WTN Worldwide Television News
WTVA Wider Television Access
YTV Yorkshire Television